The Common Insects of North America

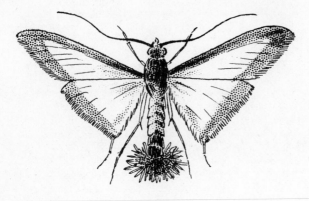

THE

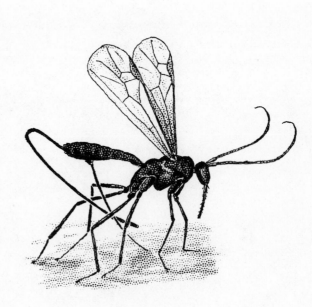

Illustrations by Charles S. Papp

COMMON INSECTS OF NORTH AMERICA

by Lester A. Swan

and Charles S. Papp

FOREWORD BY EVERT I. SCHLINGER

HARPER & ROW, PUBLISHERS
NEW YORK, EVANSTON, SAN FRANCISCO, LONDON
1817

THE COMMON INSECTS OF NORTH AMERICA. Copyright © 1972 by Lester A. Swan and Charles S. Papp. All rights reserved. Printed in the United States of America. No part of this book may be used or reproduced in any manner whatsoever without written permission except in the case of brief quotations embodied in critical articles and reviews. For information address Harper & Row, Publishers, Inc., 10 East 53rd Street, New York, N.Y. 10022. Published simultaneously in Canada by Fitzhenry & Whiteside Limited, Toronto.

FIRST EDITION

STANDARD BOOK NUMBER (TRADE): 06-014181-6

STANDARD BOOK NUMBER (HARPERCREST): 0-06-014179-4

LIBRARY OF CONGRESS CATALOG CARD NUMBER: 75-138765

Designed by Lydia Link

To that delightful quartet
Trona and Eva,
Roger and Kenneth

CONTENTS

* Minor orders—Protura, Entotrophi (Diplura), Zoraptera, Strepsiptera—are not treated in detail since these insects are less likely to be seen; they are described briefly in the Pictured Key to the Insect Orders.
For breakdown into families, with text references, see Appendix II.

FOREWORD

Over the years I have had the distinct pleasure of working and field-tripping with a good number of different groups of young people engaged in various kinds of insect studies. Some of these nature students were truly naïve, while others were quite sophisticated, yet both always seemed fascinated with their rather unknown world of insects.

Often I was able to give partially satisfying answers to some of the questions posed by them on such subjects as: How can ants carry such large seeds? Do scarab beetles really eat dung? How do those mayfly nymphs breathe under water? But without exception the most frequent question was: What kind of insect is that? To give a proper answer to this one was often difficult even for a practicing taxonomist.

Entomologists have already described nearly one million different species of insects in the world, and there may be a similar number of species awaiting study and analysis. For various reasons insects are described and figured in a manner suitable for publication in hard-to-find scientific journals, and even if these journals were available to the general public they likely would not be read by many. Therefore, to answer the question: What kind of insect is that? interested persons attempt to name the insect by referring to one or another of the available textbooks or guides in entomology, only to find the task completely frustrating, particularly when one is on a field trip without a microscope. In most books or guides I have examined, I found that the figure available for the insect was poorly reproduced, that variations of the insect were not included, that sexual dimorphism wasn't shown, or more often than not I found a statement about the insect in question that read as follows: "This family of insects is rarely encountered and its biology is not well known." I have had too many such experiences myself, particularly since as an "entomologist" I am expected to name any insect that one may choose to thrust under my nose.

Having thoroughly enjoyed reading the book and the interesting manner in which the text was prepared, and having carefully examined the excellent quality and number of insect drawings presented, it is my hope that this book will find its way into the hands of both the students of insects as well as all general reference libraries. Such a book will enhance anyone's chances of recognizing the insects in our region, the most essential fact one has to ascertain before further reading and reference can be achieved.

More particularly, in my view, this book will facilitate a better understanding of the insects around us both at home and in the field, and should serve as an excellent reference for such groups as the Boy and Girl Scouts of America, 4-H Club members, and various elementary school groups that need a quick and easy but still a rather thorough guide to the insects they encounter as they study the American environment.

I commend these authors for undertaking this task and spending the many long and arduous hours of library and laboratory research necessary to produce this truly welcome addition toward a synthesis and clarification of our knowledge of insects.

May I wish Messrs. Swan and Papp every success with their book, and I for one feel much more confident now that I shall be able to better answer that proverbial question: What kind of insect is that?

EVERT I. SCHLINGER
Professor of Entomology
University of California
Berkeley, California

PREFACE

The purpose of this book is to provide any interested person an easy way to identify the more common insects of North America north of Mexico. Detailed classification is a complex subject and not our concern here; this is properly left to the taxonomists or specialists in that field. This "guide" differs from others in scope and in the emphasis on pictures for identification. The drawings, unlike most photographs, show sufficient details to enable the user to identify his specimens more readily. The object is to minimize the text and to dispense with descriptions of structural features and maculations (markings) which can be shown and need little comment. The descriptive text—limited mostly to range, colors, size, habits, and food preference—is in most cases either opposite or below the picture, an advantage that hardly needs comment. The general features of each group—order and family—precede the descriptions of species in that group.

This book should be useful to the general naturalist and to the gardener, farmer, or grower in helping to solve his specific insect problems. We hope that it will also serve as a guide for the hobbyist or amateur collector, and as an introduction for the casual or leisure study of insects by the interested layman or more serious study by the student. We have not included instructions for collecting and preserving insects, although we would encourage this as a useful and interesting hobby. Booklets on this subject are available from federal and state agencies at little or no cost; therefore, instructions of this kind are not repeated here. *Collection and Preservation of Insects* (USDA Miscellaneous Publication No. 601) is one of the best; single copies may be obtained free from the Publications Division, Office of Information, U.S. Department of Agriculture, Washington, D.C. 20250.

Similar booklets are also available from the Agricultural Extension Service affiliated with land-grant or agricultural colleges.

The range (or distribution) of the insect in many cases can be given only in very general terms and is not always to be taken too literally. Distribution records are derived from the reports of workers in entomology and in agriculture and may undergo frequent change; the report or lack of it from a given locality does not necessarily mean that the insect in question will or will not be found there now. The range of many insects is being constantly extended where conditions are favorable, partly from natural causes but also as a result of the tremendous increase in travel and commerce. We have today "argosies of magic sails" (as foretold by Tennyson) descending daily on all our major cities with human and other cargoes from far-off countries. Every plane from a foreign land, every truck and automobile crossing the Mexican border or entering and leaving California and other states by any route, is a potential carrier of insects beyond their existing borders.

It is perhaps inevitable that the great natural barriers—the oceans, mountains, and deserts—will in effect be gradually eliminated and only climatic and biotic factors will set the limits of an insect species. This is not to say that we should give up on quarantines and inspections or that we should not exercise great individual care to avoid the spread of pest species. It means that greater effort in this direction is necessary. It is no longer as simple as watching a few ports of entry of ocean-going vessels. Pest species accidentally introduced usually come without their natural enemies, and if other conditions are favorable, they are much more troublesome than in their place of origin. Charles Elton, in *The Ecology of Invasions by Animals and Plants,* has shown the far-reaching effects of some of these introductions over the years.

A word should be said about names (in addition to what we have to say about them in the Introduction). The genus name in parentheses denotes an older name, for the convenience of those who may read about that insect elsewhere in the literature under its former name. In technical writings, the genus name in parentheses customarily denotes the subgenus, which is not necessary for our purposes. Where we show a species name in parentheses, this also denotes an older name. We have taken pains to use the common names that are well established and appear in the list of *Common Names of Insects* of the Entomological Society of America. In other instances we have sometimes taken the liberty of using names which we have come to associate with certain insects in practice and which may not be in common or wide use.

We would like to acknowledge our indebtedness to Dr. Howard E. Evans of Harvard University, who read the original draft of the manuscript, for his criticism and advice, and to Dr. Frederick Streams of the University of Connecticut, who read the manuscript later and made many valuable suggestions. We are especially grateful to Dr. Evert I. Schlinger of the University of California for his early interest and encouragement and for giving generously of his time in a critical review of the manuscript and writing the Foreword. Many others have extended us courtesies and help for which we are grateful. We had the privilege of sharing several extended collecting trips with Professor Philip H. Timberlake, thereby benefiting from his entomological experience of sixty years. Dr. Lauren D. Anderson, and Dr. John D. Pinto, Jack Hall and Saul Frommer were very helpful, being ready at all times to procure and lend us specimens from the insect collection of the University of California, Riverside. Without the advice, encouragement and inspiration provided over many years by George T. Okumura, Chief of Laboratory Services of the California State Department of Agriculture in Sacramento, Charles Papp's part in this book would hardly have been possible. The Agricultural Experiment Stations and the Agricultural Research Service in the United States, and the Canada Department of Agriculture have been liberal and prompt in providing information whenever requested. We wish also to thank Miss Margi Papp and Mrs. Magda Papp for painstakingly typing the manuscript. Needless to say, Mrs. Papp and Mrs. Helen Swan have our deep gratitude for putting up with it all. We are appreciative of Mrs. Theodore Illés Hodo's assistance in checking parts of the manuscript, and especially of the help and patience of our editor, production editor, and copy editor at Harper & Row during the long editorial process. Lester Swan especially wishes to add his personal thanks to Dr. Bernard Tesniere for his interest and helpful suggestions.

Escondido, California
Sacramento, California

L. A. S.
C. S. P.

The Common Insects of North America

"What sort of insects do you rejoice in where *you* come from?" the Gnat inquired.

"I don't *rejoice* in insects at all," Alice explained, "because I'm rather afraid of them—at least the large kinds. But I can tell you the names of some of them."

"Of course they answer to their names?" the Gnat remarked carelessly.

"I never knew them to do it."

"What's the use of their having names," the Gnat said, "if they won't answer to them?"

"No use to *them*," said Alice; "but it's useful to the people that name them, I suppose. If not, why do things have names at all?"

"I can't say," the Gnat replied. "Further on, in the woods down there, they've got no names—"

Lewis Carroll
THROUGH THE LOOKING-GLASS

Introduction to Insects

On the assumption that many of our readers will not have had a formal course in entomology (the study of insects) or more than a general course in biology, a discussion of classification and the place of insects in the animal kingdom seems appropriate. For them also, the general structural features and habits peculiar to insects together with some of the terms necessary to an understanding of the descriptions are included. These introductory topics, in outline, are as follows:

What is an insect?
How insects and other animals are named
Phyla of the Animal Kingdom
Relatives of insects: other arthropods
Other animals associated with insects
The number of species of animals
How insects are born and grow

Predators and parasites
The structure of insects
    Wings: venation
    Antennae and mouthparts
The senses of insects
    Insect vision
Respiration, circulation, digestion
Defensive mechanisms
The value of insects

What is an insect?

For some strange reason many people do not think of insects as "animals." We have seen this confusion on TV panel shows, where the panelists are otherwise quite knowledgeable. Perhaps this failure (or reluctance?) to identify them with the more familiar and higher forms of animals is because insects are so strange and alien to our world. Even though they do not always reveal themselves, they are ever present, sometimes in far greater numbers than we would wish. Briefly, an insect (in the adult stage) is a small animal with *six* legs (each with five main segments) and three distinct body parts: head, thorax, and abdomen. Nearly all insects have a pair of antennae ("feelers"), and most of them (but not all) have wings.

Two pairs of wings are the rule, with an important exception—the Diptera, or flies, which have only one pair.

Insects are the most numerous of all animals in the number of *species* (different kinds) and in total numbers as well, if we exclude certain microscopic forms of animal life. They have been (and continue to be) the most adaptive of all forms of life in their long evolutionary history, which accounts for their success as a group and their great variety. This has been demonstrated by the ability of some species to develop resistance to chemical insecticides in a short period of time. Some occupy a very small niche in the scheme of things—indeed are scarcely ever seen or felt—but they nevertheless persist as they have through countless millions of years. Many insects are very primitive and have undergone little or no change during eons of their existence, as can be seen from fossil records. The place of insects in the animal kingdom can best be seen by looking at the biological system of classification and making some comparisons.

How insects and other animals are named

It would be almost impossible to study insects or any other animals as a disorganized array of the various forms. Someone has aptly said, "It is natural for man to classify." With such a vast number of different kinds of animals it is also necessary for him to arrange them in an orderly system. The *Animal Kingdom* is divided into *Phyla*. Insects belong to the *Phylum* called *Arthropoda*, which means joint-legged, and the *Class Insecta* or *Hexapoda* (having six feet or legs). The Phylum is, thus, the broader category of classification. Below this are narrowing divisions arranged thusly:

CLASSIFICATION			EXAMPLES	
	European corn borer	American oyster	Guppy	Man
Phylum	Arthropoda	Mollusca	Vertebrata	Vertebrata
Class	Hexapoda (Insecta)	Pelecypoda	Pisces	Mammalia
Order	Lepidoptera	Prionodesmacea	Cyprinodontiformes	Primates
Family	Pyraustidae	Ostreidae	Poeciliidae	Hominidae
Genus	Ostrinia	Crassostrea	Lebistes	Homo
Species	nubilalis	virginica	reticulatus	sapiens

Starting with the biological unit, or the *species*, the individuals are so similar they can interbreed and reproduce fertile progeny capable of carrying on their own kind. Closely related species are grouped into *genera* (sing., *genus*). Various genera with similar structural and other features are grouped into *families*. Family names for animals always end in -*idae* and for plants generally in -*ceae*.* Families in turn are grouped into *orders*,

* Some exceptions: Cruciferae (mustard family), Compositae (sunflower family), Umbelliferae (parsley family), Labiatae (mint family), Gramineae (grass family).

orders into *classes,* and classes finally into *phyla.* It is not always this simple in actual practice; species with geographical variations are often broken down into subspecies with a third name. Suborders, superfamilies (ending in *-oidea*), subfamilies (ending in *-inae*), tribes (ending in *-ini*), and subgenera are also used in classification, but these complications are generally not necessary for our purposes.

Insects and other animals (like ourselves) are always given two names— *genus* and *species.* This is in accordance with the "binomial system of nomenclature" as devised by Linnaeus, the eighteenth-century Swedish naturalist and botanist. The first letter of the genus is customarily capitalized, that of the species is not. In print they are usually *italicized* (or in typing *underlined*). The scientific name for the European corn borer, for example, is *Ostrinia nubilalis.* The names are derived from Latin and Greek to make them uniform and international, regardless of language. It appears that we still must have "common names," and uniformity is desirable here also. For this purpose the Entomological Society of America publishes a list of the *Common Names of Insects,* and these recommendations are generally followed in the United States and Canada.*

The great majority of insects described and named do not have "common names"—their penalty perhaps for not being well known. The scientific names are usually followed by the name of the author who first described the species and for convenience in referring to the literature. Names change; after further study an insect may be reassigned to another or new genus. When this happens it is customary to retain the species name and that of the original author but with the latter in parentheses. In writing, the genus only followed by "sp." is sometimes given; this is done when the species is not definitely known or the exact identity is not necessary for the purpose at hand. The genus followed by "spp." denotes several undetermined species of that genus. While generic names are never duplicated within the animal kingdom, a species name may be used again in a different genus.

Phyla of the animal kingdom

The main Phyla of animals are as follows: Protozoa (single-celled microscopic forms found practically everywhere); Porifera (sponges); Coelenterata (jellyfish, sea anemones, corals); Echinodermata (starfish, sea ur-

---

* Perhaps it should be noted that in compound names the recommended practice is to separate the two parts when one of them indicates the proper group; for example, house fly, horse fly, deer fly (which are true flies, i.e., Diptera), and lace bug, bed bug, harlequin bug (which are true bugs, i.e., Hemiptera). Otherwise they are joined together, as for example, mayfly, dragonfly, sawfly, whitefly, lanternfly (which are not true "flies"), and mealybug, spittlebug (which are not true "bugs"). Names descriptive of the larval stage are often used, such as webworm, wireworm, cabbageworm, bollworm, and are written as one word. (Insect larvae may be wormlike, but they are not of course "worms" in the strict sense.)

chins, sea cucumbers); Bryozoa ("moss" animals); Annelida (segmented worms, including earthworms); Nematoda (nematodes or roundworms); various other kinds of worms; Onychophora (soft, velvety caterpillar-like animals with a pair of antennae and many pairs of legs);* Arthropoda (see below); Mollusca (shellfish, snails, sea slugs, limpets, chitons, octopuses, squids, and cuttlefish); and finally the Chordata, or Vertebrata, if we consider this a separate Phylum from some minor groups of marine animals, such as the tunicates, that have a "notochord" (which is something different from a vertebra or backbone).

The main *Classes* of vertebrates are: Reptilia (snakes, lizards, turtles); Amphibia (frogs, toads, salamanders); Mammalia (mammals, including man); Aves (birds); and Pisces (fishes).

Relatives of insects: other arthropods

The main *Classes* of arthropods or joint-legged animals are as follows:

Hexapoda (Insecta): insects
Chilopoda: centipedes ⎫
Diplopoda: millipedes ⎪
Symphyla: symphylans† ⎬ Myriapoda
Pauropoda: pauropods‡ ⎭
Arachnida: scorpions, solpugids, harvestmen (daddy longlegs), spiders, mites, ticks, king (or horseshoe) "crabs"
Crustacea: crayfish, lobsters, crabs, shrimps, sowbugs, sand hoppers, barnacles, copepods

It will be seen that some animals we often encounter in gardens, around the house, and elsewhere (centipedes, millipedes, symphylans, sowbugs, spiders, harvestmen, mites, and ticks) and normally think of as insects are not so classified (see Figs. 1 to 3). As already noted, insects have three pairs of legs and three distinct body parts. Spiders, harvestmen, mites, and ticks, on the other hand, have four pairs of legs and no more than two body parts—the abdomen and cephalothorax (a fusion of head and thorax). In mites and harvestmen, head and body are fused into one. Young mites start life with only three pairs of legs. Breathing in mites may be by means of tracheae or directly through the skin. Some mites are able to spin silk like

* They occur in rotten wood in tropical countries, and are very ancient (fossils from the oldest Cambrian); they are carnivorous and trap their prey with a jet of saliva. The group is sometimes considered as a Class under the Arthropoda.

† Symphylans are small centipede-like animals with long antennae and eleven or twelve pairs of legs; they live in humus mostly, but some eat plant roots and are garden pests.

‡ Pauropods are minute segmented animals with short two-branched antennae and eight or nine pairs of legs; they live in humus.

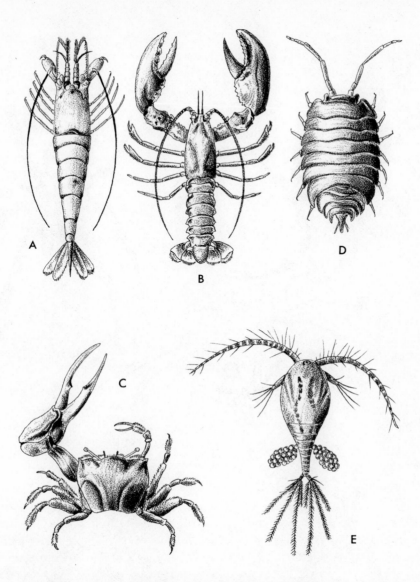

Fig. 1. Relatives of Insects: Other Arthropods

1. Crustaceans—have a hard outer shell, breathe by gills, range in size from the microscopic copepods to the eastern lobster three feet long and weighing 35 pounds. Like insects, they have antennae, an exoskeleton, and shed their skin (molt).

A, Shrimp (*Crago* sp.); *B*, Lobster (*Homarus* sp.); *C*, Fiddler crab (*Uca* sp.)—these are decapods ("ten-legged"); fresh-water crayfish (not shown), which resemble lobsters, are included in this group; *D*, Sowbug (*Onicus* sp.)—an isopod ("legs equal"), with seven pairs of legs; *E*, Copepod (*Cyclops* sp.)—having "oarlike" appendages, an important part of the microscopic life (plankton) of the ocean on which higher forms of marine animals live. Sand hoppers (not shown) also belong to the crustaceans; they are amphipods, resembling isopods, except that they are usually flattened *laterally*. It may seem strange to some that barnacles (Cirripedia) are also crustaceans.

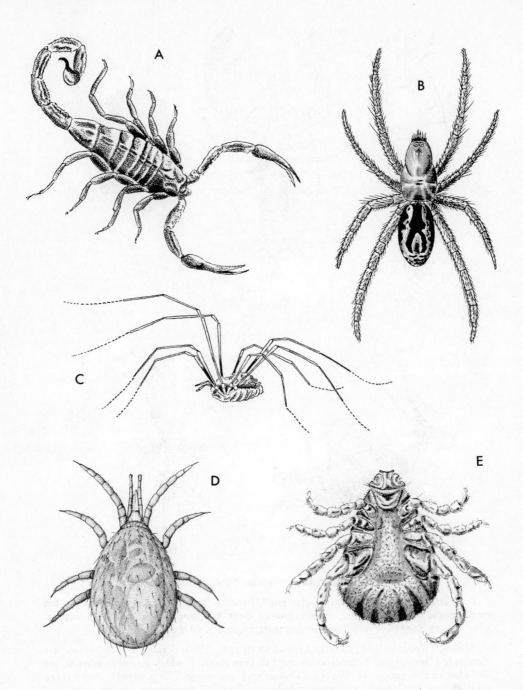

Fig. 2. Relatives of Insects: Other Arthropods

2. Arachnids—have four pairs of legs, two body parts, no antennae.

A, Scorpion; B, Spider; C, Daddy longlegs; D, Mite; E, Tick.

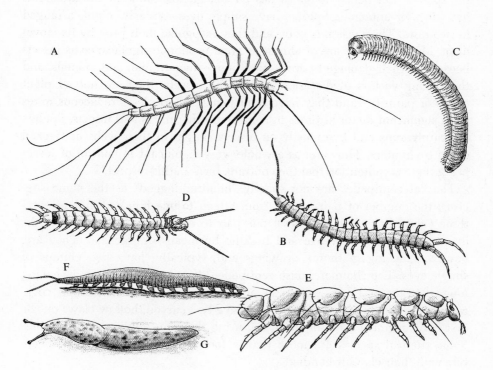

Fig. 3. Relatives of Insects: Other Arthropods

3. Chilopods (centipedes)—have a pair of legs on most body segments, a pair of claws, and a pair of antennae. Tergites, or dorsal plates, hide some segments.

4. Diplopods (millipedes)—have two pairs of legs on most body segments, a pair of antennae, no claws.

5. Symphylans—have about fifteen body segments, eleven or twelve pairs of legs, and a pair of antennae.

6. Pauropods—have eleven or twelve body segments, eight or nine pairs of legs, and (unlike other myriapods) a pair of two-branched antennae.

A, House centipede (*Scutigera forceps*); *B,* Large centipede (*Scolopendra* sp.); *C,* Millipede (*Parajulus* sp.); *D,* Symphylan (*Scutigerella immaculata*); *E,* Pauropod (*Pauropus silvaticus*).

Unrelated Forms Associated with Insects

1. Onychophorans—centipede-like, tropical.

2. Gastropods (Mollusca)—garden slugs, garden and tree snails.

F, Onychophoran (*Peripatus* sp.); *G,* Garden slug.

the spiders, but daddy longlegs has no silk glands. None of the arachnids has wings or antennae. Spiders have simple eyes—typically eight, arranged in two rows. Their vision is poor, and they recognize their prey by its movement. Mites and ticks are of almost as great economic significance as insects because of their damage to crops, their annoyance to man and animals, and their being vectors of disease. But like the insects, they are not all plant eaters or parasitic, and they have their beneficial species. Predaceous mites are a significant factor at times in control of other mites and insects. Spiders are carnivorous and live mostly on insects caught in their webs or tracked down by hunters. They are, as a whole, very significant in control of insect pests; there is contention that they outrank birds in this respect.

The flat centipedes are not exactly "hundred-legged" as the name suggests; the number of pairs varies from fifteen to one hundred and seventy-three (according to the species), one pair to each body segment. Like the insects, they have antennae and breathe by means of tracheae. They are, however, lacking in thorax or wings and, typically, have two clumps of simple eyes. The common house centipede differs from the others in having long, delicate legs (fifteen pairs) and compound eyes. It lives on cockroaches and other insects. Centipedes have two curved, hollow claws on the first body segment back of the head, using them to paralyze their victims. Some tropical species are more than a foot long and can inflict a severe bite with their claws if handled.

Millipedes ("thousand-legged") have rounded bodies and two pairs of legs for each body segment; superficially, they look like the larvae of some insects. They have two antennae like the centipedes but no poison claws and never more than one hundred and fifteen pairs of legs! These slow-moving creatures feed mostly on decaying vegetable matter but occasionally resort to plant roots. The centipedes make a fast getaway when uncovered, while the timid millipedes often curl up into a coil.

Sowbugs, or "woodlice," are land crustaceans and very successful as a group, being abundant in gardens everywhere. These animals are closely related to the marine isopods (Order: Isopoda) and, since they have gill-like breathing organs, must live in damp places—under stones or logs. They have the flexible "armor" typical of crustaceans and usually seven pairs of legs, all equal in length, which accounts for the name *iso*, meaning "equal," and *pod*, meaning "foot" or "leg." Sowbugs feed on decaying vegetable matter and often curl up into a ball when disturbed.

Other animals associated with insects

Two other "invaders" from the sea (mollusks) have successfully established themselves on land (the land snails and slugs) and have become garden pests. They are gastropods (Class: Gastropoda), which literally means "stomach foot." Gastropods, of course, are not closely related to insects, but we mention them because of their association with plants. The giant African snail is the most serious pest among the land snails. Toads are one of the few natural enemies of land slugs, whose slimy nature is distasteful to most other animals. Snails, on the other hand, have many enemies: birds, poultry, predaceous snails, ground beetles, and "fireflies" (lampyrid beetles).

Mention should also be made of nematodes, since some larger types are apt to be mistaken for immature forms of insects; also because of their close association with insects and crop plants, as well as their economic importance. They are not related, of course, to insects, as biological relationships go, but belong to the Phylum Nematoda, as indicated above. Also referred to as roundworms or eelworms,* they have some resemblance to pieces of thread; the name comes from the Greek word *nema*, meaning "thread." They range in size from microscopic forms to those of several feet in length, and though they are called "worms" and may be very small, nematodes are actually complicated animals. Their elongated cylindrical bodies, covered with a tough cuticle, are pointed, without legs or cilia and not segmented, though they are ringed (annulated). They are mostly bisexual and go through three stages of development: egg, larva, and adult, much like insects.

Nematodes are found practically everywhere; there are marine forms and others are found in fresh water and in the soil as parasites of insects or predators of their own kind. Parasitic forms are the cause of human diseases: the tropical filariasis (which is transmitted by mosquitoes), hookworm disease, and trichinosis. One of the largest nematodes invades the human intestines. The crop pests (Figs. 4A, 4B) attack all parts of the plant: stem, leaves, flowers, and roots. The relatively large "mermithid" nematodes (Fig. 4D) enter the bodies of soil-inhabiting grubs, ants, grasshoppers, aphids, and other insects, and cause fatal injury as they emerge from their hosts. Tiny forms called "neoaplectanid" nematodes (Fig. 4C) attack such crop pests as the cabbage root maggot, Colorado potato beetle, European corn borer, codling moth, Japanese beetle, and tobacco budworm and can be sprayed on plants with a high-powered spray gun like an insecticide. These nematodes, which have a very complex life cycle, harbor bacteria that are fatal to the insects. They are, thus, not true parasites but saprophytes living on dead insects. The study of these interesting animals belongs to the science of nematology.

*"Eelworm" is applied only to the plant-parasitic and free-living forms.

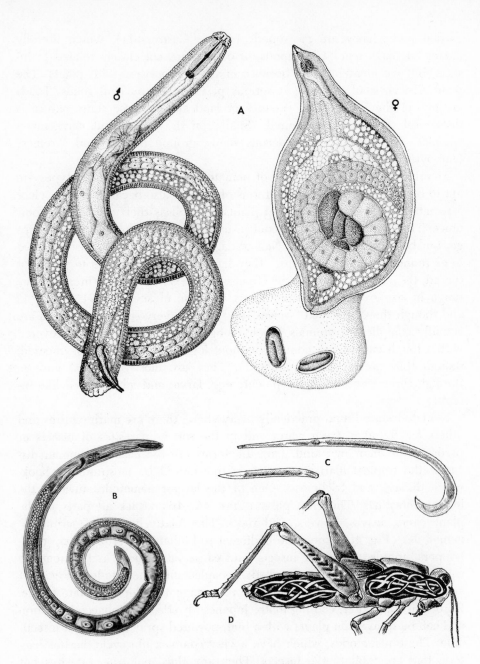

Fig. 4. Unrelated Forms Associated with Insects

3. Nematodes—unsegmented round worms, microscopic to several feet long.

A, Sugar beet nematode, *Heterodera* sp., microscopic in size; B, *Hoplolaimus* sp., plant parasitic form, microscopic in size; C, Larval and adult stages of *Neoaplectana glaseri*, semiparasitic on Japanese beetle grubs; D, Mermithid nematode, a large parasitic form in body of grasshopper, about to emerge.

The number of species of animals

How the various groups of animals compare in number of species will be seen in Fig. 5. These estimates of the number of species of insects are admittedly low. The new species being recorded grows constantly, and there are undoubtedly many thousands to be described: the British Museum (Natural History) estimates that approximately 1,000 new species of insects are being described each year. This field is scarcely touched in the tropics, where insects and other forms of life are most abundant and where remote regions remain unexplored. Estimates of the probable number of species of insects vary anywhere from 2 million to 10 million. It is believed that more than three-quarters of the butterflies and moths have been described but probably less than one-quarter of the Hymenoptera and Diptera, and no more than one-third of the Coleoptera. The number of described species of insects for North America, north of Mexico, is estimated to be about 90,000.

How insects are born and grow

One of the distinctive features of insects is the phenomenon called *metamorphosis*. The term is a combination of two Greek words: *meta,* meaning "change," and *morphē,* meaning "form." It is commonly defined as a marked or abrupt change in form, structure, and habit and refers to all stages of development. In the case of *complete* or *complex metamorphosis* (Fig. 6), the forms are: egg, *larva, pupa,* and adult; in *gradual* or *simple metamorphosis* they are: egg, *nymph,* and adult. The more primitive orders —Protura, Diplura, Thysanura, and Collembola—are "without" metamorphosis; the young grow in size but not in form.* The amphibians—frogs, toads, and salamanders—go through a metamorphosis similar to that of insects but without the pupal stage. As everyone knows, frogs and toads spend their larval stage as free-swimming tadpoles (polliwogs); the gill-breathing larval stage of salamanders is spent inside the egg. Many aquatic and marine animals also go through a "larval" period.

All life begins with a single cell; this is the egg, except in the lower forms of animals which multiply by division or other means. In some insects, fertilization of the egg by sperm is not necessary for reproduction. Generally, development within the egg is not considered when discussing metamorphosis. These changes cannot be viewed, of course, without the microscope, and they constitute another study, embryology. It should be borne in mind that the larva is perfectly formed within the mature egg and may even have acquired its color. In higher animals, the most important development takes

* See footnote, p. 39.

Estimated number of species of all animals

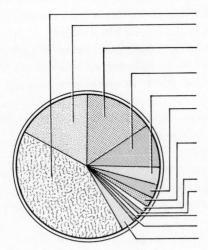

GROUP:	COMMON NAMES:	NUMBER OF SPECIES:
HEXAPODA (INSECTA)	Insects	900,000
OTHER **ARTHROPODA**	Spiders, Centipedes, Crawfish, etc.	52,000
MOLLUSCA	Clams and other Shellfish, Snails	82,000
CHORDATA	Mammals, Birds, Fish, Reptiles, etc.	38,000
ALL OTHER **ANIMALS**	Sponges, Corals, Worms, etc.	48,000
		1,120,000

Estimated number of described species of insects

COLEOPTERA:	Beetles	300,000
LEPIDOPTERA:	Butterflies, Skippers, Moths	120,000
HYMENOPTERA:	Bees, Ants, Wasps, Sawflies, Horntails	100,000
DIPTERA:	Flies, Mosquitoes, Gnats, Midges	78,000
HEMIPTERA:	The True Bugs	31,000
HOMOPTERA:	Scales, Mealybugs, Aphids, Cicadas, Leafhoppers	26,500
ORTHOPTERA:	Roaches, Crickets, Grasshoppers, Katydids, Mantids	20,000
ODONATA:	Dragonflies, Damselflies	5,000
NEUROPTERA:	Antlions, Dobsonflies, Alderflies, Lacewings	4,000
TRICHOPTERA:	Caddisflies	3,000
COLLEMBOLA:	Springtails	3,000
OTHERS:		13,000
		703,500

Fig. 5. Number of Species of Animals

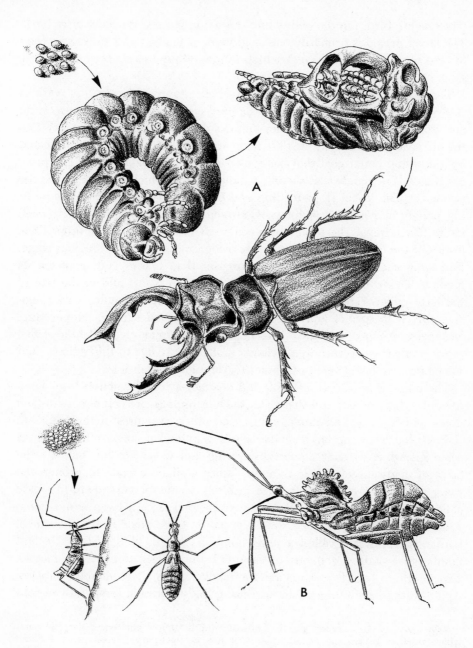

Fig. 6. Metamorphosis of Insects

A, Complete or complex metamorphosis of a stag beetle—egg, larva, pupa, adult; *B,* Gradual or incomplete metamorphosis of the wheel bug—egg, young nymph, older nymph, adult.

place *before* birth (in the embryonic stage); in insects, it occurs *after* birth. The larval period is primarily one of growth, of feeding and storing up food for the pupal and adult stages which follow. Many insects feed very little, some not at all, during their adult lives.

The young insect—larva or nymph—sheds its coat (molts) at various stages of growth, since it outgrows the hard covering or *cuticula*. Insects do not grow gradually as many other animals do. They grow by stages. When the old coat gets too tight, it splits open and the insect crawls out, protected by a new and larger coat that has grown underneath the old one. The process is aided by glands that secrete a molting fluid which helps separate the two coats. The insect stops feeding and becomes inactive during this molting period, and is quite helpless. Conspicuous changes occur during each molt. The form of the insect between each molt is called an *instar*. The instars of some Lepidoptera are hardly recognizable from the previous stage. Following each molt the insect is hungrier than before and increases its feeding. This is noticeable in the damage done to plants and in the size of the fecal pellets. The number of instars, or frequency of molts, varies considerably with species and to some extent with food supply, temperature, and moisture (thus, latitude). The Lepidoptera commonly molt four or five times, while the seventeen-year cicada molts twenty-five to thirty times, and some primitive insects molt continuously throughout their lives.

The pupal stage is one of profound change—a transformation from larva to adult. Many tissues and structures, such as prolegs, are completely broken down and true legs, antennae, wings, and other structures of the adult are formed. As when the larva molts, this is a quiescent stage.* The pupa in some species is without a special covering and is said to be "naked"; the larva of many insects, such as moths, spins a silken *cocoon* to protect the pupa, while that of the butterflies fashions a smooth parchment-like case called a *chrysalis* which is often seen suspended from a twig or overhang. Among the higher forms of Diptera, pupation takes place in the thickened, hardened larval skin called a *puparium*. The length of the life cycle and number of broods or generations vary with the species and climatic factors. Two or three broods is common for many insects in temperate climates. Aphids often have ten generations during the summer in temperate regions

---

* Discovery of the so-called juvenile hormone and the steroid hormone called ecdysone, which together govern the metamorphosis of insects, opens up interesting possibilities for control of insects through manipulation of these factors. It has been shown that the juvenile hormone "must be absent during certain stages of insect life or normal development will not occur." Contact with the substance (in infinitesimal amounts) at this time will prevent the insect from completing its normal development. Ecdysone, which regulates the molting of larvae and pupae, has been produced synthetically and used experimentally to cause immature larvae to develop precociously into pupae and to terminate diapause in pupae. This abnormal development leads to premature death of the insect.

and many more during the season in southern regions. The shortest life cycle for an insect is ten days, and the longest is seventeen years!

Hibernation takes place during the winter season. This is a period of *diapause*, when body temperature[*] is greatly reduced and metabolic processes are almost at a standstill; it may occur in the egg, larval, pupal, or adult stage, according to species. It is the insect's way of adjusting to the cold and dwindling food supply, and its response to certain cyclical changes. Recent research indicates that hibernation is triggered by the length of day. Many insects start preparing for the winter before the end of summer, and this is not primarily due to changes in temperature but to seasonal variations in the amount of daylight. Silkworms, which normally hibernate as eggs, were kept under constant but short periods of light (12 hours per day). As long as the eggs and larvae were kept under these conditions egg laying continued for generation after generation without interruption (diapause). Under long periods of light the moths laid eggs that went into diapause and remained there indefinitely. The influence of the light exerts itself on the moth and is transmitted in some subtle way to the egg. It is believed to do this through a "diapause hormone."[†]

The adult insect does not grow in the usual sense, but it does change. Many insects are not mature when they emerge from the pupa. The adult period is primarily one of reproduction and is sometimes of short duration. The mature adult is called an *imago*, the Latin for "likeness" or "image." Their food is often entirely different from that of the larval stage, as in the Lepidoptera and Hymenoptera. In the Coleoptera, the largest order, the food of both stages is commonly the same. Insects normally reproduce by means of fertilized eggs and mate by coition. The *genitalia* (external reproductive organs) are often an important means of identifying species. The

---

[*] Insects are "cold-blooded" and do not have the ability to regulate their body temperature as do the so-called warm-blooded animals (birds and mammals); the insect's body temperature varies with the environmental temperature. Collectively, bees behave like warm-blooded animals in controlling hive temperature.

[†] Diapause in the fall webworm, which hibernates in the pupal stage, is induced in the larval stage, as in many other insects. Recent studies of this insect show that humidity, quality of food, temperature, and the length of day (photoperiod) are all factors in determining *when* diapause occurs. It has been shown that diapause in the pink bollworm and in the sugarcane borer, which hibernate in the larval stage, is controlled primarily by the length of day. It is believed that these insects, like the Colorado potato beetle, are limited to certain geographical regions by their response to photoperiod. The boll weevil, on the other hand, has been able to adjust to conditions on the high plains of Texas by entering diapause considerably earlier in the fall than it does in central Texas, thus avoiding death from an earlier freeze date. See R. F. Morris, "Factors Inducing Diapause in *Hyphantria cunea*," *Canad. Ent.*, 99(5):522–529 (1967); M. J. Lukefahr et al., *Factors Inducing Diapause in the Pink Bollworm*, Tech. Bul. 1304, U.S. Dept. Agr. (1964); W. L. Sterling and P. L. Adkisson, *Differences in the Diapause Response in Boll Weevils from the High Plains and Central Texas* . . . , Bul. 1047, Agr. Exp. Sta., Texas A&M Univ., College Station (1966). The classic study of this phenomenon in the silkworm was made by Makita Kogure in Japan in 1933.

female stores the sperm in a pouch called the *spermatheca* and fertilizes the eggs as they are laid. Many female insects have an *ovipositor* protruding from the end of the abdomen for thrusting eggs into the ground (grasshoppers), into tissues of plants (sawflies), or into the bodies of other insects (parasitic wasps). In the parasitic Hymenoptera, in which the ovipositor reaches the highest development, it is a fine hollow tube resembling a hypodermic needle and is used for depositing eggs, injecting paralyzing fluid, or puncturing the host to secure body fluids as food. Beetles, moths, and flies (excepting some tachinid parasites) do not have an ovipositor. That of bees, nonparasitic wasps, and ants has lost its original function and is strictly a stinging or offensive weapon. Because of its origin the stinger is possessed by females only.

Some insects can reproduce normal offspring without fertilization of the egg—by a type of reproduction called *parthenogenesis*. Bees, ants, wasps, and aphids are notable examples. In the honey bee colony, workers (females) and potential queens are produced from fertilized eggs, the drones (males) from unfertilized eggs. In ant colonies, queens determine the sex of their progeny the same way that queen bees do. Some worker ants also reproduce by parthenogenesis; they usually produce males only and are not known to mate. Among the parasitic Hymenoptera, females are generally produced from fertilized eggs and the males from unfertilized eggs. In a few species the females can reproduce their own kind asexually (without mating) for many generations or indefinitely. Aphids commonly reproduce for several generations during the summer by means of parthenogenesis; the young are born alive and may contain within them the embryos of the next generation. In temperate climates the early generations of aphids are mostly female but later sexual winged forms of both sexes usually appear and migrate to other plants. Here they mate and the females lay fertilized eggs which go into diapause for the winter. A rare variant of parthenogenesis, called *paedogenesis*, is found in some gall midges; the larvae or pupae in this case produce living young. Another form of reproduction, called *polyembryony*, is found in some parasitic Hymenoptera. A fertilized egg divides repeatedly, forming many individual embryos. In some species several hundred adult wasps may emerge from a parasitized larvae in which a single egg was deposited.

Predators and parasites

Not all insects are phytophagous (plant eaters) or scavengers. A very considerable proportion of them, called entomophagous insects, feed on phytophagous or other types of insects. They do not resort to plants if their prey is scarce. Adult parasitic wasps live mostly on pollen but do not eat plants. A rare case of mixed feeding habits is found in the blister beetles.

The adults eat plants, but the larvae live in the soil or on plants and feed on other insects or their eggs. The phenomenon of predation (when an insect eats other insects, mostly pests), as is notable among the Coleoptera, is familiar to most people. That of parasitism is not generally understood, since it is much less obvious. The female parasite usually deposits her eggs in or on the victims, and more commonly the larval forms are attacked. The larvae hatching from the eggs consume the host *gradually* so as to keep their food supply alive until their own development is complete or nearly so. Only one host is necessary to sustain the parasitic larva in contrast to the predator, which generally requires more than one to complete its development. This parasitic trait is most prominent among the Hymenoptera, but it is also found among the Diptera, notably the family of Tachinidae. Both of these traits have great significance as factors in "natural control," or suppression of insect pests, and as agents of "biological control."

Three major kinds of biological control techniques have been developed to supplement natural control and to replace or supplement chemical control: "introduction" of potentially useful parasites and predators from other countries or areas—long-range projects which are being carried on by some universities and departments of agriculture; "inundation" or mass rearing and release over limited areas of natural enemies of insect pests, with repeated applications required as in the case of chemical insecticides—agents provided by commercial companies or growers' cooperatives; "conservation" or altering the environment to increase the number and effectiveness of native and introduced beneficial species—a field undergoing study but still in its infancy.

The structure of insects

While the adult insect's body is made up of three parts (head, thorax, and abdomen), the division is not always obvious. The head and thorax are nearly always distinct (a notable exception is in the scale insects), but the division between thorax and abdomen is obscure in several groups. This is due to an overlapping of a segment of the thorax and the first body segment, or even several. Grasshoppers, cicadas, treehoppers, and some of the Hemiptera, such as stink bugs, are examples. The *pronotum* (upper side of the prothorax) of stink bugs forms a large plate that overlaps the abdomen, giving the two parts a solid appearance and resemblance to a shield. In the treehoppers the pronotum overlaps in a similar way and is humped, sometimes giving them an odd angular shape and the appearance of the thorny stem projection on certain plants.

An insect's body is not supported by a bony skeleton but by a tough body wall or *exoskeleton*. The tough covering of skin is referred to as the *cuticle* or cuticula (some writers refer to the insect covering as the *integument,* as

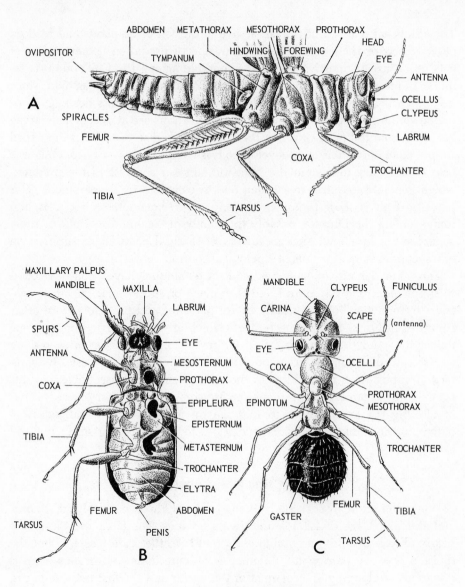

Fig. 7. Structure and Parts of Insects

A, Grasshopper (side view); *B*, Beetle (underside); *C*, Ant (top view).

in the case of plants). The cuticle contains a layer of wax which determines its permeability to water (even to insecticides) and prevents desiccation or drying up. It also contains chitin and other substances, principally a protein called arthropodin. Chitin is not soluble in water, dilute acids, alkalis, or organic solvents. The relative proportions of these substances determine the hardness and resistance of the skin. The cuticle of each segment is formed

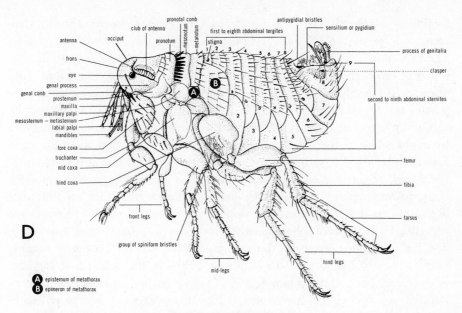

Fig. 7D. Flea (side view).

into several hardened plates called *sclerites*, separated by infolds or *sutures* which give it flexibility. The cuticle of the larva is not usually as heavily sclerotized as that of the nymphs and adults.

Segmentation is not peculiar to insects. The other arthropods are segmented, as are the annelids, such as earthworms. The difference is that the arthropods have legs. The insect's body is made up of twenty to twenty-one segments, if we count the six head segments which have become fused into a single capsule (see Fig. 7). The *head* structures will be described later. The body segments have four areas whose names are: *notum* or *tergum* (the dorsal surface or top); the *sternum* (the ventral surface or underside); the *pleura* or sides. The *thorax* is made up of three segments: *prothorax, mesothorax*, and *metathorax*.* Each of these segments bears a pair of *legs*. The *wings* are attached to the mesothorax and metathorax, never to the prothorax or first segment. The Diptera have only one pair of wings, attached to the mesothorax or middle segment; knoblike rudimentary wings called *halteres* appear on the metathorax and are believed to have a balancing function. The thickened front wings of beetles, called *elytra*, serve as a protective covering.

The insect's leg consists of five independent, movable parts (see Fig. 7).

---

* Applying this terminology to the thoracic segments, the following terms are used: *pronotum* for top of prothorax, *prosternum* for the underside; *mesonotum* for top of mesothorax, *mesosternum* for the underside; *metanotum* for top of metathorax, *metasternum* for the underside.

Beginning at the thorax, they are: *coxa* (pl., *coxae*), *trochanter* (pl., *trochanters*), *femur* (pl., *femora*), *tibia* (pl., *tibiae*), and *tarsus* (pl., *tarsi*). The tarsus or "foot" has from two to five segments; the last one usually has a pair of *tarsal claws,* with pads called *pulvilli.* The tarsi of the larvae more often have a single claw. Claws are regarded by some writers as a separate part of the leg, the *pretarsus.* Some insect legs are adapted for leaping by having enlarged femora, as in the grasshoppers and beetles; or for digging by modification of the tibia and tarsus, which act as a scoop, as in the mole cricket (see Fig. 8). Others are modified for grasping and are called *raptorial* legs. This is a predatory adaptation, as in the praying mantid, ambush bugs, waterscorpions, and mantispids. The femur and tibia on the front legs are armed with spines that oppose one another when closed. Some parasitic Hymenoptera (*Brachymeria,* for example) have hind legs modified for grasping their prey during oviposition. Honey bees and bumble bees have a "pollen brush"—rows of stiff hairs—on the broad first segment of the tarsus for gathering honey, and a "pollen basket"—a smooth concavity with fringe of long stiff hairs—on the hind tibia for carrying the pollen to the nest. The tibiae of many insects have sharply pointed spurs; true butterflies have one pair, skippers usually have two, bumble bees have one or two.

Insects use their legs in an efficient manner. In walking, they create what essentially amounts to a tripod effect, by moving the middle leg on one side of the body in unison with the front and hind legs on the other side. Thus, with three feet normally on the ground, the insect has greater stability than most other walking animals. *Prolegs* (fleshy body projections) occur only in the larvae and are used for clinging to plants (see Fig. 9). The larvae of the Diptera are usually without legs of any kind. Beetle larvae (grubs) never have a series of prolegs like caterpillars; at best, they have only one pair at the tip of the body. Prolegs of sawfly (Hymenoptera) larvae outnumber those of the Lepidoptera (six to eight pairs in all, compared with three to five); only the latter have *crochets,* or hooks.

The *abdomen* may have eleven or twelve segments, but in most cases they are difficult to distinguish. Some insects have a pair of appendages called *cerci* at the tip of the abdomen. They may be short, as in the grasshoppers, termites, and cockroaches, or extremely long, as in the mayflies (the third or middle "tail" is not a cercus), or curved, as in the earwigs. The male reproductive organs are on the ninth abdominal segment and sometimes have a pair of claspers associated with them. The female reproductive organs are between the seventh and eighth segments and sometimes associated with an ovipositor, as already described.

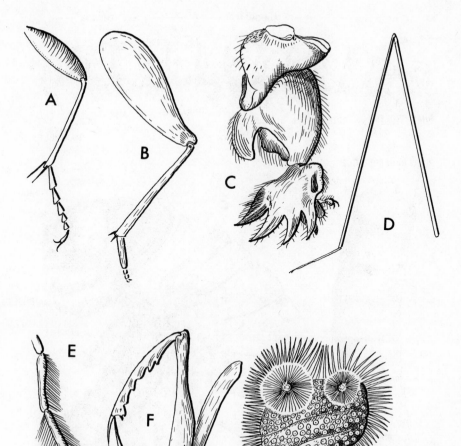

Fig. 8. Leg Adaptations of Some Insects

A, Running (ground beetle); *B*, Jumping (cricket); *C*, Digging (mole cricket); *D*, Walking on grass (walkingstick); *E*, Swimming (whirligig beetle); *F*, Grasping (praying mantid); *G*, Hanging onto hairs (louse); *H*, Suction cups (diving beetle).

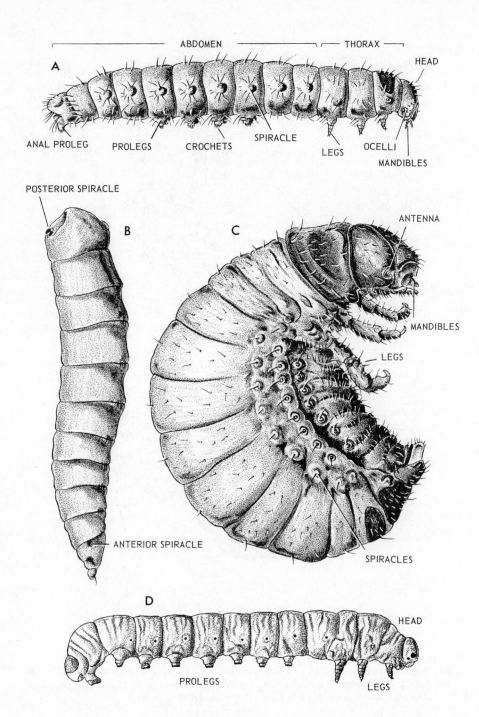

Fig. 9. Common Larval Forms of Insects

A, Caterpillar (butterfly); *B*, Maggot (fly); *C*, Grub (beetle); *D*, Larva of a sawfly.

Insects are the only invertebrates having wings. Thus, if it has wings and no backbone, it is an insect. Not all insects have wings, however. The Thysanura, Collembola, and other primitive orders, as well as the Anoplura, Mallophaga, and Siphonaptera, are entirely without them. In other orders wingless forms are scattered.* Usually it is the female that cannot fly, seldom the male. The wing actually has two surfaces, being an expansion of the body wall during its development. This can sometimes be seen when the insect is emerging from the pupa and the wings are still sacs. Gradually they are spread out and the two surfaces fuse together. The veins, which give them structural support, also contain the tracheae and nerves and, indeed, *are* veins which conduct blood, despite many contrary statements. Arnold and others have shown that blood circulates in the veins of all winged insects. In the geometrid moths, blood in the forewings circulates in a peculiar manner; the tracheae are suspended centrally in certain veins, and blood circulates in opposite directions on the two sides of the tracheae.

Venation, or arrangement of the veins of the wings, is different for each species of insect. Thus, it serves as a means of identification, and systems have been devised to designate the venation for descriptive purposes (see Fig. 10). Wing surfaces are covered with fine hairs or scales, or they may be bare. It will be noted that the names of the orders in the majority of cases end in *-ptera*, which comes from the Greek word meaning "wing." Thus, each of these names denotes some feature of the wings. Hemiptera means "half-winged"; Hymenoptera means "membrane-winged"; Diptera means "two-winged," and so on. Siphonaptera is a combination (Siphonaptera) from words meaning "siphon" or tube and "without wings." Apterous insects are those without wings; when apterous and winged forms occur in the same species, the latter are said to be *alate* (or are called alates).

ANTENNAE AND MOUTHPARTS

The main feature of the insect's head are the eyes, antennae, and mouthparts. The eyes are described later. The antennae are a prominent and distinctive feature of insects, and one pair is always present on the adult's head (except in the Protura and often in the scale insects); they are usually located between or in front of the compound eyes. Antennae are segmented, vary greatly in form and complexity (as will be seen in Fig. 11), and are often referred to as "horns" or "feelers," which is misleading. They are primarily organs of smell but serve other functions in some insects.

* The recent discovery of a winged fossil insect about as old as the oldest known wingless fossil suggests that the wingless insects may not be older, as now generally believed, than some winged forms.

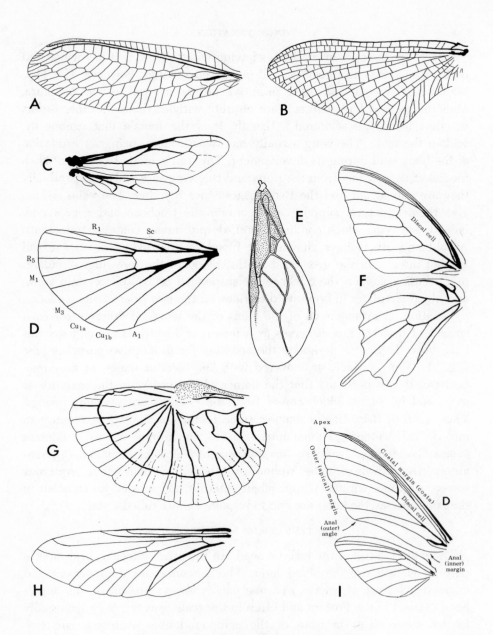

Fig. 10. Some Insect Wings Showing Venation

A, Green lacewing (*Chrysopa* sp.)—forewing; B, Mayfly—forewing; C, Hornet—forewing and hindwing; D, Tent caterpillar (*Malacosoma* sp.)—forewing, indicating subcosta (Sc), radius (R), median (M), cubitus (Cu), and anal vein (A); E, Reduviid bug—elytron (forewing); F, Swallowtail butterfly—forewing and hindwing; G, Earwig—hindwing; H, Horse fly; I, Hawk moth—forewing and hindwing, indicating margins, apex, and discal cell.

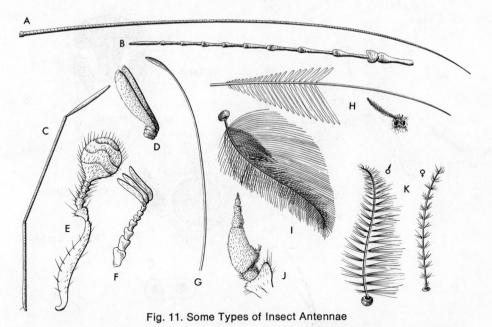

Fig. 11. Some Types of Insect Antennae

A, Cockroach (ciliate); *B*, Longhorn beetle; *C*, Assassin bug (filiform); *D*, Stylopid; *E*, Bark beetle (capitate); *F*, Scarab beetle (lamellate); *G*, Butterfly (clubbed); *H*, Moth (bipectinate); *I*, Chironomid fly (plumose); *J*, Horse fly; *K*, Mosquito: male (left) showing Johnson's organ, female (right).

Ants use them to communicate with one another, and for male mosquitoes they are sound receptors used to locate the females. The fleshy antennae of the larvae of swallowtail butterflies are believed to be tactile organs. Antennae of immature insects are generally much shorter than in the adults and are sometimes lacking.

The most remarkable structural feature of insects and the most complicated is the mouth. There are many variations in form and function of the mouthparts, but they fall into two basic types: those for chewing, or the *mandibulate* type, and those for sucking, the *haustellate* type (see Fig. 12). The chewing type is the more primitive and is better developed. While they differ considerably in appearance, the basic parts are generally found in the sucking types, though they may be greatly modified.

Typically, there are three pairs of mouthparts; one moves in a vertical plane and two move horizontally. Those operating vertically are the *labrum*, or upper lip, and the *labium*, or lower lip. The labrum, a single plate, is not really a mouthpart but a sclerite of the head. It assists in guiding the food into the mouth. Attached to the inner surface of the labrum is the *epipharynx*, which is believed to correspond to the palate

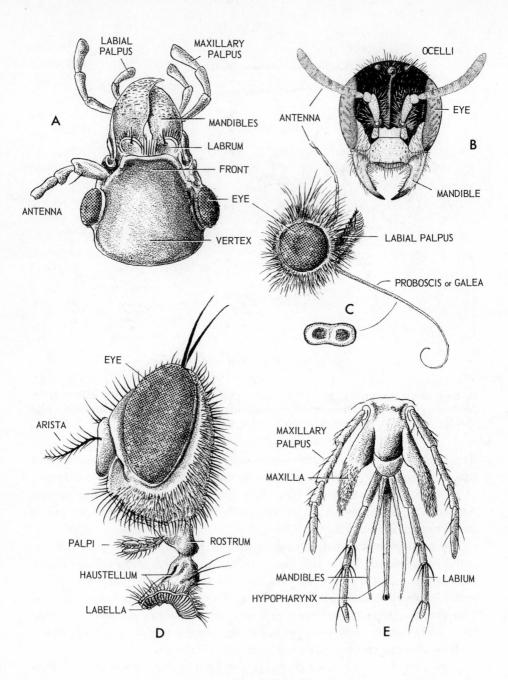

Fig. 12. Some Types of Insect Mouthparts

A, Chewing (beetle—top view); *B*, Chewing-lapping (bee—front view, with proboscis retracted); *C*, Siphoning (butterfly—side view, with proboscis uncoiled); *D*, Sponging (house fly—side view); *E*, Piercing-sucking (flea—front view).

of higher animals and to serve as an organ of taste. On the inner surface of the labium, where the salivary glands open, is the *hypopharynx;* it serves as an organ of taste and is said to be the "true tongue." Associated with the labium is a pair of appendages called the *labial palpi* (or palps), one on each side of the mouth. Operating on a horizontal plane are the upper jaws or *mandibles,* the principal organ for chewing, tearing, and biting; they usually have teeth. The lower jaws, called *maxillae,* are a more intricate structure, operating horizontally like the mandibles; they assist to some extent in chewing but are mainly for holding the food when the mandibles are open. Associated with the lower jaws are the *maxillary palpi* (or palps); they are longer than the labial palpi, both of which might be mistaken for antennae. The maxillary palpi have tactile hairs and possibly organs of taste or smell.

There are intermediate types of mouthparts: *rasping-sucking,* as found in the thrips, and *chewing-lapping,* as found in honey bees, wasps, and bumble bees. Sucking types are greatly varied. *Piercing-sucking* mouthparts are typical of the Hemiptera (bugs), Homoptera (aphids, scales, mealybugs), bloodsucking lice, fleas, mosquitoes, and the so-called biting flies. This type of mouthpart is characterized by a tubular, jointed beak, with needle-like stylets. Inside the outer tube is a smaller tube for sucking up the juices, and the stylets, used for piercing. The outer tube is the labium, and the stylets and inner tube have been formed by elongations of the mandibles and maxillae. In the *sponging* types, as found in the house fly, the labium forms a retractile proboscis with a spongelike organ on the end; there are only two stylets (incapable of piercing the skin), no mandibles or maxillae, except the maxillary palpi. In the *siphoning* types, as seen in the butterflies and moths, the mandibles are wanting and the labium and maxillary palpi greatly reduced. The retractile proboscis is formed largely by the fusing of the two maxillae into a tube. The proboscis is not a piercing instrument but is used to suck up liquids.

The mouthparts of the immature insects tend to be more varied than in the adults. In the nymphs they are similar to those of the adults. Larval forms generally have the mandibulate types regardless of the kind possessed by the adults. In the Lepidoptera the types are the two extremes: mandibulate in the caterpillars and haustellate in the adults.

The senses of insects

The nervous system of insects is made up of *ganglia* (small, white masses of nerve cells) distributed throughout the body and "wired" together, as it were, by cords called *connectives.* Nerve fibers extend from each ganglion to the sensory organs (eyes, antennae, etc.) and to other organs and appendages of the body. Typically, there are two ganglia (right and left) in each body segment. They lie in the body cavity along the underside of the body (not down the back as in vertebrates). Two larger masses of ganglia appear in the head, one below the esophagus and one above. The latter is referred to as the "brain," but it does not have the importance of the brain in higher animals. Each pair of ganglia works more or less independently of the others, controlling movements of the appendages or organs associated with it. It is a simpler, more direct process, without the elaborate associating, selecting, and coordinating involved in the mental processes of higher animals.

The organs of sound reception (hearing) in insects, where they occur, range from a simple single-celled receptor to the more elaborate *tympana* of grasshoppers and cicadas. The tympanum (see Fig. 13A) is a thin membranous area in the body wall, connected to a more or less complicated structure called the *chordotonal organs.* These organs translate the vibrations to impulses that are carried by nerves to ganglia in the thorax; appropriate tympanal organs occur in pairs. The most conspicuous are those of the grasshoppers: two oval plates, one on each side of the first abdominal segment (see Fig. 7). In katydids, crickets, and termites they are on the tibiae of the front legs; those of the katydids are oval and fairly easy to see; in the crickets and termites they appear as small slits. Tympana are fairly well developed in the Lepidoptera and generally located in the mesothorax; in the geometrids they are in the first abdominal segment. The abdominal tympana and associated structure in cicadas are among the most elaborate. The "Johnson's organs" in the antennae of male mosquitoes are an interesting form of sound receptor. The males orient themselves with dead reckoning toward the buzzing females by adjusting their direction of flight to bring the delicately vibrating antennae in phase with one another. The cerci of crickets and cockroaches are thought to have a sound reception function also.

Insects show a wide range of sound reception—about thirteen octaves— often in frequencies beyond the range of the human ear (more than 20,000 cycles per second). Moths respond quickly to the echo-locating sounds emitted by bats—in a range completely inaudible to humans—and can be repelled by a mechanical reproduction of the sound.

The insect's sense of touch comes from sense hairs or *setae* which are scattered over the surface of the body, and possibly from the palpi. Organs of taste are located in the mouthparts as one would expect, chiefly in the palpi. Some Hymenoptera taste with their antennae, or at least distinguish sweet from sour. Certain butterflies are known to have organs of "taste" on the tarsi of the hind legs; when the feet touch a flower, the promise of

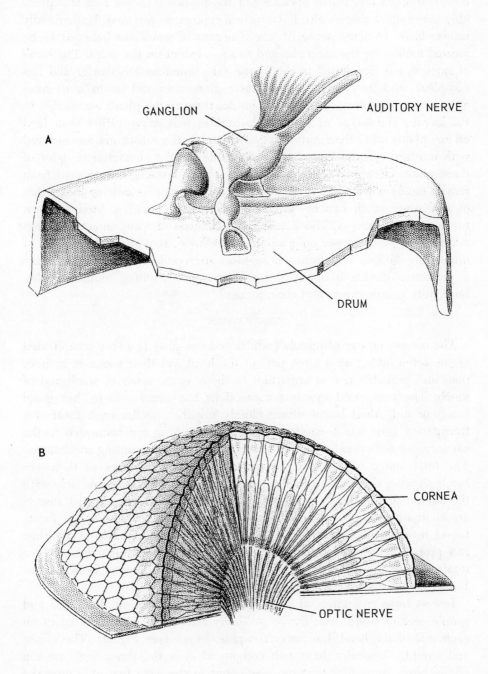

A

GANGLION

AUDITORY NERVE

DRUM

B

CORNEA

OPTIC NERVE

Fig. 13.

A, Eardrum (tympanum) of a grasshopper or locust (After E. G. Gray); *B*, Compound eye of a bee (After Karl von Frisch).

nectar prompts a reflexive uncoiling of the proboscis. Some flies and probably many other insects also have taste receptors on the tarsi. Undoubtedly insects have an acute sense of smell; organs of smell are believed to be located mainly on the antennae and to some extent on the palpi. The sense of smell is less developed in the larvae (the antennae are shorter and less complex), and they react more to taste. Butterflies and moths must have a discriminating sense of smell that guides them to the plants acceptable to the larvae. The larvae of cabbage butterflies will starve rather than feed on any plants other than certain crucifers, unless the substitutes are smeared with mustard oil—the ingredient responsible for the characteristic odor of these plants. Extremely minute amounts of sex attractants extracted from female moths will attract the male from a distance of a mile or more; that of the gypsy moth can be chemically synthesized and is used in traps for population surveys. The defensive mechanism of many insects employs odors; some worker ants emit volatile substances that warn the colony of impending danger. Secretions whose odors are repellent to their enemies are a protective device in many different groups, including moths, beetles, bugs, ants, grasshoppers, and cockroaches.

INSECT VISION

The compound eye of insects (adults and nymphs) is a very complicated organ, often taking up a large part of the head, yet their vision is, at best, poor and probably not as important to them as the sense of smell and of touch. The compound eye is not one lens but many minute, hexagonal facets or individual lenses fitting closely together. Under each facet is a transparent cone which conducts the light to a long rod connected to the surrounding retinal cells (see Fig. 13B). There is no focusing mechanism. The total image is thus a combination of as many images as there are facets sighting the object. Since the number of images varies inversely with the square of the distance from the object, it would appear that insects are short-sighted. It is believed that an insect with the number of facets found in a butterfly can see a clear image two to three feet away; they can probably see movement much farther than this. Dragonflies have a total of more than 50,000 facets in both eyes. A swallowtail butterfly has 17,000 in each eye and a house fly about 4,000.

Insects have two kinds of eyes: the *compound eye* just described and simple eyes or *ocelli*. The larvae generally have from one to six ocelli on each side of the head, but never compound eyes (see Fig. 9). The adults and nymphs, typically, have two compound eyes and three ocelli on top of the head. Butterflies have no ocelli, but moths have two—one near the base of each antenna. Many other insects have no ocelli, as in the case of the beetles. These simple eyes of insects are sensitive to light but apparently do not produce an image. Many insects perceive color, including parts of the spectrum that humans do not see. Black light, toward the ultraviolet portion of the spectrum and not visible to us, is quite effective in insect traps, especially against certain moths. The sensations recorded by insects are not to be judged in terms of those recorded by humans. What an insect sees, hears, smells, or tastes is, no doubt, entirely different from what is conveyed to us by the same sensation.

Respiration, circulation, digestion

Insects do not breathe by means of lungs. Oxygen is supplied to the tissues through a system of tubes called *tracheae* and finer branches called *tracheoles*. The tracheae open to the outside through *spiracles* or openings appearing in pairs on various segments of the body (see Fig. 9). Commonly, there is one pair each on the mesothorax, metathorax, and the first eight body segments, or ten in all. Many insects have fewer spiracles than this. Mosquitoes, for example, have only one pair on the eighth abdominal segment. In the higher animals the lungs serve to purify the blood and provide it with oxygen, which is taken up by the hemoglobin in the red corpuscles and carried to the tissues. There is no hemoglobin in the insect's blood, which ordinarily has nothing to do with respiration. Some aquatic forms—certain caddisflies, pyralid moths, and midges—are among the very few exceptions. Their blood contains hemoglobin which extracts oxygen from the water through *blood gills,* like fish. Other aquatic insects (the immature forms) breathe by means of *tracheal gills.* These are membranous expansions of the body wall with numerous tracheoles which take up oxygen from the water and distribute it to the tissues in the normal manner.

There is no elaborate circulatory system in insects. The blood, a clear fluid (sometimes with a greenish or yellowish tinge), simply flows through the open spaces in the insect body. The *heart* is a simple organ (a long tube closed at one end and open at the other) with several valves. It lies along the middle of the back just under the body wall. Muscles cause it to expand and contract, drawing the blood in and forcing it out, and circulating it through the spaces. The main function of the blood is to carry food materials to the various organs and pick up waste material for excretion.

The digestive system of insects consists of an *alimentary canal* within the main body cavity, or *haemocoele,* and extends from mouth to *anus,* which opens on the twelfth abdominal segment. Its length and complexity vary greatly, according to the food eaten. Digested materials are absorbed into the surrounding blood, and waste materials are *egested* from the anus.

Defensive mechanisms

In the body wall of many insects are glands that secrete defensive substances (such as cause the disagreeable odor of stink bugs or of earwigs) which are repellent and sometimes toxic to other insects. One or several glands may be located in the head or along the sides of the thorax and abdomen. They are located near the surface and connected to saclike reservoirs. The secretion is often forced out by muscular contractions and hydrostatic pressure or a combination of both. In some cases the reservoirs are everted (turned inside out), exposing the volatile secretion in droplets on the skin. Some lady beetles exude droplets of blood from leg joints by muscular contractions. The attacker gets trapped in the sticky secretion, which hardens on exposure to air. A similar secretion is exuded from glands on the sides of aphids (not to be confused with the anal honeydew secretions desired by ants). The larva of the swallowtail butterfly has an eversible gland—they are called *osmeteria*—that protrudes from the prothorax (just back of the head) like a two-pronged fork when the insect is alarmed. The defensive secretions are discharged in other ways. Sometimes the reservoirs are connected to the trachea and the substance discharged through the spiracles; some grasshoppers and cockroaches expel substances in this way. The bombardier beetle expels defensive secretions in audible puffs, often aimed accurately at its enemies; these are actually explosions within the reservoir resulting from the mixing of highly volatile chemicals and triggered by enzymes. Formicine ants are well known for secreting formic acid which is expelled from anal glands. Some of these ants use the formic acid in conjunction with other chemicals secreted in mandibular glands with large reservoirs located in the head.

The value of insects

Through the millions of years that insects and plants have existed together, they have evolved and adjusted themselves to a reasonable balance that permits both to survive. The relationship between plants and insects in many instances is so close that it is doubtful whether they could exist apart. Most of our fruits and many of our vegetables and field crops depend on insects for pollination. Only the bees are generally given credit for this. Other pollinators are butterflies, wasps, ants, and certain flies (especially the syrphid or hover flies). Bees have much greater economic importance in the production of fruits and seeds than they have as a source of honey and beeswax.

The part played by predatory and parasitic insects in the suppression of insect pest populations has been mentioned. Many phytophagous insects are beneficial in controlling weeds. When they have their proper place in the scheme of things, plants are said to be "wild" and we admire the annual show of flowers; when they become established where they are not wanted, they are tagged "weeds." The importance of insects here is shown dramatically sometimes when a plant is introduced, accidentally or intentionally, from one country into another without the insects that normally keep it within bounds in its native habitat. The American prickly pear cactus which overran the range in Australia—it was introduced innocently as an ornamental—and the Klamath weed in northern California (an invader from Europe) are well-known examples. Importation and release of the appropriate *insects* were required to bring about control of the plants.

Insects are the main source of food for many fish and other small animals, especially birds. Nestling birds are fed insects primarily at a time of the season when other food is not available. Insects have been an important source of food for humans too and are eaten in some parts of the world even today; they were an important part of the diet of many aboriginal tribes before they became "civilized" and adopted the taboos of the white man. Grasshoppers ("Mormon crickets") were eaten by Indians whom the Mormon settlers first encountered in Utah. Salmonflies (stoneflies) were gathered along streams for food by Indians in the Northwest. The larvae and pupae of the Pandora moth are prized as food by certain Indians in California and Oregon. The eggs of aquatic bugs (the prolific corixids or "water boatmen") are a popular food in Mexico, as are the larvae of a butterfly which bores into century plants (source of the native drink pulque). Termites and caterpillars are a delicacy in Africa, as are leaf-cutting ants in South America. But grasshoppers, prepared in various ways, seem generally to be the most favored insect food. Some scientists who have delved into the subject believe that insects might some day become an important source of protein—they are high in this food factor and said to be quite palatable, especially when mixed with other foods. This could come about, it is pointed out, as the human population explosion precludes the luxury of utilizing the limited land resources for raising cattle.

Insects are useful as scavengers, disposing of dead wastes, and they contribute substantially to soil fertility. Along with earthworms, they aerate the soil and aid in the penetration of water, as well as enrich it with their own wastes and decay. In the tropical forest where there are no earthworms, ants and termites are the chief aerators. Insects are valuable subjects for biological and medical research—they have contributed enormously to the study of heredity—and are a source of many chemical substances having potentially valuable properties. The study of insect physiology has broad technological applications. The destruction caused by some insects is, of course, very great, and it is not our intention to minimize this. We dwell on their value only because this is less evident and not generally appreciated.

Pictured Key to the Insect Orders

APTERYGOTA—The Wingless Insects
Insects "without" Metamorphosis*

Plate A

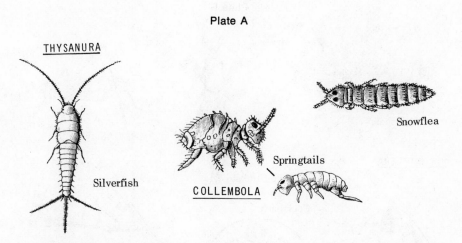

THYSANURA

Silverfish

COLLEMBOLA

Springtails

Snowflea

1. THYSANURA: Bristletails

Wings—absent. **Antennae**—long, filiform, many-segmented. **Mouthparts**—chewing type. **Tarsi**—two or three segments. **Cerci**—long, usually with tail-like appendage between the two. **Where found**—in moist places outdoors; some are household pests. **Young**—closely resemble adults.

2. COLLEMBOLA: Springtails

Wings—absent. **Antennae**—usually short; four to six segments. **Mouthparts**—chewing type. **Tarsi**—one segment fused with tibiae. **Cerci**—absent. **Where found**—in moist soil, on water, and on snow; some are pests in greenhouses. **Young**—closely resemble adults.

---

* The primitive wingless insects do not undergo metamorphosis in the sense that the winged insects do, no change in form being perceptible between hatching and maturity; they change, of course, in size and in the genitalia, as more advanced forms do. The nymphs or young of the more primitive winged insects also resemble the adults but differ in not having wings; "wing pads" appear usually after several molts and the wings are not fully developed until the insect reaches maturity.

PTERYGOTA—The Winged Insects
Insects with Gradual or Simple Metamorphosis

Plate B

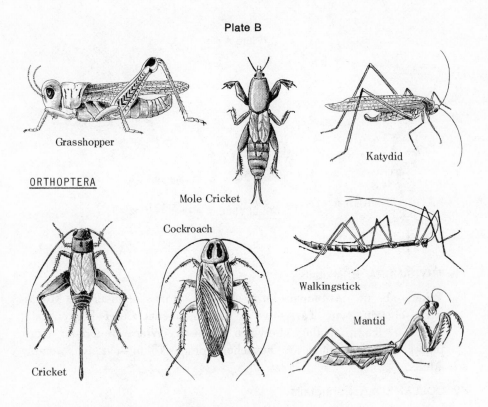

Grasshopper

ORTHOPTERA

Mole Cricket

Katydid

Cockroach

Walkingstick

Cricket

Mantid

3. ORTHOPTERA: Cockroaches, Grasshoppers, Katydids,
Crickets, Mantids, Walkingsticks

Wings—two pairs, forewings (tegmina) usually narrow and thickened, hindwings broad, membranous; absent or rudimentary in some forms. **Antennae**—short or long, filiform, many-segmented. **Mouthparts**—chewing type. **Tarsi**—one to five segments. **Cerci**—short or long. **Where found** —mostly outdoors on plants; certain cockroaches are household pests. **Young** (nymphs)—resemble adults.

Plate C

DERMAPTERA   PLECOPTERA   EMBIOPTERA

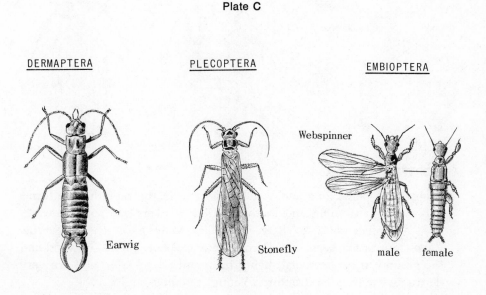

Webspinner

Earwig   Stonefly   male   female

4. DERMAPTERA: Earwigs

Wings—two pairs. **Antennae**—fairly long, filiform, many-segmented. **Mouthparts**—chewing type. **Tarsi**—three segments. **Cerci**—pincer-like. **Where found**—outdoors in moist locations, occasionally in garbage cans; mostly in the coastal regions. **Young** (nymphs)—resemble adults which guard them for a time, confining them to the nest.

5. PLECOPTERA: Stoneflies

Wings—two pairs, hindwings much larger than forewings; at rest, they are folded like a fan. **Antennae**—long, filiform, many-segmented. **Mouthparts**—chewing type, very primitive. **Tarsi**—three segments. **Cerci**—long, tail-like. **Where found**—on stones and plants near water. **Young** (naiads or nymphs)—are aquatic, do not resemble adults.

6. EMBIOPTERA: Embiids or Webspinners

Wings—two similar pairs in males, absent in females. **Antennae**—fairly long, filiform, many-segmented. **Mouthparts**—chewing type, very primitive. **Tarsi**—three segments. **Cerci**—short, one or two segments. **Where found**—in the soil or grass, under bark; they are gregarious, live in interconnected silk-lined tunnels. **Young** (nymphs)—resemble adults.

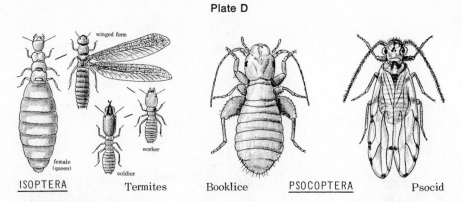

ISOPTERA   Termites   Booklice   PSOCOPTERA   Psocid

7. ISOPTERA: Termites

Wings—two pairs, long and equal in length, in the reproductive forms only during swarming and mating period; workers and soldiers wingless. **Antennae**—short to long, beadlike. **Mouthparts**—chewing type. **Tarsi**—four or five segments. **Cerci**—present. **Where found**—in wood and the ground; they are social, living in organized colonies, and are very destructive to wooden structures. **Young**—resemble adults.

8. PSOCOPTERA: Booklice and Psocids

Wings—two pairs in winged forms, forewings much larger than hindwings. **Antennae**—short to long, filiform, 13 to 50 segments. **Mouthparts**—chewing type. **Tarsi**—two or three segments. **Cerci**—none projecting. **Where found**—in dark, damp places, under bark and leaves, in rotten wood outdoors; some are pests in homes, warehouses, and museums. **Young** (nymphs)—resemble adults.

MALLOPHAGA   ANOPLURA

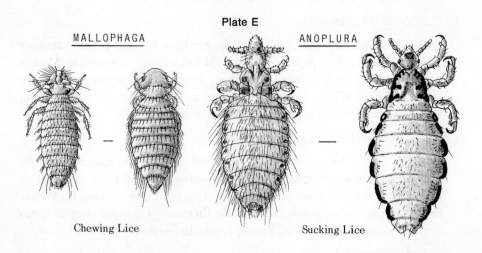

Chewing Lice   Sucking Lice

9. MALLOPHAGA: Chewing Lice

Wings—absent. **Antennae**—short, filiform or clublike, three to five segments. **Mouthparts**—chewing type, greatly reduced. **Tarsi**—one or two segments. **Cerci**—absent. **Where found**—on body of birds, domestic fowl, mammals; some are confined to one host species or to closely related ones. **Young** (nymphs)—resemble adults.

10. ANOPLURA: Sucking Lice

Wings—absent. **Antennae**—short, bristle-like, three to five segments. **Mouthparts**—piercing-sucking type; eversible set of fine stylets forming a tube. **Tarsi**—one segment. **Cerci**—absent. **Where found**—on body of mammals, often confined to one or closely related species; the human body louse rests on clothing when not feeding, is a vector of typhus. **Young** (nymphs)—very similar to adults.

Plate F

THYSANOPTERA

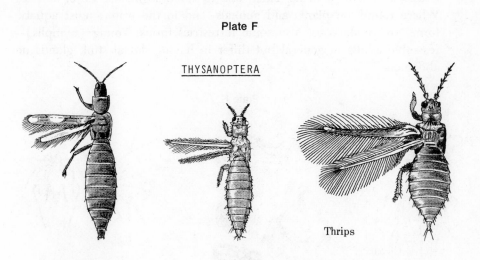

Thrips

11. THYSANOPTERA: Thrips

Wings—two pairs in winged forms; slender, fringed with long hairs. **Antennae**—fairly long, six to nine segments. **Mouthparts**—rasping-sucking type, conelike. **Tarsi**—one or two segments, ending in blunt tip with eversible pad or bladder. **Cerci**—absent. **Where found**—on flowers or leaves of plants. **Young** (nymphs or "larvae")—early stages similar to adults, fourth stage quiescent; some have fifth or "pupal" stage, forming cocoons in the soil.

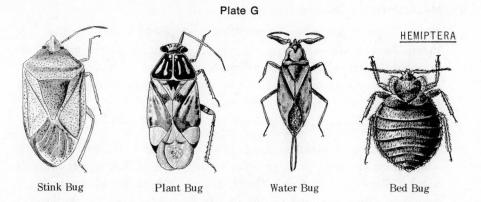

HEMIPTERA

Stink Bug   Plant Bug   Water Bug   Bed Bug

12. HEMIPTERA (Heteroptera): True Bugs

Wings—two pairs in winged forms, forewings (hemelytra) partly thickened, tips membranous, hindwings membranous; folded flat over the body. **Antennae**—usually filiform, short in aquatic forms, long in terrestrial forms; four or five segments. **Mouthparts**—piercing-sucking type, enclosed in jointed beak. **Tarsi**—one to three segments. **Cerci**—absent. **Where found**—on plants and animals, and in the water; most aquatic forms are predaceous, also some terrestrial forms. **Young** (nymphs)—resemble adults in general but differ in having dorsal stink glands on abdomen.

Plate H

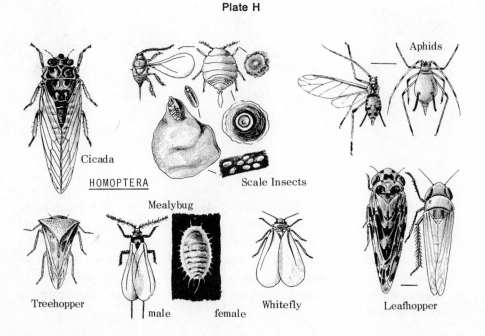

Cicada

HOMOPTERA

Aphids

Scale Insects

Mealybug

Treehopper   male   female   Whitefly   Leafhopper

13. HOMOPTERA: Scale Insects, Mealybugs, Whiteflies, Aphids, Cicadas, Leafhoppers, Planthoppers

Wings—two pairs when present, usually membranous; held rooflike over the body. **Antennae**—usually threadlike; four to ten segments. **Mouthparts**—piercing-sucking type, enclosed in jointed beak arising from back of head (it is at front of head in the Hemiptera). **Tarsi**—one to three segments. **Cerci**—absent. **Where found**—on plants; all are plant feeders. **Young** (nymphs)—generally resemble adults.

Plate I

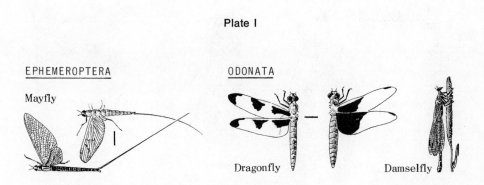

EPHEMEROPTERA — Mayfly

ODONATA — Dragonfly, Damselfly

14. EPHEMEROPTERA: Mayflies

Wings—two pairs, triangular-shaped, forewings larger than hindwings. **Antennae**—very short, few segments. **Mouthparts**—vestigial only (adults are short-lived and do not feed). **Tarsi**—one to five segments. **Cerci**—long, filamentous, often with long, tail-like appendage between the two. **Where found**—around water. **Young** (nymphs or naiads)—do not resemble adults; they are aquatic, sturdy, with well-developed mouthparts.

15. ODONATA: Dragonflies and Damselflies

Wings—two pairs, membranous, about equal in size; dragonflies hold them out to the sides at rest; damselflies hold them together vertically, like mayflies. **Antennae**—very short, hairlike, few segments. **Mouthparts**—chewing type, mandibulate, well developed. **Tarsi**—three segments. **Cerci**—present. **Where found**—normally near fresh water. **Young** (naiads)—aquatic, do not resemble adults.

PTERYGOTA—The Winged Insects
Insects with Complete or Complex Metamorphosis

Plate J

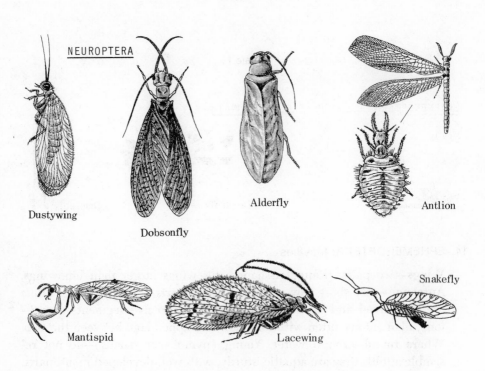

Dustywing

Dobsonfly

Alderfly

Antlion

Mantispid

Lacewing

Snakefly

16. NEUROPTERA: Dobsonflies, Alderflies, Dustywings, Lacewings, Antlions, Mantispids, Snakeflies

Wings—two similar pairs, membranous. **Antennae**—variable, often fairly long and threadlike, many-segmented. **Mouthparts**—chewing type. **Tarsi** —five segments. **Cerci**—absent. **Where found**—mostly on plants as predators. **Young** (larvae)—unlike adults; those of the dobsonflies and alderflies are aquatic.

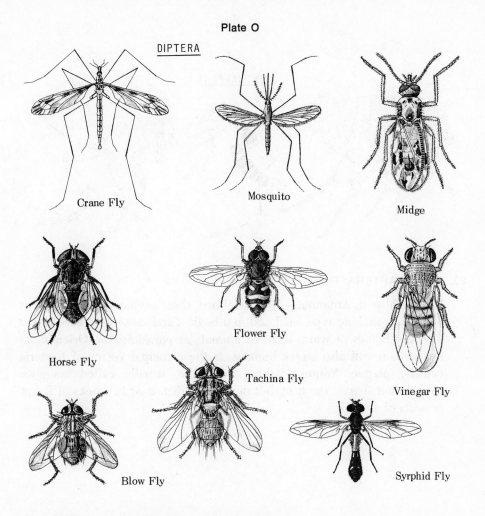

DIPTERA

Crane Fly

Mosquito

Midge

Horse Fly

Flower Fly

Vinegar Fly

Blow Fly

Tachina Fly

Syrphid Fly

22. DIPTERA: Flies, Mosquitoes, Gnats, Midges

Wings—one pair ("front"), membranous, scaly in a few species, absent in some forms; knobs or halteres occur in place of the hind pair. **Antennae**—variable, usually short and threadlike, aristate in some forms. **Mouthparts**—piercing-sucking, rasping, and lapping types; vestigial in some forms (such as the bot flies) which may not feed at all. **Tarsi**—five segments. **Cerci**—absent. **Where found**—on plants and animals, foods and waste products, and around water; most of them feed on nectar or plant exudates, many are predaceous or parasitic; the blood-sucking kinds and others include the most important vectors of organisms causing disease. **Young** (larvae)—are usually wormlike, totally unlike the adults, and called maggots; they prefer moist media, breed in meats, decaying animal and vegetable matter, and wastes; some invade the tissues of plants and animals; many are aquatic.

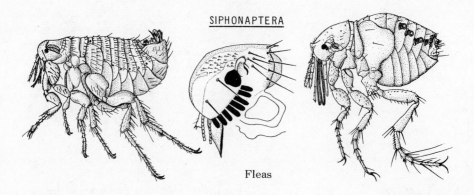

SIPHONAPTERA

Fleas

23. SIPHONAPTERA: Fleas

Wings—absent. **Antennae**—short, knobbed, three-segmented. **Mouthparts** —piercing-sucking type, enclosed in a beak. **Tarsi**—absent. **Where found** —on the bodies of warm-blooded animals as parasites; the Oriental rat flea, which will also attack humans, is the principal vector of bacteria causing plague. **Young** (larvae)—wormlike, usually called maggots; they do not live on the host, but develop in dirt, dust in cracks of floors, or nests of animals.

MINOR ORDERS

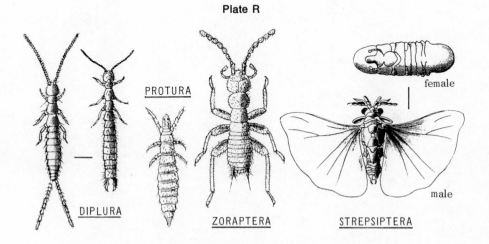

PROTURA

DIPLURA

ZORAPTERA

STREPSIPTERA

female

male

24. DIPLURA (Entotrophi), PROTURA, ZORAPTERA, STREPSIPTERA

Diplura or Entotrophi (Campodeids and Japygids)—wingless, blind, with chewing-type mouthparts hidden in ventral pouch; they are found under stones, logs, or debris, are "without" metamorphosis and are very primitive. They resemble the Thysanura, used to be included with them.

Protura (Proturans)—wingless, without eyes, with stylet-like mandibles; they are found in damp soil and decaying vegetation, are "without" metamorphosis, and are very ancient. They are unique in having no antennae; they "feel their way" with forelegs raised upward.

Zoraptera (Zorapterans)—winged and wingless forms occur, the latter being blind; they are very rare, mostly tropical (two species occur in the southern states), are found in colonies under bark and in rotten wood, and undergo gradual metamorphosis.

Strepsiptera (Twisted-wing Parasites)—the male is winged, the female is wingless and ovoviviparous, giving birth to larvae. (See footnote, p. 568.) This group is believed to be closely related to the meloid beetles, and is included with the Coleoptera in some classifications.

PART I.

APTERYGOTA

The wingless insects—Thysanura, Collembola, Protura, and Diplura—are placed in the subclass Apterygota. The others—the "winged insects"—are placed in the subclass Pterygota (Part II and Part III).

Insects "without" Metamorphosis

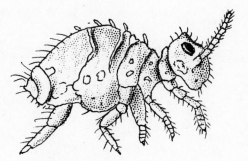

1

THYSANURA: Bristletails

These are primitive wingless insects with long filiform or threadlike antennae and exposed biting mouthparts. Some have eyes; others are without them. Their bodies are slender, ten- or eleven-segmented, clothed in glistening scales. They have two to eight pairs of styli—small, pointed appendages —on the ventral side; "these are regarded as vestiges of limbs"—a survival from ancestors having legs on many or all segments of the body. The Machilidae* have these styli on all abdominal segments; the others have them on segments two to nine only. Two long filiform, many-segmented appendages (or cerci) protrude from the posterior end, with a long many-segmented "tail" in between. Outdoor species are found under rocks, logs, in leafmold or rotten wood. When disturbed they dart out and look for another hiding place.

Indoor species are a nuisance, live on a great variety of foods and materials, are found in kitchens, bathrooms, and elsewhere. Active only at night, they eat cereals, flour, paper with glue, paste or sizing (such as wallpaper), bookbindings, starch in clothing, rayon fabrics. They are hardy creatures and can live without food for several months. Species resembling *Lepisma* are known to feed on termite eggs and nymphs. Their chief natural enemies are ants, which they are often able to elude because of their slippery scales. They are "without" metamorphosis, and continue to molt throughout their lives.

* The family Machilidae is sometimes placed in a separate order, Microcoryphia.

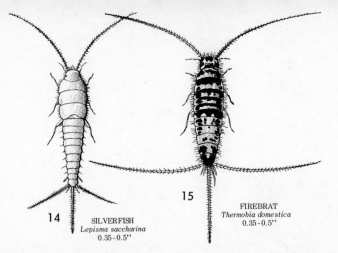

15

14   SILVERFISH
      Lepisma saccharina
      0.35 - 0.5"

FIREBRAT
Thermobia domestica
0.35 - 0.5"

(1) SILVERFISH: *Lepisma saccharina* (Lepismatidae). **Fig. 14.**

Also called the common silverfish. **Range:** Throughout North America. **Adult:** Silver gray; maxillary palpi five-segmented; two pairs of styli in both sexes; setae occur singly or in small groups, never in tufts or combs. Whitish, oval eggs are laid a few at a time over a period of several weeks and hatch in two to eight weeks, depending on temperature. **Length:** .35–.5". **The young:** Look like the adults, except that they are smaller. They become mature about two years after hatching, are active the year long in the South. **Food:** See introduction above. Silverfish prefer warm, damp places, are found in bookcases, closets, cupboards, bathrooms, behind baseboards and partitions. The **Giant Silverfish,** *Ctenolepisma urbana,* is slate gray, may be distinguished by its size (about .6" long) and long antennae (longer than the body); it has two pairs of styli, and dorsal setae arranged in combs, three rows on a side.* The **Four-lined Silverfish,** *C. quadriseriata,* may be distinguished by its violet-brown scales, four longitudinal stripes on abdomen, and three pairs of styli; it has three rows of dorsal setal combs on each side.

(2) FIREBRAT: *Thermobia domestica.* **Fig. 15.**

Range: Throughout North America. **Adult:** Brown and tan, mottled; maxillary palpi six-segmented; three pairs of styli in the female, two in the male; setae occurring in tufts or combs, not singly. Eggs—soft, white, opaque when laid, turning yellowish later—are laid 50 at a time, in several batches over a period of time. **Length:** .35–.5". **The young:** Look like adults, except that they are smaller. They mature from three to twenty-four months after hatching, depending on temperature and humidity; forty-one molts have been recorded for one firebrat. They live up to two and one-half years. **Food:** See introduction above. Firebrats are quick-moving, prefer hot, dark places, are commonly found around furnaces and heating pipes.

* See Ruth E. Slabaugh, "A New Thysanuran, and a Key to the Domestic Species of Lepismatidae (Thysanura) Found in the United States," *Ent. News,* 51:51–95 (1940).

2

COLLEMBOLA: Springtails

The springtails are primitive wingless insects, usually minute or very small. Many species are cosmopolitan in distribution and are found in both the Old and the New World. They are very ancient, fossil records going back to the Devonian period—more than 400 million years ago according to present reckoning.* The body may be a globular mass, with thorax and most of the abdominal segments fused, or elongate and distinctly segmented. Both sexes are usually very similar. They are clothed with dense hairs (setae) and brilliantly pigmented, or with scales and brightly iridescent. The mouthparts are adapted for chewing, or the mandibles and maxillae are long and sharp like stylets and adapted for sucking. The antennae are four- to six-segmented and vary greatly in length and structure. The eyes are simple (ocelli), typically sixteen in number, eight located on a large patch behind each antenna; many species have "pseudocelli" on various parts of the dorsal surface. Behind the antennae of some species is a sensory organ— the postantennal organ—which varies from a simple slit to a more elaborate raised structure.

The abdomen is six-segmented; on the first segment is a ventral tube or colliphore, which is a distinctive feature of the group. The colliphore usually ends in a bilobed structure and has filaments which can be extruded; it is believed to serve a respiratory function, or possibly as a means of clinging to surfaces. The fourth (or fifth) abdominal segment is usually equipped with a ventral "spring" called the furcula which enables the insect to jump

* According to Jeannel, "the Collembola are the most primitive of all insects."

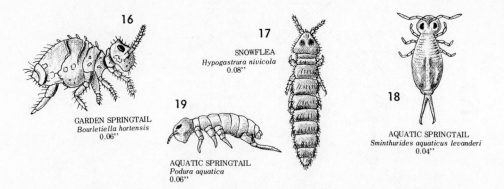

16

GARDEN SPRINGTAIL
Bourletiella hortensis
0.06"

17
SNOWFLEA
Hypogastrura nivicola
0.08"

19

AQUATIC SPRINGTAIL
Podura aquatica
0.06"

18

AQUATIC SPRINGTAIL
Sminthurides aquaticus levanderi
0.04"

some distance; on the third segment is a button-like "catch" called the tenaculum which holds the spring in place. The legs are well developed and in many species equipped with elaborate toothed claws—usually a large one (the unguis), with a smaller one (the unguiculus) opposing it—and accompanying setae (tenent hairs). Respiration in most species is cutaneous and over the entire body. Like the Thysanura, they are "without" metamorphosis. Springtails are found mostly in ground litter and on the surface of quiet waters; those living mainly on water are not truly aquatic in that they do not lay their eggs in the water. They usually feed on fungi and decaying organic matter, but some attack germinating seeds, roots, and other parts of tender shoots. A few get into houses, where they are found in damp places or in the soil of potted plants. Some can be collected by beating vegetation; a better way is to shake moss, bark, leaves, rotted wood, and other debris in a pan with a little water, on which they will float. Their chief enemies are predaceous mites.

(1) GARDEN SPRINGTAIL: *Bourletiella hortensis* (Sminthuridae). **Fig. 16.**

Range: Throughout most of North America; Europe. **Adult:** Globular; dark purple to blackish, with numerous small, irregular white spots, body often dull greenish dorsally; eyes eight on a side, on large black patches partially or completely ringed with white; legs, furcula, and ventral tube slate gray; head and body with dense short, curving setae; anal appendage broad, fanlike. **Length:** .06". This is the most common of the economic species, abundant in spring and early summer, especially on ground previously cultivated. **Food:** Seedlings; holes are chewed in leaves and stems of tender shoots, the damage sometimes being severe.

(2) AQUATIC SPRINGTAIL: *Sminthurides aquaticus levanderi.* **Fig. 18.**

Range: Throughout North America. **Adult:** Rose pink to deep purple or violet, sides of abdomen with pale spots, dorsum often with long, broad white stripe; eyes eight on a side, antennae pale basally and purplish distally, slightly longer than head; large claw on first and second feet long, slender, toothed; setae simple, moderate in length. **Length:** .04″. It is found on quiet waters, can remain submerged up to four days. The male grasps the female antennae with his own while mating, and is carried about by her for several days off the water surface. All eggs of one female develop into males or females, the number of both sexes remaining about equal until late in the season when the females predominate, since they live longer. Hibernation usually takes place in the egg stage.

(3) SNOWFLEA: *Hypogastrura (Achorutes) nivicola* (Hypogastruridae). **Fig. 17.**

Range: Maine, Massachusetts, Pennsylvania, Michigan, Iowa, Ontario; Europe. **Adult:** Elongate, distinctly segmented; dark indigo blue; head prognathous, eyes eight on a side, antennae as long as head, postantennal organ with four peripheral tubercles; large claw curved inward slightly, smaller one half its length, tenent hair with minute knob; two short anal spines, conicla and straight or curved inward slightly; dense coat of simple setae; furcula and tenaculum present. **Length:** .08″. Abundant in forests in winter and early spring on the surface of the snow after a thaw ("often sprinkled over whole acres of woodland") or floating on the water; they often show up in buckets of maple sap. *H. harveyi* is also abundant at this time and may be found crawling over bark or old stumps, sometimes on the snow; it is readily distinguished from *nivicola* by its long anal spines. *Anurida maritima* is cosmopolitan but strictly maritime, blue-black, about .12″ long; eyes five on a side, antennae shorter than head; it is distinguished by the postantennal organ, with rosette of six to ten peripheral tubercles; setae dense, furcula and anal spine absent. It is abundant on seashores, found crawling on the sand, rocks, and seaweed at low tide, feeding on dead crustaceans; at high tide it burrows in the sand or hides in rock crevices, enough air being retained by the hairs to sustain it during submerged periods.

(4) AQUATIC SPRINGTAIL: *Podura aquatica* (Poduridae). **Fig. 19.**

Range: Widely distributed throughout North America; Europe, Siberia. **Adult:** Elongate, distinctly segmented; blue-black or reddish brown; head hypognathous, eyes eight on a side, antennae reddish brown, shorter than head; legs reddish brown, large claw long, slender, and curving, smaller one only rudimentary, tenent hair long and without knob; only a few minute, curving setae on each segment; furcula very long and flattened ventrally. **Length:** .06″. It is very abundant on lakes and ponds and along the edge of streams, "often found skipping from the surface of the water to the vegetation and back again to the water as the waves wash the shore."

PART II.

PTERYGOTA

The "winged insects" are placed in the subclass Pterygota. The wingless insects are placed in the subclass Apterygota (Part I).

Insects with Gradual or Simple Metamorphosis

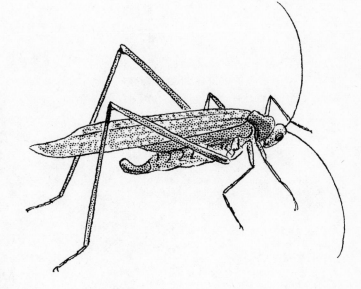

3

ORTHOPTERA: Cockroaches, Mantids, Walkingsticks, Grasshoppers, Katydids, Crickets

The name of this order comes from the Greek *orthos*, meaning "straight," and *pteron*, meaning "wing," the reference being to the forewings, which are not used for flying. Certain families are so distinctive that some authors place them in separate orders. Most of them have well-developed biting and chewing mouthparts, two pairs of wings, and a pair of cerci. The forewings —called the tegmina (sing., tegmen)—are straight and narrow, thickened, and distinctly veined; the hindwings are broad and membranous, well veined, and fold like a fan. In some species wings are lacking or rudimentary. Grasshoppers, katydids, and crickets are noted for their powerful hind legs (adapted for jumping by greatly enlarged femora) and the stridulations of the males. Crickets and katydids produce their sounds by rubbing specialized parts of the forewings together; some grasshoppers rub the forewing and hindwing together, others rub a "file" on the inner surface of the hind femur against a raised, roughened area on the outer edge of the forewing. Crickets and others in the group can be attracted and collected by reproducing their stridulatory sounds.

Metamorphosis in this group is gradual, with nymphs generally going through five developmental stages or instars, which is to say they molt five times. They feed mostly on plants, with the exception of the mantids, which are fierce predators, destructive to some useful insects but on the whole beneficial; mantids can be successfully colonized by transferring their egg cases from one place to another. A number of species of longhorn grasshoppers and crickets (*Oecanthus*) are predaceous on various kinds of Homoptera, and many families show a strong tendency toward cannibalism.

Many egg pods of grasshoppers are destroyed by predators—especially birds, rodents, and the larvae of bee flies, blister beetles, and ground beetles. Sarcophagid flies parasitize the adults and nymphs and sometimes prey on the eggs. Many grasshoppers are attacked by mermithid nematodes (long, round worms), others are susceptible to bacterial and microsporidian infections; their cannibalistic tendencies aid in the spread of disease.

Cockroaches (Blattidae)

Most of the cockroaches—about 55 species are known in the United States—are not troublesome and live outdoors. Only the five species illustrated are commonly found indoors in North America. They carry filth, may spread disease, destroy fabrics and bookbindings; and their offensive odors contaminate food. Cockroaches are broad, flattened, dark brown, reddish brown, light brown, or black. They forage at night, hide during the day; most of them have wings and can run rapidly.* The females lay their eggs in leathery capsules, called oothecae, which are formed at the end of the body.

"Cockroaches were abundant in the Paleozoic [era] and are one of the most ancient groups of living insects." According to Jeannel "there are no fundamental differences between these Paleozoic cockroaches [from the earliest deposits of the Carboniferous period] and those of the present day. It would seem that they have survived unchanged, perhaps from Devonian times; that is to say, for about 400 million years." Their fossil remains are so abundant in the strata of the Pennsylvanian that this period is often referred to as the Age of Cockroaches.

(1) AMERICAN COCKROACH: *Periplaneta americana.* **Fig. 20.**

Range: Widely distributed over the world in mild climates; a native of tropical and subtropical America. **Adult:** Reddish brown to dark brown. It forages mostly on the first floor of buildings. The female carries the egg capsule a day or two, then glues it to some object in a protected place; about 12 nymphs hatch from each capsule in about two to three months. An adult female lives a year or more and lays many capsules. **Length:** 1.5–2″. **Nymph:** Color same as adult. Undergoes thirteen molts, matures in about a year. Grows slowly under unfavorable conditions, that is, where good sanitation is practiced; develops in basements and sewers. The **Brown Cockroach,** *P. brunnea*—a tropical species found in Florida and Texas—is from 1″ to 1.3″ long, glossy brown, with sides and tip of tegmina paler, markings

---

* Investigations have shown that cockroaches have the ability to learn to avoid the insecticides and dusts commonly used as control measures (boric acid dust proved to have the least repellency). See W. Ebeling and D. A. Reierson, "The Cockroach Learns to Avoid Insecticides," *California Agriculture,* 23(2):12–15 (1969), Agr. Ext. Sta., Univ. of California, Berkeley.

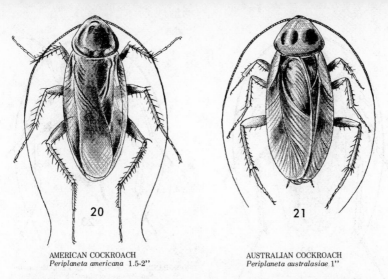

AMERICAN COCKROACH
Periplaneta americana 1.5-2"

AUSTRALIAN COCKROACH
Periplaneta australasiae 1"

on thorax darker and more diffuse than in *americana;* the female has shorter cerci, and the last abdominal segment is not so deeply notched as in *americana;* it is found in buildings and out of doors.

(2) AUSTRALIAN COCKROACH: *Periplaneta australasiae.* **Fig. 21.**

Range: World-wide. **Adult:** Reddish brown to dark brown, with yellow markings on thorax, yellow streaks at base of forewings. Forages mostly on the first floor of buildings. Eggs are laid and hatch as in (1) above. **Length:** 1". **Nymph:** Color same as adult. Matures in about a year; develops in warm, damp places in buildings or out of doors. The **Smoky-brown Cockroach,** *P. fuliginosa,* is from 1" to 1.3" long, ranges from Texas to Florida; it is common out of doors and around warehouses along the waterfront, matures in about a year.

(3) BROWN-BANDED COCKROACH: *Supella supellectilium.** **Figs. 22, 23.**

Also called tropical cockroach. **Range:** Throughout the world. **Adult:** Light brown; wings mottled, reddish brown in the female, lighter in the male. Eggs are laid and hatch as in (1) above. **Length:** .5". **Nymph:** Color as in adult. Matures in four to six months; develops and lives all over buildings.

(4) ORIENTAL COCKROACH: *Blatta orientalis.* **Figs. 24, 25.**

Also called black beetle and shad roach. **Range:** World-wide. **Adult:** Black or dark brown; wings vestigial in the female, short in the male. More sluggish than the others. Forages on the first floor of buildings. Eggs are laid and hatch as in (1) above. **Length:** 1–1.25" (female); male a little shorter. **Nymph:** Color as in adult, matures in about a year.

(5) GERMAN COCKROACH: *Blattella germanica* (Blattellidae). **Fig. 26.**

Also called croton bug and water bug. **Range:** World-wide. Most com-

* Proposed name change: *Supella longipalpa* (Blattellidae).

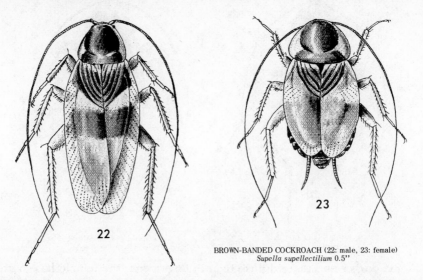

BROWN-BANDED COCKROACH (22: male, 23: female)
Supella supellectilium 0.5"

mon of the indoor species. **Adult:** Light brown, with longitudinal stripes on back. The female carries an ootheca for about a month, drops it a day or so before the eggs are ready to hatch; about 30 nymphs emerge from a capsule. **Length:** About .6". **Nymph:** Color as in adult. Undergoes six molts, matures in four to six months; develops and lives in kitchens and bathrooms. The **Surinam Cockroach,** *Pycnoscelus surinamensis,* is oval in outline, dark brown, from 1.5" to 2" long, with wings one-third as long as body; slow-moving, it is often found in greenhouses, where it feeds on young plants.

Mantids (Mantidae)

Despite the striking difference in appearance, mantids are closely related to cockroaches; the two families are sometimes placed in a separate order, Dictyoptera. Mantids are easily recognized by the greatly elongated pro-thorax and powerful front legs adapted for grasping prey; the coxa is extremely long, resembling a femur, while the large femur is grooved and spined ventrally, locking pincer-like against a tibia similarly armed. Pro-tective coloring makes them difficult to see, and though strong fliers, they are most often found in the characteristic "praying" attitude, quietly wait-ing for their unsuspecting prey. The eggs are glued to stems and twigs of plants in a sticky mass which hardens; a female may lay from three to six capsules of 50 to 400 eggs each. Northern species have one brood and hibernate in the egg stage. They make interesting captives (or pets?).

(6) EUROPEAN MANTID: *Mantis religiosa.* **Fig. 27.**

Introduced, established in North America about 1899; also called the praying mantis. **Range:** Eastern U.S., ranging northward into Ontario, where it is quite common. **Adult:** Brown or green. **Length:** 2.5" (including wings).

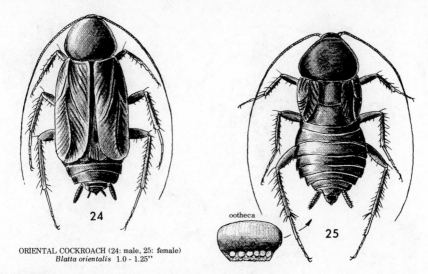

ORIENTAL COCKROACH (24: male, 25: female)
Blatta orientalis 1.0 - 1.25"

(7) CAROLINA MANTID: *Stagmomantis carolina.* **Fig. 28.**

A native species. **Range:** Southern U.S., westward to Arizona, seldom north of Maryland; sometimes found in California, Utah, and Colorado. **Adult:** Grayish brown, with green tarsi, or all green. **Length:** 2.4". *S. limbata* has smoky brown, mottled wings with dark patch at the apex, and male has opaque costal area; it is a Mexican species ranging northward into Arizona and New Mexico. The **Agile Ground Mantid,** *Litaneutria minor,* a small gray species from .6" to 1.2" long, occurs in California, Colorado, and southward; the male has full wings, the female has only stubs.

(8) CHINESE MANTID: *Tenodera aridifolia sinensis.*

Introduced into North America around 1896. **Range:** New Jersey, southern New England, New York, westward to Ohio. **Adult:** Front wings brown, with green margins. **Length:** 3". The **Narrow-winged Mantid,** *T. angustipennis,* another Asiatic species, is smaller, makes a cylindrical egg mass; it is established in some eastern states, was first reported from Delaware and Maryland in 1933.

Walkingsticks (Phasmatidae)

This family is remarkable for its mimicry of dead twigs and leaves. Some authors place it in a separate order, Cheleutoptera; others include it in an order called Dictyoptera, with the cockroaches and mantids. The prothorax is short, the mesothorax extremely long; North American species are wingless, excepting one found in Florida, *Aplopus mayeri*. They live in trees and feed on the foliage but usually are not numerous enough to be very destructive. The seedlike eggs—one to a capsule—are scattered over the ground. A few species are parthenogenetic; some, such as *Carausius morosus,* are reared as laboratory animals for research purposes.

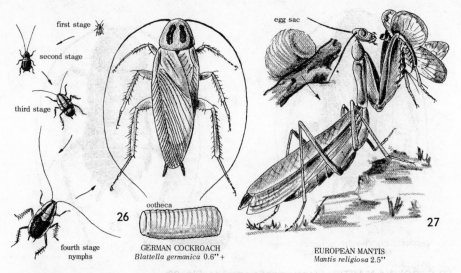

first stage

second stage

third stage

26

fourth stage
nymphs

ootheca

GERMAN COCKROACH
Blattella germanica 0.6" +

egg sac

27

EUROPEAN MANTIS
Mantis religiosa 2.5"

(9) WALKINGSTICK: *Diapheromera femorata.* **Fig. 29.**

Range: Common and widely distributed east of the Rocky Mts., northward into Canada, and southward into New Mexico. **Adult:** Brown to greenish. It inhabits deciduous trees.* Eggs are laid on the ground in the fall and hatch in the spring. **Length:** 3.7". The pale yellowish brown **Arizona Walkingstick,** *D. arizonensis,* is more slender, from 3.4" to 3.6" long, has shiny black marks on underside of mesothorax and metathorax; it is common on spiny shrubs such as mesquite and acacia.

Grasshoppers or Locusts (Acrididae = Locustidae)

The acridids or grasshoppers—sometimes called shorthorn grasshoppers —may be differentiated from the tettigoniids or longhorn grasshoppers by their short antennae (or "horns"), and by the tympanum (Fig. 7A) or ear on the side of the first abdominal segment, the three-segmented tarsi, and the female ovipositor consisting of short "valves" adapted for digging (Fig. 35C). Nearly all cultivated and wild plants are attacked by grasshoppers. Only a few species are sufficiently numerous to cause serious crop damage; 90 percent of this injury in the U.S. is caused by the five species described below in examples (10) to (14). About 600 species of grasshoppers have been identified in the U.S. and Canada.

Eggs are laid in the soil in a cemented mass or pod commonly containing from eight to twenty-five eggs (the pod of each species is distinctive); they are usually placed in uncultivated areas such as field margins, pastures, roadsides, or on the range, from late August to October, and hatch from

* In 1969 it was reported to have "severely defoliated red oak and locust on about 100,000 acres of Ouachita National Forest in eastern Oklahoma and western Arkansas." *Cooperative Economic Insect Report,* 19(47):846 (Nov. 21, 1969). Plant Prot. Div., Agr. Res. Serv., U.S. Dept. of Agr., Hyattsville, Md.

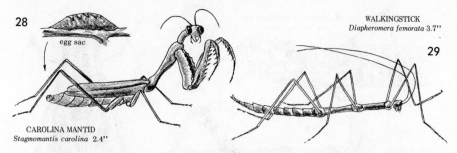

28

egg sac

CAROLINA MANTID
Stagmomantis carolina 2.4"

WALKINGSTICK
Diapheromera femorata 3.7"

29

mid-May to July.* Most species hibernate in the egg stage; a few overwinter as partly grown nymphs or as adults. Nymphs usually undergo five instars and develop to maturity in 40 to 60 days.

(10) MIGRATORY GRASSHOPPER: *Melanoplus sanguinipes (bilituratus).* **Fig. 30.**

Range: Southern Ontario, Minnesota, westward to Alberta and Montana, southward to Texas, Arizona, Mexico. **Adult:** Light brown and yellow, hind tibiae red; male with broad cerci and distinct swelling between the middle legs. A female lays up to 21 pods, each containing about 16 greenish white eggs; the pod is about 1" long, with brownish green froth and heavily cemented sides. **Length:** 1" (male) to 1.25" (female). Spectacular and extensive mass flights occurred in the Plains states from 1938 to 1940. **Food:** All crops, range plants. [See Diptera (38).]

(11) RED-LEGGED GRASSHOPPER: *Melanoplus femurrubrum.* **Fig. 31.**

Range: Arizona to central Kansas, western Oklahoma, and the prairie provinces, eastward. **Adult:** Greenish gray to olive-green or reddish brown, with yellowish underside, row of black specks on the tegmen and black bar on each side of the body under the wings; hind tibiae bright red, with black spines. Egg pods contain about 20 eggs each. Males do not stridulate. **Length:** .9" (male) to 1" (female). **Food:** Mainly grains, alfalfa, and beans in the Southwest, soybeans in the Middle West.

(12) TWO-STRIPED GRASSHOPPER: *Melanoplus bivittatus.* **Fig. 33.**

Range: Throughout North America. **Adult:** Reddish black or brownish black to dark olive-green, with two pale yellow stripes on either side of pronotum extending to base of the tegmina. A female may lay 12 pods, each with up to 100 eggs. **Length:** 1" (male) to 1.5" (female). It frequently has six instars. **Food:** Alfalfa, red clover, grasses, corn, barley, wheat, vegetables, deciduous fruit trees.

(13) DIFFERENTIAL GRASSHOPPER: *Melanoplus differentialis.* **Fig. 32.**

Range: Throughout North America. **Adult:** Very shiny, brownish yellow;

* An unusual grasshopper, the South American *Marella remipes,* feeds on the floating leaves of water lilies, which it eats from the flat upper surface. It is an air-breather but spends considerable time under water swimming about and clinging to the stems of water plants; its eggs are laid under water.

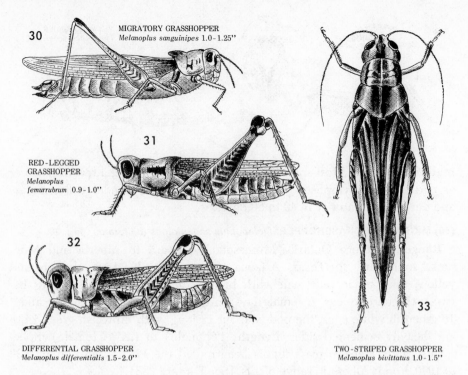

30 MIGRATORY GRASSHOPPER
Melanoplus sanguinipes 1.0-1.25"

31

RED-LEGGED
GRASSHOPPER
*Melanoplus
femurrubrum* 0.9-1.0"

32

33

DIFFERENTIAL GRASSHOPPER
Melanoplus differentialis 1.5-2.0"

TWO-STRIPED GRASSHOPPER
Melanoplus bivittatus 1.0-1.5"

hind femora yellow, with black herringbone markings on outer face; hind tibiae yellow, with prominent black teeth, tarsi yellow; antennae yellow or bright red. A female may lay 8 pods, each with up to 100 eggs. Males do not stridulate. **Length:** 1.5–2" (to wing tip); largest species in the genus. It regularly has six instars. **Food:** Grain and forage crops, cotton, sugar beets, vegetables, fruits. [See Diptera (91).]

(14) CLEAR-WINGED GRASSHOPPER: *Camnula pellucida.* **Fig. 34.**

Range: Throughout North America; one of the most common western species, particularly at higher elevations and in northern part of range. **Adult:** Pale brown, with darker markings; pronotum with median and lateral ridges, large black spot on the sides; forewing with irregular markings, hindwing transparent. They may fly in swarms but do not migrate great distances. **Length:** .75" (male) to 1" (female). **Food:** Range grasses, grains, vegetable crops.

(15) CAROLINA GRASSHOPPER: *Dissosteira carolina.* **Fig. 35A.**

Range: Widespread throughout North America. **Adult:** Brown or grayish brown, depending on habitat, with fine specks of gray and red on the tegmen; the hindwing is pitch black, with sharply contrasting narrow yellow border which serves to distinguish it from other species; high median ridge on pronotum. Egg pods contain more than 20 and up to 70 eggs each. **Length:** 1.5–1.9". One of the species most commonly seen along roadsides

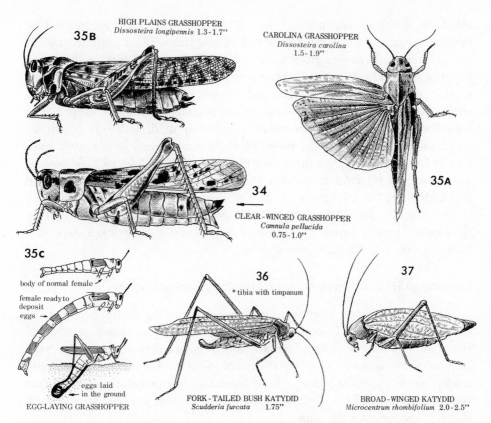

HIGH PLAINS GRASSHOPPER
Dissosteira longipennis 1.3-1.7"
35B

CAROLINA GRASSHOPPER
Dissosteira carolina
1.5-1.9"
35A

34

CLEAR-WINGED GRASSHOPPER
Camnula pellucida
0.75-1.0"

35C

body of normal female

female ready to
deposit
eggs →

eggs laid
in the ground

EGG-LAYING GRASSHOPPER

36

* tibia with timpanum

FORK-TAILED BUSH KATYDID
Scudderia furcata 1.75"

37

BROAD-WINGED KATYDID
Microcentrum rhombifolium 2.0-2.5"

and in fields; they fly away quickly when disturbed, exposing their colorful wings. The stridulations of the male are not loud. **Food:** Grasses, grains, beans, and other cultivated crops. This species is not as destructive as the others mentioned above.

(16) HIGH PLAINS GRASSHOPPER: *Dissosteira longipennis.* **Fig. 35B.**

Range: Confined to a relatively small area (though it migrates great distances at times)—southeastern Colorado, southwestern Kansas, Oklahoma and Texas Panhandles, northern New Mexico—within the "shortgrass area" of the Great Plains at elevations of 4,000 to 6,000 feet. **Adult:** Brownish yellow, with long, brown-spotted tegmen, wings blue at base, gradually becoming black toward discal area. After initial flights and dispersal (when they may be attracted to lights) they become gregarious. Egg pods are massed in limited areas ("egg beds"), mainly in the grama grass and buffalo grass range. The average number of eggs per pod is 65. **Length:** 1.3–1.7". **Nymph:** They persist in bands, start migrating a week after emerging, gradually disperse as they feed and grow. Five or six instars. One brood. **Food:** Mostly range grass, which they sometimes damage severely. [See Diptera (92).] The **Green Valley Grasshopper,** *Schistocerca* *shoshone,* is one of our largest, from 1.6″ to 2.8″ long, easily recognized by the leaf-green color, red hind legs, and yellow stripe along midline of head and pronotum; it ranges from California to Colorado, Texas, and Mexico to Nicaragua. The **Vagrant Grasshopper,** S. *vaga,* is nearly the same size, similar in range, may be recognized by the all-brown coloring and tan stripe along the midline. Both species are sometimes pests of grapes. The **American Grasshopper,** S. *americana,* is brown-spotted, with yellow midline stripe and yellow underside, migrates northward from Florida each summer as far as Ontario and the Great Plains. Several species of *Schistocerca* are commonly referred to as bird grasshoppers because of their powerful flight; they are close relatives of "two of the most destructive grasshoppers in the world": the **Desert Locust,** S. *gregaria,* which has migrated in devastating hordes across vast areas of Africa and southwestern Asia, and S. *paranensis,* a South American species.

Longhorn Grasshoppers and Katydids (Tettigoniidae)

The tettigoniids, or longhorn grasshoppers, may be distinguished from the acridids, or grasshoppers, by their long filamentous antennae, the ear slit at the base of the front tibiae, the four-segmented tarsi, and the swordlike female ovipositor. Ocelli are usually present. Katydids are predominantly green, have exceedingly long antennae, are more often heard than seen; the characteristic shrill song interspersed with an occasional "Katy-she-did" accounts for the name.

(17) NORTHERN KATYDID: *Pterophylla camellifolia.* **Fig. 38A.**

Also known as the true katydid, rough-winged katydid, or eastern katydid. **Range:** East of the Rocky Mts. **Adult:** Dark green, tinged with yellow, dark brown triangle on back where bases of the two wings overlap; the delicate gauzelike wings are relatively short and rounded, the eyes have spots like pupils. In the fall during mating season, the stridulating males can be heard in the treetops beginning at dusk. Eggs are inserted in loose bark of trees and hatch the following spring. One brood. **Length:** 1–1.4″. **Food:** Foliage of trees.

(18) FORK-TAILED BUSH KATYDID: *Scudderia furcata.* **Fig. 36.**

Range: Throughout North America, common in the West. **Adult:** Dark green, with long narrow tegmina. It gets the name from the peculiar forked appendage at the tip of the abdomen in the male. Oval, flat eggs are inserted in the edge of leaves in the fall and hatch in the spring. One brood. **Length:** 1.75″ (to tip of folded wings). **Food:** Foliage, stems, flowers, fruit of trees. The western form is sometimes destructive to oranges, eating holes in the rind of young fruit.

(19) BROAD-WINGED KATYDID: *Microcentrum rhombifolium.* **Fig. 37.**

Range: Southern states from the Atlantic to the Pacific coasts, northward to Utah, Colorado, and Kansas. **Adult:** Green, with broad angular wings. Flat, oval eggs are laid overlapping one another in a double or single row, glued to the side of a twig or edge of leaf; they are laid in the fall and hatch in the spring. **Length:** 2–2.5″ (to tip of folded wings). **Food:** Many kinds of trees; common on willow, cottonwood, citrus. Their small numbers preclude economic damage. The **Angular-winged Katydid,** *M. retinerve,* an eastern species, is a paler green, head with yellowish tinge, pronotum with middle of front margin straight.

(20) MORMON CRICKET: *Anabrus simplex.* **Fig. 38B.**

Range: West of the Missouri River to the Cascade and Sierra Nevada ranges, from the Canadian border to California and northern Arizona. **Adult:** Dark brown, black, or green; stout-bodied, wings absent. The tegmina are vestigial, extending a little beyond the pronotum in the male and largely concealed by the pronotum in the female; adult males use these wing stubs to make chirping or singing sounds. The swordlike ovipositor of the female curves upward and is about the same length as the hind femora. Dark brown (later gray) eggs are deposited in midsummer, hatch the next spring; unlike other grasshoppers' eggs, they are laid singly, as many as 100 closely grouped just below the ground surface. **Length:** 1.4–1.8″. **Nymph:** Black, with white markings on edge of shield just back of the head. Requires 60 days to reach maturity; seven instars. They have a strong tendency to migrate all their lives, moving in bands on favorable days, up to one mile a day. Cannibalism is prevalent. **Food:** A great variety of range grasses and weeds; wheat and other cereals, alfalfa, truck crops. Severe outbreaks have occurred, as when the Mormons settled Utah in 1848; a great flock of gulls saved the settlers' first crop, enabling the colony to survive. Indians used the "crickets" as food.

(21) COULEE CRICKET: *Peranabrus scabricollis.* **Fig. 38C.**

Named for the dry ravines or coulees where this grasshopper is found. **Range:** Montana, Idaho, eastern Washington. **Adult:** Reddish brown, with lighter markings. Stout-bodied; wings completely absent in the female. The short wing stubs of the male are used for chirping. Elongated eggs are inserted singly at the base of tiny grass stools, in arid sagebrush areas; they are laid in the fall and hatch early in the spring. In contrast to the closely related Mormon cricket, which continues egg laying all summer, the Coulee cricket disappears by July; like the Mormon cricket, it migrates in great hordes and is strongly cannibalistic. **Length:** 1.5″. **Food:** Weeds such as sagebrush, bitterroot, wild mustards; field and garden crops. Meadowlarks are sometimes effective in cleaning up an area of Coulee crickets.

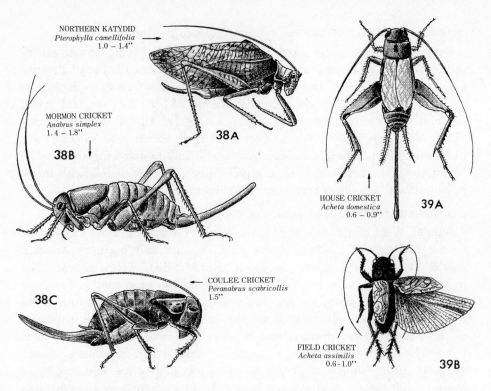

NORTHERN KATYDID
Pterophylla camellifolia
1.0 – 1.4"

38A

MORMON CRICKET
Anabrus simplex
1.4 – 1.8"

38B

HOUSE CRICKET
Acheta domestica
0.6 – 0.9"

39A

38C

COULEE CRICKET
Peranabrus scabricollis
1.5"

FIELD CRICKET
Acheta assimilis
0.6 - 1.0"

39B

Crickets (Gryllidae)

The Gryllidae, or "true crickets," are very similar to the longhorn grass-hoppers (17). They have long filiform antennae, three-segmented tarsi, pear-shaped ovipositor in the female; tympana or ears are usually located on both sides of the front tibiae. Ocelli are present in some, absent in others. Some species have fully developed wings; many are wingless or have only stubs. When fully developed and at rest, the tegmina lie flat on the back, turned down sharply at the edges. The chirping of the male is produced by stridulating organs similar to those of the katydids; they are on the dorsal surface of the tegmina, near the base, and consist of a file, scraper, and tympanum or vibrating area. The tegmina are raised above the body at a 45-degree angle, and the files are rubbed against the scrapers. Crickets lay their eggs singly in the ground, and singly in rows in plant tissues. As a whole they do little harm.

(22) HOUSE CRICKET: *Acheta (Gryllus) domestica.* **Fig. 39A.**

House crickets are difficult to distinguish from field crickets. **Range:** Throughout North America. **Adult:** Shiny black. Lives outdoors, often in garbage dumps, is fond of warmth and enters houses or bakeries mainly in the fall when the weather turns cold. Eggs are laid in the ground during the summer, hatch the following spring. **Length:** .6–.9". **Food:** Young plants outdoors are injured occasionally, and woolens and carpets indoors are sometimes damaged. They are chiefly a nuisance indoors, with their incessant chirping, which is faster than that of the field cricket, the rate depending on the temperature.

(23) FIELD CRICKET: *Gryllus* sp. *(Acheta assimilis)*. **Fig. 39B.**

Very common throughout the country in various forms difficult to distinguish. **Range:** Throughout North America. **Adult:** Black or brown. Eggs are laid in the ground, preferably in damp places. Hibernates in egg stage in the North, in nymph or adult stage in the South. **Length:** .6–1″. **Nymph:** Develops to maturity in 9 to 14 weeks, undergoes 8 to 12 instars. Adults appear in July and August. One brood; in southern climates up to three. **Food:** Cotton (seedlings), seeds of alfalfa and cereals, tomatoes, cucurbits, peas, beans; paper, woolen and fur garments indoors. Sometimes abundant enough to cause serious damage. They often get indoors and are mistaken for the house cricket.

(24) SNOWY TREE CRICKET: *Oecanthus fultoni (niveus)*. **Fig. 40.**

Range: Throughout North America. **Adult:** Whitish, shaded with pale green. Single black spot on front side of each of the first two antennal segments. Wings of the male are broad, paddle-like, lie flat on the back at rest; forewing of the female is narrow, wrapped closely about the body. Their incessant chirping is plainer at night with fewer competing noises, the rate being dependent on the temperature. The stridulating male raises its wings straight up, exposing a gland on the metathorax which emits a secretion very attractive to the female, who devours it eagerly before mating. (The male's musicianship, however, is lost to his mate—she can't hear the chirping.) Eggs are laid in pin-size holes made in the bark of trees, after a drop of excrement is placed there; they are laid in the fall, sealed over with a secretion, and hatch in the spring. **Length:** .7″. **Nymph:** Almost white. Feeds on foliage, small insects, fungi, pollen, ripe fruit; grows slowly, maturing late in summer. One brood. **Food:** Apple, peach, plum, prune, cherry, berries. Egg holes (usually in single rows) sometimes injure twigs and brambles of bush fruits. The **Four-spotted Tree Cricket,** *O. quadripunctatus,* and the **Black-horned Tree Cricket,** *O. nigricornis,* are more eastern in distribution.

Mole Crickets (Gryllotalpidae)

The mole crickets are closely related to the "true crickets" (Gryllidae) but differ greatly in appearance. They are distinguished by the broad front legs adapted for digging. The tibiae are very wide, terminated in strong sharp "teeth" or toes called dactyls; two tarsal segments, shaped

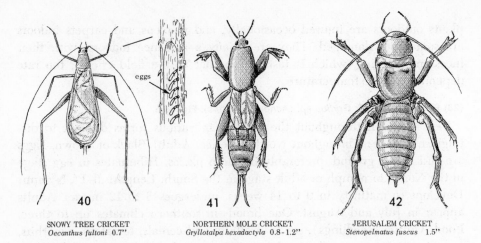

40
SNOWY TREE CRICKET
Oecanthus fultoni 0.7"

41
NORTHERN MOLE CRICKET
Gryllotalpa hexadactyla 0.8-1.2"

42
JERUSALEM CRICKET
Stenopelmatus fuscus 1.5"

like blades, move across the dactyls like the teeth of a mowing machine. In the genus *Gryllotalpa,* the front tibiae have four dactyls (Fig. 8C); in the *Scapteriscus* they have two. Though they are subterranean, mole crickets have stridulatory organs like those of the true crickets.

(25) NORTHERN MOLE CRICKET: *Gryllotalpa hexadactyla.* **Fig. 41.**

Our most common and widely distributed species. **Range:** From Canada to the southern part of South America. **Adult:** Brown, covered with fine velvety hairs. They spend most of their time in the ground digging tunnels, but are good fliers. The female deposits her eggs in a chamber at the end of the burrow; she guards the nymphs until the first or second molt is completed. **Length:** .8-1.2". The life cycle requires about one year. **Food:** Roots of various plants; frequently destructive in gardens, nurseries, and orchards. The European species, *G. gryllotalpa,* now established in some eastern states, is much larger (about 2" long), has a life cycle of three to four years. The **Southern Mole Cricket,** *Scapteriscus acletus,* and the **Changa,** *S. vicinus,* prefer dry soils, have a life cycle of one year. The latter—a native of Puerto Rico—has become a pest in Florida and other southern states. The **Pigmy Mole Cricket,** *Tridactylus minutus* (Tridactylidae), is aquatic, burrows in the sand, and hops about at the edge of lakes and streams, swims by means of modified plates (natatory lamellae) on the hind tibiae; it is blackish brown with buff-colored markings, about .2" long, the hind tarsi lacking. *T. apicalis*—plain shiny brown—*has* hind tarsi.

Cave and Camel Crickets (Gryllacrididae)

Cave and camel crickets have curved, humped bodies, are medium to large in size, and are wingless. They are usually brown and have hind legs with large femora and long spined tibiae. They are common but not often seen, since they live in caves or are nocturnal in habit.

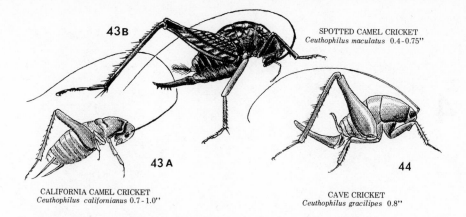

43B

SPOTTED CAMEL CRICKET
Ceuthophilus maculatus 0.4-0.75"

43 A

44

CALIFORNIA CAMEL CRICKET
Ceuthophilus californianus 0.7-1.0"

CAVE CRICKET
Ceuthophilus gracilipes 0.8"

(26) JERUSALEM CRICKET: *Stenopelmatus fuscus.* **Fig. 42.**

Also called sand cricket. Mexicans call it *Niña de la Tierra* (child of the earth) because of its large human-like head. A popular name in some places is "potato bug," possibly because they occasionally attack potato tubers. **Range:** West of the Mississippi River; common along the Pacific coast. **Adult:** Smooth, shiny, amber brown with darker stripes across the abdomen; wingless. Head large, with powerful jaws; legs spiny. They live in the soil and under rocks during the day, come out at night in search of food. Clumsy, painfully slow in movement, they are often found lumbering on the surface in the daytime in spring during mating season. The white sperm sac of the male is torn off by the female vulva in mating. Masses of oval white eggs are laid in holes in the ground which look like miniature birds' nests. **Length:** 1.5". **Food:** Principally other insects; probably beneficial but not numerous enough to be important. Some species feed on plant roots, tubers, decaying animal and vegetable matter. The females are cannibalistic and often devour their mates. They bite but are harmless. (Sometimes placed in a separate family Stenopelmatidae.)

(27) CALIFORNIA CAMEL CRICKET: *Ceuthophilus californianus.* **Fig. 43A.**

Range: Pacific coast states, British Columbia, Utah. **Adult:** Pale brown, with darker bands across the abdomen. Ovipositor of the female short, with six or seven spines at the tip. They live in the soil, under stones or logs, are often found in or at the entrance to gopher holes. **Length:** .7–1". The **Spotted Camel Cricket** (Fig. 43B), *C. maculatus*—from .4" to .75" long—ranges from Manitoba and the Dakotas east to Maryland and south to Arkansas. *C. brevipes* is darker brown and more mottled than *maculatus*—about .6" long—and ranges from eastern Canada to Kentucky, where it is more often found in caves and other recesses. The **Cave Cricket** (Fig. 44), *C. gracilipes,* is about .8" long and common in eastern North America.

4

DERMAPTERA: Earwigs

The earwigs are medium to large insects, usually brownish in color, with short, leathery forewings and longer membranous hindwings which fold straight back under the front pair. Some species are wingless. The body terminates in a pair of curved, pincers-like cerci which can exert a fairly good pinch; they use them in defense, and to grasp attacking ants from their bodies. Compound eyes are present but not ocelli; mouthparts are adapted for chewing. Eggs are laid in a nest in the soil and are brooded by the mother. The nymphs are cared for until ready to leave on their own. Metamorphosis is gradual. Feeding habits are exceedingly variable; some species are at various times plant feeders, scavengers, and predators; others are largely carnivorous.

While some species of earwigs are considered pests, it appears that their value as predators may outweigh the harm they do. Much of the objections to these insects stems from their offensive appearance and odor and their habit of showing up in garbage cans and occasionally indoors. (Long-persisting putrid odors come from secretions sprayed at enemies.) There are about twelve native species of earwigs and by last count some seven more unintentionally introduced in ship cargoes. Some may be a valuable addition to our fauna of beneficial species. They are found mostly in the warmer regions having adequate moisture, and are most common in our coastal areas.

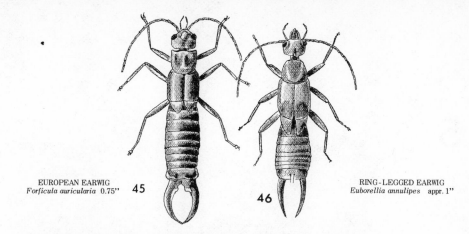

EUROPEAN EARWIG
Forficula auricularia 0.75" **45**

46

RING-LEGGED EARWIG
Euborellia annulipes appr. 1"

(1) EUROPEAN EARWIG: *Forficula auricularia* (Forficulidae). **Fig. 45.**

Range: Eastern and western coastal areas mostly. Native of Europe, first came to the Pacific Northwest in the early 1900's. **Adult:** Reddish brown. Cerci of the male curved, those of female nearly straight. It can fly only by taking off from a high place; the wings are not strong enough to take off from the ground or for sustained flight. **Length:** .75". **Nymph:** For the first few days after eggs hatch, the nest is kept closed to prevent escape. Four molts; period from hatching to maturity lasts about 68 days. Broods average 30. They hibernate in the egg stage and as adults under boards, stones, and cells in the ground. One or two broods. **Food:** Various garden flowers and vegetables, ripe fruit, garbage, a variety of insects including fleas. An introduced parasitic fly, *Bigonicheta spinipennis* (Tachinidae), contributes substantially to control.

(2) RING-LEGGED EARWIG: *Euborellia annulipes* (Labiduridae). **Fig. 46.**

Range: Throughout North America; most common in the South and Southwest. **Adult:** Brown to blackish, yellowish brown on underside. Legs yellowish, with brown rings on femur and tibia. Antennae 16-segmented, black, except third and fourth segments (from the apex), which are white. Eggs hatch in about 14 days. **Length:** Approximately 1". **Nymph:** Five molts; average time from hatching to maturity is 80 days. **Food:** Plants, other insects. It is said to feed on grain and grain insects in Nebraska, and to frequent packing plants in other parts of the country. In California it is reported to feed on the larvae of eye gnats (*Hippelates*). In corn processing plants in Indiana it is called the "starch bug." *E. cincticollis*, very similar in appearance, is a native of western and equatorial Africa and now established in California, where it is sometimes found in the company of *Labidura riparia*. The latter, also a newcomer to California, is found in irrigated alfalfa fields, and appears to be a valuable predator, primarily of soil- and litter-inhabiting insects; it will also climb alfalfa plants and feed on aphids. It occurs in the southern states from Florida to Texas.

5

PLECOPTERA: Stoneflies

This is a small order of aquatic insects, also called salmonflies. The adults are found on plants or stones along streams, or flying over the water. The nymphs, called naiads, require well-aerated moving water and usually live in streams with stony or gravelly bottoms; some northern species live in lakes. In the aquatic stage most of them live on vegetation or are omnivorous; others are carnivorous; some adults feed on green algae, but many apparently feed little or not at all. A species found in the Northwest, *Brachyptera* (*Taeniopteryx*) *pacifica* (Fig. 47A), occasionally injures the buds of peach, apricot, and plum trees in the spring, especially those near the Columbia River or other large streams. Stoneflies are important as a source of food for game fish; they are also eaten by frogs, turtles, and dragonfly nymphs. Anglers refer to the adults as "browns" and copy them in wet and dry flies.

Most of the adults have two pairs of long membranous wings; the hindwings are wider than the forewings, and are folded fanlike over the body when at rest. A few species are brachypterous or are wingless. They have primitive mouthparts, two or three ocelli, three-segmented tarsi, many-segmented tapered antennae. Some are active in daytime, others at night and are attracted to lights. The eggs are dropped or otherwise deposited in the water; some are spherical, with a sticky coating (when wet); others are elongated or of various shapes, with a plate for adhering to surfaces. Egg production is generally high, often exceeding a thousand for one female, sometimes several thousand. Metamorphosis is incomplete.

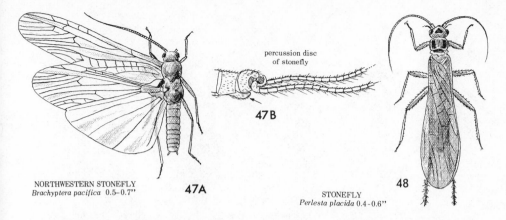

percussion disc
of stonefly

47B

NORTHWESTERN STONEFLY
Brachyptera pacifica 0.5–0.7"

47A

48

STONEFLY
Perlesta placida 0.4–0.6"

The nymphs commonly undergo twenty-two molts or more, and usually require a year to mature, in some cases two or three years. They are flat, have stout legs with well-developed claws, long tapered antennae set far apart on the head, and an extremely long pair of cerci. The tracheal gills are tufts of hairs or filaments located internally (in the rectum) or externally (along the sides of the head, thorax, the first few abdominal segments, and anus). In some species remnants of the gills are carried over into the adult stage and are useful in identification.

The so-called fall and winter stoneflies (found in the families Nemouridae, Taeniopterygidae, Capniidae, and Leuctridae) are interesting for their strange reversal of the hibernation pattern of most insects. The nymphs do not develop beyond the first instar stage until the cold weather approaches and the days grow short, when their development is quickened rather than arrested. The adults emerge in the coldest period, from November to March, and are active on the warmer days, when they feed and mate. The largest and one of the most abundant of these is *Taeniopteryx nivalis,* a dark brown species with wingspread of more than one inch, found along the flowing streams in Illinois and eastward.

(1) STONEFLY: *Perlesta placida* (Perlidae). **Fig. 48.**

Perlidae is the largest family of stoneflies, named for the roundness of the head, said to "rival a pearl in brilliance." The apical segments of the palpi and two basal segments of the tarsi are greatly reduced. Some males are brachypterous; and some (*Classenia, Acroneuria*) have a percussion disc (Fig. 47B) or "hammer" on the underside of the ninth abdominal segment which they strike against hard surfaces. **Range:** *P. placida* is widely distributed east of the Rocky Mts. **Adult:** Brown to blackish; head and costal margin of wings yellow, abdomen yellowish. **Length:** .4–.6"; wingspread: .6–1". Females are a little larger than males. *Perlesta* males do not have the "hammer."

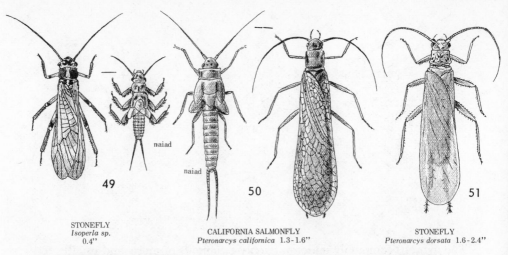

49

50

51

STONEFLY
Isoperla sp.
0.4"

CALIFORNIA SALMONFLY
Pteronarcys californica 1.3-1.6"

STONEFLY
Pteronarcys dorsata 1.6-2.4"

(2) CALIFORNIA SALMONFLY: *Pteronarcys californica* (Pteronarcidae). **Fig. 50.**

Range: Washington to California, eastward to Kansas and New Mexico. **Adult:** Dark grayish brown, with lighter or reddish and darker shades, paler on underside; head blackish, wings with blackish veins. **Length:** 1.3" (males) to 1.6" (females); wingspread: 2.9–3.4". Anglers use the adults for bait; Indians used them for food, scooping up large numbers from streams after shaking the overhanging bushes. **Naiad:** Inhabits the shallow parts of swift-flowing large and small streams; it is common in the upper reaches of the Sacramento River and its tributaries; attains a length of 3.5".

(3) STONEFLY: *Pteronarcys dorsata.* **Fig. 51.**

This is our largest species. **Range:** Labrador to Tennessee, westward to Kansas, Minnesota, Saskatchewan, and Alaska. **Adult:** Dark brown, with yellowish markings; wings somewhat clouded, veins blackish. **Length:** 1.6–2.4"; wingspread: 2.8–4.2".

(4) STONEFLY: *Isoperla patricia* (Perlodidae).

Range: British Columbia to California, east to Montana and Colorado. **Adult:** Pale yellowish brown, with darker pattern on head and pronotum; posterior half of mesothorax and metathorax dark brown, wings glassy, with costal margins pale yellowish and veins light brown. **Length:** .4". **Naiad:** Pale yellowish, with darker areas, dark longitudinal stripes on abdomen (there are light and dark forms), about .5" long. Three ocelli form an equilateral triangle on the head. The **Two-lined Stonefly,** *I. bilineata*—the common eastern species, ranging from Newfoundland and New Jersey to Saskatchewan and Colorado—is yellowish, with brown spot in ocellar triangle of head; pronotum brown, with broad, yellow median stripe; appendages darker toward the tips, wings transparent greenish, with brownish veins; from .4" to .6" long, wingspread from .7" to 1". The adult and nymph of *Isoperla* sp. are pictured in Fig. 49.

6

EMBIOPTERA: Embiids or Webspinners

This is a very small order of extraordinary insects which occur mostly in the tropics and subtropics but range into the southern part of the temperate zone. The females are always wingless. The males of most species have two similar pairs of long membranous wings with simple venation; they fold them flat over the back when at rest and at such times resemble stoneflies. They are readily distinguished by the large head and eyes, the short stout legs with three-segmented tarsi, and the one- or two-segmented cerci. The neck is flexible and the head moves about somewhat in the manner of mantids. The enlarged first segment of the front tarsi contains silk glands, and the thread is used to fashion silken tunnels. They live in colonies with interconnecting tunnels, constructed in or near their food—in dry grass, on the ground under leaves, in lichens, or under bark. The hind femora are enlarged and adapted for running rapidly backwards in the tunnels to elude their enemies; the wings of the males are adapted for this movement by being flexible, and can be bent forward over the head so they will not be caught on the sides of the tunnel.

The elongated curved eggs, with sloping lids, are laid in clusters on the walls of the tunnels; the female guards them well, as she does the young nymphs. Females are without metamorphosis, changing outwardly in size only as they mature. Winged males undergo gradual metamorphosis, and are unique in that the wing pads develop internally and do not appear as typical wing pads until the last nymphal instar, which is analogous to the pupal stage of insects having a complete metamorphosis. Some species are

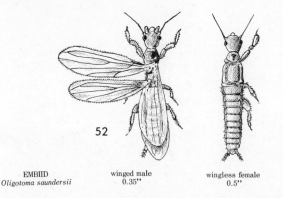

52

EMBIID
Oligotoma saundersii

winged male
0.35"

wingless female
0.5"

abundant, and the winged males are frequently attracted to lights; they often appear in houses, where they like to collect on the white walls near lamps. Most species reproduce in the normal manner, but a few are parthenogenetic and only females are known to occur.

(1) EMBIID: *Oligotoma saundersii* (Oligotomidae). **Fig. 52.**

Three species of this genus are common in tropical and subtropical America, and are often attracted to lights; they are believed to have been introduced from Asia. **Range:** *O. saundersii* occurs in Florida, Texas, the West Indies, and Cuba. **Adult:** Male light brown, head darker; female all chocolate-brown. **Length:** .35" (male) to .5" (female). Abundant in many localities. *O. nigra* is slightly smaller, dark chestnut-brown, with body, legs, and pigmented areas of wings lighter shades; head one-half longer than wide; eyes large, bulging; antennae with at least 20 segments; female all dark chocolate-brown; it is abundant in Arizona and southern California, where it is closely associated with the date palm, colonies often being found among leaf bases and offshoots of the trunks. *O. humbertiana* occurs in Mexico. *Gynembia tarsalis*, widely distributed throughout California, also found in Arizona, is often associated with *O. nigra* under stones in southern California, also found in open fields in dry grass or soil cracks; the female is light yellowish brown, head reddish brown; a little more than .5" long; head slightly longer than wide; the male is unknown. *Oligotoma* and *Gynembia* may be separated by the "sole-bladder" on the hind tarsi—in *Oligotoma* there is only one, and in *Gynembia* there are two. *Oligotoma* is widely distributed across the southern part of the U.S., with both sexes present; *Gynembia* is restricted to California and Arizona, and only females are known to occur.

7

ISOPTERA: Termites

Sometimes called white ants, the termites bear only a superficial resemblance to ants and are much more primitive in origin. Though they have evolved into social insects, they appear to be more closely related to cockroaches. They may be distinguished by the broadly joined thorax and abdomen as contrasted with the waist or petiole of ants. In the winged forms, termites have two pairs of long, narrow, opaque wings of equal size, whence comes the name of the order. Ants have two pairs of transparent wings unequal in size and much shorter in relation to the length of the body. Like ants (also the social wasps and bees), termites live in colonies and have a caste system, as described in (1) and (3) below. All castes in the termite colony are bisexual; ant colonies and those of the other social Hymenoptera are composed mostly or entirely of females, the males being short-lived and relegated to the sex role only. It is believed that caste is determined among termites after hatching—not in the egg stage as appears to be the case with ants—by hormones secreted by the soldiers and reproductives which are passed on to the nymphs in the oral and anal feeding and inhibit their growth to soldiers or reproductives.

Termites are extremely sensitive to light, and more especially air currents and humidity, excepting the winged forms (which are attracted to lights during swarming) until they have shed their wings and paired off to establish a new colony. Most termites feed entirely on wood, which is literally digested for them by protozoa living in the gut; these microscopic single-celled animals take up the tiny particles of ingested wood, making the cellulose available to their host. The relationship is mutualistic; each, in fact, is dependent on the other for survival. It is believed that the protozoa are passed from one termite to the other in the process of feeding predigested food to the young. Not all termites, however, live on wood. Some mound-building tropical species raise fungus on spongelike masses of their own excrement, in large chambers much like those of our leaf-cutting ant, *Atta texana* (Fig. 1208). In the tropical rain forests and even arid regions, where there are no earthworms, termites serve the useful purpose of hastening the conversion of dead vegetable matter into humus and aerating the soil. In parts of Africa and Australia certain species construct huge mounds twelve feet or more in height and may be so numerous as to be destructive to grasslands. The so-called magnetic termites of northern Queensland, Australia, construct slablike mounds with jagged tops, the axes pointed north and south like a magnetic compass.

(1) EASTERN SUBTERRANEAN TERMITE: *Reticulitermes flavipes* (Rhinotermitidae). Figs. 53, 54, 57d, 59d.

Range: Eastern United States; most numerous in the southeastern and Gulf states. Scattered localities in southern Ontario, particularly around Toronto. **Adult:** Four forms or castes: (1) Reproductives—fertile males and females; they are sometimes called "imagos" or "alates." They have yellowish brown to black bodies, pigmented compound eyes, whitish opaque wings of equal size which are shed after nuptial flight and pairing off to found new colonies. When the nest is well established, the queen turns to egg laying, attended by certain workers; her abdomen becomes greatly enlarged, egg production may reach a dozen or more a day for many years. The male or king remains in the nest. Flights occur in spring after a warm rain, and any time through spring and summer. King and queen remain in the old colony, which may survive for many years. (2) Secondary reproductives—pale in color, with compound eyes, and only wing buds. The females—"secondary or supplemental queens"—augment egg production of the queen if necessary, but are unable to produce winged progeny that can found new colonies. (3) Sterile workers—grayish white, eyeless, wingless. Most numerous of the forms and most destructive. (4) Sterile soldiers—like workers except for long narrow heads, long mandibles. **Nymph:** When the colony is started they are cared for by the parents, later by workers. Four instars are completed in three to four months. In older colonies development takes longer, and they grow larger. **Food:** Wood and wood products. They eat the soft spring growth of wood, hollowing out timbers along the *length* of the grain, causing more damage than nonsubterranean termites; they eat *all* the wood, leaving only brown specks of excrement. They are most destructive to second-growth lumber with large amounts of sapwood and sometimes injure living trees. They must maintain contact

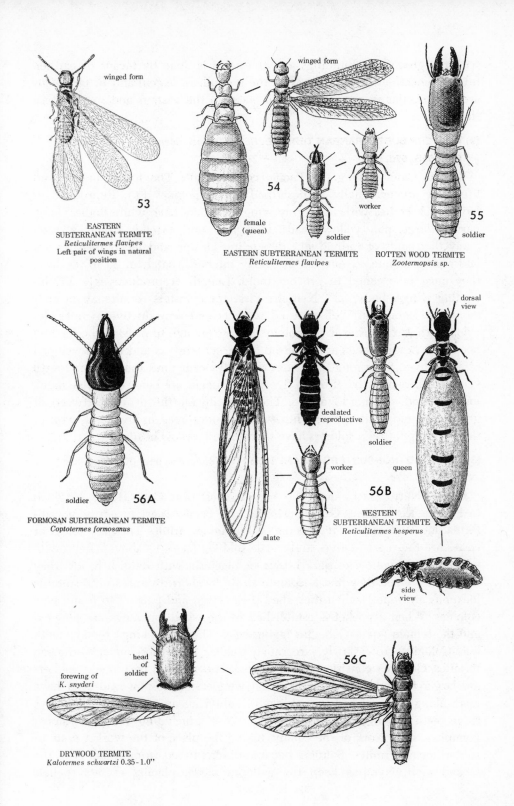

53

EASTERN
SUBTERRANEAN TERMITE
Reticulitermes flavipes
Left pair of wings in natural
position

54

EASTERN SUBTERRANEAN TERMITE
Reticulitermes flavipes

55

ROTTEN WOOD TERMITE
Zootermopsis sp.

56A

FORMOSAN SUBTERRANEAN TERMITE
Coptotermes formosanus

56B

WESTERN
SUBTERRANEAN TERMITE
Reticulitermes hesperus

56C

DRYWOOD TERMITE
Kalotermes schwartzi 0.35-1.0"

with the ground; in buildings and trees this is done by means of earthen tubes connecting the galleries with the ground. *R. hageni* and *R. virginicus* are very similar in appearance, also occur in the eastern and southeastern states.

(2) WESTERN SUBTERRANEAN TERMITE: *Reticulitermes hesperus.*
   Figs. 56B, 57d, 59d.

Range: Chiefly along the Pacific coast. **Adult:** The forms, habits, and life cycle are essentially as described for *R. flavipes* (1). Reproductives have black bodies with light gray wings. Soldiers have white bodies, long, narrow, large, pale yellow heads, long jaws, and no eyes. Colonies reach many thousands of individuals, depending on age and location. Eggs are laid one at a time, at one- to three-day intervals, hatch in 30 to 90 days; they must be cleaned to prevent mold. **Length** (reproductives): .4″, including wings; body: .2″. **Nymph:** First three instars require about four months to complete. Well-matured workers and reproductive nymphs develop in the sixth instar if the colony is large enough to provide the feeding necessary. Seven instars are required for larger workers and fully developed reproductives. Swarming does not ordinarily occur until the third or fourth year. Workers may live three to five years; queens are believed to live much longer. **Food:** As in (1) above. They also invade the dead heartwood of grapevines and weaken them. *R. tibialis* is widespread in arid lands west of the Mississippi River; soldiers have short, broad, yellow heads.

(3) DRYWOOD (NONSUBTERRANEAN) TERMITE: *Kalotermes* spp. (Kalotermitidae).
   Figs. 56C, 57c.

Range: Narrow strip along the Atlantic coast from Cape Henry, Virginia, to Florida Keys; along Gulf of Mexico and Pacific coast to northern California, local infestation at Tacoma, Washington. **Adult:** The only forms or castes are the reproductives and sterile soldiers [see (1) above]. Reproductives are light yellow to dark brown or blackish, with reddish heads; they have wings when they first become mature, fly short distances until suitable locations are found, when they shed their wings and pair off to found new colonies. When a pocket is established in the wood, openings are plugged and the female (queen) begins laying eggs. The male (king) remains with her. Colonies grow slowly, are much smaller than those of subterranean termites. Galleries consist of series of pockets connected by tunnels. Soldiers are larger than the reproductives and wingless; their powerful jaws have teeth along the inner edge (soldiers of subterranean termites do not have these teeth). **Length** (reproductives): .35–1″, including wings. **Nymphs:** Nymphs do the work of the colony, taking the place of the worker caste of subterranean termites. Soldiers become differentiated from others after the second molt. Nymphs keep the galleries clean, placing excretal pellets (made from feces and partly digested wood) in unused galleries or pushing them out through openings in the wood which are later sealed over. Holes are also made for ventilation and to control humidity; they are sealed over with a brownish or blackish secretion that hardens into a paper-thin plug. **Food:** Wood structures of homes, piled lumber, furniture. They attack soft spring growth as well as denser summer growth, cutting galleries *across* the grain. Wood is contacted directly, no connection being made with the ground. Infestations are evidenced by discarded wings, by piles of excretal pellets, which have concave surfaces (in contrast to convex surfaces of pellets discarded by powder-post beetles); by plugged openings, surface blisters, and thick narrow shelter tubes made of excretal pellets cemented together for passage from one piece of wood to another. The **Western Drywood Termite,** *Incisitermes minor* (Fig. 57a), is about as destructive to buildings in southern California as are subterranean termites. **Powder-post Termites,** *Cryptotermes* spp. (Fig. 57b), also attack sound dry wood, are found only in Florida and Louisiana.

(4) DAMPWOOD TERMITE: *Paraneotermes simplicornis.*

Range: Southern Florida, southwestern states, Pacific coast states. **Adult:** Winged reproductives are dark brown, including wings. Soldiers have flat, brown or yellowish brown heads; mandibles are short, thick near the base, with narrow tips. **Nymph:** The abdomen is usually spotted, where dark contents of intestines show through. **Food:** Wood structures, posts, poles, trees. Colonies are located in wood partially or wholly in the ground. They require moisture, enter the wood below ground, feed on sapwood and heart-wood of living trees as well as timbers. They may leave wood to burrow into the ground, have been known to attack fruit trees by burrowing between the roots. *Neotermes castaneus* also prefers damp wood, occurs in the same range. *Zootermopsis* spp. (Figs. 55, 58a) are among the largest species, occur along the Pacific coast from British Columbia to Baja California, in Idaho, Montana, Nevada, Arizona, and New Mexico; they are found in stumps, fallen trees, old logs, and may invade buildings. The **Pacific Dampwood Termite,** *Z. angusticollis,* is found in California and western Oregon; nymphs are about .5″ long, soldiers about .75″ long. The latter may be recognized by the large brownish head and long dark mandibles; winged forms are brownish, nearly 1″ long, and are often seen at lights. Carpenter ants also infest partially decayed wood; their work may be recognized by the clear chambers cut across the grain and the accumulations of fine to coarse wood fibers removed from the chambers. Wood wasps or horntails also infest timber (usually fire-killed or damaged before it is felled) and sometimes leave exit holes about one-quarter inch in diameter in walls or floors where the adults emerge. Various beetles—lyctid, anobiid, cerambycid, bostrichid—

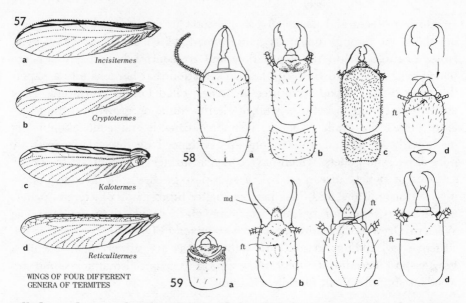

WINGS OF FOUR DIFFERENT
GENERA OF TERMITES

Incisitermes

Cryptotermes

Kalotermes

Reticulitermes

called powder-post beetles also attack wood and cause extensive damage to buildings [see reference to pellets in (3) above].

A key to termites

Imagos (sexual reproductives) and more especially the workers are difficult to identify. Imagos may be recognized by the wing venation and dentition (teeth or notching) of the mandibles; the latter is about the same for both castes but can be misleading, since it is almost identical in several genera. The wing venation of four genera—*Incisitermes, Cryptotermes, Kalotermes,* and *Reticulitermes*—is shown in Fig. 57. The wing venation of *Coptotermes* and *Heterotermes* is similar to that of *Reticulitermes*. The wing membrane of *Coptotermes*, however, has many minute hairs, while the anterior costal margin of the forewing in *Heterotermes* has a definite row of small, fine hairs; the forewing of *Reticulitermes* has very sparse or no hairs. *Coptotermes* imagos are also differentiated by the round head and large ocelli (one alongside each eye). In the genera above, ocelli are present, the antennae have less than 24 joints, and the cerci are two-jointed; ocelli are absent in *Zootermopsis*, the antennae 24-jointed, the cerci four- to eight-jointed. The **Formosan Subterranean Termite,** *Coptotermes formosanus* (Fig. 56A), was first discovered in the continental United States in 1965 at a shipyard in Houston, Texas; infestations have since been found in New Orleans and Lake Charles, Louisiana, and in Galveston, Texas. Infestations at Lake Charles were found mainly in dead trees and stumps in a swampy area. The species is very damaging to structures, is found in Asia, South Africa, and is a serious pest in Hawaii.

The soldier caste is less difficult to identify than the others. Eight genera of termite soldiers may be recognized by the following key* and associated Figs. 58 and 59:

1. Antennae with at least 23 joints, cerci long and four- to five-jointed (Fig. 58a) .... *Zootermopsis*
   Antennae with fewer than 23 joints, cerci short and two-jointed 2
2 (1). Head short, truncated in front, mandibles comparatively short (Fig. 59a) .... *Cryptotermes*
   Head long and narrow, not truncated in front 3
3 (2). Head without fontanelle (Figs. 58b, 58c) 4
   Head with fontanelle (ft—Figs. 58d, 59b, 59c, 59d) 5
4 (3). Pronotum with anterior margin deeply incised and angular (pr—Fig. 58c) .... *Incisitermes*
   Pronotum with anterior margin concave (pr—Fig. 58b) .... *Kalotermes*
5 (3). Fontanelle large and tubular (ft—Fig. 59c) .... *Coptotermes*
   Fontanelle not large and tubular, sometimes very small, circular, and distinct 6
6 (5). Head long and narrow, mandibles without prominent tooth, basal inner margin of left mandible with serrations, fontanelle small and circular, pronotum flat (Figs. 59b, 59d) 7
   Head shorter in relation to width, mandibles with prominent tooth; pronotum saddle-shaped, with front portion raised (Fig. 58d) .... *Amitermes*
7 (6). Mandibles long, slender, curved slightly inward at tips (md—Fig. 59b) .... *Heterotermes*
   Mandibles short, thick, hooked at tips (md—Fig. 59d) .... *Reticulitermes*

* The key and figures are taken from *Cooperative Economic Insect Report,* 16(47): 1091–1098 (Nov. 25, 1966); 16(50):1181 (Dec. 16, 1966), Plant Prot. Div., Agr. Res. Serv., U.S. Dept. Agr., Hyattsville, Md.

A *key* is an arrangement of characteristics, usually in couplet form, which describes species, genera, family or other grouping. Start with the first couplet, select one of the two contrasting statements which fits the specimen at hand, and proceed to the next couplet number shown at the end of the dotted line; continue until the appropriate line is reached ending with the name. The number in parentheses refers back to the couplet (not necessarily the one immediately preceding) that leads up to this point; it is useful when the key is long and one wishes to trace back from the name or other point.

8

PSOCOPTERA (CORRODENTIA): Booklice and Psocids

This is a small order of minute, frail, generally inconspicuous insects found in dark, damp places. They have large heads with prominent eyes, chewing mouthparts, 13- to 50-segmented antennae—sometimes extremely long—and a necklike prothorax. In the winged forms the forewing is much longer than the hindwing, the veins reduced and often "kinked"; at rest the wings are folded rooflike, as in the cicadas. In other forms the wings are reduced to a few scales or absent. The legs are all similar and suited to running. Species living indoors feed on paper, starch, and grain, and are sometimes quite numerous in food warehouses. The outdoors species live under bark or hidden on plants, in dead leaves and rotten wood. Psocids also feed on mildew and various microscopic molds as well as dead plant and other materials. They damage books by eating the glue or paste in bindings, and are sometimes injurious to dried specimens of insects and plants. Metamorphosis is gradual or incomplete.

(1) BOOKLOUSE: *Liposcelis divinatorius* (*terricolis*) (Liposcelidae). **Fig. 60.**

Range: Cosmopolitan. One of the most common species. **Adult:** Pale, soft-bodied, wingless, with large head, *two* distinct thoracic segments, and stout legs. Eggs are white, oval-shaped. Length of life cycle varies from 24 to 110 days. In heated buildings they reproduce continuously. Reproduction is by parthenogenesis (possibly for prolonged periods only). **Length:** .04″. **Nymph:** White, motionless on hatching; becomes darker with each molt. Hibernation is usually in the egg stage. Six to eight broods. **Food:** Dried

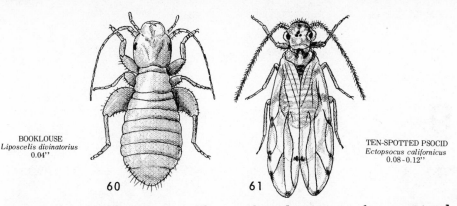

BOOKLOUSE
Liposcelis divinatorius
0.04"

60

61

TEN-SPOTTED PSOCID
Ectopsocus californicus
0.08–0.12"

paper and vegetable matter, molds, cereal products. It is often associated with the **Deathwatch,** *Trogium pulsatorium* (Trogiidae), a smaller species with head broader than wide, *three* distinct thoracic segments, and wings reduced to small scales. *T. pulsatorium* produces a ticking noise by striking a thickened knob on the ventral side of the abdomen in front of the genitalia against paper or similar thin, resonant material. The tapping is done by the female, sometimes five or six taps per second for a minute and continuing with pauses for an hour; it is believed to be a mating call. (See Deathwatch Beetles, p. 415.) *Lepinotus inquilinus* (Atropidae), a dark brown indoor psocid, also makes a tapping noise but with "a likeness to a creak"; each is three seconds long and they are emitted at intervals of 15 to 80 seconds.[*] This species and the smaller, lighter-colored *L. patruelis* are both widespread and common in warehouses and granaries; hundreds of the latter are often seen scurrying about in the light near windows.[†]

(2) TEN-SPOTTED PSOCID: *Ectopsocus (Peripsocus) californicus* (Caeciliidae).
Fig. 61.

Range: Atlantic coast states from Florida to New York; Tennessee, Kentucky, California. **Adult:** Grayish brown; top of head with tan spots, eyes bluish black. Wings clear; veins distinct, tan to light brown; forewings with ten brown spots at ends of veins, hindwings unmarked. Eggs are laid commonly in groups of six, covered with strands of silk spun from the mouth. **Length:** .08–.12". **Nymph:** It undergoes five or six instars, matures in 14 to 40 days; entire life cycle varies from 30 to 50 days. Ten to twelve broods in greenhouses. It reproduces by parthenogenesis; males are unknown. **Food:** Fungus growths on leaves of almost any tree or shrub, sooty mold when moist, fungus mites, own eggs and exuviae. They feed under a light webbing. *E. pumilis* is smaller, head and thorax grayish orange, with conspicuous hairs on dorsum, wings tan or pinkish, veins reddish brown; it is widespread, common in warehouses and granaries.

[*] J. V. Pearman, "On Sound Production in the Psocoptera and on a Presumed Stridulatory Organ," *Entomologist's Monthly Magazine*, 64:179–186 (Aug., 1928).
[†] *Trogium pulsatorium* is referred to as the "larger pale booklouse."

9

MALLOPHAGA: Chewing Lice

These are small to minute wingless insects with oval or elongated, flattened bodies, averaging about .08″ in length. They are pale or yellowish, with dark brown or black spots or bands. They have strong biting mouthparts on the underside of the head; the antennae are three- to five-segmented, the legs short. Also called biting lice or bird lice, they are external parasites of birds and fowl, less often of mammals, feeding mostly on feathers, hair or fur, and the outer layer of skin; some are known to feed on freshly drawn blood of the host. They are restricted to either single or closely related host species. Eggs are laid singly on feathers or hair of the host. Metamorphosis is incomplete. About 2,600 species are known world-wide.

(1) CHICKEN BODY LOUSE: *Menacanthus stramineus* (Menoponidae). **Fig. 62.**

Sometimes called large poultry louse. **Range:** Cosmopolitan. Pale yellow to brownish, elongated, with relatively long, stout legs; head short, triangular in outline, areas back of the eyes expanded laterally, antennae in grooves which open laterally. **Length:** .08–.1″. **Nymph:** Transparent when newly hatched, darkens after several molts. Full grown two weeks after hatching. **Host:** Chickens, turkeys, pigeons, peacocks. Found mostly around the vent and under the wings of young and old birds. It is known to feed on freshly drawn blood.

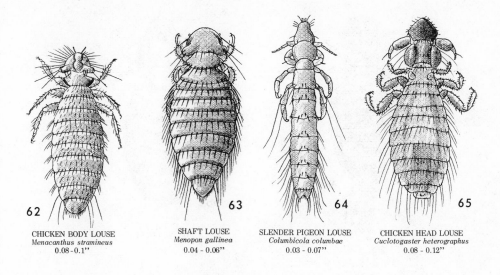

62 CHICKEN BODY LOUSE
Menacanthus stramineus
0.08-0.1"

63 SHAFT LOUSE
Menopon gallinea
0.04 - 0.06"

64 SLENDER PIGEON LOUSE
Columbicola columbae
0.03 - 0.07"

65 CHICKEN HEAD LOUSE
Cuclotogaster heterographus
0.08 - 0.12"

(2) SHAFT LOUSE: *Menopon gallinea.* **Fig. 63.**

Sometimes called common hen louse. **Range:** Cosmopolitan. Often found associated with the chicken body louse (1). **Adult:** Pale yellow; stouter, with longer head and shorter legs than the body louse, shorter overall. **Length:** .04–.06". **Host:** Chickens, turkeys, pigeons, ducks. Found around the vent as well as the back and breast; it clings to feathers rather than the skin, does not infest young chickens.

(3) SLENDER PIGEON LOUSE: *Columbicola columbae* (Philopteridae). **Fig. 64.**

Range: Throughout North America. **Adult:** Yellowish white; tarsus with two claws. **Length:** .03–.07". **Nymph:** Undergoes three molts, matures in three weeks. **Host:** Pigeons. The closely related **Chicken Head Louse,** *Cuclotogaster heterographus* (Fig. 65), is from .08" to .12" long, dark gray, clings to head at base of feathers or down; it passes from one bird to another by contact, is most injurious to young ones. The **Wing Louse,** *Lipeurus caponis,* is about .12" long, slender, grayish, with front of head hemispherical; it is found on large wing feathers of chickens and other fowl, feeds on hooklets. The **Fluff Louse** (lesser chicken louse), *Goniocotes gallinae,* is .04" long, with broad body and rounded abdomen, angulate temples; it is found on fluffy portion of feathers, feeds on barbs and barbules. The **Large Turkey Louse,** *Chelopistes meleagridis,* is .16" long, with temporal lobes extending far back and ending in long stylelike processes.[*]

[*] See Donald MacCreary and E. P. Catts, *Ectoparasites of Delaware Poultry Including a Study of Litter Fauna,* Bul. 307 (Tech.), Agr. Exp. Sta., Univ. of Delaware, Newark (1954); D. H. Brannon, *Poultry Pests and Their Control,* Ext. Cir. 330, Ext. Serv., Washington State Univ., Pullman (1962).

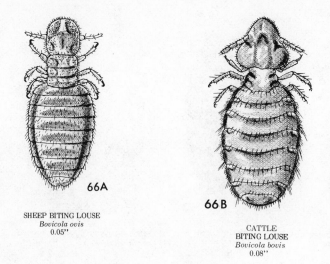

SHEEP BITING LOUSE
Bovicola ovis
0.05"

CATTLE
BITING LOUSE
Bovicola bovis
0.08"

(4) SHEEP BITING LOUSE: *Bovicola ovis* (Trichodectiae). **Fig. 66A.**

Range: Throughout North America. **Adult:** Body long and ovoid, yellowish brown, with dark bands; head rounded in front, about as wide as long and conspicuously red; legs short, tarsus with single claw. Eggs are attached to hair or wool. **Length:** .05" (male); female slightly longer. **Nymph:** Matures in about two weeks after hatching. **Host:** Sheep. The lice are most numerous and annoying in winter. The **Cattle Biting Louse** (red cattle louse), *B. bovis* (Fig. 66B), is about .08" long, a pest of cattle in winter; the red color distinguishes it from the smooth cattle sucking lice, which are blue [see Anoplura (4)]; it reproduces mainly by parthenogenesis, though males are present. The **Goat Biting Louse,** *B. caprae,* infests the common goat; the **Angora-goat Biting Louse,** *B. limbata,* is damaging to the valuable mohair. The **Horse Biting Louse,** *B. equi,* is found on horses, asses, and mules around the neck and base of tail. The **Dog Biting Louse,** *Trichodectes canis,* is .04" long, broad, clear yellow, with dusky lateral bands, abdomen *with* pleural plates; it is most troublesome to puppies, and an intermediate host of the dog tapeworm *Dipylidium caninum* [see Siphonaptera (2)]. The **Cat Louse,** *Felicola subrostrata,* is .04" long, stout, with pointed head, distinct banding, abdominal segments *without* pleural plates.

10

ANOPLURA: Sucking Lice

Like the Mallophaga, these are flat-bodied, wingless parasitic insects that live externally on their hosts and cannot live long apart from them. The bloodsucking lice, or "true lice," are all parasitic on mammals, including man; they are generally host-specific—confined to one or related host species —as are the human lice. The human body louse is unique among lice in that it rests on clothing when not feeding and does not remain in contact with the host at all times. Sucking lice are parasitic on most mammals except bats, marsupials, and carnivores other than dogs.

The mouthparts, greatly modified for piercing and sucking, consist of three stylets which are retracted within the head when not in use. Lice feed frequently, making a new puncture each time, causing irritation and itching. The antennae are three- to five-segmented, the eyes greatly reduced or lacking; the legs are short, tarsi one- or two-segmented, with sharp claws on one or more pairs for clinging to hairs of the host. The eggs of lice are unusual in having a distinct cap or operculum; they are glued to hairs of the host or to clothing fibers in the case of the human body louse. Metamorphosis is gradual or incomplete. About 225 species of sucking lice are known, according to Ferris.

There are two distinct species of human lice: the **Body and Head Louse,** *Pediculus humanus,* and the **Crab Louse,** *Phthirus pubis.* There are two racial forms or subspecies of *Pediculus humanus,* which are capable of cross-mating; the **Head Louse,** *P. h. capitis,* loses its subspecific characteristics and acquires those of the **Body Louse,** *P. h. humanus,* after four gener-

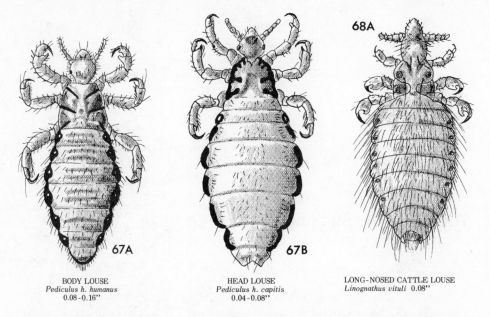

68A

BODY LOUSE
Pediculus h. humanus
0.08-0.16"

HEAD LOUSE
Pediculus h. capitis
0.04-0.08"

LONG-NOSED CATTLE LOUSE
Linognathus vituli 0.08"

67A   67B

ations. The body louse—not the head louse—is the vector or carrier of *Rickettsia prowazeki,* the cause of "epidemic or louseborne typhus." Typhus is rare in tropical countries where little clothing is worn, and where the head louse occurs. A louse contracts the disease when feeding on an infected person, dies within a week or ten days. The infection is transmitted to another person when the feces of the infected louse is deposited on the skin, and the organisms enter the body through a cut or abrasion, or possibly by scratching. The lice tend to leave a feverish host for another, thus spreading contaminated feces when the organisms are most virulent.

"Brill's disease," a mild relapsing form of louseborne typhus which recurs in persons who have had the disease before, should not be confused with "murine or endemic fleaborne typhus" (see Siphonaptera: Fleas). The body louse is also a carrier of trench fever (believed to be a rickettsial disease) and relapsing fever, caused by spirochetes—a kind of bacteria—called *Borrelia recurrentis.* The latter can be spread only when the infected louse is crushed on the skin, and the contaminated blood of the louse comes in contact with an abrasion or opening in the skin.

(1) BODY LOUSE: *Pediculus humanus humanus* (Pediculidae). **Fig. 67A.**

Range: Cosmopolitan. **Adult:** Grayish white; elongated, with legs approximately same size, no hairy processes laterally. Besides being larger than the head louse, the body louse is usually lighter in color, has more slender antennae, less pronounced constrictions between the abdominal segments. Yellowish eggs are cemented to the fibers of underclothing and hatch in about a week. **Length:** .08–.16". **Nymph:** It molts three times,

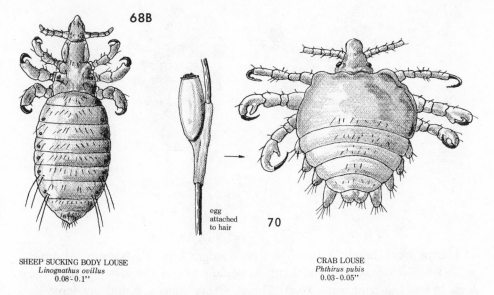

68B

egg
attached
to hair   **70**

SHEEP SUCKING BODY LOUSE
Linognathus ovillus
0.08-0.1"

CRAB LOUSE
Phthirus pubis
0.03-0.05"

matures in eight to nine days when remaining in contact with the human body, or in two to four weeks when clothing (to which it clings) is removed at night. If infested clothing is not worn for several days all the lice on it die. **Host:** Man.

(2) HEAD LOUSE: *Pediculus humanus capitis.* **Fig. 67B.**

Range: Cosmopolitan. **Adult:** Grayish white, with darker margins. (See Body Louse, above, for distinguishing features and life cycle.) Eggs are cemented to hairs on the head or back of the neck. **Length:** .04–.08". **Host:** Man.

(3) CRAB LOUSE: *Phthirus pubis.* **Fig. 70.**

Range: Cosmopolitan. **Adult:** Whitish, short, with hairy lateral processes. First pair of legs smaller than second and third pair, which have claws resembling those of a crab. **Length:** .03–.05". Life cycle similar to that of the body louse and head louse. It is usually found in the pubic region, sometimes on other hairy parts of the body—under the arms, on the eyebrows. It can survive only short periods removed from man, is usually spread by human contact, sometimes by way of infested toilets, furniture, or bedding. **Host:** Man.

(4) SHEEP SUCKING BODY LOUSE: *Linognathus ovillus* (Linognathidae). **Fig. 68B.**

This family—the smooth sucking lice—is characterized by the smooth dorsum. **Range:** Throughout North America. **Adult:** Yellowish brown, oval in shape; head larger than thorax. Legs are all approximately equal. It is found in colonies on various parts of the body and face. Eggs are attached

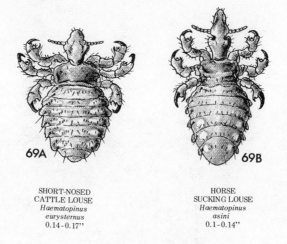

SHORT-NOSED
CATTLE LOUSE
*Haematopinus
eurysternus*
0.14-0.17"

HORSE
SUCKING LOUSE
*Haematopinus
asini*
0.1-0.14"

to hair or wool, hatch in 10 to 18 days. **Length:** .08–.1". **Nymph:** Matures in about two weeks after hatching. *L. pedalis* infests the lower part of the legs, below the true wool. **Host:** Sheep. Other species found on domestic animals include: the **Dog Sucking Louse,** *L. setosus;* the **Goat Sucking Louse,** *L. stenopsis;* the **Long-nosed Cattle Louse** (or blue louse), *L. vituli* (Fig. 68A), and the **Little Blue Louse,** *Solenoptes capillatus.* The last two are distinguished by their size, *L. vituli* being much larger (about .08" long); both are distinguished from the cattle biting louse by their colors [see Mallophaga (4)]. The family Haematopinidae—the wrinkled sucking lice—includes such pests as: the **Short-nosed Cattle Louse,** *Haematopinus eurysternus* (Fig. 69A), which is much larger (from .14" to .17" long) than the long-nosed cattle louse, and has a prominent protuberance (ocular point) on each side, behind the antenna; the **Cattle Tail Louse,** *H. quadripertusus*—hardly distinguishable from *eurysternus*—and the **Horse Sucking Louse,** *H. asini* (Fig. 69B), which is similar but smaller (from .1" to .14" long); the bluish gray or gray-brown **Hog Louse,** *H. suis*—largest of the sucking lice (the female is .25" long) and the only louse attacking hogs (it is found in the ears mostly, and in folds of the skin, mainly in winter). This family is characterized by the wrinkled dorsum, the genus by the ocular points.

11

THYSANOPTERA: Thrips

The thrips (sing. and pl.) constitute a distinctive group, named for the marginal fringe of long bristles on the wings. They are minute to small in size, usually depressed. The narrow wings have few veins and are folded flat on the back with the bristles held against the margins; in some species they are only vestigial. The head is vertical, with large compound eyes, usually three ocelli, and six- to nine-segmented antennae. The mouthparts are peculiar: a short cone or rostrum protrudes downward from the ventral side of the head and encloses stylets; the latter are poorly developed and used for rasping the plant integument, not as a food channel, the juices instead being sucked up through the cone which is applied to the wound.

The abdomen has ten segments, the basal one attached closely to the thorax; the terminal segment of some species is like a long tube. The tarsi are one- or two-segmented, with one or two claws having a bladder-like enlargement at the tip. Most thrips crawl with the abdomen curved up over the back; adults can leap and fly off quickly. Most of them reproduce in the usual manner, but parthenogenesis occurs in some species, with males scarce or unknown, examples being the greenhouse and gladiolus thrips. The female of some species lack an ovipositor and lay their eggs at random in crevices or debris; others have one consisting of four curved parts and insert their eggs in plant tissues. Some are ovoviviparous, giving birth to living young.

Metamorphosis in the group is gradual, though a "pseudopupal" or quiescent stage occurs and immature forms are often referred to as "larvae"

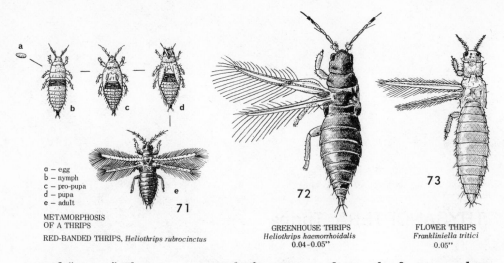

a — egg
b — nymph
c — pro-pupa
d — pupa
e — adult

METAMORPHOSIS
OF A THRIPS
RED-BANDED THRIPS, *Heliothrips rubrocinctus*

GREENHOUSE THRIPS
Heliothrips haemorrhoidalis
0.04–0.05"

FLOWER THRIPS
Frankliniella tritici
0.05"

and "pupae." There are commonly four instars: during the first two, when they are active and feed the same as the adults, there are no wing pads; in the third instar the wing pads first appear; in the last or quiescent stage feeding stops completely and transformation to the adult takes place (Fig. 71). Some species enter the soil at this stage and form a cocoon. Thrips are found on all kinds of vegetation, usually feed on leaves, flowers, stems, or fruit, and include some serious plant pests. However, some are scavengers, feeding on dead vegetable matter, sap, and fungi, and a substantial number are predaceous. The genus *Aeolothrips,* sometimes referred to as the banded-wing thrips, are all beneficial so far as known, and prey on plant-feeding thrips, other small insects, and mites; they are very active and run about on flowers, foliage, and grasses in search of other thrips larvae and small prey. Some valuable predators also occur in the family Phlaeothripidae—they are mostly dark brown or black, with tip of abdomen tubular, ovipositor lacking; among them are the widely distributed *Leptothrips mali* (the black hunter thrips), and *Haplothrips faurei.* The latter is said to be an important predator on the European red mite (*Panonychus ulmi*) and the brown mite (*Bryobia arborea*) in the Niagara Peninsula of Ontario; it also attacks eggs of the Oriental fruit moth.

Thrips (Thripidae)

The Thripidae is the largest family of thrips and includes the most injurious species. The body is depressed, the antennae usually seven- or eight-segmented. Wings may be present or absent; females usually have well-developed sawlike ovipositors, curved downward at the apex. A generation is normally completed in about two weeks, and many generations occur in a year; in warm climates they breed all year long on various plants.

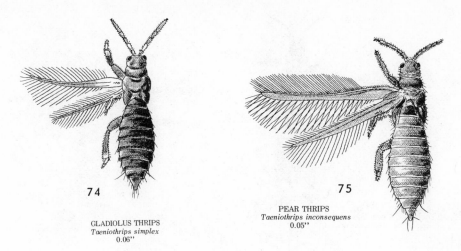

74

GLADIOLUS THRIPS
Taeniothrips simplex
0.06"

75

PEAR THRIPS
Taeniothrips inconsequens
0.05"

(1) GREENHOUSE THRIPS: *Heliothrips haemorrhoidalis.* **Fig. 72.**

A tropical and subtropical species, found outdoors only in warmer climates. **Range:** Cosmopolitan; widely distributed in greenhouses. **Adult:** Dark brown, with lighter median band and appendages. Body surface reticulated (covered with a network of lines), the antennae eight-segmented. Males are not known to occur. Eggs are inserted in leaves. **Length:** .04–.05". **Larva:** Milky white, with accumulations of dark excrement on the back. Remains on plant during quiescent stage. **Food:** Practically all greenhouse plants, especially roses, carnations, chrysanthemums, fuchsias, crotons, cinerarias, cucumbers. Adults and larvae cause stippling of leaves, the tips of which curl and wither. *H. rubrocinctus* is pictured in Fig. 71.

(2) GLADIOLUS THRIPS: *Taeniothrips simplex.* **Fig. 74.**

Range: Entire North America, wherever gladioluses are grown. **Adult:** Brownish, with white band across basal third of wings; third antennal segment white. Body very slender. They are not often seen unless plants are examined carefully. Reproduces without mating by parthenogenesis. Kidney-shaped eggs are inserted in growing tissues or corms, hatch in about a week. **Length:** .06". **Nymph:** First two instars yellowish; third and fourth instars are in quiescent stage. About six broods in greenhouses. They hibernate in corms in colder climates. **Food:** Gladioluses, iris, lilies; the worst pest of gladioluses. Infested leaf sheaths become brown, leaves "silvered"; bud sheaths dry out, become straw-colored, flowers have whitish streaks. They feed on corms in storage; feeding areas become russeted.

(3) PEAR THRIPS: *Taeniothrips inconsequens.* **Fig. 75.**

Range: Throughout North America; South America, Europe. **Adult:** Brown, with grayish wings. Eggs are laid in stems of fruit and foliage.

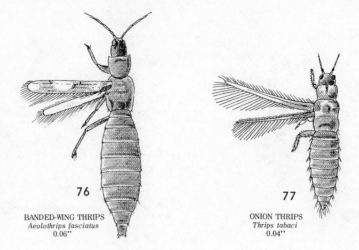

76

BANDED-WING THRIPS
Aeolothrips fasciatus
0.06"

77

ONION THRIPS
Thrips tabaci
0.04"

Length: .05". **Nymph:** Pale whitish (for which it gets the name "white thrips"). Feeding stops in May or June, when it drops to the ground and forms a cell in the soil; it remains in quiescent stage until October or November, when it pupates and goes into hibernation. Adults emerge the following spring. One brood. **Food:** Pear, cherries, prunes. Serious pest of prunes in Pacific Northwest. Buds and young flowers become brown and blasted in appearance.

(4) FLOWER THRIPS: *Frankliniella tritici.* **Fig. 73.**

Range: Throughout North America. **Adult:** Yellowish brown, thorax tinged with orange; body very slender, anterior angles of prothorax each with long, strong bristle; antennae eight-segmented; wings clear. Extremely active, flying about from plant to plant and field to field. Minute creamy white eggs are laid in tender plant tissues. **Length:** .05". **Nymph:** White to lemon-colored. In active stage it feeds about six days and molts twice, then drops to the ground, where it undergoes two more molts and pupates. **Food:** Grasses, alfalfa, vetch, clover, weeds—where it breeds; roses, peonies, many other garden flowers. Petals become flecked, flowers deformed. The **Western Flower Thrips**, *F. occidentalis,* is very similar, clear lemon yellow to dusky yellowish brown, attacks seedling cotton, safflower, onions, lettuce, potatoes, alfalfa, black grama grass, cantaloupes, and sometimes grapefruit in the western states. The **Tobacco Thrips**, *F. fusca,* is brown to yellowish brown, about .04" long, the female long-winged or brachypterous; it attacks seedling cotton, peanuts, tobacco, roses and other flowers in the eastern and southern states. *F. exigua* infests cotton all across the Cotton Belt.

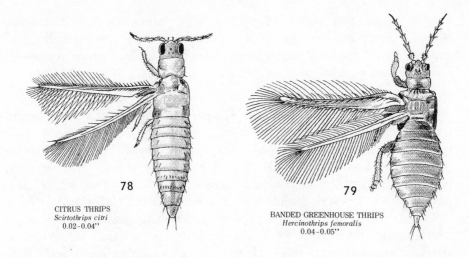

78

CITRUS THRIPS
Scirtothrips citri
0.02-0.04"

79

BANDED GREENHOUSE THRIPS
Hercinothrips femoralis
0.04-0.05"

(5) ONION THRIPS: *Thrips tabaci.* **Fig. 77.**

Range: Cosmopolitan; occurs wherever onions are grown. **Adult:** Pale yellow to dark brown; antennae seven-segmented; posterior margin of eighth abdominal segment bearing a comb of fine hairs; forewing gray. **Length:** .04". **Nymph:** Milky white at first, later turns to green or lemon yellow, with red eyes. Feeds on tender part of center leaves. Life cycle as in (4) above. Hibernates in remnants of onion crop, in crowns of alfalfa and clover. Five broods in Iowa, to ten in California. **Food:** Onions, cauliflower, cabbage, spinach, beets, turnips, cotton. Common pest of greenhouse cucumbers and tomatoes, or roses and other flowers. Rasping of leaves causes whitish or silver blotches, a condition referred to as "white blast" or "silver top"; leaves become crinkled and twisted. Transmits spotted wilt, a virus disease, from outdoor dahlias and other infected plants to greenhouse tomatoes. In California, *T. madroni*—dark brown, with light band across forewing—occurs on madroño, willow, manzanita, ceanothus, toyon, and other plants, and causes deep pits in apples where eggs are inserted in the skin.

(6) CITRUS THRIPS: *Scirtothrips citri.* **Fig. 78.**

Range: Southwestern states, and from northern California to Mexico; it does not occur in the Gulf states. **Adult:** Orange-yellow; pronotum and lateral margins of abdomen brownish, legs and antennal segments three to eight smoky brown. Kidney-shaped eggs are inserted in new leaves, leaf stems, green twigs, fruit stems, or fruit. **Length:** .02–.04". **Nymph:** Yellowish to orange, with prominent bright red eyes. Life cycle similar to that of flower thrips (4). Hibernates in egg stage. Ten to twelve broods. **Food:**

Chiefly citrus; also found on pepper trees, laurel, magnolia, chamiso, sumac, willow, live oak, fir, pecan, date palm, rose, grape, alfalfa, and cotton. One of the most serious pests of citrus in the interior regions of California and in Arizona.

(7) BANDED GREENHOUSE THRIPS: *Hercinothrips femoralis.* **Fig. 79.**

Also called the sugar-beet thrips. **Range:** Sporadically over all of North America. **Adult:** Resembles the greenhouse thrips (1); somewhat more robust, dark brown to yellowish brown; wings grayish brown, with three whitish crossbands. **Length:** .04–.05″. **Nymph:** Pale yellow, with red eyes. **Food:** Sugar beet, bean, celery, cucumber, begonia, cotton, cactus, date palm, banana. A pest in greenhouses. The **Bean Thrips**, *Caliothrips (Hercothrips) fasciatus,* is similar, dark brown, forewing grayish brown, with white transparent band at base and one at three-fourths of the length; it attacks beans and many other plants, occurs on oranges and prickly wild lettuce in California.

Banded-wing Thrips (Aeolothripidae)

This family is believed to be entirely predaceous. The body is not depressed. The antennae are nine-segmented, the apex of the third segment having one or two elongate sensory openings. The ovipositor is large, sawlike, and upturned at the apex. Colors are usually dark brown to black, wings whitish or mottled, the forewing broad, rounded at the apex, and distinctly veined.

(8) BANDED-WING THRIPS: *Aeolothrips fasciatus.* **Fig. 76.**

The most widely distributed species in the group. **Range:** Throughout North America; Europe, Africa, Asia, Hawaii. **Adult:** Yellowish to dark brown, highly variable with blackish forms often present. Forewing has two wide, dark crossbands; basal fourth of the wing is white. Some forms are brachypterous. **Length:** .06″. **Larva:** After two instars it drops to the ground and spins a cocoon. Not single-brooded as are most species of *Aeolothrips,* which resemble the pear thrips in life cycle. **Food:** Predaceous on species of *Frankliniella;* also attacks onion thrips and others. Found more often on grasses, row or field crops, wild and cultivated flowers than on trees. *A. aureus* and *A. duvali,* western species, are very similar to *fasciatus; duvali* is more brilliantly colored, the third antennal segment dark brown, often yellowish brown toward the base; *aureus* is golden yellow or light brown, with tip of abdomen dark brown. Both have broader wings than *fasciatus.*

12

HEMIPTERA (HETEROPTERA): The True Bugs

In some classifications the Hemiptera, which means "half-winged"—a reference to the partly thickened and partly membranous forewing—is divided into two suborders: the Heteroptera ("wings dissimilar") and the Homoptera ("wings the same"). We have treated them as separate orders. The Hemiptera (or Heteroptera) range in size from minute to very large, are easily recognized by the wings and shape. The basal half of the forewing or hemelytron is thickened, the apical half membranous like the hindwing. The mouthparts are the piercing-sucking type, consisting of a three- or four-segmented rostrum or beak (the labium greatly elongated) enclosing the mandibles and maxillae in the form of stylets. The maxillae form a food channel through which juices are sucked, and a salivary channel. In the Hemiptera the mouthparts arise from the front part of the head. (In the Homoptera they usually arise at the back of the head and touch the front coxae.) The beak is held below the body, between the legs, when not in use; the palpi are lacking. The compound eyes are with few exceptions large, the ocelli usually present. The antennae are short and hidden under the head in the aquatic forms, long in the terrestrial forms; they are four- or five-segmented, the individual segments often long.

The prothorax of the Hemiptera is large and free, the mesothorax and metathorax are united, with the mesothorax forming a large triangle called the scutellum. The forewings, because of their structure, are called hemelytra; the membranous tips overlap when folded, forming a fairly well-defined X. Venation of the hindwings is reduced and irregular. In some species the wings are lacking or greatly reduced. The legs are usually long, with three-segmented tarsi. Metamorphosis is incomplete; the young are called nymphs, resemble the parents, and are wingless until almost mature, when the wing pads begin forming. Many species are aquatic or semiaquatic and in this form mostly predaceous. Terrestrial forms are phytophagous with the exception of a few families or species that are carnivorous; some of the latter are important in natural control of certain insect pests. The most important predators are the assassin bugs (Reduviidae), ambush bugs (Phymatidae), damsel bugs (Nabidae), and certain species of stink bugs (Pentatomidae). The stink bugs are noted for their offensive odors and sometimes spoil the taste of berries. This is caused by highly volatile defensive substances which some species spray at their enemies.

Water Boatmen (Corixidae)

Most numerous of the aquatic Hemiptera, the Corixidae, or water boatmen, range from the tropics to the subarctic and are found in fresh, brackish, and saline water. They swim rapidly but are usually seen on the bottom in shallow water. Dorsally, they are flattened, with the head overlapping the thorax, the pronotum in most cases almost or entirely covering the scutellum. The broad rostrum or beak is cone-shaped, the stylets used more for rasping than sucking. The short front legs, with one-segmented spoon-shaped tarsi and comblike fringe of bristles, are used for scraping and straining small plant and animal life from the ooze and sweeping them into the mouth opening. The long, slender middle legs end in two claws; the hind legs are flattened and fringed with hairs for swimming. The front tarsi (called palae) of the males have "pegs" arranged in rows which are useful for identification. Certain males have pegs on the base of the front femur with which they stridulate by rubbing the head. Some species are attracted to lights and swarm around them. The corixids differ from other water bugs in breaking the surface of the water with their heads to renew air, rather than with the posterior end of the body.

(1) WATER BOATMAN: *Arctocorixa alternata.* **Fig. 80.**

A very common species. **Range:** Northern part of North America. **Adult:** Gray and black mottled. Eggs are oval, with a nipple on the top, attached by a pedestal to any support in the water; they hatch in one to two weeks. **Length:** .3–.4″. **Nymph:** Five instars. Life cycle from egg to adult requires about six weeks. Hibernates as adult in the water. *Sigara alternata*—about .25″ long, dark brown, pronotum crossed by 8 to 10 narrow brown bands—is common in roadside ditches. *Corisella tarsalis*—about .2″ long, narrow and shiny, gray, pronotum crossed by 7 to 10 similar bands—is found in

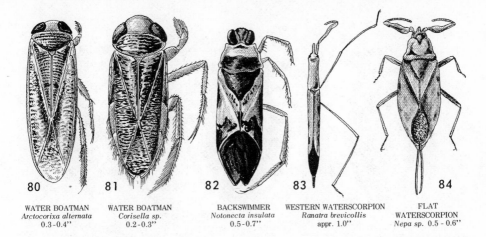

80	81	82	83	84
WATER BOATMAN	WATER BOATMAN	BACKSWIMMER	WESTERN WATERSCORPION	FLAT
Arctocorixa alternata	*Corisella sp.*	*Notonecta insulata*	*Ranatra brevicollis*	WATERSCORPION
0.3-0.4"	0.2-0.3"	0.5-0.7"	appr. 1.0"	*Nepa sp.* 0.5 - 0.6"

somewhat saline lakes and ponds. Both of these species are widespread throughout North America. *Corisella* spp. (Fig. 81) are collected in Mexico and sold for bird and fish food; reeds are placed in the water and later collected with attached eggs for human consumption.

Backswimmers (Notonectidae)

The notonectids, or backswimmers, are familiar in ponds everywhere. They are boat-shaped, convex dorsally, which helps to keep them on an even keel as they swim on their backs. Ocelli are lacking. The front and middle legs are equipped for grasping; the longer hind legs are fringed with hairs and adapted for swimming. Backswimmers are often seen on the surface of the water with their long hind legs held straight out and pointed forward, like an oarsman poised for a fast start. Air is renewed at intervals by breaking the surface with the point of the abdomen, the body pointed down at an angle. They fly long distances, sometimes in swarms, and are attracted to lights. Fierce predators, notonectids can inflict painful bites if handled. **Food:** Midges, mosquito larvae, fish fry, small crustaceans.

(2) BACKSWIMMER: *Notonecta undulata.*

A widely distributed species. **Range:** Most of North America. **Adult:** Black and white; western specimens are often pale. Elongated white eggs are attached to plant stems and hatch in about two weeks. **Length:** .4–.5". **Nymph:** Five instars. One or two broods. Adults appear in July. *N. insulata* (Fig. 82), common in the West, is from .5" to .7" long, brownish, greenish, or yellowish, with black markings.

Waterscorpions (Nepidae)

The Nepidae, or waterscorpions, may be recognized by their long scissors-like claws, the long tail-like respiratory organ, and three pairs of oval discs

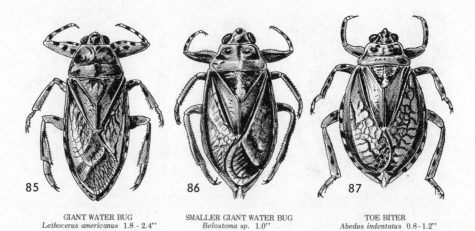

GIANT WATER BUG	SMALLER GIANT WATER BUG	TOE BITER
Lethocerus americanus 1.8 - 2.4"	*Belostoma sp.* 1.0"	*Abedus indentatus* 0.8 - 1.2"

on the underside of the abdomen. They have conspicuous compound eyes, short proboscis, long middle and hind legs adapted for walking about in vegetation rather than swimming (which they do clumsily). Some species (*Ranatra*) have long, slender, rounded bodies, others (*Nepa,* Fig. 84), are broad and flat. Eggs are inserted in plant tissues, with two slender filaments (longer than the eggs) protruding. The oval discs on the underside of the abdomen are called "false spiracles" or "static sense organs," and are believed to be a kind of depth gauge, useful for surface breathers not having the buoying effect of air carried by other aquatic insects. They are fierce predators, will bite when handled. Their prey consists of mosquito larvae, mayfly nymphs, small Crustacea. *Ranatra* stridulates by rubbing together thickened areas at the base of the front coxa against the edge of the coxal cavity. They have well-developed wings but seldom fly and only at night.

(3) WESTERN WATERSCORPION: *Ranatra brevicollis.* **Fig. 83.**

Range: Dominant species in California. **Adult:** Brown throughout. Lives at bottom of pools of shallow water. Deposits eggs in living or dead aquatic plants; they hatch in two weeks to a month. **Length:** About 1". **Nymph:** Five instars, each of about one-week duration. Hibernates as adult. *R. fusca* is dark reddish yellow to brownish black, from 1.4" to 1.7" long (breathing tube about .8" long), occurs in eastern Canada, New England, south to Florida, west to Mexico and California.

Giant Water Bugs (Belostomatidae)

Variously called fish killers, electric light bugs, toe biters, the belostomatids, or giant water bugs, are among our largest insects. They are brown, broad and flat, with short, stout rostrum. Front legs are raptorial, hind legs flattened and fringed with hairs, adapted for swimming. They are strong fliers, attracted to lights. These water bugs are vicious biters; they inject a poisonous fluid into their prey, which are captured and sucked dry. Found in ponds and quiet pools in streams, they attack a great variety of aquatic insects, tadpoles, and fish (sometimes several times their size). They feign death when removed from the water and eject a fluid from the anus. Eggs are laid in massed rows of 100 or more above the water, on cattails and other supports, and hatch in one to two weeks. Females of *Belostoma* (Fig. 86) and *Abedus* glue their eggs to the backs of the males, who carry them until hatched.

(4) GIANT WATER BUG: *Lethocerus americanus.* **Fig. 85.**

Widespread and one of the commonest species. **Range:** Throughout North America. **Adult:** Brown. Strongly attracted to electric lights, often found stranded in the street. Eggs are grayish to dark brown, striped at apical end; white ring marks edge of lid, which is opened at hatching and remains hinged. **Length:** 1.8–2.4″ (width: .8–1″).

(5) TOE BITER: *Abedus indentatus.* **Fig. 87.**

Range: California. **Adult:** Brown; smaller, flatter, and more oval than giant water bug. Pale brown, elongated eggs cemented to back of male, hatch in 10 to 12 days. Several broods. **Length:** About half the size of (4). *Abedus* is strictly western, extends from California to Panama. *Benacus* is confined to the East and, like *Lethocerus,* strongly attracted to lights.

Water Striders (Gerridae)

The gerrids, or water striders, inhabit the surface film of fresh-water ponds, lakes, and streams for the most part.* Many have long, slender bodies, others are stout and broad; they are usually dull brown on top, with a silvery white, velvety waterproof pile on the underside. A scent gland opens in the middle of the metasternum, which may explain why fish do not attack them. The front legs are short and modified for grasping. The middle and hind pair are long, like stilts, with claws located some distance from the last tarsal segment; thus, the film or skating surface is not broken. They are gregarious, and predaceous on aquatic insects that come to the surface and others falling on the water. Wing polymorphism is the rule; winged, brachypterous, and apterous forms are commonly found in a single colony of *Gerris.* They usually fly only at night.

(6) WATER STRIDER: *Gerris remigis.* **Fig. 88.**

Range: Throughout North America. **Adult:** Brown or black; mostly apterous, found in running water or pools in streams. Long, cylindrical eggs are laid during spring and summer in parallel rows, glued to objects at the

---

* They are also called "skaters" or, as in Texas, "Jesus bugs"—apparently for their ability to walk on water.

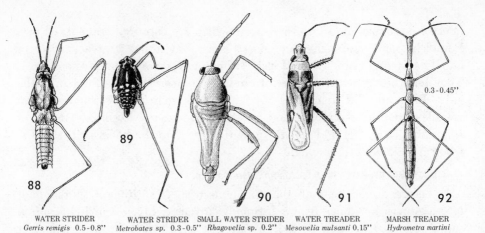

WATER STRIDER	WATER STRIDER	SMALL WATER STRIDER	WATER TREADER	MARSH TREADER
Gerris remigis 0.5–0.8"	*Metrobates* sp. 0.3–0.5"	*Rhagovelia* sp. 0.2"	*Mesovelia mulsanti* 0.15"	*Hydrometra martini*

water's edge or floating. **Length:** .5–.8". **Nymph:** Five instars, each lasting about a week. Hibernates as adult under rocks or logs at the water's edge. *G. comatus* ranges across Canada and the northern states; hemelytra purplish black, veins prominent, gold-flecked; head and pronotum with golden pubescence; from .3" to .4" long. *G. marginatus* is closely related, more southerly in range. *Metrobates* (Fig. 89) is stout-bodied, prefers larger, swift-flowing streams; *M. hesperius*, an eastern species, is grayish black, with broad dark stripe on each side of the dorsal surface in the apterous form; head and sides of abdomen brownish, antennae and legs brownish and red; from .12" to .2" long. *Halobates* is tropicopolitan, occurs in the open ocean and in the protected reefs; it is stout-bodied, with mesonotum and metanotum fused, inner margin of eyes convexly rounded, always apterous, tibia and first tarsal segment of middle leg with fringe of long hairs. *H. sericus* has been found 200 miles off the coast of Monterey, California.

(7) SMALL WATER STRIDER: *Rhagovelia obesa* (Veliidae). **Fig. 90.**

Also called riffle bug. **Range:** Ontario, Manitoba, New England, south to Florida, west to Minnesota, Colorado, Utah, California. **Adult:** Black or dark brown; pronotum with narrow lateral margin and small spot on each side reddish; abdomen with narrow margins reddish. It is apterous or has very small wings. The middle tarsus is deeply cleft, with fringed hairs attached at base of cleft, the hairs capable of being spread out like a fan. Eggs are laid on objects at the water's edge. **Length:** About .2" or less. They are predaceous, found in large numbers in the swiftest riffles of streams, and overwinter as adults.

(8) WATER TREADER: *Mesovelia mulsanti* (Mesoveliidae). **Fig. 91.**

Range: Widely distributed throughout North America. **Adult:** Mostly green; head greenish yellow, antennae brownish; hemelytra white, with black veins and brownish suffusion posteriorly; underside with fine silky,

93

| VELVET WATER BUG | VELVET WATER BUG | SHORE BUG | SHORE BUG |
| *Merragata hebroides* 0.1" | *Hebrus sobrinus* 0.09-0.1" | *Pentacora signoreti* 0.25-0.3" | *Salda buenoi* 0.25" |

white pubescence. Winged and apterous forms occur. Eggs are laid in plant stems and spongy wood. **Length:** About .15". They are very active, predaceous on mosquito larvae and pupae and on other organisms, live on the surface of ponds, hibernate as adults.

(9) MARSH TREADER: *Hydrometra martini* (Hydrometridae). **Fig. 92.**

Range: Manitoba and Ontario, south to Florida, Texas, California. **Adult:** Brown to blackish, sometimes with bluish tinge; pronotum with pale median stripe; hemelytra white, with dark veins. Eggs are laid singly on aquatic plants, above the water surface. **Length:** .3–.45". They are fragile, sluggish, predaceous on mosquito larvae and pupae and on other surface organisms. They overwinter as adults.

(10) VELVET WATER BUG: *Merragata hebroides* (Hebridae). **Fig. 93A.**

Range: Ontario, south to Florida, west to Alberta, Colorado, Mexico, California; also Hawaii, where it was originally described but is believed to have been introduced from California. **Adult:** Mostly black; legs yellow, wings gray; hemelytra with dark brown veins; head purplish, antennae four-segmented, yellow basally, brown apically; dense, silky, white or gray pubescence. Winged and apterous forms appear. Eggs, surrounded by a gelatinous mass, are laid in moss, under algae or stones, and incubate in 8 to 12 days. **Length:** Less than .1". **Nymph:** Five instars; requires 20 to 25 days to develop. Hibernates as adult. They are sluggish, found in moist places hiding in the debris, and prey on Collembola and other organisms. The closely related *Hebrus sobrinus* (Fig. 93B) is widespread in California, has five-segmented antennae.

(11) TOAD BUG: *Gelastocoris oculatus* (Gelastocoridae). **Fig. 94.**

Range: Widely distributed, from New England to Oregon, California, Mexico; Ontario and Manitoba. **Adult:** Brown to yellowish brown, often with darker mottling; legs dull yellow, with dark bands. Eggs are laid in the sand or mud and under stones, away from the water's edge. **Length:**

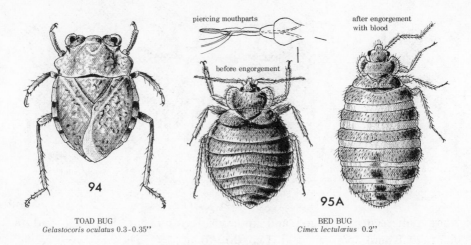

piercing mouthparts

after engorgement with blood

before engorgement

94

95A

TOAD BUG
Gelastocoris oculatus 0.3-0.35"

BED BUG
Cimex lectularius 0.2"

.3–.35". They are predaceous, run rapidly and leap, are found on the shores of lakes, ponds, and streams. They hibernate as adults.

(12) SHORE BUG: *Pentacora signoreti* (Saldidae). **Fig. 93C.**

Range: New England to Florida, Cuba, west to Texas, Mexico, California, Manitoba, and Saskatchewan. **Adult:** Pale, with black markings as shown; lateral margins of abdomen, and femora, yellow; sides of pronotum and hemelytra with row of short black setae. Eggs are laid in clumps of grass or moss near the water's edge. **Length:** .25–.3". They are predaceous, common on the shores of brackish lakes bordered with sparse vegetation. *Salda buenoi* (Fig. 93D) is dull black, with abbreviated hemelytra and pale spots apically, about .25" long; it occurs in the Pacific coast states and in British Columbia.

Bed, Bat, and Bird Bugs (Cimicidae)

Best known for the bed bugs, this is a small family of bugs that suck the blood of mammals and birds. They are broad and flat—enabling them to creep through narrow cracks—with only vestigial forewings, barely extending over the base of the abdomen. The sides of the pronotum are rounded and flangelike. The long, slender antennae are four-segmented, the beak and tarsi three-segmented; ocelli are absent.

(13) BED BUG: *Cimex lectularius.* **Figs. 95A, 95B.**

This is the common bed bug. **Range:** Throughout North America; cosmopolitan. **Adult:** Ovate, very flat, reddish brown. Gregarious and nocturnal in habit, found in houses inhabited by man. It lives at least six months; some can withstand starvation for a year or more. Eggs are laid in their hiding place—200 to 300 by each female—and hatch in about a

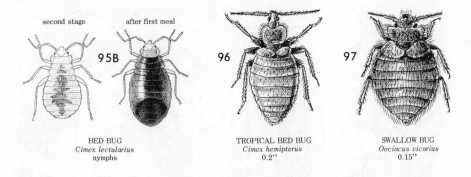

second stage   after first meal

95B   96   97

BED BUG
Cimex lectularius
nymphs

TROPICAL BED BUG
Cimex hemipterus
0.2"

SWALLOW BUG
Oeciacus vicarius
0.15"

week. **Length:** About .2". **Nymph:** Undergoes five instars, each requiring a blood meal. The life cycle from egg to adult takes from six to eight weeks under favorable conditions (to the bug) or can be greatly prolonged. **Host:** Man (bites are in a typical 1-2-3 linear pattern); chickens, mice, rats, rabbits. It is not known to be an important vector of organisms causing human disease. *C. hemipterus* (Fig. 96)—a tropical species, about the same size—has similar habits but is a more vicious biter. The **Bat Bug**, *C. pilosellus,* is about .15" long, infests bats. The closely related **Swallow Bug**, *Oeciacus vicarius* (Fig. 97), about the same size, is found in the nests of swallows, where they overwinter while the birds migrate. The **Poultry Bug**, *Haematosiphon inodorus,* attacks poultry, sometimes gets into dwellings.

Plant Bugs (Miridae = Capsidae)

The mirids comprise a fairly large family commonly called plant bugs or leaf bugs. The most common and destructive ones are species of *Lygus,* collectively referred to as lygus bugs. Most plant bugs suck plant juices, but many species are predaceous and some of them very beneficial. Mostly small, they have large compound eyes but no ocelli. The rostrum and antennae are four-jointed, the tarsi three-jointed. A distinctive feature is the well-developed cuneus (a small triangular area at the end of the hardened portion of the forewing); a few species are brachypterous.

(14) TARNISHED PLANT BUG: *Lygus lineolaris.* **Fig. 98.**

Range: Throughout North America. **Adult:** Straw green to dark brown, with yellow, reddish brown, and black markings; cuneus yellow, with black tip. Elongate, slightly curved eggs are inserted in plant tissues, especially stems and leaf petioles. **Length:** .25". **Nymph:** Pale yellow or green. It molts five times. Hibernates as adult (possibly nymph) under leaves or other shelter. Three to five broods. **Food:** Alfalfa seed and hay crops, cotton squares and bolls, vegetables, deciduous fruits, strawberries; weeds such as wild mustard, butterweed, fleabane, goldenrod, dog fennel, and aster. With *L. hesperus* and *elisus,* it is among the principal pests of cotton in

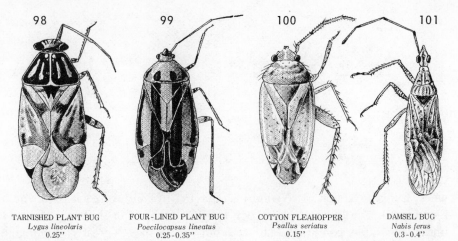

98   99   100   101

TARNISHED PLANT BUG
Lygus lineolaris
0.25"

FOUR-LINED PLANT BUG
Poecilocapsus lineatus
0.25-0.35"

COTTON FLEAHOPPER
Psallus seriatus
0.15"

DAMSEL BUG
Nabis ferus
0.3-0.4"

the Southwest.* *L. hesperus* and *elisus* are also injurious to fruit blossoms, and carrots grown for seed in the Pacific Northwest; *hesperus* damages safflower buds in California. The **Superb Plant Bug**, *Adelphocoris superbus* —a western species, about .3" long—is blood red to orange or brownish yellow, with midsections black and brownish, antennae and legs black (except coxae, which are pale brown); it feeds mostly on alfalfa; unlike *Lygus* bugs, it hibernates in the egg stage. The **Apple Red Bug**, *Lygidea mendax,* resembles *A. superbus* closely in appearance and habits, is a pest of apples in the East; it causes stippling of leaves and gnarling of fruit.

(15) FOUR-LINED PLANT BUG: *Poecilocapsus lineatus.* **Fig. 99.**

Range: East of the Rocky Mts. **Adult:** Greenish yellow, with four black stripes on thickened part of hemelytra. Eggs are inserted in slits in stems (at right angles to direction of growth with tips protruding) in the fall, hatch in May and June. **Length:** .25-.35". **Nymph:** Bright red to orange, with black dots on thorax; yellow stripe on each side of wing pads in last instar. It feeds on leaves, causing white and dark spots, or the entire leaf turns brown and falls off. Hibernates in egg stage. One brood. **Food:** Gooseberry, currant; rose, zinnia, chrysanthemum, aster, dahlia, phlox, sunflower, snapdragon, and other flowers. A chalcid wasp, *Cirrospilus ovisugosus* (Eulophidae), preys on the eggs.

(16) COTTON FLEAHOPPER: *Psallus* (= *Pseudotomoscelis*) *seriatus.* **Fig. 100.**

Range: All major cotton-producing states. **Adult:** Pale green, with small dark spots on body, four black marks near wing tips. Eggs are laid in stems of croton (goatweed), which is common throughout the Cotton Belt.

---

* Besides *Lygus* bugs, the main pests of cotton in this and other areas of the Cotton Belt are: the bollworm, cabbage looper, beet armyworm, salt-marsh caterpillar, pink bollworm, tobacco budworm, and boll weevil; the cotton aphid, cotton fleahopper, cotton leafworm, and certain thrips are important cotton pests in some areas.

Length: .15″. **Nymph:** Pale green (after feeding); eyes prominent, scarlet. Five instars. Hibernates in egg stage. Six to eight broods. **Food:** Tender weeds such as evening primrose and horsemint, where they mate; when these become less succulent, they fly to cotton (in presquaring or early fruiting stage) and feed on tender parts of the plants. They migrate to croton and other weeds to feed and lay eggs when the cotton becomes tough. The **Garden Fleahopper,** *Halticus bractatus*—blackish, about .1″ long—is a pest of alfalfa, clover, and garden crops in the eastern states; the female has two forms: one long-winged and typically hemipteran, the other short-winged, ovate, densely black, resembling a flea beetle (both have the jumping habit); the male is typically hemipteran in form, straight-sided, very active but quite able to fly, has front and middle femora of pale yellow; eggs are inserted in leaves and stems; the nymphs are green. The **Yucca Plant Bug,** *Halticotoma valida,* is bluish black, with reddish brown head and pronotum, about .12″ long; the bright scarlet nymphs are often very numerous on ornamental yucca plants, cause stippling of leaves which turn yellow; adults run rapidly to base of plant if disturbed rather than fly off.

Damsel Bugs (Nabidae), Flower Bugs or Minute Pirate Bugs (Anthocoridae), Ambush Bugs (Phymatidae)

These are small families of predaceous bugs. In the damsel bugs, the front legs are raptorial, the tarsi three-segmented; the antennae are long, four- or five-segmented, the rostrum or beak usually long and four-segmented. The flower bugs—also known as the minute pirate bugs—have a well-developed cuneus like the plant bugs; the antennae are short, four-segmented, the beak and tarsi three-segmented. The ambush bugs are characterized by their odd and sometimes grotesque shapes, suggesting the family name, Phymatidae, which means growth or swelling. The thorax is often extended into sharp angles, rounded projections, and spines. They are medium-sized, robust predators that lie in wait among flowers and foliage for their prey. A remarkable structural feature is the front legs with their greatly enlarged femora, curved at the ends and armed with teeth, and sickle-shaped tibiae also armed with teeth; the tarsi are bent back, apparently useless.

(17) DAMSEL BUG: *Nabis ferus.* **Fig. 101.**

One of the commonest species, inhabits fields and orchards. **Range:** Throughout North America. **Adult:** Grayish to brownish or almost black. **Length:** .3–.4″. **Food:** Aphids, leafhoppers, treehoppers, small caterpillars.

(18) INSIDIOUS FLOWER BUG: *Orius insidiosus.* **Fig. 102.**

Best known of our anthocorids. **Range:** Widely distributed over North

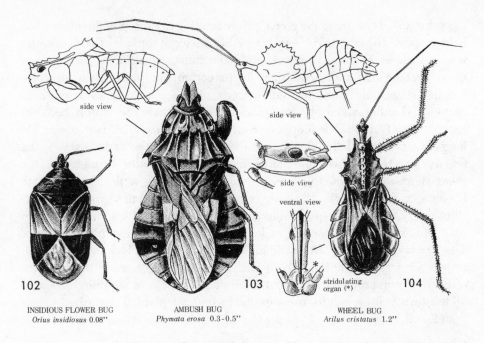

102	103	104
INSIDIOUS FLOWER BUG	AMBUSH BUG	WHEEL BUG
Orius insidiosus 0.08"	*Phymata erosa* 0.3-0.5"	*Arilus cristatus* 1.2"

America. Common on flowers. **Adult:** Black. Thickened area of hemelytra yellowish white, with large, blackish triangular spot at tip; membrane milky white. Sometimes mistaken for chinch bug (29), on which it preys. **Length:** .08". **Food:** chinch bugs, thrips, corn earworm (small larvae and eggs), walnut husk fly (eggs), other small insects and eggs, mites. Adults and nymphs insert beaks into eggs and suck them dry. They are abundant on corn in the eastern states, deposit their eggs on silks. *O. tristicolor* is an important enemy of mites in the citrus groves of California. *Anthocoris melanocenus* is important in control of the pear psylla in British Columbia.

(19) AMBUSH BUG: *Phymata erosa.* **Fig. 103.**

Range: Throughout North America; common in the West. **Adult:** Pale or yellowish brown, with darker markings, and wide brown band across middle of abdomen. Black, oval eggs are glued to plants. **Length:** .3–.5". **Food:** Bees, butterflies, and other flower visitors.

Assassin Bugs (Reduviidae)

This large family of predaceous bugs includes some bloodsucking species that attack man and other animals. They are medium to large in size, with long, narrow head, long four- or five-segmented antennae, the last segment filiform. The beak is stout, three-segmented, the tip held in a groove in the prosternum when at rest; this groove contains stridulating organs with which they make squeaking noises [see (*) Fig. 104].

(20) WHEEL BUG: *Arilus cristatus.* **Fig. 104.**

Range: The eastern and southern states. **Adult:** Brown. Named for the "cogwheel" crest on the prothorax. Eggs are laid in clusters. **Length:** 1.2″. **Nymph:** Blood red, with black markings. **Food:** Many kinds of insects; a voracious predator, it attacks large caterpillars such as hornworms and sucks them dry. It will bite humans with painful results.

(21) MASKED HUNTER: *Reduvius personatus.* **Fig. 105.**

Also called kissing bug or masked bed bug hunter. **Range:** Eastern and midwestern states. **Adult:** Black or dark brown. Beak curved, not slender and tapered as in *Triatoma* (22). They are very active and attracted to lights, enter houses in search of bed bugs. Eggs are laid singly in the dust in cracks and corners. **Length:** .7–.8″. **Nymph:** Body, legs, and antennae are covered with a viscid substance to which dust and lint adhere; thus "masked," they are visible only when moving. Hibernates as fourth or fifth (final) instar, matures the following June. **Food:** bed bugs, *Triatoma* spp., flies, and other insects. It will bite humans.

(22) BLOODSUCKING CONENOSE: *Triatoma sanguisuga.* **Fig. 106.**

Also called the big bed bug or Mexican bed bug. **Range:** New Jersey to Florida, west to Illinois and Texas; considered a common insect in parts of Kansas. **Adult:** Brownish or black, with six reddish orange spots on each side of abdomen, above and below. Beak slender and tapered, not curved as in *Reduvius* (21), and almost bare. Oval pearly white eggs are deposited singly from May to September, each batch being laid after a blood meal. When they have mated and found a host the adults no longer fly. **Length:** .75–.8″. **Nymph:** Undergoes eight instars. Entire life cycle requires about three years. They are found mostly in the nests of wood rats. **Host:** Wood rats preferred, but will feed on almost any animal including man; bites are very painful, cause a severe reaction in some persons (they are multiple— as many as 15, grouped or scattered, at a time). *T. heidemanni* is very similar, but has a hairy beak, occurs in about the same range. *T. lecticularius* and *T. gerstaeckeri* are pests in Texas. The **Western Bloodsucking Conenose,** *T. protracta,* is smaller than *sanguisuga,* all black, occurs in Utah, Arizona, California, Mexico. Species of *Triatoma* transmit Chagas' disease (caused by *Trypanosoma cruzi*) among small animals, and to man; the disease is endemic in Mexico, Central and South America.*

* The principal carrier of Chagas' disease in Argentina is *T. infestans.* This is the "great black bug of the Pampas" which attacked Darwin one night while he was on an expedition into South America. The severe "bouts of illness that afflicted him" after the voyage of the *Beagle* are now believed to have been due to Chagas' disease and traceable to this incident. The infective agent is transmitted from mammal to mammal via the feces of infected bugs, through the mucous membranes (eyes, nose, mouth) and skin abrasions; mammals commonly involved are dogs, cats, wood rats, opossums, and armadillos.

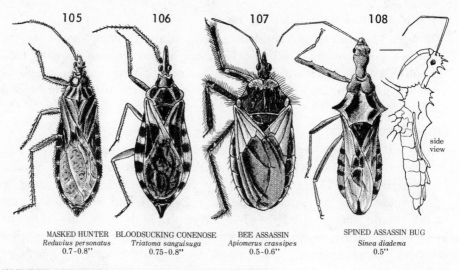

MASKED HUNTER	BLOODSUCKING CONENOSE	BEE ASSASSIN	SPINED ASSASSIN BUG
Reduvius personatus	*Triatoma sanguisuga*	*Apiomerus crassipes*	*Sinea diadema*
0.7-0.8"	0.75-0.8"	0.5-0.6"	0.5"

(23) BEE ASSASSIN: *Apiomerus crassipes.* **Fig. 107.**

Range: Throughout North America; very common in the western states. **Adult:** Black; pronotum, scutellum, and abdomen broad and fringed in deep red; thick pile on head, thorax, and legs. Front and middle tarsi very small. Commonly found on flowers. **Length:** .5–.6". **Food:** Bees and other flower visitors.

(24) SPINED ASSASSIN BUG: *Sinea diadema.* **Fig. 108.**

Range: Throughout North America. **Adult:** Dull brownish yellow to brown; head, pronotum, and front femora covered with spines; margin of female's abdomen wavy. Cylindrical white eggs with cone-shaped cap are laid in small groups, covered with a reddish yellow secretion. **Length:** .5". **Food:** Larvae of many injurious insects; all stages of the Mexican bean beetle are attacked.

(25) WESTERN CORSAIR: *Rasahus thoracicus.* **Fig. 110.**

Range: Pacific Northwest and California. One of the commonest species in California. **Adult:** Body amber, wing membrane black, with amber spot near middle of forewing. Nocturnal, attracted to lights. Makes a squeaky noise when handled, can inflict a painful bite. **Length:** .17". **Nymph:** Hibernates in fifth and last instar, matures in the spring. One brood. **Food:** Larvae and adults of various injurious insects. *R. biguttatus,* sometimes referred to as the two-spotted corsair, is similar in appearance and habits, occurs throughout the southern and southwestern states, Mexico, and the West Indies.

(26) THREAD-LEGGED BUG: *Emesa brevicoxa.* **Fig. 109.**

Range: California. **Adult:** Brown; long, slender body, prothorax not distinctly separated from mesothorax. Wings about one-third as long as

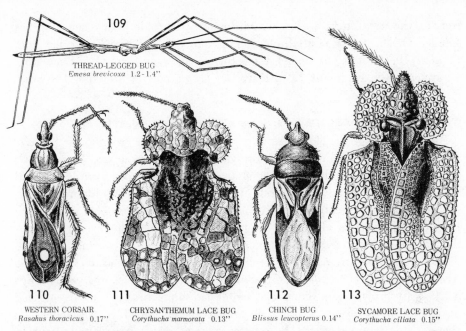

109

THREAD-LEGGED BUG
Emesa brevicoxa 1.2 - 1.4"

110

WESTERN CORSAIR
Rasahus thoracicus 0.17"

111

CHRYSANTHEMUM LACE BUG
Corythucha marmorata 0.13"

112

CHINCH BUG
Blissus leucopterus 0.14"

113

SYCAMORE LACE BUG
Corythucha ciliata 0.15"

abdomen. Threadlike middle and hind legs. Front legs heavier than others, with greatly elongated coxa (more than twice as long as the head), femur spined; tibia closes back on femur. **Length:** 1.2–1.4". **Food:** Other insects, spiders. Large numbers often congregate in trees, sheds, or barns.

Lace Bugs (Tingidae)

The Tingidae, or lace bugs, are not likely to be confused with any other group. They are small, oval or rectangular in outline, and lacelike in appearance owing to reticulated pattern on the head, thorax, and wings. The head is often covered with a hood, the pronotum extended on the sides like another pair of wings, the abdomen hidden by the wide lacelike hemelytra. The antennae are four-jointed, the tarsi two-jointed, ocelli lacking, scutellum generally wanting or vestigial. The nymphs are unlike the adults; they are much darker and covered with spines. They suck plant juices in all stages and are found on the underside of leaves, which they leave blotched and spotted with specks of excrement. Eggs are inserted in plant tissues—in groups, on underside of leaves, generally along the midrib—and are often covered with a viscid substance which forms cone-like projections on the plant surfaces when it hardens.

(27) SYCAMORE LACE BUG: *Corythucha ciliata.* **Fig. 113.**

Range: Throughout North America, mostly east of the Rocky Mts. **Adult:** Whitish above, dark beneath. Hibernates in adult stage under bark of trees. **Length:** .15". **Food:** Sycamore, ash, hickory, paper mulberry. *C. con-*

fratana, the western form, has white dorsum with pale brown markings, is smaller. Other closely related and common species are: the **Elm Lace Bug,** *C. ulmi;* the **Cotton** (or **Bean**) **Lace Bug,** *C. gossypii;* the **Oak Lace Bug,** *C. arcuata;* the **Eggplant Lace Bug,** *Gargaphia solani;* and the **Basswood Lace Bug,** *G. tiliae.*

(28) CHRYSANTHEMUM LACE BUG: *Corythucha marmorata.* **Fig. 111.**

Range: Throughout North America; also common in greenhouses. **Adult:** Body black, thorax whitish, with brown specks; antennae and legs yellow; four brown transverse lines on hemelytra. They suck juices from leaves and stems in all stages. Several broods. **Length:** .13″. **Food:** Goldenrod, chrysanthemum, asters, scabiosa, and other garden flowers. The **Hawthorn Lace Bug,** *C. cydoniae,* is about .1″ long, sucks the juice from leaves of hawthorn, cotoneaster, pyracantha, and Japanese quince, causing them to discolor and fall prematurely; it hibernates in egg or adult stage. The **Azalea Lace Bug,** *Stephanitis pyrioides,* is about .12″ long, with wings marked in brown and black and antennae light brown, causes grayish discoloration on upper surface of azalea leaves and blackish specks of excrement on underside; it hibernates in egg stage. The **Rhododendron Lace Bug,** *S. rhododendri,* is very large for a lace bug—nearly .4″ long— with black body and grayish wings, attacks rhododendron, mountain laurel, and related plants, causing white-peppered appearance on upper surface of leaves; it hibernates in egg stage.

Lygaeid Bugs (Lygaeidae)

This is a large family which feeds mostly on plants; the big-eyed bugs (*Geocoris*) are predaceous. The antennae are four-segmented and located low on the sides of the head. The forewing has four or five unbranched longitudinal veins which do not run parallel with one another as they do in the next family, Coreidae. The best-known member of the family is the chinch bug because of its damage to wheat in the midcontinent and to lawns elsewhere.

(29) CHINCH BUG: *Blissus leucopterus.* **Fig. 112.**

Range: From the East Coast to the western plains. **Adult:** Black; forewing snowy white, with black spot near middle of costal margin; legs and base of antennae reddish brown or yellowish brown. In another form, the wings cover only the base of the abdomen, are brown, with white tips. Slightly curved white eggs (later dark red) are laid in the soil, or roots, or in stems near the ground. **Length:** .14″. **Nymph:** Bright red, with white band across the back; final stage black with white spots on the middle. It sucks juices from roots and stems. Five instars. When the grain matures, they migrate on foot in vast hordes to more succulent grains, grasses, or corn. Hibernates as adult in clump of grass or under litter. Two or three broods. **Food:** Wheat, rye, barley, corn, sorghum, oats, rice, grasses, millet. When abundant, chinch bugs may destroy an entire crop. They are often attacked by fungi, the most important being *Beauveria bassiana;* diseased insects become covered with a white cottony growth, and the contagion spreads rapidly under moist conditions. A subspecies, *B. l. hirtus*—the **Hairy Chinch Bug**—is a serious pest of lawns in New England. *B. arenarius* occurs on the coastal dunes along the Atlantic seaboard; the majority are brachypterous. The **Southern Chinch Bug**, *B. insularis,* is a pest of St. Augustine grass in the Gulf states; it resembles *leucopterus* closely, has fully developed wings. The above species comprise the so-called *Leucopterus* Complex. The **Western Chinch Bug**, *B. occiduus,* is black, with short hemelytra.

(30) FALSE CHINCH BUG: *Nysius ericae.* Fig. 114.

Range: Throughout North America. **Adult:** Light or dark gray. **Length:** .12–.16″. **Nymph:** Pale gray, with reddish brown abdomen. Life cycle and habits similar to the chinch bug (29). **Food:** Alfalfa, clover, grasses, grains, sugar beets, cotton, truck crops, deciduous fruits, grapes, berries, flowers, citrus. *N. raphanus* is the more common form in southern California.

(31) LARGE MILKWEED BUG: *Oncopeltus fasciatus.* Fig. 115.

Range: Most of North America. **Adult:** Orange or bright red, with three large black areas on thorax and hemelytra; black spots on underside. Antennae and legs black. Elongated bright red eggs are laid in milkweed plants. **Length:** .4–.6″. **Nymph:** Bright red, with black legs and antennae. Hibernates as adult on trees and around buildings. Like the boxelder bug (34), they often appear in large swarms on warm winter or spring days. **Food:** Milkweed. Several species of *Lygaeus* also feed on milkweed and resemble the large milkweed bug. In the **Small Milkweed Bug**, *L. kalmi,* the black on the pronotum is separated from the black on the front of the wings, and the black area at the middle does not extend across the wings.

(32) BIG-EYED BUG: *Geocoris pallens.* Fig. 116.

Range: Western part of North America. **Adult:** Pale yellowish green; head and thorax have minute black spots, both are same width as abdomen. Scutellar area black, with posterior edges lighter. Predaceous in both adult and nymph stages. **Length:** .12–.16″. **Food:** Aphids, leafhoppers, *Lygus* bugs. *Geocoris* spp. are very numerous in cotton, alfalfa, and sugarbeet fields and a valuable part of the predator-parasite complex that helps keep the insect pests of these plants under control. *G. punctipes, G. decoratus,* and *G. atricolor* are commonly found with *G. pallens.* The European red mite is a prey of *G. punctipes* on cotton.

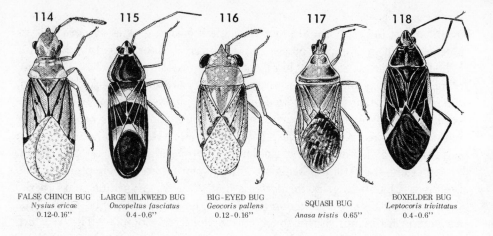

FALSE CHINCH BUG
Nysius ericae
0.12-0.16"

LARGE MILKWEED BUG
Oncopeltus fasciatus
0.4-0.6"

BIG-EYED BUG
Geocoris pallens
0.12-0.16"

SQUASH BUG
Anasa tristis 0.65"

BOXELDER BUG
Leptocoris trivittatus
0.4-0.6"

Coreid Bugs (Coreidae)

All members of this family are phytophagous. They have large eyes and four-segmented antennae. The longitudinal veins of the forewing are more numerous than in the preceding family, Lygaeidae, and run parallel to one another.

(33) SQUASH BUG: *Anasa tristis*. Fig. 117.

Range: Throughout North America. **Adult:** Brownish yellow; appears black because of dense covering of black hairs. Protruding margins of abdomen are orange or orange-and-brown striped; margins of pronotum yellow. Shiny brown eggs are laid singly or in groups of 15 to 40 on underside of leaves (or on stems). **Length:** .65". **Nymph:** Pale green, changing to blackish thorax and brownish abdomen, often covered with white powder. Young nymphs feed in clusters. Feeding causes affected part of leaves to droop, turn black and crisp. Hibernates as adult under any shelter it can find. One brood. **Food:** Squash, pumpkin, melon, gourd. [See Diptera (93).]

(34) BOXELDER BUG: *Leptocoris trivittatus*. Fig. 118.

Range: Throughout North America. **Adult:** Flat-topped, grayish brown to black, with three red stripes on thorax, red veins on hemelytra. Body bright red, head black. Red markings more pronounced on eastern forms. Eggs are laid on the bark and leaves of boxelder but preferably on seed pods of female trees. **Length:** .4–.6". **Nymph:** Bright red, marked with black when half grown. Feeds on foliage. Females hibernate as adults in buildings or other shelter. One or two broods. **Food:** Boxelder, maple; sometimes fruit of deciduous fruit trees which are punctured, causing deformities. Damage to trees is minor. In the fall and spring they appear in swarms or clusters on tree trunks and buildings in sunny locations [see Milkweed Bug (31)]; they sometimes invade houses. *L. rubrolineatus* is the western form.

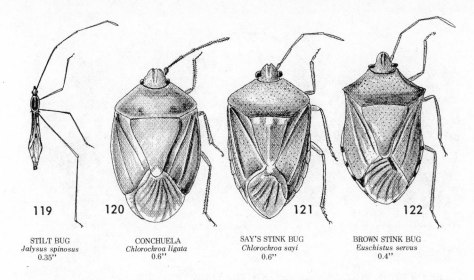

| 119 | 120 | 121 | 122 |

STILT BUG
Jalysus spinosus
0.35"

CONCHUELA
Chlorochroa ligata
0.6"

SAY'S STINK BUG
Chlorochroa sayi
0.6"

BROWN STINK BUG
Euschistus servus
0.4"

(35) STILT BUG: *Jalysus spinosus* (Berytidae = Neididae). **Fig. 119.**

Range: Eastern North America, ranging westward to Colorado. **Adult:** Pale brown. Extremely long, knobbed antennae with black terminal segments; slender body with long slender legs, resembling a mosquito. **Length:** .35". **Nymph:** Whitish and dark green. Hibernates as adult. **Food:** Potatoes, tomatoes, nettle, and other plants of the potato family; alfalfa, grasses. They puncture stems and ovaries of tomato plants; flowers and young fruit dry up or fall off. Western forms occur in California, Oregon, Utah, Arizona.

Stink Bugs (Pentatomidae)

This is a large family commonly known as stink bugs. The family name, Pentatomidae, refers to the five-jointed antennae. They have small eyes, only two ocelli. The scutellum is triangular, or U-shaped, and very large. The hind tibiae are armed with numerous thornlike spines, or with short, even spines and long hairs. They are mostly plant feeders, but there are some important predators among them.

(36) CONCHUELA: *Chlorochroa ligata.* **Fig. 120.**

Range: British Columbia, Colorado, south into Mexico. **Adult:** Olive green, with darker specks, narrow marginal border and tip of scutellum orange or reddish. Pale green cylindrical eggs are glued in masses to leaves, later turn gray or brown. **Length:** .6". **Nymph:** Gray to brownish, with orange or reddish markings. They feed in compact colonies. **Food:** Wide variety of plants; serious pest of cotton in Texas and New Mexico. They feed on the bolls, sucking juice from immature seeds; young bolls become soft, turn yellowish and fall off.

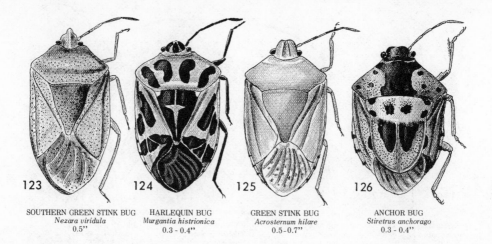

123	124	125	126
SOUTHERN GREEN STINK BUG	HARLEQUIN BUG	GREEN STINK BUG	ANCHOR BUG
Nezara viridula	*Murgantia histrionica*	*Acrosternum hilare*	*Stiretrus anchorago*
0.5"	0.3 - 0.4"	0.5-0.7"	0.3 - 0.4"

(37) SAY'S STINK BUG: *Chlorochroa sayi.* **Fig. 121.**

Range: Western states; most abundant in Texas, Oklahoma, New Mexico, Arizona, and California. **Adult:** Bright green, with minute specks, and three orange spots at base of scutellum; tip of scutellum whitish or orange. Closely resembles the conchuela (36) except for lighter color and spots. **Length:** .6". **Food:** Wide variety of plants, including alfalfa, cotton, potatoes, beans, asparagus, peas, sunflower, oats, grasses, weeds. Very damaging to cotton.

(38) BROWN STINK BUG: *Euschistus servus.* **Fig. 122.**

Also called southern brown stink bug. **Range:** Eastern U.S., westward to Texas, New Mexico, Colorado. **Adult:** Brown, speckled, with narrow checkered border. **Length:** .4". **Food:** Cotton, cabbage, corn, blackberry, peach. A serious pest of cotton, sometimes causes damage in peach orchards, along with the southern green stink bug (39), the green stink bug (41), and the tarnished plant bug (14). They puncture the skin of fruit, causing droplets of gum to appear, and deformities called "catfacing" and dimpling. The **Western Brown Stink Bug,** *E. impunctriventris,* and the **Dusky Stink Bug,** *E. tristigmus,* are very similar. The **One-spot** (or **Spined**) **Stink Bug,** *E. variolarius,* is about .5" long, light grayish brown with brown specks, yellow legs, and sharp projection on each side of pronotum; the male has a black spot on the ventral side on the genital segment; it is a pest of peach, pear, and field crops in the East, Midwest, and Far West. The **Consperse Stink Bug,** *E. conspersus,* is about .5" long, pale brown with black specks and red or yellow antennae, yellowish underneath, legs spotted with black; it is a pest of soft fruits, especially pear, from California to British Columbia.

(39) SOUTHERN GREEN STINK BUG: *Nezara viridula.* **Fig. 123.**

Range: Southeastern states and the Deep South. **Adult:** Uniformly light green, speckled. **Length:** .5". **Food:** Tomatoes, potatoes, okra, and other vegetables; cotton, peach [see Brown Stink Bug (38)]. A serious pest of cotton.

(40) HARLEQUIN BUG: *Murgantia histrionica.* **Fig. 124.**

Also called calico bug, fire bug, or collard bug. **Range:** Southern half of the U.S. **Adult:** Flashy, red and black; flat, shaped like a shield. Eggs are keg- or barrel-shaped (with black "stays"), and laid in double rows of twelve or more, glued together on the underside of leaves. **Length:** .3–.4". **Nymph:** Oval, red and black spotted like the adult. Three broods and a partial fourth. White and yellow blotches appear where they feed. **Food:** Cole crops like cabbage, broccoli, turnip, horseradish, and kale are favored. A serious garden pest.

(41) GREEN STINK BUG: *Acrosternum hilare.* **Fig. 125.**

Also called the green soldier bug. **Range:** Throughout North America, south into Central and South America. **Adult:** Bright green, with narrow yellow, orange, or reddish border. Eggs similar to those of harlequin bug (40) are glued in clusters to the underside of leaves. **Length:** .5–.7". **Food:** Peach, apple, cherry, bean, okra, pea, eggplant, cabbage, tomato, turnip, corn, cotton, orange, mesquite, mustard, soybeans.

(42) RICE STINK BUG: *Oebalus pugnax.* **Fig. 127A.**

Range: Arkansas, Mississippi, Louisiana, Texas. **Adult:** Straw-colored, with black specks. Cylindrical eggs are laid on leaves, stems, or head of rice, on grass, Mexican weed, usually in clusters of 10 to 40 arranged in two rows; they are green when laid, turn reddish before hatching. **Length:** About .5". **Nymph:** Young nymph nearly round; head, antennae, thorax, legs black; abdomen red, with two transverse black spots. In rice fields it feeds on panicles, sucking the juice from developing kernels, or empties the glumes when rice is in the milk stage, causes "peckiness" when rice is in the soft-dough stage. Hibernates as adult in clump of grass or under debris. Several broods. **Food:** Grasses and sedges, weeds, cultivated rice. Two or three generations feed on grasses before migrating to rice fields. Two wasps—*Ooencyrtus anasae* (Encyrtidae), and *Telenomus podisi* (Scelionidae)—parasitize the eggs, sometimes reducing the population of later broods significantly.

(43) ANCHOR BUG: *Stiretrus anchorago.* **Fig. 126.**

Range: Throughout most of North America. **Adult:** A predaceous bug with varied color patterns; metallic blue to greenish black or violet, marked with orange or red. **Length:** .3–.4". **Nymph:** All black, or black with red spot. **Food:** Caterpillars, larvae of beetles; larvae, pupae, and adults of Mexican bean beetle.

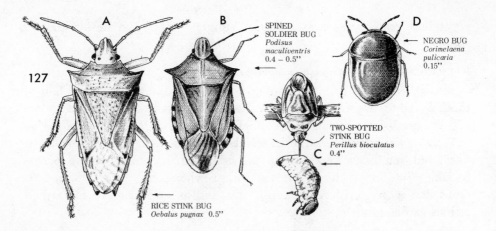

A

B

SPINED
SOLDIER BUG
*Podisus
maculiventris*
0.4 – 0.5"

D

NEGRO BUG
*Corimelaena
pulicaria*
0.15"

127

TWO-SPOTTED
STINK BUG
Perillus bioculatus
0.4"

C

RICE STINK BUG
Oebalus pugnax 0.5"

(44) TWO-SPOTTED STINK BUG: *Perillus bioculatus.* **Fig. 127C.**

Range: Throughout North America. **Adult:** As striking as the harlequin bug, with several color forms. Generally yellow or red, with black Y on pronotum and two black spots on thorax. Each female deposits up to 260 eggs in the vicinity of hosts. **Length:** .4". **Nymph:** Young nymphs feed on host's eggs, each eating an average of 452 eggs during this period. Both nymphs and adults attack caterpillars and the larvae of beetles, holding them in the air as they suck out the body juices. An individual bug destroys 150 to 200 larvae in its lifetime. **Food:** Cutworms and other caterpillars; larvae of beetles, especially the Colorado potato beetle. Introduced into France from North America in the 1930's, along with the spined soldier bug (45), for biological control of the Colorado potato beetle.

(45) SPINED SOLDIER BUG: *Podisus maculiventris.* **Fig. 127B.**

Range: Throughout North America. **Adult:** Yellowish or pale brownish, covered with black specks; shoulder sharply pointed, short black line near apexes of wings extending beyond end of abdomen; conspicuous spine on middle (underside) of front tibia. Metallic bronze eggs are laid in batches of 20 to 30; each female is capable of laying up to 1,000 eggs in a period of 5 to 8 weeks. **Length:** .4–.5". **Nymph:** The nymphs cluster about eggshells for a few days, until the first molt, and may feed on plant juices to some extent at this time, but are strictly predaceous after the first molt. Hibernates as adult. Two broods. **Food:** Hairless caterpillars, larvae of sawflies and leaf-eating beetles. Said to be "the most useful of the American predaceous Hemiptera" and to "rank next to *Calosoma* (ground beetles) as an enemy of the fall armyworm." Used in biological control of the Colorado potato beetle, as noted above (44). *P. serviventris* has similar habits. *P. sagitta* is abundant in Mexico wherever the Mexican bean beetle and other species of *Epilachna* occur.

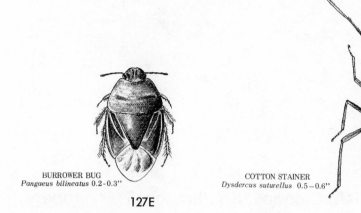

127F

BURROWER BUG
Pangaeus bilineatus 0.2-0.3"

127E

COTTON STAINER
Dysdercus suturellus 0.5–0.6"

(46) NEGRO BUG. *Corimelaena (Thyrecoris) pulicaria* (Cydnidae). **Fig. 127D.**

The negro bug and closely related species generally called negro bugs are sometimes placed in the family Corimelaenidae, or Thyrecoridae. **Range:** *C. pulicaria* is widely distributed. **Adult:** Strongly convex, shiny black, with stripe on each side of body, very large U-shaped scutellum which covers most of the abdomen and wings; the tibiae are armed with several rows of stout spines. **Length:** .15". **Food:** Berries and other fruits, celery, flowers, grasses and weeds; sometimes a serious pest of celery, often found in large numbers on flowers in early spring. The closely related **Burrower Bug,** *Pangaeus bilineatus* (Fig. 127E) (Cydnidae), is dull blackish, from .2" to .3" long, with tibiae expanded and armed with long spines; it is found under stones and burrowing in the soil around the roots of grasses and other plants.

(47) COTTON STAINER: *Dysdercus suturellus* (Pyrrhocoridae). **Fig. 127F.**

Range: Southern states; Florida to Texas, Mexico, the West Indies. **Adult:** Head and pronotum bright red, remainder of body dark brown, crossed with pale yellow lines. Eggs are laid on cotton plants or dropped on the ground. **Length:** .5–.6". **Food:** Cotton, citrus; rose mallow appears to be its natural food. It feeds on immature cotton bolls, puncturing the seeds; exudations during feeding stain the lint. It sometimes invades citrus orchards in Florida, puncturing the fruit and sucking out the juice. The crushed insect makes a rich orange-yellow dye. The **Arizona Cotton Stainer,** *D. mimulus*—brilliant black, with red on head and shoulders, yellowish hemelytra—is sometimes a serious pest in Arizona.*

---

* Buxton, George M., "Cotton Stainers, *Dysdercus* spp., a potential threat to cotton and citrus," *National Pest Control Oper. News,* May 1972.

13

HOMOPTERA: Cicadas, Spittlebugs, Treehoppers, Leafhoppers, Planthoppers, Psyllids, Aphids, Whiteflies, Scale Insects and Mealybugs

This is a very large order of exceedingly diverse groups of insects which do not lend themselves to generalization. Superficially, some of them look more like plant outgrowths or exudations, or accumulations of nonliving material, than insects. The mouthparts have been characterized in the introduction to the Hemiptera (see p. 109). The wings are usually membranous in the winged forms and held rooflike or sloping at the sides in resting position. Many species and forms are apterous. Metamorphosis is incomplete or simple, excepting the whiteflies and male scale insects which have a pupal stage.

Cicadas (Cicadidae)

Also known as harvestflies, cicadas are medium to large in size, and easily recognized by the wide blunt head with prominent eyes and the long, clear membranous forewings which are held peaked over the body at rest. They have three distinct ocelli, short bristle-like antennae, plainly visible proboscis, stout front legs, and three-jointed tarsi. The monotonous, sustained chorus of the male cicadas on a hot summer's day is a familiar and pleasant sound. The music is produced by complex sound organs—actually a pair of drums located in the metathorax and vibrated by muscles, not by beating. An air cavity acts as resonator and is connected to the outside through the thoracic spiracles.

The nymphs, which spend their developmental period in the ground, have

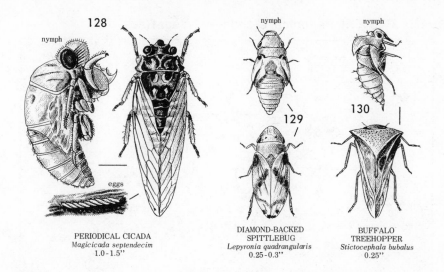

128

nymph

eggs

PERIODICAL CICADA
Magicicada septendecim
1.0 - 1.5"

nymph

129

nymph

130

DIAMOND-BACKED
SPITTLEBUG
Lepyronia quadrangularis
0.25 - 0.3"

BUFFALO
TREEHOPPER
Stictocephala bubalus
0.25"

front legs modified for digging, with elongated coxa and scooplike femur; when development is completed they leave the ground and climb up on plants, where the adults emerge through a split in the top of the last nymphal skin. The adults suck juices from the limbs and twigs of trees, while the subterranean nymphs feed on the roots, neither of which has a noticeable effect on the trees; the damage comes from the egg slits, which cause dieback of the twigs. There are many species of cicadas, with life cycles from one year upward, but the best known is the periodical cicada; the eastern form (often called the "seventeen-year locust") has a 17-year life cycle, while the southern form has a 13-year life cycle. Western species of cicadas are generally smaller and, while common, are not so numerous.

(1) PERIODICAL CICADA: *Magicicada (Tibicina) septendecim.* **Fig. 128.**

Range: The 17-year or eastern form—Massachusetts and Vermont westward to Michigan, Wisconsin, and Kansas, southward to Texas, northern Alabama and Georgia; the 13-year or southern form—Virginia to southern Iowa, Oklahoma, southward to the Gulf coast, eastward to the Atlantic coast. **Adult:** Body wedge-shaped, black or brownish black; legs mostly orange, wings transparent with reddish orange tinge and orange margins. Eggs are laid from late May to early July in slits made in the bark and wood of twigs, 12 to 20 in each puncture—several punctures may be made in a row—and hatch in six to seven weeks. **Length** (including wings): 1–1.5". **Nymph:** The young nymph is antlike, drops to the ground, enters the soil, and constructs a cell alongside rootlets from which it sucks sap. They remain in the ground for 17 or 13 years, then burrow to the surface, where they sometimes construct mud "chimneys" two or three inches high. All emerge at one time, climb the trunks of trees or stems of plants, shed their skins, and

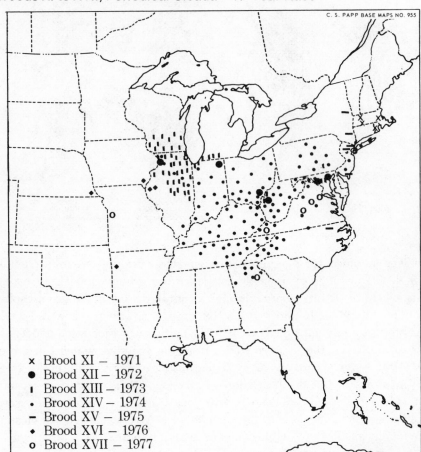

C. S. PAPP BASE MAPS NO. 955

x Brood XI – 1971
● Brood XII – 1972
ı Brood XIII – 1973
. Brood XIV – 1974
– Brood XV – 1975
♦ Brood XVI – 1976
o Brood XVII – 1977

Economic Insect Survey and Detection   U.S. Department of Agriculture

NOTE: Brood X of the periodical cicada (17-year race), which appeared in 1970, is the most widespread and abundant of the broods. According to the *Cooperative Economic Insect Report* (USDA) of April 17, 1970, it was scheduled to appear suddenly in large numbers "over most of the northeast quarter of the Nation . . . about the last week of May. . . . For approximately 6 weeks it will fill the countryside with its remarkable song, mate, lay its eggs in twigs, and pass away as suddenly as it appeared." This brood will not appear again until 1987 (see table). Brood XXX of the 13-year race was also scheduled to appear (in northeast Louisiana) in 1970, but "there is some doubt as to the existence of this brood." The two races are difficult to separate except by an expert. (Specimens should be submitted to the Department of Entomology, U.S. National Museum of Natural History, Smithsonian Institution, Washington, D.C. 20560.)

Table of Coincidence of Broods of Periodical Cicadas

	I	II	III	IV	V	VI	VII	VIII	IX	X	XI	XII	XIII	XIV	XV	XVI	XVII
XVIII		1945				1932				1919				1906			
XIX			1946				1933				1920				1907		
XX				1947				1934				1921				1908	
XXI	1961				1948				1935				1922				1909
XXII		1962				1949				1936				1923			
XXIII			1963				1950				1937				1924		
XXIV				1964				1951				1938				1925	
XXV	1978				1965				1952				1939				1926
XXVI		1979				1966				1953				1940			
XXVII			1980				1967				1954				1941		
XXVIII				1981				1968				1955				1942	
XXIX	1995				1982				1969				1956				1943
XXX		1996				1983				1970				1957			
XVIII			1997				1984				1971				1958		
XIX				1998				1985				1972				1959	
XX	2012				1999				1986				1973				1960
XXI		2013				2000				1987				1974			
XXII			2014				2001				1988				1975		
XXIII				2015				2002				1989				1976	
XXIV	2029				2016				2003				1990				1977
XXV		2030				2017				2004				1991			
XXVI			2031				2018				2005				1992		
XXVII				2032				2019				2006				1993	
XXVIII	2046				2033				2020				2007				1994
XXIX		2047				2034				2021				2008			
XXX			2048				2035				2022				2009		

Agricultural Research Service, U.S. Department of Agriculture

Horizontal (across top): Broods of 17-year race. Read down for year due.
Vertical (side): Broods of 13-year race. Read from right to left for year due.
Diagonal (intersection of horizontal and vertical rows): Broods of both races due to appear each year.

emerge as adults. As many as 20,000 to 40,000 may emerge under one tree. Egg slits can cause severe damage to young fruit trees. Adults live for 30 to 40 days. The general location of the various broods and year of emergence are known and mapped (see p. 134). The "harvestman" or **Dogday Cicada,** *Tibicen linnei,* is sometimes mistaken for the periodical cicada; it is much larger, with lighter markings on the body and greenish margins on the wings; it is believed to have a two-year cycle. *Tibicen dorsata* is brown and black, with powdery areas above and brownish wings; it is one of the largest cicadas, about 2.25" long to tip of folded wings, appears each year in the Midwest but is not abundant. *T. pruinosa* is green and black, with greenish legs and wings, about 1.75" long, very common in the Midwest. A common western species is the **Orchard Cicada,** *Platypedia areolata;* it is dark bronze, with yellow and amber markings, greenish yellow band across the thorax, about .75" long.

Spittlebugs (Cercopidae)

Spittlebugs are named for the sticky, bubbly mass of froth with which the nymphs surround themselves. They are also called froghoppers for their slight resemblance to frogs and the adults' habit of hopping about on grasses and shrubs. Bristle-like antennae arise *between* the eyes (see Fulgorid Plant-hoppers, p. 141). The tarsi are three-segmented; the apex of the tibiae have circlets of spines (see Leafhoppers, p. 137).

(2) DIAMOND-BACKED SPITTLEBUG: *Lepyronia quadrangularis.* **Fig. 129.**
Range: Widely distributed and common. **Adult:** Brownish, with two oblique brown bands across the hemelytra, which are tipped with a small blackish curve. Eggs are laid between leaf and main stem of grasses, from midsummer to late fall. **Length:** .25–.3". **Nymph:** Feeds and molts under spittle, though it moves about. Hibernates in egg stage. One brood. **Food:** Grasses, shrubs, herbs. The **Meadow Spittlebug,** *Philaenus spumarius*—gray or brown and spotted, about .37" long—is destructive to alfalfa and clover in the northeastern and north central states, also found in the South and Far West. The **Saratoga Spittlebug,** *Aphrophora saratogensis,* is brown, with light median dorsal stripe on head and pronotum, tan and silvery mottled transverse bands on hemelytra, from .35" to .4" long; it is widely distributed and a major pest of red pine plantations in the Lake states. *Prosapia bicincta* adults are white when they first emerge, except for two narrow orange bands across the wings and one on the thorax, but the white quickly turns to black (the orange remains); they are .35" long, have two broods, overwinter as eggs, cause severe damage to Bermuda grass along the southern Atlantic Coastal Plain; they are also found on other grasses, grains, and ornamentals.

Treehoppers (Membracidae)

Treehoppers are distinguished by the enlarged prothorax projecting above the head and back over the abdomen; the shapes in some species are bizarre in the extreme. They are adept at both flying and jumping, are found in grasses and shrubs as well as trees. Eggs are inserted in the bark of trees, in straight or crescent-shaped slits, sometimes causing serious injury to young trees.

(3) BUFFALO TREEHOPPER: *Stictocephala (Ceresa) bubalus.* **Fig. 130.**

Range: Throughout North America. **Adult:** Bright green or yellowish. Eggs are inserted in a series of crescent slits in the bark, in pairs forming parentheses, 6 to 12 eggs in each puncture; they are laid in the fall and hatch in the spring. **Length:** .25″. **Nymph:** Green, with large spiny processes on back. It drops to the ground, feeds on alfalfa, grasses, or other succulent plants. Five instars. It matures in about six weeks, returns to trees (as adult) to feed, mate, and oviposit. Hibernates in egg stage. One brood. **Food:** Alfalfa, clover, grasses, wild plants, deciduous fruits, ornamentals, tomatoes, potatoes. Egg slits are injurious to young fruit trees and nursery stock.

Leafhoppers (Cicadellidae)

The Cicadellidae, or leafhoppers, comprise a large family of small, slender insects with short bristle-like antennae, and usually two ocelli. They may be distinguished by the double row of spines lengthwise on the curved hind tibiae (see Spittlebugs). The forewings are slightly thickened, and often marked with color patterns matching those of the head and thorax. Adults and nymphs have the peculiar habit of running sidewise; when the plants are disturbed, the adults hop or fly away, and the long-legged nymphs run sidewise for cover. They cause injury to plants by sucking the juices and injecting salivary substances which obstruct the conductive system; discoloring of leaves and stunting of growth results. Both spittlebugs and leafhoppers are vectors of viruses causing Pierce's disease, aster yellows, curly top, and other plant diseases.

(4) BEET LEAFHOPPER: *Circulifer tenellus.* **Fig. 131.**

Range: Western U.S. and Canada, Mexico, eastward to Missouri, Illinois, and Texas. **Adult:** Pale greenish or yellow to dark brownish, usually with darker blotches. Eggs are inserted in veins, leaf petioles, or stems of plants. Adult females hibernate around wild desert host plants such as sagebrush. First-generation adults fly many miles, often in huge swarms, to summer breeding areas of Russian thistle and sugar beets; fall movements take them back to the desert hibernating areas. In transferring from one breeding ground to another they often travel 200 to 300 miles or more. **Length:** .15″.

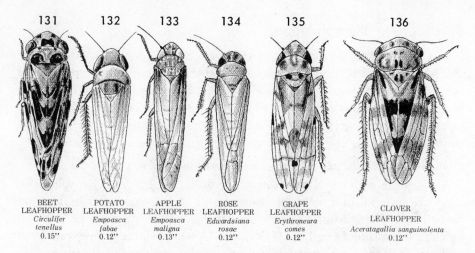

131	132	133	134	135	136
BEET LEAFHOPPER *Circulifer tenellus* 0.15"	POTATO LEAFHOPPER *Empoasca fabae* 0.12"	APPLE LEAFHOPPER *Empoasca maligna* 0.13"	ROSE LEAFHOPPER *Edwardsiana rosae* 0.12"	GRAPE LEAFHOPPER *Erythroneura comes* 0.12"	CLOVER LEAFHOPPER *Aceratagallia sanguinolenta* 0.12"

Nymph: Pale-colored, matures in three to eight weeks. One to three or more broods. **Food:** Sugar beets, table beets, tomatoes, potatoes; wild plants such as sagebrush, saltbush, greasewood, wild mustard, Russian thistle, afilaria, sea blite. Their greatest harm comes from the transmission of curly top, a plant virus disease that stunts and kills beet plants and reduces their sugar content. They also transmit tomato yellows. They are commonly parasitized by certain wasps (Dryinidae), big-eyed flies (Pipunculidae), and twisted-wings or Strepsiptera (Stylopidae).

(5) POTATO LEAFHOPPER: *Empoasca fabae.* **Fig. 132.**

Range: Eastern half of North America, westward to Colorado and Wyoming. **Adult:** Greenish, with faint white spots on head and thorax, and six round white spots along anterior margin of prothorax. It sucks plant juices from veins on underside of leaf, leaving blotched appearance. Eggs are laid in main veins or petioles on underside of leaves. **Nymph:** Similar to adult but paler in color. Five instars; matures in two weeks. Not known to hibernate in northern part of range, believed to migrate from warmer regions each spring. Two to four broods. **Food:** Potatoes, beans, eggplant, rhubarb, celery, dahlias, alfalfa, soybeans, clover, peanuts, apple nursery stock, wild plants. One of the most serious pests of potatoes in eastern half of the U.S. It produces hopperburn, a virus-like disease resulting in extensive stippling, curling, and browning of leaves of potato, eggplant, rhubarb, and dahlia. The **Southern Garden Leafhopper,** *E. solana,* also causes hopperburn. The pale green **Intermountain Leafhopper,** *E. filamenta,* is a serious pest of potatoes in the Pacific Northwest and transmits viruses. The greenish **Western Potato Leafhopper** (or melon leafhopper), *E. abrupta,* occurs in Oregon, Texas, and California, where it causes serious injury to melons, potatoes, and celery. The **Arid Leafhopper,** *E. arida,* occurs in the lower elevations of California, Arizona, and Utah, is a serious pest of potatoes.

These species are all very similar in appearance, attack a wide variety of crops, and are very prolific.

(6) APPLE LEAFHOPPER: *Empoasca maligna.* **Fig. 133.**

Range: Throughout North America. **Adult:** Greenish in color. Appears in May and June. Eggs are inserted in midrib of larger veins and stems of leaves where more than one generation occurs. In the Pacific Northwest, where one brood occurs, most eggs are deposited in June and July under the bark of trees and remain unhatched until the following spring. **Length:** .13″. **Nymph:** Pale green to greenish white; very active. One or two broods. Where two broods occur, the first generation matures by midsummer, lays eggs for the second brood. Hibernates in egg stage in bark of trees. **Food:** Apple, rose, currant, gooseberry, raspberry, potato, sugar beet, bean, celery, grains, grasses, weeds. The foliage of apple becomes pale, with greenish white specks as the result of leafhopper feeding; margins of new foliage become slightly curled. Specks of excrement may appear on fruit. The **White Apple Leafhopper,** *Typhlocyba pomaria,* has similar habits, is sometimes associated with the apple leafhopper and rose leafhopper (7).

(7) ROSE LEAFHOPPER: *Edwardsiana (Typhlocyba) rosae.* **Fig. 134.**

Range: Throughout North America. **Adult:** Slender, whitish or greenish yellow. Lives two months or more. Eggs are laid in tissues of leaves, from June to August, hatch in about a month. Overwintering eggs are laid in bark of trees, cause small blisters. **Length:** .12″. **Nymph:** Like adult, sucks plant juices from underside of leaves. Hibernates in egg stage in bark of trees. Two broods. **Food:** Same as apple leafhopper (6); injury also the same. On roses they cause leaves to appear mottled and pale and to drop prematurely. Attacks chiefly rose and apple.

(8) GRAPE LEAFHOPPER: *Erythroneura comes.* **Fig. 135.**

Range: Throughout North America, wherever grapes are grown. **Adult:** White or pale yellowish with red or yellow markings; in early spring and summer forms, pale yellow predominates; hibernating forms are almost entirely red. Eggs are laid in leaf tissues. **Length** .12″. **Nymph:** Sucks plant juices from underside of leaves. Hibernates as adult in trash on the ground. Two or three broods. **Food:** Chiefly grape; also blackberry, Virginia creeper, Boston ivy. Feeding causes leaves to become mottled or blotched with white, to turn brown and drop prematurely. *E. elegantula* occurs in all grape-growing areas of California, is pale yellow, with reddish and brown markings.

(9) CLOVER LEAFHOPPER: *Aceratagallia sanguinolenta.* **Fig. 136.**

Range: Throughout North America. **Adult:** Short, wide, flattened; light gray, with brown markings. **Length:** .12″. **Nymph:** Appears in early spring;

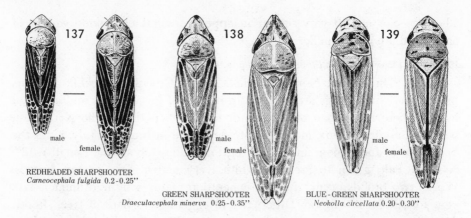

137

male

female

REDHEADED SHARPSHOOTER
Carneocephala fulgida 0.2-0.25"

138

male

female

GREEN SHARPSHOOTER
Draeculacephala minerva 0.25-0.35"

139

male

female

BLUE - GREEN SHARPSHOOTER
Neokolla circellata 0.20-0.30"

sucks plant juices from leaves and stems. Hibernates as adult. Two broods in northern states, more in the South. **Food:** Clover, alfalfa, lupines, peas, vetch, grains, corn, beets, cabbage, weeds, potatoes. Especially damaging to clovers, during seedling stage and following cutting. Reported to carry potato yellows (dwarf) virus.

(10) SIX-SPOTTED LEAFHOPPER: *Macrosteles fascifrons (divisus)*.

Range: Throughout North America. **Adult:** Yellow or yellowish green to dark greenish brown, with six very small black spots on the head. **Length:** .12". **Nymph:** Grayish. Feeds on underside of leaves. Hibernation is not believed to occur in northern part of range, except for some eggs laid in fall-sown rye, wheat, and barley; it migrates from the lower Mississippi Valley each spring. Two or more broods. **Food:** Grains, grasses, lettuce, celery, carrots, potatoes; asters (it is sometimes referred to as the aster leaf-hopper), marigold, chrysanthemum, dahlia, and other flowers; strawberries, wild plants. Very damaging to lettuce in Ontario. Vector of aster yellows or purple-top wilt disease organisms. The **Painted Leafhopper,** *Endria (Poly-amia) inimica* (Fig. 140B), is abundant on grasses and other vegetation throughout North America; it is about .15" long, grayish yellow, with three pairs of large black spots on front of head and pronotum, and on scutellum.

(11) REDHEADED SHARPSHOOTER: *Carneocephala fulgida.* **Fig. 137.**

Range: California. **Adult:** Green, with reddish head. **Length:** .2-.25". Hibernates in adult stage. Four broods. **Food:** Grasses preferred; succulent growth of grapevines in spring. Vector of virus causing Pierce's disease; scalding and drying of leaves, wilting of fruit, and death of vines result.

(12) GREEN SHARPSHOOTER: *Draeculacephala minerva.* **Fig. 138.**

Range: Oregon, Utah, south to Texas, Mexico, and Central America. **Adult:** Green; in some regions it turns brown late in the fall. The head is more pointed than in the redheaded sharpshooter (11). Eggs are laid in grasses in early spring. **Length:** .25–.35″. Hibernates in adult stage. Three broods. **Food:** Prefers grasses but feeds on many other plants, including grapevines. Vector of Pierce's disease. The **Blue-green Sharpshooter,** *Neokolla* (*Hordnia*) *circellata* (Fig. 139), is slightly smaller, green to bright blue on top, yellow underneath; it carries the disease to grapevines in the coastal region.

Fulgorid Planthoppers (Fulgoridae), Psyllids or Jumping Plantlice (Psyllidae), Bark Aphids, Gall Aphids, or Phylloxeras (Phylloxeridae)

The fulgorid planthoppers comprise a family of greatly varied little insects with long, slender legs and odd prolongations of the head. Larger, more bizarre forms occur in the tropics, one of the best known being the **Lantern-fly,** *Lanternaria phosphorea* (Fig. 140A)—which, incidentally, is not luminous as suggested by the name. The wings of the fulgorids are long or sometimes quite short, or lacking in some species. The antennae arise on the sides of the face beneath the eyes (see Spittlebugs). Two ocelli are usually present, and the tarsi are three-segmented. They feed on native plants but are not important economically; many are found only in arid regions. The psyllids, or jumping plantlice, are tiny insects which have been likened to "miniature cicadas." They are more solid than aphids, have stronger legs, with the hind pair adapted for jumping. The antennae are usually ten-jointed. Both sexes have wings, the front pair thicker than the hind pair. The bark aphids and gall aphids or phylloxeras include two very troublesome species: the grape phylloxera—one of the most destructive grape pests, especially in some western states—and the balsam woolly aphid, a serious forest pest.

(13) PALLID SCOLOPS: *Scolops pallidus* (Fulgoridae). **Fig. 143.**

Range: Southwestern states. **Adult:** Yellowish and gray. **Length:** .25″. **Food:** Dry grasses, weeds, shrubbery. *S. sulcipes* is brown to grayish, from .25″ to .4″ long, has long horn with upward sweep and wedge shape viewed from above; forewing with numerous small areoles or squares apically; it is one of the most common fulgorids in the central states and eastward, where it feeds on weeds and grasses. The stubby little **Cranberry Toad Bug,** *Phylloscelis atra* (Fig. 141), is common in cranberry bogs, where it causes some damage to plants. *Asarcopus palmarum* is an Egyptian species found in southern California date groves; it attacks crown leaves and fruit stems, depositing much honeydew on plants; nymphs are reddish brown; adults are dark brown, with rudimentary wings, and hop about like the nymphs.

(14) PEAR PSYLLA: *Psylla pyricola* (Psyllidae). **Fig. 142.**

A major pest of pears, first introduced into the East in 1832. **Range:** Pear-growing areas of eastern U.S. and Canada, the Pacific Northwest, and Cali-

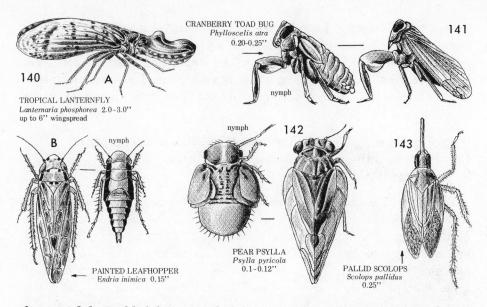

TROPICAL LANTERNFLY
Lanternaria phosphorea 2.0-3.0"
up to 6" wingspread

CRANBERRY TOAD BUG
Phylloscelis atra
0.20-0.25"

nymph

141

nymph 142

143

PAINTED LEAFHOPPER
Endria inimica 0.15"

PEAR PSYLLA
Psylla pyricola
0.1-0.12"

PALLID SCOLOPS
Scolops pallidus
0.25"

fornia. **Adult:** Reddish brown to dark brown, with reddish or green markings. The transparent wings fold rooflike at rest. Pear-shaped orange-yellow eggs are laid at base of fruit spurs, on leaf buds or twigs in early spring, and on leaves in late spring, summer, and fall, mostly on the upper surface along the midvein, a few on the underside or along edge of leaves. **Length:** .1–.12". **Nymph:** Flat and broad, yellowish green to reddish brown, with bright red eyes. It has five instars, immerses itself in droplet of honeydew during second to fourth stages, moves about during the last or "hardshell" stage, when wing pads become prominent and color brownish black. Hibernates as adult in bark or litter. Three to five broods. **Food:** Primarily pear, preferably Bartlett and D'Anjou varieties; occasionally quince. Feeding has toxic effect, causing leaves to turn yellow. Honeydew harbors black fungus (sooty mold) which drips on the fruit. The psyllid is said to be the vector of a virus causing "pear decline." The **Apple Sucker,** *P. mali*—pale green, about .15" long—is a pest of apples in eastern Canada; it deposits large quantities of honeydew on foliage, has one brood, overwinters in the egg stage; the yellowish green nymphs feed on opening buds, stems and leaves. The **Hackberry-nipple-gall Maker,** *Pachypsylla celtidismamma,* causes mammiform galls on underside of hackberry leaves. *P. celtidisvesicula* causes blister-like galls on underside of hackberry leaves; these are smaller and more numerous than the nipple galls. Adults of both species may appear in swarms in the fall, and get stuck in fresh paint or invade homes in a desperate search for a place to hibernate.

(15) POTATO (TOMATO) PSYLLID: *Paratrioza cockerelli.* **Fig. 144.**

Range: North and South Dakota, Nebraska, Kansas, Oklahoma, Texas,

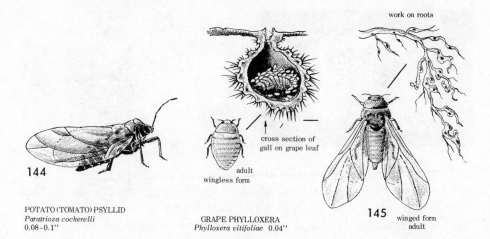

work on roots

cross section of
gall on grape leaf

144

adult
wingless form

145 winged form
adult

POTATO (TOMATO) PSYLLID
Paratrioza cockerelli
0.08–0.1"

GRAPE PHYLLOXERA
Phylloxera vitifoliae 0.04"

and westward, excepting Washington and Oregon; British Columbia to Saskatchewan; Mexico. The number of psyllids diminishes eastward from the 100th meridian, as the humidity increases. **Adult:** Black, with white transverse band on first abdominal segment, white inverted Y on the last segment. Very active, it jumps quickly and takes to flight readily. Eggs are yellow to orange, attached to leaves, usually on the underside, by a stalk. **Length:** .08–.1". **Nymph:** Pale green; flat, scalelike, with row of short hairs or spines around the margin. Five instars. They do not hibernate in the North; they migrate in May and June from their southern wintering and breeding grounds in Texas and New Mexico (where they live chiefly on *Lycium*) to the potato-growing areas. **Food:** Potatoes, tomatoes; wild plants of the Solanaceae family, chiefly wild ground cherry and Chinese lantern, horsenettle and buffalo bur, matrimony vine. Toxic injections of the nymphs (not adults) cause "psyllid yellows"; the leaves curl and turn yellow or purplish, yield and quality are reduced.

(16) GRAPE PHYLLOXERA: *Phylloxera vitifoliae* (Phylloxeridae). **Fig. 145.**

Range: Throughout the grape-growing areas of North America. Mostly leaf-infesting forms in the East, root-infesting forms in the West. **Adult:** Minute, oval- or pear-shaped, yellowish green or yellowish brown. **Length:** .04". Besides the immature forms, there are four forms of adult. They hibernate as eggs on grape canes or as aphids in the nodules of galls on grape roots. The eggs hatch in the spring; the aphids migrate to leaves, where they feed and form galls. Those on the roots resume feeding in the spring. Mature aphids (parthenogenetic females) give birth to living young in the galls; these in turn form new galls, and other generations follow. Leaf-inhabiting aphids drop to the ground, reach the roots through cracks in the soil; here they form galls and produce other generations. In the fall,

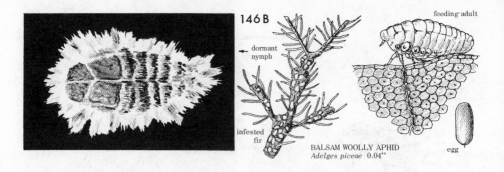

146B

feeding adult

dormant nymph

infested fir

BALSAM WOOLLY APHID
Adelges piceae 0.04"

egg

winged forms appear, leave the ground, and lay eggs on the vines; these eggs hatch into males and true females. Mating takes place on the cane, where each female lays one egg which remains here for the winter. Only parthenogenetic females are normally found in California. Winged forms appear in coastal areas, but they are unable to found new colonies; infestations are spread by crawling aphids. Five to eight generations.

(17) BALSAM WOOLLY APHID: *Adelges (Chermes) piceae.** **Fig. 146B.**

Range: New England, Maritime Provinces of Canada, Appalachian region, Oregon, Washington, and British Columbia. Introduced from Europe about 1900. **Adult:** Black, hemispherical in form; covers itself with white, curly wax threads. There are females only, reproduction being by parthenogenesis. Some of the first to hatch in spring may have wings, but this is rare; they are spread chiefly by the crawling nymphs and the wind. Pale yellow eggs are laid in a mass on twigs, under the wax. **Length:** .04". **Nymph:** Brown at first. They move about until a suitable place is found on the bark, where they remain to suck plant juice. Hibernates as nymph, in a form having white fringe and "backbone" of wax, remains at base of bud, on the twig or stem. Two or three broods. **Food:** Only true fir (*Abies*). Trees turn rust red and die as the result of severe attack on stems. Injection of a substance before feeding causes "gout disease," resulting in swelling of buds and nodes and gnarling of branches. No known parasites attack this aphid. The **Eastern Spruce Gall Aphid,** A. (*Chermes*) *abietis,* causes a pineapple-shaped growth at the base of current or one-year-old shoots of Norway and white spruces. The **Cooley Spruce Gall Aphid,** A. (*Chermes*) *cooleyi,* attacks blue spruces. A. *lariciatus* forms a globular, pineapple-like gall on twigs of white, black, and blue spruces; the alternate host is larch; the complete life cycle of the various forms requires two years. Nymphs of the **Pine Leaf Chermid** (or pine leaf aphid), *Pineus pinifoliae,* cause browning and death of new branch tips of white pine; they overwinter in the wax-fringed form, and winged adults migrate the following spring to the

* Genus recently changed to *Adelges.*

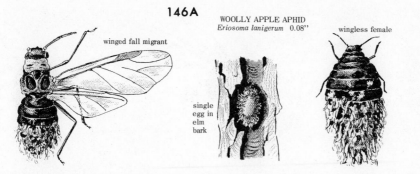

WOOLLY APPLE APHID
Eriosoma lanigerum 0.08"

winged fall migrant

wingless female

single
egg in
elm
bark

alternate (primary) host, red and black spruces, where they form conelike galls and remain until the next year; the damage to pine depends on the number of spruce in the stand. The **Pine Bark Aphid,** *P. strobi,* attacks white pine, is easily recognized by the white cottony patches on the trunks and limbs.

Aphids (Aphididae)

Aphids, or plantlice, are minute—most of them less than one-tenth of an inch long—soft-bodied insects that suck plant juices. The beak or rostrum arises well back on the underside of the head, the antennae are long, situated on the front of the head between the eyes; winged forms usually have three ocelli. The tarsi are two-jointed and terminate in two claws. Many aphids have a pair of tubules called cornicles—one on each side near the posterior end—which exude a sticky protective substance when they are attacked. Though minuscule, they live in such large colonies and reproduce so rapidly (having numerous generations or broods) that they may be very destructive. They are found on all kinds of plants—on leaves, bark, roots, blossoms, and fruit; many produce galls and other deformities, or cause curling of leaves, which affords them protection. Their anal excretions (honeydew) are desired by ants, which protect them from enemies. Many aphids are vectors of fungus and virus diseases of plants. Most species have winged as well as apterous forms, but some do not. Males are sometimes rare or do not appear. Mating usually takes place in the fall, at which time oviparous females lay overwintering eggs; reproduction is by parthenogenesis at other times. Unmated females are ovoviviparous, giving birth to living young. They live on an average about one month, and each female produces from 80 to 100 offspring.

(18) WOOLLY APPLE APHID: *Eriosoma lanigerum.* **Fig. 146A.**

Range: World-wide. **Adult:** Reddish or purplish under the bluish white, cottony wax. **Length:** .08". They suck juices from bark of branches and roots (are more destructive on latter); twigs become swollen and knotted,

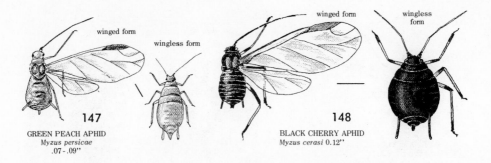

GREEN PEACH APHID
Myzus persicae
.07-.09"

148

BLACK CHERRY APHID
Myzus cerasi 0.12"

and roots develop nodules. In northern regions they hibernate as egg or nymph, in the latter case on the roots of apple. In warm regions mature females may hibernate on trunks or branches. Eggs are usually deposited in the bark of elm trees in the fall; aphids hatching in the spring feed here for two generations and produce winged forms that migrate to apple, hawthorn, or mountain ash, where some work their way down to the roots. Wingless males appear in the fall and mate with wingless females, each of which lays a single egg in the bark on a branch or the trunk of a tree. Some winged females are active during most of the summer. **Food:** Apple, pear, quince, elm, hawthorn, mountain ash. These aphids are parasitized by a chalcid wasp which leaves many black "mummified" aphid remains stuck to the branches [see Hymenoptera (39)]. The **Woolly Elm Aphid,** *E. americanum,* is very similar to *lanigerum* in appearance and life history; it is purplish and often covered with a white cottony wax, causes curling of *individual* leaves of elm (*lanigerum* causes *groups* of leaves to curl); its alternate host is shadbush (also known as Juneberry or serviceberry).

(19) GREEN PEACH APHID: *Myzus persicae.* **Fig. 147.**

Also known as spinach aphid or tobacco aphid. **Range:** World-wide. **Adult:** Wingless forms vary from light to dark green, with yellowish tint, during summer; in late summer and fall they vary from pink to red; winged forms are usually dark brown, except for yellowish abdomen, partly covered dorsally with irregular dark patch. **Length:** .07–.09". In the southern states, Arizona, California, extreme western Oregon and Washington, practically all are females. Continuous reproduction by unmated (parthenogenetic) females also occurs as far north as warmer parts of New Jersey, Maryland, Virginia, Tennessee, Arkansas, Oklahoma, Washington; hibernation takes place here in the adult stage. In colder areas, males and egg-laying females appear in the fall; hibernation takes place in the egg stage, other stages being unable to survive 0 degrees F. or lower temperatures. Eggs are laid on peach, plum, and cherry; nymphs hatching in the spring feed here, later broods spread to vegetables and other plants. As many as 30 broods occur in the extreme South and parthenogenetic reproduction is

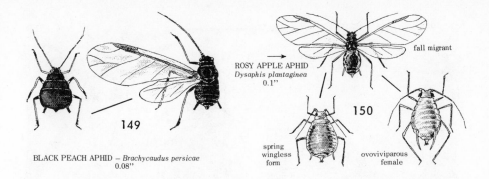

ROSY APPLE APHID
Dysaphis plantaginea
0.1"

fall migrant

150

spring
wingless
form

ovoviviparous
female

149

BLACK PEACH APHID – *Brachycaudus persicae*
0.08"

continuous. **Food:** Peach, plum, prune, cherry, citrus; cabbage and related cole crops, dandelion, endive, mustard greens, parsley, turnip, tobacco, potato, rose; most destructive to spinach, peppers, beets, celery, lettuce, and chard, it spreads virus diseases such as potato leaf roll, beet mosaic, beet yellows (which also occurs in spinach), and lettuce mosaic. [See Hymenoptera (26).]

(20) BLACK CHERRY APHID: *Myzus cerasi.* **Fig. 148.**

Range: Pacific Northwest, California, Colorado, Utah, Nevada. **Adult:** Shiny black. Winter eggs are tucked in among buds, and hatch when the buds burst. A large species. **Length:** .12". The young develop on new growth, curling the leaves and feeding from within; leaves become covered with sticky honeydew. Winged adults appear in midsummer, migrate to peppergrass, watercress, and other wild crucifers. Later broods return to cherry in the fall and produce wingless males and oviparous females which lay overwintering eggs. Several broods. **Food:** Sweet cherry, sour cherry (less often); peppergrass, watercress, and other Cruciferae.

(21) BLACK PEACH APHID: *Brachycaudus persicae (persicaecola).* **Fig. 149.**

Range: Most abundant in the Middle Atlantic states; also occurs in the Pacific Northwest, California, Utah, New Mexico. **Adult:** Shiny black. **Length:** .08". **Nymph:** Reddish brown. Adult wingless forms live on roots throughout the year; in late winter and early spring some migrate to new growth above ground, start new colonies on twigs and young shoots. Winged forms develop, presumably in response to overcrowding, and migrate to other trees. Migrants return to peach in late fall; males and oviparous females appear and mate, and the latter lay overwintering eggs. **Food:** Peach, apricot, almond. The **Thistle Aphid,** *B. cardui,* is greenish or reddish, curls leaves of plum and prune in the Pacific Northwest.

(22) ROSY APPLE APHID: *Dysaphis plantaginea (Anuraphis rosea).* **Fig. 150.**

Range: Throughout North America, wherever apples are grown. **Adult:** May be distinguished by the long cornicles flanged at the tips and by the

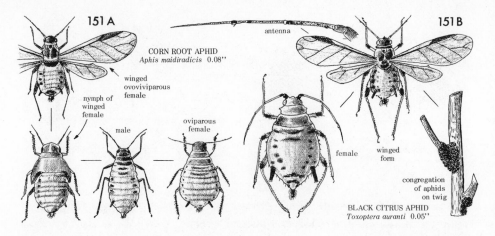

151 A

CORN ROOT APHID
Aphis maidiradicis 0.08"

winged
ovoviviparous
female

nymph of
winged
female

male

oviparous
female

antenna

151 B

female   winged
form

congregation
of aphids
on twig

BLACK CITRUS APHID
Toxoptera auranti 0.05"

long antennae [see (25) below]; spring form on apple purplish or rosy brown, winged form migrating to plantain brownish green. Pale green eggs—they turn shiny black later—are deposited on twigs, in bud axils, and in bark crevices of apple trees, and hatch in spring when the buds begin to swell. **Length:** .1". Several broods feed on fruit spurs and young fruit, causing leaves to curl and fruit to become gnarled and stunted. Winged forms appear in July, abandon apple for the alternate host, plantain; several broods develop here, winged forms appear again and fly to apple trees. Wingless male and oviparous female forms now appear and mate, and the latter lay overwintering eggs. **Food:** Apple, plantain.

(23) CORN ROOT APHID: *Aphis (Anuraphis) maidiradicis.* **Fig. 151A.**

Range: Throughout the corn- and cotton-growing areas east of the Rocky Mts. **Adult:** Blue-green. The overwintering eggs, which are shiny dark green, are collected by the cornfield ant [see Hymenoptera (77)], and stored in its nest. **Length:** .08". The ant carries the young to certain wild plants, especially smartweed, in the spring and later to roots of corn and cotton. Winged forms appear in the summer months, and some fly to other fields, thus aiding in the spread of infestations. They are, however, almost entirely dependent on the ants and are rarely found on roots unless attended by the ants. **Food:** Corn, cotton, various grasses and weeds, particularly smartweed. The closely related **Clover Aphid,** *Nearctaphis (Anuraphis) bakeri,* is sometimes very injurious to red and alsike clover grown for seed.

(24) RUSTY PLUM APHID: *Hysteroneura setariae.* **Fig. 152A.**

Range: Eastern North America, westward into Colorado. **Adult:** Rusty brown. Eggs are deposited on small twigs in the fall, and hatch when the buds open. **Length:** .06–.08". The first few broods are wingless, feed on foliage of plum and peach; new growth is distorted and crumpled, terminal buds stunted. Winged forms appear later, migrate to several varieties of grasses, where they breed during remainder of summer. Winged forms reappear in the fall, migrate to plum and young peach trees; males and oviparous females mate, and the latter lay overwintering eggs. **Food:** Plum, peach, corn, grasses, sugar cane, Virginia creeper.

(25) APPLE APHID: *Aphis pomi.* **Fig. 154.**

Also called green apple aphid. **Range:** Throughout North America, wherever apples are grown. **Adult:** Green, with black legs, sometimes with a yellowish head; cornicles and antennae are short [see (22) above]. Eggs are green at first, later turn black like those of the rosy apple aphid; they are laid in the fall, usually on water sprouts, and hatch in the spring when the buds begin to swell.* They feed mostly on succulent terminal twigs, stunting their growth and curling the leaves. Winged forms migrate to other apple trees. Male and oviparous female forms appear in the fall and mate, and the latter lay overwintering eggs. Nine to seventeen broods. **Food:** Apple throughout the year, occasionally pear. The **Apple Grain Aphid,** *Rhopalosiphum fitchii,* is often found on apple and pear in the early season; it is light green, with dark green bands on abdomen, may be distinguished from *A. pomi* by the green legs. [*R. maidis* is described in (34) below.]

(26) SPIREA APHID: *Aphis spiraecola.*

Also known as the citrus aphid. **Range:** Throughout North America; the most important aphid on citrus in California and Florida. **Adult:** Apple-green like tender citrus leaf, with black cornicles and cauda; winged form has dark brown or almost black thorax. **Length:** .07″. **Nymph:** Thorax changes to pink as wing pads begin to form, darkens as they develop. So far as known, no males or oviparous females appear on citrus. Eggs have been found on spirea in Florida in the winter. Winged forms appear when they are ready to migrate to other trees; in Florida they are present throughout the year but are most numerous in April and May. **Food:** Citrus, spirea. Injury consists mostly of stunted terminal growth and curled leaves; sooty mold forms on honeydew and drips on fruit, which then must be washed. The **Black Citrus Aphid,** *Toxoptera auranti* (Fig. 151B), is black or mahogany brown, often with purplish cast, about .05″ long, winged form with short black band along costal margin of forewing near apex; it is widespread, attacks citrus, holly, and camellia. This species and the spirea aphid are confined mostly to citrus in California.

(27) COTTON (MELON) APHID: *Aphis gossypii.* **Fig. 152B, 152C.**

Range: Throughout North America; world-wide. **Adult:** Light yellow in spring, changing to dark green in summer, blackish in the fall. **Length:** .04–

---

* The aphids emerging from overwintering eggs are all females and are known as "stem mothers," as in the case of other species of similar habit.

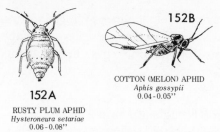

152B

COTTON (MELON) APHID
Aphis gossypii
0.04-0.05"

152A

RUSTY PLUM APHID
Hysteroneura setariae
0.06-0.08"

152C

PARASITE – a
small braconid
wasp, *Lysiphlebus*,
ovipositing in a
cotton aphid

.05". Only parthenogenetic, ovoviviparous females occur in the Cotton Belt. They may hibernate part of the winter as winged adults in the ground litter; development occurs on weed hosts in the spring, winged forms appear when conditions prompt a change of plants. New generations may occur every five days in summer. Males and oviparous females appear in the North on melons and other plants, and they hibernate in the egg stage. **Food:** Cotton, citrus, melon, cantaloupe, cucumber; many kinds of garden flowers and shrubs, including aster, chrysanthemum, lilac, rose; various vegetables and wild plants. Feeding stunts growth of seedling cotton, causes leaves to shed and bolls to open prematurely in the fall. It "seldom increases to damaging numbers unless its natural enemies are killed or retarded," as in the use of insecticides. [See Hymenoptera (26).]

(28) BEAN APHID: *Aphis fabae.* Fig. 155.

Also known as the black bean aphid. **Range:** Throughout North America. **Adult:** Dark olive-green to black; resembles the cabbage aphid (33). In the North, overwintering eggs are laid on euonymus, to a limited extent on snowball and deutzia. It is believed to produce only females in the South, where the favored host plant is dock. **Food:** Chiefly bean, beet, chard, and weeds as indicated; it is injurious to safflower in California. The minute **Buckthorn Aphid,** A. *nasturtii,* a pest of potatoes in the northeastern states, is less than .04" long, yellowish green to dark green or almost black, the winged form with darker head and thorax. The **Grapevine Aphid,** A. *illinoisensis,* is deep reddish brown, attacks succulent terminal shoots of grape.

(29) STRAWBERRY ROOT APHID: *Aphis forbesi.* Fig. 153.

Range: East of the Rocky Mts., California. **Adult:** Dark bluish green to almost black; coloring and egg-shaped form distinguish it from the strawberry aphid (30). It attacks crowns and roots, stems, new leaf buds, and underside of leaves. In the North, males and oviparous females appear in the fall and mate, and the latter lay overwintering eggs. In the South and milder areas of California, only females appear and parthenogenetic reproduction is continuous. Most females are wingless; winged forms appear in the summer and fall. **Food:** Strawberry. In the eastern states, the cornfield

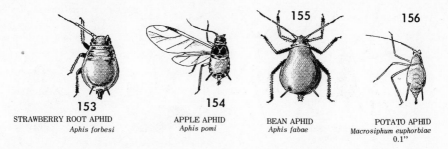

153
STRAWBERRY ROOT APHID
Aphis forbesi

154
APPLE APHID
Aphis pomi

155
BEAN APHID
Aphis fabae

156
POTATO APHID
Macrosiphum euphorbiae
0.1"

ant [see Hymenoptera (77)] carries many of the aphids to the crowns and roots of strawberry plants, where they are more injurious.

(30) STRAWBERRY APHID: *Chaetosiphon (Pentatrichopus) fragaefolii.*

Range: Throughout North America. **Adult:** Wingless form pale green or whitish to yellowish, dull in appearance, with relatively long antennae; winged form shiny yellowish green, with extensive black markings on head and body [compare with (29) above]. **Length:** .06". Occurs on new shoots and buds in crown of plant. They thrive in cool weather, are most numerous in spring and fall; the peak population in California is in March, when the proportion of winged forms is high. **Food:** Strawberries. They transmit several virus diseases of strawberries.

(31) POTATO APHID: *Macrosiphum euphorbiae.* **Fig. 156.**

Also called pink and green tomato aphid. **Range:** Throughout North America. **Adult:** Light to medium green, pink, or mottled green and pink; many green individuals have a darker longitudinal stripe on the back. **Length:** .1". In colder areas the usual sexual forms occur in the fall, with overwintering eggs laid on stems of roses mainly; spring forms migrate to other plants. In warmer areas the usual parthenogenetic females occur, without egg stage; winter and summer in this case are spent on similar plants. **Food:** Potato, rose, bean, tomato, pea, asparagus, eggplant, turnip, clover, buckwheat, corn, aster, iris, gladiolus, ground cherry, lamb's-quarter, nightshade, pigweed, shepherd's-purse, wild lettuce, apple, citrus. The **Rose Aphid,** *M. rosae* (Fig. 157), is similar in appearance, may occur on roses throughout the season; it overwinters in the egg stage on rose plants.

(32) PEA APHID: *Acyrthosiphon (Macrosiphum) pisum.* **Fig. 160.**

Range: Throughout North America, where peas and alfalfa are grown. **Adult:** Green. A large species. **Length:** .12–.17". Overwinters as egg or ovoviviparous female, usually on alfalfa or clover. In spring these fertilized eggs hatch into ovoviviparous "stem mothers" which give rise to new infestations. In the Midwest they begin migrating to other plants early in May and are most injurious to peas in June. Winged forms appear when overcrowding takes place. Seven to twenty broods. **Food:** Peas, sweet peas,

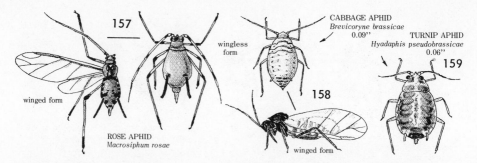

157

winged form

ROSE APHID
Macrosiphum rosae

wingless
form

CABBAGE APHID
Brevicoryne brassicae
0.09"

TURNIP APHID
Hyadaphis pseudobrassicae
0.06"

159

158

winged form

alfalfa, clover, wild leguminous plants. A very prolific species under favorable conditions and one of the worst pests of peas. In severe infestations the plants wilt and large bronze-colored patches appear in the field. [See Hymenoptera (26).] The **Foxglove Aphid,** A. solani—a pest of potatoes in the Northwest, Pacific Northwest, and Southwest—is apple green (wingless form) or brownish yellow (winged form) with dark ring around cornicles.

(33) CABBAGE APHID: *Brevicoryne brassicae.* **Fig. 158.**

Range: Throughout North America. **Adult:** Grayish green; distinguished from other aphids by the powdery, waxy covering. **Length:** .09". Thirty or more generations of females, winged or wingless, are produced throughout the year in the South. In the North, males and oviparous females occur in the fall; overwintering eggs are laid on residues of host crops left in the field. **Food:** Chiefly cabbage, cauliflower, broccoli, collards, and kale; seldom mustard or turnip.

(34) TURNIP APHID: *Hyadaphis (Rhopalosiphum) pseudobrassicae.* **Fig. 159.**

Also called false cabbage aphid. **Range:** Widely distributed, most destructive in the South. **Adult:** Pale green, resembling cabbage aphid (33) without the waxy covering; winged forms with head, spots, and wing veins black. **Length:** .06". Similar in habits to cabbage aphid, except that oviparous females are rare. In the Gulf states, 46 generations have been observed in a year. **Food:** Mainly turnip, mustard, radish; other crucifers in seedling stage. The **Corn Leaf Aphid,** *Rhopalosiphum maidis,* is similar—green, with dark green head—occurs in the heart leaves, young tassels, and heads of corn and sorghum, and sometimes damages late-seeded barley in the Plains states. [*R. fitchii* is described in (25) above.]

(35) SPOTTED ALFALFA APHID: *Therioaphis maculata.* **Fig. 162.**

Range: Southern two-thirds of the U.S.; especially destructive to alfalfa grown west of the Mississippi River. **Adult:** Pale yellow, with *six* or more rows of black spots along the back; the winged form has smoky areas along the wing veins. **Length:** .06". Males are rare. Females are mostly wingless, parthenogenetic forms which give birth to living young without mating. Sexual forms occur in colder climates, where they overwinter in the egg

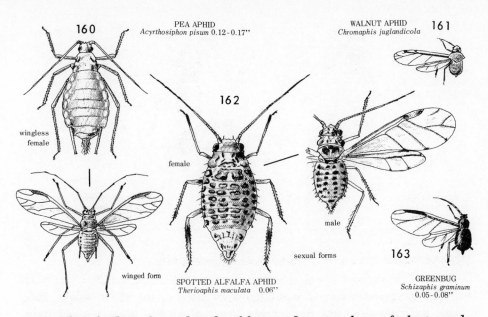

PEA APHID
Acyrthosiphon pisum 0.12-0.17"

160

WALNUT APHID
Chromaphis juglandicola

161

162

wingless
female

female

male

sexual forms

163

winged form

SPOTTED ALFALFA APHID
Therioaphis maculata   0.06"

GREENBUG
Schizaphis graminum
0.05-0.08"

stage. They feed on the underside of leaves, first near base of plant, gradually work their way up and feed on the stems. Twenty-nine or more broods. **Food:** Alfalfa mainly; clovers to some extent. In Arizona peak populations occur in April, July, October, sometimes in January. The **Sweetclover Aphid,** *T. riehmi,* is yellow, may be distinguished by the *four* rows of spots and its association with sweet clover (*Melilotus*); also, the ovoviviparous, parthenogenetic female is always winged. The **Yellow Clover Aphid,** *T. trifolii,* resembles *T. maculata* more closely (may be the same species)—has *six* rows of spots—differs only in that its true host is red clover (*Trifolium*). These are introduced species. Two braconid wasps—*Praon palitans* and *Trioxys utilis*—imported from the Mediterranean area, help control *T. maculata.*

(36) WALNUT APHID: *Chromaphis juglandicola.* **Fig. 161.**

Range: California, Oregon, British Columbia. **Adult:** Pale yellow. Wingless females which appear in the fall to lay eggs have two dark bands across the back. Feeding is from underside of leaves; nymphs tend to settle along midrib of leaf and branches radiating from it. The usual ovoviviparous, parthenogenetic females—some winged—occur throughout the summer. Tiny males with smoky wings, and difficult to see, appear in the fall, as do oviparous females which lay overwintering eggs. **Food:** English walnuts. "Most important insect attacking walnut." *Trioxys pallidus*—a braconid wasp first imported in 1959—is an effective parasite of this aphid. The **Black Pecan Aphid,** *Finocallis caryaefoliae,* feeds on both sides of pecan leaves, causing premature drop; it prefers inner shaded part of tree; nymphs are pale green, gradually become darker and finally black.

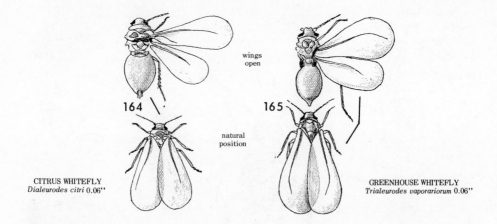

164

wings
open

165

natural
position

CITRUS WHITEFLY
Dialeurodes citri 0.06''

GREENHOUSE WHITEFLY
Trialeurodes vaporariorum 0.06''

(37) GREENBUG: *Schizaphis (Toxoptera) graminum.* **Fig. 163.**

Range: Throughout North America, first recorded in the U.S. (Virginia) in 1882; most destructive in wheat-growing areas of the West. **Adult:** Pale green; wingless form has dark green stripe down the back, winged form has brownish yellow head and blackish cornicles. **Length:** .05–.08''. In the North, shiny black eggs are laid on plants in the fall, and hatch into females in the spring. Succeeding generations, excepting the last, are parthenogenetic females (winged and wingless) which give birth to living young. Males and oviparous females appear in the fall. In the South all are females, the egg stage is omitted, and they hibernate as nymphs or adults. Five to fourteen broods. **Food:** Grains, alfalfa, wild and cultivated grasses.

Whiteflies (Aleyrodidae)

Whiteflies are tiny mothlike insects, from .04'' to .12'' long. Adults of both sexes have two pairs of membranous wings coated with a white powdery wax. Like scale insects, nymphs lose their legs after the first molt; unlike scale insects, females get their legs back in the adult stage. The nymphs resemble early instars of unarmored scales; they are flattened, oval, often with a marginal fringe of white waxy filaments or "plates" around the outer margin, not unlike mealybugs. A pupal stage precedes the adult, which emerges through a T-shaped slit in the pupal case.

(38) CITRUS WHITEFLY: *Dialeurodes citri.* **Fig. 164.**

Range: Florida and other Gulf states. **Adult:** It lives ten days on an average. Pale yellow eggs are scattered on underside of leaf, each attached by a stalk. Unfertilized eggs hatch into males. **Length:** .06''. **Nymph:** Moves about for a few hours, inserts its beak into leaf, settles down to sucking sap; legs become vestigial after first molt. Trees are weakened, honeydew and sooty mold drop on fruit. A pupal stage occurs after the third molt, resembles the last nymphal instar. **Food:** Citrus, chinaberry and umbrella trees, gardenias, privet, prickly ash, Japanese persimmon; a serious pest of citrus in Florida. The **Cloudy-winged Whitefly,** *D. citrifolii,* is similar, except for darkened area in middle of wings; also, its eggs are black, and it is attacked by yellow aschersonia fungus (*citri* is not). The **Woolly Whitefly,** *Aleurothrixus floccosus,* is a pest of citrus in southern California. The **Citrus Black-fly,** *Aleurocanthus woglumi,* a native of India and south Asia, attacks citrus, mango, persimmon, pear, quince, coffee, myrtle, cherimoya, and sapote in Mexico, is normally controlled by parasites and lady beetles; the wings are slaty blue with colorless spots forming a white band across the middle when folded, abdomen red, antennae yellowish white. Curved elliptical eggs are attached by pedicels to underside of leaves in a spiral formation; the ovate larva is shiny black, with green spot on anterior part of abdomen when full grown, and numerous spines; the pupa is ovate, black with waxy marginal band. It is a potential threat to citrus in the U.S. [See Diptera (59) Mexican fruit fly.]

(39) GREENHOUSE WHITEFLY: *Trialeurodes vaporariorum.* **Fig. 165.**

Range: Common in gardens in warm climates, a serious pest in greenhouses; they do not survive outdoors in cold climates. **Adult:** "Like snowflakes" when flushed from heavily infested plants. Yellowish green cone-shaped eggs are attached to the underside of leaves by a short pedicel. **Length:** .06″. **Nymph:** Thin, flat, oval-shaped; semitransparent, pale green with waxy filaments. Hibernates in immature stage. **Food:** Fruits, vegetables, flowers. Troublesome to greenhouse cucumbers, tomatoes, ornamental plants. Breeds on vegetables such as squash, potatoes, tomatoes during summer, reinfests greenhouses in the fall. The **Banded-wing Whitefly,** *T. abutilonea,* occurs on mallow, is sometimes a serious pest of cotton, beans, and crucifers.

Soft Scales, Armored Scales, and Mealybugs (Coccoidea)

Soft or unarmored scales (Coccidae and others), armored scales (Diaspididae), and mealybugs (Pseudococcidae) are grouped in the superfamily Coccoidea and are known collectively as coccids. They include a great many species of small insects that suck plant juices, occur in vast aggregations, and are very destructive, especially to fruit or other trees. Some are exposed, with an epidermal covering; others are covered with a waxy secretion, or with a hard shell. All are motile at some time—during the so-called crawler stage—but the majority remain fixed for the greater part of their existence. The antennae may be well developed or lacking; the beak arises between the front legs; the tarsi are one-jointed ending with a single claw. Like aphids, the young are born alive or from eggs. Adult females are wingless. Males very soon pass through a "quiescent metamorphosis" in a cocoon, or under a shell, usually emerge as delicate two-winged adults with two long tails, and disappear soon after mating. Many species, particularly unarmored scales, produce great amounts of honeydew, which attracts ants and sooty mold. The coccids are more severely parasitized than any other group of insect pests, and largely by chalcid wasps. The commercial value of some scales was recognized many centuries ago. The **Lac Insect,** *Laccifer lacca,* of India and southeast Asia, is used to make shellac. Other species are used for wax and dyes.

(40) BLACK SCALE: *Saissetia oleae* (Coccidae). **Fig. 166.**

Range: Nearly all citrus areas of the world. A minor pest in the Gulf states. **Adult:** Full-grown females are dark brown to black, hemispherical, with ridges on the back forming an H. Up to 2,000 eggs are laid under a scale; they hatch in about 20 days. **Size:** .2" across, .04–.12" thick. **Nymph:** First instar or "crawler" light brown with black eyes, six-segmented antennae. Remains under scale for a few hours after hatching, then migrates to twigs and branches. Once settled, they remain stationary. The H forms on female after second molt. Development requires 8 to 10 months in single-brooded forms in the interior regions of California. They are double-brooded in the coastal areas. Males have a pupal stage, emerge as two-winged adults, but are rare; females reproduce mostly by parthenogenesis. **Food:** Citrus, almond, walnut, olive, fig, apple, pear, apricot, grape, oleander, rose. In Florida, mostly oleander, rarely citrus. [See Hymenoptera (42).] The **Hemispherical Scale,** *S. coffeae* (*hemisphaerica*)—convex, strongly elliptical, smooth, dark brown—is found on ferns and other greenhouse and house plants. The **Brown Soft** (or "tortoise") **Scale,** *Coccus hesperidium*—oval and flattened, fleshy, much like the black scale except for color, light brown or greenish brown (male elongated, white, conspicuous)—attacks many ornamentals, is a minor pest of citrus in Florida; females give birth to living young. The **Azalea Bark Scale,** *Eriococcus azalea*—elongate and oval, white, accompanied by sooty mold which grows on the honeydew secretions of the scale—is found on rhododendron and Japanese andromeda as well as azalea. The **Tuliptree Scale,** *Toumeyella liriodendri*—a large scale, about .35" in diameter, hemispherical, wrinkled, accompanied by sooty mold—hibernates as first instar nymph, females give birth to living young; heavy infestations on twigs and branches of tulip tree may cause its death.

(41) EUROPEAN FRUIT LECANIUM: *Lecanium corni.*

Also called brown apricot scale in Pacific Northwest. **Range:** Throughout North America. **Adult:** The female is brown, often covered with a brown or cottony substance, oval in shape. The delicate male is smaller, flatter, transparent, with well-marked ridges. The soft covering of the female hardens, forming a shell to protect the eggs. Each female lays several hundred to a few thousand pearly white eggs by the end of June. **Length** (female): .1–.2″. **Nymph:** Yellow to pale brown. Young nymphs hatching in spring migrate to the underside of leaves, settle along the midrib and veins to feed, remain here until late summer, when they migrate back to the limbs. Hibernates half-grown on bark, on the underside of a limb. One brood. **Food:** More than fifty species of shade trees, ornamentals, fruit trees; sometimes conifers, highbush blueberries (in New England), and grape. [See Hymenoptera (42).] The **Terrapin Scale**, *L. nigrofasciatum*—shiny, nearly hemispherical, reddish brown, about .12″ in diameter—occurs on fruit and shade trees in the eastern and southern states; it is marked by the presence of sooty mold formed on honeydew deposits of the scale. The **Calico Scale**, *L. cerasorum*—conspicuous, almost globular, brown and white, about .4″ in diameter—is common on walnut, pear, and stone fruits in California. **Fletcher's** (or "arborvitae") **Scale**, *L. fletcheri*—more or less hemispherical, dark brown (nymphs are amber to pale orange-yellow)—is found on arborvitae, yew, pachysandra, and many other plants in greenhouses and outdoors. The **Magnolia Scale**, *Neolecanium cornuparvum*—one of the largest scales, about .5″ long, convex, smooth, shiny dark brown, covered with a white waxy bloom—coats twigs and branches of magnolia if unchecked; it hibernates as first-instar nymph.

(42) COTTONY MAPLE SCALE: *Pulvinaria innumerabilis.*

Range: Throughout the U.S. and Canada. Native of Europe. **Adult:** The female is brown, convex, easily recognized by the cottony sac, several times her size, which she produces on twigs in the summer. The sac is formed gradually and the body of the scale is raised above the twig, often at an angle. About 3,000 eggs are laid in a sac; the female dies after oviposition. The eggs hatch during June and July. **Length** (female): About .12″. **Nymph:** Partly grown female is greenish or yellowish, oval, flattened. Young scales crawl from twigs to the underside of leaves, settle along the midrib and veins, where they suck the sap. They mature by August or September, mate, and the males die. The partly grown females migrate to the twigs where they hibernate; growth is completed in the spring. One brood. **Food:** Maple, pear; many other forest, shade, and fruit trees; grape, gooseberry, currant, Virginia creeper. Most destructive in northern part of range. A closely related species, *P. acericola*, called the **Maple Leaf Scale**, is destructive at times to maples in Arkansas (also found on sassafras). The female of the **Cottony Taxus Scale**, *P. floccifera*, is hemispherical, flattened, light brownish, about .12″ long when it hibernates in the fall; it forms a long, white, fluted egg sac (three to four times as long as the insect itself) after maturing in the spring, is a pest of camellia, abutilon, and acalypha. The **Cottony Peach Scale**, *P. amygdali*, is also similar.

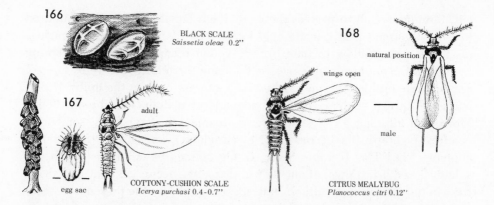

166

BLACK SCALE
Saissetia oleae 0.2''

168

natural position

wings open

167

adult

male

egg sac

COTTONY-CUSHION SCALE
Icerya purchasi 0.4-0.7''

CITRUS MEALYBUG
Planococcus citri 0.12''

(43) COTTONY-CUSHION SCALE: *Icerya purchasi* (Margarodidae). **Fig. 167.**

Range: South Carolina, Georgia, Florida, westward to California. A native of Australia. **Adult:** Mature female reddish brown, with black legs and antennae, distinguished by the large, white, cottony, fluted egg sac. From 600 to 1,000 bright red eggs are laid in each sac, hatch in a few days in the summer and in about two months in the winter. **Length** (including egg sac): .4–.7'' (sac about two and one-half times as long as the body). **Nymph:** When newly hatched it is bright red, with darker six-jointed antennae and thin brown legs. First and second instars feed on leaves, along the midrib and veins, later migrate to twigs and finally the trunk. In the third instar, it is broadly oval, reddish brown (obscured by cottony wax secretion), the antennae are nine-jointed. After the third molt, the body is freed of its waxy secretion, the antennae are eleven-jointed, the egg sac begins to form, and egg laying starts. The male is smaller, more elongate; it stops feeding after the second molt, seeks shelter on the tree or ground, and forms a flimsy cocoon, transforming to the delicate two-winged adult. **Food:** Great variety of trees and other plants. Once a serious threat to citrus in California, it was controlled by introduction of Vedalia [see Coleoptera (199)] in one of the most spectacular successes in biological control; an imported parasitic fly, *Cryptochetum iceryae,* also helps to control it. The **Red-pine Scale,** *Matsucoccus resinosae,* is similar in appearance and life history; it is a serious pest of red (Norway) and several imported species of pines in New England, New York, New Jersey, and Pennsylvania; it has two broods, hibernates as first-instar crawler under bark. The **Sycamore Scale,** *Stomacoccus plantani,* is one of the smallest scale insects (the cottony-cushion scale by contrast is one of the largest); unlike other margarodids, the male is wingless and has wax glands, spends part of its life in a cottony cocoon. *Margarodes* spp.—called **Ground Pearls**—are pests of Bermuda grass in the South and Southwest; the female deposits about 100 eggs in a white waxy sac, the preadult nymph forms a hard, faintly yellowish green globular shell (about .12″ in diameter) resembling a pearl on the rootlets; they are found on grapevine roots in southern California.

(44) EUROPEAN ELM SCALE: *Gossyparia spuria* (Eriococcidae).

Range: New Jersey, Maryland, Pennsylvania, midwestern and far western states. **Adult:** The female is reddish purple, surrounded by white cottony fringe, oval in shape. The minute male forms a conspicuous white cocoon early in the spring, emerges as a reddish, winged or wingless adult in April or May. The female deposits eggs on a twig, beneath herself; they hatch in an hour or less. **Length** (female): .2–.4″. **Nymph:** Lemon yellow; migrates to leaf to feed, returns to limb or trunk in the fall. Hibernates as second-instar nymph in bark crevice. **Food:** Elms. *Dactylopius opuntiae* and *confusus* (Dactylopiidae) are "cochineal" coccids (producing a red dye of that name), feed on prickly pear cactus (*Opuntia*); *D. opuntiae* occurs from Texas to California and Mexico, *confusus* occurs in Florida, Kansas, and from Colorado to Wyoming and Utah.* *Dactylopius* sp. has been used successfully to control prickly pear cactus on rangeland.†

Mealybugs (Pseudococcidae)

Mealybugs are oval, flattened, soft-bodied insects, distinctly segmented but without a clear definition between head, thorax, and abdomen. They are covered with a white, powdery or mealy wax, and have many lateral and anal filaments which are useful in identification. Mealybugs congregate in large compact colonies amid masses of eggs, and suck the sap from roots, leaves, and fruit stems; they cause premature fruit drop and deposits of honeydew with sooty mold. Potentially destructive, they are normally controlled by parasites and predators.

(45) CITRUS MEALYBUG: *Planococcus (Pseudococcus) citri.* **Fig. 168.**

Range: Throughout the world on citrus. **Adult:** Pale yellow to brownish orange, with short, irregular anal filaments [see Pictured Key to the Insect Orders: (*) Homoptera]. Oval yellow eggs are laid in a loose cottony mass of 200 to 350 eggs. The female dies when oviposition is completed. **Length:** .12″. **Nymph:** Pale yellow, free of wax when newly hatched. The female has three instars, then transforms to an adult. The male (Fig. 168) goes through three instars, forms a flimsy, cottony cocoon; it emerges as an adult with a

---

* The cochineal insect, *Dactylopius coccus,* is the cochineal of commerce (the "true cochineal"), being the largest of the genus and thus producing the most dye; it is believed to be a native of Mexico or Central America, has become established in many countries. Aniline dyes have largely replaced cochineal today, but the industry still flourishes in the Canary Islands.

† Richard D. Goeden, Charles A. Fleschner, and Donald W. Ricker, "Biological Control of Prickly Pear on Santa Cruz Island, California," *Hilgardia,* 38(16):579–606 (1967), Univ. of California, Berkeley.

pair of wings and a pair of halteres with hooks, and long white filaments protruding from the tip of the abdomen. Hibernates mostly in the egg stage. Two or three broods. **Food:** Citrus; also found on avocados, cotton, potatoes, and ornamentals in California. Sometimes a serious pest of cucumbers and tomatoes in greenhouses. It is the principal mealybug pest of citrus in Florida, and persists in the coastal areas of California, where it is controlled by periodic releases of insectary-reared lady beetles (*Cryptolaemus montrouzieri*) and chalcid wasps (*Leptomastix dactylopii*). The **Rhodes-grass Scale,** *Antonina graminis*—probably introduced from the Orient, first discovered in 1942—is a pest of lawns, infests Bermuda grass and St. Augustine grass in southern Texas, also occurs in Louisiana and Florida; the adult is about .12″ in diameter, dark purplish brown, covered with a cottony secretion, the first-instar nymph or crawler cream-colored; the sedentary second instar forms a waxy covering (only females are present, and they give birth to living young); its principal parasite is an encyrtid wasp, *Anagyrus antoninae.*

(46) CITROPHILUS MEALYBUG: *Pseudococcus fragilis (gahni).* **Fig. 169.**

Range: Widespread—outdoors or in greenhouses. **Adult:** Similar to citrus mealybug (45) except that lateral filaments are wedge-shaped and anal filaments are longer. **Food:** Deciduous fruits, various ornamental trees and flowers, as well as citrus. [See Hymenoptera (40).]

(47) LONG-TAILED MEALYBUG: *Pseudococcus adonidum (longispinus).*

Range: Widespread—outdoors or in greenhouses. **Adult:** Distinguished from others by longer lateral filaments and long, tapering anal filaments, also by absence of egg masses; the female is ovoviviparous, giving birth to living young. **Food:** Many different kinds of plants, *Dracaena* (an ornamental) being a preferred host; this was a serious pest of avocados in California prior to the introduction of two chalcids, *Hungariella peregrina* and *Anarhopus sydneyensis.*

(48) BAKER'S (GRAPE) MEALYBUG: *Pseudococcus obscurus (maritimus).** **Fig. 170.**

Range: Widely distributed; occurs farther north than the other mealybugs described above. **Adult:** Pearly to gray, more elongated than the others; the powdery wax covering is thin, the lateral filaments are short and slender, and the anal filaments one-fourth to one-half as long as the body. The female deposits 400 to 600 eggs in a cottony white sac which she carries around. **Length:** About .2″. **Nymph:** Spring brood feeds on new shoots. Hibernates

---

* These are actually two different species but very difficult to separate. See H. L. McKenzie, "Taxonomic Study of California Mealybugs, with Additional Species from North and South America," *Hilgardia,* 29(15):681–770 (1960); 31(2):15–52 (1961); 32(14):644–645 (1962); 35(10):255–256 (1964). The name "grape mealybug" applies to *P. maritimus,* according to *Common Names of Insects.*

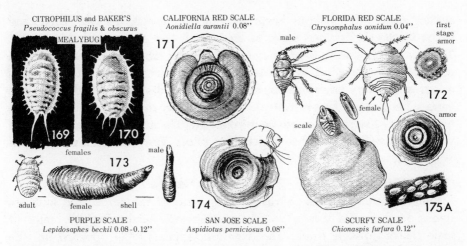

in egg stage or as very young nymph in egg sac. Two broods. **Food:** Grapes, citrus, walnut, pear, apple. Attacks grapes as far north as Michigan. Usually controlled by native parasites and predators. The chalcid parasite *Acerophagus notativentris* attacks the mealybug on grapes in California but not on citrus. The wingless female of the **Comstock Mealybug,** *P. comstocki,* is elliptical in shape, about .25" long, with fringe of short soft spines and covering of white wax (the winged male is seldom seen); it hibernates as egg under bark, infests apples, peaches, pears, grapes, and catalpa in the eastern states, is a pest of citrus in some parts of the world.

Armored Scales (Diaspididae)

The most conspicuous difference between the armored scales and the soft scales is the thick armor made of wax and exuviae (skins cast off after molting). The armor is not attached to the body and is enlarged by adding threads of wax around the outer edge; the circular scales do this by rotating their bodies, while the elongate or oystershell type do it by moving their bodies from side to side. The head and thorax are fused into one, and body segmentation is obscured. The posterior end of the body, called the pygidium, is greatly compressed and contains the wax glands and tubes, which open at the tip. Species are differentiated by the structure and markings of the lobes and plates making up the fringe of the pygidium. Their life cycle is similar to that of soft scales. The young (crawlers) are born alive or from eggs; they are minute, motile, and very active prior to the first instar settled stage, and look much like soft scales do at this time. During the first molt females lose their legs, antennae, and anal filaments. After the first molt the more elongated males may be differentiated; when they emerge from under the scale, they are minute, fragile two-winged insects with well-developed antennae, eyes, and an anal appendage called the stylus. Shortly after settling down, the nymphs begin forming the armor; the size, color, and shape of the shells often serve to distinguish the genera and the species. There are from one to four, or even six, broods; hibernation may be in the egg, immature, or adult stages. Armored scales are generally more injurious to citrus than any other insects; they are widely distributed and difficult to control because of the protective covering. Injury to plants results from the toxic effect of salivary secretions injected into them (causing leaves to turn yellow) as well as from the extraction of juices. They do not leave deposits of honeydew.

(49) CALIFORNIA RED SCALE: *Aonidiella aurantii.* **Fig. 171.**

Range: Throughout the citrus-growing areas of the world; the southern states, southern California, and in greenhouses. **Adult:** Reddish color of scale shows through armor. The pygidium appears to be "pushed in," the lateral margins extending as far as the tip; there are three pairs of well-defined lobes on the pygidial fringe (the two inner ones are notched) with deeply fringed plates between the lobes. The inner sclerotized structures, visible on the ventral side of the pygidium, distinguish this scale from the **Yellow Scale,** *A. citrina,* which for many years was thought to be the same species. Females give birth to living young. **Length:** (female): .08"; width: .07". **Nymph:** Upon emerging from under the mother's armor, it crawls about actively, settles down after wandering around for a few hours. Three instars. Fertilization of the female, necessary for reproduction, occurs in the so-called gray adult stage after the second molt, when the wax rim is soft and light gray. The fully mature female is protected by a ventral membrane as well as the hard covering. **Food:** Citrus chiefly; occurs on many other plants including castor beans, nightshade, rose, carob, eucalyptus, walnut, laurel, sumac. The most serious pest of citrus; attacks all parts of trees: trunk, branches, twigs, leaves, fruit. [See Hymenoptera (37) and (43).]

(50) FLORIDA RED SCALE: *Chrysomphalus aonidum.* **Fig. 172.**

Range: Gulf states, Cuba, Mexico; only greenhouses in California. The Mediterranean region. **Adult:** The armor (darker and more convex than in the California red scale) is made up of three rings: a brown inner ring near the center, a reddish brown middle ring, a wider reddish brown (or black) outer ring with gray margin. Unlike California red scale, the armor can be lifted from the body. Their life histories and habits are similar except that the female in this case lays eggs; they are deposited and hatched under the scale covering, but must first be fertilized. **Length:** About .04". **Food:** Citrus; also rose, palms, many tropical and subtropical plants in greenhouses. An important potential pest of citrus in Florida; normally controlled by a native parasite, *Pseudomalopoda prima.* The **Obscure Scale,** *Melanas-*

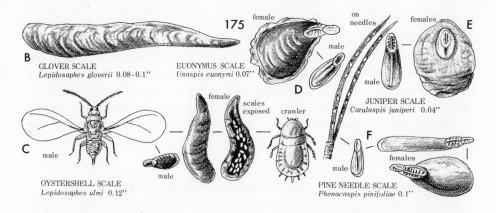

175

B GLOVER SCALE
Lepidosaphes gloverii 0.08-0.1"

EUONYMUS SCALE
Unaspis euonymi 0.07"

female

male

D
female / scales exposed / crawler

on needles

females E

male

JUNIPER SCALE
Carulaspis juniperi 0.04"

male

C male

OYSTERSHELL SCALE
Lepidosaphes ulmi 0.12"

male

F

male

females

PINE NEEDLE SCALE
Phenacaspis pinifoliae 0.1"

pis obscurus—dark gray, about .1″ in diameter—is a pest of oak, pecan, elm, hickory, and hackberry; it overwinters as a nymph. The **Chaff Scale,** *Parlatoria pergandii,* is present in the majority of citrus groves in Florida.

(51) PURPLE SCALE: *Lepidosaphes beckii.* **Fig. 173.**

Range: Throughout the citrus-growing areas of the world. **Adult:** The armor is purplish or purplish brown, shaped like an oyster shell, often curved like a comma. Pearly white oval eggs are laid under the shell, 40 to 80 by a single female; she contracts her body to make room. Length: .08–.12″. **Nymph:** The crawler is whitish with brown posterior tip. It wanders about for a few hours or a day or two, settles down on branch, twig, leaf, or fruit. It spins a cottony wax thread for protection while the shell is being formed. The mass of threads remain until the nymphs are half grown; this is referred to as the "fuzz stage." The male may be recognized after the first molt by its covering, which is much shorter and narrower than that of the female; it emerges from under the scale about 50 days after birth. Fertilization of the female occurs after the second molt. Two broods and a partial third, or a fourth if the winter is mild. **Food:** Citrus. Ranks next to California red scale as a pest of citrus, is more widely distributed. [See Hymenoptera (37).] The **Glover Scale,** *L. gloverii* (Fig. 175B)—very similar to the purple scale, but narrower and straighter—is prevalent in the citrus groves of Florida.

(52) OYSTERSHELL SCALE: *Lepidosaphes ulmi.* **Fig. 175C.**

Range: Northern two-thirds of the U.S., westward to Wyoming; the Pacific Northwest; southern part of Canada. **Adult:** The armor is light to dark brown, shaped like a tiny oyster shell. **Length** (female): .12″, about one-third as wide. **Nymph:** Crawlers are whitish, move about over the bark for a few hours or a day or two, later change to polished brown. The spring brood is full grown by the middle of July, when winged males emerge from the shell. Hibernates in the egg stage. One or two broods. **Food:**

Apple, pear, plum, quince, raspberry, currant, grape, apricot, almond, walnut, lilac, ash, willow, poplar, elm, soft maple, sycamore, dogwood, horse chestnut, linden.* They sometimes spread over the entire bark, causing cracking and scaling. The **Camellia Scale**, *L. camelliae*—the female is pear-shaped, flattened, dark brown—occurs on camellia, mainly cuttings, in the South and in greenhouses. The **Tea Scale**, *Fiorinia theae*—the female is oval or boat-shaped, brown—is the most important scale on camellia and Chinese holly in the South. *F. floriniae* also occurs on camellia, is common in greenhouses.

(53) SAN JOSE SCALE: *Aspidiotus (Quadraspidiotus) perniciosus.* Fig. 174.

Range: All fruit-growing areas of the U.S. and Canada. **Adult:** The female's armor is grayish with a dark nipple-like projection in the center; that of the male is smaller, darker, elongated, with the nipple near the larger end. The delicate saclike bodies underneath are bright yellow. Males emerge from shells as delicate, yellowish, winged adults. Females give birth to living young. **Length** (female): .08″. **Nymph:** Yellowish; the active young nymph crawls about for a few hours, settles on bark, leaves, or fruit to feed and form wax cover. It causes small reddish spots on fruit and leaves. Hibernates as second instar one-third grown, in what is called "sooty-black stage"; it becomes full grown when the trees are in bloom. Two to six broods. **Food:** Apple, pear; also sweet cherry, peach, prune, other deciduous fruits, nuts, berries, shade trees, ornamental shrubs. It can become a serious orchard pest. The **Forbes Scale**, *A. forbesi*, is very similar, occurs alone on deciduous fruit trees or in association with San Jose scale. The **Walnut Scale**, *A. juglansregiae*, also similar, attacks apple, apricot, cherry, peach, pear, plum, and walnut in California. The **White Peach Scale** (or "West Indian peach scale"), *Pseudaulacaspis pentagona*, is found on peach, cherry, and other stone fruits, chiefly in the South; the males— usually found in clusters—are elongated and pure white, while the females are circular and brownish white. The **European Fruit Scale**, *A. ostreaeformis*, is often confused with *perniciosus;* crawlers of both species generally settle in the calyx or stem of apples, but males and females of the single-brooded *ostreaeformis* are not found on the fruit; the shells of immature forms of *ostreaeformis* are more grayish (they are blackish in *perniciosus*) and the nipples lighter in color, and infestations are never as severe.† The **Putnam Scale**, *A. ancylus*—the circular female is dark gray or black, with brick-red nipple (the male shell is oblong)—may cause

* Essig names 97 host plants in widely divergent families.

† See C. V. G. Morgan and B. J. Angle, "Notes on the Habits of the San Jose Scale and the European Fruit Scale on Harvested Apples in British Columbia," *Canad. Ent.,* 100(5):499–503 (1968). Change of *Aspidiotus* to *Quadraspidiotus* has been proposed.

serious damage to elms and other trees; it hibernates as a yellow, partly grown nymph. The **Oleander Scale,** *A. nerii (hederae),* is a pest of oleander in nursery stock, occasionally attacks hemlock in the East; the female is buff-colored, the smaller male is white.

(54) SCURFY SCALE: *Chionaspis furfura.* **Fig. 175A.**

Range: Idaho and Utah, eastward. **Adult:** The female's armor is grayish white, pear-shaped; that of the male is narrower, about one-fourth as long, straight-edged, very white, with three ridges along the top. The male emerges from the shell as a delicate winged adult, dies soon after mating. The stationary female lays about 40 reddish purple eggs under its armor. **Length** (female): .12″. **Nymph:** Newly hatched purplish nymphs crawl about for a few hours, usually settle on the bark; a few settle on fruit, and cause spotting. Hibernates in egg stage. One or two broods. **Food:** Pear, apple, quince, gooseberry, currant, black raspberry, mountain ash, elm and other deciduous trees. Closely related species include the **Elm Scurfy Scale,** *C. americana,* which also attacks hackberry, and the **Dogwood Scale,** *C. corni.* The **Pine Needle Scale,** *Phenacaspis pinifoliae* (Fig. 175F)—the female scale is elongated oval, convex, snowy white, about .1″ long—sucks the sap from needles of pine and spruce; it is widespread over North America, overwinters in the egg stage. The **Juniper Scale,** *Carulaspis juniperi* (Fig. 175E), attacks juniper, cypress, red cedar, arborvitae. The **Euonymus Scale,** *Unaspis euonymi* (Fig. 175D)—the females are pear-shaped, flattened, grayish brown (males are white, conspicuous), about .07″ long—is the main pest of euonymus and pachysandra, also occurs on bittersweet; it overwinters as a mature female.* *U. citri* is sometimes a serious pest on citrus in nurseries in Florida.

---

* See J. C. Schread, *Scale Insects and Their Control,* Bul. 578, and *Scales and Mealybugs,* Circ. 216, Connecticut Agr. Exp. Sta., New Haven (1954, 1961); W. C. Nettles et al. (in Bibliography—General); C. D. Guldner, Jr., and H. E. Thompson, *Some Important Insect Pests of Shade Trees and Shrubs in Kansas,* Bul. 500, Agr. Exp. Sta., Kansas State Univ., Manhattan (1967); C. H. Hill, *The Biology and Control of the Scurfy Scale on Apples in Virginia,* Tech. Bul. 119, Agr. Exp. Sta., Virginia Polytech. Inst., Blacksburg (1952).

14

EPHEMEROPTERA: Mayflies

In some localities where mayflies appear earlier than May they are (more appropriately perhaps) called shadflies. The order is named for the ephemeral nature of the adults, which live only a few hours or at most a few days. They eat nothing in this stage, having only vestigial and non-functional mouthparts. They are fragile creatures with long soft bodies ending in two long filamentous cerci, and often with a third median appendage to match. The large triangular, membranous forewings have numerous longitudinal veins and cross-veins which often serve to identify the species; the hindwings are quite small, and in a few species they are lacking. At rest the wings are held together and erect. Mass emergence often occurs, more especially in the genera *Hexagenia* and *Ephemera*. Huge swarming flights are a familiar sight in the vicinity of streams and lakes; the swarms consist mostly of males, with individual females flying in to mate. They are often attracted to lights, and pile up in windrows on beaches and elsewhere. Eggs are deposited on the surface of the water and sink to the bottom.

The nymphs (Fig. 176) are aquatic and by no means ephemeral—they are sturdy creatures, well adapted for living at the bottom of quiet bodies of water or rapidly flowing streams, and usually live for a year or more. Some species crawl about on the vegetation or bottom, some burrow into the bottom, others live in the silt or moss. From four to seven of the ten abdominal segments have gills. They live on algae, diatoms, and aquatic vegetation, and to some extent on other aquatic insects. While metamorphosis is incomplete, the nymphs are unique in having a brief period during which they fly and are referred to as "subimagos" or "subadults." When mature the nymphs swim to the surface or climb up plant stems or rocks, where they break the nymphal skin, wait briefly for the wings to dry, and fly off. The subimago period lasts a few minutes to forty-eight hours, depending on the species, when the final molt occurs and the mature adult emerges. The subimagos are dull in appearance; the true adults are shiny, have longer tails and legs. Parthenogenesis occurs in various groups, and some species are known to be ovoviviparous. Mayflies are important in the food chain, serving as food for fish, dragonfly nymphs, and birds. Anglers imitate the adults in dry flies—they refer to them as "duns" or "spinners"—and pattern wet flies after the nymphs.

(1) MAYFLY: *Hexagenia limbata* (Ephemeridae). **Fig. 177A.**

This family contains some of the more common and abundant mayflies. The naiads have long cylindrical bodies, tapered at both ends, sharp mandibles that curve outward like tusks, long fringed antennae, three cauda (anal filaments), and legs adapted for digging in the mud. The adults have two or three caudal filaments. The hind tarsi are four-jointed. **Range:** *H. limbata* is widely distributed throughout eastern North America. **Adult:** Pale yellow, with brownish stripes; thorax reddish brown. **Length:** .6–.9″. **Naiad:** Yellowish white, with markings as in the adult; about 1″ long full grown. It burrows in the bottom of quiet bodies of water. A subspecies, *H. l. californica,* is the only member of the family found in California, being common along the Sacramento and San Joaquin rivers; it is yellow, tinged with red on the abdomen, and has extensive dark color patterns; the costal margin of the forewing is light brown, the cross-veins from the costa to M_1 have heavy black margins. *Ephemera simulans* occurs in Ontario, the Appalachian region, Florida, the Midwest, and parts of the Northwest; head and thorax are dark reddish brown, antennae pale brown with whitish tips, abdomen yellowish to light brown, tails dark yellowish brown, wings with blotches and margined cross-veins; forewing about .6″ long. This species and *Ephoron album* are important in the food chain supporting fish in Lake Erie* (at least prior to its advanced state of pollution).

(2) MAYFLY: *Ephemerella spinifera* (Baetidae, Ephemerellinae).†

This is a large family of small, widely distributed mayflies. The hind-wings of the adults are greatly reduced or lacking. The hind tarsi have three or four freely movable joints. The naiads have long antennae and legs, three

---

* See N. W. Britt, "Biology of Two Species of Lake Erie Mayflies, *Ephoron album* (Say) and *Ephemera simulans* (Walker)," *Bul. Ohio Biol. Survey,* New Series, 1(5):1–70 (1962), Ohio State Univ., Columbus.

† Burks treats the Ephemerellinae as a separate family, Ephemerellidae.

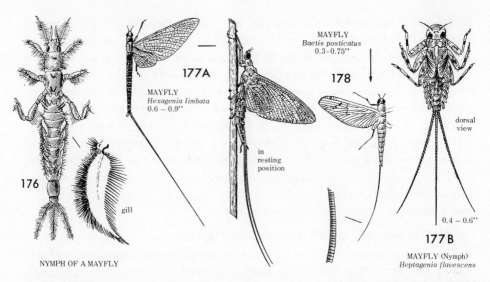

MAYFLY
Hexagenia limbata
0.6 – 0.9"

177A

MAYFLY
Baetis posticatus
0.3 - 0.75"

178

in
resting
position

176

gill

dorsal
view

0.4 – 0.6"

177 B

NYMPH OF A MAYFLY

MAYFLY (Nymph)
Heptagenia flavescens

caudal filaments, the median shorter than the cerci; the head is not strongly depressed, and the eyes are lateral. They live in the mud of still waters, under falls and in rapids, seldom exceed more than .3" in length. **Range:** *E. spinifera* occurs in California, where the genus is one of the most common. **Adult:** Head pale yellow, thorax yellowish brown; abdomen bright purple-rose, with wide stripes on sides and white posterior margins on segments 4 to 10. **Length** (forewing): .6". The naiads live in the silt and debris of slow-moving streams. *E. glacialis carsona,* also a western species, is similar, with forewing about .7" long, head and thorax of adult dark yellowish brown; abdomen deep reddish, with margins of each segment creamy, and a creamy line along the sides. *E. trilineata* occurs in the southeastern states; the body of the adult is light yellowish brown, abdomen with three dark, longitudinal dorsal stripes; the naiads are found at the margin of lakes or in slow-moving streams.

(3) MAYFLY: *Baetis posticatus* (Baetinae). **Fig. 178.**

Range: The upper Mississippi Valley. **Adult:** Head and thorax dark brown to black, abdomen greenish white; legs and tails white, wings transparent. Swarming takes place in May. The female crawls under the water to lay her eggs in a gelatinous mass of 80 to 300. **Length:** .3–.75". **Naiad:** It is found in swift-flowing streams, molts twenty-seven times, the adult emerging in October; eggs are laid at this time and hatch in August of the following year. Three broods in two years. *B. spinosus* is very similar, occurs in the entire eastern portion of the U.S. and parts of Canada; the adult female has a large U-shaped mark on the head, the naiads are found in still waters as well as the rapids of swift-flowing streams. *B. insignificans* and *B. bicaudatus* are western species; in both cases the hindwings have only two longitudinal veins, the third being entirely lacking; a short, pointed projection of the costal margin near the base is present in *bicaudatus* but lacking in *insignificans*.

(4) MAYFLY: *Heptagenia flavescens* (Heptageniidae). **Fig. 177B.**

In the adults of this family the hind tarsus is five-jointed (each joint freely movable), and the eyes of the male are simple. In the naiads the head is flattened, and the eyes are located dorsally. They have two or three caudal filaments. *H. flavescens* is a large yellowish species. **Range:** Georgia to Texas, north to Minnesota and Manitoba. **Adult:** Yellowish; top of head tan, top of thorax reddish brown, abdomen with broad, reddish brown median band; caudal filaments white or pale yellowish; in the female (larger than the male) head and thorax are all yellowish. **Length** (forewing): .4–.6″. **Naiad:** Dark brown, with irregular white spots on body; each side of head has a triangular white spot; about .6″ long full grown. Three caudal filaments. *H. hebe,* similar but smaller—about .3″ long— occurs in eastern U.S. and Canada; the adult male is yellowish brown, the female yellowish with brown markings; the naiad is dark brown, about .3″ long full grown, with prominent white spots on the abdomen, smaller white spots on the thorax and three white spots in front of the eyes, three caudal filaments.

15

ODONATA: Dragonflies and Damselflies

The name of this order comes from the Greek meaning "tooth"—a reference to the toothed mouthparts. Distinguishing features of the group are the two pairs of membranous, netlike wings of nearly equal size, the large compound eyes, and well-developed mouthparts. They are graceful in flight and often beautifully colored. The legs of the adults are adapted for clinging to plants, being too close together for walking. The femora and tibiae have rows of spines increasing in size from front to hind legs; with the legs held close to the body, they form a "basket"—an efficient device for catching small insects on the wing. The lower lip (labium) of the naiads or nymphs is an unusual instrument; it is greatly elongated and hinged, with a pair of toothed and spined scoops on the end that cover the mouth when retracted, which is why it is often referred to as a "mask" (Fig. 179).

The copulatory mechanism is a unique feature of the group (Fig. 180A). The abdomen has ten segments with the genital openings of both sexes being near the tip. Males have genital appendages at the anal end with which they grasp the female by the head (in the case of dragonflies) or by the prothorax (in the case of damselflies) during copulation. The male copulatory organs are located on the underside of the second and third abdominal segments; they consist of claspers, penis, and vesicle to which the sperm is transferred before mating by bending the tip of the abdomen forward and ejecting the sperm capsule from the anal genital organs to the vesicle. The female receives the sperm by bending her abdomen under the male and placing the tip against the second body segment. This curious act

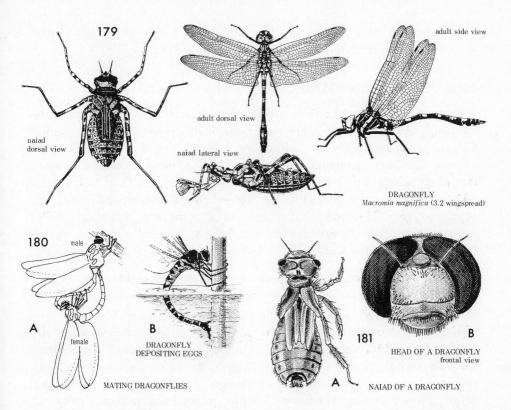

179

adult side view

naiad
dorsal view

adult dorsal view

naiad lateral view

DRAGONFLY
Macromia magnifica (3.2 wingspread)

180 male

A

female

B

DRAGONFLY
DEPOSITING EGGS

MATING DRAGONFLIES

181

A

NAIAD OF A DRAGONFLY

B

HEAD OF A DRAGONFLY
frontal view

of mating, in flight or on plants, can hardly escape one's notice around lakes, streams, and ponds.

Most of the Odonata breed around permanent bodies of fresh water, but a few develop in saline water or are semiaquatic and develop in bogs. Eggs are laid in the water or nearby (Fig. 180B). Damselflies and some dragonflies have well-developed ovipositors and insert their eggs in plant stems, or in the mud and other damp places; unlike most other insects, the males of many species appear to be much concerned in this matter and accompany the females consistently while they oviposit. Metamorphosis in the group is incomplete or gradual. The naiads or nymphs (Fig. 181A) are important predators of mosquito larvae and are themselves eaten by fish, frogs, birds, and other aquatic insects. They undergo from ten to fifteen molts, complete their development in one year or less and up to five years. When fully grown the nymph crawls up a plant and out of the water, where the adult quickly emerges through a slit at the top of the last nymphal skin; the latter remains attached to the plant as a ghostlike reminder of its earlier aquatic life. As adults they are active predators and on the whole beneficial, since they destroy many kinds of noxious insects, more especially gnats and mosquitoes. The order is divided into two suborders: Anisoptera—

the dragonflies—and Zygoptera—the damselflies—which are easily differentiated in either the adult or the immature (aquatic) stages.

ANISOPTERA: DRAGONFLIES

Dragonflies are stout-bodied, with hindwings wider at the base than the forewings. At rest the wings are held horizontally, sometimes tilted slightly forward and down. They are skillful, swift fliers, very alert and difficult to capture. The huge compound eyes (Fig. 181B) are never separated by more than their width. The naiads are stout and often short. They lack caudal gills, breathe through anal gills, the water being drawn in and ejected to supply oxygen; the ejections also serve as a propellant. Adult dragonflies are sometimes very destructive to honey bees, especially around queen-rearing apiaries in the South, where they have earned the name "bee hawks."

(1) TEN-SPOT DRAGONFLY: *Libellula pulchella* (Libellulidae). **Fig. 182A.**

This is a large family of brightly colored dragonflies, varying from one to four inches long, with stout tapered bodies. The females do not have well-developed ovipositors and either drop their eggs in the water or deposit them around aquatic plants. The well-disguised naiads are bottom crawlers and inhabit quiet, shallow waters. *L. pulchella* is a large conspicuous species, widely distributed. **Range:** Throughout North America. **Adult:** Mostly brown. Thorax with white pubescence and two yellowish stripes on the sides; abdomen with two interrupted stripes the entire length. Each wing has three large brown spots. The mature male has ten chalky white spots between these—two on each forewing and three on each hindwing. They range widely and are seldom seen resting. The female drops her eggs over submerged vegetation, unaccompanied by the male. They are often seen hovering over quiet streams, ponds, canals, and reservoirs. **Length:** 2–2.3″. **Naiad:** Hairy; crawls on the bottom in silt and debris. Somewhat elongated, tapering gradually toward the posterior end, eyes very prominent. *L. (Belonia) saturata,* a beautiful bright red species, common in the West and Southwest, is found around ponds and slow-moving streams; it is slightly larger than *pulchella,* all red including the wings, excepting the eyes and back of the head, which are brown.

(2) WIDOW DRAGONFLY: *Libellula luctuosa.* **Fig. 182B.**

Range: Eastern Canada, the U.S. east of the Rocky Mts., Mexico. **Adult:** Dark brown, head mostly black; legs black; abdomen with five black longitudinal stripes and intervening yellow stripes on the sides. Basal third of wings has broad, dark patch. **Length:** 1.7–2″; wingspread: 3.4″. Common around ponds, they are steady fliers and are frequently seen resting on plants. *Macromia magnifica* (Fig. 179) occurs from British Columbia to

California and Arizona; it is a large brownish species with yellow markings, very fast in flight; the female oviposits by striking the end of the abdomen on the surface of the water at frequent, short intervals.

(3) DRAGONFLY: *Celithemis eponina.* **Fig. 183.**

Range: The U.S. east of the Rocky Mts., Ontario. **Adult:** Wings yellowish with brown markings; veins yellow, turning red at full maturity. Top front part of head yellow, hind part brown; thorax yellowish, with black stripes. Abdomen long and slender, blackish, with broad, yellow band on each side. The male holds the female as she oviposits. **Length:** 1.4–1.7″; wingspread: 3″. The largest species in the genus, common around ponds, flutters somewhat like a butterfly in flight. **Naiad:** Rather delicate, like the adult. Swims poorly, climbs over vegetation.

(4) BIG GREEN DARNER: *Anax junius* (Aeshnidae). **Fig. 184.**

This family includes some of the largest, fastest dragonflies that often wander far from the water. They breed in quiet water; females have well-developed ovipositors and insert their eggs in the stems of aquatic plants. The naiads are elongate, with long legs, and climb actively over submerged plants and debris; their prey includes small fish. *A. junius* is widespread and very common. **Range:** Throughout North America. **Adult:** Thorax green, abdomen bluish. Femora reddish brown, remainder of legs black. Wings transparent, with amber tinge. Females insert their eggs in stems of aquatic plants below the water surface. **Length:** 2.7–3.2″; wingspread: 4.2″. **Naiad:** Body long and smooth, with protective coloration. The eyes cover most of the sides of the head. *A. walsinghami,* the largest species in North America, is from 3.5″ to 4.6″ long, occurs in the southwestern states, Mexico, south to Guatemala; the coloration is similar to that of *junius.*

(5) DRAGONFLY: *Hagenius brevistylus* (Gomphidae). **Fig. 185.**

This family is readily distinguished from the others. The eyes are widely separated on the top of the head, the thorax is only slightly skewed, placing the wings farther back and the legs forward. The female has no ovipositor, drops her eggs in the water during flight, striking the surface with the tip of the abdomen; she is unattended by the male. They are black, with yellowish and greenish stripes, clear-winged, rest much of the time but are very hard to see. They are found around the larger, muddy streams. The naiads may be recognized by the depressed, wedge-shaped head, four-segmented antennae (the fourth segment minute or vestigial), the two-segmented front and middle tibiae. They are stiff-legged, slow-moving; most of them burrow in the mud. *H. brevistylus* is a large, black, robust species. **Range:** Widely distributed over eastern U.S. and Canada. **Adult:** Black; sides of thorax yellow, with broad black lateral stripes. Abdomen

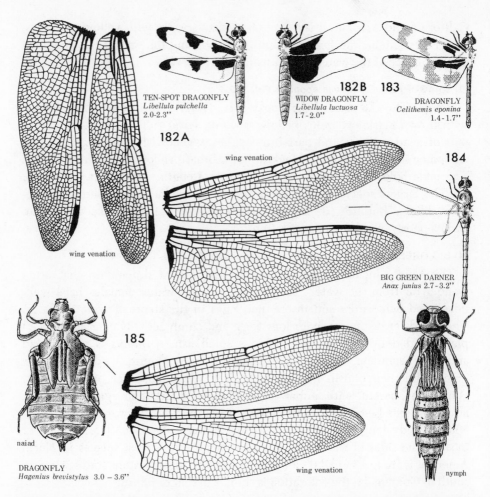

TEN-SPOT DRAGONFLY
Libellula pulchella
2.0-2.3"

182A

182B

WIDOW DRAGONFLY
Libellula luctuosa
1.7-2.0"

183

DRAGONFLY
Celithemis eponina
1.4-1.7"

wing venation

184

wing venation

BIG GREEN DARNER
Anax junius 2.7-3.2"

185

naiad

DRAGONFLY
Hagenius brevistylus 3.0 – 3.6"

wing venation

nymph

very stout, legs and wings very long and powerful. Wings transparent, with blackish veins; narrow edge of costa yellowish. They are fearless, attack butterflies and other large insects. They fly back and forth over open woodland streams and roads, using the same perches repeatedly. **Length:** 3–3.6". **Naiad:** Very flat and wide, slow-moving. Unlike most gomphids, it crawls on the bottom, does not burrow; it is found at the edge of large streams with muddy bottoms.

ZYGOPTERA: DAMSELFLIES

The dainty damselfly is something of a contrast to the least robust of the dragonflies. The body is more slender, and both pairs of wings are similar in size and shape, narrowing toward the base and usually petiolate or stalked. The wings are held together vertically and parallel to the body or tilted slightly when at rest. They are weak fliers and frequently come to rest on plants. The eyes are separated by more than their width. The long, slender naiads or nymphs have three conspicuous feather-like caudal lamellae or gills which protrude from the posterior end of the abdomen.

(6) BLACK-WING (BROAD-WINGED) DAMSELFLY: *Agrion maculatum* (Agrionidae). **Fig. 188A.**

This family is sometimes designated as Agriidae, or Calopterygidae. The members have broad wings, which are not petiolate, with close venation and numerous cross-veins, and bodies of metallic sheen. They flutter in flight, somewhat like butterflies, and are found along streams. The naiads have protective coloration and cling to submerged vegetation. *A. maculatum* is a forest species. **Range:** Eastern U.S. and Canada; Nevada and California. **Adult:** Body brilliant metallic green with bluish cast. Entire wings black, those of female paler and with white stigma. The female inserts her eggs in plant stems below the surface of the water with the male usually nearby. **Length:** 1.7″; wingspread: 2.6″. **Naiad:** Light brown, with darker markings. The antennal segment nearest the head is shorter than the head. Clings to plants in streams of medium size. In *A. aequabile* (Fig. 188C) only the distal half or less of the hindwing and a third of the forewing are black. The naiads occur in larger streams than is the case with *maculatum;* the antennal segment nearest the head is longer than the head. The **Ruby Spot,** *Hetaerina americana* (Fig. 187A), is a slender species found throughout most of North America. In the male, the abdomen is greenish bronze or dark brown; the basal fourth of both pairs of wings are bright red, the remainder has a slight brownish tint. In the female, the abdomen is greenish, and the wings are without the red basal areas.

(7) NARROW-WINGED DAMSELFLY: *Lestes unguiculatus* (Lestidae). **Fig. 188B.**

The damselflies in this family are long and slim with narrow, clear wings constricted at the base to form a petiole or stalk. They are found around the edge of ponds, marshes, and sheltered lakes. Eggs are deposited in aquatic plants above the water line, with the male and female usually in tandem during oviposition. The caudae (gills) of the naiads are long and feather-like. *L. unguiculatus* is common around temporary and semipermanent ponds in the open. **Range:** Nova Scotia to British Columbia, south to New Jersey, Tennessee, Missouri, Oklahoma, Utah, and California. **Adult:** Top front of head and thorax brownish, hind part of head blackish, with yellow transverse bar. Coxae and femora pale yellow. Abdomen dark metallic green, pale yellow on sides. Colors and markings of the male and female are similar. Mature females have blue eyes. **Length:** 1.4–1.5″. **Naiad:** Medium to dark brown; caudal lamellae light to dark brown, darkened along the tracheae and margins. *L. disjunctus* is widely distributed, dark bronze, head and thorax nearly black, latter yellow on sides and

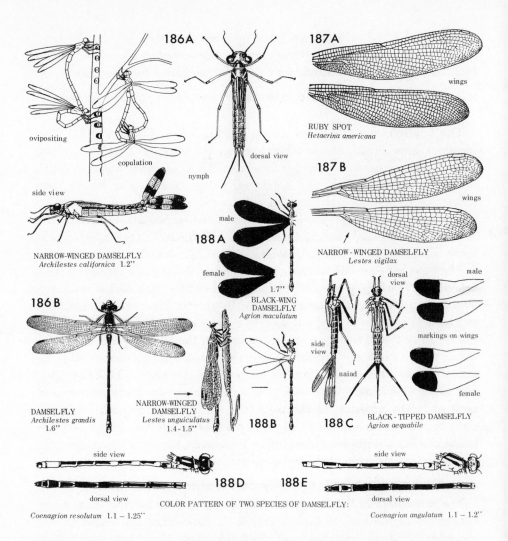

186A

ovipositing

copulation

nymph

side view

dorsal view

NARROW-WINGED DAMSELFLY
Archilestes californica 1.2"

male

188A

female

1.7"

BLACK-WING DAMSELFLY
Agrion maculatum

187A

wings

RUBY SPOT
Hetaerina americana

187B

wings

NARROW-WINGED DAMSELFLY
Lestes vigilax

186B

DAMSELFLY
Archilestes grandis
1.6"

NARROW-WINGED DAMSELFLY
Lestes unguiculatus
1.4-1.5"

188B

side view

naiad

dorsal view

male

markings on wings

female

188C

BLACK-TIPPED DAMSELFLY
Agrion aequabile

side view

dorsal view

188D

188E

side view

dorsal view

COLOR PATTERN OF TWO SPECIES OF DAMSELFLY:

Coenagrion resolutum 1.1 – 1.25"

Coenagrion angulatum 1.1 – 1.2"

underneath with light yellow middorsal line posteriorly; the naiad is variegated yellow and dark brown, with three pale bars (often broken) crossing caudal lamellae, about 1" long (including the lamellae), abundant in marshes and slow-moving weedy streams. *Archilestes grandis* (Fig. 186B)—a western species, about 1.6" long—is yellowish to greenish brown, with a double row of irregular markings on each side of the abdomen; the eggs are inserted in vertical willow stems near the water's edge and above the surface, with the male holding the female; six eggs are placed in a puncture, where they overwinter, hatching in the spring. *A. californica* (Fig. 186A) is pale in color, about 1.2" long, with a white stripe on each side of the abdomen and a black and white spot on the side of each segment, commonly mistaken for *Lestes vigilax* (Fig. 187B).

(8) COMMON BLUET: *Enallagma ebrium* (Coenagrionidae). **Color plate 6h.**

This family consists of small brightly colored damselflies, usually black and blue, with narrow, clear petiolate wings. Most of the small blue damselflies belong to the genus *Enallagma*. The naiads are green or brown and climb about on plants in lakes and ponds. *E. ebrium* is most common around marshy areas in alkaline soils. **Range:** Newfoundland to British Columbia, south to Maryland, west to Utah. **Adult:** Male blue and black, blue predominating. Most females are greenish or brownish yellow, with extensive black markings; abdomen black on top, greenish yellow or bluish on sides. **Length:** 1–1.2″. **Naiad:** Light and dark brown. Caudal lamellae pale, with dark patches and darkened tracheal "twigs." *E. hageni* is very similar, ranges from Nova Scotia to British Columbia, south to North Carolina, west to Kansas.

(9) DAMSELFLY: *Coenagrion angulatum.* **Fig. 188E.**

Range: Western Ontario to Alberta, south to Iowa, Minnesota; British Columbia. **Adult:** Male blue and black; top of head black, face pale greenish; top of prothorax black, with narrow anterior margin and broad lateral margin blue; abdomen blue, with black spots. Female stouter and sometimes larger than the male, yellowish green to pale brown, with black markings as in the male. **Length:** 1.1–1.2″. **Naiad:** Shaped like that of *E. ebrium* (8), antennae longer than the head. Brown, abdomen with pale middorsal line; caudal lamellae broad, pear-shaped, uniformly pale brown, with slightly darkened tracheae; about .75″ long. Found in prairie sloughs, ponds, slow-moving streams, permanent or semipermanent. Abundant on the Canadian prairies.

(10) DAMSELFLY: *Coenagrion resolutum.* **Fig. 188D.**

Range: Newfoundland to Alaska, south to New York, west to Iowa, Saskatchewan, and higher altitudes of Nevada and California. **Adult:** Pale blue, passing to yellowish green below; head black above, with a pair of blue spots and transverse line between also blue, face and rear of head yellowish green; prothorax black above, sides pale green; abdomen pale blue, yellowish green ventrally, with bluish markings. Female generally paler. **Length:** 1.1–1.25″. **Naiad:** Pale yellowish brown, with numerous small dark spots; abdomen brown, with pale middorsal line, paler at lateral margins and with dark dots; caudal lamellae pale, without markings; about .9″ long. Found in marshy waters.

PART III.

PTERYGOTA

The "winged insects" are placed in the subclass Pterygota. The wingless insects
are placed in the subclass Apterygota (Part I).

Insects with Complete or Complex Metamorphosis

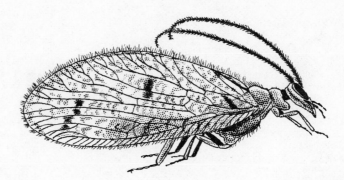

16

NEUROPTERA: Dobsonflies, Alderflies, Dustywings, Lacewings, Antlions, Mantispids, and Snakeflies

The name of this order, from the Greek meaning "nerve" and "wing," refers to the venation. Commonly referred to as the nerve-winged insects, the group is generally divided into three suborders: (1) Sialodea—dobsonflies and alderflies; (2) Planipennia—lacewings, dustywings, mantispids, antlions, and spongillaflies; (3) Raphidiodea—snakeflies. These are sometimes considered as separate orders, with dobsonflies and alderflies classified as Megaloptera. The Neuroptera are characterized by two pairs of very similar membranous wings having many veins and cross-veins. Adults have biting mouthparts, large eyes, long many-segmented antennae, and five-segmented tarsi. Cerci are absent in both adults and larvae.

The larvae of the Sialodea are aquatic and have biting mouthparts. Those of the Planipennia are terrestrial, excepting the spongillaflies (Sisyridae), which are parasitic on fresh-water sponges. They have suctorial mouthparts (biting-type but modified for piercing and sucking); the mandibles and maxillae are long, curved, sickle-like blades opposed to one another, with the maxillae underneath, and each grooved, forming a canal that opens into the mouth. Victims are pierced by the blades, and the body juices are drawn in through the canals. The larvae of both groups are largely carnivorous, those of the lacewings being important predators of aphids and other small injurious insects. Metamorphosis is complete. Most of the larvae spin cocoons; some pupate in earthen cells.

SIALODEA: DOBSONFLIES AND ALDERFLIES

Adult dobsonflies (Corydalidae) are large insects with three ocelli and all tarsal segments cylindrical, which serves to distinguish them from the alderflies. The eggs are laid in rounded masses of several thousand on branches, rocks, and other objects overhanging the water, and are covered with a whitish waxy secretion. The larvae are called dobsons, hellgrammites, toebiters, and other names, and are used as bait by anglers. They live under stones in swift- and slow-flowing streams, and feed on all kinds of aquatic insects. They have paired lateral filaments on the first eight abdominal segments, and a pair of short hooked appendages at the tip of the abdomen. Pupation takes place in cells under stones and logs along the riverbank above the water line; adults emerge in early summer. The complete life cycle requires two to three years.

Alderflies (Sialidae) are small, smoky brown to black insects found around streams and lakes bordered with alder. They are weak fliers, active only at twilight. Ocelli are lacking, and the fourth tarsal segment is dilated and bilobed. The eggs—cylindrical, with rounded top and curved process—are laid in single-layered masses on plants and other overhanging objects along the water's edge; a female lays several thousand eggs. The larvae may be found under stones in lakes and streams, and are used as bait by anglers. They are brownish yellow to white, may be easily recognized by the seven pairs of five-segmented lateral filaments and the long, pointed (unsegmented) terminal process containing tracheae.

(1) DOBSONFLY OR HELLGRAMMITE: *Corydalus cornutus* (Corydalidae). **Fig. 189.**

One of the largest species, common but not abundant. **Range:** Eastern North America. **Adult:** Dull grayish; wings held rooflike over the body when at rest. Mandibles half as long as the body in the male (they are used mostly for clasping the female during mating), greatly reduced in the female. They are not active during the day, live only briefly, and are not believed to feed. **Length:** 4–5.2″. **Larva:** Found under stones in fast-flowing, shallow water; it can swim but usually crawls. Mature larvae are well over 3″ long. The first seven abdominal segments have short tracheal tufts (gills) at the base of the lateral filaments, except in the first instar.

(2) ALDERFLY: *Sialis infumata* (Sialidae). **Fig. 190.**

Range: Eastern North America. **Adult:** Dark brownish. Mouthparts poorly developed. Adults are short-lived. **Length:** .4–.6″; wingspread: 1″. **Larva:** Predaceous, very active, most often found in swift-flowing streams; also occurs in lakes and ponds. Pupates in cell made in the soil above the water line. *S. mohri,* widely distributed over eastern U.S. and Canada, is similar,

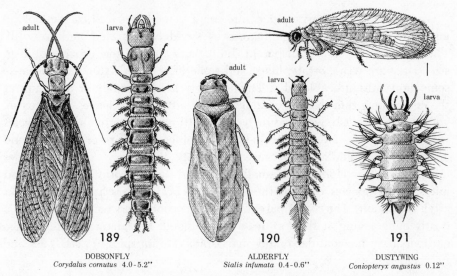

189
DOBSONFLY
Corydalus cornutus 4.0-5.2"

190
ALDERFLY
Sialis infumata 0.4-0.6"

191
DUSTYWING
Coniopteryx angustus 0.12"

may be distinguished by its velvety appearance owing to fine hairs, the concave median area of the vertex, and six to ten accessory cross-veins in the costal area of the forewing. *S. joppa* is the same size, black, except for yellowish raised lines and dots on the head and narrow ring around the eyes; wings dark brown, lighter toward the apex; occurs in the Northeast. *S. hamata* has the same coloring and general structure as *joppa,* occurs in the Pacific Northwest, Alberta, Montana, and Wyoming. *S. californica* occurs from British Columbia to California. The males of *S. rotunda,* another Far West species, have a prominent "horn" on the front of the head between the eyes; the wings are almost black, the apical half sometimes clear; resembles the eastern *S. vagans.*

PLANIPENNIA: DUSTYWINGS, BROWN LACEWINGS, GREEN LACEWINGS, ANTLIONS, AND MANTISPIDS

The dustywings (Coniopterygidae) comprise a small group of minute insects with long antennae, large eyes, and no ocelli. They are named for the powdery wax covering on the wings and body. The larvae prey on small homopterous insects and on mites. Brown lacewings (Hemerobiidae) are small fragile-appearing insects, usually brown, with proportionately large wings, both pairs similar. The body and wings are covered with hairs. They are slow-moving, erratic in flight. The oval, white eggs are laid singly or in groups, usually attached on their sides to the underside of leaves. The larvae are oval, smooth, with short curved mandibles. Both adults and larvae are predaceous; the latter are sometimes referred to as "aphidwolves," and are valuable enemies of aphids and mealybugs. They hibernate as larvae or pupae.

Green lacewings (Chrysopidae) are small to medium-sized fragile-looking insects with small head, large lustrous eyes widely separated, and no ocelli. The wings are usually transparent, both pairs similar and nearly equal in size. They are weak, erratic fliers, and sometimes called "stinkflies" because some emit a disagreeable odor. The oval eggs are laid singly or in groups at the end of long thin threads or stalks usually attached to leaves. The larvae, yellowish to grayish, with red or brown mottling, have prominent tubercles with tufts of hair, and long sickle-like mandibles. They are predaceous, and often called "aphidlions" because they stalk their prey, which includes aphids and other Homoptera, small larvae and the eggs of other insects, and mites. Some species conceal themselves by heaping trash on their backs after each molt. They spin parchment-like cocoons and emerge through a neatly cut opening at the top, leaving it hinged. For characteristics of the antlions (Myrmeleontidae) and mantispids (Mantispidae), see (6) and (7) below.

(3) DUSTYWING: *Coniopteryx angustus* (Coniopterygidae). **Fig. 191.**

Common in citrus groves of the Southwest. **Range:** Southern California, Arizona. **Adult:** Yellow or dark green, with mealy wings. Oval, flattened eggs are laid on leaves of trees and shrubs. **Length:** .12″. **Larva:** Smooth, broad, tapering to a point at tip of abdomen, with long legs and needle-like mandibles. Spins thin, flat cocoon resembling cover of scale insect. **Food:** Chiefly citrus red mite. This, and three other dustywings—*Parasemidalis flaviceps, Conwentzia nigrans* and *hageni*—are said to be "among the most effective enemies of the citrus red mite in southern California."

(4) BROWN LACEWING: *Hemerobius pacificus* (Hemerobiidae). **Fig. 192.**

Range: Alaska to California; Arizona, New Mexico. **Adult:** Light to dark brown. **Length:** .4″. **Larva:** Pale white when newly hatched, dull amber after second molt; third and final molt occurs in pupal case. **Food:** Mealybugs, aphids, and other homopterans, mites. *Sympherobius angustus,* similar in appearance and color but smaller and more slender, is a valuable predator in the Southwest.

(5) GREEN LACEWING: *Chrysopa oculata* (Chrysopidae). **Fig. 193.**

Most common of the eastern species, generally known as golden-eye lacewing. **Range:** East of the Rocky Mts. **Adult:** Pale green, with black crescent under each eye, broad blackish band under antennal sockets, vertex with two submedian black spots; wings transparent, wing venation green, with many cross-veinlets black.* **Length:** .6–.8″. **Larva:** Three instars; final molt occurs in cocoon. Life cycle requires from five to six weeks. Hibernates as pupa. Three to four broods. **Food:** Aphids and other small

* See R. A. Bram and W. E. Bickley, *The Green Lacewings of the Genus* Chrysopa *in Maryland,* Bul. A-124. Agr. Exp. Sta., Univ. of Maryland, College Park (1963).

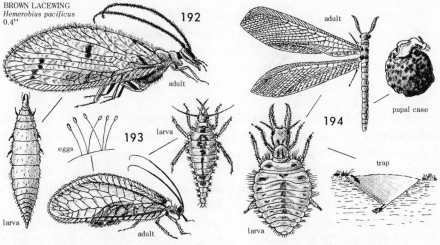

BROWN LACEWING
Hemerobius pacificus
0.4"

192

adult

adult

pupal case

193

eggs

larva

larva

194

adult

trap

larva

larva

GREEN LACEWING *Chrysopa oculata* 0.6-0.8"

ANTLION *Hesperoleon abdominalis* 1.6"

insects, mites. *C. californica* is pale green, with yellow longitudinal stripe on thorax, from .4" to .6" long, occurs west of the Rocky Mts., lays stalked eggs singly or in *small* groups; it has been mass-produced and used effectively against the grape mealybug in California and in greenhouses for control of the citrus mealybug on gardenias. *C. majuscula*—also a western species—is pale yellow, with reddish head and blackish abdomen, lays stalked eggs in *large* groups. *Polystoechotes punctatus* (Polystoechotidae), largest of our lacewings, has long forewings, short forklike veins in the discal area, short antennae, occurs throughout North America.

(6) ANTLION: *Hesperoleon abdominalis* (Myrmeleontidae). **Fig. 194.**

Antlions are named for their predation on ants. They have attracted attention for the interesting form and habits of the larvae, which dig cone-shaped pits in the sand and lie buried at the bottom, waiting for an ant or other small insect to fall into the trap; the caving sand prevents their escape. The larvae—called "doodlebugs"—have an oval, plump abdomen, narrow thorax and head with large sickle-like mandibles. A larva builds many pits, enlarges them as it grows. It walks backwards, uses its head and mandibles as a shovel to toss away the sand. Cocoons resembling those of the lacewings are made in the sand. The adults are long delicate creatures —in contrast to the stubby, ferocious-looking larvae—and resemble a damselfly; they have short knobbed or clubbed antennae, long membranous wings, both pairs similar in size and shape and often spotted. They fly slowly and infrequently, and lay their eggs in the sand. In captivity the eggs take up to two years to develop. Not all antlions build cone-shaped pits; some simply lie buried in the sand or hidden in debris waiting for their prey. **Range:** *H. abdominalis* is widespread, occurs as far west as Arizona and

195

FALSE MANTID
Mantispa brunnea 0.9"

New Mexico. **Adult:** Pronotum yellow, with brown lateral bands; abdomen dark. **Length:** 1.6"; wingspread: 1.8". **Larva:** Brownish. **Food:** Chiefly ants. In *Myrmeleon immaculatus,* common throughout North America, the pronotum is grayish, with a pair of faint yellowish spots in front; forewings with row of elongated light and dark areas along the radius, cells R_1 and Rs very long; the larva is pinkish gray, head with two pairs of brown spots, body with three pairs of long black spots, covered with short, stout setae; overwinters as partly grown larva, spins cocoon in early spring or summer. Both of these species dig pits. Some antlions are heavily parasitized by bee flies.

(7) FALSE MANTID: *Mantispa brunnea* (Mantispidae). **Fig. 195.**

Mantispids are generally rare, but interesting for their striking resemblance to praying mantids. They have no ocelli; the short antennae are circled with rows of bristles. The raptorial-type front legs have long, broad coxae and femora, the latter armed with spines and locking against the tibiae, as in the mantids; they are attached at the front of an extremely long prothorax. The two pairs of similar wings, resembling those of lacewings, are folded flat over the body when at rest. The white or reddish oval eggs, attached to short stalks, are laid in clusters on vegetation. The adults and larvae are predaceous. The young larvae have three- to five-segmented antennae, short mandibles, legs equipped with two claws. They are active predators until they find spider egg sacs or wasp nests with sufficient food to sustain them, when they become parasitic and take on the characteristics of a grub. They spin cocoons, pupate in the last larval skin. *M. brunnea* is very common. **Range:** The eastern and southern states. **Adult:** Yellow and black or brownish. **Length:** .9". **Food:** Insects and spiders. *M. b. occidentis,* very similar, is the western form or subspecies.

RAPHIDIODEA: SNAKEFLIES

The snakeflies have long transparent wings and probably come nearest to the dobsonflies in general appearance. These strange insects may be distinguished by the "long serpentine neck" (the prothorax)—the front legs are attached to the base—and the large head tapered abruptly at both ends. Adults and the long, flattened larvae are both predaceous. The latter are found under the loose bark of trees or in the surrounding litter, and do not spin cocoons to pupate; they are active predators and attack wood-boring insects mostly. The few North American species occur only in the West.

(8) SNAKEFLY: *Agulla unicolor* (Raphidiidae).

Range: Pacific Northwest, California, Nevada, Utah, Colorado, probably the other mountain states. **Adult:** Brown and black; head more reddish brown (which serves to distinguish it from the others); remainder of thorax and abdomen dark brown or black, posterior margins of abdominal segments yellowish. Female similar in structure and colors to the male, but slightly larger. **Length** (forewing): .3–.4″. The larva of *Raphidia bicolor*—mottled gray to brown, with shiny black head—is a valuable predator of overwintering codling moth larvae in Colorado orchards, according to Quist. (For *Raphidia* sp. see Pictured Key to the Insect Orders: Neuroptera.)

17

MECOPTERA: Scorpionflies

The name of this small group of strange, little-known insects refers to the unusually long wings; they are narrow, membranous, with numerous cross-veins, both pairs similar in size and appearance. In some species the wings are conspicuously marked, in others vestigial or lacking. The common name, scorpionfly, refers to the enlarged genital capsule of the male but is more properly applied to the Panorpidae, since this peculiarity is lacking in the others. The chewing-type mouthparts are often located at the end of a snout or prolongation of the head. Eggs are laid singly or in clusters, in the soil or on the ground. The larvae resemble caterpillars or grubs, have numerous spines, three pairs of true legs, well-developed prolegs (from four to ten pairs), large head with short, thick antennae, and usually live in moss, under stones, in rotten wood or decaying organic matter. Many species are predaceous in the adult and larval stages, others feed largely on dead or injured insects, and a few adults feed on nectar and fruits. They are single-brooded, and metamorphosis is complete.

(1) SCORPIONFLY: *Panorpa nebulosa* (Panorpidae). **Fig. 196.**

   In this family—the "true scorpionflies"—the tarsi have two claws and are not fitted for grasping, and the last abdominal segment of the male is greatly enlarged. The wings are spotted with black. Adults sit on the foliage when at rest. Courting males vibrate their wings rapidly in front of the females and provide them with small pellets of saliva which they devour eagerly—a ritual suggestive of the snowy tree cricket and the dance

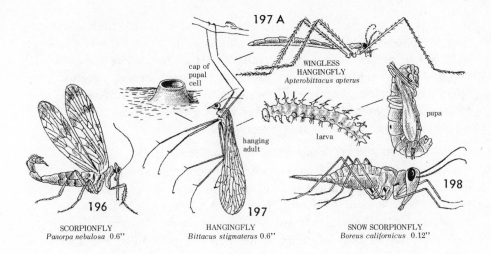

197 A

cap of
pupal
cell

WINGLESS
HANGINGFLY
Apterobittacus apterus

pupa

hanging
adult

larva

SCORPIONFLY
Panorpa nebulosa 0.6"

HANGINGFLY
Bittacus stigmaterus 0.6"

SNOW SCORPIONFLY
Boreus californicus 0.12"

196   197   198

flies. Eggs are laid in clusters in the soil. The larvae are caterpillar-like, have from four to eight pairs of prolegs and twenty to twenty-eight simple eyes on both sides of the head. They make short burrows in the soil but leave them to feed on the surface; some are predaceous or cannibalistic. They pupate in cells in the soil and overwinter in the last instar. *P. nebulosa* is very common. **Range:** Ontario, Quebec, Maine to Georgia and Tennessee, west to Illinois and Wisconsin. **Adult:** Brownish, with black wing markings. **Length:** .6". It is tolerant to a wide range of ecological situations and is most often found in wooded areas.

(2) HANGINGFLY: *Bittacus stigmaterus* (Bittacidae). **Fig. 197.**

This group is easily differentiated from the Panorpidae. The tarsi have a single claw fitted for grasping; the last abdominal segment is not enlarged and the wings are not spotted. They hang suspended by the front legs most of the time and feed on live insects, reaching for them as they come by. Their ability to capture prey equals or exceeds that of the praying mantid. *B. stigmaterus* is a shady woodland species. **Range:** New York to Georgia, westward to Kansas. **Adult:** Light brown, eyes black, wing veins prominent; long-legged, slender, resembling a crane fly. It seldom flies, unless disturbed or when a struggling prey pulls it away from the twig. They mate from a hanging position. The eggs are rectangular, with a depression in the center of each side; they are laid in marshy or wet ground during summer, and hatch the following spring. **Length:** .6"; wingspread: 1.2". **Larva:** Dark brown, cylindrical, with rounded head, three rows of spines on each side; stout antennae, large mandibles. Feeds mostly on dead moss roots. Young larvae stand up on their hind ends when disturbed and feign death by curling the head and thorax. **Food** (adult): Mostly flies. In *B. apicalis* the tips of wings are dark brown; it occurs from New York to North Carolina, west to Oklahoma, in deciduous forests. *B. chlorostigma* is a large dark brown species with a bright yellow spot on the costal margin of the wings, found in Michigan and California. *Apterobittacus apterus* (Fig. 197A) is wingless, occurs only in central California, around the San Francisco Bay area; it climbs monkey-fashion, swings from twig to twig, often holding on by one tarsus as it reaches out for prey.

(3) SNOW SCORPIONFLY: *Boreus californicus* (Boreidae). **Fig. 198.**

These are very small insects, wingless or with only vestigial wings. They are sometimes called snowfleas, and are often seen hopping about in great numbers on the snow in winter. They have no ocelli; females have a long ovipositor, sometimes nearly as long as the body. The bare larvae are C-shaped, live in moss or under stones and debris. Adults feed mostly on moss and small insects or other animals living among the moss roots. *B. californicus* occurs in the Pacific coast states. **Range:** California, Oregon, Washington. **Adult:** Black. Wings of the male are half as long as the abdomen. They mate on the snow in winter. **Length:** .12″. A subspecies, *B. c. fuscous*, brownish black or reddish black, occurs in British Columbia and Alberta. [See Collembola (3).]

18

TRICHOPTERA: Caddisflies

This is the "largest single group of predominantly aquatic insects." "Trichoptera" refers to the hairy wings, which are held roof-like over the body when at rest (Fig. 199). (The popular name is "sedgeflies" in Britain and "rails" in Ireland.) Many are covered with a dense coating of fine hairs and resemble moths slightly; others are nearly bare. The wings of *Lepidostoma* males have numerous black scales mixed with the hairs. Venation is simple, the hindwings shorter and usually wider than the forewings. The adults are mostly small, some shade of brown or gray, have long legs, very long many-segmented antennae, large compound eyes, and vestigial mouthparts excepting the two pairs of palpi. The wings are generally well developed but reduced in some females. Most of them are nocturnal in habit and often attracted to lights. Diurnal species sometimes visit flowers. Their flight is somewhat erratic, with frequent circular movements; they run rapidly on alighting and take off with a jump. Metamorphosis is complete. Eggs are laid in gelatinous masses or strings on rocks and plants, usually under the water but sometimes above it on overhanging objects. A female lays from 300 to 1,000 eggs. The life cycle usually requires a year, most of this being in the larval stage; the adult stage lasts about a month, whereas the egg and pupal stages are much shorter.

The larvae of caddisflies are called caddisworms, the "caddis" being a reference to their silk-spinning habits; they are found mostly in fresh-water streams and lakes. They have small one-segmented antennae, well-developed mouthparts adapted for chewing, usually a pair of anal appendages with

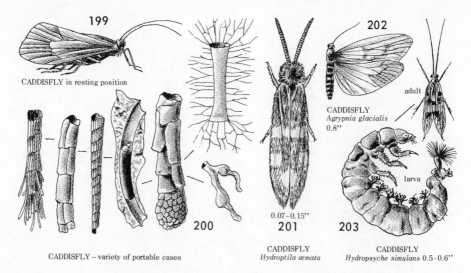

199 CADDISFLY in resting position

200 CADDISFLY – variety of portable cases

201 CADDISFLY *Hydroptila armata* 0.07 - 0.15"

202 CADDISFLY *Agrypnia glacialis* 0.8" — adult

203 CADDISFLY *Hydropsyche simulans* 0.5 - 0.6" — larva

hooks, and finger-like gill tufts. The majority construct portable cases—in an amazing variety of shapes—from stones, leaves, sticks, or sand, held together with a cement-like secretion and silk threads (Fig. 200). They move about with head and legs protruding, and remind one of their amusing distant relatives of the tide pools—the hermit crabs—in their protective coverings of snail shells. The case makers pupate inside their shelters, lining them with silk and closing the ends. Some larvae, however, are free-living —either completely, as are the *Rhyacophila,* or in the early instars, as are some hydroptilids; these forms are adapted for crawling about in fast-flowing streams, have strong anal hooks and widely separated legs. Free-living forms pupate in cocoons constructed of silk and sand, stones, or debris and attached to some object. Some species remain fixed in a net of fine mesh spun of silk during the developmental period. The open end of the net is anchored to a plant or other object, and the structure trails with the current; some are finger-shaped (as in the Philopotamidae), others are long silken funnels (as in the Psychomyiidae). The hydropsychids erect a net in front of tubelike retreats under rocks and logs. Some psychomyiid larvae, mainly in the genus *Phylocentropus,* make burrows in the sand at the bottom of streams and line them with cement. Net-spinning and tube-making forms pupate in their shelters. The larvae of all species spin cocoons before pupating. The mature pupa—actually the adult minus wings, still in the pupal case—cuts its way out of the cocoon, swims to the surface, and settles on a log or rock, where the transformation to adult is soon completed and emergence takes place.

The majority of caddisworms live on diatoms, algae, and higher forms of aquatic plants, but a large number are carnivorous and live on other aquatic larvae and crustaceans. The *Hydropsyche* and *Rhyacophila* are

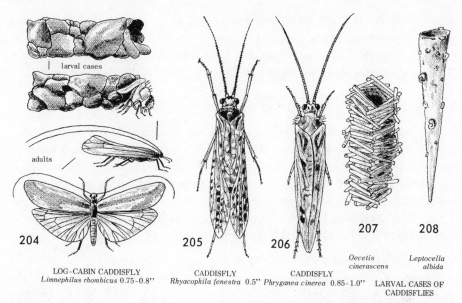

larval cases

adults

204

LOG-CABIN CADDISFLY
Limnephilus rhombicus 0.75-0.8"

205

CADDISFLY
Rhyacophila fenestra 0.5"

206

CADDISFLY
Phryganea cinerea 0.85-1.0"

207

*Oecetis
cinerascens*

208

*Leptocella
albida*

LARVAL CASES OF
CADDISFLIES

important predators of black fly (Simuliidae) larvae. Caddisflies are an important source of food for fish and constitute a large percentage of the food of eastern brook trout. Anglers imitate the adults in dry flies and the larvae in wet flies.

(1) CADDISFLY: *Rhyacophila fenestra* (Rhyacophilidae). **Fig. 205.**

Range: Eastern North America. **Adult:** Head and abdomen straw-colored; thorax brown dorsally, yellowish on the sides. Antennae and palpi dark brown, eyes black. Forewings gray, with white markings, hindwings bluish. **Length:** .5". **Larva:** Free-living. Constructs stone case for pupating. **Food:** Other aquatic larvae.

Microcaddisflies (Hydroptilidae)

The smallest caddisflies—averaging less than .25" long—are found in this family. The adults are very hairy and usually have mottled color patterns. They gather in large numbers on plants near the water, and are sometimes attracted to lights. Males have a brush or pair of brushes on the ninth abdominal segment. The minute first-instar larvae are free-living; the last-instar larvae build portable cases, often purselike in form. The abdomen is swollen, much wider than the thorax, and without gills. They are found in the borders of ponds, lakes, and streams, where they feed on algae; at times they are quite abundant.

(2) CADDISFLY: *Hydroptila armata*. **Fig. 201.**

Range: North Central states. **Adult:** Mottled gray and brown; the underside, legs, and palpi whitish. The head and wings have patches of white hairs. **Length:** .07–.15″. *H. hamata* is very similar, except that the femora are dark brown or black (they are brownish yellow in other species of the genus); it occurs in mountainous and hilly regions from Mexico to Arizona, Wyoming, Washington, eastward to Ontario, New York, North Carolina.

Large Caddisflies (Phryganeidae)

This family includes the largest caddisflies. They are found around lakes and marshes or quiet streams. The maxillary palpi are four-segmented in the males, five-segmented in the females. The front tibiae have two or more spurs; the middle tibiae have four or more spurs. The larvae construct long, cylindrical portable cases with leaves and sticks often arranged in spiral form. The mesonotum and metanotum are membranous, each having a tuft of long setae laterally; the lateral gills are covered with hair.

(3) CADDISFLY: *Phryganea cinerea.* **Fig. 206.**

Range: Northern states and Canada, from the Eastern seaboard to the Rocky Mts. **Adult:** Mostly gray or brown; forewings with brown or gray patches along the posterior margins forming triangular markings when folded, hindwings gray. They are crepuscular in habit, flying just before and after dark, and difficult to see during the day resting on plants at the edge of lakes and in marshes. **Length:** .85–1″. **Larva:** Head, pronotum, and legs yellow; front part of head with dark line in the middle, oblique line on each side. Front and back margins of the pronotum black. About 1.4″ long full grown; the case is spiral in form, 1.8″ long. *Agrypnia glacialis* (Fig. 202) is one of three species of this family known to occur in California; it is light brown to pale yellow, with straw-colored pubescence, about .8″ long. *A. dextra,* very similar, is reported from Idaho; both probably occur in other western states.

(4) CADDISFLY: *Hydropsyche simulans* (Hydropsychidae). **Fig. 203.**

Adults in this family have two spurs on the front tibiae, four on each of the others. They are found mostly along swift-flowing streams. The larvae are very active, often gregarious. They erect elaborate nets to catch food, are believed to be omnivorous. They have numerous branched tracheal gills on the underside of the abdomen, and do not construct portable cases. *H. simulans* occurs in the North Central states, south to Texas and west to Colorado. **Adult:** Head and body various shades of brown, with varicolored patches of hair. Forewings with cream, gray, and brown mottling and patches of lighter and darker hairs; hindwings uniformly gray. **Length:** .5–.6″. **Larva:** Head dark brown, with light patches; thorax and legs straw-colored to brownish yellow. Abdominal segments with numerous flattened setae; about .7″ long full grown. *H. cornutu* is very similar; the adults have less contrast in colors and smaller eyes. *H. occidentalis* is a widespread species, also occurs in California. *H. orris*—together with *Cheumatopsyche campyla* and *Potamyia flava*—is so abundant in cities along the upper Mississippi River in Iowa that the insects are a nuisance and create something of a health hazard; they swarm around lights and windows, and dart into the faces of people (the minute setae dislodged from the wings and bodies are irritating to the eyes and lungs).*

Log-cabin Caddisflies (Limnephilidae)

This is one of the largest families of caddisflies. The maxillary palpi are three-segmented in the males, five-segmented in the females. The front tibiae have one spur; the middle tibiae have two or three spurs. The larvae build portable cases of various shapes; some live in quiet waters, others in swift-flowing, cold streams. The larva has a median dorsal hump on the first abdominal segment; the antennae are usually midway between the eyes and base of the mandibles, or if not, they are closer to the latter. Some build cases with sticks and stones in arrangements suggestive of a log cabin; later stages fashion another case of bark and shells.

(5) CADDISFLY: *Limnephilus rhombicus.* **Fig. 204.**

Range: Greenland to Newfoundland, New York, westward to Wisconsin, Saskatchewan. **Adult:** Brownish yellow; wings with cream-colored and chocolate brown oblique stripes, head and thorax armed with long silvery or brownish yellow bristles. **Length:** .75–.8″. **Larva:** Yellowish brown, with scattered brown dots, front of head with distinctly brown central area, brown lines on each side; about .8″ long full grown. *L. submonilifer* is a slender brown species about .6″ long, with variegated spots on the wings, brushlike black spines on underside of the front femur and tibia, occurs in the northeastern states and eastern Canada; the larva is brownish, about .7″ long, constructs a neat cylindrical case about .8″ long from stems of leaves and grass on the bottom of ponds and marshes which dry up in early summer; this species has a second brood after water is restored with late summer rains.

(6) CADDISFLY: *Oecetis inconspicua* (Leptoceridae).

The adults in this family are slender, have long, slender antennae arising at the base of the mandibles, and five-segmented maxillary palpi in both sexes. In the larvae the mesonotum is membranous except for a pair of parentheses-like sclerotized bars. *O. inconspicua* is one of the most common and widely distributed caddisflies. **Range:** Throughout North America.

---

* See C. R. Fremling, *Biology and Possible Control of Nuisance Caddisflies of the Upper Mississippi River,* Res. Bul. 483, Agr. and Home Econ. Exp. Sta., Iowa State Univ., Ames (1960). Cities on Niagara and St. Lawrence Rivers have this trouble.

Adult: Brown, with reddish cast and no conspicuous markings. **Length:** .4–.5″. **Larva:** Sclerotized or hardened plates are pale yellowish to yellowish brown, head plain or with brownish patterns; first abdominal segment has large dorsal hump without setae. It is about .3″ long, constructs a case from grains of sand and small stones, often irregular in shape and with large stones on the sides, about .4″ long. *O. cinerascens,* which does not occur as far west, is interesting for its "log cabin" case, .4″ long, made of sticks (Fig. 207); the adult—about .5″ long—is brownish, with pale hairs giving it a hoary effect; the larva is brown, pronotum and legs speckled with light brown. *Leptocella albida*—a little larger than *O. cinerascens,* with about the same range—is more artistic and fashions a smooth, slender hornlike case with grains of sand and tiny bits of shells (Fig. 208), about .8″ long (a quarter of it, at the small end, is cut off prior to pupation); the head, thorax, and wings of the adult are covered with white hairs, the larva is yellowish brown, with dark brown middle and hind legs and V on the forehead.

19

LEPIDOPTERA: Butterflies, Skippers, and Moths

The name of this order means "scaly wings," being a combination of two Greek words: *lepidos,* or "scale," and *pteros,* or "wing." The wings, and the legs and body as well, of most butterflies and moths are covered with minute flat scales arranged like the shingles on a roof. On some moths the scales are so dense they come off as a cloud of dust when the insects are caught and struggle to free themselves. Because they are also slippery and come off easily the scales often enable the moths to elude their enemies. Unlike most insects they manage to escape unscathed from the sticky traps set by the orb-spinning spiders, except for a few expendable scales left behind.

Along with the protective scales many male butterflies have specialized scent scales called androconia which serve to attract the females. These may be scattered among the other scales (as in the Pieridae), or they may occur in patches or in folds of the wings. Those of the monarch butterfly appear as a small, black patch on one of the veins on the upper side of each hindwing. In the skippers they occur in a small fold of the costal or front margin of the forewings; among the swallowtails they appear in a fold of the anal or inner margin of the hindwings. These and the regular scales can sometimes be useful for identification. The scales have many shapes and colors and are very beautiful when viewed under a microscope. They are an interesting subject for color microphotography. The chemical (pigmentary) and physical properties (together with the arrangement) of the scales give the butterflies and moths their distinctive colors. Scales built up in layers produce colors owing to the refraction of light, while surfaces bearing ridges produce colors by diffraction of light. Blues and metallic greens are produced in this way.

Aside from their scaly wings, the adults of the Lepidoptera are characterized by large compound eyes, long antennae, and well-developed legs. Butterflies lack ocelli, but moths usually have one near the base of each antenna. Mouthparts are greatly reduced except for the maxillae, which are well developed in most cases; these are elongated and fused to form a tube—the proboscis or "tongue"—which is extended to suck up liquid foods and coiled up when not in use. Tympana or ears are fairly well developed and usually located in the mesothorax. The Lepidoptera reproduce in the usual manner, parthenogenesis being rare. Metamorphosis is complete.

The larvae, called caterpillars, are marked by the presence of prolegs (or larvapods)—the short fleshy projections on the underside of the abdomen which they use in walking and for clinging to plants. The number varies from two or three pairs in the "loopers" to five pairs, including those on the hind end; a few have none. Typically, they have four pairs on abdominal segments three to six and a pair of anal prolegs on the last segment. The crochets or hooks on the bottom of the prolegs occur only in the lepidopterous larvae and are sometimes useful in identification; they may be arranged in single to triple rows, often in a circle or semicircle. The head of the caterpillar is usually distinct from the rest of the body, equipped with one pair of minute three-segmented antennae, and one to six pairs of ocelli on each side. The thoracic segments bear three pairs of segmented legs, each with a claw. Spiracles or breathing pores occur on each side of the prothorax and segments one to eight of the abdomen—nine pairs in all. The body is usually covered with hairs or setae which may be long or short, dense or sparse; their arrangement is often useful in identification.

Butterfly or moth?

Basically, there is little difference between butterflies and moths, but it is not difficult to distinguish between them. Two structural features will serve to differentiate most butterflies from moths: (1) the manner in which the two wings on either side are coupled together during flight and (2) the type of antennae.

In flight the anal margin of the forewing covers the costal margin of the hindwing and the two beat together in perfect union. The most primitive living Lepidoptera—moths that are sometimes classified in a separate suborder, Jugatae—lock their wings together by means of a flap or lobe called the jugum. This is located at the inner margin of the forewing near its base, and overlaps the hindwings. Only a few species of this kind occur in

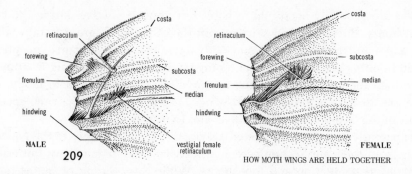

209

HOW MOTH WINGS ARE HELD TOGETHER

North America, the best known being the "swifts" (Hepialidae)—rather large, fast-flying moths whose larvae are wood borers or root feeders. Most moths have a well-developed "hook and eye" mechanism in the form of one or more bristles, called the frenulum, which project from the base of the costal margin of the hindwing and hook into a flap (in the males) or a row of bristles (in the females) on the underside of the forewing. The latter is called the retinaculum (see Fig. 209). The frenulum of the male is usually a single bristle or spine; that of the female consists of two or more bristles. In many species this difference between the locking mechanisms serves to distinguish the male from the female. Only a few moths lack this coupling device. The wings of butterflies and skippers, on the other hand, are kept together in flight by the humeral lobe—a process on the hindwings which grips the underside of the forewings. Butterflies and skippers never have a frenulum, with probably only one exception—some Australian skippers in the genus *Euschemon.*

The usual division of the Lepidoptera into two suborders is based on the second feature that serves to distinguish moths from butterflies. The Rhopalocera (from the Greek meaning "clubbed horns or antennae") comprises the butterflies and skippers; the Heterocera (meaning, in effect, the "others") comprises the moths. This distinction holds good generally for the North American species. (A few species of moths found in tropical America and in Australasia have clubbed antennae.) The antennae of moths are tapered and either hairlike or feathery; there are a few exceptions to this rule, but these always have the frenulum. Butterflies never have feathered antennae. Frenatae—meaning "having a frenulum"—is also used as a suborder designation for moths not having a jugum.

Some differences in other features or in behavior often serve to distinguish butterflies from moths, but they are less reliable clues than the structural differences mentioned. The majority of moths rest with their wings spread out flat to the sides or over the back, but some do not, notably the geometrids. Butterflies and skippers, on the other hand, generally hold their wings upright when resting, but there are again some exceptions.

Moths are generally nocturnal in habit but some are active during the twilight hours. Butterflies and skippers, on the other hand, are generally diurnal, and many remain inactive on cloudy days. There are of course exceptions again in either case. The diurnal moths tend to be more striking in appearance than the crepuscular or the nocturnal moths. Some female moths are apterous, or brachypterous (short-winged), and unable to fly; such species occur in the Geometridae, Psychidae, and Lymantriidae.

Apparently butterflies suggested a seductive kind of beauty to the Navaho Indians. In their purification ceremonies stylized dry paintings of butterflies symbolized temptation and foolishness. While the coloring of moths tends to be dull, many are as brightly colored or as gaudy as the most provocative butterflies. The general features of the early stages of the butterflies and moths are similar. There are conspicuous differences in the pupal stage, however. The pupae of moths are nearly always enclosed in silken cocoons; butterfly pupae (chrysalises) are naked and often hung from a limb or other structure, head down. The attachment is made by the cremaster, a short projection at the tip of the abdomen which is fastened to the surface by a silken "button" spun by the larva. The chrysalises of the large family of Nymphalidae are suspended in this manner. Many butterflies (the Papilionidae and Pieridae) also secure themselves by means of a girdle of silk wrapped around the thorax, holding the pupa head upwards, suggestive of an Indian papoose. (Indeed, the term *pupa*, meaning "baby," was proposed by Linnaeus for this reason.) This silken thread is believed to be a vestige of the cocoon-spinning habit of the butterfly's moth ancestors.

The skippers spin a loose cocoon and more closely resemble the moths in this respect. Many moths (and some butterflies) pupate in the ground—in a chamber formed by a hardened larval secretion and sometimes lined with silk. Some moths spin cocoons of silk and leaves, while others that bore into plant stems and trees may pupate in the chambers or mines excavated by the larvae.

Butterfly or skipper?

In common usage, skippers are lumped with butterflies because of their close similarities. Their ancestry, however, is different, and in both cases their closest relatives are among the moths. In taxonomic classifications they are separated into two superfamilies: the Papilionoidea or "true butterflies" and Hesperioidea or "skippers." The bodies of the true butterflies are relatively more slender and the wings proportionately larger than those of the skippers. The antennae of skippers arise far apart on the head, and while also clubbed, are usually hooked at the tip and often end in a short tapered portion (the apiculus) extending beyond the club. Butterflies have a single spur on the tibiae of the hind legs, whereas skippers usually have two. Generally small, with stout bodies, the skippers are named for the habit of swift flight punctuated by rapid stops and starts. Many of them hold the front wings out at a slightly different angle from the hind pair when at rest. The larvae of skippers are smooth, generally plainer than those of the true butterflies, with large globular head and slender prothorax suggestive of a neck. The chrysalis of the skipper is suspended by silk and protected by a loosely spun cocoon formed from leaves and held together with a few strands of silk.

Adult butterflies and moths are harmless. They feed on the nectar of flowers primarily and to a lesser extent on the honeydew of other insects or on decaying fruit. As a group they are important pollinators. The caterpillars of the Lepidoptera are nearly all plant feeders, and many of them are notorious pests. Only a few are entomophagous, feeding as predators (or more rarely as parasites) on aphids, leafhoppers, scales, mealybugs, and the larvae or pupae of ants. Among the butterflies the habit occurs only in the family Lycaenidae. Certain noctuid moths are predators of scale insects on *Eucalyptus*. They feed from beneath a shelter formed from the victim's remains and woven together with silk; this later becomes a cocoon. The larvae of the phycitid moth *Laetilia coccidivora* are predators of certain cochineal scale insects (*Dactylopius*) which feed on prickly pear cactus in the Southwest. Larvae of epipyropid and cyclotornid moths found in Australia and India are external parasites of leafhoppers, and some species in the family Cosmopterigidae are predatory on scale insects.

Adaptive coloration—color patterns which give protection against enemies —is exhibited to a striking degree in the Lepidoptera. These include markings and coloring for concealment—"cryptic" colors, as they are called—and conspicuous patterns or "warning coloration" by which distasteful or inedible species are said to "advertise their distastefulness."* Some hairy or brightly colored caterpillars feed openly, with impunity, apparently because they look unpalatable or suggest some distasteful species to predators. The bright and sometimes dazzling "eye spots" on certain caterpillars, such as the spice-bush swallowtail, and on the wings of some saturnid, nymphalid, and noctuid moths are believed to "terrify potential predators."†

Migration is not an uncommon trait among butterflies. The best-known example is the monarch butterfly. Monarchs often congregate in great masses, enveloping an entire tree, and are commonly seen on the Monterey Peninsula in California during the winter months. Some fly as far as a thousand miles, from Canada to Texas, and back again in the spring. It has

* See E. B. Ford, *Moths;* and Hugh B. Cott, Bibliography: General.
† *Ibid.*

been demonstrated that many of those returning are the same ones that left in the fall as well as progeny of the original migrants. Other migratory butterflies include the buckeye, little sulphur, painted lady, and the purple wing.

RHOPALOCERA: BUTTERFLIES

Swallowtails and Parnassians (Papilionidae)

This is a large world-wide group, most of the New World species being tropical. They are medium to large in size and include some of the largest and prettiest butterflies. Most of our swallowtails have the characteristic tail-like projection of the hindwing, and a movable projection, called the epiphysis, on the tibiae of the front legs. The eggs are globular, with flattened base. The larvae are smooth and the anterior end larger than the posterior; they all have the forked osmeterium which protrudes from the prothorax and emits a repugnant odor when they are disturbed. The swallowtail chrysalises are attached at the anal end by the cremaster, and the head is usually held upward by a silk girdle.

The parnassians, which do not have the tail on the hindwing, are mostly alpine and confined to the mountains of the western region. The eggs are turban-shaped, with a rough granular-like surface. The larvae are flattened, resembling a leech, and have small heads; they are black or dark brown, with light spots. Pupation takes place on the ground, in leaves held together loosely by strands of silk. An interesting feature of the parnassians is the presence of an appendage (or "pouch") on the underside of the female after mating. It is formed during copulation from a waxy secretion of the male and prevents later mating attempts by other males. The Papilionidae lack the tympana that appear on the thorax or first body segment of many butterflies and moths.

(1) TIGER SWALLOWTAIL: *Papilio glaucus.* **Color plate 3e, f.**

One of the most common of the eastern butterflies. Often attracted to decaying animal matter and to smoke; frequently seen around mud puddles. **Range:** Hudson Bay to Florida, west to Texas. **Wings:** Yellow, with black streaks and black border. Spots on outer margin of border are yellow, those on inside are blue. Southern females are dimorphic: the background color in half of them may be dark brown instead of yellow. Dwarf forms, *canadensis,* are found in Canada; other dwarf forms are common among the spring broods in eastern parts of the range. **Spread:** Male, 3–4"; female, 3.5–5". **Larva:** Green, with large orange and black eyespots on the metathorax. Lives in a nest formed by folding a single leaf. Two or three broods. Hibernates as pupa. **Food:** Birch, ash, poplar, wild cherry.

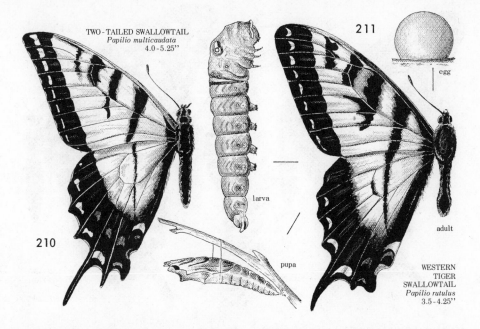

TWO-TAILED SWALLOWTAIL
Papilio multicaudata
4.0-5.25"

211

egg

larva

pupa

adult

210

WESTERN
TIGER
SWALLOWTAIL
Papilio rutulus
3.5-4.25"

(2) TWO-TAILED SWALLOWTAIL: *Papilio multicaudata (daunus).* **Fig. 210.**

One of the largest western species and a strong flier. **Range:** Common in the eastern valleys of the Rocky Mts., Arizona, central mountain range of California, and the Sierra Nevada Mts. as far north as Mt. Shasta. **Wings:** Similar to *P. glaucus* (1) in coloring and markings. Distinctive features are two long tails and extended lobe (or short tail) at the anal angle of the hindwing. (It is sometimes called the three-tailed swallowtail.) **Spread:** 4–5.25". **Larva:** Apple green except for brownish head. Two broods, May to August. **Food:** Ash, wild cherry, laurel, willow.

(3) WESTERN TIGER SWALLOWTAIL: *Papilio rutulus.* **Fig. 211.**

Pacific coast counterpart of the eastern tiger swallowtail but does not exhibit dimorphism. A slow flier, found at all but higher altitudes. **Range:** Common throughout the western states. **Wings:** Similar in color and markings to the two preceding species. A dwarf form, *arcticus,* occurs in Alaska. **Spread:** 3.5–4.25". **Larva:** Deep green; conceals itself by folding a leaf. **Food:** Poplar, willow, alder, hop.

(4) PALE SWALLOWTAIL: *Papilio eurymedon.* **Fig. 212.**

Range: Valleys of the Coast Range, from northern Mexico to the Fraser River canyon of British Columbia, east to Colorado. **Wings:** Markings quite similar to the tiger swallowtail, but the background color is very pale yellow or white, and the tails are more slender. **Spread:** 3.5–4". **Larva:** Pale green; head purplish brown. **Food:** California coffee (*Rhamnus californica*) and various other wild plants.

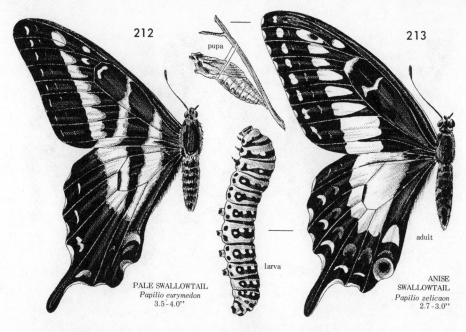

212

pupa

213

larva

adult

PALE SWALLOWTAIL
Papilio eurymedon
3.5-4.0"

ANISE
SWALLOWTAIL
Papilio zelicaon
2.7-3.0"

(5) ANISE SWALLOWTAIL: *Papilio zelicaon.* **Fig. 213.**

Very common in the valleys of California. **Range:** From Arizona to Vancouver Island, east to Colorado. **Wings:** Background color yellow. Spots on wing border yellow, hindwings with row of pink spots on inner edge of black border. Resembles Bruce's swallowtail (8) except that the wings are shorter. **Spread:** 2.7–3". **Larva:** Early stage black with orange spots; mature stage green with black bands on each segment and a series of orange spots on the bands. Very similar to the black swallowtail (6). Broods occur throughout most of the year. **Food:** Anise, fennel, and other plants of the carrot family. Occasionally found on related cultivated plants in gardens, and on lemon and orange trees. *P. oregonius*, which ranges over Oregon, southern British Columbia, east to Colorado, is similar, except that the black is lighter and the wing border narrower.

(6) BLACK SWALLOWTAIL (PARSLEYWORM): *Papilio polyxenes asterius* (*ajax*). **Fig. 214.**

Also variously called common American swallowtail, celeryworm, carrotworm. Most common of our swallowtails. **Range:** North America, to the Hudson Bay. **Wings:** Black, with two rows of yellow spots near margin and yellow spotband on the inside; intermediate pink spots on hindwings. Early brood smaller, with the yellow spots larger. **Spread:** 2.75–3". **Larva:** Early stage brown, with a white "saddle"; older larva green, with black band and yellow spots on each body segment as in the anise swallowtail (5). Two to four broods. Hibernates as pupa. **Food:** Carrot family. Sometimes a pest of garden and field crops.

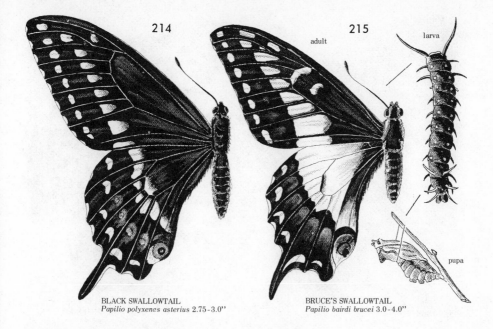

214

215

adult

larva

pupa

BLACK SWALLOWTAIL
Papilio polyxenes asterius 2.75-3.0"

BRUCE'S SWALLOWTAIL
Papilio bairdi brucei 3.0-4.0"

(7) BAIRD'S SWALLOWTAIL: *Papilio bairdi bairdi.* **Color plate 1 (top).**

Range: Arizona, throughout the Sierra Nevada Mts., Mojave Desert. **Wings:** Similar to black swallowtail (6) except that the yellow spotband is wider. **Spread:** 3.5".

(8) BRUCE'S SWALLOWTAIL: *Papilio bairdi brucei.* **Fig. 215.**

Range: Western slopes of the Rocky Mts. **Wings:** Similar to (6) and (7), except that yellow has now become dominant over the black. **Spread:** 3–4".

(9) POLYDAMUS SWALLOWTAIL: *Battus (Papilio) polydamus.* **Fig. 216.**

Swift, erratic fliers, rather common in the Deep South. **Range:** Florida, Georgia, Gulf states, Texas, and into Mexico. **Wings:** Hindwings lack tails. Forewing black, with pale yellow spots; hindwing shiny green, with greenish yellow spots. Prominent scent glands (androconia) in fold along anal margin of hindwing. **Spread:** 3–4". **Larva:** Black or various shades of brown, with two prominent reddish orange tentacles protruding from prothorax. Four rows of short fleshy projections (papillae) orange-colored. **Food:** Dutchman's-pipe and other vines of the birthwort family.

(10) PIPEVINE SWALLOWTAIL: *Battus (Papilio) philenor.* **Color plate 3c.**

Range: Middle Atlantic states to Massachusetts, west to Arizona and California, into Mexico. Most common in the South. **Wings:** Similar in coloring to *polydamus* (9). Forewing greenish black, hindwing iridescent blue-green. Early spring form is smaller and hairy, similar to the form *hirsuta* found in California. Prominent scent glands as in *polydamus*.

216

217

POLYDAMUS SWALLOWTAIL
Battus polydamus 3.0-4.0"

PALAMEDES SWALLOWTAIL
Papilio palamedes 3.5-4.25"

Spread: 3.75–4.25". **Larva:** Light to dark brown. Long paired tentacles at anterior end, shorter pairs at posterior end, short paired rows along sides. Two to three broods. Hibernates as pupa. **Food:** Pipevine and closely related plants.

(11) SPICE-BUSH SWALLOWTAIL: *Papilio troilus.* **Color plate 3d.**

A common species of eastern Canada and U.S. Also called the green-clouded or green-spotted swallowtail. **Range:** Canada to the Gulf states, east of the Mississippi. **Wings:** Forewing dark green or almost black, with yellow marginal spots; hindwing bluish green, marginal spots tinted with green. Orange spots on costal and anal margins of hindwings. **Spread:** 3.75–4.25". **Larva:** Smooth, green and yellow when full grown. Large orange and black eyespots, with second similar pair behind these. Lives in a folded leaf. Two or three broods, depending on climate. Hibernates as pupa. **Food:** Spice bush, sassafras, sweet bay.

(12) PALAMEDES SWALLOWTAIL: *Papilio palamedes.* **Fig. 217.**

A slow flier. **Range:** Virginia to Key West, west to Missouri, Texas, and south into Mexico. **Wings:** Mostly black, with yellow spots and markings. **Spread:** 3.5–4.25" or more. **Larva:** Speckled green, with pair of black and orange eyespots on metathorax. Two or three broods. Hibernates as pupa. **Food:** Sweet bay, and other plants and trees of the laurel family.

(13) ORANGE-DOG: *Papilio cresphontes.* **Fig. 218.**

Our largest butterfly, often referred to as the giant swallowtail. The caterpillar is also called orange puppy. **Range:** Ontario, eastern and

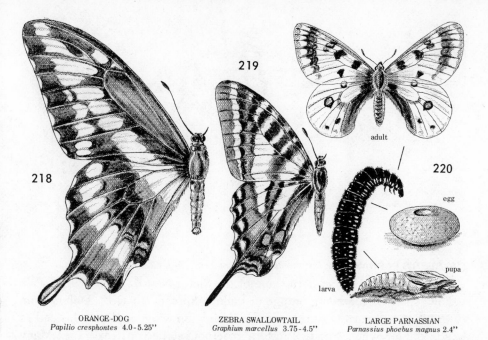

218

220

adult

egg

pupa

larva

ORANGE-DOG	ZEBRA SWALLOWTAIL	LARGE PARNASSIAN
Papilio cresphontes 4.0-5.25"	*Graphium marcellus* 3.75-4.5"	*Parnassius phoebus magnus* 2.4"

central states, Florida. **Wings:** Dark green, with broken yellow band starting at the apex, and two branches in the discal area. Yellow band runs across central part and apex of hindwing. **Spread:** 4–5.25". **Larva:** Dark brown, with cream-colored markings. Two to three broods. **Food:** Prickly ash, hop tree, citrus. Sometimes a pest of citrus in Florida. The **Orange** or **Citrus Swallowtail,** *P. thoas,* a tropical species often seen in citrus groves of the Southwest, is very similar.

(14) ZEBRA SWALLOWTAIL: *Graphium (Papilio) marcellus.* **Fig. 219.**

Highly variable but easily distinguished from other species. **Range:** Canada to the Gulf, east of the Mississippi River. **Wings:** Dark brown, with white stripes; long, white-tipped tail. Early forms are smaller, with more white than forms emerging later from the same brood. The summer and later broods are much larger, have relatively more white area and longer tails with more white on the tips. **Spread:** 3.75–4.5". **Larva:** Light green, with yellow and black transverse bands. Two to four broods. Hibernates as pupa. **Food:** Papaw.

(15) BALDUR PARNASSIAN: *Parnassius clodius baldur.* **Fig. 221.**

Range: Northern California (Sierra Nevada Mts.), Oregon, British Columbia. **Wings:** Pale yellow; pair of black spots, or red spots ringed with black, on hindwing. Broad translucent border on forewing. **Spread:** 2.5–3". **Larva:** Flattened, with small head; dark brown or black, covered with white spots. **Food:** Stonecrop, bilberry, saxifrage. There are many subspecies and varieties of *clodius* (Fig. 223).

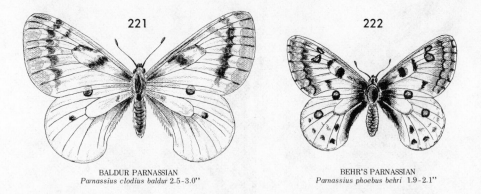

BALDUR PARNASSIAN
Parnassius clodius baldur 2.5-3.0"

BEHR'S PARNASSIAN
Parnassius phoebus behri 1.9-2.1"

(16) BEHR'S PARNASSIAN: *Parnassius phoebus behri.* **Fig. 222.**

Range: Rocky Mts., eastern slope of the Sierra Nevada Mts. **Wings:** Pale yellow; spots on hindwing orange or yellow. **Spread:** 1.9–2.1". Differs from *clodius* (15) in narrow translucent border of forewing; red spots instead of bands on forewing. **Food:** Western roseroot, stonecrop, saxifrage. Many subspecies and varieties of *phoebus* also occur.

(17) LARGE PARNASSIAN: *Parnassius phoebus magnus.* **Fig. 220.**

Range: Rocky Mts. **Wings:** Similar to Behr's parnassian (16) but larger, with spots on hindwing darker and more heavily margined. **Spread:** 2.4".

Butterflies: Sulphurs and Whites (Pieridae)

The Pieridae is a very large world-wide family, ranging on this continent from the subarctic to the tropics. They are small or medium in size. The wings are nicely rounded as a rule, white or yellow (sometimes orange), with dark borders and simple markings. Their coloring gives rise to the name "butterfly." Sexual dimorphism is common, the females often differing from the males in color and markings.

The eggs are elongated, spindle-shaped, with ridges and fine cross-lines. The larvae are usually slender and smooth; the coloring is green, with longitudinal stripes. The osmeteria characteristic of the Papilionidae larvae are lacking. They commonly feed on plants of the mustard family, notably crucifers such as cabbage, cauliflower, and broccoli, and on legumes such as alfalfa, clover, vetch, lupines, peas, beans, and peanuts. This family includes some well-known pests of agriculture and one of the forests. The chrysalises are pointed and attached by a cremaster and girdle. Several broods in a season is the general rule. Some species are migratory but not consistently.

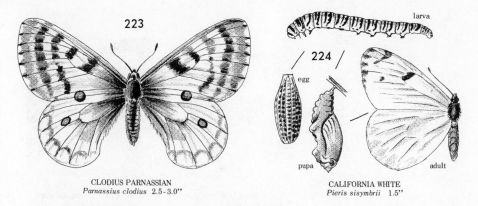

CLODIUS PARNASSIAN
Parnassius clodius 2.5-3.0"

CALIFORNIA WHITE
Pieris sisymbrii 1.5"

(18) CALIFORNIA WHITE: *Pieris sisymbrii.* **Fig. 224.**

Found mainly in the mountainous regions and the foothills. **Range:** Rocky Mts. to California. **Wings:** White background, with sharply defined black markings on upper side of forewing; narrow black bar at end of discal cell. Underside of hindwing with veins broadly outlined in blackish brown. Female similar to male, except that markings are darker. Occasional yellow variants appear. **Spread:** 1.5". **Larva:** Pale yellow, with black cross-stripes. Two broods. **Food:** Wild crucifers such as rock cress, rock cabbage, jewel flower. The Colorado subspecies, *P. s. elivata,* is found on both sides of the Continental Divide; the veins on the underside of the hindwing are narrowly edged with blackish brown. **Becker's White,** *P. beckerii,* is very similar to *sisymbrii,* ranges from the eastern slopes of the Sierras to the western slopes of the Rockies; the markings are heavier, the black bar at end of discal cell is nearly square, with white center; markings along veins on underside of hindwing are broader, and greenish.

(19) SOUTHERN CABBAGEWORM: *Pieris protodice.* **Fig. 225, 226.**

Also called common white, or checkered white. **Range:** Found in practically every state and in Canada, but less common along the East coast. **Wings:** Males resemble those of the California white (18) except that underside of hindwing is plain; markings of both sexes are more diffuse than in *sisymbrii* and the bar at end of discal cell is broader. Pronounced brown scaling along veins in the female. Spring forms are dwarfed, with brown markings of the male reduced on upper side, and veins on underside marked with brown as in *sisymbrii,* markings on upper side of female's wings deeper in color. **Spread:** 1.3–1.6". **Larva:** Varies from pale to deep green or bluish, with four yellow and green stripes, and numerous black dots. Three or more broods, depending on geographical area. Hibernates as pupa. **Food:** Crucifers. A pest of cabbages but less serious than *P. rapae* (20).

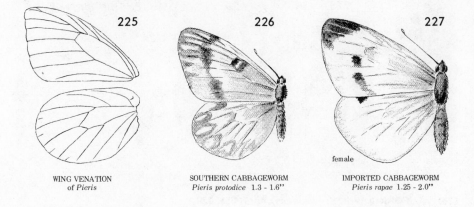

225

226

227

female

WING VENATION
of *Pieris*

SOUTHERN CABBAGEWORM
Pieris protodice 1.3 - 1.6"

IMPORTED CABBAGEWORM
Pieris rapae 1.25 - 2.0"

(20) IMPORTED CABBAGEWORM: *Pieris rapae.* **Figs. 227, 228.**

Also called European cabbageworm or cabbage butterfly. **Range:** Most of the continent, wherever agricultural crops are grown; it came here accidentally in ship cargoes, by way of Quebec, about 1860. **Wings:** White, with apex of forewing tipped black, black spot on forewing of male, two on that of female; black spot on costal margin of hindwing. Yellowish eggs are laid singly on outer leaves. **Spread:** 1.25–2". **Larva:** Pale green, with numerous spots, body covered with fine hairs; dark stripe down the back, row of yellow spots (forming a broken line) along each side; about 1.25" long full grown. Five instars; the first two feed on outer leaves, the others on head and wrapper leaves. Three broods, according to climate; life cycle requires about 31 days in summer. Hibernates as pupa in crop debris, emerges in May in northern part of range. It feeds with larva of the cabbage looper (213), which it resembles, and the diamondback moth (284). **Food:** Turnip, cabbage and other crucifers. This is one of our worst agricultural pests. Larval mortality from a granulosis virus is high in some places, varies with the moisture; it is also susceptible to the bacterium *Bacillus thüringiensis.*

(21) GREAT SOUTHERN WHITE: *Ascia monuste.* **Figs. 229, 230, 231.**

Also called Gulf white cabbageworm. **Range:** Florida, the Gulf states, tropical America, ranging northward to Virginia and Kansas; sometimes migratory, at which times large swarms may be seen flying along the Florida coast. **Wings:** White, with black margins in the male; the female is dimorphic: one form being white, the other—a migratory form named *phileta*—varying from pale gray to brownish gray. **Spread:** 2–2.5". **Larva:** Yellow, with four dark green or purplish stripes; about 1.5" long full grown. **Food:** Wild and cultivated crucifers. A pest of cabbage.

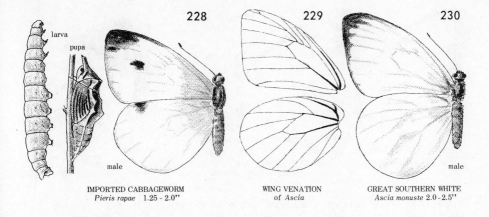

228   229   230

larva

pupa

male

male

IMPORTED CABBAGEWORM
Pieris rapae   1.25 - 2.0"

WING VENATION
of *Ascia*

GREAT SOUTHERN WHITE
Ascia monuste 2.0 - 2.5"

(22) PINE BUTTERFLY: *Neophasia menapia.* **Fig. 232.**

Range: Oregon, Washington, Idaho, British Columbia. **Wings:** White, with black markings; the female has a yellowish cast, with same markings on forewing as the male but with heavier markings on hindwing; many females have orange spots along apical margin of hindwing. They fly above the forest trees, sometimes in large swarms. Emerald green eggs are laid in rows near tip of needles at the tops of the trees. **Spread:** 1.75". **Larva:** Pale green in early stage, with shiny head; they feed in clusters, encircling a needle, with heads pointed toward the tip. Older larva dark green, with two lateral stripes, pale green head, covered with fine hairs; it feeds singly. They drop by silk thread to lower vegetation or trunk to pupate. One brood. Hibernates in egg stage. **Food:** Ponderosa pine; western white pine and lodgepole pine when intermixed with the favored host tree. "Potentially one of the most dangerous enemies of ponderosa pine in the northwestern states," according to Keen. Severe outbreaks occurred in 1893 and 1922. [See Hymenoptera (28).]

(23) LARGE MARBLE: *Euchloë ausonides.* **Fig. 233.**

Erratic in flight, keeps close to the ground. **Range:** Arizona to Alaska, east to Colorado. In California it is found in the lowlands and mountains of the coastal region, in the central and northern parts of the state. **Wings:** Forewing white, with black markings. Hindwing yellowish, darker in the female. Characterized by green marbling on underside of hindwing. **Spread:** 1.65–1.9". **Larva:** Dark green, with a stripe down the back and three on each side. **Food:** Rock cress and other crucifers.

(24) SARA ORANGE TIP: *Anthocharis sara.* **Fig. 234.**

Not abundant but fairly common in the mountainous regions of the Far West. **Range:** New Mexico to Wyoming, west to the Pacific coast. **Wings:**

female

231	232	233 234
GREAT SOUTHERN WHITE *Ascia monuste* 2.0-2.5"	PINE BUTTERFLY *Neophasia menapia* 1.75"	LARGE MARBLE *Euchloë ausonides* 1.65-1.9" SARA ORANGE TIP *Anthocharis sara* 1.25-1.75"

White, tipped with deep orange; black markings. Orange of female less vivid, with broken black borders. The smaller spring form has dark spots on border of hindwing, underside of hindwing with patches of dark green scales. **Spread:** 1.25–1.75". **Food:** Wild radish and mustard, rock cress, and related plants. In the subspecies *A. s. stella* the dimorphic females are yellow on the upper surface of both wings. The **Falcate Orange Tip,** *A. genutia,* occurs east of the Rockies, from Texas to southern New England, is easily recognized by the falcate forewing—"sickle-like" at the apical margin; the apex is reddish orange in the male, white in the female; underside of hindwing is mottled yellow and brown; wingspread from 1.3" to 1.5".

(25) DWARF YELLOW: *Nathalis iole.* **Fig. 235.**

Also called dainty sulphur. Very small but a fast flier. **Range:** The Deep South, through the North Central states, west to Arizona and southern California. **Wings:** Yellow, with black wing tips and markings. Bars across the base of the forewing and costal margin of the hindwing are much wider in the female. Androconia plainly visible in the male as a patch near the base of the costal margin of the hindwing. **Spread:** .75–1.25". **Larva:** Dark green, with purple band down the back, and yellow and black stripe along the sides. Stiff hairs over the body. Pair of cone-shaped tubercles or short projections with reddish bristles on the prothorax. Two or more broods. **Food:** Fetid marigold, sneezeweed, chickweed, alfilaria, cultivated marigolds. [See Fairy Yellow (41).]

(26) ALFALFA CATERPILLAR: *Colias eurytheme.* **Figs. 236, 237.**

Also called alfalfa butterfly or orange sulphur. A polymorphic species occurring in many different forms, according to seasonal and climatic factors. **Range:** Canada and the entire U.S. but uncommon or rare in northern Canada and southern Florida and Texas. **Wings:** The typical *eurytheme* is the spring form: deep orange or orange-red on upper side, with black markings; underside yellow. Albinism is quite common in the females. The albinos of *eurytheme* have wider black borders than those of the closely

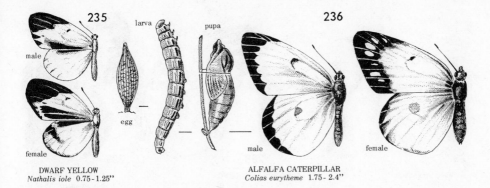

235

male

larva

egg

pupa

236

female

male

female

DWARF YELLOW
Nathalis iole 0.75-1.25"

ALFALFA CATERPILLAR
Colias eurytheme 1.75- 2.4"

related *philodice* (27). **Spread:** 1.75–2.4″. **Larva:** Dark green, covered with short, fine hairs. Faint white stripe down the back, with a narrow black or red line inside the white. A similar line on each side. Three or more broods. Hibernates as pupa and possibly sometimes as adult. **Food:** Alfalfa, clover, and other legumes. At times a serious pest of alfalfa in the southwestern states. [See Hymenoptera (23).] It is susceptible to a virus and the bacterium *Bacillus thüringiensis*.

(27) CLOUDED (COMMON) SULPHUR: *Colias philodice*. **Color plate 4a.**

Swarms are common around mud puddles and in open fields. **Range:** From Canada to the Gulf of Mexico, east of the Rocky Mts. **Wings:** Yellow, otherwise very similar to *eurytheme*. **Spread:** 1.25–2″. **Larva:** Green, with a faint stripe down the back and on each side. Three broods. Hibernates as pupa, possibly as larva or as adult. **Food:** Clover or other legumes.

(28) PINK-EDGED SULPHUR: *Colias interior*. **Fig. 238.**

Range: Canada and the border states, west to the Rocky Mts. **Wings:** Yellow in the male, resembling *philodice* (27) except that border markings are pink. Pale yellow or white in the female, with forewings tipped lightly in black and no markings on hindwing. **Spread:** 1.3–2″. **Larva:** Green, covered with fine, short hairs. Darker green stripe along the back. Spiracles show up white, with red line between. One brood. Hibernates as first-instar larva. **Food:** Several species of blueberries.

(29) SCUDDER'S SULPHUR: *Colias scudderii*. **Figs. 239, 240.**

Range: Montana, Utah, Colorado, British Columbia. **Wings:** Similar to *philodice* (27) in the male except that black borders are wider, with rosy fringes. Usually white (sometimes pale yellow) in the female with borders lightly black or lacking. **Spread:** 1.8–2″. **Food:** Huckleberry and willow.

(30) LABRADOR SULPHUR: *Colias pelidne*. **Fig. 241.**

An Arctic species, possibly a variant of *interior* (28). **Range:** Labrador, northern Quebec to east coast of Hudson Bay, Baffin Island. **Wings:** Fore-

WING VENATION
of *Colias*

male

238

PINK - EDGED SULPHUR
Colias interior 1.3-2.0"

female

239

male

SCUDDER'S SULPHUR
Colias scudderii 1.8-2.0"

wing pale yellow, hindwing greenish in the male (borders are narrower than those of *interior*); only a trace of black at the apex, or none at all, in the female. The female of a variant, *labradorensis,* has no black markings. **Spread:** 1.25–1.5".

(31) ARCTIC SULPHUR: *Colias nastes.* Fig. 242.

A small Arctic species, with numerous variations. **Range:** Northern region of Canada, including Labrador. **Wings:** Dull white to greenish yellow. May be distinguished by the series of light spots on border of forewing and hindwing of both sexes. **Spread:** 1.2–1.6".

(32) BEHR'S SULPHUR: *Colias behrii.* Color plate 4b.

An Arctic-Alpine species. Thought to be rare until an area called Tuolumne Meadows in the High Sierras was made accessible by the opening of Tioga Road in 1915. Large swarms occur in July in its favored haunts. **Range:** Sierra Nevada Mts. above Yosemite. **Wings:** Dark greenish, with darker borders, light spot on hindwing; underside lighter green. A darker shade follows the veins in the female. An albino form of the female, *canescens,* is also common; a light greenish tinge obscures the white. **Spread:** 1.5". **Food:** Dwarf bilberry.

(33) CALIFORNIA DOG FACE: *Colias eurydice.* Fig. 243, Color plate 4f.

Swift in flight. Marking on forewing of male forms a "dog's head." **Range:** Lower mountain ranges and foothills of California, where it is the state butterfly. **Wings:** Darkened area of forewing of the male has a beautiful purplish iridescence. Wings of female are plain yellow except for black spots on the forewing. Pointed apex of the forewing is also a distinctive feature of this species. **Spread:** 1.8–2". **Larva:** Green, with combination white and orange line along sides, black spot or band on each segment between the two lateral lines. Covered with tiny black tubercles and hairs. Highly variable. Two broods. **Food:** Legumes, more especially false indigo or lead bush.

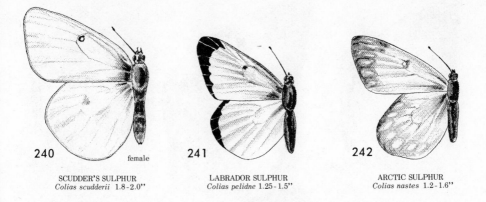

240 female
SCUDDER'S SULPHUR
Colias scudderii 1.8-2.0"

241
LABRADOR SULPHUR
Colias pelidne 1.25-1.5"

242
ARCTIC SULPHUR
Colias nastes 1.2-1.6"

(34) SOUTHERN DOG FACE: *Colias caesonia.* **Fig. 244.**

Unlike *eurydice* (33), the sexes are very similar in color and markings. **Range:** Common in southern states and south into Mexico; ranges northward into North Central states, as far as southern Ontario. **Wings:** Yellow, with black markings. Lighter fringe markings on hindwing of female. **Spread:** 1.8–2.4". **Larva:** Similar to *eurydice.* Also very variable. Three or more broods. Hibernates as pupa. **Food:** False indigo or lead bush, and other legumes such as clover.

(35) LARGE SULPHUR: *Phoebis philea.* **Fig. 245.**

Also called orange-barred or red-barred sulphur. **Range:** Gulf states, Mexico, Central America. **Wings:** Deep yellow, with orange bar across discal cell of forewing and orange shading on outer margin of hindwing. **Spread:** 2.75–3.5". **Larva:** Variable. Typically green, with tiny black spines. Black and yellow band along sides, the latter near ventral surface. Two or more broods. **Food:** Wild senna, partridge pea, and possibly clover.

(36) LARGE ORANGE SULPHUR: *Phoebis agarithe.* **Fig. 246.**

Sometimes called dark sulphur. **Range:** Common in the Gulf states, Mexico and Central America. **Wings:** Bright orange in the male, with no markings; orange or delicate pink in the female, with brown spot and marginal markings on forewing. **Spread:** 2.25–2.5". **Food:** Various species of *Cassia.*

(37) CLOUDLESS SULPHUR: *Phoebis sennae eubule.* **Fig. 247, Color plate 4c.**

Sometimes called pale sulphur. They are often migratory. **Range:** Eastern North America, from Canada to the Gulf. **Wings:** Bright yellow, unmarked. Tend to be darker in the female, with a dark spot on discal area of forewing and a row of brown spots along the border. **Spread:** 2.2–2.75". **Larva:** Pale green, black spots in bands across each segment, yellow stripe along sides. Two or more broods. **Food:** Legumes, more especially senna and partridge pea.

245

LARGE SULPHUR
Phoebis philea 2.75-3.5"

male

243

244

male

female

female

CALIFORNIA DOG FACE
Colias eurydice 1.8-2.0"

SOUTHERN DOG FACE
Colias caesonia 1.8-2.4"

246

LARGE ORANGE SULPHUR
Phoebis agarithe 2.25-2.5"

(38) MEXICAN SULPHUR: *Eurema mexicana.* **Figs. 248, 249.**

Species of *Eurema* are sometimes referred to collectively as "little sulphurs"; they are typically yellow or bright orange with black margins. **Range:** This species is abundant in Central America, ranges northward through Mexico and our central states. Common in Arizona and Texas. **Wings:** Distinctive features are pointed hindwing and broad, black border on forewing. Forewing of male white, hindwing pale yellow, with orange tinge along coastal margin; borders of forewing much lighter shade in the female. Hindwing of summer form reddish. **Spread:** 1.25–1.75". **Food:** Wild senna.

(39) NICIPPE YELLOW: *Eurema nicippe.* **Fig. 250.**

Also called sleepy orange, often seen in large swarms. **Range:** Central America, northward. Common as far north as southern Illinois and Indiana. Rare in New England. **Wings:** Quite variable. Various shades of orange and yellow. Typically orange in the male, with wide black borders; lighter coloring and borders in the female. The latter are occasionally albino. A variant, *flava,* is bright yellow in place of the orange. **Spread:** 1.5–2". **Larva:** Velvety green, with white and yellow lateral stripe. **Food:** Senna, clover.

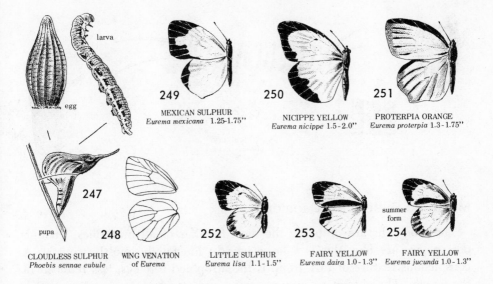

249 MEXICAN SULPHUR *Eurema mexicana* 1.25-1.75"

250 NICIPPE YELLOW *Eurema nicippe* 1.5-2.0"

251 PROTERPIA ORANGE *Eurema proterpia* 1.3-1.75"

247

248 CLOUDLESS SULPHUR *Phoebis sennae eubule*

WING VENATION of *Eurema*

252 LITTLE SULPHUR *Eurema lisa* 1.1-1.5"

253 FAIRY YELLOW *Eurema daira* 1.0-1.3"

summer form 254 FAIRY YELLOW *Eurema jucunda* 1.0-1.3"

larva

egg

pupa

(40) LITTLE SULPHUR: *Eurema lisa*. Fig. 252.

A migratory species, known to travel from the mainland to Bermuda, 600 miles distant. Swarms are commonly seen around mud puddles. **Range:** New England, west to the Rocky Mts. and south into Mexico. **Wings:** Yellow, with black borders. **Spread:** 1.1–1.5". **Larva:** Green, covered with fine, soft hairs, giving it a fuzzy appearance. White lateral line along spiracles. Two to three or more broods. **Food:** Wild senna and partridge pea, clover and probably other legumes.

(41) FAIRY YELLOW: *Eurema daira* (*delia*). Fig. 253.

Range: Gulf states, north to North Carolina and Arkansas. **Wings:** Yellow, with black borders and black bar across inner margin of forewing; black bar not present in the female. Underside, hindwing and tip of forewing are red. In the summer form, *jucunda* (Fig. 254), the black border of the hindwing extends around the entire margin, and the underside of the wings are a lighter red. **Spread:** 1–1.3". The resemblance to *Nathalis iole* (25) will be noted; the dwarf yellow is smaller and has an additional black bar along the costal margin of the hindwing.

(42) PROTERPIA ORANGE: *Eurema proterpia*. Fig. 251.

Range: Cuba, Mexico, Texas and Arizona. **Wings:** Deep, bright orange, with solid black border on costal margin of forewing in the male; dull orange, with wider and diffused borders in the female. In the spring form, or *gundlachia*, the background coloring is not as deep and the tip of the hindwing is longer and sharply pointed. **Spread:** 1.3–1.75".

egg

larva

surface of egg

255

MONARCH
Danaus plexippus

pupa

Milkweed Butterflies (Danaidae)

The monarchs are probably the most familiar of all butterflies. Though our species are few, this is a large world-wide family, mostly tropical. A conspicuous feature of this group is the atrophied forelegs, less reduced in the female. The antennae, without scales, and attached on top of the head, have only a moderate swelling to form the club. The androconia of the male, located alongside a vein (Cu_2) near the outer angle of the hindwing, are conspicuous.

The cone-shaped eggs are cut off at the top and have longitudinal ridges with cross-lines. The smooth larvae, which feed on milkweed, are very colorful, with dark crossbands and a pair of tentacles near the anterior and posterior ends. The chrysalises are attached by a cremaster only, and often gleam with golden or silvery spots.

The monarchs have few predatory enemies because their choice of food makes them distasteful and sometimes poisonous. For this reason they have their mimics, such as the viceroy, and have a life span that is much longer than most butterflies. Their migratory habit has been commented on elsewhere (see p. 201). They are true migrators, like birds, massing in impressive swarms as they make the flight southward in the fall. Swarming does not take place on the return flights in the spring, and the movement northward is not noticeable.

(43) MONARCH: *Danaus plexippus.* **Fig. 255, Color plate 2a.**

Also called milkweed butterfly. Slow-flying. **Range:** Most of North America. **Wings:** Background coloring light brown. Black borders, with white dots; blackened veins. Black borders are wider in the female than in the male; the latter may be distinguished by the androconia. **Spread:** 3.5–4″. **Larva:** Background color bright green, with black and yellow bands. The monarch produces successive broods throughout the year. **Food:** Milkweeds. It is mimicked by the viceroy (100), which is strikingly similar.

(44) QUEEN: *Danaus gilippus berenice.* **Color plate 3a.**

This is not a migratory species. **Range:** Florida, Gulf states, Georgia, Mississippi Valley to Kansas. **Wings:** Brownish ground color obscures veins. Narrow black apical and marginal marking on the forewing. Spots whiter and black border of hindwing relatively wider than in the monarch. **Spread:** 3.1–3.3″. **Larva:** Brownish, with brown and yellow bands. **Food:** Milkweeds.

Butterflies: Satyrs and Wood Nymphs (Satyridae)

This is a small group in the temperate zone of North America but common and widely distributed. Weak fliers, they are commonly seen flitting about the grass and shrubs, and quickly hide near the ground when alarmed. A few species are Arctic or Alpine and live in remote, bleak regions. Some tropical species are crepuscular. They are medium in size, dull in coloration, usually some shade of brown, with conspicuous eyespots on both sides of the wings. Typically, some of the main veins of the forewing are swollen at the base. As in the Danaidae the forelegs are reduced in size, more especially in the males. When at rest the forewings are folded inside the hindwings. The less colorful "Arctics" are hairy, enabling them to withstand cold temperatures.

The eggs have a variety of shapes but are usually ovoid or spherical, with ridges and cross-lines, flattened at the top and rounded at the bottom. The mature larvae are generally pale green or brown and are striped. They are tapered at both ends and enlarged at the middle, and are differentiated from most other larvae by having the anal end forked. The newly emerged larvae have large heads, larger around than the rest of the body. The larvae of the Arctics tend to be stouter and are covered with hairs which help protect them from extremely cold temperatures. Their food is mostly grasses. The unornamented chrysalises are green or brown and usually hung by the cremaster, but pupation also takes place on the ground.

(45) SOUTHERN WOOD NYMPH: *Cercyonis (Minois) pegala pegala.* **Fig. 258.**

This is the largest species of the genus. **Range:** Gulf states to Canada. **Wings:** Brown, with large orange patch on forewing. Lower eyespot on forewing is sometimes smaller or absent. Underside with orange patch and two large, bright eyespots on forewing and four similar spots on hindwing. **Spread:** 2.75″. **Larva:** Pale green, with red anal fork and four stripes. Covered with short, fine hairs. One or two broods. Hibernates as first instar larva. **Food:** Grasses. *C. p. alope* is generally smaller but otherwise very similar, occurs along the Atlantic coast, has two prominent eyespots on forewing; wingspread from 2″ to 2.8″. The orange patch tends to become obscured as one goes farther north.

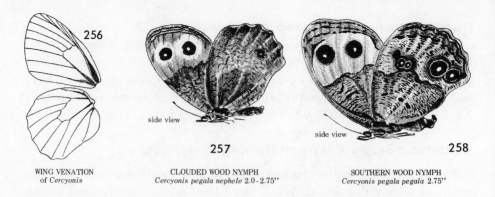

256

side view

257

side view

258

WING VENATION
of *Cercyonis*

CLOUDED WOOD NYMPH
Cercyonis pegala nephele 2.0-2.75''

SOUTHERN WOOD NYMPH
Cercyonis pegala pegala 2.75''

(46) CLOUDED WOOD NYMPH: *Cercyonis pegala nephele.* **Figs. 256, 257.**

Range: Mountains of New York, New England, and Pennsylvania, ranging westward and north into Canada. **Wings:** Similar to *pegala* and *alope* (45) but orange patch lacking. Eyespots are smaller, have a faint orange ring. **Spread:** 2–2.75''. *C. p. ariane* is abundant in grasses of northern California, Oregon, Nevada, and Utah, has series of six spots on underside of hindwing in groups of three, the center one larger; ring around spots on forewing of female are more yellow in place of orange.

(47) SYLVAN SATYR: *Cercyonis silvestris.* **Fig. 259.**

Common on semiarid, grassy slopes. Erratic in flight. **Range:** Southern California. **Wings:** Dark brown in male, lighter in female; two prominent eyespots on forewing, eyespots on both surfaces of hindwing generally obscured or missing. Androconia of male prominent. **Spread:** 1.5''. The subspecies *C. s. paula* or **Little Satyr,** found in northern California, Oregon, and Washington, has two eyespots on the forewing (the one near inner angle obscured in the male), two prominent eyespots on underside of forewing, and a series of small ones on the hindwing.

(48) LITTLE WOOD SATYR: *Euptychia (Megisto) cymela (euryta).* **Fig. 260.**

A common species in partially wooded areas; very elusive. **Range:** Canada, eastern U.S. to Kansas and Nebraska, Texas. **Wings:** Brown, with two eyespots on forewing and on hindwing, upper and lower sides. **Spread:** 1.75''. **Larva:** Light brown, with white head; tubercles and fine hairs covering body also white. Finely striped. One or more broods. Hibernates as larva. **Food:** Grasses. A closely related species, *E. mitchelli*, found in southern Michigan and in Ohio, has a series of four eyespots on underside of forewing and six on hindwing.

(49) CAROLINA SATYR: *Euptychia hermes.* **Fig. 261.**

Range: New Jersey, south to Florida, west to Texas and lower Mississippi Valley. **Wings:** Dark gray, without eyespots on upper surface. Series of

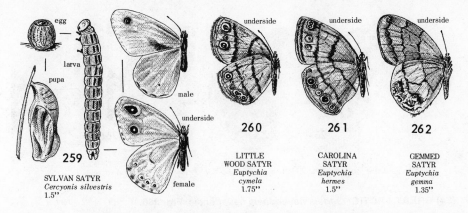

egg

larva

pupa

259

SYLVAN SATYR
Cercyonis silvestris
1.5"

male

underside

female

underside **260**

LITTLE
WOOD SATYR
*Euptychia
cymela*
1.75"

underside **261**

CAROLINA
SATYR
*Euptychia
hermes*
1.5"

underside **262**

GEMMED
SATYR
*Euptychia
gemma*
1.35"

eyespots on underside, all somewhat obscured excepting one on forewing and two on hindwing. **Spread:** 1.5". **Larva:** Similar to *cymela* (48) except for yellowish tubercles and hairs. **Food:** Grasses.

(50) GEMMED SATYR: *Euptychia (Neonympha) gemma.* **Fig. 262.**

Range: West Virginia to the Gulf and into Mexico. **Wings:** Brownish gray, with two black spots on outer margin of hindwing. Violet patch with series of silvered spots near outer margin on underside of hindwing. **Spread:** 1.35". **Larva:** Spring brood green, fall brood brown, with darker stripes. Paired tentacles on head and near anus. Two broods. **Food:** Grasses.

(51) CALIFORNIA RINGLET: *Coenonympha tullia california.* **Fig. 263.**

Range: California to British Columbia, east to Nevada. **Wings:** White. Underside greenish, with several eyespots on hindwing. Exhibits seasonal dimorphism. Summer form yellowish on upper side, brownish on underside; a northern form is similar except that eyespots on underside of hindwing are lacking. **Spread:** 1.2–1.8". **Larva:** Green or brown, slender, with rounded head and two short projections on anal segment. **Food:** Grasses.

(52) PEARLY EYE: *Lethe (Enodia) portlandia.* **Fig. 264.**

A woodland species. Male "pearly eyes" perch on trees, establishing a "territory" like birds often do, and fight off any other males intruding. **Range:** Canada to the Gulf of Mexico, east of the Rocky Mts. **Wings:** Fawn-colored, with several darker eyespots on both wings. Both sexes very similar. **Spread:** 1.75–2". **Larva:** Pale green, with a pair of short, red projections from the head. Forked tip of abdomen red. One or two broods. Hibernates as larva. **Food:** Grasses.

(53) CREOLE PEARLY EYE: *Lethe creola.* **Fig. 265.**

Range: Manitoba to the Gulf of Mexico. **Wings:** Very similar to *portlandia* (52) but more yellowish. May be readily distinguished by the raised scales between the veins of the forewing, extending in from the eyespots. **Spread:** 1.6–2".

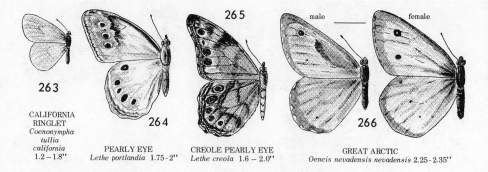

265

263

CALIFORNIA
RINGLET
*Coenonympha
tullia
california*
1.2 – 1.8"

264

PEARLY EYE
Lethe portlandia 1.75-2"

CREOLE PEARLY EYE
Lethe creola 1.6 – 2.0"

male ———— female

266

GREAT ARCTIC
Oeneis nevadensis nevadensis 2.25-2.35"

(54) GREAT ARCTIC: *Oeneis nevadensis nevadensis.* **Fig. 266.**

Also called Felder's Arctic. Found in open spaces among the coniferous forests and grassy foothills of lower altitudes. **Range:** Northern California, Nevada, British Columbia. **Wings:** Light brown, with darker margins. Eyespots sometimes white-pupiled. Gray-brown marbling on underside of hindwing. Coloration and patterns serve as an effective camouflage against the bark of trees while they rest. **Spread:** 2.25–2.35". **Larva:** Brown, with black median stripe; additional stripes of black, brown, or white. Rounded head, body tapering sharply at anal end. Single brood. Hibernates as larva. **Food:** Grasses.

(55) GREATER ARCTIC: *Oeneis nevadensis gigas.* **Fig. 267.**

Range: Higher altitudes of British Columbia and to the south. **Wings:** Bold eyespot near apex on underside of hindwing. Dark grayish marbling on underside of hindwing. **Spread:** 2.1–2.5".

(56) LABRADOR ARCTIC: *Oeneis taygete.* **Fig. 268.**

Also called white-veined Arctic. Found in Arctic region and higher altitudes to the south. **Range:** Labrador, the Canadian Arctic, Alaska, Rocky Mts. **Wings:** Light brown, plain (no eyespots). Contrasting mottled areas and white veins on underside of hindwing characterize this species. **Spread:** 1.75".

(57) NOVA SCOTIAN ARCTIC: *Oeneis jutta.* **Fig. 269.**

Also called the jutta Arctic. One of the most southerly-ranging of the Arctics. Also occurs in Europe. **Range:** Maine, New Hampshire, Nova Scotia, Newfoundland, to the Canadian Arctic. **Wings:** Light brown background. Orange band near margin of both wings contains black, white-pupiled eyespots. **Spread:** 1.8–1.9".

(58) COMMON ALPINE: *Erebia epipsodea.* **Fig. 270.**

Found in the Far North or high altitudes of the West. **Range:** New Mexico to Alaska, eastward to Manitoba. **Wings:** Dark brown, with three or four black eyespots, white-pupiled and ringed with red. **Spread:** 1.5–1.8".

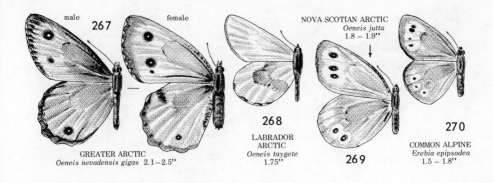

male 267 female

NOVA SCOTIAN ARCTIC
Oeneis jutta
1.8 – 1.9"

268

LABRADOR
ARCTIC
Oeneis taygete
1.75"

269

270

COMMON ALPINE
Erebia epipsodea
1.5 – 1.8"

GREATER ARCTIC
Oeneis nevadensis gigas 2.1–2.5"

Butterflies: Heliconians (Heliconiidae)

Distasteful and even poisonous to predators, the heliconians are relatively free from attack and are mimicked by many butterflies and some moths. They are primarily tropical and range from the southern states to the tropical Americas. The forewings are typically long and narrow, extending well beyond the rounded hindwings. The front legs are greatly reduced as in the Danaidae and Satyridae. The eggs are barrel-shaped, with ridges and cross-lines. The young larva is covered with hairs which later become branched spines. The hairs rise from button-like projections arranged in rows or geometrical patterns. Their food consists of various species of passionflower (*Passiflora*). The chrysalises are very irregular in shape, with numerous sharp projections, and suspended by a cremaster. Extreme variations of color and markings are common among these colorful butterflies. This group is sometimes classified as a subfamily (Heliconiinae) of the Nymphalidae, our next family.

(59) ZEBRA: *Heliconius charitonius.* **Fig. 271.**

A distinctive species. **Range:** Southern states, tropical America. **Wings:** Black background, with yellow bands. **Spread:** 3–3.4". **Larva:** White, with six rows of branched spines, brown and black spots. Two or more broods. **Food:** Passionflower.

(60) JULIA: *Dryas (Colaenis) julia.* **Fig. 272.**

Mimics other heliconians. **Range:** Southern Florida and Texas, tropical America. **Wings:** Deep orange, with black borders and markings. The female is brownish. **Spread:** 3.25–3.7".

(61) GULF FRITILLARY: *Agraulis (Dione) vanillae.* **Fig. 273.**

Range: Virginia, south to the Gulf, west to Arizona, California and Mexico. **Wings:** Brownish orange, with black spots and markings. Spots on discal cell have white pupils; those along margin of hindwing are the basic orange with black outline. **Spread:** 2.5–3". **Larva:** Yellowish brown,

271

272

ZEBRA
Heliconius charitonius 3.0-3.4"

JULIA
Dryas julia 3.25-3.7"

GULF
FRITILLARY
Agraulis vanillae 2.5-3"

with darker stripes. Six rows of black, branched spines, two longer ones on the head. **Food:** Passionflower.

Brush-footed Butterflies (Nymphalidae)

This is the largest family of "true butterflies" and includes many of the most common ones. They are generally medium to large in size and strong fliers. Some tropical species are among the largest butterflies. Background coloring is commonly orange or brown and intermediate shades, with black markings. In the females the background colors are usually lighter and the markings heavier. The front legs are much smaller than the others (like the Danaidae, Satyridae, and Heliconiidae) and have numerous hairs which give them the appearance of a brush. The antennae are usually as long as the body or sometimes longer, and heavily coated with scales. The discal cell of the hindwing is almost always open to the outer margin, as is sometimes the case with the forewing. They are usually most active at midday.

The eggs are spherical and cut off at one end, or barrel-shaped, with vertical ribs or hexagonal ridges (like a net). The larvae are usually brown, or black; they are covered with hairs in the early stages and later with branched spines, both arranged symmetrically. Many species feed only at night and hide during the day. The chrysalises are suspended by a cremaster only, and have numerous projections; the head is usually split or forked.

Some members of this group are noted for their migratory habit, as for example the painted lady, the buckeye, and the California tortoise shell. They are not "true migrators," as in the case of the monarch, since their migrations occur only at intervals and they are normally capable of over-wintering as an adult or pupa. Populations of the California tortoise shell and the painted lady are subject to wide fluctuations, and sometimes the larvae build up to a point where they consume all the available food plants, culminating in massive swarming of the adults and their migratory flight. Apparently prompted by overcrowding and shortage of food, the phenomenon is suggestive of the periodic mass migrations of lemmings and their

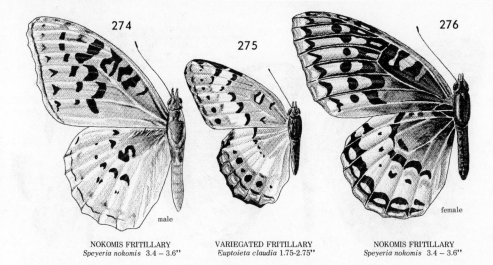

274

275

276

male

female

NOKOMIS FRITILLARY
Speyeria nokomis 3.4 – 3.6"

VARIEGATED FRITILLARY
Euptoieta claudia 1.75-2.75"

NOKOMIS FRITILLARY
Speyeria nokomis 3.4 – 3.6"

march to the sea. Spectacular flights of both of these butterflies have been witnessed in California. These flights are northward in the summer, to the Rocky Mountains of Colorado and Canada, where they are apparently able to overwinter as adults. A mass flight southward late in the fall has been witnessed in the case of the tortoise shell. The painted lady has been known to travel 1,400 miles, from California to the Canadian Rockies, at an average speed estimated to be eight miles an hour.

(62) VARIEGATED FRITILLARY: *Euptoieta claudia* (Argynninae). **Fig. 275.**

The name FRITILLARY refers to the checkered or spotted pattern of coloration. This genus lacks the silver spots on the underside of the hindwings that characterize the next group, *Speyeria*. **Range:** Eastern U.S., uncommon in the northern half; the Southwest, into Mexico and Central America. **Wings:** Area inside the black zigzag line through the median is brownish; outside this line, a light orange band and darker orange near the outer margin. Other markings black. **Spread:** 1.75–2.75". **Larva:** Reddish yellow, with two brown stripes on each side and a series of white spots on the back, between the bands. Six rows of black, branching spines; the pair on top of the first segment longer than the rest and pointed forward. Two or three broods. Hibernates as adult. **Food:** Violets, pansies, passionflower, purslane, stonecrop, mandrake or May apple.

(63) NOKOMIS FRITILLARY: *Speyeria (Argynnis) nokomis.* **Figs. 274, 276.**

This genus is commonly referred to as the GREATER FRITILLARIES or SILVER-SPOTS. They are readily distinguished by the large silver spots on the underside of the hindwings. The larvae are all brown or black, with six rows of black, branched spines; they are nocturnal feeders, remain hidden during the day. *S. nokomis* is one of the largest and most beautiful species in the

HESPERIS FRITILLARY
Speyeria atlantis hesperis
2.25 - 2.4"

279

277

278   280

REGAL FRITILLARY
Speyeria idalia
2.75 - 4.0"

LETO FRITILLARY
Speyeria cybele leto
3.5"

SILVER-BORDERED FRITILLARY
Boloria selene myrina
1.5"

genus. **Range:** Western slopes of the Rocky Mts., from Colorado to Mexico, including the Great Basin of California. **Wings:** Brownish orange, with black markings, base of forewing reddish, hindwing with pale reddish or straw-colored submarginal band; dark background color of female is brown, the lighter area yellow. Silver spots on underside of hindwing are very bright, edged with black. **Spread:** 3.4–3.6". In the **Apache Fritillary,** *S. n. apacheana,* the disc on the underside of the hindwing is the same color as the surrounding submarginal area; in *nokomis* it is darker.

(64) REGAL FRITILLARY: *Speyeria idalia.* Fig. 277.

Markings and coloration are distinctive. At times quite common in its territory. **Range:** Georgia to Maine, west to the Plains states. Mountainous regions of New York and Pennsylvania. **Wings:** Background coloring of forewing yellowish brown, that of the hindwing shiny blue-black. Marginal spots on hindwing of male yellowish brown; in the female they are creamy, as are those in the submarginal row. Underside of forewing yellowish brown, with marginal row of silvery crescents; hindwing darker brown, with three rows of silvery spots. **Spread:** 2.75–4". **Food:** Violets.

(65) LETO FRITILLARY: *Speyeria cybele leto.* Fig. 278.

Prefers wooded mountainous areas. **Range:** Southern California (common in Yosemite) to British Columbia, east to Colorado, Wyoming, and North Dakota. **Wings:** Similar to *nokomis* (63). Background coloring darker in general. Dark area in the male is more reddish; in the female it is velvety brown, with marginal areas straw-colored. **Spread:** 3.5". **Food:** Violets.

(66) SPANGLED FRITILLARY: *Speyeria cybele cybele.* **Color plates 5a, b, c.**

Range: Eastern Canada and the U.S., excepting the extreme southern states. **Wings:** Similar to *leto* (65). Markings of the male are darker, dark coloring of the female is chocolate brown; the lighter marginal areas are light brown rather than yellowish as in *nokomis* (63) and *leto.* Numerous silver spots in discal area on underside of hindwing. Varies considerably throughout its range; tends to be darker in northern region and lighter in western part of range, with smaller silver spots. **Spread:** 3–4″.

(67) HESPERIS FRITILLARY: *Speyeria atlantis hesperis.* **Fig. 279.**

A mountain species, common in forested zones. **Range:** Colorado and westward. **Wings:** Yellowish brown, with slightly darker areas toward the base, female lighter than the male, with heavier black markings; underside reddish-brown, darker in the female, spots on underside yellowish white rather than silver. **Spread:** 2.25–2.4″. The **Atlantis Fritillary** (or mountain silver-spot), *S. a. atlantis,* is the eastern form, ranging from the Maritime Provinces south along the Appalachian Range to Virginia, west to Wisconsin; the background coloring is deep brown, underside darker, with yellowish submarginal band well defined and always present; spots on underside of hindwing usually but not always silvered, outer border of forewing almost solid black on upper side, spread 2.5″.

(68) SILVER-BORDERED FRITILLARY: *Boloria selene myrina.* **Fig. 280.**

This genus belongs to a group commonly called the LESSER FRITILLARIES. They are small copies of *Speyeria,* resembling them in all but size; many are northern or Arctic in range. Immature forms of both groups are also similar; the larvae feed at night, commonly on violets, and hibernate as early instars. This species prefers marshy areas. **Range:** Eastern Canada and the U.S. to North Carolina, westward to Alberta and Montana. **Wings:** Brownish orange, with black markings. Underside of forewing light brown, with slightly silvered marginal spots; hindwing has red and yellow mottling, with silvery spots. **Spread:** 1.5″. **Larva:** Olive-brown, with green mottling. Covered with sharp spines protruding from tubercles; those on the sides of the prothorax are much longer than the others. One to three broods. Hibernates in early larval stage. **Food:** Violets.

(69) WHITE MOUNTAIN FRITILLARY: *Boloria titania montina.* **Fig. 281.**

Unique because of its restricted range. **Range:** White Mountains of New Hampshire. **Wings:** Orange-brown, with black markings. Female a little lighter but otherwise closely resembles the male. Underside of hindwing reddish brown; area toward the base and costal margin is more reddish, with white patches; row of well-defined silver spots along the outer margin. **Spread:** 1.5–1.7″.

281	282	283	284
WHITE MOUNTAIN FRITILLARY	PURPLE LESSER FRITILLARY	POLAR FRITILLARY	CHALCEDON CHECKERSPOT
Boloria titania montina	*Boloria titania grandis*	*Boloria polaris*	*Euphydryas chalcedona*
1.5 – 1.7''	1.6''	1.25''	2.0 – 3.0''

underside

(70) PURPLE LESSER FRITILLARY: *Boloria titania grandis.* **Fig. 282.**

Range: Canada and the northernmost states; throughout the Rockies at higher elevations. **Wings:** Closely resemble *montina* (69) but a little darker on the underside, with less white and more dark spots showing in the marginal area of the hindwing. **Spread:** 1.6''.

(71) POLAR FRITILLARY: *Boloria polaris.* **Fig. 283.**

Range: Arctic America, Norway, Greenland. Circumpolar. **Wings:** Color and markings similar to *titania* (69) and (70). A greater amount of white shows through the brown and reddish areas. Marginal and submarginal spots are conspicuous against the white background coloring. **Spread:** 1.25''.

(72) CHALCEDON CHECKERSPOT: *Euphydryas chalcedona* (Melitaeinae). **Fig. 284.**

This genus and the one following (*Melitaea*) are called CHECKERSPOTS. Most of them are western species. Small to medium in size, they resemble the lesser fritillaries in general appearance, except in the light markings that sometimes form a striking checkered pattern. There is usually little or no silvering of spots on the underside of the wings. The larvae are covered with short spines having hairs or sharp points. The chrysalises are rounded at the anterior end, gray or white, with orange or black spots, and adorned with pointed tubercles. Some species of checkerspots have such a great variety of forms that it is almost impossible to identify them. This is especially true of *chalcedona*. **Range:** California and adjoining states; very common. **Wings:** Background color is brown, with yellowish spots. Some forms are lighter, others much darker (almost black); in some the spots are fused and elongated, in others almost entirely lacking. **Spread:** 2–3''. **Larva:** Mature larva mostly black, with white stripe down the back, and a broken lateral band, also white. Covered with white tubercles which support white spines and hairs. **Food:** Monkey flowers, bee plant, and other figworts.

285   286   287   288

EDWARDS' CHECKERSPOT
Euphydryas chalcedona colon
1.75-2.5"

LEANIRA CHECKERSPOT
Melitaea leanira
1.25 – 1.75"

ACASTA CHECKERSPOT
Melitaea acastus
1.5 – 1.75"

HARRIS' CHECKERSPOT
Melitaea harrisii
1.4 – 1.7"

(73) EDWARDS' CHECKERSPOT: *Euphydryas chalcedona colon.* **Fig. 285.**

Range: Northern California, along the Sierra Nevada and Cascade Mts. to British Columbia; Idaho and Montana. **Wings:** Area just inside the submarginal row of light spots on the hindwing is black. Red spots around margin are well defined. **Spread:** 1.75–2.5".

(74) LEANIRA CHECKERSPOT: *Melitaea (Chlosyne) leanira.* **Fig. 286.**

Lives in isolated colonies. **Range:** Foothills of mountain ranges, Arizona to British Columbia. **Wings:** Reddish brown, with yellowish white spots in submarginal and median areas, red spots along margins. Underside of forewing is pale yellowish brown, hindwing pale yellow. Extremely variable. **Spread:** 1.25–1.75". **Food:** Bird's-beak.

(75) ACASTA CHECKERSPOT: *Melitaea (Chlosyne) acastus.* **Fig. 287.**

Found in arid regions of the West. **Range:** Arizona, Nevada, Utah, Mojave Desert, southeastern Oregon to western Nebraska. **Wings:** Yellowish brown, with black markings. Underside of forewing is much paler, with white margins and light brown borders; hindwing pearly white, with light brown submarginal band and borders. **Spread:** 1.5–1.75". One brood. **Food:** Aster.

(76) HARRIS' CHECKERSPOT: *Melitaea (Chlosyne) harrisii.* **Fig. 288.**

Occurs in small colonies, as do many other checkerspots. **Range:** Eastern U.S. and Canada, west to Manitoba and the Rocky Mts., south to Mexico. **Wings:** Reddish brown, black or dark brown along outer margins. Underside of forewing bright orange; hindwing has whitish areas and submarginal row of crescents. **Spread:** 1.4–1.7". **Larva:** Brownish yellow, with black band along the back; black spines. One brood. **Food:** White aster.

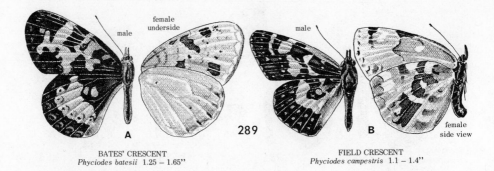

BATES' CRESCENT
Phyciodes batesii 1.25 – 1.65"

289

FIELD CRESCENT
Phyciodes campestris 1.1 – 1.4"

(77) PEARL CRESCENT: *Phyciodes tharos.* Fig. 290.

This genus belongs to a group called CRESCENT-SPOTS or CRESCENTS, characterized by one or more pearly white or silvery crescent-shaped spots in the outer margin, on underside of hindwings. Small in size, they resemble the checkerspots in general appearance and in wing venation. As in the checkerspots, the discal cell of the hindwing is open to the outer margin. Background coloring of the wings is orange or yellowish brown, with darkened or black markings and margins; the latter are repeated on the underside, on a paler background. There may be one to several broods, depending on location; those appearing in the spring are often much darker than the summer broods. The immature forms are similar to those of the closely related checkerspots. *P. tharos* is very common in open fields and along roadsides. **Range:** All of North America from southern Canada to Mexico, excepting the Pacific coast. **Wings:** Light brown, with extensive black borders. The dark areas are larger in the spring broods. **Spread:** 1.25–1.65". **Larva:** Dark brown, with yellow spots, yellow stripe on each side, and short spines. One to several broods. Hibernates as larva. **Food:** Aster.

(78) BATES' CRESCENT: *Phyciodes batesii.* Fig. 289A.

Also called the tawny crescent. **Range:** Eastern Canada and the U.S. **Wings:** Similar to *tharos* (77). Dark areas are heavier and more extensive. **Spread:** 1.25–1.65".

(79) FIELD CRESCENT: *Phyciodes campestris.* Fig. 289B.

Also called the meadow crescent. Prefers moist areas at low altitudes. **Range:** Arizona; Pacific coast, California to Alaska, east of the Rocky Mts. to the Great Plains. **Wings:** Orange, with extensive black markings. Underside paler, with yellowish spots. The female is larger than the male, has heavier dark areas and yellow spots; underside darker also. **Spread:** 1.1–1.4". Two broods. **Food:** Aster.

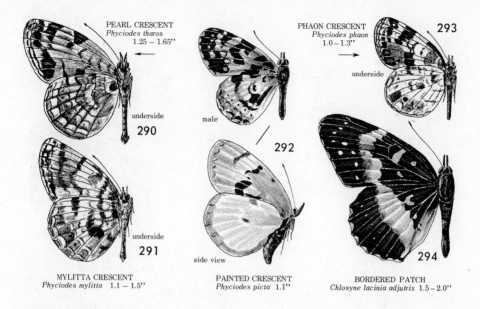

PEARL CRESCENT
Phyciodes tharos
1.25 – 1.65"

underside
290

underside
291

MYLITTA CRESCENT
Phyciodes mylitta   1.1 – 1.5"

male

side view

PAINTED CRESCENT
Phyciodes picta   1.1"

292

PHAON CRESCENT
Phyciodes phaon
1.0 – 1.3"

underside

293

294

BORDERED PATCH
Chlosyne lacinia adjutrix   1.5 – 2.0"

(80) MYLITTA CRESCENT: *Phyciodes mylitta*. Fig. 291.

Abundant throughout the West Coast in grassy areas and along streams. **Range:** Mexico to British Columbia, west of the Rocky Mts. **Wings:** A distinguishing feature is the relatively large area of bright yellowish brown and reduced dark areas. Female larger and a little paler but otherwise similar to male. **Spread:** 1.1–1.5". **Larva:** Velvety black, yellow underneath, with a pair of faint yellow lines down the back and each side. Head black, covered with hairs. Six rows of spines; these are black except on the fourth to sixth segments, where they are yellow. Several broods. **Food:** Thistle.

(81) PAINTED CRESCENT: *Phyciodes picta*. Fig. 292.

Range: Nebraska, Kansas, Oklahoma, Colorado, New Mexico, Arizona, Mexico. **Wings:** Yellowish brown, with black markings. Underside pale yellow, except median area of forewing, which is reddish with black markings. **Spread:** 1.1". **Larva:** Mottled with yellowish or greenish tints. Seven rows of spines which are light brown in spring broods and greenish yellow in fall broods. Several broods. **Food:** Aster.

(82) PHAON CRESCENT: *Phyciodes phaon*. Fig. 293.

Common in the Southeast. **Range:** Georgia, south to Cuba and Guatemala, west to Kansas and southeastern California. **Wings:** Similar to *picta* (81). Underside of forewing has more black, with large orange-brown areas; underside of hindwing has more distinct pattern of pale yellow, white, and brown or black. **Spread:** 1–1.3". Several broods. **Food:** Lemon verbena, matgrass.

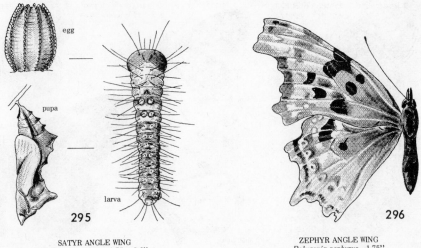

egg

pupa

larva

295

296

SATYR ANGLE WING
Polygonia satyrus   1.75 – 2.0"

ZEPHYR ANGLE WING
Polygonia zephyrus   1.75"

(83) BORDERED PATCH: *Chlosyne lacinia adjutrix.* **Fig. 294.**

This genus, closely related to *Phyciodes,* is a small group called PATCHED BUTTERFLIES. *C. l. adjutrix,* like most of the patched butterflies, is a southwestern species but more abundant in Central America; other common names for it are adjutrix patch or Scudder's patched butterfly. **Range:** Mostly Texas, Mexico. **Wings:** Brownish black, with small but prominent white spots in submarginal area of forewing. Both wings have an orange patch in median area and orange spots between patch and base. **Spread:** 1.5–2". **Larva:** Color highly variable. Black to reddish brown, with varied stripes and blotches. Several broods. **Food:** Sunflower.

(84) SATYR ANGLE WING: *Polygonia satyrus* (Nymphalinae). **Fig. 295, Color plate 5g.**

This genus comprises a group called the ANGLE WINGS, distinguished by the concave curvature and deeply indented margin of the forewing and a tail-like projection of the hindwing. They are small to medium in size, and have brownish yellow wings with black spots. The underside is mottled brown or gray, often with a silvery C or "comma" on the hindwing. Because of the color patterns and ragged edges of the wings, when held upright at rest, the angle wings look like leaves and are difficult to detect. They seldom visit flowers, preferring sap and fruit juices, and are deceptive in flight owing to the flashing effect of the color patterns. Hibernation occurs in the adult stage. The eggs are pale green, with ribs spaced well apart; they are laid on the underside of leaves, usually in strings of three or four, one on top of the other. The chrysalids are soft-brown or greenish. The larvae are usually brown, and have rows of spines. Seasonal and sexual dimorphism are common. Like the tortoise shells, the next group (90), they hibernate as adults. *P. satyrus* ranges wide but is mostly a western species. **Range:** Can-

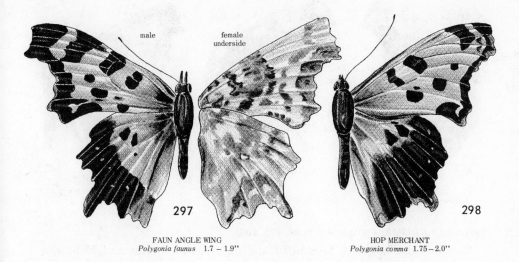

male   female
underside

297   298

FAUN ANGLE WING   HOP MERCHANT
Polygonia faunus 1.7 – 1.9″   *Polygonia comma* 1.75 – 2.0″

ada and the western half of the U.S. **Wings:** Brownish, with dark spots and border; dark border very narrow or almost absent on the hindwing. **Spread:** 1.75–2″. **Larva:** Black; wide greenish white stripe down the back, with a black V at each segment. Seven rows of spines, some rows green, others black. **Food:** Nettle, possibly elm.

(85) ZEPHYR ANGLE WING: *Polygonia zephyrus.* **Fig. 296.**

Inhabits mountainous regions. **Range:** Rocky Mts. and westward. **Wings:** Similar to *satyrus* (84) above. More yellowish, black markings not as heavy. Distinguished by pale gray underside. Replaces *satyrus* at higher altitudes. **Spread:** 1.75″. **Food:** Azalea, currant, gooseberry.

(86) FAUN ANGLE WING: *Polygonia faunus.* **Fig. 297.**

Also called the green comma. Found mostly at moderate elevations, in forested mountainous areas. **Range:** Canada, south to Georgia, and westward. **Wings:** Brownish yellow, with slightly greenish sheen. Heavy black borders, deeply indented hindwings. Underside mottled brown, with two submarginal rows of green spots. **Spread:** 1.7–1.9″.

(87) HOP MERCHANT: *Polygonia comma.* **Fig. 298.**

Sometimes called the comma. **Range:** Eastern Canada, south to Carolina, west to the Great Plains. **Wings:** Brownish yellow, with black markings and margins. The "comma line" on underside of hindwing is enlarged or hooked at both ends. In the summer form (*dryas*) black is diffused over most of the hindwing. **Spread:** 1.75–2″. **Larva:** Variable, from brown to almost white, with numerous branched spines (one pair on the head). Two or three broods. **Food:** Nettle, elm, hops. Sometimes a pest.

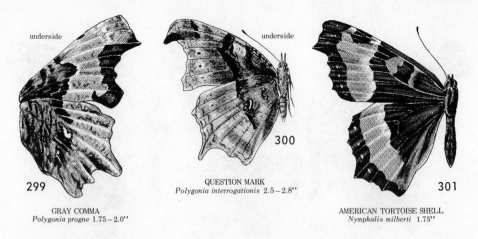

underside

underside

300

299

QUESTION MARK
Polygonia interrogationis 2.5 – 2.8"

301

GRAY COMMA
Polygonia progne 1.75 – 2.0"

AMERICAN TORTOISE SHELL
Nymphalis milberti 1.75"

(88) GRAY COMMA: *Polygonia progne.* Fig. 299.

Range: Canada, south to South Carolina, west to Kansas and Nebraska. **Wings:** Brownish yellow, with black markings and margins. Yellow along inner edge of black margin on the forewing. Broad dark margin of hindwing broken by yellow spots. Underside dark and light shades of gray. **Spread:** 1.75–2". **Food:** Gooseberry, currant. *P. gracilis* is darker on upper side, lighter gray on underside; the "comma line" is angular, not hooked or enlarged at either end.

(89) QUESTION MARK: *Polygonia interrogationis.* Fig. 300.

Range: Southern Canada; the entire U.S. east of the Rocky Mts., south into Mexico. **Wings:** This species differs from others in the genus by its large size, the more uniform wing margins, and the dark spot in cell M_2 of the forewing; also the "comma" in this case consists of a silvery curved line with a separate dot forming a "question mark." **Spread:** 2.5–2.8". **Larva:** Reddish brown, with lighter markings and numerous branched spines (one pair on the head). Two to five broods. **Food:** Nettles, elm, hackberry, hops, false nettle, basswood.

(90) MILBERT'S OR AMERICAN TORTOISE SHELL: *Nymphalis milberti.* Fig. 301.

The TORTOISE SHELLS resemble their close relatives, the angle wings (84). The margins of the wings are smoothly notched; the outer margin of the forewing is not deeply concave as in *Polygonia.* They are medium in size, black or brown, with black, orange or yellow markings. Like the angle wings they hibernate as adults. Like other tortoise shells *milberti* is largely confined to the northern regions or to the mountains farther south. **Range:** Canada, south to Virginia, Colorado, and California. **Wings:** Brown, with broad submarginal band of yellow shading into red toward the margin; black margin, with row of blue submarginal crescents. **Spread:** 1.75". **Larva:** Black above, yellow below, with yellow lateral line and an orange one above this; nu-

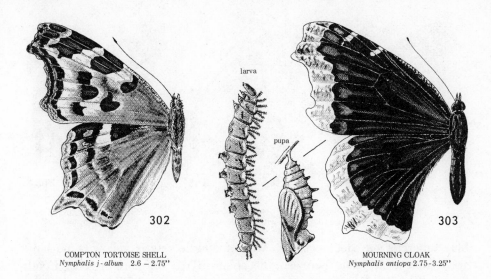

larva

pupa

302

303

COMPTON TORTOISE SHELL
Nymphalis j-album 2.6 – 2.75"

MOURNING CLOAK
Nymphalis antiopa 2.75-3.25"

merous white dots, hairs and branched spines. Three broods. Hibernates as adult or pupa. The larvae feed in colonies. **Food:** Nettle, sometimes willow and sunflower.

(91) COMPTON TORTOISE SHELL: *Nymphalis j-album (vau-album).* **Fig. 302.**

Range: Alaska, Canada, south to the Carolinas, west to Colorado. **Wings:** Yellowish brown to orange-brown, with black markings and white patch along costal margin near apex of each wing. They mimic dead leaves with their wings folded, exposing marbled gray underside. **Spread:** 2.6–2.75". **Food:** Willow, poplar, white birch. Never numerous, populations fluctuate widely from year to year.

(92) MOURNING CLOAK: *Nymphalis antiopa.* **Fig. 303.**

One of the best-known and most beautiful butterflies. In some places it goes by the name "spiny elm caterpillar." **Range:** Distributed widely over Canada and the U.S., temperate zone of Europe and Asia; common in California. It is called the "Camberwell beauty" in Great Britain, where it occasionally migrates from the Continent. **Wings:** Velvety, purplish brown, with wide yellow margin and submarginal row of blue spots. **Spread:** 2.75–3.25". **Larva:** Black, with the usual rows of branched spines characteristic of the Nymphalidae. Spotted with tiny white dots, rows of red spots on back. Two broods. Hibernates as adult. **Food:** Willow, poplar, elm.

(93) CALIFORNIA TORTOISE SHELL: *Nymphalis californica.* **Fig. 304.**

Occurs in the western mountains. **Range:** Western states, west of the Rocky Mts. **Wings:** Brownish yellow, with yellow mottling, black borders and spots. Single black spot and row of purple marginal spots on hindwing. **Spread:** 1.6–2.2". **Larva:** Black, with five branched spines rising from

CALIFORNIA TORTOISE SHELL
Nymphalis californica 1.6–2.2"

WHITE PEACOCK
Anartia jatrophae 1.75 – 2.0"

PAINTED LADY
Vanessa cardui 2.0–2.25"

blue tubercles on each segment. Area of middle row is yellow. Fine hairs arise from white dots between the spines. Two or more broods. **Food:** Snowbush, and other species of *Ceanothus*. The larvae will completely defoliate large areas at times, giving rise to massive swarming and migration of the adults when they emerge.

(94) BUCKEYE: *Junonia (Precis) coenia (lavinia)* (Vanessinae). **Color plate 5f.**

This striking species and the one following are referred to as PEACOCKS. A pugnacious butterfly, the male buckeye quickly pounces on others passing by. **Range:** From Colombia in South America to the southern part of Canada. **Wings:** Grayish brown to darker shades; patch of lighter shade on forewing, pale oval area along inner margin of hindwing. Eyespots ringed with reddish orange. Pair of black-bordered orange markings on forewing near costal margin and base. **Spread:** 2.25–2.5". **Larva:** Gray-green, with pale yellow longitudinal stripes, and branched spines (two on the head). Two or more broods. Hibernates as adult. **Food:** Plantain, stone-crop, snapdragon, monkey flower.

(95) WHITE PEACOCK: *Anartia jatrophae.* **Fig. 305.**

A delicate tropical species. **Range:** Tropical America, Florida, Texas. **Wings:** White, with pinkish tints, yellowish orange margins, and grayish orange markings along costal margin of forewing. **Spread:** 1.75–2".

(96) PAINTED LADY: *Vanessa cardui.* **Fig. 306.**

This genus, collectively known as THISTLE BUTTERFLIES, includes the most universal of all species—the painted lady. **Range:** *V. cardui* is found practically everywhere that thistle grows. **Wings:** Brownish yellow, with a rosy tint and black markings; black postmedian spots on hindwing, white spots on apex of forewing; spots (with pupils) on hindwing repeated on underside. **Spread:** 2–2.25". **Larva:** Pink, with yellow lateral stripes between segments, scattered black spots, seven rows of tubercles with spines; head

| 307 | 308 | 309 |

WEST COAST LADY
Vanessa carye 2.0''

PAINTED BEAUTY
Vanessa virginiensis 2.0''

RED ADMIRAL
Vanessa atalanta 1.75–2.0''

black, covered with hairs. Two or more broods. Usually hibernates as adult, sometimes as pupa. **Food:** Nettle, thistle, mallow, borage, burdock, hollyhock, cocklebur. In grazing country, where it feeds primarily on thistle (as in Montana, for example), it is considered beneficial, but in some areas it has taken to a host of cultivated plants; in California it is now considered a pest of cotton. For migratory habits, see introduction to this family (Nymphalidae). (Proposed genus: *Cynthia.*)

(97) WEST COAST LADY: *Vanessa carye.* **Fig. 307.**

The most common species of this group on the West Coast. **Range:** From Argentina to British Columbia, east to Utah and Colorado. **Wings:** Similar to *cardui* (96). Background coloring darker, without the rosy tint; a complete black band crosses middle of discal cell of forewing. The spots on the hindwing are repeated on the underside, as they are in *cardui.* **Spread:** 2''. **Larva:** Black, with orange blotches and black spines, and scattered white hairs. Several broods. Hibernates as adult or pupa. **Food:** Various species of mallow. (Proposed genus: *Cynthia.*)

(98) PAINTED BEAUTY: *Vanessa virginiensis.* **Fig. 308.**

Also called American painted lady and Virginia lady. **Range:** Central America to southern Canada. Less common in the Far West than in the East. **Wings:** Similar to *carye* (97) and *cardui* (96). May be distinguished readily by the two large eyespots on the underside of the hindwing. The other species, as noted above, have all the spots repeated on the underside of the hindwing. **Spread:** 2''. **Larva:** Similar to *cardui* except for the row of silvery spots on each side, on the posterior half of the body. Two or more broods. Hibernates as adult or pupa. **Food:** Cudweed, everlasting, mugwort, burdock. (Proposed genus: *Cynthia.*)

(99) RED ADMIRAL: *Vanessa atalanta.* **Fig. 309.**

Not to be confused with the "admirals" in the next group (*Limenitis*).

The older name "alderman" was suggested to the British years ago by similarities to the aldermanic robes worn as a mark of this office. **Range:** World-wide. **Wings:** This species is easily identified by the rich black background, the bright orange band and white apical spots on the fore-wing, the orange margins of the hindwing, and the blue-tipped apex. Two or more broods. Hibernates as adult or pupa. **Spread:** 1.75–2″. **Food:** Hops, nettle, false nettle, pellitory.

(100) VICEROY: *Limenitis archippus* (Limenitidinae). **Color plate 2b.**

This genus is commonly referred to as VICEROYS and ADMIRALS and includes several mimics. Their flight is marked by quick wing beats followed by a glide. Eggs are laid singly on the tips of leaves. The larvae are humped, greenish or brown, with patches of various colors. The head has two lobes with short projections, and behind it are two clublike tubercles. Fleshy projections appear on the fifth, ninth, and tenth segments. Hibernation takes place in a tube formed by folding the uneaten portion of a leaf remaining on both sides of the rib and lining it with silk; the stem is secured to the twig with silk to keep the leaf from falling off. The viceroy is also called the "mimic" because of its close resemblance to the monarch (43). They may be distinguished in flight by the way they glide; the viceroy holds the wings straight out to the sides (as do the admirals), while the monarch holds them at a slight upward angle. **Range:** Canada to Mexico, east of the Rocky Mts.; southwestern states. **Wings:** Reddish brown, with black veins. **Spread:** 2.5–2.8″. **Larva:** See above. Two or more broods. **Food:** Poplar, aspen, willow, apple, plum, cherry. *L. a. floridensis* (Color plate 3b) occurs in the Southeast.

(101) WHITE ADMIRAL: *Limenitis arthemis.* **Fig. 310, Color plate 2f.**

Also called banded purple. **Range:** Wooded areas of Canada, Great Lakes region (Minnesota, Wisconsin, northern Michigan, Ontario), higher elevations of New England, New York, and Pennsylvania. **Wings:** Black, with broad white median band, and white and blue crescents along margins; submarginal row of red spots with blue borders on hindwing. **Spread:** 2.5″. **Larva:** Olive-brown, with white patches. One to two broods. Hibernates as larva in early instar. **Food:** Willow, poplar, aspen, hawthorn, birch.

(102) RED-SPOTTED PURPLE: *Limenitis astyanax.* **Fig. 311, Color plate 2e.**

Starts where white admiral drops off, taking to the lower elevations and going southward. **Range:** Southern Canada; the U.S. east of the Rocky Mts. **Wings:** Purplish, with white marginal crescents. Submarginal row of red spots on underside. **Spread:** 3.1–3.4″. Similar to *arthemis* (101). Two or three broods. **Food:** Willow, poplar, linden, currant, huckleberry, apple, quince, plum, cherry. It is very pugnacious, will chase after any small object thrown into the air.

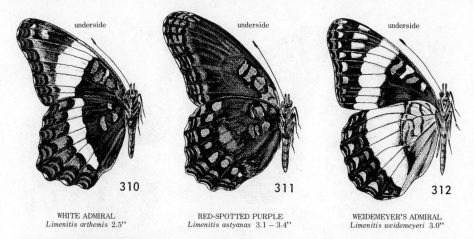

underside	underside	underside
310	311	312

WHITE ADMIRAL	RED-SPOTTED PURPLE	WEIDEMEYER'S ADMIRAL
Limenitis arthemis 2.5"	*Limenitis astyanax* 3.1 – 3.4"	*Limenitis weidemeyeri* 3.0"

(103) WEIDEMEYER'S ADMIRAL: *Limenitis weidemeyeri.* **Fig. 312, Color plate 2c.**

Confined largely to the Rocky Mts. **Range:** Mostly Colorado, also found in Mono Lake area of California, on eastern side of the Sierra Nevada Mts. **Wings:** Resembles white admiral superficially. Easily distinguished by white marginal spots and absence of crescents on the forewing. **Spread:** 3". **Larva:** Similar to other admirals. **Food:** Willow, aspen, cottonwood, poplar.

(104) LORQUIN'S ADMIRAL: *Limenitis lorquini.* **Fig. 313, Color plate 2d.**

Western species, found in moist wooded areas and along streams. **Range:** West Coast from California to British Columbia, east to Colorado. **Wings:** Easily identified by orange-red apical patch on forewing and yellowish tint of band across both wings. **Spread:** 2.25–2.75". **Food:** Willow, aspen, cottonwood, chokecherry.

(105) CALIFORNIA SISTER: *Limenitis bredowii californica.* **Fig. 314.**

Range: Nevada, California, Arizona, New Mexico, western Texas, Mexico. **Wings:** Resembles Lorquin's admiral (104). May be readily distinguished by the blue marginal band along the margins and the blue markings toward the base of the hindwings. **Spread:** 2.5–3". **Food:** Oaks.

(106) RUDDY DAGGER WING: *Marpesia petreus* (Eurytelinae). **Color plate 4h.**

The DAGGER WINGS are readily identified by the extended apex of the forewing and the long tail-like extension of the third median vein (M_3) of the hindwing and a shorter tail at the apex. The discal cell is open to the margin in both the forewing and the hindwing. They are mostly tropical. **Range:** This species (*petreus*) is found in southern Florida and Texas. **Wings:** Orange-brown, with black or dark brown margins and markings. **Spread:** 2.6–2.8". **Larva:** Brown on upper surface, yellow on underside, with black spots on the sides and slanted black and white lines toward the posterior end. Head yellow, with black spot, two black streaks, and a pair of horns. Two or three broods. **Food:** Figs.

underside

313

LORQUIN'S ADMIRAL
Limenitis lorquini 2.25 – 2.75"

CALIFORNIA SISTER
Limenitis bredowii californica 2.5-3.0"

314

(107) FLORIDA PURPLE WING: *Eunica tatila.* **Fig. 315.**

The PURPLE WINGS (or violet wings) are tropical, common in the shady hardwood forests—the hummocks—of the tropics. **Range:** This species (*tatila*) is found in southern Florida. **Wings:** Dark brown or black, with purple iridescence, large white spots on forewing. **Spread:** 1.2–1.4".

(108) GOATWEED: *Anaea andria* (Charaxinae). **Color plate 5d.**

The LEAF WINGS are mostly tropical, and are named for their mimicry of dead leaves when resting with wings folded. They are characterized by the concave outer margin and pointed apex of the forewing, the tail at the end of the third median vein (M_3) of the hindwing, and their beautiful orange-red or orange-brown coloring. **Range:** This species (*andria*) is found in the Gulf states, north to Georgia, Tennessee, Ohio, Illinois, southern Michigan, Nebraska. **Wings:** Reddish orange, with black markings. The female is lighter than the male, with wide, irregular postmedian band of pale yellowish brown and darker margins. Similar to the **Florida Leaf Wing,** A. *aidea floridalis,* and to A. *portia,* which have slightly serrated margins and are more reddish. **Spread:** 2.5–3". **Larva:** Gray-green, tapering to a point at the anal end; head rounded, same color. Covered with many tiny, pointed tubercles. Later instars feed from folded leaf. Two broods. Hibernates as adult. **Food:** Goatweed. **Morrison's Leaf Wing,** A. *aidea morrisoni* (Color plate 5e), is found in Florida.

(109) HACKBERRY BUTTERFLY: *Asterocampa celtis* (Apaturinae). **Fig. 316.**

This genus comprises the HACKBERRY BUTTERFLIES or EMPERORS. They are noted for their swift, dodging flight and the habit of alighting on tree trunks or occasionally on people. The forewings are moderately concave at the outer margins. In the male the forewing is narrower and more pointed

315

FLORIDA
PURPLE WING
Eunica tatila
1.2 – 1.4"

316

HACKBERRY BUTTERFLY
Asterocampa celtis 1.9 – 2.4"

317

SOUTHERN SNOUT BUTTERFLY
Libythea carinenta 1.8"

at the apex than in the female. Background coloring is varying shades of brownish yellow, with black margins and markings, and white spots. Eyespots are a conspicuous feature of the hindwing and appear in some species on the forewing. The larvae are yellow, green and white striped. Like the satyrs they are tapered from the middle toward both ends, and the anal end is forked. A pair of heavy, branched horns protrude conspicuously from the head. These butterflies are not found along the Pacific coast. *A. celtis* is found wherever hackberry trees grow. **Range:** North Central states, southern New England, New York, Pennsylvania, and southward. **Wings:** Brownish yellow, with six red-ringed eyespots on the hindwing and one on the forewing; female lighter than the male. Underside purplish gray, with the spots repeated. **Spread:** 1.9–2.4". **Larva:** See above. One or more broods. Hibernates as larva. **Food:** Hackberry. The **Mountain Emperor,** *A. c. montis,* is very similar except that it has two eyespots on the forewing, the front one not being as well developed as the other; its range is the mountainous areas of Colorado and Arizona. **Alicia,** *A. c. alicia,* is paler or fawn-colored, with darker and heavier margins, darker eyespots (one on the forewing); its range is the Gulf states.

(110) TAWNY EMPEROR: *Asterocampa clyton.* **Color plate 5i, j.**

Range: Southern New England to Nebraska, southward. **Wings:** Dark brown, all but obscuring the eyespots on the hindwing. Forewing lacks eyespots. Female much larger and paler than the male, with eyespots on hindwing very conspicuous. **Spread:** 2.5–2.6". **Larva:** See *celtis* (109). **Food:** Hackberry.

(111) LEILIA: *Asterocampa leilia.* **Color plate 5h.**

Range: Southern Texas, New Mexico, Arizona, Mexico. **Wings:** Similar to *celtis,* but larger, with two eyespots on the forewing. **Spread:** 2.15–2.65".

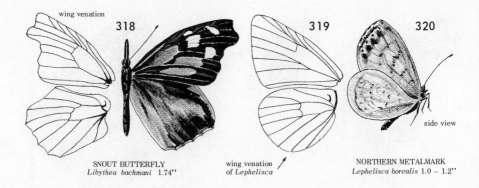

wing venation

318

319

320

side view

SNOUT BUTTERFLY
Libythea bachmani 1.74"

wing venation
of *Lephelisca*

NORTHERN METALMARK
Lephelisca borealis 1.0 – 1.2"

Snout Butterflies (Libytheidae)

The smallest family of butterflies, this group is easily recognized by the conspicuous projection of the labial palpi; the last joint is extended and the pair come together to form the "snout." The front pair of legs are reduced in the male, as they are in the metalmarks (the next group); all three pairs are normal and functional in the female. The eggs are ovoid, with vertical ridges thickening as they approach the apex; every other ridge is higher than the next one, which does not run the full length. The larva is humped, being greatly enlarged at the anterior end and having a small head. The chrysalis is usually green, with yellow lines across the abdomen, wedge-shaped and pointed toward the head, and ridged on both sides. Snout butterflies sometimes migrate in great swarms.

(112) SNOUT BUTTERFLY: *Libythea (Libytheana) bachmani.* **Fig. 318.**

There is only one genus of this butterfly in North America north of Mexico, and only one species in the eastern part of the country. **Range:** Southern Ontario, New England, south and west to southern California. **Wings:** Brown, darker or almost black in the marginal and apical areas, with orange patches on both wings, white spots near the apex of the forewing. The costal margin of the hindwing is concave. A similar species, *L. carinenta* (Fig. 317), paler in coloration, occurs in the Southwest. **Spread:** 1.74". **Larva:** Velvety green, with yellow stripes. The thorax is humped and has two black, fleshy projections with a yellow ring. Three or more broods. Hibernates as pupa. **Food:** Hackberry.

Butterflies: Metalmarks (Riodinidae)

In some classifications the metalmarks are placed with the next family, the Lycaenidae. Not many species of the smaller group occur in our area; many, brilliantly colored, are found in the tropics. In contrast, the coloring of our specimens is subdued. The name comes from the metallic appearance

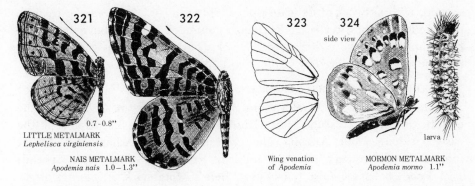

321   322   323   324

side view

0.7-0.8''

LITTLE METALMARK
Lephelisca virginiensis

NAIS METALMARK
Apodemia nais 1.0−1.3''

Wing venation
of *Apodemia*

MORMON METALMARK
Apodemia mormo 1.1''

larva

of the spots and other markings which often have a steel-blue sheen. As in the snout butterflies, the front legs of the males are greatly reduced but those of the females are functional. In some genera a branch of the precostal vein at the base of the hindwing projects out along the inner margin, forming a short frenulum or "hook," as occurs in moths. Small, and weak fliers, they often come to rest on the underside of leaves and spread their wings out flat in the manner of moths. Their antennae are generally long and slender. Eggs are turban-shaped, with many short spines or network of fine ridges. The larvae are comparatively short and broad, covered with many hairs. The chrysalises are short and fat, slightly curved and flattened on one side, hairy on the other, usually speckled brown in color, with numerous projections and two orange spots on the thorax; they are supported by a cremaster and silk girdle but hang with the head down.

(113) NORTHERN METALMARK: *Lephelisca (Calephelis) borealis.* **Figs. 319, 320.**

Found in dry hilly or partially wooded areas. **Range:** New England, south to Virginia, west to Ohio. **Wings:** Dark brown, with black spots and streaks, marginal and submarginal spots. **Spread:** 1–1.2''. **Food:** Groundsel (ragwort).

(114) LITTLE METALMARK: *Lephelisca (Calephelis) virginiensis.* **Fig. 321.**

Common in the South, found in moist meadows and grassy fields. **Range:** Florida to Virginia, Gulf coast to Louisiana and probably Texas, north to Ohio. **Wings:** Light orange-brown dorsally, underside more reddish, fringes not checkered; both sides have steel-blue metallic spots arranged in rows. **Spread:** .7–.8''. Believed to have three broods.

(115) NAIS METALMARK: *Apodemia nais.* **Fig. 322.**

Range: From central Colorado, on both sides of the Rocky Mts., south into Mexico. **Wings:** Coppery red, with grayish brown dusting, dark brown and white markings, fringes checkered; forewing has white spot near costal margin and apex; underside of hindwing has coppery submarginal band with row of black spots along the outer edge. Eggs are green, turban-shaped, with network of delicate ridges. **Spread:** 1–1.3″. **Larva:** Gray, with pale yellow stripes, tufted hairs. Thorax humped, rising above the head, body tapered toward posterior end. Blackish brown chrysalis suspended by posterior end and secured by girdle. **Food:** Wild plum.

(116) MORMON METALMARK: *Apodemia mormo.* **Figs. 323, 324.**

Found mostly in higher elevations. **Range:** Washington, Oregon, on the eastern side of the Cascade Mts.; California, in the High Sierras and on the eastern slopes; Utah, Arizona, New Mexico. **Wings:** Ashen gray, except inner area of forewing, which is reddish. Numerous white spots. **Spread:** 1.1″. **Food:** Saltbush, false buckwheat. **Behr's Metalmark**, *A. mormo virgulati,* is similar except that the ground color is reddish brown. It is the most abundant of the metalmarks in California, ranging from the Sacramento Valley southward into Mexico.

Gossamer-winged Butterflies: Hairstreaks, Harvesters, Coppers, Blues (Lycaenidae)

This is a large, world-wide group consisting of four subfamilies: Theclinae (hairstreaks), Gerydinae (harvesters), Lycaeninae (coppers), and Plebejinae (blues). They are mostly small butterflies (including one of the smallest), with three pairs of legs adapted for walking, although the front pair in the males is slightly reduced and lacks the tarsal claws and pads. The term "gossamer-winged" is more imaginative (and poetic perhaps) than descriptive. The colors of the wings are iridescent purple, blue, brown, green, or copper. John Adams Comstock has described the pretty little blues as "lustrous azure, like bits of summer sky that have dropped to earth." Most of the hairstreaks have one or two short, slender tails on the hindwing (Fig. 325); the "hairstreaks" are irregular lines on the underside of the wings. Some species have a habit of rubbing the hindwings back and fourth when folded upward and at rest.

The eggs of the Lycaenidae are turban-shaped, with many minute prominences. The caterpillars are short and broad, flattened at the posterior end, and covered with short, fine hairs. Dorsally they resemble the mossy sea chitons somewhat. The harvesters feed on aphids, and some hairstreaks and blues secrete honeydew which is attractive to ants; the latter in return give them protection from insect predators and parasites. The chrysalises are short, flat on one side, curved on the other, and covered with short hairs. They are fastened by a cremaster and silk girdle.

(117) GREAT PURPLE HAIRSTREAK: *Atlides halesus* (Theclinae). **Fig. 326.**

Mostly tropical, probably the flashiest of the butterflies coming within our range. **Range:** Gulf states, north to New Jersey and Illinois, west to

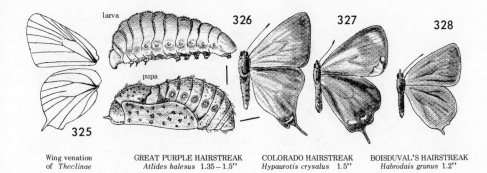

| Wing venation of *Theclinae* | GREAT PURPLE HAIRSTREAK *Atlides halesus* 1.35–1.5" | COLORADO HAIRSTREAK *Hypaurotis crysalus* 1.5" | BOISDUVAL'S HAIRSTREAK *Habrodais grunus* 1.2" |

California. **Wings:** Iridescent purple, bluish green toward the base, with black borders; black spot on forewing, greenish spots at anal angle of hindwing. Underside plain brownish gray, with green, blue, and red spots at the anal angle of the hindwing. **Spread:** 1.35–1.5". **Larva:** Green, with short, fine hairs giving a velvety appearance. Two broods. **Food:** Mistletoe growing on live oak, possibly the oak and some associated plants.

(118) COLORADO HAIRSTREAK: *Hypaurotis crysalus.* **Fig. 327.**

Very colorful, similar to *A. halesus* (117). **Range:** Colorado, Utah, Arizona, California (uncommon). **Wings:** Deep, vivid reddish purple, with black margins, submarginal orange spots at anal angle of forewing and hindwing. Underside pale yellowish brown, with white streaks edged with brown; submarginal spots repeated as red and black eyespots with green margins. **Spread:** 1.5".

(119) BOISDUVAL'S HAIRSTREAK: *Habrodais grunus.* **Fig. 328.**

Unusual in its crepuscular habit. John Garth and J. W. Tilden report it as flying in the semidarkness of dawn and hovering "above the live oaks after the last rays of sunlight have left the canyon walls" of Yosemite. **Range:** Nevada, California. They sometimes swarm in the foothills of the San Bernardino Mts. **Wings:** Brown, sometimes with orange tinge in the female. Underside pale yellowish brown, with series of small marginal and submarginal spots. **Spread:** 1.2". **Food:** Live oak.

(120) COTTON SQUARE BORER: *Strymon melinus.* **Fig. 329.**

Also called the common hairstreak or gray hairstreak. **Range:** Common throughout North America. **Wings:** Slate or blue-gray, with two large black spots and large and smaller reddish orange spots above these at the anal angle of the hindwing. Lighter gray on underside, with white streak edged in black; spots at anal angle repeated on the underside; short and long tail. Rubs hindwings back and forth when resting. Pale green eggs laid singly on plants. **Spread:** 1.1–1.2". **Larva:** Velvety, bright green, with very fine, short yellowish hairs and small hypognathous head, about .5" long

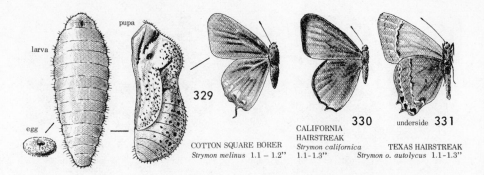

pupa

larva

329

egg

330 CALIFORNIA
HAIRSTREAK

underside 331

COTTON SQUARE BORER
Strymon melinus 1.1 – 1.2"

Strymon californica
1.1-1.3"

TEXAS HAIRSTREAK
Strymon o. autolycus 1.1-1.3"

full grown; succession of oblique lines on body segments. Color and markings variable. Two or more broods. Succession of broods throughout the year in warm climates. **Food:** Hops, mallow, knotweed, hawthorn, St.-John's-wort, cotton, lima beans and other cultivated field crops; apple, lemon. A minor pest of cotton; the young larvae feed on the leaves, and as they mature they attack the squares and bolls, one larva sometimes destroying twenty to thirty squares.

(121) CALIFORNIA HAIRSTREAK: *Strymon (Satyrium) californica.* **Fig. 330.**

Inhabitant of the western mountains and foothills. **Range:** British Columbia to California, east to Colorado. **Wings:** Brown, reddish toward the outer margins. Underside is gray, with three or four submarginal red spots on the hindwing near the anal angle. **Spread:** 1.1–1.3". **Food:** Buckbrush, mountain mahogany, possibly oak. The **Sylvan Hairstreak,** *S. sylvinus,* is similar, occurs in the canyons where willows grow, has lighter ground color on the underside and a single red spot near the anal angle.

(122) TEXAS HAIRSTREAK: *Strymon (Euristrymon) ontario autolycus.* **Fig. 331.**

Range: Texas, Oklahoma, Missouri, Kansas. **Wings:** Brown, with orange-red patches on forewing, and orange-red spots at anal angle of hindwing. Underside of forewing has submarginal and median white streak edged with brown; streaks continue on hindwing, with bright red marginal spots near anal angle. **Spread:** 1.1–1.3". **Food:** Oak, hawthorn.

(123) SOUTHERN HAIRSTREAK: *Strymon (Euristrymon) favonius.* **Fig. 332.**

Range: Florida, Georgia; occasionally strays to West Virginia and New Jersey. **Wings:** Similar to *S. ontario* (122). The orange patches on the forewing are smaller, and the marginal red spots on the underside of the hindwing between the margin and submarginal streak are replaced by a broad, red band. **Spread:** 1.1". **Food:** Oaks.

(124) ALCESTIS HAIRSTREAK: *Strymon (Phaeostrymon) alcestis.* **Fig. 333.**

Range: Texas, Oklahoma, Kansas, Arizona. **Wings:** Plain brownish gray dorsally. Underside somewhat reddish, with white streaks crossing both

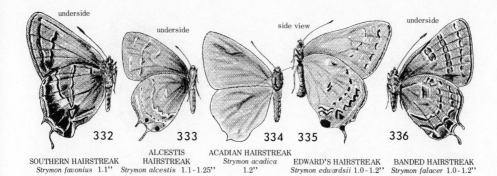

332	333	334 335	336
SOUTHERN HAIRSTREAK	ALCESTIS HAIRSTREAK	ACADIAN HAIRSTREAK *Strymon acadica*	EDWARD'S HAIRSTREAK	BANDED HAIRSTREAK

SOUTHERN HAIRSTREAK *Strymon favonius* 1.1"   ALCESTIS HAIRSTREAK *Strymon alcestis* 1.1-1.25"   ACADIAN HAIRSTREAK *Strymon acadica* 1.2"   EDWARD'S HAIRSTREAK *Strymon edwardsii* 1.0-1.2"   BANDED HAIRSTREAK *Strymon falacer* 1.0-1.2"

wings; black spot and blue patch at anal angle of hindwing, another black spot just above this, with submarginal row of reddish orange crescents edged on the inside in black. **Spread:** 1.1–1.25". **Food:** Chinaberry.

(125) ACADIAN HAIRSTREAK: *Strymon (Satyrium) acadica.* **Fig. 334.**

Range: A narrow strip from Nova Scotia to New Jersey, west to the Rocky Mts. **Wings:** Slate gray, with orange spot at anal angle of hindwing. Female similar except that the orange spot is larger. Underside pale brown, with postmedian line broken into black, distinctly separated spots ringed with white; row of reddish orange submarginal crescents. **Spread:** 1.2". **Food:** Willow.

(126) EDWARD'S HAIRSTREAK: *Strymon (Satyrium) edwardsii.* **Fig. 335.**

Fast-flying, pugnacious. **Range:** Eastern Canada to Manitoba; New England to Georgia, west to Colorado. **Wings:** Grayish brown. Underside paler, with postmedian line broken into black oval spots ringed with white; the spots are not well separated as in *acadica* (125), with which it is easily confused. Blue-green patch near the anal angle of the hindwing, with row of submarginal reddish orange crescents similar to *alcestis* (124) and *acadica.* **Spread:** 1–1.2". **Food:** Oak.

(127) BANDED HAIRSTREAK: *Strymon (Satyrium) falacer (calanus).* **Fig. 336.**

Quite common and widespread. **Range:** Eastern Canada to Manitoba, south to Georgia, Texas, Colorado. **Wings:** Similar to *edwardsii* (126). Can be distinguished by solid dark line edged with white in place of separate spots. **Spread:** 1–1.2". **Larva:** Variable. Green or brown, with contrasting stripes and oblique lines across body segments. One brood. Hibernates in egg stage. **Food:** Oak, butternut.

(128) STRIPED HAIRSTREAK: *Strymon (Satyrium) liparops.* **Fig. 337.**

Wide-ranging but not numerous. **Range:** Eastern Canada to Manitoba; New England, south to Georgia, west to Kansas, Colorado. **Wings:** Similar to *edwardsii* (126) but easily distinguished by white broken lines widely

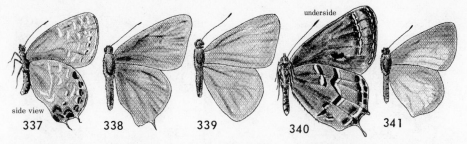

underside

side view

337   338   339   340   341

STRIPED HAIRSTREAK	HEDGEROW HAIRSTREAK	CORAL HAIRSTREAK	OLIVE HAIRSTREAK	BEHR'S HAIRSTREAK
Strymon liparops 1.2"	*Strymon saepium* 1.2"	*Strymon titus* 1.3"	*Mitoura gryneus* 0.9 – 1.0"	*Callipsyche behrii* 1.1"

separated with lighter bands between. **Spread:** 1.2". **Food:** Oak, willow, wild apple, plum, blackberry, blueberry, shadbush.

(129) HEDGEROW HAIRSTREAK: *Strymon (Satyrium) saepium.* **Fig. 338.**

Found in mountainous areas. **Range:** West Coast to the Rocky Mts. Common in California. **Wings:** Distinguished by dark reddish brown (sepia) coloring. Underside lighter, with series of narrow dark lines edged on both sides with fine white lines. Black spots and blue patch near anal angle of hindwing, submarginal orange crescents. **Spread:** 1.2". **Food:** Buckbrush, mountain mahogany.

(130) CORAL HAIRSTREAK: *Strymon (Chrysophanus) titus.* **Fig. 339.**

The only common one of this group without tails on hindwing. **Range:** Ontario to the Gulf states and westward. **Wings:** Grayish brown. Some females have red spots near anal angle of hindwing. Easily distinguished by the submarginal band of coral red spots on the underside of hindwing. **Spread:** 1.3". **Larva:** Green, with numerous fine, short brown hairs giving downy appearance. Rosy patches on anterior and posterior segments. Hibernates in egg stage. One brood. **Food:** Wild cherry and plum, oak, mistflower.

(131) OLIVE HAIRSTREAK: *Mitoura (Callophrys) gryneus.* **Fig. 340.**

Widespread but often scarce. **Range:** Eastern half of the U.S.; Ontario. **Wings:** Brownish yellow, with the marginal and apical areas darker or nearly black. Underside yellowish green (olive); brown lines edged with white, two short white lines near base of hindwing; orange submarginal crescents. **Spread:** .9–1". Two to three broods. **Food:** Red cedar.

(132) BEHR'S HAIRSTREAK: *Callipsyche (Satyrium) behrii.* **Fig. 341.**

Western mountain species. **Range:** Scattered colonies in southern British Columbia, Washington; northern California, Wyoming to New Mexico. **Wings:** Yellowish brown, with darkened outer margins and apical area. Underside paler, grayish toward margins, with the usual white streaks and

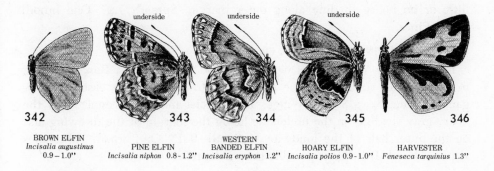

| 342 | 343 | 344 | 345 | 346 |

BROWN ELFIN
Incisalia augustinus
0.9 – 1.0"

PINE ELFIN
Incisalia niphon 0.8 - 1.2"

WESTERN
BANDED ELFIN
Incisalia eryphon 1.2"

HOARY ELFIN
Incisalia polios 0.9 - 1.0"

HARVESTER
Feneseca tarquinius 1.3"

submarginal spots. It is tailless. **Spread:** 1.1". **Food:** Lupines, locoweed, antelope bush.

(133) BROWN ELFIN: *Incisalia augustinus.* Fig. 342.

This genus is referred to as the "elfins." They do not have tails on the hindwings as do most of the hairstreaks; like them, they have a habit of rubbing the hindwings together when perched on a plant. Outer margins of the hindwings tend to be scalloped. **Range:** The brown elfin ranges through southern Canada, from Newfoundland to Manitoba, south to Virginia, Illinois and Michigan, New Mexico and California. **Wings:** Plain brown. Underside paler, orange-brown or yellowish brown toward the margins; broken median band on forewing, irregular curved band on hindwing; row of darker submarginal spots. **Spread:** .9–1". **Larva:** Green, with yellowish stripes on back and along sides, oblique stripes on the segments. One brood. Hibernates as pupa. **Food:** Sheep laurel, blueberries. The **Western Elfin,** *I. iroides,* is sometimes treated as a subspecies of *augustinus;* it has a pronounced lobe on the hindwing, and the larva feeds on sedum.

(134) PINE ELFIN: *Incisalia niphon.* Fig. 343.

Also called eastern pine elfin or banded elfin. **Range:** Eastern Canada to Florida, west to Manitoba and Colorado. **Wings:** Shiny dark brown dorsally, yellowish brown on underside. Outer margin of hindwing deeply scalloped. Underside crossed by cinnamon-brown bands, some of them bordered with white. Underside of forewing has two transverse bars in discal cell; underside of hindwing with narrow hoary shading on outer margin. **Spread:** .8–1.2". **Larva:** Body green, with two white stripes on each side, thin coating of short yellowish brown hairs; head yellow. One brood. Hibernates as pupa. **Food:** Pines.

(135) WESTERN BANDED ELFIN: *Incisalia eryphon.* Fig. 344.

Range: Rocky Mts., west to the Pacific coast. **Wings:** Larger than the pine elfin (134) but otherwise similar. Underside of forewing has only one distinct transverse bar at end of discal cell; underside of hindwing with little or no hoary shading along outer margin. **Spread:** 1.2″. One brood. **Food:** Pines.

(136) HOARY ELFIN: *Incisalia polios.* **Fig. 345.**

Found in dry, open places where the food plant thrives. **Range:** Alaska to Washington, south along the Rocky Mts. to New Mexico, east to Michigan, Virginia, Nova Scotia. **Wings:** This species may be recognized by the hoary gray shading on the underside along the margin of the forewing and on the outer half of the hindwing. **Spread:** .9–1″. One brood. Hibernates as pupa. **Food:** Bearberry, possibly ground laurel.

(137) HARVESTER: *Feneseca tarquinius* (Gerydinae). **Fig. 346.**

A predator in the larval form. **Range:** Canada, Nova Scotia to Ontario, south to the Gulf states. **Wings:** Orange-brown, with black spots and markings. Underside paler, with brown mottling; spots reappear as on upper surface. **Spread:** 1.3″. **Larva:** Brown, with green stripes. Unusually short larval period; only three molts, period ten or eleven days. Two broods. Hibernates as pupa, and possibly as adult. The pupa resembles a monkey's head. **Food:** Aphids found on alder, ash, witch hazel, beech, hawthorn, wild currant. The harvester larva feeds under a protective covering of aphid skins held together with silk web.

(138) BRONZE COPPER: *Lycaena thoe* (Lycaeninae). **Fig. 347.**

The coppers are generally small, stout-bodied, fast fliers. They are reddish brown or orange, with black margins and spots. The males are usually darker than the females, with iridescent purplish overcast. Their habitation is mostly in the northern temperate regions—in marshes and grassy, open spaces, less often in wooded areas. **Range:** The bronze copper ranges across Canada from Quebec to Saskatchewan, and in the U.S. from Maine to New Jersey, west to Colorado. **Wings:** The male is bronze, with black markings and spots, reddish orange submarginal band on hindwing. Background coloring of the female is orange. Underside of forewing in both sexes is more reddish, with gray in the apical area. **Spread:** 1.3–1.4″. **Larva:** Short, fat, tapered toward both ends; light green with dark stripe down the back. Two broods. Hibernates in egg stage. **Food:** Yellow dock.

(139) AMERICAN COPPER: *Lycaena phleas* (*phlaeas*). **Fig. 348.**

One of our most common butterflies, abundant in all but the southern and far western states. Very pugnacious; will attack larger butterflies and even birds and other animals. **Range:** Nova Scotia to Alberta, south to Georgia, west to Kansas; isolated colonies occur farther west. **Wings:** Forewing bright metallic copper, with darker margin and black spots; hindwing dark brown, with brownish yellow submarginal band indented. Early spring specimens are lighter in color, with smaller spots. **Spread:** 1″. One to four or possibly

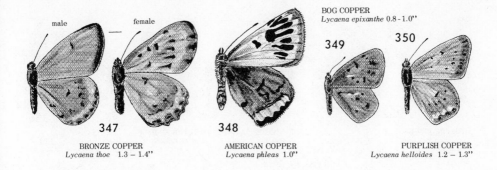

male female

BOG COPPER
Lycaena epixanthe 0.8 - 1.0''

349 350

347 348

BRONZE COPPER
Lycaena thoe   1.3 – 1.4''

AMERICAN COPPER
Lycaena phleas 1.0''

PURPLISH COPPER
Lycaena helloides 1.2 – 1.3''

more broods. Hibernates as pupa. **Food:** Sheep sorrel and closely related plants. This is considered a subspecies of the European *L. phleas* but differs little from it; the full name is *L. p. americana.* Various aberrant forms have been given subspecies names.

(140) BOG COPPER: *Lycaena epixanthe.* **Fig. 349.**

One of our smallest coppers, found in cranberry bogs. **Range:** Eastern Canada, northeastern U.S., west to Wisconsin and Manitoba. **Wings:** Brownish yellow with purplish overcast and black spots; the female is paler than the male. Orange submarginal crescents at least near anal angle of hindwing. Underside much paler and reddish. **Spread:** .8–1''. **Food:** Cranberry.

(141) PURPLISH COPPER: *Lycaena helloides.* **Fig. 350.**

Range: Western U.S. and Canada, east to northern Illinois, Michigan, and Ontario. Common along the Pacific coast. **Wings:** Brownish orange, with purplish iridescence and black markings, submarginal band of orange crescents on hindwing. Background coloring of the female is more orange. Underside of forewing pale red, hindwing with purplish cast and sub-marginal row of red crescents. **Spread:** 1.2–1.3''. **Larva:** Green, broken lines along the back and yellowish stripe on each side, a series of short oblique yellow lines on each body segment along the sides. Head small, pale amber; body covered with fine colorless hairs. Two broods in northern areas, six or seven in southern California. **Food:** Yellow dock, sheep sorrel, knotgrass, knotweed, baby's breath. The **Dorcas Copper,** *L. dorcas,* a northern species ranging southward to Michigan, Ohio, and Maine, is very similar. It is dark brown, purple tinged, with the submarginal band of crescents reduced, sometimes to thin lines.

(142) GREAT COPPER: *Lycaena xanthoides.* **Fig. 351.**

One of our largest coppers. **Range:** Upper Mississippi Valley and westward. Common in the foothills of California during summer. **Wings:** Shiny brownish gray, with orange patches in discal area and near angle of fore-

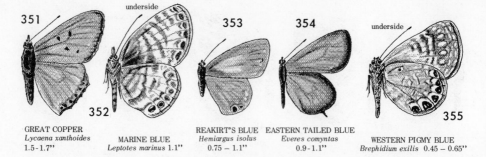

351

underside

353   354

underside

352

355

GREAT COPPER
Lycaena xanthoides
1.5–1.7"

MARINE BLUE
Leptotes marinus 1.1"

REAKIRT'S BLUE
Hemiargus isolus
0.75 – 1.1"

EASTERN TAILED BLUE
Everes comyntas
0.9–1.1"

WESTERN PIGMY BLUE
Brephidium exilis 0.45 – 0.65"

wing; row of submarginal orange crescents on hindwing. Underside creamy white, with black spots and orange patch near outer margin of hindwing. **Spread:** 1.5–1.7". One brood. **Food:** Bitter dock.

(143) MARINE BLUE: *Leptotes marinus* (Plebejinae). **Fig. 352.**

The blues are predominantly temperate in range, but there are many tropical species. They are small, generally blue on the upper side but sometimes brown or white. The underside is usually white or brown, marked with numerous lines and spots. The females are normally darker than the males, with wider dark borders, sometimes with little or no blue. The larvae of many blues are closely associated with ants, which protect them for their honeydew. The fluid is secreted by a gland visible in the last (or next to the last) body segment when present. **Range:** Western Illinois, Nebraska, west to Nevada, and California, south to Arizona, Texas, and Central America. **Wings:** Male pale blue, female brown, with a diffusion of blue toward the base; some of the underside markings show through. Underside white, with numerous wavy lines and series of pale blue, iridescent submarginal spots on hindwing. **Spread:** 1.1". **Larva:** Highly variable in color and markings. Varies from pale green, with little or no markings, to brown, with heavy chocolate markings. **Food:** Mainly buds and blossoms of wisteria, alfalfa, sweet pea, and other vetches.

(144) REAKIRT'S BLUE: *Hemiargus isolus.* **Fig. 353.**

Range: Most of the region west of the Mississippi River, eastward to Illinois, southern Michigan, western Ohio, Mississippi. **Wings:** Male pale blue; female brownish gray, with diffusion of blue toward the base. Black spot on outer margin of hindwing near anal angle. Underside pale grayish brown, with row of white-circled black spots across postmedian area of forewing. **Spread:** .75–1.1". **Food:** Rattleweed, locoweed, mesquite, clover.

(145) EASTERN TAILED BLUE: *Everes comyntas.* **Fig. 354.**

Very widespread and common species, the only one in the subfamily having hindwings with tails. **Range:** Southern Canada, practically all of the U.S., south into Central America. **Wings:** Male pale blue, with dusky margins; female all brown. Sharply defined markings on underside including bar across discal cell. In spring forms the females are darker brown, the males have narrower dark margins, undersides of both are lighter. Not all specimens have tails. **Spread:** .9–1.1″. **Larva:** Green, with darker stripe down the back, faint oblique stripes along sides. Black papillae, with short white hairs, cover the body; head dark brown. Two or more broods. Hibernates as full-grown larva. **Food:** Legumes. The **Western Tailed Blue,** *E. amyntula,* is very similar, found west of the Rocky Mts., from California to British Columbia and east to Michigan and Ontario. Markings on the underside are smaller and fewer, less distinct, than in *comyntas;* and bar across the discal cell is very faint or not visible. Some authorities do not consider this a separate species.

(146) WESTERN PIGMY BLUE: *Brephidium exilis.* **Fig. 355.**

This is the smallest of our butterflies. For this reason and because of its weak flight it is generally unnoticed. **Range:** Southern California, Texas, tropical America. Some strays reach Nebraska and Oregon. Very common in southern California. **Wings:** Light brown, with diffusion of blue toward the base, row of dark spots along outer margin of hindwing. Underside pale brown, with light spots and numerous short, white streaks. Row of shiny round spots on outer margin of hindwing. **Spread:** .45–.65″. **Larva:** Yellowish green, covered with brown tubercles having short clublike processes. The latter give it a frosty appearance and resemblance to the underside of leaves of the food plant. Markings variable; a yellow line may appear along the back and a bright yellow lateral line below the spiracles. Many broods. **Food:** Saltbush, lamb's-quarters, *Petunia.* The **Eastern Pigmy Blue,** *B. isophthalma pseudofea,* a resident of Georgia, Florida, and the Gulf coast, is very similar, feeds on glasswort; *exilis* may be distinguished by the white spot at inner angle of hindwing and the white fringe near apex.

(147) ORANGE MARGINED BLUE: *Lycaeides (Plebejus) melissa.* **Fig. 356.**

Also called Melissa blue. Congregates in swarms around wet places. **Range:** New England to North Carolina, west to Manitoba, Kansas, Colorado, California. **Wings:** Male pale blue; female brown or gray, with marginal row of orange crescents on both wings. Underside gray; outer margin of hindwing decorated with row of elongated orange spots, banded with shiny green toward the margin and with black opposite this. **Spread:** .9–1.15″. **Food:** Locoweed, mints, lupine, alfalfa and other legumes.

(148) GREENISH BLUE: *Plebejus saepiolus.* **Fig. 357.**

Also called Saepiolus blue. **Range:** Across central and southern Canada, south to Maine, the Great Lakes states, New Mexico and California. **Wings:** The male is blue, with dusky margins and a diffusion of green. The dimorphic female is more often brown, with reddish marginal markings. In

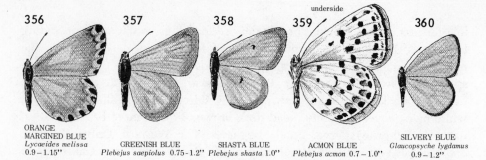

underside

356   357   358   359   360

ORANGE
MARGINED BLUE
Lycaeides melissa
0.9 – 1.15"

GREENISH BLUE
Plebejus saepiolus 0.75-1.2"

SHASTA BLUE
Plebejus shasta 1.0"

ACMON BLUE
Plebejus acmon 0.7 – 1.0"

SILVERY BLUE
Glaucopsyche lygdamus
0.9 – 1.2"

another form, the female is greenish blue, diffusing into brown in the area of the forewing, with reddish orange crescents on the outer margin of the hindwing. Underside pale bluish green, with prominent submarginal spots on both wings (bright orange on hindwing). The blue female form is common in west-central and southwestern Colorado but not in California, where the more normal brown female prevails; the latter occurs in northwestern Colorado. **Spread:** .75–1.2". **Food:** Alsike.

(149) SHASTA OR ALPINE BLUE: *Plebejus (Icaricia) shasta.* **Fig. 358.**

Small but rugged Alpine species. **Range:** Near the stormy summits of the High Sierras and Rockies—California, Colorado, Idaho, Washington, Montana. **Wings:** The male is blue, with dusky margins; the female is brown, bluish toward the base, usually with row of black-centered orange crescents on outer margin of hindwing. Underside dark gray, with black spots; hindwing with submarginal row of complex spots, the outer part metallic bluish green. **Spread:** 1". **Food:** Lupines.

(150) ACMON BLUE: *Plebejus (Icaricia) acmon.* **Fig. 359.**

Range: Western U.S., east to the Great Plains; rarely, Minnesota, Nebraska, Kansas. **Wings:** The male is blue on the upper side, with orange marginal band on forewing; the female is brownish; both sexes have orange border with black spots on hindwing. Underside grayish, with black submarginal spots on forewing and prominent orange and black marginal spots on hindwing. **Spread:** .7–1". **Larva:** Yellowish, with black spots, narrow green stripe along the back; clothed with short, fine hairs. Two or more broods. **Food:** Locoweed, rattleweed, bird's-foot trefoil, umbrella plant, lupine, buckwheat.

(151) SILVERY BLUE: *Glaucopsyche lygdamus.* **Fig. 360.**

Widespread northern species. **Range:** Alaska, east to Nova Scotia, south to Georgia, west to Colorado and New Mexico. **Wings:** The male is silvery blue, with narrow black borders; the female is darker blue, dusky at the borders and toward the base, with dark spot in discal cell on forewing. Underside brownish gray, with postmedian row of white-margined black spots. **Spread:** .9–1.2". **Food:** Everlasting peas, lupine, vetch, greasewood.

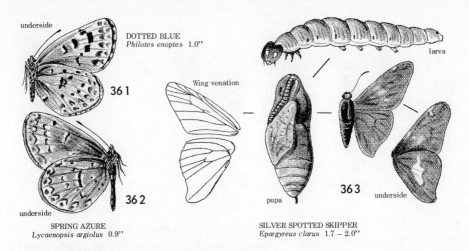

underside

DOTTED BLUE
Philotes enoptes 1.0"

361

Wing venation

larva

363

underside

pupa

underside

SPRING AZURE
Lycaenopsis argiolus 0.9"

SILVER SPOTTED SKIPPER
Epargyreus clarus 1.7 – 2.0"

362

(152) DOTTED BLUE: *Philotes enoptes.* **Fig. 361.**

A mountain species. **Range:** Western states. **Wings:** The male is pale purplish blue, with dusky margins; the female is darker or brownish, sometimes with red submarginal crescents on hindwing. Underside dull gray, with many small, black spots; two rows of black spots appear along the margins, with a narrow orange band between them on the hindwing. **Spread:** 1". **Food:** Blossoms of buckwheat.

(153) SPRING AZURE: *Lycaenopsis (Celastrina) argiolus.* **Fig. 362.**

One of the first butterflies to fly in the spring, and one of the most variable. Males are often seen around mud puddles or moist places. **Range:** Throughout North America, from the subarctic to Central America, with the possible exception of Florida. **Wings:** Very fragile. Upper side pale lilac blue, with little or no dark fringes; the female is sometimes nearly white or may have darker apical areas. Underside white to gray or brownish, with vague spots or sometimes with spots clearly defined; some broods in the North have large dark brown patches, others have wide dark borders. Spring brood smaller in size than others. Eggs are laid in flower buds. **Spread:** .9" and less. **Larva:** Body white, with reddish tinge, darker dorsal stripe, faint greenish, oblique stripes on sides; clothed in short, fine hairs; head brownish. One to three broods. Hibernates as pupa. **Food:** Dogwood, blueberry, sumac, meadowsweet, New Jersey tea, crownbeard.

Skippers (Hesperiidae)

The skippers comprise a large world-wide family characterized by stout bodies, relatively small wings, and fast, darting flight. In general they lack the great variety of gaudy coloring and patterns of the true butterflies and are less conspicuous. Colors vary from dark to lighter or yellowish browns, with simple white or yellow markings. The larvae of "true" skip-

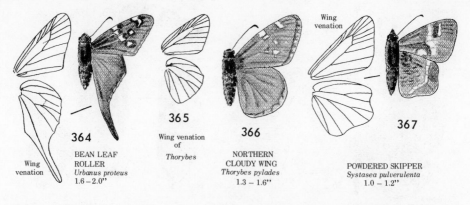

Wing venation

364
BEAN LEAF
ROLLER
Wing venation
Urbanus proteus
1.6-2.0"

365
Wing venation of
Thorybes

366
NORTHERN
CLOUDY WING
Thorybes pylades
1.3 - 1.6"

Wing venation

367
POWDERED SKIPPER
Systasea pulverulenta
1.0 - 1.2"

pers usually live in nests made of folded leaves; some caterpillars chew off the leaf tip and return with it, others leave at night for feeding outside the shelter. The larvae of giant skippers are stem borers and pupate in their tunnels; their smaller relatives pupate on the ground in loosely constructed cocoons made of leaves. (For other distinguishing features of the skippers see introduction to the Lepidoptera, p. 200.)

Our Hesperiidae are divided into two main subfamilies: the Pyrginae (formerly the Hesperiinae), numbered here (154) to (164); and the Hesperiinae (formerly the Pamphilinae), numbered (165) to (175). (A third subfamily, Pyrrhopyginae, represented by only one genus, occurs mostly in southern Arizona and in Mexico.) Our giant skippers comprise a small family (three genera), the Megathymidae (176); with the Hesperiidae, they form the superfamily Hesperioidea.

(154) SILVER SPOTTED SKIPPER: *Epargyreus clarus* (Pyrginae). **Fig. 363.**

Common and widespread. **Range:** Southern Canada to South America, Atlantic to Pacific coasts. **Wings:** Dark brown, with white and yellow spots on forewing. White or silvery patch on underside of hindwing distinguishes it from any other skipper. **Spread:** 1.7–2″. **Larva:** Lemon green, with transverse darker green bands. Head large, rusty brown, with two large orange spots. Lives in nest during the day, crawls out at night to feed. One or more broods. Hibernates as pupa on the ground. Chrysalis reddish brown, with darker markings. **Food:** Locust, wisteria, wild beans and other legumes.

(155) BEAN LEAF ROLLER: *Urbanus proteus.* **Fig. 364.**

Also called the long-tailed skipper. Tropical, but extending northward. **Range:** South and Central America, southern states and along Atlantic seaboard as far as Connecticut; occasionally common in southern California. **Wings:** Dark brown, with greenish overcast near base of both wings. White spots on forewing carry through to underside; female same as male. Underside paler, except anal angle and tail, which are dark brown. **Spread:** 1.6–2″. **Larva:** Green, similar to *clarus* (154). Darker green stripe down middle of back; yellowish subdorsal stripe and pale green stripe lower down on side. Black and yellow spots on body segments between lines. Three or more broods. Hibernates as pupa in leaves on the ground. Chrysalis covered with white powder. **Food:** Mesquite, wisteria, butterfly-pea, wild bean. A minor pest of cultivated beans and peas.

(156) GOLDEN-BANDED SKIPPER: *Autochton cellus.* **Color plate 6j.**

One of the prettiest skippers, a slow flier. **Range:** The Virginias, north to New York and Ohio, south to Texas, Arizona, and South America. **Wings:** Dark brown, with distinctive yellow band across forewing; scalloped fringe same color as band. **Spread:** 1.6–2″. **Food:** Hog peanut.

(157) NORTHERN CLOUDY WING: *Thorybes pylades.* **Figs. 365, 366.**

Very pugnacious. **Range:** Southern Canada and the entire U.S. east of the Rocky Mts. **Wings:** Dark brown, with small white spots on forewing; the translucent spots appear on underside. **Spread:** 1.3–1.6″. **Larva:** Green, with darker stripe down the back, two lighter stripes along the side; head black. Two or more broods. Hibernates as pupa. **Food:** Bush clover, yellow clover, tick trefoil. The **Southern Cloudy Wing,** *T. bathyllus,* is similar, occupies much of the same range, has larger white spots.

(158) CHECKERED SKIPPER: *Pyrgus communis.* **Color plate 6b.**

Very pugnacious, fast flier. **Range:** Canada to Mexico. **Wings:** Dark brown, with extensive white markings which are reproduced on the underside. Varies from very dark brown (in the females) to lighter browns (in the males). Underside lighter, with hindwing creamy white. **Spread:** 1.1–1.25″. **Larva:** Yellowish or greenish, with indistinct stripes on the back and side. Head black, covered with white hairs. Several broods in the South. Hibernates as pupa or larva. **Food:** Mallow, checkerbloom, wild hollyhock. The **Grizzled Skipper,** *P. centaureae* (Color plate 6e), a circumpolar species, resembles *communis,* has checkered pattern with large white spots, lighter brown background; underside of hindwing brown, with green scaling; spread from .8″ to 1.1″.

(159) POWDERED SKIPPER: *Systasea (Antigonus) pulverulenta.* **Fig. 367.**

Easily distinguished by shape and coloration. **Range:** Southern Texas, west to California, south to Central America. Rare in southern California. **Wings:** Brownish yellow, mottled with brown. White spots and markings on forewing; white submarginal line on hindwing. Underside pale, with similar but less distinct markings. **Spread:** 1–1.2″. Two broods.

(160) LARGE WHITE SKIPPER: *Heliopetes ericetorum.* **Fig. 368.**

Swift flier, difficult to capture. **Range:** Southern Texas, Arizona, and California; lower elevations of the southern Sierra Nevada Mts. **Wings:** White,

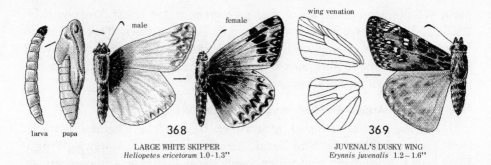

male female wing venation

larva pupa 368 369

LARGE WHITE SKIPPER
Heliopetes ericetorum 1.0 - 1.3"

JUVENAL'S DUSKY WING
Erynnis juvenalis 1.2 – 1.6"

with light brown marginal markings. Female more heavily marked, resembling the checkered skipper (158). **Spread:** 1–1.3". **Larva:** Greenish yellow, with green stripe on back, two along the side. Several broods. **Food:** Mallow, pigweed. *H. domicella* (Color plate 4i) occurs in Arizona.

(161) COMMON SOOTY WING: *Pholisora catullus.* **Color plate 6c.**

Very widespread. **Range:** Practically all of temperate North America from Canada southward, with the possible exception of Florida. **Wings:** Black both sides. Typically, series of small white marginal and submarginal spots on forewing of male; faint series of marginal spots on hindwing of females. Conspicuous white spots on forewing repeated on underside. Number of spots variable. Some western males have a series of well-defined spots on hindwing also. **Spread:** 1–1.2". **Larva:** Yellowish green; tiny white tubercles with hairs cover body. Faint stripe along the side. Two or more broods in southern areas. Hibernates as larva. **Food:** Lamb's-quarters, pigweed, ragweed.

(162) JUVENAL'S DUSKY WING: *Erynnis juvenalis.* **Fig. 369.**

The dusky wings (*Erynnis*) are difficult to tell apart by colors and patterns since the specimens are so variable. The genitalia are the only reliable clue in many cases. *E. juvenalis* and most of the skippers in this group rest with the wings held out to the side at a slight upward angle. They are sometimes confused with *Thorybes* (157) but have a shorter discal cell in the forewing and a heavier antennal club without the apiculus. **Range:** Southern Canada and eastern U.S. **Wings:** Dark brown or nearly black background on forewing, with lighter marking between the veins and translucent spots. Brown hairy covering of hindwing obscures markings. Spots are larger and more distinct in the female, the underside paler, with numerous spots. **Spread:** 1.2–1.6". **Food:** Oaks. **Horace's Dusky Wing,** *E. horatius,* is similar, occurs from the Atlantic seaboard to the Rocky Mts.; the distinction can sometimes be made by the presence of two spots near the apex on underside of the hindwing in *juvenalis,* which is most often the case in females and in the north. *E. propertius,* largest and most common

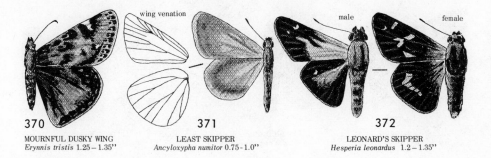

wing venation   male   female

370   371   372
MOURNFUL DUSKY WING   LEAST SKIPPER   LEONARD'S SKIPPER
Erynnis tristis 1.25–1.35"   Ancyloxypha numitor 0.75-1.0"   Hesperia leonardus 1.2–1.35"

of the dusky wings on the West Coast, is also similar; *propertius* has gray hairy covering, *juvenalis* is more brownish.

(163) FUNERAL DUSKY WING: *Erynnis funeralis.* **Color plate 6d.**

Range: Kansas to Texas, and westward; occasionally in southern Colorado. Most abundant of the dusky wings in southern California. **Wings:** Dark brown, with white fringe on hindwing. Fringe and long, narrow forewing are distinctive features. Indistinct brownish patch on outer portion of forewing serves to distinguish it from the mournful dusky wing (164). **Spread:** 1.35". **Food:** Alfalfa and other legumes.

(164) MOURNFUL DUSKY WING: *Erynnis tristis.* **Fig. 370.**

Range: Arizona, California, Mexico. **Wings:** Similar to *E. funeralis* (163), having white fringe on hindwing. It has a shorter forewing, with more distinct markings, than *funeralis.* **Spread:** 1.25–1.35". **Food:** Oak.

(165) LEAST SKIPPER: *Ancyloxypha numitor* (Hesperiinae). **Fig. 371.**

Low, weak flier. **Range:** Quebec to Florida, westward to Texas and the Rocky Mts. **Wings:** Orange, with brown shading of margins and across forewing. In some cases the forewing is almost all brownish. **Spread:** .75–1". **Larva:** Grass green, with short hairs. Head dark brown, mottled. Two or more broods. **Food:** Grasses. The tropical *A. arene,* found in southern Texas and Arizona, and in Mexico, is much smaller; the forewing of *arene* has more orange, and the underside of both wings is all pale reddish brown.

(166) LEONARD'S SKIPPER: *Hesperia leonardus.* **Fig. 372.**

Flies only in the fall. **Range:** Ontario, Maine to Florida, west to Iowa and Kansas. **Wings:** Dark brown; darker or almost black on outer half of forewing and marginal areas of hindwing. Female has more dark area, with lighter markings, on both wings. Underside dark red, with prominent cream-colored spots on median area of hindwing. **Spread:** 1.2–1.35". **Larva:** Brown, with red and green shading, mottled. Single brood. Hibernates as young larva. **Food:** Grasses.

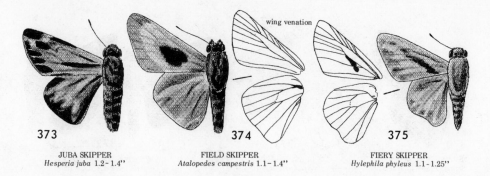

wing venation

373   374   375

JUBA SKIPPER
Hesperia juba 1.2 - 1.4"

FIELD SKIPPER
Atalopedes campestris 1.1 - 1.4"

FIERY SKIPPER
Hylephila phyleus 1.1 - 1.25"

(167) JUBA SKIPPER: *Hesperia juba.* **Fig. 373.**

A powerful flier. **Range:** Northern California, east to Colorado, north to Oregon, Washington, and Idaho, in high and low elevations. **Wings:** Reddish brown, with dark brown, sharply defined borders on the forewing. Colors diffused in the female. Underside brown; hindwing uniformly colored, with conspicuous white patches. **Spread:** 1.2–1.4". Two broods. **Food:** Grasses. The **Nevada Skipper,** *H. nevada,* which ranges from Arizona and California northward into Canada, is similar; the ground color and borders of *nevada* are darker, the underside has larger white patches.

(168) COBWEB SKIPPER: *Hesperia metea.* **Color plate 4j.**

Begins flying in early spring. **Range:** New England to Florida, west to Wisconsin, Missouri, Texas. **Wings:** Brown, with lighter markings; female is larger and darker. Markings of upper side appear on underside in white, with white submarginal band along outer margin of hindwing. In some cases the white markings on underside are lacking. **Spread:** 1–1.3". **Food:** Grasses.

(169) FIELD SKIPPER: *Atalopedes campestris.* **Fig. 374.**

Also called the sachem. **Range:** New York to the Gulf coast, west to the Pacific coast. **Wings:** Yellowish brown, with dark margins, well diffused with ground color in the female. The male is easily identified by the patch of raised black scales in the discal area of the forewing. Light spots appear faintly on the underside, which has a lighter ground color. Underside of hindwing grayish brown, with yellowish brown mottling. **Spread:** 1.1–1.4". **Food:** Grasses. The sachem caterpillar constructs a shelter (or "tent") at base of leaf, chews off tip, and returns with it for feeding.

(170) FIERY SKIPPER: *Hylephila phyleus.* **Fig. 375.**

Widespread, abundant in southern California. **Range:** Connecticut, west to Nebraska, south to Argentina; southern California, north to San Francisco Bay area. **Wings:** Male yellowish orange, with dark triangular markings along margins, and shading of veins to suggest rays. Underside paler yellow, with black spots and streaks. In sharp contrast the female is drab brown,

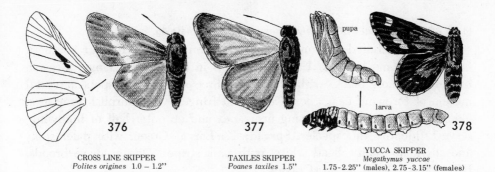

376

377

pupa

larva

378

CROSS LINE SKIPPER
Polites origines 1.0 – 1.2"

TAXILES SKIPPER
Poanes taxiles 1.5"

YUCCA SKIPPER
Megathymus yuccae
1.75-2.25" (males), 2.75-3.15" (females)

with lighter patches. **Spread:** 1.1–1.25". **Larva:** Green, yellowish brown, or grayish brown, with brown stripes on back and on side; head dark brown or black, with reddish stripes. Two or more broods. **Food:** Grasses.

(171) TAWNY-EDGED SKIPPER: *Polites themistocles (taumas).* **Color plate 4d.**

This and Peck's skipper (173) are the most numerous of the eastern skippers. **Range:** Central Canada, southward east of the Mississippi River to the Gulf states, west to Colorado. **Wings:** Brown, with large orange area along the costal margin of the forewing. Females are darker and more variable, with smaller and more orange patches. Underside paler, with orange area greatly enlarged. **Spread:** .8–1.1". One or two broods. **Food:** Grasses.

(172) CROSS LINE SKIPPER: *Polites origines (manataaqua).* **Fig. 376.**

Range: Same general area as *themistocles* (171) except that it does not extend as far south and includes all of New England. **Wings:** Similar to *themistocles* but darker, more greenish brown, with black streak through discal cell. In some forms this cross line is obscured and the species distinguished only by examination of the genitalia. It is less common than *themistocles.* **Spread:** 1–1.2". **Food:** Grasses.

(173) PECK'S SKIPPER: *Polites coras (peckius).* **Color plate 4e.**

Small but numerous. **Range:** Maritime Provinces of Canada, south to Central America, west to the Rocky Mts. **Wings:** Dark brown, with yellow patches on both wings. Yellow areas repeated on underside, more boldly on hindwing; the latter markings serve to identify the species readily. **Spread:** .9–1". **Larva:** Reddish brown, with lighter mottling; head black. Two or more broods. Hibernates as larva or pupa. **Food:** Grasses.

(174) TAXILES SKIPPER: *Poanes taxiles.* **Fig. 377.**

Range: Western Nebraska, and Colorado, south to Arizona and Mexico. **Wings:** The male is golden yellow, with narrow brown margins; the female is darker, with broader dark margins and dusky toward the base. Underside of male bright yellow, with shading of brown along outer margins; female is mottled brown, with lighter spots. **Spread:** 1.5". **Food:** Grasses.

(175) ROADSIDE SKIPPER: *Amblyscirtes vialis.* **Color plate 4g.**

Widely distributed and one of most common skippers. **Range:** Southern Canada, and most of the U.S. Uncommon in California. **Wings:** Grayish brown to almost black, with cluster of three small, light spots at costal margin of forewing near apex. Checkered fringes, with purplish gray area along outer margins of forewing near apex and on outer half of hindwing, serve to identify this species. **Spread:** 1″. **Larva:** Green, with paler spots and short, fine hairs; head white, with pink stripes. One or more broods. Hibernates as pupa. **Food:** Grasses.

(176) YUCCA SKIPPER: *Megathymus yuccae* (Megathymidae). **Fig. 378.**

One of the "giant skippers"—fast-flying; very stout, hairy body, resembling a moth. **Range:** Southern states, from Florida to Texas. **Wings:** Dark brown, with orange patches (some may be white) on forewing; orange fringes, usually solid on the hindwing and checkered on the forewing. **Spread:** 1.75–2.25″ (male); 2.75–3.15″ (female). In the **Navaho Giant Skipper,** *M. y. navaho,* found in the mountains and deserts of the Southwest, the ground color is black; the costal spots on the forewing are white, the others yellow; border of hindwing yellow, with submarginal row of small yellow spots on hindwing of the female but absent in the male. In the **Colorado Giant Skipper,** *M. y. coloradensis,* occurring in eastern Colorado, western Kansas and Oklahoma, and northeastern New Mexico, the band of spots on the forewing is wider, the area between this and the margin is accented by light scales, as are the veins; submarginal row of large yellow spots on hindwing of the female, usually the same but fainter in the male. The **Texas Giant Skipper,** *M. texana,* is similar to *yuccae,* occurs from South Dakota to New Mexico and Texas; the ground color is almost black in the male and brown in the female, with the spots near the apex of forewing white, those below yellow, submarginal row of irregular orange spots on hindwing of female. The larvae of these skippers bore into the base of leaves or stems of yucca and agave. Pupation takes place in the tunnels.

HETEROCERA (FRENATAE): MOTHS

Sphinx Moths (Sphingidae)

Our moths in this family are mostly medium to large in size, with rounded, pointed bodies protruding well beyond the hindwings and frequently with tufts of hair on the end or along the sides. They are sometimes referred to as hawk moths or hummingbird moths. The forewings are extremely long and narrow in comparison to the hindwings, with eleven or twelve veins in the former and eight in the latter, and relatively small discal cells. In contrast to butterflies, which have large wing areas and fly leisurely—ten to twelve beats per second on the average—the heavy-bodied hawk moths have relatively small wing areas and must beat their wings with great rapidity. Thus they hover over flowers sipping nectar through their long uncoiled probosces, looking more like hummingbirds, for which they are often mistaken. They fly at night and are generally most active at dusk. These moths lack the tympana possessed by most other moths. Their antennae are thickened from the base outward, usually to the middle only, and curved or hooked at the end.

The larvae are mostly large, some shade of green, with oblique stripes along the sides and an appendage resembling a horn on the anal segment. The anterior segments are retractile and are telescoped in when the larva is at rest or alarmed. Their habit of humping the body suggested an Egyptian sphinx and accounts for the name Sphingidae. They pupate in the soil (in silk-lined cells) or in litter. In some species the proboscis extends well beyond the body in the pupal stage, like a long horn.

(177) TOMATO HORNWORM: *Manduca (Protoparce) quinquemaculata.* **Fig. 379.**

Sometimes called the five-spotted hawk moth, a reference to the orange spots along each side of the adult's body. **Range:** Southern U.S. and north into Canada. **Adult:** Grayish or brownish wings, with white and dark streaking and mottling. May be distinguished from the tobacco hornworm (178) by the well-defined zigzag streaks on the hindwing, and the five pairs of orange spots on the body (instead of six as in *sexta*). Eggs are laid on the underside of leaves. **Spread:** 4–5″. **Larva:** Light or dark green, with black caudal horn and eight oblique white stripes hooked backward at the base, forming an L or V; from 3″ to 4″ long full grown. (In the tobacco hornworm larva, the horn is red and there are seven oblique stripes without the base.) Two broods in the South. Hibernates in the ground as pupa. **Food:** Tomatoes, tobacco, potatoes, peppers, eggplant, and related weeds. Eats the leaves mostly, sometimes the fruit.

(178) TOBACCO HORNWORM: *Manduca (Protoparce) sexta.* **Fig. 380.**

Range: Northern states, south into South America. **Adult:** Very similar to *M. quinquemaculata;* see (177). Both species are often found in the same garden. **Larva:** See (177). These hornworms and the catalpa sphinx (179) are frequently seen covered with a mass of tiny larvae or cocoons of a parasitic wasp, *Apanteles congregatus.*

(179) CATALPA SPHINX: *Ceratomia catalpae.* **Fig. 381.**

Range: New Jersey to Colorado and southward. **Adult:** Mostly grayish or brownish. Eggs are white, laid in masses on underside of leaves, as many as a thousand in one mass. **Spread:** 2.5–3″. **Larva:** Green or black, with dark

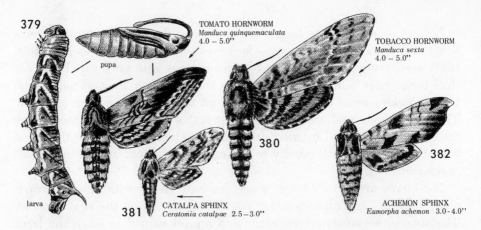

379

TOMATO HORNWORM
Manduca quinquemaculata
4.0 – 5.0"

pupa

TOBACCO HORNWORM
Manduca sexta
4.0 – 5.0"

380

382

larva

381   CATALPA SPHINX
Ceratomia catalpæ 2.5 – 3.0"

ACHEMON SPHINX
Eumorpha achemon 3.0 - 4.0"

green markings and black caudal horn; highly variable in coloring and markings; about 3″ long full grown. Gregarious in early stages. Two broods in southern part of range. Overwinters in the ground as naked brown pupa. **Food:** Catalpa. The most serious pest of this tree.

(180) ACHEMON SPHINX: *Eumorpha (Pholus) achemon.* **Fig. 382.**

Range: Southern Canada, the entire U.S., northern Mexico. **Adult:** Marbled brownish gray ground color, with deep red patches on forewing and one on each side of body where wings attach. Hindwing pink, with dark spots and brown border. Large green, spherical eggs are laid singly on upper side of outer leaves. **Spread:** 3–4″. **Larva:** May remain green (as in early instars) or turn pink, tan, or brick red; six to eight white or pale yellow oblique stripes extend across the spiracles; from 3″ to 4″ long full grown. Early instar has black caudal horn longer than body; the horn is lost when the caterpillar is about one-half inch long and is replaced by a black spot. Feeds on leaves. Pupates in smooth cell constructed in the soil. Two broods in warm climates. Hibernates as pupa. **Food:** Grapes; at times a serious pest.

(181) WHITE-LINED SPHINX: *Hyles (Celerio) lineata.* **Fig. 383.**

Most common of the sphinx moths. **Range:** Southern Canada, the entire U.S., south into Central America. **Adult:** Brown, with white stripes on thorax and head, broad white line across middle of forewing, white veins and border. Hindwing dark brown, with broad rosy white band across the middle. Flies in bright sunlight, attracted to lights at night. **Spread:** 3.1″. **Larva:** Generally green, with two black subdorsal lines, yellowish or orange head and caudal horn; a black form sometimes occurs, with yellow subdorsal and middorsal stripes; about 3.2″ long full grown. Two broods. Hibernates as pupa. **Food:** Portulaca, chickweed, purslane, azalea, evening primrose, elm; at times of overcrowding it takes to cultivated fruits, more especially grapes, and to some field crops.

Giant Silkworm Moths (Saturniidae)

A world-wide group of medium to very large moths; the family name is taken from Roman mythology and was probably suggested by the markings of these beautiful moths. The common name refers to the dense silken cocoons in which the large caterpillars pupate. The enormous Atlas moth, a saturniid of northern India in the foothills of the Himalaya Mountains, resembles our Cynthia moth but is several times larger and one of the largest moths in the world. The hindwings of some large African and South American species have extremely long tails; to Alfred Werner and Joseph Bijok these tropical butterflies "with their long hindwing tails trailing behind them [look] like ghosts [gliding] through the forests in the late hours of the night." Large feathery antennae are a conspicuous feature of the saturniids, more especially in the males. The frenulum is lacking, as are the spurs on the hind tibiae in many cases. Like hawk moths, saturniids lack tympana or ears. The proboscis is absent or rudimentary. The larvae are pale green to brownish, black in the early stages, armed with spines or tubercles. Long paired spines appear on the upper side of the mesothorax and metathorax and a single middorsal spine on the ninth abdominal segment in some species; others have paired but less prominent spines, often branched, on other segments. Several species of saturniids are important defoliators in pine forests of the West. Efforts to reel or card the fibers of the Prometheus and Polyphemus cocoons have not been successful.

(182) PANDORA MOTH: *Coloradia pandora.* **Fig. 384.**

Range: Eastern foothills of the Rocky Mts., Montana to Mexico, west to the Cascade Mts. and the Sierra Nevada Mts. **Adult:** Dark, heavy body, with grayish brown wings and small, dark discal spot on each wing. Base and interior margin of hindwing covered with fine pinkish hairs shading to wine color in the male. Antennae of males are feathery and heavy; females have slender antennae and heavier body. Globular eggs are laid in clusters on branches or trunks of trees, or in litter. Adults appear in June or July, eggs hatch in August. **Spread:** 3–4″. **Larva:** Black or brownish newly hatched, covered with short hairs; head shiny, black. Mature larva from 2.5″ to 3″ long, brown to yellowish green, with short, stout branched spines on each body segment. Young larvae feed in groups on new pine shoots, hibernate during the first winter in clusters at base of needles; they descend to the ground the following June to form silk-lined cells and pupate, and remain in this stage for a year. **Food:** Ponderosa pine, Jeffrey pine, lodge-pole pine. Serious epidemics of this species and *C. doris* have occurred in the past. Adults of *doris* are less heavily scaled, discal spots are oblong, and hindwings translucent; the larvae have long branched spines on two thoracic segments and on the anal segment only. *C. pandora* is not normally destructive because it does not feed on terminal buds, and a two-year life cycle permits trees to recover; also, it is held in check by a "wilt" disease and parasites, by screech owls that seize the moths on the wing, and by ground squirrels and chipmunks that dig up and eat the pupae. Mono Indians of California are said to eat the caterpillars, which are smoked out of the trees; the Klamath Indians of Oregon are said to prefer the pupae.

(183) CEANOTHIS SILK MOTH: *Platysamia (Hyalophora) euryalus.* **Fig. 385.**

Range: Pacific coast, east to Wyoming and Utah. **Adult:** Densely covered with reddish hairs; white and black markings. **Spread:** 4–5″. **Larva:** Large, pale green to blue, with golden tubercles on back and bluish tubercles on sides. **Food:** Willow, manzanita, *Ceanothus. P. gloveri* is similar, ranges from Arizona, in the Rocky Mts., north into Alberta. Inner half of wings are much darker red; outer band has purplish cast. The larva, 3″ long full grown, is yellowish green, with black spines, two rows of tubercles on the back, black spots between.

(184) CECROPIA MOTH: *Platysamia (Hyalophora) cecropia.* **Fig. 386, Color plate 1 (bottom).**

The largest of our moths. **Range:** Atlantic seaboard, west to Montana and the Great Plains. **Adult:** Body deep red, with white bands. Wings reddish brown, with reddish crescent on forewing and white one on hindwing, both bordered with black; black submarginal spots on forewing, with purplish shading along costa near apex. Prominent white band on both wings with reddish border along the outside; wing margins clay-colored. **Spread:** 5.5″. **Larva:** Pea green, with bluish tint; about 4″ long. Two pairs of coral red tubercles on thorax; fifteen yellow tubercles along the back, two rows of blue ones along each side. Hibernates as pupa; the large spindle-shaped cocoon attached to the side of a twig is prominently displayed when leaves have fallen. **Food:** Many forest trees, including birch, maple, ash, willow, elm, lilac, apple, wild cherry, and others.

(185) POLYPHEMUS MOTH: *Antheraea polyphemus.* **Figs. 388, 389, Color plate 6a.**

Range: Entire continent of North America. **Adult:** Ocher or purplish to reddish brown, with an occasional melanic specimen (upper side of wings almost wholly black), or the other extreme, albino. Large eyespot on both wings, margined with yellow on forewing and blue and black on hindwing. **Spread:** 4–5″. **Larva:** Apple green, with pale yellow oblique stripes on each side, orange tubercles arising from red spots. Two broods in the South, one in the North. Hibernates as pupa; the cocoon is spun with leaves, generally falls to the ground. **Food:** Oak, birch, hickory, maple, elm, poplar, willow, alder, madroño, basswood, chestnut, sycamore, and others.

(186) CYNTHIA MOTH: *Samia cynthia.* **Fig. 390, Color plate 6f.**

Brought here from Europe in 1861 for its silk possibilities. **Range:** Atlantic

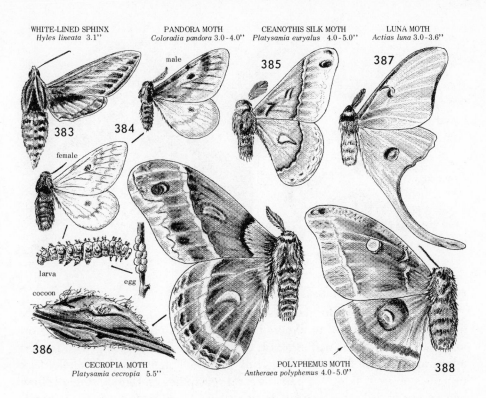

WHITE-LINED SPHINX
Hyles lineata 3.1"

PANDORA MOTH
Coloradia pandora 3.0-4.0"

CEANOTHIS SILK MOTH
Platysamia euryalus 4.0-5.0"

LUNA MOTH
Actias luna 3.0-3.6"

male

385

387

383   384

female

larva

egg

cocoon

386

CECROPIA MOTH
Platysamia cecropia 5.5"

POLYPHEMUS MOTH
Antheraea polyphemus 4.0-5.0"

388

seaboard. **Adult:** Easily recognized by white spots or tufts on abdomen. Greenish brown wings, with white markings; white crescents bordered on lower side with yellow. Black apical spot on forewing, with bluish shading above. **Spread:** 4–5". **Larva:** Green when fully grown, with black spots. Dorsal and lateral tubercles are shades of blue; those in row below spiracles are black-bordered. Head and anal shield yellow. Hibernates as pupa. The cocoon is spun in folded leaf, and stem fastened to twig by silk; hanging from a tree, it is not an easy mark for birds. **Food:** *Ailanthus,* which they sometimes strip of leaves; to lesser extent lilac, sycamore, linden, wild cherry, and other trees and shrubs.

(187) LUNA MOTH: *Actias luna.* **Fig. 387.**

Range: Canada to Florida, west to the Great Plains. **Adult:** Wings are delicate green, shading to pale gray, with transparent eyespots. Narrow border along outer margins except on those emerging later in the season. **Spread:** 3–3.6". **Larva:** Variable colors; resembles the Polyphemus larva (185) except that Luna has yellow lateral stripe and no oblique stripes on the side. One or two broods. Hibernates as pupa. Pupates in loosely fitting, papery cocoon which falls to the ground. **Food:** Walnut, hickory, persimmon, sweet gum.

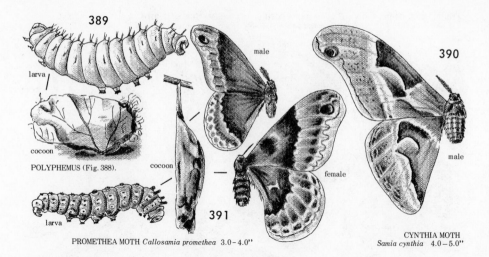

389

male

390

larva

cocoon

POLYPHEMUS (Fig. 388).

cocoon

male

larva

391

male

female

PROMETHEA MOTH *Callosamia promethea* 3.0–4.0"

CYNTHIA MOTH
Samia cynthia 4.0–5.0"

(188) PROMETHEA MOTH: *Callosamia promethea.* **Fig. 391.**

Range: Canada to Florida, west to the Great Plains. **Adult:** Ground color of male is dark maroon to almost black, markings obscured; female lighter and markings well defined. **Spread:** 3–4". **Larva:** Bluish green, with frosted appearance; from 2" to 3" long full grown. Three rows of small black tubercles on each side; two pairs of coral red spines on thorax, single yellow one on eleventh body segment; head yellowish green. Hibernates as pupa. The cocoon hangs down from a branch, like that of Cynthia (186). **Food:** Spice bush, sassafras, wild cherry, other trees and shrubs.

(189) IO MOTH: *Automeris io.* **Fig. 392.**

Range: Canada to Florida, west to the Great Plains. **Adult:** Forewing bright yellow in the male, reddish brown in the larger female. Background coloring of hindwing yellow, with reddish orange submarginal band and shading of same color along inner margin, black eyespot partially circled by black line. **Spread:** 2.5–2.8". **Larva:** Green, with pink and white lateral stripe and numerous clusters of urticating hairs. [See (193), (223), (228), (238) to (241) for other "stinging caterpillars."] Hibernates as pupa; spins thin cocoon in leaves on the ground. **Food:** A great variety of trees, shrubs, and plants such as corn.

(190) BUCK MOTH: *Hemileuca maia.* **Fig. 393.**

The common name is said to come from this moth's habit of flying when the deer is hunted. **Range:** Nova Scotia and Maine to Florida, west to Colorado and New Mexico. **Adult:** Abdomen brown or black, with red tuft on tip. Wings reddish brown, with broad, white median band, semitranslucent, almost devoid of scales. Diurnal or semicrepuscular. Eggs are laid in clusters around twigs, hatch in April or May. **Spread:** 2–3" (almost). **Larva:**

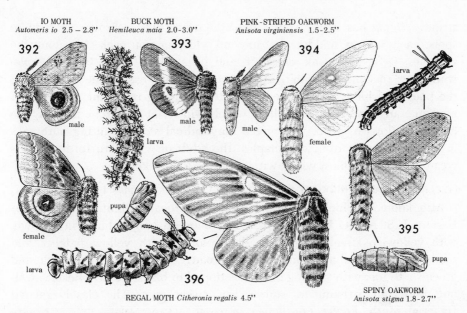

IO MOTH
Automeris io 2.5 – 2.8"

BUCK MOTH
Hemileuca maia 2.0-3.0"

PINK-STRIPED OAKWORM
Anisota virginiensis 1.5-2.5"

392

393

394

larva

male

male

male

female

larva

pupa

395

female

larva

pupa

396

REGAL MOTH *Citheronia regalis* 4.5"

SPINY OAKWORM
Anisota stigma 1.8-2.7"

Dark brown to black, with yellowish dots and branched, stinging spines arising from tubercles; about 2.5" long full grown. They feed gregariously, molt five times, and pupate in the ground. Adults emerge in the fall (a few remain in the ground and overwinter as pupae). Single-brooded. Hibernates in egg stage. **Food:** Oak, willow.

Royal Moths (Citheroniidae)

This is a small family of medium to large moths, also called regal moths. The proboscis is feebly developed, and they may not feed at all as adults. The moths lack a frenulum but have tibial spurs. They pupate in the ground without forming a cocoon. Our examples are sometimes shown as Saturniidae.

(191) PINK-STRIPED OAKWORM: *Anisota virginiensis.* **Fig. 394.**

Range: Canada to the Carolinas, west to Kansas and Missouri. **Adult:** Male reddish brown excepting central area of forewing, which is almost transparent; body of female yellow, wings brownish yellow, with wide pink band along outer margin of forewing. **Spread:** 1.5–2.5". **Larva:** Dark green, with white speckling; two purplish red stripes, three rows of short black spines on each side and a pair of long spines on prothorax; about 2" long full grown. **Food:** Oaks. Two similar species occur in the same general area: the **Orange-striped Oakworm,** A. *senatoria,* and the **Spiny Oakworm,** A. *stigma* (Fig. 395). The males of *senatoria* closely resemble the females of *stigma,* the forewings of both being profusely spotted with black; *stigma* is a little larger and more heavily scaled. The larva of *senatoria* is black, with orange stripes; *stigma* has longer spines than the other two, is brown with white spots, has a narrow middorsal stripe and wider one on each side along the spiracles; it feeds on oak, chestnut, and hazel. The **Green-striped Maple-worm** (or "rosy maple-moth"), *A. rubicunda,* occurs east of the Rocky Mts.; the body and hindwing are yellowish, the forewing rosy pink, with broad median transverse band of yellow which widens at the front margin; the larva is yellowish green, with green longitudinal stripes; during outbreaks it strips the leaves from soft maples; the adult of a western form, *alba,* is cream-colored to almost white, with faint pinkish markings.

(192) REGAL MOTH (HICKORY HORNED DEVIL): *Citheronia regalis.* **Fig. 396.**

Also known as hickory horned moth and royal walnut moth. **Range:** Eastern U.S. **Adult:** Body brown, with yellowish bands; hindwing brown, with yellowish streaks, forewing slate-colored, with yellow patches and reddish brown scaling along veins. **Spread:** 4.5″. **Larva:** Brownish and greenish, thorax with six long red horns tipped in black. **Food:** Hickory, walnut, butternut, persimmon, sumac, ash, sweet gum. The closely related **Imperial Moth,** *Eacles* (*Basilona*) *imperialis* (Color plate 7a), is canary yellow, with purplish markings, wingspread from 4″ to 5″; the brown or green larvae feed on a great variety of trees and shrubs.

Tiger Moths (Arctiidae)

A large family of medium-sized to large moths, the tiger moths are named for their color patterns of white, brown, and orange, with contrasting stripes and spots. Some are plain, however, with few or no markings. The antennae are pectinate, the proboscis well developed in some genera and not in others. Eggs are usually laid on leaves of host plants in large clusters, often covered with hairs from the female moth's body. The caterpillars are robust and hairy; some are called "woollybears," and they are often seen scurrying along the ground or other surface. Their cocoons are made of silk and hairs shed during spinning.

(193) SILVER-SPOTTED TIGER MOTH: *Halisidota argentata.* **Fig. 397.**

Range: From the Atlantic to the Pacific in coniferous forests. **Adult:** Body covered with yellowish hairs; forewing reddish brown with silvery white spots, hindwing light tan or almost white with a few brown markings; female larger and darker than male. They emerge in July and August, deposit eggs on needles or twigs. **Spread:** 1.5–2″. **Larva:** Densely covered with tufts of brown or black poisonous hairs, about 1.5″ long full grown. They feed in clusters under web formed with dead needles, hibernate here in early stage; they disperse the following spring and feed singly until mature, when they spin cocoons of silk and body hairs, attaching them to twigs, needles, or trunk of tree, or to litter on forest floor. **Food:** Douglas fir, true fir, Sitka

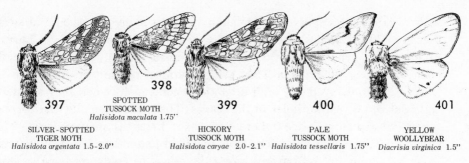

397   398 SPOTTED TUSSOCK MOTH *Halisidota maculata* 1.75''   399   400   401

SILVER-SPOTTED TIGER MOTH *Halisidota argentata* 1.5-2.0''   HICKORY TUSSOCK MOTH *Halisidota caryae* 2.0-2.1''   PALE TUSSOCK MOTH *Halisidota tessellaris* 1.75''   YELLOW WOOLLYBEAR *Diacrisia virginica* 1.5''

spruce, shore pine, Monterey pine, and other conifers.

(194) SPOTTED TUSSOCK MOTH: *Halisidota maculata.* **Fig. 398.**

Not to be confused with the Lymantriidae or tussock moths. **Range:** Northern Atlantic coast region, west to California. **Adult:** Forewing brownish yellow, with brown splotches; hindwing paler. **Spread:** 1.75''. **Larva:** Dense covering of black hairs, with intermixing of white and yellow hairs; wide row of shorter, tufted black hairs (or "tussocks") down the back. **Food:** Oak, maple, alder, willow, poplar, other trees and shrubs. An eastern species, *H. caryae,* the **Hickory Tussock Moth** (Fig. 399), closely resembles *H. argentata* (193); the adult is a little larger and lighter in coloring; the larva prefers hickory when available, spins a very thin cocoon that looks as though the hairs were growing through the covering. Adults of *H. tessellaris* (Fig. 400), the **Pale Tussock Moth,** are pale straw in coloring, with bluish green lines on thorax; the larvae have black hairs, without the short-tufted row, and may be tinted yellow, brown, or orange. These species feed on a great variety of trees and shrubs.

(195) YELLOW WOOLLYBEAR: *Diacrisia virginica.* **Fig. 401.**

Range: Throughout North America. **Adult:** Wings white, with small black spot on each. **Spread:** About 1.5''. **Larva:** Covered with long, dense hairs, usually pale yellow, sometimes brownish yellow, red, or white. Feeds on flowers, leaves, and tender stems. Several broods. Hibernates as pupa in cocoons made of silk and hairs, in the leaves, clods of dirt, or trash; 20 to 30 cocoons are often found under the same shelter. **Food:** A great variety of field and garden crops, including many ornamentals. The **Banded Woolly-bear,** *Isia isabella,* is similar, except that both wings are yellow; the larva is black at both ends, with a wide reddish brown band in the middle, and has an "evenly clipped appearance" in contrast to other woollybears; it feigns death by curling into a ball; hibernates as a larva.

(196) SALT-MARSH CATERPILLAR: *Estigmene acrea.* **Fig. 402.**

Range: Throughout North America. **Adult:** Abdomen orange-yellow, with row of six black spots down the middle and white tip; forewing of male white, hindwing orange, both wings of female white; wing surfaces marked with variable black spots. The female may lay as many as a thousand eggs; they are laid in patches on the underside of leaves. **Spread:** 1.8–2.5″. **Larva:** Another "woollybear"; gray or black to reddish brown, with black head; about 2.2″ long full grown. They are gregarious, feeding on underside of leaves, skeletonizing them, until after the second molt, when they separate and feed singly, eating holes in the leaves. There are six molts, the cocoon woven mostly from hairs; hibernation takes place as mature larva. The larvae sometimes move en masse from one field to another, like armyworms; paper or aluminum foil barriers along the edges of fields are used to stop them. **Food:** Alfalfa, clover, corn, cotton, asparagus, beans, beets, cabbage, carrots, celery, onions, lettuce, strawberries, plum, apple, many ornamentals. Fungus and bacterial diseases often take a heavy toll, leaving dead caterpillars clinging to the plants. Its principal parasites are *Exorista mella* and *Leschenaultia adusta* (Tachinidae).

(197) FALL WEBWORM: *Hyphantria cunea.* **Fig. 403.**

Range: Southern Canada and the entire U.S. **Adult:** Wings satiny white, usually with some brown or black spots. Abdomen yellow or orange, with black dorsal and lateral spots; antennae black and white, front coxae and femora reddish orange. Eggs are laid in masses on underside of leaves, partially covered with hairs. **Spread:** 2–2.4″. **Larva:** Densely covered with pale yellow to brown hairs; long white hairs rising from black and orange tubercles tend to give it a grayish appearance; about 1″ long full grown. They feed in colonies from within large webs constructed around the end of branches, are most active at night. [The eastern tent caterpillar (229) builds its nest in a crotch, crawls outside to feed.] They are subject to severe parasitization, which tends to make colonies compact. Five to eight (typically six) instars. One to four broods. Hibernates as pupa in cocoon attached to trunk of tree or in the ground litter. **Food:** More than a hundred different kinds of broad-leaved woodland, shade, and fruit trees, and bushes are said to be hosts; common ones are wild cherry, walnut, madroño, alder, willow, cottonwood, honeysuckle, blueberry. The **Spotless Webworm,** *H. textor,* is very similar; the adult is all white, has no markings.

Owlet Moths and Underwings (Noctuidae)

This is much the largest family of Lepidoptera, formerly known as the Phalaenidae and one of the most destructive to cultivated plants. The adults are medium to large in size, fly mostly at night or dusk, and are strongly attracted to lights; they are sometimes called millers. The forewings are mottled dull brown or gray; the hindwings are paler and plain, and the frenulum is always present. The proboscis is usually well developed, and so are the tympana or ears, befitting the insects' nocturnal habit.

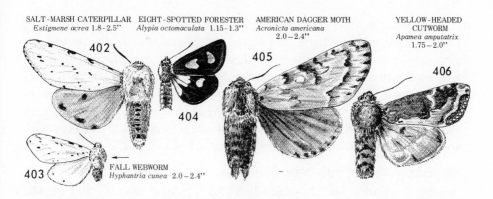

SALT-MARSH CATERPILLAR   EIGHT-SPOTTED FORESTER   AMERICAN DAGGER MOTH   YELLOW-HEADED
   Estigmene acrea 1.8-2.5"   *Alypia octomaculata* 1.15-1.3"   *Acronicta americana*   CUTWORM
                                                                  2.0-2.4"   *Apamea amputatrix*
                                                                              1.75-2.0"

402

405

404

406

FALL WEBWORM
Hyphantria cunea 2.0-2.4"

403

The larvae of many noctuids are called cutworms, and others are called armyworms. The latter are so named because of their habit of migrating en masse from one field to another when overcrowding takes place; they normally feed at night like other noctuids but may travel by day during severe outbreaks, eating as they go. Cutworms are named for their habit of chewing off the tender shoots just above or below the ground line. They are usually grayish or brownish, curl up when at rest or when disturbed. Most of them remain in the soil by day, including those that eat the upper parts of the plant, and emerge at night to feed. Some are entirely subterranean, feeding on plant roots and stems below the ground; others bore into the plant stems. Caterpillars of the underwings feed mostly on leaves of broad-leaved forest and shade trees. Some noctuid larvae are "semiloopers," having three pairs of prolegs (including the anal pair) instead of the usual five; in bringing the prolegs forward the body is humped as in the "true loopers" or geometrids. Noctuid larvae are naked or clothed at best in fine, soft hairs. Many of them pupate in the ground without a cocoon, in a smoothly formed cell. They are normally attacked by many predators, parasites, and disease organisms.

(198) EIGHT-SPOTTED FORESTER: *Alypia octomaculata.* **Fig. 404.**

Range: Northeastern Atlantic seaboard. **Adult:** Black, with eight large spots—two on the forewing yellow, two on the hindwing usually white. Often flies during the day. **Spread:** 1.15–1.3". **Larva:** Humped near the hind end. Bluish white; each segment with orange (or reddish) and black cross-stripes, and black dots; about 1.5" long full grown. One or two broods. Hibernates in the ground or litter in thin cocoon made of chips and silk, or in a tunnel made in wood. **Food:** Virginia creeper, Boston ivy, grapes.

(199) AMERICAN DAGGER MOTH: *Acronicta americana.* **Fig. 405.**

Range: Quebec to Georgia, west to Manitoba and Utah, south to the Gulf states. **Adult:** Forewing marbled fawn, hindwing brownish, lighter in male. It takes a little imagination to see the "dagger" near the outer angle of the forewing in some specimens. **Spread:** 2–2.4″. **Larva:** Pale yellow, very hairy; the hairs are scattered, not tufted on tubercles as in the arctiids. It often rests curled up near the stem of a leaf, "plays possum" and drops to the ground when disturbed. Hibernates as pupa in loose cocoon in bark of tree or ground litter. **Food:** Maple, oak, elm, and other trees. The **Cottonwood Dagger Moth,** A. *lepusculina,* somewhat smaller, is white, with gray dusting; the **Smeared Dagger Moth,** A. *oblinita,* is similar but streaked with gray on and between the veins, has a wingspread of 1.5″ to 2″.

(200) YELLOW-HEADED CUTWORM: *Apamea amputatrix.* **Fig. 406.**

Range: Labrador and Newfoundland to Virginia, west to the Pacific, southwest to New Mexico. **Adult:** Forewing dark brown or blackish, with gray marbling and crimson middle band; hindwing dull gray, with outer third and veins dark brownish. **Spread:** 1.75–2″. **Larva:** Immaculate, pale smoky, with brownish yellow head. **Food:** Wide variety of field and garden crops. The closely related and common **Glassy Cutworm,** *Crymodes devastator,* ranges from the Gaspé (Quebec) to New Jersey, west to the Pacific; it is somewhat smaller, dark brownish, with light gray and blackish areas; the larva is immaculate, dirty white, or green, with darker mouthparts, large head and cervical shield, prominent tubercles; it is a general feeder like the yellow-headed cutworm.

(201) IRIS BORER: *Macronoctua onusta.* **Fig. 407.**

Range: Quebec and Maine to Virginia and Georgia, west to Iowa. **Adult:** Forewing dark brownish, with black markings; hindwing lighter. Eggs are laid on dead and dying leaves in the fall. **Spread:** 1.6–2″. **Larva:** Green at first, turns pinkish, with brown head, lighter stripe down the back, rows of black spots on the sides; from 1.5″ to 2″ long full grown. Pupates in the soil. Single-brooded. Hibernates in egg stage. **Food:** Roots and crowns of iris. The **Stalk Borer,** *Papaipema nebris,* is a somewhat smaller moth, grayish brown, forewing sometimes wich white spots along the discal area and the costal and apical margins; eggs are laid on burdock, ragweed, and other plants in late summer; the young larva is brownish, with dark purple band around the middle and brown or purple longitudinal stripes; the mature larva is creamy white to light purple, without markings, about 1.25″ long; the larvae bore into stalks and stems of dahlia, aster, and other thick-stemmed plants. See (251).

(202) ARMYWORM: *Pseudaletia unipuncta.* **Fig. 408.**

Range: Canada to the Gulf states, east of the Rocky Mts.; occasionally found in New Mexico, Arizona, and California. **Adult:** Brownish gray, with white spot near center of forewing; hindwing lighter brown, with darker scaling along the veins. Eggs are laid at night in masses or rows in folded

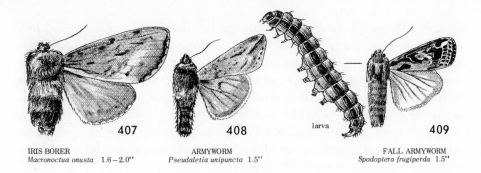

407
IRIS BORER
Macronoctua onusta 1.6 – 2.0''

408
ARMYWORM
Pseudaletia unipuncta 1.5''

larva

409
FALL ARMYWORM
Spodoptera frugiperda 1.5''

leaves or leaf sheaths of grains and grasses. **Spread:** 1.5''. **Larva:** Early stage pale green; it is a looper until half-grown. The mature larva is smooth, almost hairless, greenish brown to reddish brown, with darker or nearly black stripe down the back and along the side, broad dorsal stripe with fine, lighter line in the center, about 1.5'' long [compare with Fall Armyworm (203)]. Pupates in cell in the ground. Three broods. Hibernates as larva, sometimes as pupa. All stages are present during winter in the extreme south. They feed mostly at night; older larvae will eat the entire plant. The eggs of a parasitic fly, *Winthemia quadripustulata,* are often noticeable on the larva's back near its head [see Diptera (96), also (94).] The larvae are also heavily parasitized by braconid wasps, *Apanteles* spp. **Food:** Corn, sugar beets, alfalfa, clover, timothy, and various other grasses, sorghums, flax, millet, rice, and various small grains. Outbreaks are more apt to occur after a mild, dry winter followed by a cold, wet spring.

(203) FALL ARMYWORM: *Spodoptera (Laphygma) frugiperda.* **Fig. 409.**

Range: Extreme southern parts of Florida and Texas; tropical North, Central, and South America. It migrates northward each summer as far as Montana, Michigan, and Maine. Appearance in late summer or fall accounts for the name. Confined mostly to southern counties in Nevada and California. **Adult:** Forewing dark brownish gray, mottled, with oblique mark near center and irregular white or gray patch near apex; female darker than male; hindwing white, with pearly or pinkish luster and brownish border. It becomes active just after sunset. Eggs are laid in masses of 50 to 100 and covered with hairs. **Spread:** 1.5''. **Larva:** Difficult to distinguish from armyworm (202). White, with black head when newly hatched, getting darker as it feeds. Young ones are often seen curled up in leaf sheath or suspended from thread. Greenish to brownish, with white subdorsal line, brownish black stripe above spiracles and pale stripe with reddish brown mottling below when mature; about 1.5'' long full grown. Usually has prominent inverted Y on front of head. Scattered hairs are longer and black tubercles from which they arise are more prominent than in the armyworm. Up to six broods in the Gulf states, one in the North from migrants. Pupates in cell in the soil. It feeds mostly at night but does not leave plant and hide in the soil as do the armyworm and climbing cutworms. **Food:** Corn, cabbage, cotton, tobacco, peanuts, and many other field crops; grasses, rice, cereals, legumes. Injury varies with plants; on corn they chew leaves, bore into stalks and ears. The **Beet Armyworm,** S. *exigua,* is very similar in appearance and habits, slightly smaller, forewing grayish brown, with two yellowish spots near center, hindwing translucent white, with narrow brownish border; it is a pest of cotton in the Southwest, also attacks a great variety of field and garden crops, fruits, grains.

(204) YELLOW-STRIPED ARMYWORM: *Prodenia ornithogalli.* **Fig. 410.**

Range: Throughout the U.S., most abundant in the South. **Adult:** Forewing blackish brown, reddish brown, or yellowish brown, with oblique yellowish mark near center and blue-gray markings at the apex and hind angle, veins lighter shade; hindwing iridescent white, border with narrow brown line. Eggs are laid in masses on leaves and covered with hairs. **Spread:** 1.4–1.6″. **Larva:** Smooth, with scattered short hairs, inverted Y on front of head, body with broad yellowish brown dorsal stripe with white lines, pair of black triangular subdorsal spots on each segment except prothorax; yellowish, brown and orange stripes with white lines along sides. It feeds during the day. Several broods; hibernates as pupa. **Food:** A great variety of field crops. Very destructive to cotton; it eats young plant or bores into squares and bolls. The **Western Yellow-striped Armyworm,** P. *praefica,* is very similar, more common in the West; the adult is not as contrastingly marked as in *ornithogalli* and the hindwing is darker; in the larva, the stripe above the spiracles is brown instead of creamy or yellow as in *ornithogalli,* and the stripe along the spiracles is black rather than bright orange. The **Southern Armyworm,** P. *eridania,* a pest of vegetables in the South, is a "climbing cutworm," feeding on upper parts of plants; the full-grown larva is gray or nearly black, with median, subdorsal, and lateral stripes yellow.

(205) VARIEGATED CUTWORM: *Peridroma saucia.* **Fig. 411.**

One of the most common noctuids, occurring almost everywhere that crops are grown. **Range:** Throughout North America, Europe, and elsewhere. **Adult:** Forewing brownish gray, outer margin darker, front margin reddish; four wavy transverse lines, kidney-shaped spot and smaller circular one near center. Hindwing gray, with lightly shaded brownish border and pinkish cast. Eggs are dome-shaped, with a flat surface; they are laid in rows in batches of 60 or more—in trees, on fences or buildings. **Spread:** 1.5–2″. **Larva:** Smooth, gray to dark mottled brown, with intermixing of red and yellow; distinct pale yellow to orange middorsal spot on metathorax and next four or more abdominal segments, with a black W on the eighth

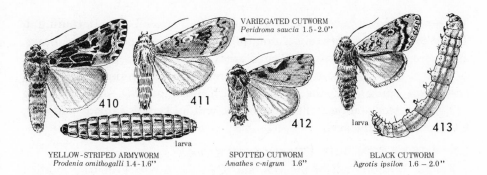

VARIEGATED CUTWORM
Peridroma saucia 1.5-2.0"

410

411

412

larva

413

larva

YELLOW-STRIPED ARMYWORM
Prodenia ornithogalli 1.4-1.6"

SPOTTED CUTWORM
Amathes c-nigrum 1.6"

BLACK CUTWORM
Agrotis ipsilon 1.6 – 2.0"

segment; two alternate black and orange stripes along the sides; 1.6" long full grown. A "climbing cutworm"; it hides by day, feeds at night. Three or more broods. Usually hibernates as pupa. **Food:** A great variety of field crops, ornamentals, and greenhouse plants; alfalfa, grasses, grains, cotton; foliage, buds, and fruit of fruit trees and vines. [See Diptera (94).]

(206) SPOTTED CUTWORM: *Amathes c-nigrum.* **Fig. 412.**

Range: Throughout North America; world-wide. **Adult:** Forewing dull gray or brownish, with basal spot near costa forming a V, filled with ocher-yellow; hindwing clear except for brownish suffusion along margins. Strongly attracted to lights. Eggs are laid singly, or in rows or batches, on leaves of trees or on cover crop in the fall. **Spread:** 1.6". **Larva:** Smooth, grayish brown or blackish (summer brood sometimes olive), with faint dorsal and subdorsal lines, very faint broad substigmatal stripe; a pair of wedge-shaped black spots on upper side of each body segment on posterior half of body, increasing in size and closer together posteriorly. A "climbing cutworm"; it hides in grass or litter during the day, feeds at night. One to three broods. Hibernates as young larva. **Food:** Vegetables, legumes, grasses, grains; leaves and buds of fruit trees.

(207) BLACK CUTWORM: *Agrotis ipsilon.* **Fig. 413.**

Sometimes called greasy cutworm. **Range:** Throughout North America; Europe. **Adult:** Forewing clay-colored, with reddish median area, suffused with black in the female; hindwing clear, with darkened margins and veins. Eggs are laid singly, a few in clusters, on leaves and stems. **Spread:** 1.6–2". **Larva:** Gray or brown to nearly black, greasy-appearing, covered with *rounded* granules of varying sizes; from 1.2" to 2.8" long full grown. A solitary "surface cutworm" (cuts seedlings off just above the ground line). Two to four broods. Hibernates as larva or pupa. **Food:** Wide variety of vegetable crops; alfalfa, clover, cotton, corn, tobacco; strawberry, grape, and other fruits. The **Clay-backed Cutworm,** *A. gladiaria,* is similar; the forewing is dark brownish, with grayish costa, the hindwing is clay-colored between the veins; spread about 1.4"; the larva—pale grayish, with reddish dorsum and grayish brown head—drags the plant to its burrow before devouring it. The **Western Bean Cutworm,** *Loxagrotis albicosta,* is a pest of field beans, and more especially corn in irrigated areas of the western plains; the adult is light to dark brown, forewing with creamy white stripes and spots in costal area, spread 1.5″; the larva is pinkish brown or gray, about 1.5″ long full grown, feeds on buds, flowers, leaves, and on kernels in pods and ears, hibernates in earthen cell.

(208) PALE WESTERN CUTWORM: *Agrotis orthogonia.*

Range: Western half of the U.S. and Canada. **Adult:** Forewing mottled grayish, hindwing clear white. They fly in the fall and are readily attracted to lights. The white spherical eggs are laid singly or several together in the soil. **Spread:** 1.2–1.5″. **Larva:** Glassy slate gray, without spots or stripes, covered with fine hairs and *flat* granules; head yellowish, with two short bluish dashes forming an H or inverted V on the front; from 1.5″ to 2″ long full grown. A subterranean feeder, it cuts off stem one or two inches below surface. Single-brooded. It usually hibernates in egg stage, hatching during warm days in winter or early spring. **Food:** Grasses, alfalfa, grains, corn sorghums, potatoes, beets, Russian thistle. The **Pale-sided Cutworm,** *A. malefida,* is gray to blackish, except in the male the hindwing is pure white, wingspread from 1.6″ to 1.8″; the larva is gray, with obscure mottling, *without* the granules, has the first prolegs reduced. The **Army Cutworm,** *Euxoa (Chorizagrotis) auxiliaris* (Color plate 7c), damages wheat and alfalfa in the spring in the Plains states; in the subspecies *introferens* the forewing is grayish brown and dark brown intermixed, with a kidney-shaped and an oblong spot, grayish border and lateral band near outer margin; hindwing brownish gray; spread from 1.5″ to 1.75″; in the subspecies *agrestis*—smaller of the two—the coloring is more subdued; the larva is greenish gray to dark brown, with two thin cream-colored stripes.

(209) DINGY CUTWORM: *Feltia subgothica.* **Fig. 414.**

Range: Canada and the northern states. **Adult:** Forewing grayish, with black markings and grayish or brownish borders; hindwing clear except for grayish or brownish borders. Eggs are laid singly or several together in the soil. **Spread:** 1.1–1.5″. **Larvae:** Dull brown, with wide grayish buff dorsal stripe divided into many triangular areas and bordered on each side by a narrow darker stripe; surface marked with isolated coarse, rounded granules. A "climbing cutworm" some of the time, feeding during the night or cooler part of day. Single brood. Hibernates as partially grown larva. **Food:** Corn, cabbage, beans, potatoes, tomatoes, cover crops, greenhouse plants. The **Granulate Cutworm,** *F. subterranea*—a subterranean feeder—is common in the middle states and the South, and a major pest of dichondra lawn in California; the forewing is mostly black, with wide grayish band along the

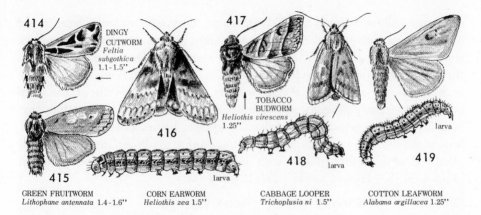

414 DINGY CUTWORM *Feltia subgothica* 1.1-1.5"

417

TOBACCO BUDWORM *Heliothis virescens* 1.25"

416

415 larva

418 larva

419 larva

GREEN FRUITWORM
Lithophane antennata 1.4-1.6"

CORN EARWORM
Heliothis zea 1.5"

CABBAGE LOOPER
Trichoplusia ni 1.5"

COTTON LEAFWORM
Alabama argillacea 1.25"

outer border, spread of 1.5"; the larva is dark gray, covered with many black, bluntly conical granules of varying sizes.

(210) GREEN FRUITWORM: *Lithophane antennata.* **Fig. 415.**

Range: Ontario to Virginia, west to the Pacific; Missouri. **Adult:** Forewing light grayish, with slight blue-violet cast, mottled with darker gray; hindwing clear. Eggs are laid singly on twigs and branches of trees. **Spread:** 1.4–1.6". **Larva:** Stubby, smooth, light green, with broken dorsal stripe, fragmentary subdorsal and lateral lines, wide stigmatal stripe, tubercles surrounded by white dots; head brownish; from 1" to 1.25" long full grown. Single brood. Pupates in flimsy cocoon in the soil. Hibernates as adult (a few as pupa) in woods or other shelter near orchards. **Food:** Common tree fruits and field crops, deciduous trees. The larvae bore into small green apples; the shallow cavities heal over without much harm, or may cause fruit drop, with possible benefits from thinning. Where the population is high they may defoliate woodland trees or invade corn fields.

(211) CORN EARWORM: *Heliothis (Helicoverpa, Chloridea) zea.* **Fig. 416.**

Also known as the bollworm (on cotton) and the tomato fruitworm (on tomato). **Range:** World-wide in distribution. (The Old World form is *armigera.*) **Adult:** Olive-green to grayish brown and reddish brown; forewing with obscure black spot near middle of front margin, scattered indistinct black markings, dark submarginal band; hindwing pale, with dark transverse streak near center, and wide dark submarginal band interrupted by light patch near middle. It flies on cloudy days but mostly in early evening; not strongly attracted to lights. Eggs—pale yellow or brown on hatching—are laid singly on leaves of young plants and later on the silk. **Spread:** 1.5". **Larva:** Whitish, with black head, when newly hatched; four pairs of abdominal prolegs, about 1.5" long full grown, skin set with little spines. The color varies: some have stripes of varying shades of cream, yellow, slate, and black; others have no stripes and may be pink, green, cream, or yellow. It feeds in whorl of leaves, on tassels and silk, later in ear of corn, and rolls up when disturbed. They are cannibalistic—when two come in contact they fight and one eats the other. One to seven broods, three or four in the Corn Belt. Hibernates as a pupa in cell made in the soil. It is unable to withstand severe winters; populations north of the 40th parallel are mostly progeny of migrants. **Food:** Corn, tomatoes, cotton, soybeans, vetch, cowpeas, alfalfa, tobacco, squash, beans, peanuts, peas, potatoes, abutilon, chrysanthemum, geranium, dahlia, gladiolus, and many other plants. It is the worst pest of corn and the second most destructive insect in the U.S. [See Hymenoptera (35); Diptera (96).]

(212) TOBACCO BUDWORM: *Heliothis virescens.* **Fig. 417.**

Range: Widespread in the southern states, California. **Adult:** Forewing light green, crossed with three oblique stripes of lighter shade; hindwing silvery, with brownish fringe. Active at night only. Eggs are laid singly on underside of leaves. **Spread:** 1.25″. **Larva:** Usually green, varying to yellowish brown or dark reddish brown, skin set with small spinules; about 1.5″ long full grown. The young larva eats sparingly of its own eggshell, then the leaf, moving gradually to the bud. Four or five broods. Hibernates as pupa in the ground. **Food:** Almost entirely tobacco for the early broods; later, they feed on beggarweed, and to a limited extent cotton, tomatoes, garden peas, sweet peas; the preferred host in California is geranium. It is parasitized by a braconid wasp (*Cardiochiles nigriceps*) and a sarcophagid fly; *Polistes* wasps and a green spider destroy many.

(213) CABBAGE LOOPER: *Trichoplusia ni.* **Fig. 418.**

Range: Throughout the U.S. and southern Canada. **Adult:** Forewing grayish brown, mottled, with silvery spot resembling a V or figure 8 near the center; hindwing paler, darker toward outer margin. It flies at night, eggs are laid singly on upper side of leaves. **Spread:** 1.5″. **Larva:** Green, with two white dorsal stripes and white lateral stripe along the spiracles, about 1.2″ full grown; body tapered toward the head, covered with spinules. It has *two* pairs of abdominal prolegs (excluding anal pair), walks with a looping motion. Often found with the imported cabbageworm (20), which it resembles, and diamondback moth (284) on cole crops. Three or more broods. Hibernates as pupa in thin cocoon attached to leaf; northern populations are believed to start with migrations from the South. **Food:** Cabbage and related plants, cotton, potatoes, spinach, lettuce, celery, parsley, tomatoes, soybeans, chrysanthemums, geraniums, carnations, and other plants. It is an important pest of cotton in the Southwest.* The **Alfalfa Looper,**

* Epizootics of polyhedrosis virus disease often occur in populations of the cabbage looper throughout the U.S.; like the imported cabbageworm, it is susceptible to the bacterium *Bacillus thüringiensis*. The alfalfa looper is often infected by a polyhedrosis or a granulosis virus in California, Washington, and British Columbia.

Autographa californica (Color plate 7b), is similar, somewhat larger, forewing glossy and more distinctly marked, with white spot resembling Y at the center and a transverse zigzag line extending from foot of Y; the larva has prolegs as in the cabbage looper, can be distinguished by the much smaller spinules; it attacks legumes, grains, cotton, many garden plants and fruits. The **Green Cloverworm**, *Plathypena scabra,* is a pest of soybeans and cowpeas in the South; the forewings, which are dark brown with lighter areas and black spots, and the grayish brown hindwings form a triangle at rest; the palpi are conspicuous, pointed forward; the larva—light green with four faint whitish stripes down the back, about 1″ long—is a looper, with *three* pairs of abdominal prolegs (excluding anal pair), drops to the ground when disturbed.

(214) COTTON LEAFWORM: *Alabama argillacea.* **Fig. 419.**

Range: Eastern Canada, eastern half of the U.S., southwestern states (excepting California); a tropical species, widespread seasonally by migrations. **Adult:** Tan or olive-gray, sometimes with purple tinge. It flies after sunset, breeds only where cotton grows; wheel-like eggs are laid singly on underside of cotton leaves. Strong fliers, they range far from cotton fields. The proboscis is adapted for piercing skin of ripe fruit on which they feed in northernmost part of range. **Spread:** 1.25″. **Larva:** Smooth, light to dark yellowish green; in dark forms the back is velvet black with white or pale yellow middorsal line; white subdorsal and lateral stripes, four black dots forming square on top of each body segment; about 1.5″ long full grown. It feeds on underside of leaves between veins, sometimes skeletonizing them. It forms a web in folded leaf to pupate; other larvae may eat the surrounding leaf, leaving the pupa hanging by threads attached only to the veins. Two to eight broods. Overwinters in Central and South America. **Food:** Cotton, rose mallow, *Hibiscus,* Portia tree; adults sometimes damage cantaloupes, grapes, peaches, and plums. It was the worst pest of cotton prior to advent of the boll weevil. The **Brown Cotton Leafworm**, *Acontia dacia,* occurs in Texas, Arkansas, and Louisiana; the larva is smooth, reddish brown, with white mottling, has a dorsal hump on the eighth abdominal segment.

(215) FLORIDA FERN CATERPILLAR: *Callopistria floridensis.* **Fig. 420.**

Range: Southern states; a pest of ferneries and greenhouses. **Adult:** Forewings brownish, with V-shaped darker patches along costal margin; hindwing plain, pale brownish. **Spread:** 1″. **Larva:** Two color phases when mature: pale green, velvety black; with two wavy, white lateral stripes; about 1.5″ long full grown. Pupates in cocoon in the soil. Several broods. **Food:** Ferns.

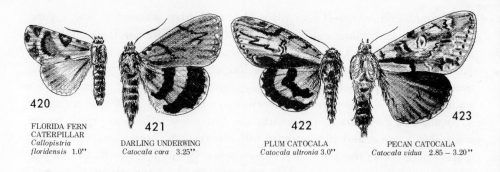

420

FLORIDA FERN
CATERPILLAR
*Callopistria
floridensis* 1.0"

421

DARLING UNDERWING
Catocala cara 3.25"

422

PLUM CATOCALA
Catocala ultronia 3.0"

423

PECAN CATOCALA
Catocala vidua 2.85 – 3.20"

(216) DARLING UNDERWING: *Catocala cara.* **Fig. 421.**

The genus *Catocala,* or "underwings," includes a great many species, most of which are not of economic importance but are prized by collectors. They are medium to fairly large moths, usually with brownish or grayish mottled forewings; the distinctive hindwings have solid bands of black and red, yellow, or white, or plain dark brown or black. Though often brilliantly colored the hindwings are displayed only in flight. Their eggs are usually laid on trees in the crevices of the bark; pupation takes place in litter. *C. cara,* a beautiful eastern species widely distributed, is fairly typical of the group. **Adult:** Forewing brownish, with short, black streaks along costal margin; hindwing red, with black bands and yellowish margin. **Spread:** 3.25". **Larva:** Clay-colored to brownish, with darker dorsal stripe and broken lateral stripe on each side, reddish or yellowish "warts" on the back, purplish head. **Food:** Poplar, willow. The larva of the closely related **Plum Catocala,** *C. ultronia* (Fig. 422), feeds on plum, apple, wild cherry; the adult is very similar to *cara* except that the forewing is more grayish.

(217) PECAN CATOCALA: *Catocala vidua.* **Fig. 423.**

Also known as the widowed underwing because of its somber colors. **Range:** Canada to Florida and westward. **Adult:** Forewing grayish with brownish streaks; hindwing black. **Spread:** 2.85–3.2". **Larva:** Grayish, the pale violet coloring obscured by stripes consisting of black spots; from 2.3" to 3" long full grown. Hibernates in egg stage. **Food:** Oak, hickory, butternut, walnut, pecan. The **Oak Catocala,** *C. maestosa* (Fig. 424), is similar, except that the hindwing is more brownish. Both are pests of pecan in California and sometimes strip the foliage.

(218) VELVETBEAN CATERPILLAR: *Anticarsia gemmatalis.* **Fig. 425.**

A tropical species, migrating northward each season. **Range:** The Gulf states, north as far as Wisconsin. **Adult:** Both wings grayish brown, crossed with brown or black zigzag lines. **Spread:** 1.5". **Larva:** Black or green, with narrow lighter stripes along the back and sides; about 1.5" long full grown. It is very active, springs into the air and wriggles violently when disturbed,

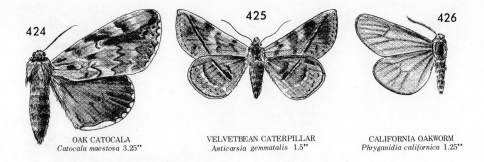

424 OAK CATOCALA
Catocala maestosa 3.25"

425 VELVETBEAN CATERPILLAR
Anticarsia gemmatalis 1.5"

426 CALIFORNIA OAKWORM
Phryganidia californica 1.25"

and "spits" a brownish substance; it feeds from the top of the plant downward. Eggs, laid singly on underside of leaf, are white and turn orange prior to hatching. Pupates in cell in the soil. Like the cotton leafworm (214) they overwinter in the tropics; local populations die in the fall and are built up again from eggs laid by swarms arriving in June and July. **Food:** Velvetbean, peanut, and soybean primarily in Florida, South Carolina, Georgia, Alabama, and Louisiana; also cotton, kudzu, alfalfa, cowpeas, horse beans, snap beans, lima beans, coffeeweeds.

(219) CALIFORNIA OAKWORM: *Phryganidia californica* (Dioptidae). **Fig. 426.**

We include one species of this small group occurring only in California. **Adult:** Pale brown, with darker veins. The male may be distinguished by a yellowish patch on the forehead and by the broader and more feathery antennae. Eggs are laid on underside of leaves, on tree trunks or other places, from two to forty in a group. **Spread:** 1.25". **Larva:** Dark olive-green, with conspicuous yellow and black stripes down the back and sides; about 1" long full grown. Hibernates in egg stage or as young larva. **Food:** Oaks, which are periodically defoliated. Other surrounding trees are sometimes attacked. Normally controlled by its enemies, such as the spined soldier bug, tachinid and wasp parasites, and by a "wilt" disease.

Notodontid Moths (Notodontidae)

The notodontid moths resemble the noctuids except that the hindwing is smaller and the body is hairier. They are often referred to as the "prominents" because of the projecting lobes on the inner margins of the front wings. In Britain some of the more fuzzy species are called puss moths and kitten moths. They are strongly attracted to lights, rest with the wings held rooflike over the body. The middle tibiae have one pair of spurs, the hind ones have two pairs; the tarsi are short and hairy, and the femora are covered with long hairs. Anal prolegs are lacking in the larvae, and they often raise their hind ends into the air, swaying them to and fro. The pupae are usually naked.

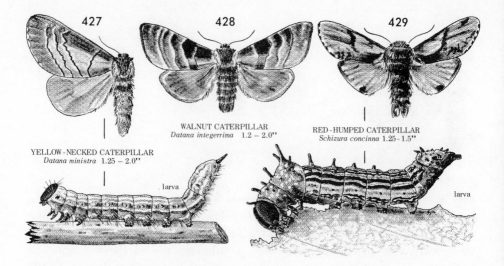

427

428

429

WALNUT CATERPILLAR
Datana integerrima 1.2 – 2.0"

RED-HUMPED CATERPILLAR
Schizura concinna 1.25 - 1.5"

YELLOW-NECKED CATERPILLAR
Datana ministra 1.25 – 2.0"

larva

larva

(220) YELLOW-NECKED CATERPILLAR: *Datana ministra.* **Fig. 427.**

Range: Canada and the entire U.S. **Adult:** Forewing reddish brown, with darker wavy lines crossing them; hindwings pale yellowish. Eggs are laid in flat masses on underside of leaf. **Spread:** 1.25–2". **Larva:** Black and yellow striped, with yellow ring around the neck; about 1.5" long full grown. They feed in colonies, many crowded together on a single leaf at first, with all heads pointed toward the edge; later they spread over more leaves and migrate to other parts of the tree. When at rest or disturbed they raise both ends of their bodies, clinging to the leaf with prolegs only, and bend as if trying to touch heads with tails. Hibernates as pupa in the ground. **Food:** Apple, pear, cherry, quince, blueberry, oak, hickory, beech, birch, elm, basswood.

(221) WALNUT CATERPILLAR: *Datana integerrima.* **Fig. 428.**

Range: Eastern and southern U.S., west to Kansas. **Adult:** Forewing brown, crossed with darker wavy lines; it may be distinguished from *ministra* (220) by the white borders along these transverse lines; hindwing plain, pale brownish. Thorax covered with tuft of dark brown hairs posteriorly. Eggs are laid in masses on underside of leaf. **Spread:** 1.2–2". **Larva:** Reddish brown, with white stripes and black head; the body changes later to brown and finally to black. Covered with long, soft, white hairs; about 2" long full grown. They feed in colonies; the larger caterpillars crawl en masse to the trunk and larger limbs to molt, after which they return to the leaves to feed. One or two broods. Hibernates as pupa in the ground. **Food:** Walnut, butternut, pecan, hickory; occasionally apple, peach, oak, beech, willow, honey locust, sumac.

(222) RED-HUMPED CATERPILLAR: *Schizura concinna.* **Fig. 429.**

Range: Canada and the entire U.S. **Adult:** Forewing reddish brown to gray, with curved brown line across the center; hindwing light gray, with brownish markings along fringe; female with few or no markings on the wings. **Spread:** 1.25–1.5″. **Larva:** Yellow, with reddish and white stripes, bright red head; fourth segment bright red, with pronounced hump and two black spines, shorter spines on other segments; about 1.5″ long full grown. Habits similar to those of the yellow-necked caterpillar (220) except that pupation takes place in a loose silken cocoon in the ground or litter. Hibernates as full-grown larva in cocoon, pupating the following summer. **Food:** Apple, pear, cherry, walnut, rose, blackberry, and other plants. The **Variable Oak Leaf Caterpillar,** *Heterocampa manteo,* attacks hardwoods in the Midwest and eastward; the moth is ash gray, with narrow wavy band near outer margin of forewing, spread about 1.5″; the larva is yellowish green with pale yellow median stripe bordered by reddish brown. The **Saddled Prominent,** *H. guttivitta,* sometimes defoliates forest and shade trees (hardwoods) in the East; base color of moth is green, with mottled-gray dusting.

Tussock Moths (Lymantriidae)

Tussock moths are named for their tufted hairs or tussocks. The family name of Lymantriidae means "injurious." The moth's body is covered with long hairs; females use the tuft at the tip of the abdomen to cover their eggs. They have a frenulum but are without a proboscis, and lack the two ocelli which most moths have near the base of the antennae. In some species the females have only rudimentary wings and are unable to fly. The antennae of the males are bushy, doubly branched on one side. The larvae are covered with dense hairs, some arranged in tufts; certain species have hairs with poisonous barbs that can cause an irritating skin rash.

(223) WHITE-MARKED TUSSOCK MOTH: *Hemerocampa leucostigma.* **Fig. 430.**

Range: Eastern Canada and the U.S., west to British Columbia and Colorado. **Adult:** Wings of male gray, forewing with darker wavy crossbands. The female is wingless and dies soon after laying her eggs; they are laid in batches on the cocoon from which she emerged, and are covered with hairs and a frothy substance that hardens. **Spread** (male): 1.25″. **Larva:** Light brown, with yellow and black stripes, a bright red spot on the sixth and seventh abdominal segments; head bright red. Three long pencil-like tufts of black hairs, one on each side of the head, and one near the posterior end; four short tufts of white stinging hairs on first four abdominal segments; about 1.25″ full grown. It skeletonizes the leaves. Larvae are often seen hanging from trees by silken threads. Pupates in cocoon made of

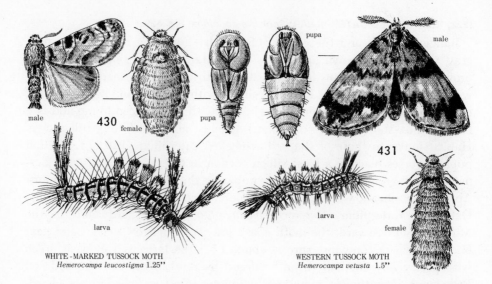

WHITE-MARKED TUSSOCK MOTH
Hemerocampa leucostigma 1.25"

WESTERN TUSSOCK MOTH
Hemerocampa vetusta 1.5"

hairs and silk, spun on tree trunk, branch, or elsewhere. One to three broods. Hibernates in egg stage. **Food:** Almost any deciduous tree; sometimes a pest of apple, pear, quince, plum. The **Western Tussock Moth,** *H. vetusta* (Fig. 431), found along the Pacific coast, is very similar in appearance and habits; it is a pest of deciduous trees, and is found in orchards of the Pacific Northwest and on apricots in California.

(224) DOUGLAS-FIR TUSSOCK MOTH: *Hemerocampa pseudotsugata*. Fig. 432.

Range: Nevada, Colorado, Idaho, Washington, eastern Oregon, British Columbia. **Adult:** Dull brownish gray. The female is wingless; she lays her eggs in a mass on top of the abandoned cocoon, covering them with froth and hairs. **Spread** (male): 1". **Larva:** Gray to light brown, with numerous red spots and shiny black head; broken narrow orange stripes along the sides; from .75" to 1" full grown. Two pencil-like tufts of black hairs behind the head, a longer one at the posterior end; short brushlike tuft of light brown or cream-colored hairs on top of the first four and last abdominal segments. They travel in search of food, dropping from the trees by threads. One brood. Hibernates in egg stage. **Food:** Douglas fir and true fir. A major defoliator of fir; serious outbreaks have occurred in the West.

(225) RUSTY TUSSOCK MOTH: *Orgyia antiqua*. Fig. 433.

A cosmopolitan species. **Range:** Over most of North America and Europe; in the West, from California to British Columbia, east to Montana. **Adult:** Wings of male rusty brown, with white spot near anal angle of forewing. The female is practically wingless, and deposits her eggs in a mass on the remains of her own cocoon. **Spread** (male): 1.2". **Larva:** Blackish or bluish, with four short tufts of white hairs on the back; two long pencil-like tufts of

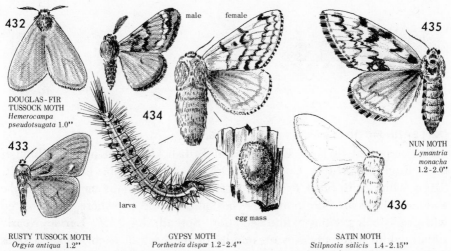

432

DOUGLAS-FIR
TUSSOCK MOTH
*Hemerocampa
pseudotsugata* 1.0"

433

RUSTY TUSSOCK MOTH
Orgyia antiqua 1.2"

male   female

434

larva

egg mass

GYPSY MOTH
Porthetria dispar 1.2-2.4"

435

NUN MOTH
*Lymantria
monacha*
1.2-2.0"

436

SATIN MOTH
Stilpnotia salicis 1.4-2.15"

black hairs in front, and one on each side near the middle; about 1" long full grown. Two broods. Hibernates in egg stage. **Food:** Alder, aspen, willow, poplar, ash, oak, various conifers and fruit trees. In orchards the fruit is sometimes damaged as well as the leaves.

(226) GYPSY MOTH: *Porthetria (Lymantria) dispar.* **Fig. 434.**

Accidentally introduced into Massachusetts from Europe in 1869. **Range:** Southeastern Canada, New England, New York, New Jersey, Pennsylvania; since 1958 it has spread southward along the Appalachian Range and westward to Minnesota and Texas. **Adult:** Male dull brownish gray, female creamy white; both with darker wavy lines crossing forewing, somewhat diffuse in the male. The male is light-bodied and a strong flier; the female is much larger, heavy-bodied, and able only to flutter along the ground. Eggs are laid on tree trunks, under stones, on fences and buildings, in masses of about 400 and covered with hairs. **Spread:** 1.2–2.4". **Larva:** Somewhat flattened; sooty gray, with long stiff tufts of brown hairs along the sides, two rows of tubercles down the back (the first five pairs blue, the next six pairs red); about 2" long full grown. It feeds mostly at night, descends to the ground and hides in the litter by day. Five (male) and six (female) instars; some strains undergo an additional instar. One brood. Hibernates in egg stage. **Food:** Oak and other deciduous trees; white pine, hemlock, and other evergreens.* [See Coleoptera (16); Hymenoptera (21) and (42); Diptera (98).] The closely related **Black Arches** or **Nun Moth,** *Lymantria monacha* (Fig. 435), of Europe is also very injurious to trees but not known to occur in North America; both sexes are similar; the forewings have black wavy lines on white background; a potential pest of eastern forests.

* See Henry A. Bess, *Population Ecology of the Gypsy Moth* Porthetria dispar L. (*Lepidoptera: Lymantriidae*), Bul. 646, Connecticut Agr. Exp. Sta., New Haven (1961).

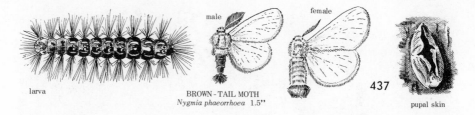

male   female

larva

437

BROWN-TAIL MOTH
Nygmia phaeorrhoea 1.5"

pupal skin

(227) SATIN MOTH: *Stilpnotia salicis.* **Fig. 436.**

First reported in New England and British Columbia in 1920. **Range:** New England, British Columbia, Oregon, Washington; Europe. **Adult:** Satiny white, with black eyes and legs, and a tuft of hairs on the tip of the abdomen. Eggs are laid in oval batches on trees and elsewhere, and covered with a glistening satiny white secretion. **Spread:** 1.4–2.15". **Larva:** Nearly square white patch of hairs on middle of the back of each segment; sides with irregular black and clay-colored markings, dense covering of short hairs arising from small warts, a few long hairs but no pencils; about 2" long full grown. One brood. Hibernates as young larva in cocoon spun in crevice of bark. **Food:** Poplar, willow, and other roadside and shade trees. [See Hymenoptera (21).]

(228) BROWN-TAIL MOTH: *Nygmia phaeorrhoea.* **Fig. 437.**

Introduced into Massachusetts in 1897. **Range:** New England, Nova Scotia, New Brunswick; Europe. **Adult:** Wings and thorax white; abdomen mostly brown, with a brownish red tuft of hairs on the tip. Strongly attracted to light. Moths emerge in July. Eggs are laid in masses on underside of leaves and covered with brown hairs, hatch in August or September. **Spread:** 1.5". **Larva:** Dark brown, nearly black, with bright red tubercle on top of two posterior segments; broken white stripe along the sides; about 1.5" long full grown. The body is covered with brown hairs except for a row of nearly white tufts on the sides; these hairs are barbed and contain poison, can cause an irritating rash and are dangerous if breathed into the lungs. [See (189), (193), (223), (238) to (241) for other "stinging caterpillars."] One brood. Hibernates as young larva in web shelter. **Food:** Willow, oak, cherry, apple, pear, and other deciduous trees. The fungus *Entomophthora aulicae,* which attacks this and many other caterpillars, has been mass-produced utilizing brown-tail moth caterpillars and used successfully to control them in Massachusetts. [See Diptera (98).]

Tent Caterpillar Moths (Lasiocampidae)

The moths in this family are medium in size, hairy, brownish yellow to brown, with a wide band of lighter or darker shade (sometimes bordered in white) across the forewing. Both sexes have bipectinate antennae. The

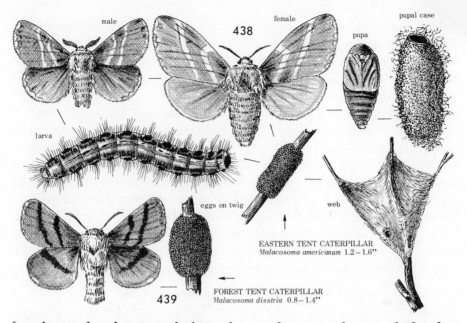

male
female
438
pupa
pupal case

larva

eggs on twig
web

EASTERN TENT CATERPILLAR
Malacosoma americanum 1.2–1.6"

439
FOREST TENT CATERPILLAR
Malacosoma disstria 0.8–1.4"

frenulum and proboscis are lacking, the mouthparts nonfunctional; they live only a few days, since no food is consumed. Eggs are banded around small twigs and covered with a foamy brown substance which hardens. They hibernate in the egg stage, which is usually of about nine months' duration; the overwintering form is actually a tiny unhatched larva within the egg. The larvae are yellowish to brown or black, with blue or orange spots and lines, and covered with long, fine hairs, those along the sides usually tufted and directed downward. They live in colonies and generally construct tents in forks or crotches of trees for shelter; trails of silk are spun by the larvae as they move away from the nest to feed. [See Fall Webworm (197).] There are usually five or six instars. Single-brooded, they pupate in cocoons constructed of silk and hair, and sometimes enclose them in folded leaves. The larvae often defoliate shade and forest trees.

(229) EASTERN TENT CATERPILLAR: *Malacosoma americanum.* **Fig. 438.**

Range: Throughout the southern part of eastern Canada and eastern half of the U.S. **Adult:** Yellowish brown to dark chocolate brown, female often lighter than male; pair of oblique whitish lines across forewing; hindwing uniform dark brown, sometimes with faint white area. **Spread:** 1.2–1.6". **Larva:** Head black, with many black setae; body with continuous, even, yellowish white median dorsal stripe bordered with reddish brown and black wavy lines; blue spots, reddish brown and yellow stripes along the sides. **Food:** Cherry, prune, peach, apple, hawthorn, many other deciduous trees and shrubs. The **California Tent Caterpillar,** *M. californicum,* and subspecies occur throughout most of western North America; adults are

own to yellowish or grayish, with the lines on the forewings usu-
at; the larva is reddish orange to brown above, pale brown below,
blue line on the sides; it feeds on willow, oak, madroño, ash, and
her trees. The **Great Basin Tent Caterpillar,** *M. c. fragile,* is less
variable than the others, usually grayish, with chocolate brown markings,
often with faint yellowish tint; the larva is brownish to black, with whitish
hairs, wide blue median dorsal stripe bordered by a fine orange line, two
blue spots on sides of each segment; it occurs in the desert areas, feeds on
poplar, aspen, willow, oak, bitterbush and other shrubs. The **Western Tent
Caterpillar,** *M. c. pluviale,* ranges from the Pacific Northwest and coastal
region eastward through Canada; adults are orange-brown to dark orange-
brown or reddish brown, lines on forewing yellowish; the larva is brown,
with a row of elliptical blue spots down the middle of the back and two
orange spots on each segment, orange spots and lines on the sides; it feeds
on alder, willow, poplar, apple, plum, cherry, and many other trees.

(230) FOREST TENT CATERPILLAR: *Malacosoma disstria.* **Fig. 439.**

This is the only species of *Malacosoma* that does not form a tent. **Range:**
Widely distributed throughout North America. **Adult:** Light yellow to dull
brownish; male quite variable, with band across forewing; female with two
brown lines across forewing. **Spread:** .8–1.4″. **Larva:** Head light blue, with
black mottling, sparse whitish orange setae; body pale bluish to dusky
brown or black, with row of diamond- or keyhole-shaped white spots down
the middle, reddish brown irregular broken subdorsal stripe. Small larvae
form clusters on leaves and twigs, larger ones on branches and trunks when
not feeding. Spins cocoon in one or more leaves webbed together with
silk (other species usually form less conspicuous cocoons in litter). **Food:**
Alder, aspen, poplar, willow, birch, oak, sweet gum, tupilo gum, black gum,
ash, and other broad-leaved trees. [See Diptera (91).]

Silkworm Moths (Bombycidae)

This is a small family of only a few genera, occurring in Asia. We include
the species used in sericulture because of its historic importance and wide
use in research. The culture of silkworms, as old as recorded history, was
for 2,000 years a secret known only to the Chinese. In A.D. 555 two monks
smuggled eggs out of China to Constantinople and set the stage for the development of an important industry in France, Italy, and Spain. The industry probably reached its greatest subsequent development in Japan. The
culture of silkworms never flourished in the U.S. as did the manufacture of
silk fabrics, owing mainly to high labor costs. The development of syn-
thetic fibers has of course reduced the importance of silk. Silkworms suffer

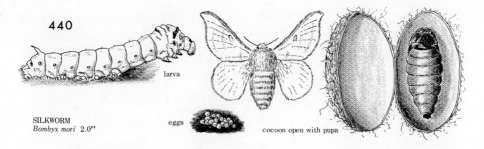

440

larva

SILKWORM
Bombyx mori 2.0"

eggs

cocoon open with pupa

from fungus, virus, and microsporidian (protozoan) infections.* The Microsporidia—transmitted from one generation to the next via the egg—ravaged the silk menageries of France a century ago, nearly putting an end to the silk industry. Louis Pasteur's research on the disease paved the way for his important contributions to bacteriology.

(231) SILKWORM: *Bombyx mori.* **Fig. 440.**

Adult: Creamy white, with darker band, wider and more pronounced in the male, across both wings. It has a relatively large body, feeble wings, hardly ever flies; it imbibes no food, lives only two or three days—long enough to lay 300 to 400 eggs. **Spread:** 2". **Larva:** Reaches full size, about 3" long, in three to four weeks. Silk for spinning the cocoon is secreted by true salivary glands opening into the mouth. The saliva hardens on exposure to the air, forming a delicate but strong flexible thread. It takes three days to form the cocoon, constructed from a single thread that averages 1,000 feet long. **Food:** Mulberry. It takes about a ton of leaves and 25,000 cocoons to make one pound of silk. In menageries only a relatively few moths, to produce the necessary eggs, are allowed to develop and emerge from the cocoons, since in doing so they secrete a fluid which softens the end of the cocoon and splits the silken thread into many pieces.

Geometrid Moths (Geometridae)

The geometrid moths comprise a large world-wide family, occurring practically everywhere that vegetation grows. They are often referred to as measuringworms, spanworms, loopers, or inchworms. The adults are medium in size, with relatively large wings and slight bodies, and are often weak fliers; the females of some species are wingless or have only rudimentary wings and are unable to fly. They usually have a frenulum and proboscis,

* The virus disease is called "jaundice" for the yellowish color of infected larvae, or *grasserie* (in France) for the swelling of the body. The microsporidian disease is called *pebrine* in France for the telltale spots on the infected silkworm; the fungus (*Beauveria bassiana*) infection is called *muscardine* for the resemblance of the white fruiting bodies covering the silkworm to a powdered-sugar confection.

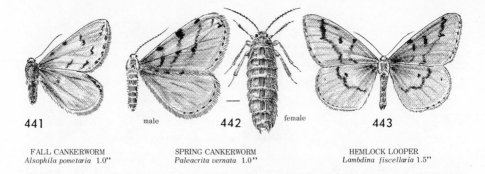

441

FALL CANKERWORM
Alsophila pometaria 1.0"

male   442   female

SPRING CANKERWORM
Paleacrita vernata 1.0"

443

HEMLOCK LOOPER
Lambdina fiscellaria 1.5"

and are semidiurnal or crepuscular. The tympana are located in the first abdominal segment (in most moths they are in the mesothorax). The caterpillars are smooth, almost hairless, with only two or three pairs of prolegs on the posterior segments of the abdomen, which accounts for their characteristic looping form of locomotion. They reach out with the forelegs while resting on the prolegs, grasp with their claws, and bring the rear forward by humping the back.

(232) FALL CANKERWORM: *Alsophila pometaria.* **Fig. 441.**

So called because the adults emerge in November or December. **Range:** Southeastern Canada, northeastern U.S. to North Carolina, west to Missouri and Manitoba; also found in Utah, Colorado, California. **Adult:** Brownish gray; forewing of male with two diffuse, white transverse bands; vestigial wings of female do not extend beyond the mesothorax. Brownish eggs, shaped like tiny flowerpots, are laid in neat clusters around twigs or branches. **Spread** (male): About 1". **Larva:** Light to dark green, with dark stripes on the back, three narrow whitish stripes on the side above the spiracles and a yellowish one below; about 1" long full grown. It has three pairs of prolegs, those on the fifth abdominal segment shorter than the others and useless. It closely resembles the spring cankerworm (233); both are often found in the same tree or hanging close together by silken threads. Pupates in cocoon in the ground. One brood. Hibernates in egg stage. Larval populations usually reach their peak in June. **Food:** Apple and elm preferably; also several oaks, cherry, hickory, maple, ash, linden. The **Winter Moth,** *Operophtera* (*Cheimatobia*) *brumata,* the **Bruce Spanworm,** *O. bruceata,* and the **Linden Looper,** *Erannis tiliaria,* are similar to the fall cankerworm and often found in the same areas in its eastern range. Forewings of male *Operophtera* are gray to dusky brown, with numerous scalloped, pale brown transverse lines. Vestigial forewings of females are longer than the width of the thorax and extend to the third abdominal segment in the winter moth; they are shorter than the width of the thorax and extend to the first abdominal segment in the Bruce spanworm. Hind tibial spurs are also shorter and mouthparts smaller in the fall cankerworm.

Females of the linden looper may be distinguished by the two rows of black spots on the back. The Bruce spanworm is a major defoliator of aspen in the prairie regions.

(233) SPRING CANKERWORM: *Paleacrita vernata.* **Fig. 442.**

So called because the moths emerge in March and April. **Range:** Same as fall cankerworm (232) except that it extends southwestward to eastern Texas. **Adult:** Closely resembles the fall cankerworm except that the wings are lighter in color, semitransparent, and more delicate. The wingless females may be distinguished by the dark middorsal stripe and the double row of reddish spines across the top of the abdominal segments. Oval, pearly eggs are laid in clusters in bark crevices. **Spread** (male): About 1". **Larva:** Light green to brown or black, with whitish lines down the back, about .75" long full grown; it has two pairs of prolegs. One brood. Hibernates in naked pupa in the ground. **Food:** See (232) above. Birds take a heavy toll of cankerworms; the English sparrow was originally imported to combat them.

(234) HEMLOCK LOOPER: *Lambdina fiscellaria.* **Fig. 443.**

Range: Northeastern states, through Canada, the Great Lakes states, northwestern coastal region. **Adult:** Light brownish yellow, with two darker wavy lines across forewing, one across hindwing. Tiny blue to gray-green or brown eggs, with a characteristic impression, are laid on trunks, twigs, and branches. **Spread:** 1.5". **Larva:** Green to brown, with diamond-shaped markings on back, about 1.5" fully grown. It has two pairs of prolegs. The larvae drop to the ground on silken threads, which are sometimes so numerous that "the whole forest looks and feels like one big cobweb." Pupates in naked pupa in bark crevice or ground litter. One brood. Hibernates in egg stage. **Food:** Hemlock, Douglas fir, western red cedar, balsam fir, spruce. The eastern form, *L. f. fiscellaria,* is a major forest pest in Newfoundland. Several serious outbreaks (each lasting about three years) of the **Western Hemlock Looper,** *L. f. lugubrosa,* have occurred in the Northwest. The **Western Oak Looper,** *L. f. somniaria,* another subspecies, infests oak. The **False Hemlock Looper,** *Nepytia canosaria,* attacks hemlock, spruce, Douglas fir, and larch in the same area.

(235) WALNUT SPANWORM: *Coniodes plumigeraria.* **Fig. 444.**

Range: Southern California. **Adult:** Male forewing silvery gray, with four wavy, brown crossbands; hindwing has brown spot near center. Wingless female is brownish gray, tinged with bronze. Eggs—oval, iridescent bronze, turning to light blue before hatching—are laid in masses on twigs and branches. **Spread** (male): 1.5". **Larva:** Black, with white patches when newly hatched; light pinkish gray to purplish, with black or brown spots, about 1" long when full grown. It has two pairs of prolegs. One brood.

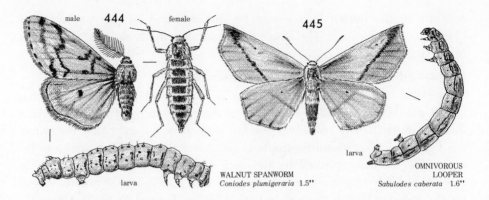

male 444   female   445

larva

WALNUT SPANWORM
Coniodes plumigeraria 1.5"

larva

OMNIVOROUS
LOOPER
Sabulodes caberata 1.6"

Hibernates as naked pupa in the soil. **Food:** Live oak normally; occasionally walnut, apple, prune, and other fruits.

(236) OMNIVOROUS LOOPER: *Sabulodes caberata*. Fig. 445.

Range: Throughout California. **Adult:** Dull brown or yellow, with two irregular darker crossbands; underside nearly white. Flies only at night. Eggs —ovoid, metallic green to chocolate brown—are laid in clusters on underside of leaves. **Spread:** 1.6". **Larva:** Yellow to pale green or pink, with yellow, brown, or green stripes on the back and sides, and some black markings. Up to 2" long full grown. It has two pairs of prolegs, retreats to webbed leaves when not feeding, and may drop on a thread when alarmed. Pupates in folded, webbed leaves. Five or six broods. **Food:** Many plants; often found in avocado groves where it normally causes little damage.

Bagworm Moths (Psychidae)

The adults of these strange insects are small to medium in size; the females are wingless, most of them little more than a "maggot." The larva is protected by a case made of silk and covered with leaves, twigs, or grass, and for this reason is sometimes called "basketworm." Pupation takes place within the case or basket. The eggs are laid here also, in some instances within the larval skin. Mating is accomplished by the female protruding her body from the case or the male inserting his abdomen. Parthenogenesis, rare among moths, occurs in some species.* The male's wings are covered with loosely attached scales, most of which are lost during his initial flight, giving the wings the appearance of semitransparency.

* An interesting example is *Solenobia triquetrella,* which is widespread in Europe and occasionally found in North America; bisexual forms also occur in Europe. The moth (female) is oval viewed from above, wingless, about .15" long, mouse gray, with whitish scales along the sides, and a brush of long wavy hairs across the posterior end upon emergence.

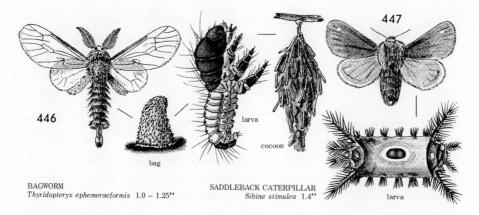

446

447

BAGWORM
Thyridopteryx ephemeraeformis 1.0 – 1.25"

bag

larva

cocoon

SADDLEBACK CATERPILLAR
Sibine stimulea 1.4"

larva

(237) BAGWORM: *Thyridopteryx ephemeraeformis.* **Fig. 446.**

Range: Atlantic states to the Great Plains. Related species occur westward and southwestward to Texas. **Adult:** Body of male is black, wings pale. Female wingless; remains within the bag, protruding the abdomen when mating. Eggs are deposited within the bag, after which female emerges, drops to the ground, and dies. **Spread** (male): 1–1.25". **Larva:** Starts constructing a bag right after hatching; this is carried about as it feeds. Head and thorax, which protrude as it feeds, are dark brown, shiny; the part within the bag is lighter and softer, from 1" to 1.25" long full grown; the bag is 1.5" to 2.5" long. Pupates within the bag. One brood. Hibernates in egg stage. The yellowish eggs may be found during winter in bags abandoned by the females. **Food:** Most deciduous or evergreen trees, which are sometimes defoliated. The bags protect the insects from birds, but not from parasites.

Slug Caterpillar Moths (Limacodidae)

The curious larvae of this group are sluglike, usually with stinging spines or hairs that can cause painful skin irritation. Some are smooth, segmented or unsegmented, head and legs sometimes retractile or barely visible. The cocoons are hard and smooth; spines imbedded within them retain their stinging properties.

(238) SADDLEBACK CATERPILLAR: *Sibine stimulea.* **Fig. 447.**

Range: Eastern and southern states. **Adult:** Velvety; forewing dark reddish brown, with two white dots near apex, hindwing lighter in shade. **Spread:** 1.4". **Larva:** Brown at the ends, including spine-covered horns, green around the middle, with purple or brown "saddle" edged with white; about 1" long full grown. Their stiff hairs or spines are mildly poisonous; children sometimes find them around shrubs and "get stung" by handling

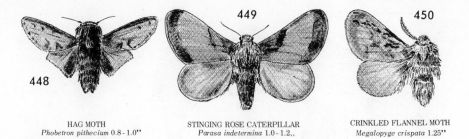

HAG MOTH
Phobetron pithecium 0.8-1.0"

STINGING ROSE CATERPILLAR
Parasa indetermina 1.0-1.2,,

CRINKLED FLANNEL MOTH
Megalopyge crispata 1.25"

them. One brood. Hibernates in smooth cocoon with stinging hairs imbedded. **Food:** Apple, cherry, corn, many ornamental plants; citrus in Florida. [See (240) for other "stinging caterpillars."]

(239) HAG MOTH: *Phobetron pithecium.* **Fig. 448.**

Range: Eastern and southern states. **Adult:** Male and female are quite different. The male is much smaller, somewhat resembles the bagworm (237); the body is black, wings reddish brown. Forewing of the female yellowish, with brown mottling; hindwing brownish. **Spread:** .8–1". **Larva:** Brownish, about .75" long full grown. Four pairs of long, plumelike processes on the back, projecting out to the sides, suggestive of the disarranged hairs of a hag; among the brown hairs of the plumes are longer black stinging hairs. One brood. Hibernates in cocoon attached to leaf or twig. **Food:** Various trees and ornamental shrubs; citrus in Florida. The **Stinging Rose Caterpillar,** *Parasa indetermina* (Fig. 449), is similar in habits and appearance to the hag moth; the larva has seven pairs of shorter processes bearing poisonous spines, is strikingly marked with red, white, and violet stripes, hibernates in cocoon on the ground; the adult is pale cinnamon brown, with green and brown markings, about 1" wingspread; the larvae feed on the underside of rose leaves, are also found on other low-growing shrubs.

(240) CRINKLED FLANNEL MOTH: *Megalopyge crispata* (Megalopygidae). **Fig. 450.**

The larvae of flannel moths usually have poisonous spines which can inflict very painful "stings." They may be distinguished from other "stinging caterpillars" by the seven pairs of prolegs (instead of five). **Range:** Atlantic coast and northern states. **Adult:** Cream-colored, with brownish and black markings. **Spread:** 1.25". **Larva:** Fleshy, covered with long, brown, silky hairs forming a pointed crest along the middle of the back; from .75" to 1" long full grown. **Food:** Apple, cherry, birch, locust, oak, bayberry, raspberry, sweet fern. The **Puss Caterpillar,** *M. opercularis* (Fig. 451), is similar; the larva is broad and flat, covered with long, silky reddish brown hairs, about 1" long full grown; beneath the long hairs are tubercles with poisonous spines, which apparently are the most severe of the stinging caterpillar hairs in their reaction; it is found in Virginia, west to Texas, feeds on hackberry, oak, maple, sycamore, English ivy, rose. The larva of *M. laceyi* can also inflict a painful sting; it feeds on oak, has defoliated large stands of shin oak in Texas. The larva of *Norape cretata* is spotted, sparsely clothed in tufts of hair, feeds on redbud, mimosa, honey locust; the reaction from its sting is usually mild. [See (189), (193), (223), (228), (238), (239), and (241) for other "stinging caterpillars."]

(241) WESTERN GRAPE LEAF SKELETONIZER: *Harrisina brillians* (Zygaenidae). Fig. 452.

Range: Southern California, Utah, Arizona, Mexico. **Adult:** Metallic blue to greenish black, wings long and narrow. Eggs are laid in clusters on underside of leaf. **Spread:** 1″. **Larva:** Yellow, with black crossbands on two thoracic segments, tufts of long black stinging hairs on each body segment, .5″ long full grown. They feed gregariously on the underside of leaves, in neat rows like soldiers on parade. Young larvae feed only on the surface tissue on underside of leaf, leaving a network of small veins; older ones skeletonize the leaf, leaving only the main veins. Where infestations are severe they may eat the fruit. Three broods. Hibernates as pupa in cocoon in bark or ground debris. **Food:** Wild and cultivated grapes; Virginia creeper and Boston ivy. The **Grape Leaf Skeletonizer,** *H. americana,* is very similar, occurs in the eastern half of the country. *H. metallica* is found in Arizona; the larvae may be distinguished by the orange-red collar on the prothorax.

Carpenterworm Moths (Cossidae)

These are fairly large, striking moths, with long, narrow forewings. They lack a proboscis, and are generally weak fliers. The larvae are wood borers, constructing galleries deep in the trunk or branches of trees; they form cocoons with silk and wood chips.

(242) CARPENTERWORM: *Prionoxystus robiniae.* **Fig. 453.**

Range: Throughout the U.S. and southern Canada. **Adult:** Forewing grayish, with darker mottling; hindwing of male has wide orange band along outer margin, that of female is tinged lightly with yellow. They fly at night and are strongly attracted to lights. Eggs are laid in crevices of bark. **Spread:** 3″. **Larva:** White, with pinkish cast and dark brown head, body covered with dark brown tubercles; up to 2″ long full grown. They bore into the trunks of trees, constructing galleries that weaken the structure. The burrows are about .5″ in diameter; occasional holes and "sawdust" are evidence of attack. Hibernates as larva in tunnel; takes three years to complete development. Pupates in cell constructed in the wood. **Food:** Oaks preferably; also locust, elm, maple, ash, poplar, willow, and others.

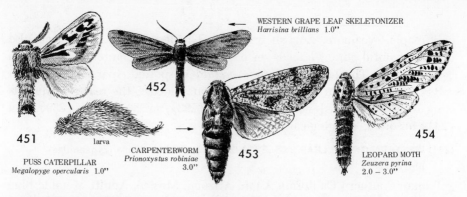

WESTERN GRAPE LEAF SKELETONIZER
Harrisina brillians 1.0"

452

451

larva

PUSS CATERPILLAR
Megalopyge opercularis 1.0"

CARPENTERWORM
Prionoxystus robiniae
3.0"

453

454

LEOPARD MOTH
Zeuzera pyrina
2.0 – 3.0"

(243) LEOPARD MOTH: *Zeuzera pyrina.* **Fig. 454.**

Range: Atlantic coast states, from Delaware northward. A native of Europe. **Adult:** White, with blue spots over entire wings, black spots along the margins. Eggs—oval, salmon-colored—are laid in bark crevices. **Spread:** 2–3". **Larva:** Pinkish white, with dark brown spots; about 1.5" long full grown. They bore into the heartwood of top branches, causing them to die and break off; holes and "sawdust" along the branches are evidence of infestation. Hibernates as partly grown larva; takes two to three years to complete development. Pupates in tunnel. **Food:** Elm, maple preferably; also other deciduous trees.

Clearwing Moths (Aegeriidae)

The wings of this group, as suggested by the name, are transparent to some degree and glassy in appearance. They are very active moths, swift fliers, diurnal in habit, and can be easily mistaken for wasps. The legs often have thick tufts of hair, and the wings have a frenulum. The larvae are borers; the pupae have projections enabling them to move outward in the tunnels.

(244) LILAC BORER: *Podosesia syringae syringae.* **Fig. 455.**

Range: Southern Canada; eastern U.S., west to Montana and Utah. **Adult:** Forewing brown or chocolate color; hindwing clear, with brown borders and veins; the legs are brown and yellow. Eggs are laid in bark, near base of tree. **Spread:** 1.5". **Larva:** Creamy white, with brown head; about .75" full grown. Borings are into the wood and under the bark; the point of attack is indicated by holes, swelling and cracking of bark. One brood. Hibernates as partly grown larva in tunnel; pupates here also. **Food:** Lilac, ash, mountain ash, occasionally privet. The **Ash Borer,** *P. s. fraxini,* a subspecies, attacks ash and mountain ash.

(245) SQUASH VINE BORER: *Melittia cucurbitae.* **Fig. 456.**

Range: Southern Canada, and the U.S., east of the Rocky Mts., south to South America. **Adult:** Forewing covered with metallic olive-brown scales;

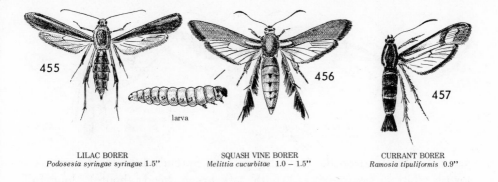

LILAC BORER
Podosesia syringae syringae 1.5"

SQUASH VINE BORER
Melittia cucurbitae 1.0 – 1.5"

CURRANT BORER
Ramosia tipuliformis 0.9"

hindwing clear, with brown border. Abdomen bright red, ringed with black stripes. Eggs are glued singly along the stems. **Spread:** 1–1.5". **Larva:** White, with brown head; about 1" long full grown. They bore into vines near the base, eating out the inner tissues and causing them to wilt; the stems are sometimes girdled and rot at the ring prior to wilting. One or two broods. Hibernates in cocoon in the ground, as larva or pupa. **Food:** Squash (especially Hubbard), pumpkin; also cucumber, muskmelon, gourds. [See Pickleworm, Melonworm (250).]

(246) CURRANT BORER: *Ramosia tipuliformis.* **Fig. 457.**

Range: Throughout North America; Europe and Asia. **Adult:** Clear wings; black and yellow markings. Eggs are laid on the canes. **Spread:** .9". **Larva:** Yellowish, about .5" long full grown. They bore into canes near the ground and feed on the pith and wood. One brood. Hibernates in the cane as nearly full-grown larva. Pupates in tunnel, near exit hole previously cut in the cane and covered with a web. **Food:** Currant, gooseberry, black elder, sumac. The **Rhododendron Borer**, *R. rhododendri,* is similar in appearance and habits; also attacks azaleas and mountain laurel if near rhododendron. The **Strawberry Crown Moth,** *R. bibionipennis,* attacks strawberries and blackberries; the larvae—white to pinkish, with brown heads—hollow the crowns; eggs are laid in midsummer on underside of leaves. Larvae of the **Raspberry Crown Borer,** *Bembecia marginata,* mine the crown and lower part of canes of raspberry, loganberry, boysenberry, and blackberry in Canada and the northern states; adults are black with yellow markings, each segment of abdomen except the last is ringed with a black and yellow band, spread from .75" to 1.5". [See (288) below.]

(247) PEACH TREE BORER: *Sanninoidea exitiosa.* **Fig. 458.**

Range: Throughout Canada and the U.S. **Adult:** Body of male light steel blue, with yellow stripes around the abdomen; both wings clear, with dark and yellowish borders. Female darker steel blue, with wide orange band around the abdomen; forewing opaque (same general coloring as the body), hindwing clear except margins. Eggs are laid singly or in groups on tree

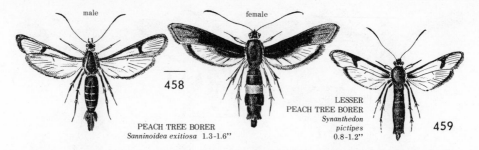

male

female

458

PEACH TREE BORER
Sanninoidea exitiosa 1.3-1.6"

LESSER
PEACH TREE BORER
*Synanthedon
pictipes*
0.8-1.2"

459

trunks mainly, and on the leaves or ground. **Spread:** 1.3–1.6". **Larva:** Yellowish white or cream-colored, with dark brown head, about 1" long full grown. They feed on the cambium and inner bark of trees, usually at the base or just below the ground line, sometimes girdling the tree; masses of gum, particles of bark and frass on the trunk indicate their presence. One brood. Hibernates as larva, the older ones within the burrow. Pupates in cocoon made of silk and bark, near entrance to burrow, on the inside or outside. **Food:** Chiefly peach; also plum, prune, apricot. In the **Western Peach Tree Borer,** *S. e. graefi,* adults lack the yellow and orange markings. The **Lesser Peach Tree Borer** (sometimes called plum tree borer), *Synanthedon pictipes* (Fig. 459), attacks mainly trunks and limbs that have been injured; both sexes of the adults resemble the male of *Sanninoidea exitiosa,* with the tufted antennae serving to distinguish the male; the larvae closely resemble those of *exitiosa.* The **Apple Bark Borer,** *Thamnosphecia pyri,* feeds in the bark and cambium of apple, pear, hawthorn, and mountain ash, kills and loosens the bark; the moth is bluish black, with yellow palpi, narrow yellow line on top of abdomen expanding to patch at base, clear wings with black veins, borders and tips (yellow-scaled below), spread of .45" to .7"; the larva is creamy white with brown head, from .5" to .75" long.

(248) DOUGLAS-FIR PITCH MOTH: *Vespamima novaroensis.* Fig. 460.

Range: Pacific Northwest and the northern Rocky Mt. states. **Adult:** Wings clear, abdomen shiny black, with red spot on the underside; orange-red spot on thorax and bands of the same color on the abdomen. Eggs are laid in bark crevices or wounds. **Spread:** 1.25". **Larva:** Slender, white, with transparent skin and brown head; 1.5" long full grown. They attack tree wounds, boring galleries through inner bark and outer layers of wood; large masses of pitch and borings at tunnel entrances indicate their presence. Hibernates as larva in gallery. Takes four years to complete development. Pupates in pitch mass. **Food:** Douglas fir, larch, Sitka spruce. The **Sequoia Pitch Moth,** *V. sequoiae* (Fig. 461), is similar in appearance and habits, except that the adult's body is striped with yellow; the larva is not transparent, and the life cycle requires two years instead of four; it attacks pines, Douglas fir, and other conifers in the same area. The damage of both moths is mostly to lumber.

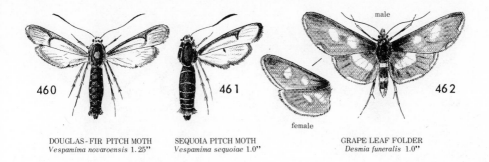

460

DOUGLAS-FIR PITCH MOTH
Vespamima novaroensis 1.25"

461

SEQUOIA PITCH MOTH
Vespamima sequoiae 1.0"

male

female

462

GRAPE LEAF FOLDER
Desmia funeralis 1.0"

Pyraustid Moths (Pyraustidae = Pyralidae)

Originally subfamilies of a large complex known as the Pyralidae, this group and the next four have generally been separated into families. They are small moths, the larvae of which usually live in webs, whence come such names as leaf roller, leaf tier, and webworm. The wings of the pyraustid moths are usually thinly scaled, often white and yellow in color. Many of them have the labial palpi enlarged and curved upward, forming a conspicuous "snout."

(249) GRAPE LEAF FOLDER: *Desmia funeralis.* **Fig. 462.**

Range: Canada to the Gulf states, west to the Great Plains; also California and British Columbia. **Adult:** Dark brown, with two white patches on forewing; one white patch on hindwing of the male, two on that of the female. Two white bands around the abdomen, male antennae thickened in the middle. They fly just before dark and at night. Eggs are laid on underside of protected leaves and on water sprouts or "suckers." **Spread:** 1". **Larva:** Bright green, with brown head and brown spot on each side of the first two body segments; about 1" long fully grown. They first feed in small groups, on leaves loosely webbed together; later feed singly in pencil-sized leaf roll. The silk strands contract as they dry, bending the leaf. Two or more rolls are made during the feeding period. Pupates in leaf fold. Three broods. Hibernates as pupa in dry leaf fold that breaks off and falls to the ground. **Food:** Grape. This species is not injurious in the East; in some years, it is the worst pest of grapes in California. It is susceptible to the bacterium *Bacillus thüringiensis.*

(250) PICKLEWORM: *Diaphania nitidalis.* **Fig. 463.**

Range: South Atlantic and Gulf states; occasionally north to Connecticut and Iowa, west to Oklahoma and Nebraska. **Adult:** Body yellowish brown, with purplish sheen; both wings with broad, yellowish brown borders; median spot on forewing and basal two-thirds of hindwing yellowish, transparent. Dark, brushlike hairs on tip of abdomen are waved in the air when the moth is resting. It flies mostly at night. Yellowish white eggs, hardly

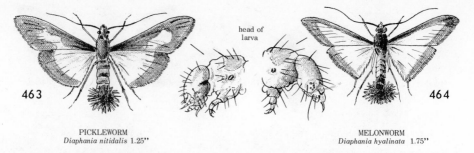

PICKLEWORM		MELONWORM
Diaphania nitidalis 1.25"	head of larva	*Diaphania hyalinata* 1.75"

463   464

visible, are laid singly or in clusters on flower and leaf buds, leaves, stalks, and young fruit. **Spread:** 1.25". **Larva:** Young larva yellowish white, with numerous, scattered dark spots; mature larva yellowish green or coppery, with brown head, about .75" long, without the spots. It feeds on flower and leaf buds, young fruit, later tunnels vines and fruit. Pupates under webbing, attached to a leaf on the plant or ground. One brood in the North, several in the South, where it overwinters; adults migrate northward in spring and summer. **Food:** Chiefly cucumber, summer squash, cantaloupe; also other cucurbits. The **Melonworm,** *D. hyalinata* (Fig. 464), in contrast to the pickleworm, feeds mainly on foliage; the adult is pearly white, with narrow, brown borders along costa and outer margin of forewing and outer margin of hindwing, the latter with two long, white appendages trailing from the base; spread about 1.75"; body in front of wings dark brown, abdomen with dark, brushlike hairs on tip; the larva is greenish, with white longitudinal subdorsal stripe, no spots. Melonworm larvae are sometimes mistaken for the corn earworm (211), which feeds only on the flowers of cucurbits, and for the squash vine borer (245), which feeds inside the stalk.

(251) EUROPEAN CORN BORER: *Ostrinia (Pyrausta) nubilalis.* **Fig. 465.**

Range: Corn-producing areas from southern Canada to North Carolina, westward to eastern Montana and Colorado; first discovered in America in 1917.* **Adult:** Yellowish brown or tan, with darker wavy bands across the wings, male much darker than the female; they are strong fliers, active at night. Eggs are laid in clusters on underside of leaves. **Spread:** 1". **Larva:** Grayish pink, with dark head, and two small brown spots on the top of each abdominal segment, from .75" to 1" long full grown. Young larvae feed on leaves, tassels, and stems between ear and stalk; when half grown they bore into stalk, leaf stem, or ear. Hibernates as full-grown larva in burrow, pupates here in loose cocoon in the spring. One to three

* It is believed to have entered this country on broomcorn shipped from Hungary or Italy between 1909 and 1914. The estimated losses caused by this insect to grain corn in the U.S. in 1969 was 163 million bushels valued at $183 million. An interesting possibility for control is suggested by the report that this insect (and the codling moth) do not survive the winter when subjected to artificial lights to extend the length of day in the fall. Because of the light, the larvae fail to go into diapause and to hibernate (see p. 17).

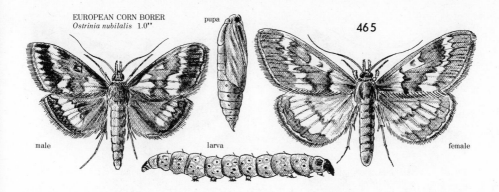

EUROPEAN CORN BORER
Ostrinia nubilalis 1.0"

pupa

465

male

larva

female

broods. Multigeneration strain predominates in southern and eastern part of range, single-generation strain in the North Central states. **Food:** More than 200 kinds of plants; mostly corn in single-generation area of the Middle West and Ontario, garden crops and ornamentals extensively as well as corn in New England, where there are usually two broods (excepting Maine). The larva of *nubilalis* may be distinguished from that of the native **Smartweed Borer,** *O. obumbratalis,* by the smaller dorsal spots, more widely distributed. For other larvae attacking corn see Stalk Borer (201), Armyworm (202), Fall Armyworm (203), Corn Earworm (211), Southern Cornstalk Borer (257), Sugarcane Borer (258), Lesser Cornstalk Borer (266). [See Hymenoptera (30) and Diptera (96) for parasites.]

(252) GARDEN WEBWORM: *Loxostege similalis (rantalis).* Fig. 466.

Range: Throughout southern Canada, the entire U.S., Mexico. **Adult:** Dull brownish yellow, with grayish markings and a few brown spots; hindwing paler, thorax often with slight greenish tinge. Eggs are laid in small or large clusters on leaves. **Spread:** .75". **Larva:** Greenish, with light middorsal stripe; three small darker spots, with one to three stiff hairs, on side of each segment; about 1" long full grown. It feeds under a light webbing, drops to the ground or retreats to tubular portion of web when disturbed. Two to five broods. Hibernates as pupa in the soil. **Food:** Alfalfa, clover, sugar beets, peas, beans, soybeans, cowpeas, cotton, corn, various weeds and grasses, scarlet verbena, castor bean. The **Morning Glory Cutter,** *L. obliteralis,* cuts the leaf stalk of morning glory, dahlia, zinnia, violets, sunflower, peppermint, wandering-Jew, hogweed, causing the leaf to hang down and wilt.

(253) BEET WEBWORM: *Loxostege sticticalis.* Fig. 467.

Range: Mississippi Valley, westward. **Adult:** Dull brown, with yellowish wavy lines and spots; dark marginal band on underside of hindwing near apex. Eggs are laid end to end in single rows, mostly on underside of leaves. **Spread:** 1–1.25". **Larva:** Yellowish to dark green, with black middorsal

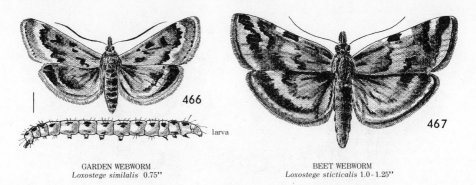

GARDEN WEBWORM
Loxostege similalis 0.75"

BEET WEBWORM
Loxostege sticticalis 1.0 - 1.25"

stripe; three small black spots with stiff hairs on the side of each segment; about 1" to 1.25" long full grown. It feeds under webbing, skeletonizes the leaves; forms tube by drawing leaves together, where it hides if disturbed. Three broods. Hibernates as pupa, possibly larva, in silk-lined cell in the soil. **Food:** Sugar beet, alfalfa, many field and garden crops, weeds. The **Alfalfa Webworm,** *L. commixtalis,* is similar, the adult buff-colored, with gray markings; it may be distinguished from *sticticalis* by the row of marginal spots near the apex on underside of hindwing; eggs are laid in overlapping clusters; the larva has light-colored middorsal stripe. The **Cabbage Webworm,** *Hellula rogatalis,* is a pest of cole crops in the South; it has a brownish yellow forewing with dark mottling, pale gray hindwing, spread of .6"; the larva is dull grayish yellow, with brownish purple stripe down the back and two along each side, black head with V marking, about .5" long full grown, feeds on inner leaves under protective webbing. The cabbage webworm is often found with the **Cross-striped Cabbageworm,** *Evergestis rimosalis,* which has a mottled yellowish brown forewing with brown zigzag lines, paler hindwing with five or six indistinct spots near the darker outer fringe, spread of about 1"; the larva is bluish gray, with fine black cross-stripes, black and bright yellow stripe along each side, about .6" long full grown, riddles tender parts of plant with holes.

Pyralid Moths (Pyralidae)

The pyralids are small attractive moths with a proboscis. They are also called snout moths because of the prominent labial and maxillary palpi. The larvae are usually smooth, grayish or brownish, with reddish brown or black heads; they feed on both live plants and stored grains and seeds. The family is noteworthy for the success of some species in adapting to an aquatic life. The preceding and following two families are sometimes included here.

(254) MEAL MOTH: *Pyralis farinalis.* **Fig. 468.**

Sometimes called meal snout moth. **Range:** Widely distributed, where

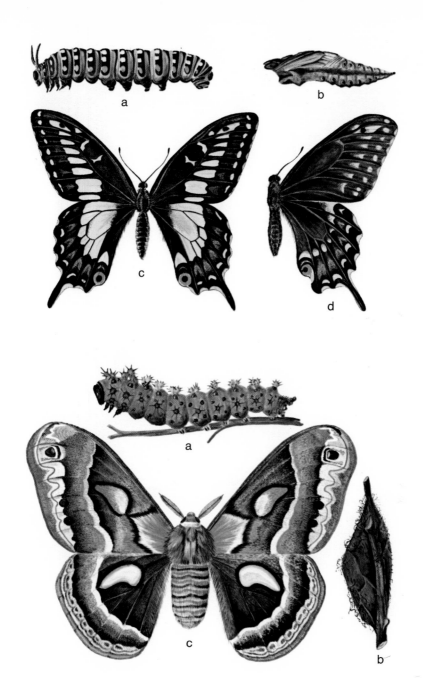

Plate 1

BUTTERFLY (Top)—Baird's Swallowtail
(*Papilio bairdi bairdi*), *a*—larva, *b*—pupa,
c—male, *d*—female

MOTH (Bottom)—Cecropia (*Platysamia
cecropia*), *a*—larva, *b*—cocoon, *c*—adult

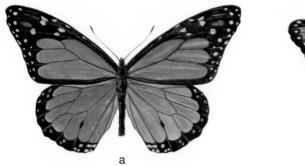

a

b

c

d

e

f

Plate 2

a—Monarch Butterfly (*Danaus plexippus*)
b—Viceroy (*Limenitis archippus*)
c—Weidemeyer's Admiral (*L. weidemeyeri*)

d—Lorquin's Admiral (*L. lorquini*)
e—Red-spotted Purple (*L. astyanax*)
f—White Admiral (*L. arthemis*)

Plate 3

a—Queen (*Danaus gilippus berenice*)
b—Florida Viceroy (*Limenitis archippus floridensis*)
c—Pipevine Swallowtail (*Battus philenor*)

d—Spice-bush Swallowtail (*Papilio troilus*)
e—Tiger Swallowtail (*P. glaucus*)
f—Tiger Swallowtail, dark form

Plate 4

a—Clouded (Common) Sulphur (*Colias philodice*), male, female
b—Behr's Sulphur (*C. behrii*)
c—Cloudless Sulphur (*Phoebis sennae eubule*), male, female
d—Tawny-edged Skipper (*Polites themistocles*)

e—Peck's Skipper (*Polites coras*)
f—California Dog Face (*Colias eurydice*), male, female
g—Roadside Skipper (*Amblyscirtes vialis*)
h—Ruddy Dagger Wing (*Marpesia petreus*)
i—Skipper (*Heliopetes domicella*)
j—Cobweb Skipper (*Hesperia metea*)

Plate 5

Plate 6

a—Polyphemus Moth (*Antheraea polyphe-mus*), female
b—Checkered Skipper (*Pyrgus communis*), male
c—Common Sooty Wing (*Pholisora catullus*), male
d—Funeral Dusky Wing (*Erynnis funeralis*), male

e—Grizzled Skipper (*Pyrgus centaureae*), male
f—Cynthia Moth (*Samia cynthia*), male
g—Filbertworm (*Melissopus latiferreanus*)
h—Common Bluet (*Enallagma ebrium*)
i—Navel Orangeworm (*Paramyelois transitella*)
j—Golden-Banded Skipper (*Autochton cellus*)

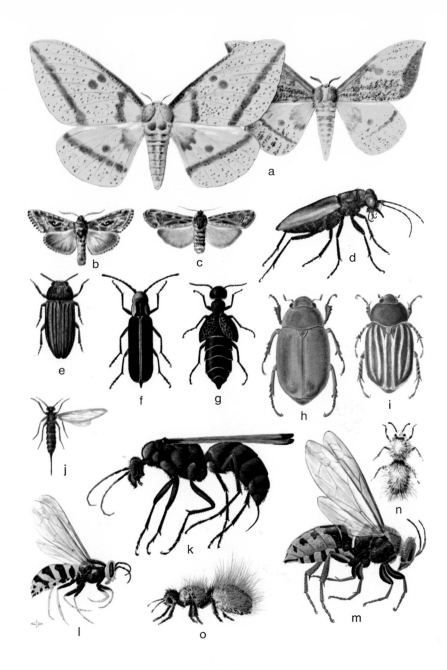

Plate 7

a—Imperial Moth (*Eacles imperialis*), male,
   female
b—Alfalfa Looper (*Autographa californica*)
c—Army Cutworm (*Euxoa auxiliaris*)
d—Tiger Beetle (*Cicindela scutellaris*)
e—Golden Buprestid (*Buprestis aurulenta*)
f—Clematis Blister Beetle (*Epicauta cinerea*)
g—Oil Beetle (*Megetra cancellata*)

h—Oak Beetle (*Plusiotis beyeri*)
i—Cypress Beetle (*Plusiotis gloriosa*)
j—Blue Horntail (*Sirex cyaneus*)
k—Tarantula Hawk (*Pepsis mildei*)
l—Yellow Jacket (*Vespula pennsylvanica*)
m—Giant Hornet (*Vespa crabro germana*)
n—Sacken's Velvet-ant (*Dasymutilla sackenii*)
o—Red Velvet-ant (*D. occidentalis*)

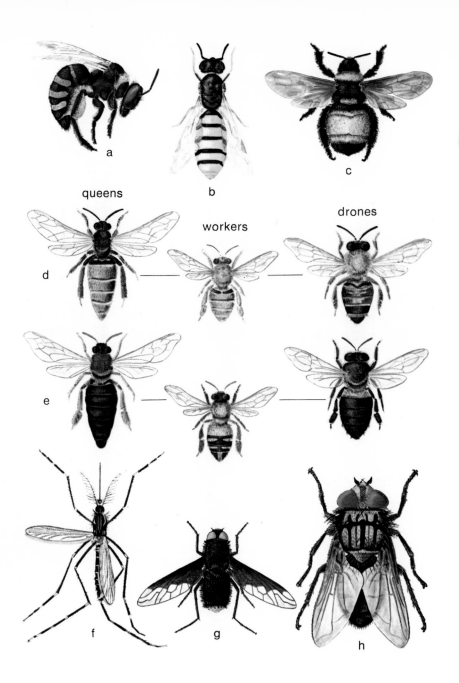

queens

workers

drones

Plate 8

a—Alkali Bee (*Nomia melanderi*)
b—Andrenid Bee (*Perdita zebrata*)
c—Pennsylvania Bumble Bee (*Megabombus pennsylvanicus*)
d—Honey Bee (*Apis mellifera*), Caucasian race

e—Honey Bee, Italian race
f—Yellow Fever Mosquito (*Aedes aegypti*), female
g—Bee Fly (*Anthrax analis*)
h—Screw-worm (*Cochliomyia hominivorax*)

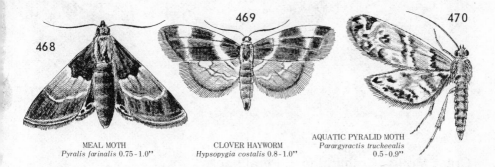

MEAL MOTH
Pyralis farinalis 0.75 - 1.0"

469

CLOVER HAYWORM
Hypsopygia costalis 0.8 - 1.0"

470

AQUATIC PYRALID MOTH
Parargyractis truckeealis
0.5 - 0.9"

damp or spoiled grain, bran, meal, or hay is stored. **Adult:** Brownish ground color; forewing light brown, with dark brown patches at the base and tip, and two wavy, white transverse lines. **Spread:** .75–1". **Larva:** Whitish; head and first body segment black; often tinted with orange toward each end, about 1" long full grown. Noted for its ability to web and bind together various kinds of seeds. It fashions a tube of silk and feeds from the open end, leaves the tube to spin a cocoon and pupate, usually cutting through the food sac to do so, causing contents to flow out. The period from egg to adult is six to eight weeks. **Food:** Cereals, dried vegetables, hay, sound wheat in damp places. The adult of the **Clover Hayworm,** *Hypsopygia costalis* (Fig. 469), is a pretty little moth with brown wings traversed by two yellowish stripes and fringed in yellow; the larva weaves webs, mixed with excrement, throughout a clover hay stack, and feeds on the leaves. It is European in origin and widely distributed.

(255) AQUATIC PYRALID MOTH: *Parargyractis truckeealis.* **Fig. 470.**

A common western species, truly aquatic. They collect at lights but may be found during the day under bridges or on trees and in bushes along streams. **Range:** British Columbia to California; Nevada to Wyoming and Montana. **Adult:** Grayish red to brownish red, with white and brownish markings; row of black spots along the outer margin of the hindwing. The female is similar to the male, except larger, with fewer ciliate antennae, has swimming hairs on the mesothoracic and metathoracic legs. Flat, yellowish to tan eggs are laid under water, glued to the surface of rocks, usually overlapping slightly, most often on rocks covered with algae. **Spread:** .5–.9". **Larva:** Greenish to blackish, with dark brown head and thoracic shield; about .6" long full grown. It lives on rocks in fast-flowing streams and in shallow riffles, breathes by means of filamentous blood gills arranged in groups on the sides of the abdominal segments. Pupates under a webbing, in cocoon attached to a rock by silk; holes are left in the sides of the web and cocoon to permit circulation of the water. Hibernates in the larval stage. **Food:** Algae, diatoms.

472

471

STRIPED SOD WEBWORM
Crambus mutabilis 1.0"

BLUEGRASS WEBWORM
Crambus teterrellus 0.6-0.85"

CORN ROOT WEBWORM
Crambus caliginosellus 0.5 – 1.0"

473

Grass Moths (Crambidae = Pyralidae)

This is a group of small moths with long, narrow forewings, often sharply pointed at the apex; the hindwings are broad, usually without markings. The wings are wrapped tightly around the body when at rest. The coloring is subdued—whitish or silvery to brownish. A prominent feature is the "snout"; the labial palpi are close together and extend well beyond the head, being about as long as the head and thorax combined. The larvae usually construct silken tubes in the ground and feed on leaves, roots, or stems.

(256) STRIPED SOD WEBWORM: *Crambus mutabilis*. Fig. 471.

Several species of *Crambus* occur in various parts of the U.S. and Canada to which this description would generally apply. **Adult:** Forewing dull ash gray, with whitish streak from base through discal cell to outer margin, outer streaking of gray and scattered black scales; fringe and hindwing brownish; they are folded close to the body when at rest. Eggs are laid near the base of grass stems or dropped at random. **Spread:** About 1". **Larva:** Light brown, with coarse hairs; paired, dark dorsal and lateral spots on each segment; about .75" long full grown. It works at night, under a loose silk webbing near the surface. The young larvae feed on leaves, the older ones form tunnels of silk and grass close to the surface; they cut off blades of grass and drag them into their tunnels to eat. Two to three broods. Hibernates as larva in the tunnel. **Food:** Pasture and field grasses, especially bluegrass and timothy, young corn, new lawns. Irregular brown spots denote the presence of sod webworms in lawns. The **Bluegrass Webworm,** *C. teterrellus* (Fig. 472), has dark reddish brown wings, with violet-gray shading between the veins and shiny lead-gray fringe; spread from .6" to .85"; the larva is yellowish white or greenish white, with dull light brown head, large tubercles and setae. Adults of the **Corn Root Webworm,** *C. caliginosellus* (Fig. 473), are smoky brown and somewhat powdery; spread from .5" to 1"; the larva is pinkish white tinged with brown, head dark brown or black; it attacks young corn, often girdling the plant and eating out the point, is also found in grass. The **Cranberry Girdler,** *C. topiarius,* is also a major pest in the grass-seed fields in Washington.

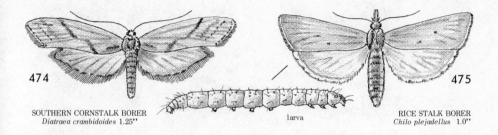

474

SOUTHERN CORNSTALK BORER
Diatraea crambidoides 1.25"

larva

RICE STALK BORER
Chilo plejadellus 1.0"

475

(257) SOUTHERN CORNSTALK BORER: *Diatraea crambidoides.* **Fig. 474.**

Range: Georgia, South Carolina, North Carolina, Virginia, Maryland. **Adult:** Female's wings whitish to smoky yellow, faintly marked; male darker. Eggs are flat, creamy white to reddish brown, laid in rows or clusters overlapping like shingles, usually on underside of lower leaves. **Spread:** 1.25". **Larva:** Dirty white, covered with spots, each having a short dark bristle; head and thoracic shield brownish yellow; about 1" long full grown. Young larvae feed in leaf whorls, later cut holes in the stalks and cover them with webbing; the first-brood larvae pupate here. The second-brood larvae feed in the stalks for a while and then make their way to the roots, where they hibernate, at which time the spots disappear; pupation takes place early in the spring. They do not bore into the ears. Two broods. **Food:** Chiefly corn; also sorghum, Johnson grass, guinea corn, grama grass. The damage to corn is sometimes severe. The **Southwestern Corn Borer,** *D.* (*Zeadiatraea*) *grandiosella,* is similar, occurs in Arizona, New Mexico, and Texas, north to Nebraska, east to Mississippi, Tennessee, and Kentucky, overlapping the European corn borer in the central states. The **Rice Stalk Borer,** *Chilo plejadellus* (Fig. 475), occurs in the rice fields of Louisiana and Texas, weakens plants by tunneling the stems.

(258) SUGARCANE BORER: *Diatraea saccharalis.* **Fig. 476.**

Range: Gulf Coast, from Florida to Texas. **Adult:** Straw-colored, forewing marked with black dots arranged in the form of a **V.** Eggs—flat, oval, creamy white, changing to yellowish and finally orange—are laid in clusters on leaves and stems. **Spread:** 1". **Larva:** Yellowish white, with brown spots. The color changes to yellow in winter, and the spots often disappear. The young larvae feed on leaves, in the whorls or leaf sheaths, later bore into the stalk to feed. Summer broods pupate in the tunnels. The fall brood hibernates in the larval stage in trash or stubble, and pupates early in the spring. Four or five broods. **Food:** Sugarcane, corn, rice, sorghums, several wild grasses. It is parasitized by the egg parasite *Trichogramma minutum* and the introduced tachinid fly *Lixophaga diatraeae* [see Diptera (98)] and braconid wasp *Agathis stigmaterus.*

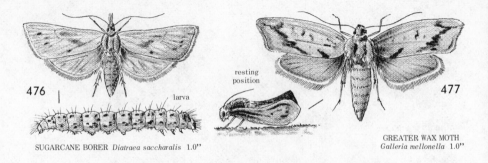

476   larva

SUGARCANE BORER *Diatraea saccharalis* 1.0"

resting
position

477

GREATER WAX MOTH
Galleria mellonella 1.0"

(259) GREATER WAX MOTH: *Galleria mellonella* (Galleriidae). **Fig. 477.**

Range: Practically anywhere that honey bees are kept. **Adult:** Forewing purplish brown, hindwing pale brown or yellowish; wings with black markings and lead-colored tips. Eggs are laid in the beehive at night when the bees are inactive. **Spread:** 1". **Larva:** White when it emerges, turning to yellow, brown, or black on the upper side. It feeds at night on the wax of the combs, forming a silken burrow. Seven and possibly eight instars. Pupates in a tough cocoon on the side of the hive. While the larvae do not attack the bees, they can destroy a hive.

Phycitid Moths (Phycitidae = Pyralidae)

The moths in this group are small, generally silky gray, and have filiform antennae. The palpi are often curved upward but are not nearly as prominent as in the Crambidae. The proboscis is often but not always present. The larvae are protected by silken "cases" or webbing, and are often surrounded by leaves bound together with silken threads. At least one species in the group has been turned to good use: the gregarious, tunneling larvae of *Cactoblastis cactorum*—a South American species—which is very destructive to prickly pear cactus (*Opuntia*) and has been used successfully to control this plant after it was introduced (as an ornamental) and became a pest in Australia and Hawaii. As already noted, the larvae of some phycitid moths are predaceous on scale insects that feed on prickly pear cactus—and thus, of course, have the effect of encouraging these plants (see p. 201).

(260) PECAN LEAF CASEBEARER: *Acrobasis juglandis.* **Fig. 478.**

Range: Mainly Georgia, Florida, and the Gulf states. **Adult:** Body white or dusky gray, wings gray. Eggs are laid in clusters on the underside of leaves. **Spread:** .75". **Larva:** Early stage brown, later stage dark green; about .5" long full grown. The young larvae feed on underside of leaves, spinning small winding cases for protection. In the fall they spin oval cases called "hibernacula," secured to twigs near the buds, where they hibernate as larvae. These larvae feed on buds and leaves in early spring, at which

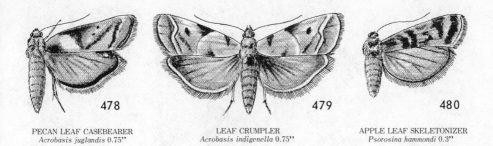

478   PECAN LEAF CASEBEARER
Acrobasis juglandis 0.75"

479   LEAF CRUMPLER
Acrobasis indigenella 0.75"

480   APPLE LEAF SKELETONIZER
Psorosina hammondi 0.3"

time they are most harmful. Pupation takes place in "pupal cases." One brood. **Food:** Pecan, hickory, walnut. The **Pecan Nut Casebearer,** *A. caryae,* is also a serious pest of pecan in the Gulf states; the adult is dark gray, with a ridge of long, dark scales across the middle of the forewing; the larva is olive-green; eggs are laid singly on the end of nuts. The first and second broods of *caryae* bore into the nuts, the third brood gnaws into the shucks without doing much harm; hibernation takes place in the larval stage in a case similar to that of *juglandis,* but pupation occurs in tunnels bored into young shoots. [See (272); also Coleoptera (425).]

(261) LEAF CRUMPLER: *Acrobasis indigenella.* **Fig. 479.**

Range: Upper Mississippi Valley, the northern states, Ontario. **Adult:** Brownish, with silvery mottling. Eggs are laid on new leaves. **Spread:** .75". **Larva:** Dark brown, hairy; about .5" long full grown. It constructs a tough hornlike case of silk and debris; this is carried along as it feeds and is attached to a twig in the fall before hibernating. It hibernates as larva inside the case surrounded by dead crumpled leaves; in the spring it feeds on buds. Pupation takes place in May or June. The new-brood larvae feed on shoots and leaves. One brood. **Food:** Apple, plum, quince, pear, wild plum and cherry. The **Destructive Pruneworm,** *A. scitulella,* is similar, attacks prune and cherry in Idaho and Oregon; the overwintering larvae feed on buds in the spring, the first brood bores into the green fruit, the second brood feeds on maturing fruit or leaves. The **Cranberry Fruitworm,** *A. vaccinii,* attacks cranberries and blueberries in the eastern states and Canada; the yellowish green larva feeds inside the berry, causing it to shrivel, and forms a webbing around the clusters. *A. comptoniella* occurs on sweet fern, *A. betulella* on birch, *A. rubrifasciella* on alder.

(262) APPLE LEAF SKELETONIZER: *Psorosina hammondi.* **Fig. 480.**

Range: Ontario, New England, west to the Mississippi Valley, south to Texas. **Adult:** Dark brown, with silvery bands on forewing. Eggs are laid on leaves. **Spread:** .3". **Larva:** Brownish green, with four black tubercles on the prothorax; about .5" long full grown. Leaves at the ends of branches

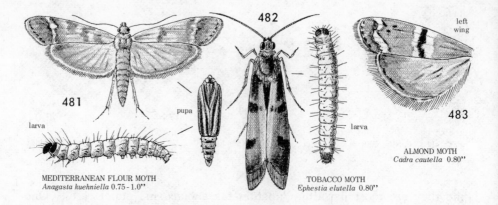

482

left wing

481

pupa

larva

483

larva

ALMOND MOTH
Cadra cautella 0.80"

MEDITERRANEAN FLOUR MOTH
Anagasta kuehniella 0.75 - 1.0"

TOBACCO MOTH
Ephestia elutella 0.80"

and the tops of trees are loosely webbed. Feeding takes place on upper surface of leaves, which are skeletonized. Pupates in web on leaves. Hibernates as pupa in leaves on the ground. Two broods. **Food:** Apple, occasionally quince and plum.

(263) MEDITERRANEAN FLOUR MOTH: *Anagasta (Ephestia) kuehniella.* **Fig. 481.**

Range: Widespread in granaries and flour mills, native of Europe. **Adult:** Forewing pale brownish gray, with wavy black transverse bands; hindwing dirty white. It has a curious habit of raising forepart of body when at rest. White eggs are laid in accumulations of flour, meal, or waste grain. **Spread:** .75–1". **Larva:** Creamy white, with pinkish cast and a few small black spots; about .5" long full grown. The food is webbed together in mats. Pupates in silken cocoon. Complete life cycle takes eight to ten weeks. **Food:** Flour and meal preferably; also bran, grain, cereals, and other foodstuffs. *Bacillus thüringiensis,* a pathogen causing disease in many moths (and important in biological control), was discovered in larvae of this moth in Germany. Closely related species attacking grains, cereals, nuts, beans, peas, raisins, tobacco, and other stored food products are the **Tobacco Moth,** *Ephestia elutella* (Fig. 482), the **Almond Moth,** *Cadra cautella* (Fig. 483), and the **Raisin Moth,** *C. figulilella.*

(264) INDIAN MEAL MOTH: *Plodia interpunctella.* **Fig. 484.**

Range: Widespread in granaries, kitchens, stores. **Adult:** Easily distinguished from other grain moths by its forewings; about one-third of the wing (toward the base) is whitish gray, the outer two-thirds reddish brown, with a coppery luster. Hindwing grayish. Eggs are laid singly or in masses on food. **Spread:** .8". **Larva:** Dull white, sometimes with greenish or pinkish cast. It spins a silken web as it feeds, leaving conspicuous masses of loose webbing clinging to the food. Pupates in silken cocoon. The complete life cycle takes about six to eight weeks. **Food:** Grains, cereals, a variety of foodstuffs including dried fruits, nuts, peas, and beans. This and the preced-

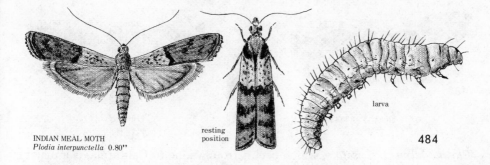

INDIAN MEAL MOTH
Plodia interpunctella 0.80"

resting
position

larva

484

ing species (263) are not major pests of stored grain, since they feed mostly on dust or broken kernels and can be removed by cleaning operations; the webbing sometimes clogs milling machinery.

(265) NAVEL ORANGEWORM: *Paramyelois transitella.* **Color plate 6i.**

Range: Southwestern states, Mexico; Oklahoma, Georgia. **Adult:** Forewing pale grayish, with darker mottling and zigzag lines; hindwing much lighter, plain except for darkened marginal fringe and veins. Eggs are laid in the navel of injured oranges, and in the cracked husks of nuts. **Spread:** .75". **Larva:** Reddish orange when hatched, changing to pinkish orange or cream; head dark reddish brown; about .75" long full grown. The two opposing crescent-shaped pigmented areas on top of the mesothorax serve to distinguish it from larvae of the codling moth (268) and filbertworm (276). It is a scavenger on injured or diseased oranges, leftover figs or other fruits. It bores into a nut after the husk has cracked, or enters a hole made by the codling moth larva, and spins a web inside. It pupates in cocoon inside the nut, constructing a tube to the exit hole where the moth escapes. Hibernates as larva in mummified fruit or nuts left on the trees or ground. **Food:** Many fruits and nuts, especially walnut and almond.

(266) LIMA BEAN POD BORER: *Etiella zinckenella.*

Range: Throughout North America. **Adult:** Mottled gray, forewing marked with orange band across the inner third, and white stripe along the outer margin from base to apex; male antenna enlarged at base. A strong flier, capable of migrating long distances. Elliptical, glistening white eggs are laid singly on lupines or pods of lima bean. **Spread:** .95–1". **Larva:** Reddish pink or tan, from .45" to .6" long when mature. It bores into the pod to feed; the entrance hole heals quickly, leaving no trace. The mature larva eats a hole in the pod, drops to the ground to pupate in cocoon. Hibernates as mature larva in cocoon in the soil. One brood on annual wild plants, two to four on lima bean and perennial lupine. **Food:** Common (annual) lupine, wild peas, locoweed, tree (perennial) lupine, lima bean. A serious pest of lima bean in California. The **Lesser Cornstalk Borer,**

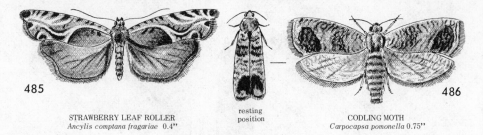

485

STRAWBERRY LEAF ROLLER
Ancylis comptana fragariae 0.4"

resting
position

CODLING MOTH
Carpocapsa pomonella 0.75"

486

Elasmopalpus lignosellus, which occurs from Maine to California, is a major pest of peanuts in Texas, especially under dry conditions, and of legumes and grasses in Georgia; the front wings are brownish yellow, with grayish margins—the wing is nearly all black in the female—spread about 1"; the larva is bluish green, with brown bands, from .5" to .75" long full grown, lives in the soil in a silken tube, tunnels the roots and stem.

Olethreutid Moths (Olethreutidae)

The moths in this family are small, generally brown or gray, with darker or lighter bands across the forewings. They resemble the preceding family (Phycitidae) but, unlike them, usually have long, pointed labial palpi. Their flying habit is mostly crepuscular or nocturnal. Many destructive species are included in the family, as denoted by its name. Some authors include them with the Tortricidae. Many are leaf rollers; others tunnel or mine twigs and needles, or bore into buds, seeds, and fruits. The wriggling larva of an olethreutid moth, *Laspeyresia saltitans,* is the activator of the Mexican "jumping bean," the seed of a species of *Croton.*

(267) STRAWBERRY LEAF ROLLER: *Ancylis comptana fragariae.* **Fig. 485.**

Range: Southern Canada and throughout the U.S. except the Southwest; California to some extent. **Adult:** Grayish or brownish, with darker and lighter bands. Eggs are laid singly on the underside of leaves. **Spread:** .4". **Larva:** Greenish or bronze; about .5" long full grown. It feeds at first on the underside of leaves under silken webbing, later on the upper surface, folding the leaf over itself with silken thread. Pupates in cocoon in folded leaf. Hibernates as pupa or as larva in folded leaf on the ground. Two broods and a partial third. **Food:** Wild and cultivated strawberry, blackberry, raspberry, dewberry. Like the Oriental fruit moth (269) it is most consistently parasitized by the braconid wasp *Macrocentrus ancylivorus.* The **Ragweed Borer,** *Epiblema strenuanum,* is similar but somewhat larger, the forewing smoky brown, with light even dusting of clay-colored scales, whitish to light brown transverse lines, and broad fringe; the larvae bore into ragweed and serve as reservoirs of *Macrocentrus ancylivorus* parasites around peach orchards.

(268) CODLING MOTH: *Carpocapsa pomonella.** **Fig. 486.**

Range: All apple-growing areas of the world. Adult: Brownish gray, forewing with chocolate-colored patch at the tip, and coppery or golden, wavy transverse markings. White, flattened eggs are laid singly on the upper surface of leaves, on twigs or fruit spurs. Spread: .75". Larva: White, with pinkish tinge, dark brown head; about .75" long full grown. It feeds briefly on leaves, then enters a young apple at the calyx cup on the blossom end, works its way to the core, often eating the seeds. It tunnels out of the fruit when full grown, pupates in a cocoon under loose bark scales or on the ground. Hibernates as full-grown larva in cocoon. One or two broods, or a partial third. Associated species are (265), (269), (276); also see Hymenoptera (8). Food: Apple, pear, quince, walnut. This moth is the worst single pest of apple and walnut, and one of the most difficult to control. Woodpeckers destroy many larvae in winter. It is often parasitized by the braconid *Ascogaster quadridentata.* The moth larvae are also susceptible to a fungus, *Beauveria bassiana,* and to the bacterium *Bacillus thüringiensis.*

(269) ORIENTAL FRUIT MOTH: *Grapholitha molesta.* **Fig. 487.**

Range: Eastern U.S., Ontario, and the Pacific Northwest; world-wide, a native of east Asia. Adult: Dark gray, with chocolate brown markings on forewing. Usually crepuscular but often flies at midday in spring. Flat, white eggs are laid on underside of leaves near the tip of young twigs, or on the latter. Spread: .5". Larva: Dirty white or gray to pink or almost red; head, thoracic and anal shields brown; about .5" long full grown. It spins a protective web. Early broods bore into tender twigs, causing dieback; later broods penetrate fruit, leaving no sign of entry. It leaves the fruit or twig to pupate in cocoon in bark or on the ground. Hibernates as full-grown larva in cocoon. Four or five broods in the North, six or seven in Georgia. Sometimes confused with the peach twig borer (287) and codling moth (268). Food: Peach, apricot, nectarine, almond, apple, quince, pear, plum, cherry, ornamentals. This moth is the worst single pest of peaches. It is attacked by many parasites, the most important being a braconid wasp, *Macrocentrus ancylivorus,* used extensively in biological control before 1954. The **Lesser Appleworm,** *G. prunivora,* is similar to the corresponding stages of the codling moth (268) but smaller; the injury to apples is the same, but the holes in the fruit are smaller, usually go in from the side and seldom penetrate to the core; it also injures prunes in the Pacific Northwest. The **Cherry Fruitworm,** *G. packardi,* is a serious pest of cherry and blueberry in some growing areas.

* The genus was recently changed to *Laspeyresia:* Donald R. Davis, Smithsonian Institution. It has also been known as *Cydia.*

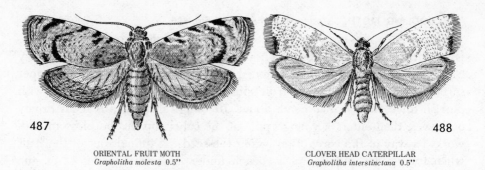

487

488

ORIENTAL FRUIT MOTH
Grapholitha molesta 0.5"

CLOVER HEAD CATERPILLAR
Grapholitha interstinctana 0.5"

(270) CLOVER HEAD CATERPILLAR: *Grapholitha interstinctana.* **Fig. 488.**

Range: Southern Canada and eastern U.S. **Adult:** Dark brownish, with six or seven short, silvery white dashes along the margin; markings on the inner margins form a double crescent when the wings are folded. Eggs are laid on stems, leaves, or heads of clover. **Spread:** About .5". **Larva:** Light green or whitish, about .25" long full grown. It usually feeds in the clover head on the soft green seeds at the base of the florets; infested flowers are stunted, open on one side only, with one side of the head often pink and the other side green. The injury is similar to that of the clover seed midge [see Diptera (19)]. The caterpillar eats leaves when there are no heads. Pupates in silken cocoon in the clover head or at the base of the plant. Two to three broods. Hibernates as larva or pupa in the litter. **Food:** Red clover mostly; also alsike and mammoth clovers.

(271) EUROPEAN PINE SHOOT MOTH: *Rhyacionia buoliana.* **Fig. 489A.**

Range: Southern Canada, New England to Virginia, west to Michigan and Illinois; coastal Washington and British Columbia. **Adult:** Rusty orange-red with silvery markings and whitish legs. Eggs are flat, yellowish, turning to reddish brown, laid singly or in small clusters on needles, twigs, or buds. **Spread:** .75". **Larva:** Brownish, with black head; about .6" long full grown. It bores into the base of the needles after spinning a protective web, causing flow of pitch and hardening of bud. Pupates inside the shoot. Hibernates as larva in the bud. One brood. **Food:** Two- and three-needled pines: red pine preferred, Scotch and mugho pines also attacked; now occurs on ornamental pines on the Pacific coast, threatens ponderosa pine. Young trees are deformed by stunting of twigs and branches. The **Nantucket Pine Tip Moth,** *R. frustrana,* common in the East and South, hollows out the tip of new pine shoots, including the buds; the moth is grayish, with brick-red patches, wingspread from .4" to .65"; first instar larvae are pinkish red to cream-colored, with dark head and thoracic shield, mature larvae light brown to orange, about .35" long; it hibernates as pupa in the tips. The **Pitch-pine Tip Moth,** *R. rigidana,* is similar and causes the same kind of

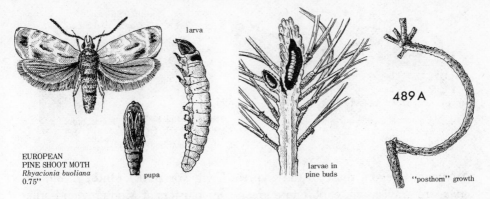

EUROPEAN
PINE SHOOT MOTH
Rhyacionia buoliana
0.75"

larva

pupa

larvae in
pine buds

489 A

"posthorn" growth

damage to pines in the South. The **Western Pine Tip Moth,** R. f. *bushnelli*
(Fig. 489B), is found among native seedlings and plantations in South
Dakota, Nebraska, and the Lake states; the forewing of the moth is yellowish
gray and reddish brown, the larva yellowish. The **Southwestern Pine Tip
Moth,** R. *neomexicana,* does similar damage to ponderosa pine in New
Mexico, Arizona, and Colorado; a little larger than *frustrana,* the basal two-
thirds of the moth's forewing is dark gray, the outer third reddish orange;
the larva is reddish. Pitch nodule or twig moths, *Petrova* spp.—with brown
and gray speckled markings, wingspread .75"—lay their eggs at the base of
pine needle sheaths; the larvae bore into new and old growth of the twigs,
branches, and stems, causing nodules or lumps of pitch and frass to form
over the feeding sites.

(272) PEA MOTH: *Laspeyresia (Grapholitha) nigricana.* **Fig. 490.**

Range: Eastern Canada; New York, Michigan, Wisconsin, Washington,
British Columbia. **Adult:** Brownish, with short black and white oblique
lines along costal margin of forewing. A weak flier, active in late afternoon.
Flat white eggs are laid singly anywhere on plant. **Spread:** .5". **Larva:** Yellowish white, darker at both ends, covered with small spots and hairs. It
bores into the pod, spins a web, and chews partway into the seeds. Pupates
in cocoon in the soil; hibernates as larva in cocoon. One or two broods.
Food: Garden and field peas, vetch, sweet peas. Several closely related
species feed on evergreen seeds. The **Pine Cone Moth** (or "pine seed-
worm"), L. *piperana,* attacks ponderosa pine in British Columbia, the
northwestern states, and California (where it also attacks Jeffrey pine); the
larva is dirty white, bores between two scales into central axis of cone, from
here into the seed, where it feeds on the endosperm; it hibernates in the
cone axis, emerges one or more years later; as many as 50 percent of the
cones may be infested. Larvae of the **Fir Seed Moth,** L. *bracteatana,* are
pinkish, bore into seeds of white, red, and other firs in Oregon, Colorado,
California. Larvae of the **Spruce Seed Moth,** L. *youngana,* are white, with
black heads, attack seeds of Engelmann, Sitka, and blue spruces in Colorado,

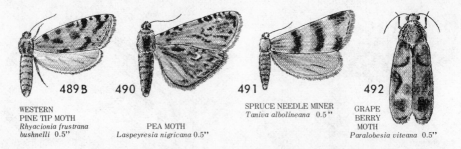

489B
WESTERN
PINE TIP MOTH
*Rhyacionia frustrana
bushnelli* 0.5"

490
PEA MOTH
Laspeyresia nigricana 0.5"

491
SPRUCE NEEDLE MINER
Taniva albolineana 0.5"

492
GRAPE
BERRY
MOTH
Paralobesia viteana 0.5"

Montana, Oregon; those of the **Cypress Twig** (or **Bark**) **Moth,** *L. cupressana,* are grayish white, bore into green cone clusters of Monterey and other cypress in California. The **Hickory Shuckworm,** *L. caryana,* is a serious pest of pecans in the South, also attacks hickory; the small white larva—about .35" long—tunnels in the shuck, preventing its normal separation from the nut; it overwinters as full-grown larva in the shuck.

(273) SPRUCE NEEDLE MINER: *Taniva albolineana.* **Fig. 491.**

Range: Eastern U.S. and Canada; Colorado, Idaho, Oregon, Washington, British Columbia. **Adult:** Dark brown. Eggs are laid in small groups, overlapping like shingles, on needles. **Spread:** .5". **Larva:** Reddish brown or greenish brown; about .3" long full grown. It mines current year's growth, bores into needle at base; many needles are cut off after mining and held to twig by funnel-shaped web. Hibernates as larva in hollow needle, continues to mine other leaves in the spring; pupates under frass within web in early summer. Two or more larvae live within a web. One brood. **Food:** Blue, Norway, Engelmann, and Sitka spruces, ornamentals in particular. [See *Pulicalvaria piceaella* (289).]

(274) GRAPE BERRY MOTH: *Paralobesia viteana.* **Fig. 492.**

Range: Eastern states, especially north of the Ohio River, east of the Mississippi River; New England and southeastern Canada. **Adult:** Brownish purple or grayish purple. It flies at dusk. Flat cream-colored eggs are laid on grape stems or berries. **Spread:** .5". **Larva:** Greenish, with brown head; about .4" long full grown. It feeds on flowers and berries, forming webbing about fruit clusters; grapes turn purplish and fall off. Pupates in cocoon after folding over part of a leaf. Hibernates as pupa in cocoon and folded leaf on the ground. Two to three broods. **Food:** Wild and cultivated grapes. This moth is a major pest in grape vineyards.

(275) EYE-SPOTTED BUD MOTH: *Spilonota ocellana.* **Fig. 493.**

Range: Northern apple-growing areas, Pacific Northwest, California; Europe. **Adult:** Ashy gray, with wide white band across forewing. Eggs are transparent discs, laid on underside of leaves. **Spread:** .5". **Larva:** Dark brown; shiny black head, thoracic and anal shields. It feeds under silk web-

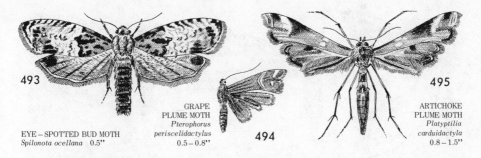

493

EYE-SPOTTED BUD MOTH
Spilonota ocellana 0.5"

GRAPE
PLUME MOTH
*Pterophorus
periscelidactylus*
0.5-0.8"

494

495

ARTICHOKE
PLUME MOTH
*Platyptilia
carduidactyla*
0.8-1.5"

bing, which is reconstructed as feeding sites are changed. Overwintered larva attacks blossoms; summer brood feeds on underside of leaves webbed together and often drawn against fruit, which is pitted with small skin perforations. Hibernates as larva in cocoon and silken case attached to twig. One brood. **Food:** Mainly apple in the North, prune and plum in the Pacific Northwest and California; pear, cherry, and other fruits.

(276) FILBERTWORM: *Melissopus latiferreanus.* Color plate 6g.

Range: California, Oregon. **Adult:** Forewing reddish brown or dusky bronze, with brilliant coppery band across the middle and narrower one toward the outer margin. Eggs, waxy white, transparent when ready to hatch, are laid singly near host fruit. **Spread:** .75". **Larva:** Creamy white, with clear amber head and thoracic shield; about .5" long full grown. It feeds within an oak apple gall or nut. Pupates in cocoon, usually where feeding. Hibernates as mature larva in cocoon within host fruit, in trash or rolled leaf. Two and partial third broods; like the codling moth, some larvae hibernate through two and even three winters. Sometimes confused with codling moth (268). **Food:** "Green apples" or oak galls of the California gallfly, acorns, fruit of Catalina cherry, wild hazelnuts, filberts, walnuts, almonds. Moths migrate from oak, enter nuts after husks crack. [See Hymenoptera (19) and (25).] A serious pest of walnuts in California and of filberts in Oregon.

(277) GRAPE PLUME MOTH: *Pterophorus periscelidactylus* (Pterophoridae). Fig. 494.

This moth belongs to a small family called plume moths because the wings are split into feather-like parts. The larvae are hairy, like woollybears (195), but unlike them, live in webs or tubes. **Range:** The grape plume moth is an eastern species. **Adult:** Forewing brownish yellow, with white and brown markings; hindwing with first two "feathers" chocolate brown, the third nearly white. It rests with wings closed, stretched out at right angles to the body to form a T. Eggs are laid on branches. **Spread:** .5–.8". **Larva:** Greenish, covered with white hairs. It feeds on terminal leaves drawn together with silk. The posterior end of the puparium is attached to a plant or other object with silk, hangs down at an angle, like some butterfly chrysalises.

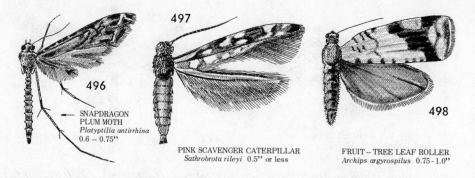

497

496

SNAPDRAGON
PLUM MOTH
Platyptilia antirrhina
0.6 – 0.75"

498

PINK SCAVENGER CATERPILLAR
Sathrobrota rileyi 0.5" or less

FRUIT – TREE LEAF ROLLER
Archips argyrospilus 0.75 - 1.0"

Food: Grapevines. The **Artichoke Plume Moth,** *Platyptilia carduidactyla* (Fig. 495), feeds on thistle, globe artichoke, and cardoon, is found in Canada from Quebec to British Columbia and in the U.S. from Connecticut to South Carolina, west to the Pacific; the forewing is brown to buff, with light spot on front margin near apex; the hindwing is grayish; spread from .8" to 1.5"; the larva is yellowish, with head, thoracic and anal shields shiny black. *Platyptilia antirrhina* (Fig. 496), a grayish brown moth, with rough scaling on the body, spread from .6" to .75", is a pest of snapdragon in greenhouses and gardens on the Pacific coast; the young larvae mine leaves; older ones bore into stems and other parts of the plant. A tropical species, the **Lantana Plume Moth,** *Platyptilia pusillodactyla,* attacks ornamental lantana in the South.

(278) PINK SCAVENGER CATERPILLAR: *Sathrobrota (Pyroderces) rileyi* (Cosmopterigidae). **Fig. 497.**

Also called pink cornworm. **Range:** Southern states, Florida to California; Oklahoma, Arkansas. **Adult:** Reddish brown; forewing with whitish or straw-colored streaks; hindwing pale grayish, with long fringe. Pearly white eggs are laid singly, sometimes in twos or threes. **Spread:** .5" or less. **Larva:** Pinkish, with pale brown head and thoracic shield. It forms webbing of frass and silk in spaces as it feeds; eats husk, kernels, and cob of corn. **Food:** Corn in field and cribs in the South; also cotton, almonds, walnuts, oranges and various other fruits, usually in a state of decay.

Leaf Roller Moths (Tortricidae)

This is a large family of small, broad moths with wide, somewhat rectangular forewings, usually grayish or brownish and strikingly marked. Because the folded wings often give the moth a distinctive bell shape, they are sometimes referred to as the "bell moths." They fly at night, and usually lay their eggs in overlapping masses like shingles, often covering them with a cement. The family name, Tortricidae—meaning "twist"—refers to its leaf-rolling tendency, a trait which is not confined to this group or charac-

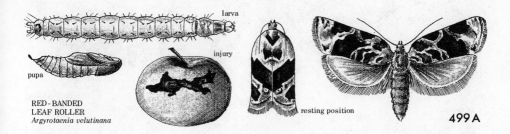

larva

injury

pupa

RED - BANDED
LEAF ROLLER
Argyrotaenia velutinana

resting position

499 A

teristic of all tortricids. The larvae—usually greenish, with scattered hairs—have a habit generally of rolling or folding leaves and forming webbed "nests." Many of them mine leaves in the early instars, but they are not true borers. When disturbed they wriggle backwards or sideways, and often drop to the ground and return on the same thread.

(279) FRUIT-TREE LEAF ROLLER: *Archips argyrospilus*. Fig. 498.

Range: Widespread throughout Canada and the U.S. **Adult:** Forewing mottled with brownish and yellowish shades; two creamy spots on costal margin. Hindwing pale yellowish brown. Eggs are laid in a single mass on trunk or branch and covered with a cement. **Spread:** .75–1″. **Larva:** Green, with dense, minute spinules; head dark reddish brown; from .8″ to 1″ long full grown. It feeds on leaves and fruit which are webbed together, and eats large holes in apples, causing characteristic rough scars. Pupates in rolled leaf and webbing. Hibernates in egg stage. One brood. **Food:** Most deciduous fruit trees, but primarily apple; also walnut in California. The **Ugly-nest Caterpillar,** *A. cerasivoranus,* best known as a defoliator of chokecherry, feeds inside a webbing or "tent" which may extend more than 100 feet along a hedgerow.

(280) SPRUCE BUDWORM: *Choristoneura fumiferana*. Fig. 499B.

Range: Labrador to Virginia, the northern states, across Canada; the Pacific coast range. **Adult:** Grayish, with indistinct yellowish brown or reddish brown splotches on forewing; hindwing uniformly dark brown, with white fringe. Green eggs are laid on underside of needles in a mass, overlapping like shingles. **Spread:** .85–1.2″. **Larva:** Dark brown, sides lighter, two rows of white dots on back, numerous short dark spinules; anal and thoracic shields brownish yellow, latter with V marks posteriorly; about .9″ long full grown. Spins cocoon soon after hatching and goes into hibernation until the following spring. Early instars mine one to four old needles, entering at the middle; later instars attack young bud, cone, or twig. Needles are cut off and incorporated into a web, which is restricted to a short tube or tent with little or no frass. Pupation takes place within the webbing. One brood. Eastern and western forms are now recognized. **Food:** Balsam fir and spruce principally; also Douglas fir, hemlock, larch, pines. This has been

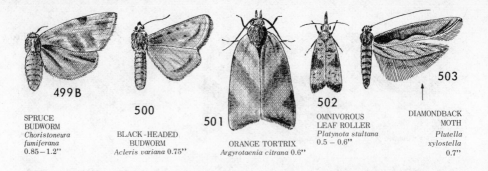

499B
SPRUCE
BUDWORM
Choristoneura
fumiferana
0.85–1.2"

500
BLACK-HEADED
BUDWORM
Acleris variana 0.75"

501
ORANGE TORTRIX
Argyrotaenia citrana 0.6"

502
OMNIVOROUS
LEAF ROLLER
Platynota stultana
0.5 – 0.6"

503
DIAMONDBACK
MOTH
Plutella
xylostella
0.7"

rated the third most destructive insect (next to the cotton boll weevil and corn earworm) in North America. Severe outbreaks have occurred in the pulpwood forests of eastern North America—from Minnesota to Maine, Quebec and New Brunswick—in stands where balsam fir is predominant. While primarily a forest pest, it is found widely in nurseries and ornamental plantings.* The **Jack-pine Budworm,** *C. pinus*—common in Christmas tree plantations and a major defoliator of jack pine in eastern Canada—is similar; it prefers jack and red pine but will attack Scotch and white pines, white and Norway spruces; it is smaller than *fumiferana,* yellowish brown, with distinct median band and basal patch separated by silvery or whitish areas on forewing; hindwing smoky, with white fringe; larva reddish brown, with thoracic and anal shields brownish yellow. The **Large Aspen Tortrix,** *C. conflictana,* is a major defoliator of aspen throughout Canada.

(281) BLACK-HEADED BUDWORM: *Acleris variana*. Fig. 500.

Range: Northern states, Canada, Pacific Northwest to Alaska. There are probably eastern and western forms. **Adult:** Grayish, with brown, black, orange, or white mottling; color patterns extremely variable. Eggs are laid singly on needles, usually the underside. **Spread:** .75". **Larva:** Body bright green, thoracic shield yellowish to brown, with middorsal groove; head reddish brown or dark brown; about .5" long full grown. First instar requires new foliage, enters shoot at base of needles; last instar often webs two current shoots together. Four or five instars. Pupates at last feeding site, adult emerges in August. One brood. Hibernates in egg stage. **Food:** Hemlock, fir, spruce; most severely attacked are balsam fir in the East and hemlock in the West. Outbreaks have occurred in the Olympic Peninsula of Washington, in British Columbia, and eastern Canada.

(282) ORANGE TORTRIX: *Argyrotaenia citrana*. Fig. 501.

Range: Southwestern states, California northward (in greenhouses) to British Columbia. **Adult:** Forewing tan to rusty brown; black diagonal band across forewings forms a V when wings are folded. Eggs—cream-

---

* Control by release of sterilized males is under investigation (see footnote, p. 630); adult males can be successfully sterilized by irradiation.

colored, turning to yellow—are laid in masses overlapping like shingles. **Spread**: .6″. **Larva**: Straw-colored or light green, with light brown head and thoracic shield; about .5″ long full grown. It makes nests in tips of twigs or among buds and blossoms, feeding on leaves, buds, and blossoms. Usually pupates in the last nest; hibernates as larva. Two to four broods. **Food**: Citrus, prunes, apricot, pear, apple, and a great variety of other plants. The **Pine Tube Moth**, *A. pinatubana*, occurs throughout the U.S. and Canada, attacks pines and western hemlock chiefly; the larva binds several needles together, forming a tube lined with papery white web, and feeds within, drops to the ground on thread to pupate, hibernates as pupa; the moth is brownish gray, with darker patches and bands on forewing, spread about .5″. The **Red-banded Leaf Roller**, *A. velutinana* (Fig. 499A), is a pest of grape and other common fruits, flowers, vegetables, many deciduous trees (with the codling moth, it is often a major pest of apples) in eastern U.S. and Canada; ground color of adult is nearly white, forewing with wide reddish band from costa to inner margin, covering about one-third of the wing, spread from .5″ to .65″; the larva is slender, pale green, with yellowish or brownish yellow thoracic shield and head, about .7″ long; it hibernates as pupa in litter, has two or three broods; eggs yellowish, laid in masses on bark.

(283) OMNIVOROUS LEAF ROLLER: *Platynota stultana.* **Fig. 502.**

Range: Mexico, California, Arizona, Texas, Arkansas; Virginia and elsewhere in greenhouses. **Adult**: Forewing dark brown, outer half yellowish brown; labial palpi and antennae comparatively long. Eggs are laid in flat clusters, overlapping one another. **Spread**: .5–.6″. **Larva**: Yellowish white, with brown head and thoracic shield when newly hatched; body coloring changes to yellowish green or brownish green; broad, transparent middorsal stripe; about .6″ long full grown. Habits similar to those of orange tortrix (282). **Food**: Alfalfa, cotton, celery, lettuce, and other field crops; walnut, orange, and a wide variety of subtropical fruits and flowers. A major pest of cotton in southern California.

(284) DIAMONDBACK MOTH: *Plutella xylostella* (*maculipennis*) (Yponomeutidae). **Fig. 503.**

This moth belongs to a family called ermine (or plutellid) moths. They have narrow forewings, pointed, fringed hindwings; these are held close against the body when at rest, and the antennae are held straight out in front. **Range**: The diamondback moth occurs wherever cole crops are grown and in greenhouses; it is a native of Europe. **Adult**: Grayish, with inner margin of male's forewing marked to form four diamond-shaped, pale yellowish areas in the middle when folded. Yellowish white eggs are laid singly and in twos and threes. **Spread**: About .7″. **Larva**: Light greenish, with

504

larva

505 ANGOUMOIS GRAIN MOTH
Sitotroga cerealella 0.5"

506

← POTATO TUBERWORM
Phthorimaea operculella 0.5"

PEACH TWIG BORER
Anarsia lineatella 0.3 – 0.5"

black head, body pointed at both ends, about .3" long full grown. Like tortricid larvae (preceding family), it wriggles violently when disturbed and often drops on a thread, a habit which serves to distinguish it from other cabbageworms [see (20), (213), and (253) above]. It feeds on all parts of the plant, often perforating leaves. Pupates in gauzy, loose saclike cocoon. Hibernates as adult in debris. Two to six or more broods. **Food:** Cole crops: cabbage, broccoli, cauliflower, collards, kale, Brussels sprouts, kohlrabi; in greenhouses, such ornamentals as sweet alyssum, candytuft, wallflower, stock. The **Arborvitae Leaf Miner,** *Argyresthia thuiella,* is common on arborvitae in the eastern states and Canada, causes tips of twigs to turn brown and die; the moth is grayish, forewing with brownish bands and silvery spots, spread about .4"; the greenish eggs are laid between the leaf scales; it overwinters as larva in the mine, has one brood.

Gelechiid Moths (Gelechiidae)

The gelechiids are small moths, usually with narrow forewings; the hindwings are pointed and have a broad fringe of hairs. The labial palpi are generally prominent and pointed upward. The family includes borers, leaf miners, and leaf rollers that form protective webs and feed on live plants and stored grains.

(285) POTATO TUBERWORM: *Phthorimaea (Gnorimoschema) operculella.* **Fig. 504.**

Range: Florida to California, and northward. **Adult:** Grayish, with minute brownish or black markings. It is active only at night. Pearly white eggs are laid singly on leaves and stems or eyes of exposed tubers. **Spread:** .5". **Larva:** Pinkish green or white, with brownish head, sometimes with reddish purple middorsal stripe; about .5" long full grown. It mines stems and leaves which are webbed together, and tunnels in stored potatoes, filling the spaces with webbing and excrement. In the field it hibernates as larva or pupa in the soil. Two to six broods. **Food:** Potato, tomato, tobacco, eggplant and other plants of the nightshade family. Most damage to potatoes occurs in storage, little in the field except in Virginia and California. *P. gallaesolidaginis* forms galls or swellings in the stems of goldenrod (Fig. 517) [see Diptera (21) and (60) for other gallmakers on goldenrod]; the moth's forewing is brownish with darker spot in the middle, the lighter areas forming an elliptical pattern when the wings are closed. The closely related **Tomato Pinworm**, *Keiferia (Phthorimaea) lycopersicella,* bores pinholes in the developing buds, green and ripening fruit of tomatoes, usually near the stems; the larva is yellowish, gray or green, with purple spots, about .25″ long.

(286) ANGOUMOIS GRAIN MOTH: *Sitotroga cerealella.* **Fig. 505.**

Range: World-wide, most commonly found in stored grains. **Adult:** Forewing buff or yellowish brown, silky, with fringe slightly darker; sometimes with small black spot near center of wing. Eggs—first white, turning reddish —are laid on wheat heads and exposed tips of corn in the field, and on kernels in storage. **Spread:** .5″. **Larva:** White, with yellowish head, about .2″ full grown. It is usually found curled up in a kernel, seldom seen elsewhere. It spins a small web and bores into the kernel to feed, cuts a circular hole partway around to form a flap before pupating to facilitate escape later as a moth. Pupates in cocoon in cavity eaten out of kernel. Hibernates as larva in kernel (in storage) or in field litter and straw bales. Five or six broods in storage houses in the North. The combine harvester destroys many. **Food:** All cereal grains. Most injurious in the South to field and stored grains. [See Coleoptera (442).] Eggs of the moth are used as host for rearing *Trichogramma* egg parasites for biological control of various pests.

(287) PEACH TWIG BORER: *Anarsia lineatella.* **Fig. 506.**

Range: Widespread throughout North America, known in the U.S. since 1860; Europe, Asia Minor. **Adult:** Brownish gray, forewing stippled with black, lightly fringed around most of the margin, hindwing lighter gray and more heavily fringed; antennae long, tapering, ringed dorsally with alternate black and silvery gray; palpi large, proximal halves plumose; wings held rooflike over body at rest. Eggs are laid on bark, leaves, or fruit. **Spread:** .3–.5″. **Larva:** Yellowish brown at first, with black head; changes to reddish brown or chocolate brown, about .5″ long full grown; a lighter band encircles the body between each segment. It bores into tender shoots, later into the fruit. Pupates in loose dark brown cocoon on rough bark. Hibernates as immature larva in hibernaculum or cell constructed of fecal pellets, bits of bark, and silk in crevice of bark, usually in a crotch; a small chimney projects above the burrow. One to four or more broods. **Food:** Peach, apricot, nectarine, plum, prune, almond. Sometimes confused with the Oriental fruit moth (269).

(288) STRAWBERRY CROWN MINER: *Aristotelia fragariae.* **Fig. 507.**

Range: North Central states, Pacific Northwest. **Adult:** Brownish black, with yellowish scales on forewing. **Spread:** .5″. **Larva:** Carmine pink, about

507
508
509

STRAWBERRY CROWN MINER
Aristotelia fragariae 0.5"

PINE NEEDLE MINER
Exoteleia pinifoliella 0.35"

PINK BOLLWORM
Pectinophora gossypiella 0.75"

.5" long full grown. It tunnels in crown just below base of leaves, causing stunting or death of plant. Hibernates as mature larva in silken hibernaculum in crown of plant. **Food:** Strawberries. [See (246) above, and Coleoptera (439).]

(289) PINE NEEDLE MINER: *Exoteleia pinifoliella*. Fig. 508.

Range: Ontario and Quebec to Georgia and Texas. **Adult:** Forewing reddish brown to golden brown, with three narrow grayish transverse bands; hindwing darker and wider. Creamy white eggs are inserted through exit holes in mines just abandoned. **Spread:** .35". **Larva:** Pinkish brown, about .2" long, with sclerotized areas of head, thoracic and anal shields brown. It mines current year's foliage, which turns yellow and dies. Pupates in needle after mining it completely, sealing entrance hole with silk. One brood. Hibernates as larva. **Food:** Pines. The **Lodgepole Needle Miner,** *Coleotechnites (Recurvaria) milleri,* has caused severe damage to lodgepole pine in Yosemite National Park; *C. starki* has done the same in Rocky Mountain National Park. *Pulicalvaria piceaella* mines needles of spruce in eastern North America, westward to Manitoba and Colorado; the forewing is dull yellowish brown near base, darker at apex, with conspicuous white transverse band near apex; hindwing silvery gray, with brownish yellow fringe; spread about .5"; the larva is orange-yellow, with shiny brown head and thoracic shield, colors later becoming bright red and black. It usually mines previous year's growth, sometimes new growth, under a loose webbing. [See spruce needle miner (273).]

(290) PINK BOLLWORM: *Pectinophora gossypiella*. Fig. 509.

Range: Arizona, California, New Mexico, Texas, Oklahoma, Arkansas, Louisiana, Florida; world-wide, believed to be a native of India. **Adult:** Grayish brown, wings folded close to body; forewing with two or more poorly defined black median spots, blackish tip and broad band near tip; tips of wings fringed, appendages ringed with black. Greenish white eggs are laid on stems, squares, terminal buds. **Spread:** .75". **Larva:** Pinkish, with darker transverse bands, yellowish brown head, about .5" long full grown. The young larva bores into a square, feeding on developing flower, or into a boll, feeding on lint and seed; the mature larva emerges and pupates in the soil or trash. Hibernates as larva in cocoon, in boll left in the field, in the soil, or in stored seed; some hibernate up to two and one-half years. **Food:** Cotton, *Abutilon,* hibiscus, okra, hollyhock, *Thespesia.* This has been rated the sixth most destructive insect in the world; it causes the most severe damage to cotton in Argentina, Brazil, China, Egypt, and India. Larvae of the **Cotton Stem Moth,** *Platyedra villella,* resemble pink bollworms, are creamy white in late stage; they feed on ovaries, flower buds, stems, leaf tips, and capsules of cotton, hollyhock, and several mallows in New York, New Jersey, Connecticut, Massachusetts, and Rhode Island.

(291) ASPEN BLOTCH MINER: *Lithocolletis tremuloidiella* (Gracillariidae). **Fig. 521A.**

Range: California, Utah, Idaho, British Columbia. **Adult:** Dull yellowish brown. **Spread:** .3″. **Larva:** Constructs irregularly shaped, winding tunnels between leaf surfaces, with blotches here and there, causing premature drop. Pupates within mine. **Food:** Aspen, poplar. *Lithocolletis* spp. mine leaves of live oak; their eggs are laid along midrib, and the young larva mines an irregular path which is later expanded to form a large blister. In the East and Pacific Northwest, the **Spotted Tentiform Leaf Miner,** *L. crataegella,* mines blotches in the leaves of apple, plum, quince, wild cherry, and hawthorn; head of larva is wedge-shaped, body flat in early stage, body yellow and cylindrical in late stage; it pupates and overwinters in the mine; the adult is golden brown with silver streaks margined in black, about .15″ long with wings folded. The **Unspotted Leaf Miner,** *Callisto geminatella,* mines apple leaves in a similar way (see Fig. 521E); the mature larva is slate gray, with greenish gray, black-spotted head; the adult is steel gray, about .2″ long with wings folded. The yellowish larva of the **Azalea Leaf Miner,** *Gracillaria azaleella,* mines the leaf, then folds edge and feeds on surface; the head, thorax, and basal half of forewing in the adult are purplish, costal half of forewing golden, margin brownish yellow, spread .4″; it hibernates as larva, has three broods. The **Lilac Leaf Miner,** *G. syringella,* is similar, attacks lilac, ash, deutzia, or euonymus, rolling the leaves or webbing several together after mining. The **Solitary** and **Gregarious Oak Leaf Miner,** *Cameraria hamadryadella* and *C. cincinnatiella,* attack oaks—red, white, and black, mostly white in the case of the solitary species —in the Northeast; one larva of *hamadryadella* occurs in a mine (see Fig. 521B), several in the case of *cincinnatiella;* there are two or three broods; hibernation is in the larval and pupal stages respectively.

(292) WEBBING CLOTHES MOTH: *Tineola bisilliella* (Tineidae). **Fig. 510.**

Range: Throughout the world. **Adult:** Yellowish or buff-colored, with satiny sheen, stiff reddish hairs on head. A weak flier, it prefers darkness but may fly about lazily in darkened corners or margins of lighted areas.

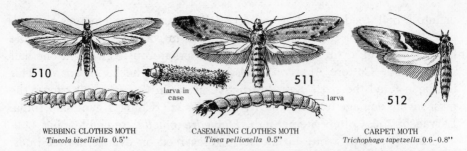

510

larva in case

511

larva

512

WEBBING CLOTHES MOTH
Tineola biselliella 0.5"

CASEMAKING CLOTHES MOTH
Tinea pellionella 0.5"

CARPET MOTH
Trichophaga tapetzella 0.6-0.8"

They fly from house to house, or infestations may be carried in articles of clothing containing wool or other animal fibers. Eggs are laid singly on food material. **Spread:** .5". **Larva:** Pearly white, with dark head, practically hairless, about .5" long full grown. On some fibers it is nearly translucent, on others opaque. It spins a silken webbing, incorporating some of the fabric, to form a feeding tube. Pupates in enlarged tube. Length of life cycle varies with temperature and humidity, may be from three months to one year. **Food:** The natural food is unprocessed materials such as pollen, feathers, hair, fur, wool, insect remains.

(293) CASEMAKING CLOTHES MOTH: *Tinea pellionella.* **Fig. 511.**

Range: Throughout the world. **Adult:** Similar to webbing clothes moth, except that the forewing is dimly spotted in a darker shade, and hairs on the head are lighter in color; the hindwing is whitish. Habits are as described above in (292). **Spread:** About .5". **Larva:** Similar to webbing clothes moth larva in size and appearance. It constructs a silken parchment-like case which is dragged about as it feeds; when disturbed it retreats into the case. Pupates in cocoon formed by enlarging the case. The pupa is white, later changing to brown. The life cycle may take from two months to four years, with long periods as dormant larva. **Food:** Same as in (292) above. The **Carpet Moth,** *Trichophaga tapetzella* (Fig. 512), also worldwide in distribution—with spread from .6" to .8"—is a pest of carpets in homes and neglected storage rooms. The closely related **European Grain Moth,** *Nemapogon granella* (Fig. 513)—with spread of .5"—is a widespread pest of grains in storage.

(294) PISTOL CASEBEARER: *Coleophora malivorella* (Coleophoridae). **Fig. 514.**

Range: Apple-growing areas from Virginia to Kansas, north into Canada. **Adult:** Mottled gray. (Most species of this family are not readily distinguished from one another in the adult stage.) Eggs are laid on underside of leaves. **Spread:** .5". **Larva:** Light brown, with darker head. It feeds from end of silken case, which is about .25" long, brownish gray, bent and broad at base, giving it some resemblance to a pistol. The case is moved about by the larva as it feeds and is maintained upright on the plant's surface. Hibernates as larva in case attached to bark. One brood. **Food:** Apple, pear,

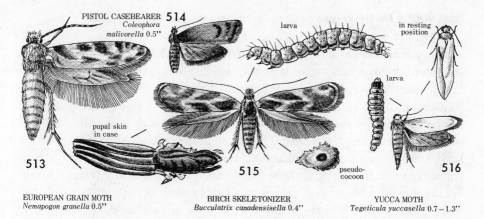

PISTOL CASEBEARER **514**
Coleophora malivorella 0.5"

larva

in resting position

larva

pupal skin in case

513

515

pseudo-cocoon

516

EUROPEAN GRAIN MOTH
Nemapogon granella 0.5"

BIRCH SKELETONIZER
Bucculatrix canadensisella 0.4"

YUCCA MOTH
Tegeticula yuccasella 0.7–1.3"

cherry, plum, quince. The **Cigar Casebearer**, *C. serratella* (Fig. 521D), is similar in appearance and habits, except that it is cigar-shaped; the case is rebuilt in the spring when feeding is resumed, looks like a little stick with the end stuck to the leaf; the larva mines leaves by protruding its head from the case, leaving telltale blotches on the leaf; it ranges farther westward than *malivorella*—to New Mexico, Montana, British Columbia. The cigar casebearer *C. occidentis* is common on apple in the East. Some other common casebearers are: the **Pecan Cigar Casebearer**, *C. caryaefoliella;* **Larch Casebearer**, *C. laricella;* **Elm Casebearer**, *C. limosipenella;* **Birch Casebearer**, *C. salmani;* and **Cherry Casebearer**, *C. pruniella.* Casebearer moths are not usually numerous enough to cause much damage, but almost every plant supports one or more species; some live on grasses, others in seed heads.

(295) BIRCH SKELETONIZER: *Bucculatrix canadensisella* (Lyonetiidae). **Fig. 515.**

Range: Northeastern U.S., west to Illinois; eastern Canada. **Adult:** Bright brown, wings crossed with silvery white bars. Eggs are laid singly on leaves. **Spread:** .4". **Larva:** Light green or yellowish green, with short hairs on small tubercles and pale brown head; about .25" long full grown. Young larvae make serpentine mines in leaves; later stages feed externally on underside of leaves and often skeletonize them. Larva drops to ground to pupate, hibernates as pupa in white, ribbed cocoon which turns dark with age; pupal stage normally extends from September to July, may be delayed three or four weeks by cool weather. One brood. **Food:** Birch, alder, oak. It is most destructive in August, is sometimes subject to severe predation and parasitism; periodic outbreaks have occurred. After the first molt, the larva of the **Cotton Leaf Perforator**, *B. thurberiella,* eats holes in cotton leaves in the Southwest; the adult is white, with black markings on the forewings; the larva is dull amber green, with two dorsal rows of black spots and distinct white tubercles; the first-instar larva makes a winding mine in the leaf.

(296) YUCCA MOTH: *Tegeticula* (*Pronuba*) *yuccasella* (Incurvariidae, Prodoxinae).
   Fig. 516.

The yucca moths are responsible for the fertilization of yucca plants, commonly known as Spanish bayonet, which are not self-pollinating. Modified mouthparts enable the female moth to scrape sticky pollen from the stamens of one plant, form it into a ball, and carry it to another plant. She inserts her eggs in the ovary and thrusts the pollen ball down the stigmatic tube, thus fertilizing the flower. The developing larvae feed on the seeds but leave some in the capsule unharmed. **Range:** Southern Ontario and Massachusetts, west to North Dakota, south to California and Mexico. **Adult:** White, immaculate; ventral surface and basal fourth of costal margin of forewing brown, with bronzy iridescence, fringes all white; dorsal surface of hindwing darker owing to thinner scaling. **Spread:** .7–1.3″. **Larva:** Body stout, fusiform, strongly tapered posteriorly, head small, not depressed, thoracic legs small, prolegs on third and sixth abdominal segments have no hooks. **Food:** Nearly every species of *Yucca*, which accounts for its wide distribution. *T. maculata maculata* is similar, somewhat smaller, white, thorax sometimes with dark spot in center of dorsum, apical third of forewing with brown spotting, hindwings considerably darker; it occurs in California, pollinates *Yucca whipplei* (the "Quixote plant").

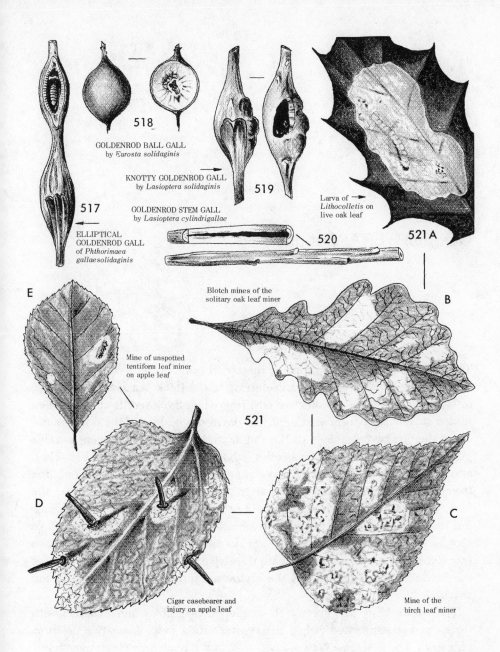

518

GOLDENROD BALL GALL
by *Eurosta solidaginis*

KNOTTY GOLDENROD GALL
by *Lasioptera solidaginis*

519

GOLDENROD STEM GALL
by *Lasioptera cylindrigallae*

517

ELLIPTICAL
GOLDENROD GALL
of *Phthorimaea
gallaesolidaginis*

520

Larva of →
Lithocolletis on
live oak leaf

521 A

E

Mine of unspotted
tentiform leaf miner
on apple leaf

Blotch mines of the
solitary oak leaf miner

B

521

D

Cigar casebearer and
injury on apple leaf

C

Mine of the
birch leaf miner

20

COLEOPTERA: Beetles

This is by far the largest order of insects in number of species described. The name means "sheath wings," referring to the thick, hardened forewings, or elytra, which form a protective covering when the insect is at rest. When folded they usually form a straight line down the middle of the dorsum and cover the last two segments of the thorax and the entire abdomen; notable exceptions occur in the blister beetles, whose elytra do not meet (Fig. 711), and in the rove beetles, which have shortened wings (Fig. 631). The line formed by the juncture of the two elytra in repose is called the suture; the front edge (as viewed when spread out or flying) is termed the lateral margin in descriptions, and the basal exterior angle becomes the humerus or shoulder. The membranous hindwings do most of the work of flying, have few veins, and are folded underneath the elytra when at rest. In some species the hindwings are missing and the elytra are fused; more rarely, both are absent.

The adults (Fig. 7B) have well-developed mandibles and maxillae with three- to five-segmented palpi; the shorter labial palpi have two or three segments. The antennae take many forms, and have ten or eleven segments. Ocelli are generally absent. The abdomen usually has nine segments—the ninth is often hidden or absent—with a pair of spiracles on the pleura (or sides) of each. The eighth sternite (underside of segment) is usually visible but partly drawn into the seventh. The last visible abdominal tergite (top of segment) is called the pygidium, the next to the last is referred to as the propygidium; either or both are exposed in some groups, as in the

Histeridae (Fig. 658). Cerci are absent. The tarsi normally have five segments, sometimes four or three. Beetles vary from minute to large in size, are usually some shade of black, brown, yellow, or red, but are often brightly colored and decorated with attractive patterns. Metamorphosis is complete (Fig. 6A).

The larvae are called grubs. Some are long, and flat or rounded; others are short and stout, and they may be smooth or wrinkled, with or without hairs or spines. They have a well-developed head, which sometimes telescopes into the thorax, especially among the borers and leaf miners. The first-instar larva of many species is armed with "egg bursters"—straight or curved spines, sometimes relatively long, that appear on the thorax and abdomen, or the head, and usually disappear during the first molt; their function is to facilitate hatching. The abdomen normally consists of nine or ten segments. The thoracic legs (three pairs) are generally well developed, but are sometimes reduced or lacking, as in the wood borers and leaf miners; abdominal prolegs are never present. The thoracic spiracles are usually located on the mesothorax. The larvae pupate in cells formed in the ground, in tunnels, or hidden in other places, and seldom form cocoons.

Adults and larvae in most cases eat the same food; the majority feed on plants, but many are scavengers or are predaceous, with some of our worst plant pests and many valuable predators included. The order is divided into two main suborders: Adephaga and Polyphaga. In the former, the underside of the first abdominal segment is split by the hind coxal or leg cavities, with much of the segment concealed by the coxal plates; and there is a distinct suture between the pronotum and episternum (anterior sclerite on the side of the thorax). These features are not present in the Polyphaga. The Adephaga are mostly predaceous in both the adult and the larval stages; they are represented here in examples (1) to (58). The remainder represent the Polyphaga.

ADEPHAGA

Tiger Beetles (Cicindelidae)

The name "tiger beetles" apparently stems from their fierce predatory habits and stealth, and possibly the color patterns. Most of them have long legs, run rapidly, take to flight quickly when disturbed, and are difficult to capture in the daytime. Sun-loving species are very sensitive to light, the mere shadow of a cloud being sufficient to curtail their activity. Both adults and larvae eat all kinds of insects, but the family is considered beneficial as a whole. Adults in the genus *Cicindela* are mostly diurnal, often attracted to lights, and may be captured easily at night with a light. They are usually found in sandy places with scattered vegetation and along the shores of

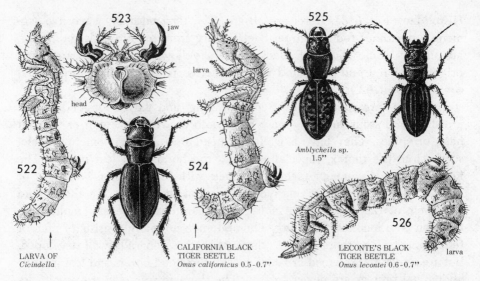

523 jaw

head

522

LARVA OF
Cicindella

524

CALIFORNIA BLACK
TIGER BEETLE
Omus californicus 0.5-0.7"

525

Amblycheila sp.
1.5"

526

LECONTE'S BLACK
TIGER BEETLE
Omus lecontei 0.6-0.7"

larva

streams and lakes; various species show a decided preference for certain types of soils. Some species are solitary, others are gregarious and are sometimes found in large colonies. Adults dig short burrows for shelter at night or on hot days, and excavate deep tunnels for overwintering.

Eggs are deposited singly, each in a separate burrow which is enlarged by the larva prior to each molt and before hibernating; overwintering tunnels are often dug to several feet in depth. There are three instars, the entire life cycle being from two to three years. They hibernate in either the adult or the late larval stage; pupation takes place in a chamber dug off the main tunnel, after the latter is filled. The larva is S-shaped (Fig. 522), has strong spines, a hump with curved hooks on the fifth abdominal segment, and powerful jaws (Fig. 523). It hides in a vertical tunnel ending in a smooth circular opening on the surface, waits with head protruding, reaches out and snatches other insects without losing its grip on the tunnel.

Considerable variations in size, color, and pattern of the adults occur within many species. They are usually identified by the presence or absence of hairs on the face and top of the head, by serrulations or minute notching of the apexes of the elytra, or by markings on the wing covers. Two genera, *Omus* and *Amblycheila* (Fig. 525), are confined to the West. *Omus* adults are black, have elytra but no wings. The principal enemies of tiger beetles are bee flies (Bombyliidae), which parasitize the larva. [See Diptera (36).] Some tiger beetles emit strong "fruity" odors when handled.

(1) CALIFORNIA BLACK TIGER BEETLE: *Omus californicus* (Cicindelinae, Megacephalini). **Fig. 524.**

Range: Along the Pacific coast, from southwestern British Columbia to southern California. **Adult:** Black. Side of elytra narrowly inflexed, thorax

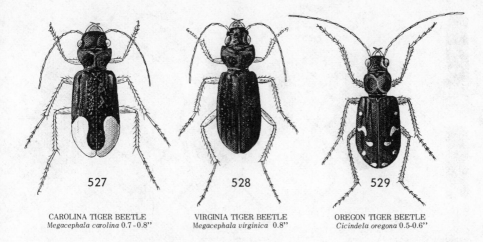

527
CAROLINA TIGER BEETLE
Megacephala carolina 0.7-0.8"

528
VIRGINIA TIGER BEETLE
Megacephala virginica 0.8"

529
OREGON TIGER BEETLE
Cicindela oregona 0.5-0.6"

distinctly margined. **Length:** .5–.7". Scarce; found under stones in dry places.

(2) LECONTE'S BLACK TIGER BEETLE: *Omus lecontei.* **Fig. 526.**

Range: Coastal Range Mts. of California. **Adult:** Shiny, black. Elytra and thorax as in (1) above. **Length:** .6–.7". *O. dejeani,* black or brownish black, may be identified by the foveae on the elytra; it is found in rotten stumps and trees in western British Columbia, Washington, Montana, and in California, and will bite viciously.

(3) CAROLINA TIGER BEETLE: *Megacephala (Tetracha) carolina.* **Fig. 527.**

Range: Southern states. **Adult:** Greenish gold, tip of abdomen and elytra orange-yellow or reddish yellow; elytra with dense punctures, irregular in size, pronotum smooth. They are nocturnal, also active at dusk, hide during the day. Eggs are laid in burrows near fresh water. **Length:** .7–.8". **Larva:** Yellowish, head and thorax blackish or bronze. *M. virginica* (Fig. 528) is dark golden green without the yellow markings, ranges westward.

(4) OREGON TIGER BEETLE: *Cicindela oregona* (Cicindelini). **Fig. 529.**

Range: New Mexico, north to Alaska. **Adult:** Blue or green, with white markings; elytra punctate, edge of apexes sometimes serrulate. Short labrum. Small patch of long hairs near inner edge of each eye distinguishes it from all other species [other than the female of *scutellaris* (7), which has similar hairs]. Markings similar to *C. repanda* (8) and *duodecimguttata* (8) except that they break down more often in *oregona.* Gregarious. **Length:** .5–.6". Found along the edge of lakes and streams. There are several subspecies variously distributed. *C. o. oregona* (*scapularis*) is bright blue. *C. o. guttifera* (Fig. 530A) is olive green with metallic luster; *C. o. depressula* (Fig. 530B) is also green, with less pronounced markings.

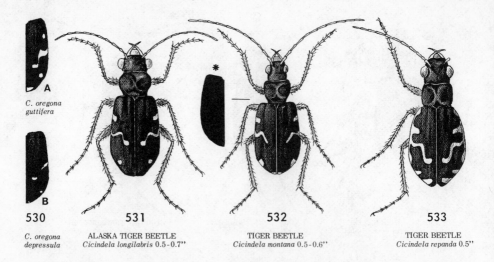

A

C. oregona
guttifera

B

530
C. oregona
depressula

531
ALASKA TIGER BEETLE
Cicindela longilabris 0.5-0.7"

532
TIGER BEETLE
Cicindela montana 0.5-0.6"

533
TIGER BEETLE
Cicindela repanda 0.5"

(5) ALASKA TIGER BEETLE: *Cicindela longilabris*. **Fig. 531.**

Range: Widely distributed throughout Canada and the western states. Distinctly a northern species. **Adult:** Black to various shades of brown and green, with white markings. Head bald, depressed, with long labrum. Similar to *C. montana* (6). May be distinguished by the many fine granules on the elytra. Gregarious. **Length:** .5–.7". Found in sandy areas around coniferous forests.

(6) TIGER BEETLE: *Cicindela montana*. **Fig. 532.**

Range: Western Canada and the U.S. **Adult:** Highly variable. Black to brown, bronze, and bright green, with white markings. Similar to *C. longilabris* (5) except that the elytra are smooth, shiny. *C. m. montana* is black. *C. m. perviridis* is a beautiful bright iridescent green. **Length:** .5–.6". Prefers clay soils, found on plains and in mountains.

(7) TIGER BEETLE: *Cicindela scutellaris*. **Color plate 7d.**

Range: East of the Rocky Mts. **Adult:** One of the most variable species. Head and thorax brilliant metallic blue or green, elytra brilliant green or coppery. Some forms are dull or bright red, olive green, or black, with or without markings. Markings are on the margin of the elytra, may be a complete or broken band with irregular edges. Front of male is hairy, labrum white. Female has small tuft of hair on inner edge of eyes, as in *C. oregona* (4), and black labrum. Gregarious. **Length:** .4–.5". Found in sandy places with scattered vegetation.

(8) TIGER BEETLE: *Cicindela repanda*. **Fig. 533.**

One of the most widely distributed species, found in Texas and Hudson Bay. **Range:** East of the Rocky Mts.; British Columbia. One of the most common species in Canada and some northern states. **Adult:** Quite uniform.

Brownish or bronze, with white markings, greenish or coppery cast. Similar to *C. oregona* (4). Short labrum. Gregarious. **Length:** .5″. Found along riverbanks; prefers moist soil. It is unique in being reported from a tundra locality. *C. duodecimguttata* is similar, except that the marginal band is continuous in *repanda*, incomplete (broken) in *duodecimguttata;* the latter is colored a more subdued brownish, and occurs in the same areas.

(9) TIGER BEETLE: *Cicindela punctulata.* **Fig. 534.**

Range: Throughout Canada and the U.S., except for the Far West. Very common. **Adult:** Slender, brownish black or slightly greenish, with bronze sheen. Row of small bluish or greenish fovea on each elytron, sometimes indistinct; white markings may or may not be present. **Length:** .4–.5″. Prefers sandy, hard-packed soils, found along paths and roads, often attracted to lights; larval burrows found among patches of grass, near edges. Emits odor resembling apples when handled. Hibernates as larva, requires only a year for life cycle. *C. pusilla* (Fig. 535), a western species, is similar but lacks the punctures; the *imperfecta* form of *pusilla* occurs among the alkali bee nesting sites in Washington but does not attack these important pollinators of alfalfa.*

(10) TIGER BEETLE: *Cicindela sexguttata.* **Fig. 536.**

Range: Abundant in eastern United States and Canada. **Adult:** Shiny, brilliant green, blue-green, or violaceous, with or without six to ten white spots (commonly with eight). Color and markings vary geographically. Elytra punctate. Solitary. **Length:** .5″. Prefers partially shaded paths in woods and nearby meadows. *C. s. denikei,* a form found in western part of range, is usually without markings.

(11) TIGER BEETLE: *Cicindela tranquebarica.* **Fig. 537.**

Range: Throughout North America. **Adult:** Black to brownish or deep blue, with distinctive oblique white markings. Eastern forms tend toward brown to bronze, with broad markings, become more greenish westward; some Far Western forms are bluish. **Length:** .5–.6″. Found on various types of soil. Like *repanda* (8), this species is found in such diverse geographical areas as Texas and the Hudson Bay.

(12) TIGER BEETLE: *Cicindela formosa.* **Fig. 538.**

Range: Throughout Canada and the U.S. **Adult:** Pigmented areas blackish red, brown, or green, with large white areas, bright metallic on the underside; in *C. f. gibsoni* the pigmented areas are purplish and greatly reduced. In *C. f. formosa,* intermediate geographically, the pigmented area is brilliant reddish purple, sharply outlined as in *generosa.* Gregarious. **Length:**

---

* See Kenneth E. Frick, "Biology and Control of Tiger Beetles in Alkali Bee Nesting Sites," *Jour. Econ. Ent.,* 50(4):503 (1957).

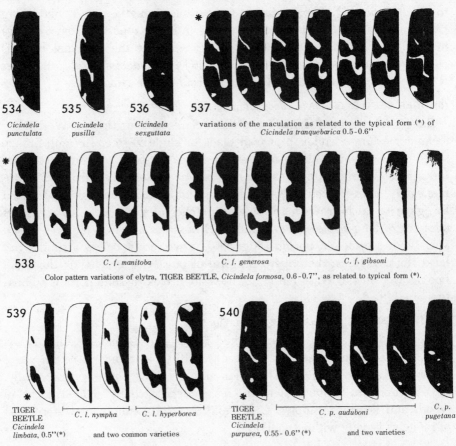

534 — Cicindela punctulata

535 — Cicindela pusilla

536 — Cicindela sexguttata

537 — variations of the maculation as related to the typical form (*) of Cicindela tranquebarica 0.5-0.6"

538 — C. f. manitoba — C. f. generosa — C. f. gibsoni

Color pattern variations of elytra, TIGER BEETLE, Cicindela formosa, 0.6-0.7", as related to typical form (*).

539 — TIGER BEETLE Cicindela limbata, 0.5"(*) — C. l. nympha — C. l. hyperborea — and two common varieties

540 — TIGER BEETLE Cicindela purpurea, 0.55-0.6" (*) — C. p. auduboni — C. p. pugetana — and two varieties

.6–.7". They are found in sandy areas with scattered vegetation. The markings of *C. lengi* are similar, the marginal band less expanded.

(13) TIGER BEETLE: *Cicindela limbata*. Fig. 539.

Range: Western Canada and U.S. **Adult:** Pigmented areas brown- or green-bronze, with much of the elytra white. *C. l. nympha,* a form found in the Prairie Provinces of Canada, is mostly yellowish white, pigmented area brown-bronze; *C. l. limbata,* found farther south, has green-bronze markings. They are similar to some forms of *C. formosa* (12) but are smaller. Gregarious. **Length:** .5". Found in sandy areas not close to water.

(14) TIGER BEETLE: *Cicindela purpurea*. Fig. 540.

Range: East of the Rocky Mts. **Adult:** Pronotum bronze, elytra cupreous, with green margins, white markings. This applies to the eastern form, *C. p. purpurea.* In the western form, *C. p. auduboni,* the elytra have various shades of green to blue-purple or black. Solitary. **Length:** .55–.6". Found on

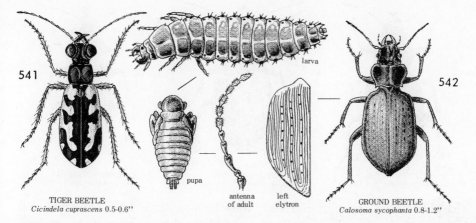

541

larva

pupa

TIGER BEETLE
Cicindela cuprascens 0.5-0.6"

antenna
of adult

left
elytron

542

GROUND BEETLE
Calosoma sycophanta 0.8-1.2"

riverbanks and nearby, in dry sandy soil. *C. limbalis* is similar, may be distinguished by the white dot near the humerus or shoulder (this dot is usually missing in *purpurea,* or less prominent) and by the middle white band, which is shorter and more oblique in *purpurea. C. limbalis* is also solitary, usually found in heavy clay soil, and at high altitudes.

(15) TIGER BEETLE: *Cicindela cuprascens.* **Fig. 541.**

Range: East of the Rocky Mts. **Adult:** Easily distinguished by the color and markings. Pigmented area coppery or greenish bronze. Similar to *C. hirticollis* and *nevadica;* it may be distinguished by the large white spot near the shoulder. **Length:** .5–.6". Found in sandy areas along rivers.

Ground Beetles (Carabidae)

This is a very large family, there being close to 20,000 known species world-wide, with 2,500 in North America. These are active insects, usually somber—predominently black—in color, though many are brilliantly colored or enhanced by iridescent hues. Mostly nocturnal, they may be found under rocks or debris in the daytime. They have long legs, with tibial spurs, and run rapidly; the tarsi are five-segmented. They are usually flattened, sometimes convex, with a prominent thorax, usually narrower than the abdomen. The eyes are large (excepting a few that live in caves), the antennae are long and eleven-segmented, the mandibles strong and sickle-like. They have prominent maxillae with five-segmented palpi, three-segmented labial palpi, and well-developed wings (excepting a few that are brachypterous or apterous). Eggs are normally laid singly in the soil; some species (in the genera *Brachinus, Galeritula, Chlaenius, Pterostichus, Craspedonotus, Carabus,* and *Calosoma*) place them in mud cells attached to plants or stones.

The larvae are usually elongate, with ten well-defined segments, tapered

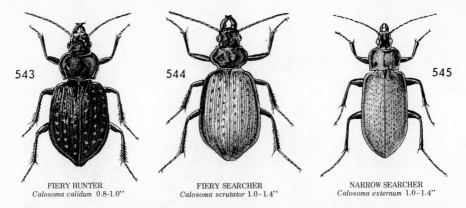

543   544   545

FIERY HUNTER
Calosoma calidum 0.8-1.0''

FIERY SEARCHER
Calosoma scrutator 1.0-1.4''

NARROW SEARCHER
Calosoma externum 1.0-1.4''

sharply toward the posterior end; they have a large head, exserted and directed forward, with six ocelli on each side, and four-segmented antennae. There is a pair of cerci, and an anal tube on the ninth abdominal segment. Pupation takes place in the soil. The life cycle of the known species is usually completed in a year, though the adults live two to three years, or longer. As implied in the name, they are terrestrial for the most part. Both adults and larvae are predaceous; a few feed on pollen, berries, seeds, and tender shoots of plants. Some species of *Harpalus* have mixed feeding habits, with much of their diet being seeds and other vegetable matter.

The family includes many important enemies of insect pests, snails, and slugs, and is considered highly beneficial as a whole. Their habits are exceedingly varied. Fierce predators, many species of *Calosoma* and *Lebia*— either the adults, or both larvae and adults—actively pursue their prey in trees. Members of the genus *Scarites* are subterranean searchers, those of the genera *Chlaenius* and *Galeritula* are active surface searchers. A few species—as in *Brachinus* and *Lebia*—are parasitic in the larval stage on the pupae of their hosts. Some adults are attracted to lights and will feed on the crushed remains of their own kind that have been stepped on or run over. At least one species, *Thalassotrechus barbarae,* inhabits the intertidal zone, in southern California. Many ground beetles exude or expel foul-smelling secretions that are used to repel their enemies.

(16) GROUND BEETLE: *Calosoma sycophanta* (Carabinae; Carabini). **Fig. 542.**

Imported from Europe for control of the gypsy moth. **Range:** Mostly New England. **Adult:** Brilliant golden green; thorax dark blue. It normally lives several years. Eggs are laid in the soil in spring. **Length:** .8–1.2''. **Larva:** Shiny black, with brown markings on underside; mouthparts, antennae, and legs dark brown; about 1'' long full grown. It matures in two to three weeks, is active by day and night during this period, consuming up to 50 caterpillars. Pupates in the soil; the adult remains in pupal cell, hibernates until the following spring. Older adults enter the soil to hibernate, females

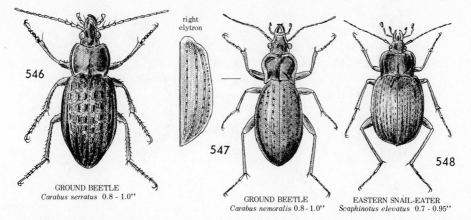

546 GROUND BEETLE
Carabus serratus 0.8 - 1.0''

right elytron

547

GROUND BEETLE
Carabus nemoralis 0.8 - 1.0''

548

EASTERN SNAIL-EATER
Scaphinotus elevatus 0.7 - 0.95''

continue to lay eggs each spring. **Food:** Caterpillars, more especially those of the gypsy moth. An adult will normally consume considerably more than a larva does in a year's time, since it is active longer periods. Both adults and larvae climb trees in search of prey.

(17) FIERY HUNTER: *Calosoma calidum.* **Fig. 543.**

Range: Throughout the U.S. and Canada. **Adult:** Black, sometimes with green margins; elytra deeply striated, with six rows of golden or red spots in deep punctures. **Length:** .8–1''. Neither adults nor larvae climb trees. **Food:** Cutworms, armyworms. *C. cancellatum,* which occurs in the West, is similar except that the deep punctures are not colored. *C. frigidum* is more slender, has green punctures and margins, ranges over forest areas of Canada and northern U.S.; only the adults climb trees.

(18) FIERY SEARCHER: *Calosoma scrutator.* **Fig. 544.**

Range: Throughout the U.S. and southern Canada. **Adult:** Black, with violaceous luster; sides of head and prothorax golden green, femora with bluish luster; elytra iridescent green, with purple margins, finely striated and punctate. Inside of curved middle tibiae and hind tibiae with dense reddish hairbrush at apex in the male. **Length:** 1–1.4''. **Larva:** Dirty yellow; head and plates on abdominal segments brown. **Food:** Tent caterpillars, bollworms, cotton leafworm, and other larvae. *C. wilcoxi*—less widely distributed, more eastern (both are reported in California)—is a small copy of *scrutator,* from .7'' to 1'' long, with reflexed margin of pronotum much narrower. *C. externum* (Fig. 545)—predominantly southern and eastern—is about the same size as *scrutator,* black and violaceous, elytra with strong coarsely punctured striae, intervals convex. The adults of all three climb trees in search of prey; the larvae do not. They are strongly attracted to lights, hibernate as adults. These voracious species, with *C. sayi,* are reported to be active in southern cotton fields even during the day where heavy applications of insecticides have not been made.

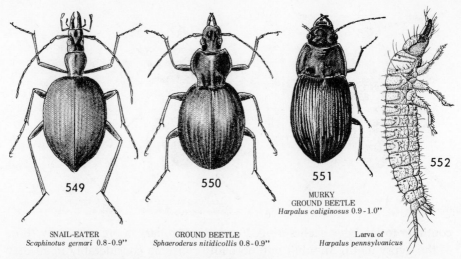

549 SNAIL-EATER
Scaphinotus germari 0.8-0.9″

550 GROUND BEETLE
Sphaeroderus nitidicollis 0.8-0.9″

551 MURKY
GROUND BEETLE
Harpalus caliginosus 0.9-1.0″

552 Larva of
Harpalus pennsylvanicus

(19) GROUND BEETLE: *Carabus serratus.* **Fig. 546.**

Range: Southern parts of the boreal forests in Canada and the northern plains, south to Oregon, New Mexico, Georgia. **Adult:** Black; outer margins of pronotum and elytra violaceous. Elytra each with three rows of oblong tubercles separated by sparsely punctate striae. **Length:** .8–1″. Usually found in open gravelly soil with sparse vegetation. *C. nemoralis* (Fig. 547), an introduced species and an important enemy of slugs, ranges from the Northeast coast to the Wisconsin side of Lake Michigan, and on the Pacific coast from Oregon to British Columbia; slightly larger than *serratus,* it is black, upper side greenish bronze or violaceous, elytra each with three rows of foveae. Both of these species hibernate as adults. Some species of *Carabus* are said to perform a courtship ritual.

(20) EASTERN SNAIL-EATER: *Scaphinotus elevatus* (Cychrini). **Fig. 548.**

Range: Massachusetts and New York, south to Florida and Texas, west to Manitoba, Nebraska, Colorado, New Mexico. **Adult:** Black, with pale bluish, purplish, or violaceous reflections. Outer margins of pronotum and elytra reflexed, usually more strongly in the latter toward the shoulders; elytral striae punctate. It is found in moist places, under leaves, rocks and logs—in forests, and on the coastal plains and lowlands. **Length:** .7–.95″. **Food:** Snails and slugs. Like most species in the genus, it is nocturnal in habit, and believed to hibernate as larva. *S. germari* (Fig. 549) is blackish, with violet iridescence, from .8″ to .9″ long, occurs eastward from Indiana.

(21) GROUND BEETLE: *Sphaeroderus nitidicollis.* **Fig. 550.**

Range: Northeastern U.S. and Canada. **Adult:** Black, with bronze iridescence. **Length:** .8–.9″. *S. fissicollis* is violaceous, about half as long as *nitidicollis,* ranges farther south. *S. lecontei* is a woodland species, shiny black, with strong violaceous reflections, about .6″ long.

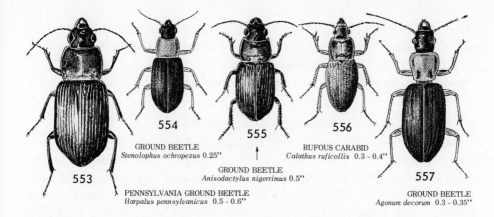

554

555

556

GROUND BEETLE
Stenolophus ochropezus 0.25"

RUFOUS CARABID
Calathus ruficollis 0.3 - 0.4"

GROUND BEETLE
Anisodactylus nigerrimus 0.5"

553

557

PENNSYLVANIA GROUND BEETLE
Harpalus pennsylvanicus 0.5 - 0.6"

GROUND BEETLE
Agonum decorum 0.3 - 0.35"

(22) MURKY GROUND BEETLE: *Harpalus caliginosus* (Harpalini). **Fig. 551.**

Range: Throughout North America. **Adult:** Dull or reddish black; elytra deeply striated. **Length:** .9–1". **Larva:** Black, with reddish head; body tapered, with two long cerci. **Food:** Larvae of the Colorado potato beetle and others, also seeds of ragweed; occasionally damages ripening strawberries —probably its way of getting needed water. Very common, often attracted to lights.

(23) PENNSYLVANIA GROUND BEETLE: *Harpalus pennsylvanicus.* **Figs. 552, 553.**

Range: Throughout North America. **Adult:** Convex, shiny black; underside reddish brown; elytra deeply striated, intervals with numerous small punctures in the female, sparse punctures in the male. **Length:** .5–.6". **Food:** Caterpillars, seeds. Very common in cotton fields of the South where insecticide treatments have not been heavy. Most numerous in pastures. *H. compar* is shiny black, with yellowish brown antennae and legs, about .6" long; it is common in fields planted to cereals and forage.

(24) GROUND BEETLE: *Stenolophus ochropezus.* **Fig. 554.**

Range: Most of the U.S., southern Canada. **Adult:** Black, elytra often iridescent, legs and base of antennae yellowish; elytra deeply striated, intervals flat, with fine punctures. **Length:** .25". Common in southern cotton fields.

(25) CORN-SEED BEETLE: *Agonoderus lecontei.*

Range: Central and southern states. **Adult:** Slender, convex; head black, pronotum brownish with large spot in the middle; elytra yellowish brown, each with long black mark near the suture, deeply striated. **Length:** .2–.25". It is attracted to lights and sometimes found in damp seed beds, where it becomes a pest. *A. comma* is very similar in appearance and habits.

(26) GROUND BEETLE: *Anisodactylus nigerrimus.* **Fig. 555.**

Range: Throughout the U.S. and southern Canada. **Adult:** Moderately shiny, black. Elytra deeply striated, with convex intervals sparsely punctate; pronotum finely punctate, base with shallow impressions, margins slightly depressed at the middle. **Length:** .5″. It is found in open fields, usually on damp clay soil rich in organic matter. Common in cotton fields of the South.

(27) RUFOUS CARABID: *Calathus ruficollis* (Agonini). Fig. 556.

Range: California, Oregon, Washington. **Adult:** Rufous (pale red); pronotum brighter red; elytra finely striated. **Length:** .3–.4″. **Food:** Codling moth and other small caterpillars. Very common; may be found under stones. *C. opaculus* is very similar, same size, reddish brown, antennae and legs paler; lateral margins of pronotum depressed; elytra darker, with fine striations; it is widely distributed. *C. gregarius* is also similar, somewhat larger, dark reddish brown; pronotum slightly depressed at lateral margins; elytral striations shallow; it hibernates in aggregations.

(28) GROUND BEETLE: *Agonum (Platynus) decorum*. Fig. 557.

Range: East of the Rocky Mts. **Adult:** Elytra blue or black, often tinged with green near margins; head green, sometimes bronzed; pronotum, scutellum, base of antennae, and legs reddish yellow. Striae shallow, without punctures; pronotum with distinct impressions at base. **Length:** .3–.35″. This species is more or less typical of this large genus of small ground beetles. *A. sinuatum* occurs from Labrador to District of Columbia, west to New Mexico, Washington, and Oregon; it is convex, shiny, black, with pronotum nearly as long as wide, elytral striae punctate; about .5″ long.

(29) GROUND BEETLE: *Pterostichus adoxus* (Pterostichini). Fig. 558.

Range: Throughout eastern U.S. and Canada. **Adult:** Black, shiny, moderately convex; margins of elytra reddish brown. Elytra striated but not punctate, deflexed at margins; pronotum has fine median line, deep impressions on each side near base. **Length:** .5–.6″. Found in wooded areas, under logs, stones, and debris. *P. melanarius* is common in fields planted to cereals and forage.

(30) GROUND BEETLE: *Amara impuncticollis* (Amarini). Fig. 559.

Range: Throughout the U.S. and Canada. **Adult:** Black or dark bronze, with greenish tinge. The elytra have shallow striae but no punctures, except for a single large puncture in the scutellar area, where the first two striae meet. **Length:** .35″.

(31) BOMBARDIER BEETLE: *Brachinus americanus* (Brachinini). Fig. 560.

Range: From the central states westward. **Adult:** Elytra dark blue, somewhat metallic, truncate at the apex, striated; head and pronotum brownish yellow. It gives off a foul-smelling gaseous secretion in audible puffs when

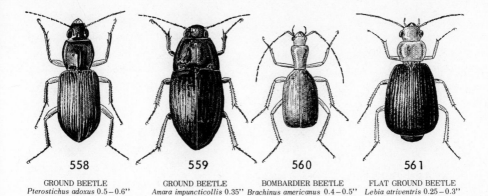

| 558 | 559 | 560 | 561 |

GROUND BEETLE
Pterostichus adoxus 0.5 – 0.6''

GROUND BEETLE
Amara impuncticollis 0.35''

BOMBARDIER BEETLE
Brachinus americanus 0.4 – 0.5''

FLAT GROUND BEETLE
Lebia atriventris 0.25 – 0.3''

attacked. (See Introduction, p. 34.) Very common. **Length:** .4–.5''. The 40-some North American species of *Brachinus* are all similar, vary from .2'' to .6'' long, are usually found in moist places, near lakes and streams. In the known species of *Brachinus*, eggs are laid singly in mud cells constructed on plants and stones. The larvae are parasitic on the pupae of their hosts, with the legs becoming reduced as in other parasitic forms; pupation takes place in the host remains. Some species parasitize whirligig beetles.

(32) FLAT GROUND BEETLE: *Lebia atriventris* (Lebiini). **Fig. 561.**

Range: Throughout North America. **Adult:** Elytra dark blue, with shallow striations and dense punctures; head and pronotum reddish yellow. **Length:** .25–.3''. The known larvae of *Lebia* become parasitic, as described in (31) above, and feed on chrysomelid beetles. Most adults are diurnal, and climb trees. *L. grandis* resembles *atriventris* closely; it is a little larger, with deeply striated elytra, without punctures, occurs throughout eastern U.S. and southern Canada. *L. scapularis* adults feed on all immature forms of the elm leaf beetle, while larvae attack the pupae; the elytra are black, each with wide, yellow median stripe and narrow, yellow outer margin; pronotum and head reddish yellow; about .25'' long.

(33) STRIATED FLAT GROUND BEETLE: *Lebia analis*. **Fig. 562.**

Range: Throughout North America. **Adult:** Elytra black, with yellow markings; pronotum reddish yellow, with paler margins; head black, constricted to form a short neck. **Length:** .2''.

(34) GROUND BEETLE: *Lebia vittata*. **Fig. 563.**

Range: East of the Rocky Mts. **Adult:** Elytra yellowish, with black stripes; pronotum reddish yellow, with paler margin; head reddish yellow. **Length:** .25''.

(35) GROUND BEETLE: *Lebia fuscata*. **Fig. 564.**

Range: Throughout North America east of the Rocky Mts. **Adult:** Elytra

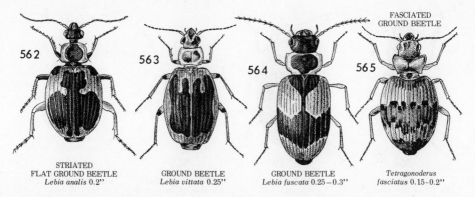

562

563

564

565

STRIATED
FLAT GROUND BEETLE
Lebia analis 0.2"

GROUND BEETLE
Lebia vittata 0.25"

GROUND BEETLE
Lebia fuscata 0.25–0.3"

*Tetragonoderus
fasciatus* 0.15-0.2"

blackish and brownish yellow; pronotum blackish, with brownish yellow margin; head black. **Length:** .25–.3".

(36) FASCIATED GROUND BEETLE: *Tetragonoderus fasciatus* (Masoreini). **Fig. 565.**

Range: Eastern U.S. and southern Ontario; in the Southwest to southern California, south to Baja California. **Adult:** Elytra gray to yellowish, with darker bands; pronotum and head dark, with bronze tinge. **Length:** .15–.2". This genus is largely tropical. *T. intersectus* occurs in the southeastern states, north to Kentucky.

(37) SLENDER GROUND BEETLE: *Galeritula* (*Galerita*) *janus* (Dryptini). **Fig. 566.**

Range: Central and eastern states, southeastern Ontario. **Adult:** Elytra bluish black; pronotum, legs, base of antennae reddish brown; dense covering of short, pale yellow hairs. **Length:** .6–.9". *G. bicolor* is very similar, with the pronotum more elongate and less rounded; females deposit their eggs in purse-shaped mud cells constructed on the underside of smooth leaves. Most of the species in this genus are confined to the southern states; only *G. lecontei* extends to California. They are strongly attracted to lights.

(38) LONG-NECKED GROUND BEETLE: *Colliuris pennsylvanicus* (Odacanthini). **Fig. 567.**

Range: Eastern U.S.; southeastern Canada. **Adult:** Elytra dull red or orange, with three black spots forming interrupted or constricted transverse band, apexes black; basal half of elytra coarsely punctate. Head and long, slender pronotum black. Found on vegetation growing in moist ground near marshes; attracted to lights. **Length:** .3".

(39) PUBESCENT GROUND BEETLE: *Chlaenius tricolor* (Chlaeniini). **Fig. 568.**

Range: Throughout the U.S. and southern Canada. **Adult:** Elytra bluish black, with fairly deep, fine striations and punctures. Head and pronotum green, legs and antennae brownish orange. Middle tibiae of male have brush of dense hairs on outer surface near tip. **Length:** .4–.5". Most species in this genus are covered with a fine pubescence; they are noted for their

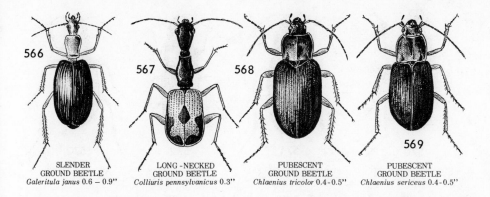

SLENDER	LONG-NECKED	PUBESCENT	PUBESCENT
GROUND BEETLE	GROUND BEETLE	GROUND BEETLE	GROUND BEETLE
Galeritula janus 0.6 – 0.9"	*Colliuris pennsylvanicus* 0.3"	*Chlaenius tricolor* 0.4-0.5"	*Chlaenius sericeus* 0.4-0.5"

foul odors. Several species are known to construct mud egg cells, each distinct in size, form, and location for the species. The cells of *tricolor* are tiny (less than .1" long), smooth, rounded and oblong, with a fold or flap on one side, attached to stems of slender grasses and sedges near streams. *C. sericeus* (Fig. 569)—a bright green species with bluish tinge—makes a smooth purse-shaped cell (.15" long) attached to stones or lower part of dead plant stems, near surface of moist soil. The cells of *C. aestivus*—a bluish black and green species with yellow pubescence—are convex and oval (a little less than .2" long) with rough surface and lidlike flap on one side for exit of larva; they are attached to dead twigs, plant stems, and the bark of trees and shrubs. Those of *C. impunctifrons*, a similar species, are smoothly convex and oblong (less than .15" long), attached to the underside of smooth leaves, a few inches to several feet above ground. *C. tomentosus* (Fig. 570)—black, with bronze sheen and fine yellowish pubescence—differs in habitat, being found in dry, grassy fields.

(40) FRAGILE GROUND BEETLE: *Badister pulchellus* (Licinini). **Fig. 571.**

Range: East of the Rocky Mts. **Adult:** Brownish yellow; elytra with black markings, shallow striae and sparse punctures; head black. **Length:** .2–.3". Members of this genus and some close relatives have a peculiarity in the mandibles, either the right or the left one being deeply notched.

(41) GROUND BEETLE: *Dicaelus purpuratus.* **Fig. 572.**

Range: Eastern and central states. **Adult:** Purplish; head black. Elytra deeply striated, intervals convex; pronotum with margins strongly reflexed. **Length:** .8–1". Common in the cotton fields of the South, and in the Southeast, where it is said to feed on snails.

(42) AMERICAN BEMBID: *Bembidion americanum* (Bembidiini). **Fig. 573.**

Range: Eastern North America, west to Kansas and south to Texas. **Adult:** Shiny, black, with bronze tinge; one segment of antennae, most of palpi, base and apex of femora reddish; elytra with fine striae and large

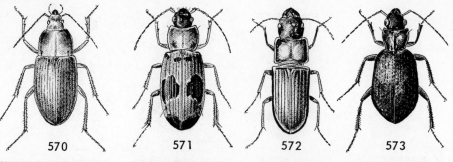

570	571	572	573
TOMENTOSE GROUND BEETLE	FRAGILE GROUND BEETLE	GROUND BEETLE	AMERICAN BEMBID
Chlaenius tomentosus 0.5"	*Radister pulchellus* 0.2 – 0.3"	*Dicaelus purpuratus* 0.8 – 1.0"	*Bembidion americanum* 0.2–0.25"

deep punctures. **Length:** .2–.25". Found in gravelly soil with sparse vegetation.

(43) PATRUUS BEMBID: *Bembidion patruele.* **Fig. 574.**

Range: Across Canada and the U.S., south to Georgia. **Adult:** Black, with greenish luster, elytra iridescent with orange markings; head, pronotum, and legs reddish brown. **Length:** .15–.2". Very common in the East, attracted to lights; found at the edge of still waters, in organic soils with sparse vegetation, takes to cultivated areas. Hibernates as adult.

(44) FOUR-SPOTTED BEMBID: *Bembidion quadrimaculatum.* **Fig. 575.**

Range: East of the Rocky Mts., south to Texas; Alberta to Newfoundland. **Adult:** Black, with bronze tinge, two to four basal segments of antennae and entire legs brownish yellow; elytra each with two yellow lateral spots, striae feeble, punctures variable. **Length:** .1–.15". According to Lindroth this is a subspecies, *oppositum*. It is one of the most abundant species of ground beetles. *B. q. dubitans,* the western form, occurs from British Columbia south to California and north to Alaska and the Yukon; all the appendages are darker than in *oppositum,* the subhumeral spot pale but well defined, the posterior one fully developed or absent. They are found in sandy or clayey soils, near lakeshores or riverbanks, in cultivated fields and gardens. They have been observed preying on the eggs of *Hylemya.* Hibernates as adult.

(45) GROUND BEETLE: *Tachys scitulus.* **Fig. 576.**

This is a large genus of minute beetles, found mostly in bare sandy and clayey areas near water, or in drier places, in warmer parts of the temperate regions or in the tropics. **Range:** *T. scitulus* occurs throughout the U.S. and southern Canada. **Adult:** Brownish yellow, head and pronotum darker, elytra with darker wide transverse band near the middle; basal segments of antennae, and legs, yellowish. **Length:** .1–.12". Found in moist places and mud flats.

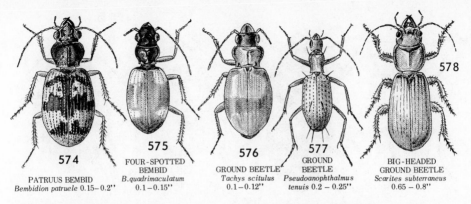

574
PATRUUS BEMBID
Bembidion patruele 0.15–0.2"

575
FOUR-SPOTTED
BEMBID
B.quadrimaculatum
0.1–0.15"

576
GROUND BEETLE
Tachys scitulus
0.1–0.12"

577
GROUND
BEETLE
*Pseudoanophthalmus
tenuis* 0.2 – 0.25"

578
BIG-HEADED
GROUND BEETLE
Scarites subterraneus
0.65 – 0.8"

(46) GROUND BEETLE: *Pseudoanophthalmus tenuis* (Trechini). **Fig. 577.**

This genus contains 50 or more species of mountain and cave dwellers restricted to the eastern U.S. **Range:** *P. tenuis* is found in caves of the eastern U.S. **Adult:** Pale brownish yellow, without eyes. **Length:** .2–.25". A few species in closely related genera are found in caves in Kentucky, Tennessee, and Virginia.

(47) BIG-HEADED GROUND BEETLE: *Scarites subterraneus* (Scaritini). **Fig. 578.**

Range: Most of the U.S. except the Northwest, south to Texas; southern Ontario. **Adult:** Shiny, pitch black; antennae and tarsi slightly paler. Head very broad, front with pair of large foveae. Elytra rather deeply striated, scarcely punctate, intervals convex. **Length:** .65–.8". They are commonly found in burrows, under stones, in damp soil. The few American forms of *Scarites* may all be subspecies of *subterraneus*.

(48) GROUND BEETLE: *Clivina bipustulata.* **Fig. 579.**

Range: Eastern U.S. **Adult:** Black; large, red spot near base and apex of each elytron; legs and antennae reddish brown. Elytra striated and deeply punctate. **Length:** .25". Strongly attracted to lights. *C. impressifrons,* similar in shape, plain reddish brown, sometimes feeds on seed corn after planting. Members of the genus are usually found near streams, in woods, or in other moist places.

(49) GROUND BEETLE: *Pasimachus depressus.* **Fig. 580.**

Range: Widely distributed east of the Rocky Mts. **Adult:** Flat, black; male shiny, female dull; elytra and pronotum with blue margins. Elytra smooth, each with a short humeral ridge. **Length:** 1–1.2". Abundant, found in daytime under stones, feeds on armyworms and cutworms at night. Species of this genus are predominantly southern, found in dry, sparsely wooded areas, prairies and fields with scattered vegetation.

(50) SPOTTED SAVAGE BEETLE: *Omophron tessellatum* (Omophronini). **Fig. 581.**

This beetle belongs to a small group called the hemispherical savage

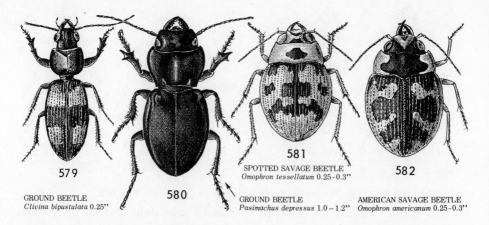

579

GROUND BEETLE
Clivina bipustulata 0.25"

580

581
SPOTTED SAVAGE BEETLE
Omophron tessellatum 0.25-0.3"

GROUND BEETLE
Pasimachus depressus 1.0 – 1.2"

582

AMERICAN SAVAGE BEETLE
Omophron americanum 0.25-0.3"

beetles which are sometimes placed in a separate family, Omophronidae. They are predaceous, live in aggregations, burrowing in the moist sand bordering lakes and streams. Collectors force them out of their burrows by pouring water over the sand; they run rapidly but make no effort to fly. They are nocturnal and often attracted to lights. **Range:** *O. tessellatum* ranges across southern Canada (not quite to the Pacific), south to Texas and Arizona. **Adult:** Yellowish brown, with green markings on head, pronotum, and elytra. Elytra with fine striae and punctures. **Length:** .25–.3". *O. americanum* (Fig. 582) is abundant in eastern U.S. and Canada, ranges west to Idaho, Utah, Arizona, Mexico; it is broad, rather flat, reddish brown with iridescent greenish markings, pale margins and green head, deep elytral striae and close punctures. *O. ovale* ranges from California to British Columbia, east to Wyoming, Montana, Alberta, is colored like *americanum,* but is more elongate, with elytral striae shallower. All are approximately the same size, and are believed to hibernate as adults.

Crawling Water Beetles (Haliplidae)

This is a small family of aquatic beetles—brownish yellow or reddish yellow, with black spots—abundant among the submerged weeds of ponds and streams. A distinctive feature is the extremely long hind coxal plates. Adults feed on vegetation; the larvae are predaceous. Some species fly. They are not proficient in swimming, the only adaptation for this activity being the long hairs on the tibiae and tarsi of the middle and hind legs. They breathe by going to the surface, storing bubbles of air under the elytra next to the spiracles. Eggs are laid in dead cells of algal filaments in the case of *Haliplus* spp. and on aquatic plants in the case of *Peltodytes* spp. The larvae are elongate (*Haliplus*), or cone-shaped (*Peltodytes*); in *Haliplus* the tenth or terminal segment of the abdomen is much longer than the

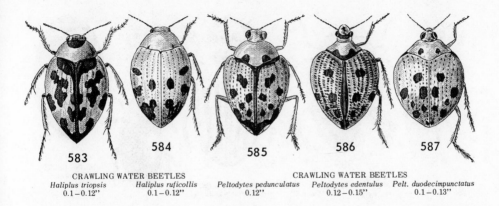

CRAWLING WATER BEETLES
Haliplus triopsis   Haliplus ruficollis
0.1–0.12"   0.1–0.12"

CRAWLING WATER BEETLES
Peltodytes pedunculatus   Peltodytes edentulus   Pelt. duodecimpunctatus
0.12"   0.12–0.15"   0.1–0.13"

others and ends in a prong; in *Peltodytes* there are two or more segmented hairs (as long as the body) on each body segment. They pupate in the soil on shore, overwinter as adults, some remaining active.

(51) CRAWLING WATER BEETLE: *Haliplus triopsis.* **Fig. 583.**

Range: Ontario, Maine to Wisconsin and Colorado, south to New Mexico and Georgia. **Adult:** Reddish yellow, with variable black spots. **Length:** .1–.12".

(52) CRAWLING WATER BEETLE: *Haliplus ruficollis.* **Fig. 584.**

Range: Throughout most of North America. **Adult:** Reddish yellow, with black spots. May be recognized by the pair of impressions on the base of the pronotum, and by the deep groove on the prosternum. **Length:** .1–.12". *Peltodytes pedunculatus* (Fig. 585) is pale yellow, with black markings, yellow legs excepting hind femora which are black; *P. edentulus* (Fig. 586) is similar, with black between the eyes; *P. duodecimpunctatus* (Fig. 587) is yellow, with twelve black elytral spots; the three species occur in the eastern half of the U.S. and Canada.

Predaceous Diving Beetles (Dytiscidae)

Diving beetles are found in almost any kind of water not excessively hot or saline. The hind tibiae have long hairs and other adaptations for swimming. Unlike other water beetles, which move their legs alternately when swimming, the dytiscids stroke them together like oars, as the backswimmers (Hemiptera) do. Aside from their manner of locomotion they may be distinguished from the Hydrophilidae by their short palpi, and from the Gyrinidae by the latter's peculiarly divided eyes. Most of them are good fliers, often migrate from one pond to another, and are sometimes attracted to lights. They break the water surface with the tip of the abdomen at intervals to get air, which is stored in bubbles under the elytra. Eggs are

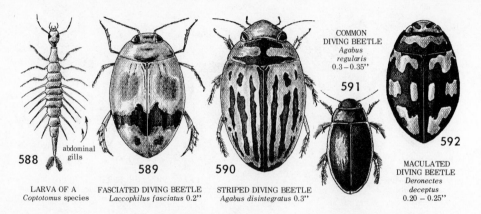

COMMON
DIVING BEETLE
Agabus
regularis
0.3—0.35"

591

588 abdominal gills

MACULATED
DIVING BEETLE
Deronectes
deceptus
0.20 – 0.25"

592

589 590

LARVA OF A
Coptotomus species

FASCIATED DIVING BEETLE
Laccophilus fasciatus 0.2"

STRIPED DIVING BEETLE
Agabus disintegratus 0.3"

laid singly on floating leaves, in plant tissues, or in masses near the shore. Both adults and larvae are predaceous.*

The larvae have eight abdominal segments (the ninth is not visible dorsally), are tapered toward both ends—to a point at the anal end. Most species have a pair of long cerci—they are lacking in *Cybister*—with rows or tufts of hair. The head is large, often depressed and tapered posteriorly. The sharp, hollow mandibles are used to inject paralyzing and digestive fluids into their prey and to draw out the liquid contents (adults tear their victims apart and chew the pieces). Fierce predators and sometimes cannibalistic, they attack all kinds of aquatic insects and other small animals including fish, and have the ability to regenerate lost parts to some extent.

The larvae of most species do not have tracheal gills and, like the adults, must come to the surface at intervals for air, breaking the film with the tip of the abdomen; or they may utilize the air trapped in bubbles on the plants. Some, however, obtain oxygen directly from the water; *Coptotomus* spp. (Fig. 588) have lateral gills about as long as the legs—a pair on each of the first six abdominal segments. Mature larvae crawl ashore to pupate in the ground. Hibernation takes place in the water, usually as adult, sometimes as larva. There is normally one generation a year, but adults often live several years. They are attacked by numerous small animals as well as fish, and certain wasps enter the water to parasitize the eggs.

(53) FASCIATED DIVING BEETLE: *Laccophilus fasciatus.* **Fig. 589.**

Range: Widely distributed in central and eastern North America. **Adult:** Greenish yellow, with broad blackish band across elytra; surface of pronotum and elytra has a wrinkled appearance. **Length:** .2".

(54) STRIPED DIVING BEETLE: *Agabus disintegratus.* **Fig. 590.**

Range: Throughout North America. **Adult:** Head and pronotum reddish

---

* They take a heavy toll of young toads in the tadpole or polliwog stage, and have been detrimental in areas where the giant toad (*Bufo marinus*) was introduced for biological control of the white grubs of May beetles.

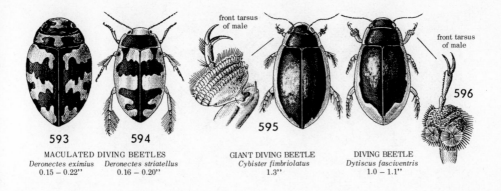

front tarsus of male

596

595

593 594

MACULATED DIVING BEETLES
Deronectes eximius   *Deronectes striatellus*
0.15 – 0.22"   0.16 – 0.20"

GIANT DIVING BEETLE
Cybister fimbriolatus
1.3"

DIVING BEETLE
Dytiscus fasciventris
1.0 – 1.1"

yellow, latter with black transverse bands; elytra brownish yellow, with three or four black longitudinal stripes on each side. **Length:** .3". *A. regularis* (Fig. 591) is blackish green, with brassy tinge and yellow margins, about .3" long; *Deronectes deceptus* (Fig. 592) is black, with yellowish orange spots, about .25" long; both are strongly convex, occur in the Southwest. *D. eximius* (Fig. 593) and *striatellus* (Fig. 594) are smaller, black, with yellow striae instead of spots, occur in the Southwest and Pacific coast states respectively.

(55) GIANT DIVING BEETLE: *Cybister fimbriolatus.* **Fig. 595.**

Range: East of the Rocky Mts. **Adult:** Brown, with greenish tinge; pronotum and elytra with yellow margins; five furrows on pronotum of female. Hind tarsi in the males have only one claw at apex; this is usually so in the females but not always. **Length:** 1.3".

(56) DIVING BEETLE: *Dytiscus fasciventris.* **Fig. 596.**

Range: Central and eastern North America. **Adult:** Greenish black; sides of pronotum and elytra margined with yellow. Basal two-thirds of female elytra deeply grooved, each with ten grooves. Labrum notched in the middle. **Length:** 1–1.1". *D. harrisii* (Fig. 597) is similar but larger—about 1.5" long—sides, base and apex of pronotum with yellow margins, elytra with narrow crossband near the apex, where the marginal band tapers; the labrum is squarely cut (without notching). *D. verticalis* (Fig. 598) and *D. hybridus* (Fig. 599) are eastern species, black, with yellow margins; *verticalis* is from 1.2" to 1.25" long; *hybridus* is slightly smaller.

Whirligig Beetles (Gyrinidae)

This small group may be distinguished from any other by the divided eyes and their circular "skating" movements on the surface of the water. Each eye is divided into two widely separated parts (Fig. 600A): the upper one is oval, remains above the water; the lower one, somewhat

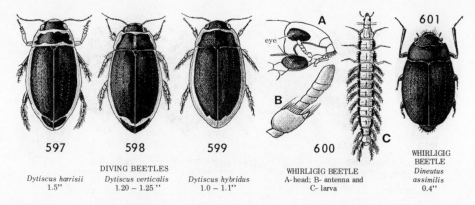

597	598	599	600		601
	DIVING BEETLES		WHIRLIGIG BEETLE		WHIRLIGIG BEETLE
Dytiscus harrisii	Dytiscus verticalis	Dytiscus hybridus	A- head; B- antenna and		Dineutus
1.5"	1.20 – 1.25 "	1.0 – 1.1"	C- larva		assimilis
					0.4"

smaller, remains below the surface of the water. They are found in lakes, ponds, and streams. The adults are oval, flattened, black or dark metallic green, have peculiar eight-segmented antennae: the first two segments scoop-shaped, the others forming a club (Fig. 600B). Held parallel with the water's surface, the highly sensitive antennae respond to air currents emanating from the wavelets created by the skating movements, and thus serve as a guidance system. Whirligigs are not only accomplished skaters, they are good fliers, and can dive and swim under water with ease. The last two pairs of legs are flattened and have an outer fringe of hairs, providing sufficient area for support on the water.

Gyrinids exude a milky secretion, usually from a region in the prothorax, for which they have been given such names as "apple smellers" and "mallow bugs." Eggs are laid in masses, often in rows, on submerged foliage. The adults are scavengers, the larvae strictly predaceous and aquatic. The latter are elongate, of nearly uniform width, have two free tarsal claws on each leg. They breathe with feather-like tracheal gills of nearly the same length as the legs—a pair on each of the first eight abdominal segments and two pairs on the ninth (Fig. 600C). Mature larvae leave the water to pupate on bordering plants, constructing a pupal case from various materials. They have a single generation each year, hibernate as adults in the mud or on plants. The species within a genus are often difficult to separate except on the basis of the male genitalia.

(57) WHIRLIGIG BEETLE: *Dineutus assimilis.* **Fig. 601.**

Range: Common and widely distributed throughout much of the U.S. and Canada. **Adult:** Shiny, all black, with bronze sheen, brownish yellow legs. **Length:** .4". *D. ciliatus* (Fig. 602), an eastern species, is slightly larger, ovate, black, with bronzed, curved stripe along sides of elytra and pronotum, dark brown legs. *D. americanus* (Fig. 603), about .5" long, is more oblong in shape, black and bronzed, occurs in streams and ponds; outer margins of elytra are curved inward near the apex, more predominantly in

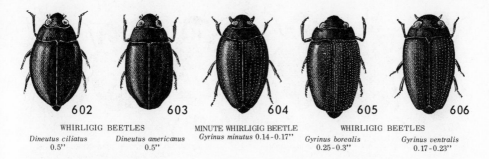

602	603	604	605	606
WHIRLIGIG BEETLES		MINUTE WHIRLIGIG BEETLE	WHIRLIGIG BEETLES	
Dineutus ciliatus 0.5"	*Dineutus americanus* 0.5"	*Gyrinus minutus* 0.14-0.17"	*Gyrinus borealis* 0.25-0.3"	*Gyrinus ventralis* 0.17-0.23"

the female. *D. solitarius* is the only species of the genus recorded from California.

(58) MINUTE WHIRLIGIG BEETLE: *Gyrinus minutus.* **Fig. 604.**

Range: Throughout most of the U.S. and Canada, Alaska. **Adult:** More flattened (less convex) than *Dineutus.* Black on top, bronzed on sides, yellowish brown on underside. Scutellum exposed (in *Dineutus* it is not). Each elytron has eleven striae, with punctures (in *Dineutus* the striae are without punctures). **Length:** .14–.17". *G. borealis* (Fig. 605), a common and widespread species, is larger—from .25" to .3" long—sides nearly parallel, shiny, black, with margins of elytra bronzed, legs brownish yellow; each elytron has eleven rows of punctures, the outer ones coarse and deep. *G. ventralis* (Fig. 606) is highly polished, black with bluish tinge and sides bronzed, from .17" to .23" long, occurs across Canada and northern U.S.

POLYPHAGA

Water Scavenger Beetles (Hydrophilidae)

These are minute to large aquatic or terrestrial beetles. The majority are aquatic, found in stagnant pools, at the edge of lakes and quiet streams where there is an abundance of vegetation, in fresh, brackish, and saline waters. The adults feed mostly on dead or decaying vegetation, a few are predaceous; some terrestrial species feed on dung, decaying animal matter, and maggots associated with these media. They are black, dull green, brownish, occasionally yellowish; the elytra may be smooth and shiny, or striated and punctured. Many are good fliers, attracted to lights in great numbers in the spring. They may be recognized by the long, four-segmented maxillary palpi which are longer than the antennae; the latter are short, inserted in front of the eyes, often concealed, seven- to ten-segmented, terminating in a hairy or pubescent club. Unlike the Dytiscidae, the hind legs move alternately when swimming, and they break the surface film headfirst for air. (Their main spiracles are on the thorax; in the dytiscids they are on the abdomen.) Eggs are laid singly (*Hydrochus* spp.) or in a mass enclosed in a

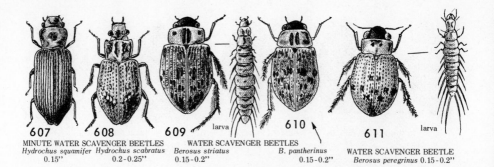

607
MINUTE WATER SCAVENGER BEETLES
Hydrochus squamifer Hydrochus scabratus
0.15"   0.2-0.25"

608

609
WATER SCAVENGER BEETLES
Berosus striatus
0.15-0.2"

larva

610
B. pantherinus
0.15-0.2"

611
WATER SCAVENGER BEETLE
Berosus peregrinus 0.15-0.2"

larva

transparent bag and carried by the female under her abdomen (*Helochares* and *Epimetopus*) or in a silken case attached to plants (or rarely set afloat). Aquatic species lay them in the water; terrestrial species lay them in damp or wet places.

The larvae are long and slender, or spindle-shaped, gray or yellowish brown, with eight to ten well-defined abdominal segments, prominent head and mandibles, long legs; a few are legless. Excepting *Helophorus* and *Berosus*, only the posterior pair of spiracles are open and functional; thus, they come to the surface tailfirst for air. The larvae of *Berosus* have long lateral gills, have no need to come to the surface for air, and can live in deeper water. There are three instars; when maturing, the larvae crawl on shore to pupate in cells formed in the mud. Some species hibernate in the larval as well as adult stage. So far as known, there are one or two generations a year. The larvae are predaceous and cannibalistic.

(59) MINUTE WATER SCAVENGER BEETLE: *Hydrochus squamifer.* Fig. 607.

Range: Northern U.S. and adjacent Canada. **Adult:** Elytra bronzy or coppery; head and pronotum dark greenish, with numerous small, rounded, flat scales. Elytra striated, intervals flat and wider than striae. **Length:** About .15". *H. scabratus* (Fig. 608), an eastern species, is very similar, from .2" to .25" long.

(60) WATER SCAVENGER BEETLE: *Berosus striatus.* Fig. 609.

Range: California to British Columbia, east to North Carolina. **Adult:** Convex, greenish yellow; head black; pronotum with pair of small black spots and black median stripes; elytra with indistinct black spots. Elytral striae with fine punctures, intervals flat, with coarse punctures. **Length:** .15–.2". *B. pantherinus* (Fig. 610) is very similar, the same size, each elytron with ten distinct black spots. In *B. peregrinus* (Fig. 611) each elytron has four or five indistinct black spots, and the pronotal stripes are lacking.

(61) GIANT WATER SCAVENGER BEETLE: *Hydrophilus triangularis.* Fig. 612.

Range: Widely distributed throughout North America. **Adult:** Elytra

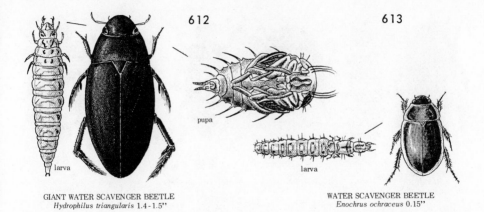

612 613

pupa

larva

larva

GIANT WATER SCAVENGER BEETLE
Hydrophilus triangularis 1.4-1.5"

WATER SCAVENGER BEETLE
Enochrus ochraceus 0.15"

shiny black with greenish tinge. Triangular yellowish markings on sides of abdomen. Prominent ridge in middle of underside of thorax. Carnivorous in adult as well as larval stage, feeds on live and dead aquatic animals. Attracted to lights. Eggs are laid in brownish egg case, attached to plant or allowed to float on the water. **Length:** 1.4–1.5". **Larva:** Spindle-shaped, head subspherical, mandibles with single inner tooth; without gills.

(62) WATER SCAVENGER BEETLE: *Enochrus ochraceus.* **Fig. 613.**

Range: Eastern U.S. **Adult:** Shiny, pale yellowish brown; head darker, pronotum and elytra with paler borders, densely punctate. **Length:** .15". **Larva:** Narrow, not spindle-shaped; right mandible with two or more teeth, left mandible with only one. Found at the edge of ponds and lakes.

(63) NARROW WATER SCAVENGER BEETLE: *Tropisternus lateralis.* **Fig. 614.**

Range: Throughout North and South America. **Adult:** Convex, shiny, greenish black; head, prothorax, and elytra margined with yellow; underside black, legs yellowish; elytra with dense fine punctures and a few scattered coarse ones. **Length:** .3–.4". Common in slow-moving streams and in lakes.

(64) DUNG SCAVENGER: *Sphaeridium scarabaeoides.* **Fig. 615.**

Range: Widely distributed in North America; introduced from Europe. **Adult:** Convex, shiny black; elytra with reddish spot near base, yellowish at apex, punctured but not striated. **Length:** About .25". The larvae feed on fresh dung and maggots associated with it.

Carrion Beetles (Silphidae)

The beetles in this family are fairly large, commonly feed on decaying animal matter and other insect larvae associated with it, particularly blow flies. A few species are found in the nests of ants; some feed on decaying fungi, others are predaceous. They are usually black, with orange, red, or

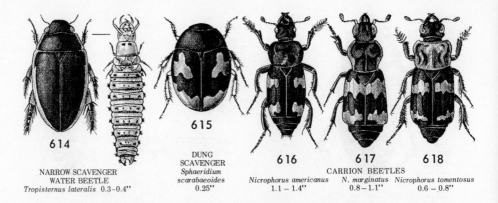

614

NARROW SCAVENGER
WATER BEETLE
Tropisternus lateralis 0.3-0.4"

615

DUNG
SCAVENGER
*Sphaeridium
scarabaeoides*
0.25"

616

Nicrophorus americanus
1.1 – 1.4"

617

CARRION BEETLES
N. marginatus
0.8 – 1.1"

618

Nicrophorus tomentosus
0.6 – 0.8"

yellow markings, may be distinguished by the clubbed antennae (ten- or eleven-segmented) and the prominent front coxae; the pronotum is much larger than the head. Species of *Nicrophorus* bury small carrion such as mice and snakes, on which the young develop. The ten-segmented larvae are dark brown or blackish, slightly depressed, broad, oval, spined or armored, and have stout legs.

(65) GIANT CARRION BEETLE: *Nicrophorus americanus.* **Fig. 616.**

Range: East of the Rocky Mts. **Adult:** Shiny black; elytra short, with two wide orange-red bands; antennal club, top of head, and pronotum orange-red. **Length:** 1.1–1.4". *N. marginatus* (Fig. 617) is very similar—from .8" to 1.1" long—with pronotum black; it occurs over much of Canada and the U.S., especially the western states.

(66) GOLD-NECKED CARRION BEETLE: *Nicrophorus tomentosus.* **Fig. 618.**

Range: East of the Rocky Mts. **Adult:** Shiny black; pronotum covered with dense coat of golden hairs; elytra with orange-red bands and epipleural fold. **Length:** .6–.8". *N. vespilloides* (Fig. 619) is black, with orange-red markings, from .4" to 1.15" long, occurs in the Pacific Northwest and in Europe; the left elytra of two subspecies are pictured (Fig. 619A, B). *N. investigator* (Fig. 620) is very similar, black, with orange-red markings, from .4" to .8" long, common in the West from Alaska to Mexico; the right elytra of two subspecies are pictured (Fig. 620A, B).

(67) PUSTULATED CARRION BEETLE: *Nicrophorus pustulatus.* **Fig. 621.**

Range: Widely distributed over most of North America. **Adult:** Shiny black, with orange-red elytral spots and antennal club (excepting the first segment). **Length:** .7". *N. p. melsheimeri*—a form occurring from Alaska to California, Colorado, Arizona, and New Mexico—has red spots and epipleural fold. *N. p. nigritus*—a subspecies found in the Southwest—is all black excepting the red-tipped antennal club.

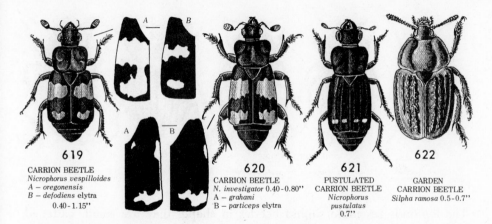

619
CARRION BEETLE
Nicrophorus vespilloides
A – *oregonensis*
B – *defodiens* elytra
0.40 - 1.15"

620
CARRION BEETLE
N. investigator 0.40 - 0.80"
A – *grahami*
B – *particeps* elytra

621
PUSTULATED
CARRION BEETLE
Nicrophorus
pustulatus
0.7"

622
GARDEN
CARRION BEETLE
Silpha ramosa 0.5 - 0.7"

(68) GARDEN CARRION BEETLE: *Silpha ramosa*. **Fig. 622.**

Range: Western North America. **Adult:** Velvety black, head and pronotum finely punctate, elytra each with three branched costae (raised ribs). **Length:** .5–.7". The shiny black larvae feed mostly on decaying vegetable matter but also attack garden and field crops, grasses, and other plants. The **Beet Carrion Beetle,** *S. opaca* (Fig. 623)—a widely scattered species (also found in Europe)—has dense orange to reddish pubescence, feeds on roots of young beets. *S. lapponica* (Fig. 624)—dull black, head and prothorax with dense coat of yellowish hairs, from .32" to .55" long—is very common on carrion from the Rocky Mts. to the Missouri River. The **Spinach Carrion Beetle,** *S. bituberosa* (Fig. 625), is dull black, from .4" to .5" long, also found on squash, pumpkin, and beets; it is more common east of the Continental Divide, and occasionally found in the Pacific Northwest.

(69) AMERICAN CARRION BEETLE: *Silpha americana*. **Fig. 626.**

Range: East of the Rocky Mts. **Adult:** Elytra brownish, with darker cross-ridges and three indistinct costae on each; pronotum yellow, with black or brownish spot. **Length:** .6"–.8". It is found in carrion and decaying fungi.

(70) RED-LINED CARRION BEETLE: *Necrodes (Silpha) surinamensis*. **Fig. 627.**

Range: East of the Rocky Mts. Becoming established west of the Rocky Mts. **Adult:** Black, sometimes with reddish tinge; elytra with solid or broken reddish crossband (sometimes missing) near apex. **Length:** .6–1". Found in carrion.

Mammal Nest Beetles (Leptinidae)

This family comprises almost the only beetles that are not free-living. They are true ectoparasites on several kinds of rodents and insectivores.

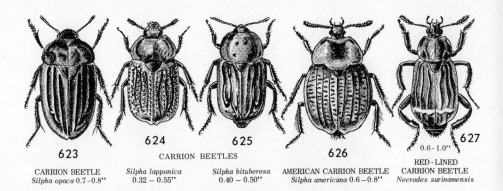

623 CARRION BEETLE Silpha opaca 0.7 - 0.8"

624 Silpha lapponica 0.32 – 0.55"

CARRION BEETLES

625 Silpha bituberosa 0.40 – 0.50"

626 AMERICAN CARRION BEETLE Silpha americana 0.6 – 0.8"

627 0.6 - 1.0" RED - LINED CARRION BEETLE Necrodes surinamensis

The leptinids resemble silphid beetles in shape; they are small, flattened, brownish yellow or reddish brown, with short recumbent hairs, eyes and wings greatly reduced consistent with their parasitic trait. They do not fly, the elytra being fused medially. The antennae are eleven-segmented and usually clubbed. The larvae resemble those of the staphylinids; they are depressed, have three-segmented antennae and no eyes, pass through three instars resembling one another except in proportions.

(71) MOUSE NEST BEETLE: *Leptinus americanus.* **Fig. 628.**

Found in the nests of field mice, moles, and shrews. **Range:** Widely distributed in North America. **Adult:** Brownish yellow, with short recumbent pubescence of same color; head without eyes, antennae only slightly or hardly enlarged apically; elytra flat, broad, inner margins fused. **Length:** .08–.1". The **Beaver Nest Beetle,** *Leptinillus validus,* is similar, about twice as long, reddish brown. The **Beaver Parasite Beetle,** *Platypsyllus castoris* (Fig. 629), is less than .1" long, with short elytra like the staphylinid beetles. Both of these species parasitize beavers in the northern states and Canada. The larvae rasp the beaver's skin, causing lesions, and feed on the exudations. The adults are believed to feed on these exudations also, and on oily skin secretions. They lay their eggs on the floor of the lodge and pupate in the earthen roof; otherwise the entire life cycle is spent on the host.

(72) HORSE-SHOE CRAB BEETLE: *Limulodes paradoxus* (Limulodidae). **Fig. 630.**

This is a member of a very small family of minute beetles, with four species widely distributed over North America. They are all associated with ants—in rotten logs and stumps, in sand, on plants—and ride on the backs of their hosts wherever they go, even on the pupae as they are carried about by the ants. The adults feed on the body exudations of their hosts; they are not carnivores or scavengers. Eyes and wings are absent; the antennae are eight- to ten-segmented, the first two segments large, the last three forming a club. The short elytra are locked together at the median

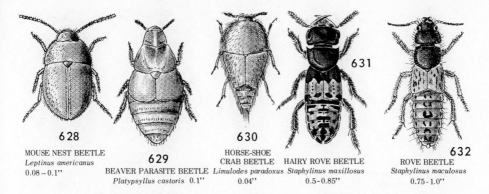

628
MOUSE NEST BEETLE
Leptinus americanus
0.08 – 0.1''

629
BEAVER PARASITE BEETLE
Platypsyllus castoris 0.1''

630
HORSE-SHOE
CRAB BEETLE
Limulodes paradoxus
0.04''

HAIRY ROVE BEETLE
Staphylinus maxillosus
0.5-0.85''

631

632
ROVE BEETLE
Staphylinus maculosus
0.75-1.0''

line, with four abdominal segments usually exposed. **Range:** *L. paradoxus* is widely distributed throughout North America. **Adult:** Shiny, reddish yellow or darker, with grayish pubescence; legs armed with spines. **Length:** .04''.

Rove Beetles (Staphylinidae)

This is a very large family—more than 26,000 species have been described world-wide, about 2,800 in North America—easily recognized by the short elytra and wings which usually cover only the first three abdominal segments when folded. Some species are apterous, and have only the elytra. They are commonly small insects, slender, uniform in width, have a habit of turning the tip of the abdomen up and over the back when alarmed, as if to strike. Colors are usually black, brown, or yellow, with spots often present; the vesture varies from bare to dense pubescence. The pronotum is commonly larger than the head, the antennae nine- to eleven-segmented (most often the latter) and usually clubbed. The mandibles are prominent and curved.

The larvae generally resemble the adults and are usually found in the same places. The abdomen is ten-segmented, the legs have five or fewer segments, and the tarsi have one claw. So far as known, the great majority of species are scavengers, and found in almost every type of habitat. A very considerable number are predaceous, and a few are parasitic; many are associated with ants, either as true guests or as active predators. The best-known parasitic species are in the genera *Aleochara* and *Maseochara*. Like the blister beetles, species of *Paederus* found in the tropics contain a vesicant or blistering agent.

(73) HAIRY ROVE BEETLE: *Staphylinus (Creophilus) maxillosus* (Staphylininae).
   Fig. 631.

Range: Widely distributed throughout North America and Europe. **Adult:** Shiny, black; band of yellowish gray hairs across the elytra. Second and

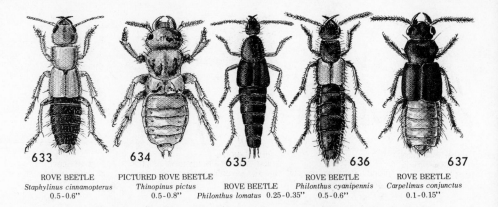

633
ROVE BEETLE
Staphylinus cinnamopterus
0.5-0.6"

634
PICTURED ROVE BEETLE
Thinopinus pictus
0.5-0.8"

635
ROVE BEETLE
Philonthus lomatus 0.25-0.35"

636
ROVE BEETLE
Philonthus cyanipennis
0.5-0.6"

637
ROVE BEETLE
Carpelimus conjunctus
0.1-0.15"

third (sometimes the fourth) abdominal segments have dense coating of yellowish gray hairs. **Length:** .5–.85". In *S. m. villosus,* the western form, the grayish pubescence on the elytra and abdomen are heavily infused with yellowish brown hairs. They are found in carrion, also feed on maggots present in the same medium.

(74) ROVE BEETLE: *Staphylinus maculosus.* **Fig. 632.**

Range: Eastern U.S. **Adult:** Brown, with reddish brown spots, top of abdomen pale; pronotum, elytra, and abdomen densely punctate, elytra sparsely pubescent. **Length:** .75–1". It is found in dung and carrion.

(75) ROVE BEETLE: *Staphylinus cinnamopterus.* **Fig. 633.**

Range: Most of the U.S. Not found in the Pacific Northwest. **Adult:** Shiny, dark brown, elytra reddish; abdomen black except for reddish tip. The pronotum has coarse, dense punctures; the elytra have dense punctures, and sparse pubescence. **Length:** .5–.6". It is found under bark and on fungi.

(76) PICTURED ROVE BEETLE: *Thinopinus pictus.* **Fig. 634.**

Range: California to Alaska. **Adult:** Amber, with black spots, circles, and curves on dorsum; elytra sparsely punctate and pubescent, wings absent, abdomen more coarsely, closely punctate and pubescent. **Length:** .5–.8". **Larva:** Resembles adult, except mandibles have no teeth, second and third thoracic segments are largely black. **Food:** Sand fleas. Both stages are found on sandy beaches, just above the water line, along the Pacific coast.

(77) ROVE BEETLE: *Philonthus lomatus.* **Fig. 635.**

Range: Throughout most of the U.S. and Canada. **Adult:** Shiny, black, with reddish tinge; pronotum and elytra sometimes brownish yellow; legs yellowish. Elytra and abdomen have moderately close, fine punctures. **Length:** .25–.35". Members of this genus occur in carrion, dung (where they prey on fly eggs and maggots), decaying vegetable matter, fungi, and are often found under stones and boards.

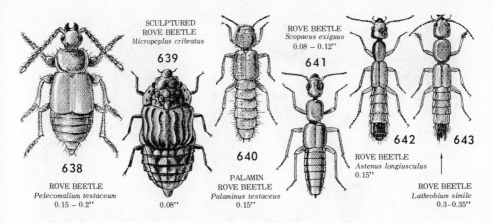

SCULPTURED
ROVE BEETLE
Micropeplus cribratus

639

ROVE BEETLE
Scopaeus exiguus
0.08 – 0.12"

641

640

642   **643**

ROVE BEETLE
Astenus longiusculus
0.15"

638

ROVE BEETLE
Pelecomalium testaceum
0.15 – 0.2"

0.08"

PALAMIN
ROVE BEETLE
Palaminus testaceus
0.15"

ROVE BEETLE
Lathrobium simile
0.3 - 0.35"

(78) ROVE BEETLE: *Philonthus cyanipennis.* **Fig. 636.**

Range: Throughout North America. **Adult:** Shiny, black; elytra metallic blue or purple. Elytra closely, abdomen finely and sparsely, punctate. **Length:** .5–.6". It is found on fleshy fungi.

(79) ROVE BEETLE: *Carpelimus conjunctus* (Oxytelinae). **Fig. 637.**

Range: Pacific Northwest. **Adult:** Shiny, black; abdomen pale brown. Head, pronotum, and elytra finely and distinctly punctate. **Length:** .1–.15". It is found near the water.

(80) ROVE BEETLE: *Pelecomalium testaceum* (Omalinae). **Fig. 638.**

Range: Alaska to California, Idaho, Nevada. Very common. **Adult:** Head black; pronotum, elytra, and epipleurae brownish yellow; abdomen nearly black. Pronotum and elytra sparsely punctate, with short decumbent setae in punctures. **Length:** .15–.2". It is found in skunk cabbage and other flowers.

(81) SCULPTURED ROVE BEETLE: *Micropeplus cribratus* (Micropelinae). **Fig. 639.**

Range: Southeastern states. **Adult:** Black, orange-brown on sides of thorax. Elytra punctate between costae; antennae nine-segmented, on sides of head; tarsi three-segmented. **Length:** .08".

(82) PALAMIN ROVE BEETLE: *Palaminus testaceus* (Paederinae). **Fig. 640.**

Range: Eastern U.S. **Adult:** Shiny, reddish yellow; abdomen reddish brown, coarsely and regularly imbricate (resembling scales or shingles); antennae and legs yellow. **Length:** About .15". It is found on leaves of various trees.

(83) ROVE BEETLE: *Scopaeus exiguus.* **Fig. 641.**

Range: Eastern U.S. **Adult:** Black, with reddish tinge; pronotum dark yellowish, antennae and legs pale yellowish. **Length:** .08–.12". *S. concavus,* a western species, is similar in shape, size, and color; the pronotum is

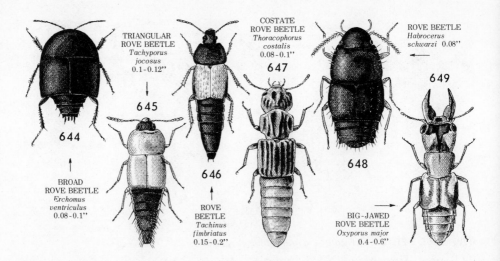

TRIANGULAR
ROVE BEETLE
*Tachyporus
jocosus*
0.1-0.12"

645

644

BROAD
ROVE BEETLE
*Erchomus
ventriculus*
0.08-0.1"

646

ROVE
BEETLE
*Tachinus
fimbriatus*
0.15-0.2"

COSTATE
ROVE BEETLE
*Thoracophorus
costalis*
0.08-0.1"

647

648

ROVE BEETLE
*Habrocerus
schwarzi* 0.08"

649

BIG-JAWED
ROVE BEETLE
Oxyporus major
0.4-0.6"

reddish, appendages brownish yellow; it may be distinguished by the prominent, deep, U-shaped concavity on the fifth abdominal sternite of the male.

(84) ROVE BEETLE: *Lathrobium simile.* **Fig. 643.**

Range: Central and eastern U.S. **Adult:** Shiny, black; elytra and most of abdomen with reddish tinge, legs paler; antennae and tip of abdomen reddish brown. Pronotum has sparse coarse punctures; abdomen has dense fine punctures. **Length:** .3–.35". *L. divisum* is very similar, slightly smaller, elytra reddish brown, with moderately coarse punctures, occurs in the Pacific coastal states and British Columbia.

(85) ROVE BEETLE: *Astenus longiusculus.* **Fig. 642.**

Range: Western North America. **Adult:** Brownish yellow, abdomen blackish; each elytron sometimes with very faint spot near the middle. Head and pronotum densely and coarsely punctate, elytra more sparsely punctate, abdomen coarsely punctate. **Length:** About .15".

(86) BROAD ROVE BEETLE: *Erchomus* (*Coproporus*) *ventriculus* (Tachyporinae).
Fig. 644.

Range: Throughout North America. **Adult:** Shiny, black, with reddish tinge; antennae and legs reddish brown; pronotum, elytra, and abdomen finely punctate. **Length:** .08–.1". It feeds on fungi under bark and in decaying vegetation.

(87) TRIANGULAR ROVE BEETLE: *Tachyporus jocosus.* **Fig. 645.**

Range: Throughout North America. **Adult:** Shiny, black, with reddish tinge; elytra, legs, and pronotum reddish yellow. **Length:** .1–.12". Members of this genus are commonly found by sweeping herbage.

(88) ROVE BEETLE: *Tachinus fimbriatus.* **Fig. 646.**

Range: Widely distributed throughout much of North America. **Adult:** Shiny, reddish brown, with pale pubescence. Pronotum and elytra finely punctate. **Length:** .15–.2″. *T. maculicollis,* a Pacific coast species, is similar, the head blackish, pronotum and abdomen smoky gray-brown, elytra pale brown, from .25″ to .35″ long. Males of this large genus may be distinguished by the dilated basal segments of the front tarsus, the division of the last abdominal sternite into two lobes, and certain differences in the next to the last abdominal sternite. They are found on fleshy fungi, decayed plant materials, dung, and carrion.

(89) COSTATE ROVE BEETLE: *Thoracophorus costalis* (Osoriinae). **Fig. 647.**

Range: Throughout most of North America. **Adult:** Dull, dark brown. Pronotum and elytra costate (four costae or raised ribs on each elytron). Head margined, abdomen not; tarsi three-segmented. **Length:** .08–.1″.

(90) ROVE BEETLE: *Habrocerus schwarzi* (Habrocerinae). **Fig. 648.**

Range: Michigan. **Adult:** Reddish brown; apex of abdomen more reddish, legs yellowish. Bristles along sides of pronotum and elytra, and on the abdomen. Elytra about as long as pronotum. **Length:** .08″ or less. *H. magnus* is very similar, found along the shore of Lake Superior. *H. capillaricornis* is also very similar, bright red-brown, with head blackish and apex of abdomen brownish yellow, from .1″ to .12″ long; a European species, it is found only in British Columbia and Oregon.

(91) BIG-JAWED ROVE BEETLE: *Oxyporus major* (Oxyporinae). **Fig. 649.**

Range: Widespread throughout much of the U.S. **Adult:** Shiny, black, with yellowish markings on elytra; tarsi yellow. Antennae inserted at sides of head, near eyes; prominent mandibles. Middle coxae widely separated. **Length:** .4–.6″. All species of Oxyporinae are found in fleshy fungi. The larvae develop very rapidly and are completely mycetophagous; the adults are believed to be predaceous, possibly partly mycetophagous.

(92) PAINTED ROVE BEETLE: *Megalopinus* (*Megalopsidia*) *caelatus*
(Megalopsidiinae). **Fig. 650.**

Range: Eastern U.S. **Adult:** Shiny, black, with oblique reddish bands on elytra. Extremely large eyes, head wider than thorax; a pair of short elongate processes extend from the labrum. **Length:** .15–.2″.

(93) ROVE BEETLE: *Stenus bipunctata* (Steninae). **Fig. 651.**

Range: Throughout North America. **Adult:** Shiny, black; tarsi and all but basal segments of antennae dark brown; each elytron with red or orange spot near the middle. Pronotum has close fine punctures; elytra more coarsely punctate. **Length:** .15–.2″. *S. juno,* also widely distributed, is very similar, about .25″ long, with body, elytra, and legs all black; a distinctive

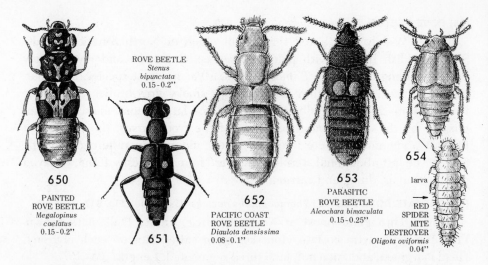

ROVE BEETLE
Stenus
bipunctata
0.15-0.2"

650

PAINTED
ROVE BEETLE
Megalopinus
caelatus
0.15-0.2"

651

652

PACIFIC COAST
ROVE BEETLE
Diaulota densissima
0.08-0.1"

653

PARASITIC
ROVE BEETLE
Aleochara bimaculata
0.15-0.25"

654

larva

RED
SPIDER
MITE
DESTROYER
Oligota oviformis
0.04"

feature is the dense hairs on the mesosternum between the coxae and on the inner side of the hind femur in the male.

(94) PACIFIC COAST ROVE BEETLE: *Diaulota densissima* (Aleocharinae). **Fig. 652.**

Range: In the tidal belt, from Alaska to central California (Monterey Peninsula). **Adult:** Black, with reddish tinge; granulose, not shiny, with dense pubescence. They are wingless. **Length:** .08–.1". **Larva:** Gray, with eighth abdominal segment black, ninth reddish. Head nearly round, body nearly uniform width, smooth, except for numerous setae; a little shorter than the adult. All stages are found in cracks in rocks, below the high-water mark; active forms often crawl about in the tide pools. *D. fulviventris* is very similar, a little smaller, occurs in southern California; in contrast, the elytra are much shorter than wide, less than half as long as the pronotum, the latter as wide as long, head as wide as pronotum, antennae as long as head and pronotum.

(95) RED SPIDER MITE DESTROYER: *Oligota* (*Somatium*) *oviformis*. **Fig. 654.**

Range: California. **Adult:** Shiny, black. Minute orange eggs are laid singly on leaves. **Length:** .04". **Larva:** Yellow, about .1" long. Found on deciduous and citrus fruit trees. **Food:** Two-spotted spider mite, European red mite.

(96) PARASITIC ROVE BEETLE: *Aleochara bimaculata*. **Fig. 653.**

Range: Throughout North America. **Adult:** Shiny, black; legs and tip of abdomen reddish brown. Adults are colonial—live in the soil in interconnected tunnels. Eggs are laid in the soil near the host, and hatch in about ten days. **Length:** .15–.25". **Larva:** Brown, horny. It seeks out the pupa of a cabbage maggot or other fly larva, gnaws a hole in it, enters and seals over the hole, and settles near the maggot's head to feed. Three instars. It consumes the entire host pupa, and pupates outside the host puparium. Two broods. Hibernates as young larva in host puparium. **Food:** Cabbage maggot or larvae of other flies. *A. bilineata* is similar, black, elytra sometimes with trace of reddish toward the apical margin; it is considered a potential biological control agent for the cabbage maggot. The adults of *Aleochara* are predators, while the larvae are true parasites. They are said to sometimes destroy 80 percent of the cabbage maggots in a field, and are also reported to attack certain muscid and sarcophagid flies in the same manner. [See Diptera (69).]

Ant-loving Beetles (Pselaphidae)

This is a fairly large family—about 5,000 species occur world-wide, 600 in North America—sometimes referred to as "short-winged mold beetles." They resemble the staphylinids, have a broader abdomen, narrow thorax, still narrower head, and a hard coating on the body. They have bulging eyes laterally placed, with a few granular facets; some are without eyes. They are mostly minute in size, yellowish or reddish, and are found in ants' nests, under loose bark, moss, or stones, in the forest litter, caves—wherever there is mold. Some feed on mites; those living with ants sometimes feed on the larvae of their hosts. The myrmecophagous species have trichomes or modified hairs that emit secretions very attractive to ants. The larval stages and life histories are not well known.

(97) ANTHILL BEETLE: *Cylindrarctus longipalpis.* **Fig. 655.**

Range: Widely distributed over North America, with possible exception of the Pacific Northwest. **Adult:** Reddish brown, abdomen with paler margin; scattered stiff hairs; palpi fairly long, with "hatchet-like" tips. **Length:** .08″.

(98) BRANDEL'S ANTHILL BEETLE: *Decarthron brandeli.* **Fig. 656.**

Range: Central and southwestern states; Pacific Northwest (?). **Adult:** Dark brownish, legs and antennae paler; elytra reddish. Ventral surface of head with large oval fossa or pit in the middle. Antennae ten-segmented. **Length:** .06″.

(99) SHINY FUNGUS BEETLE: *Baeocera falsata* (Scaphidiidae). **Fig. 657.**

This is a small group of tiny oval, convex beetles with large pronotum, small head, and long legs; they are highly polished, black, and sometimes have red markings. The elytra are truncate, exposing the tip of the abdomen. They live on fungi, and are found in rotten wood, in decaying leaves, or under the loose bark of logs. **Range:** *B. falsata* is widely distributed over the eastern and central states. **Adult:** Black, very shiny; tips of elytra and abdomen, legs, and antennae dark brown; pronotum with dense fine

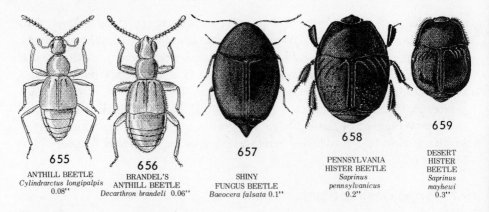

655
ANTHILL BEETLE
Cylindrarctus longipalpis
0.08"

656
BRANDEL'S
ANTHILL BEETLE
Decarthron brandeli 0.06"

657
SHINY
FUNGUS BEETLE
Baeocera falsata 0.1"

658
PENNSYLVANIA
HISTER BEETLE
*Saprinus
pennsylvanicus*
0.2"

659
DESERT
HISTER
BEETLE
*Saprinus
mayhewi*
0.3"

punctures, elytra with scattered coarse punctures and deep line along base and suture. **Length:** .1".

Hister Beetles (Histeridae)

Hister beetles comprise a small distinctive group—small in size, flattened, ovoid or hemispherical in shape. They are usually shiny black, bronze, or green, with a hard covering and short elytra which leave the last one or two abdominal segments exposed. The pronotum is large, the head small and commonly deflexed, with large curved mandibles, and eleven-segmented retractile antennae ending in a club. The larvae are elongate, nearly uniform in width, rounded or slightly depressed, with nine or ten abdominal segments and numerous setae; they are usually longer than the adults, and may have one pair of ocelli or none. Most of the adults and larvae are carnivorous; but many are scavengers, found in dung and carrion (where they also prey on fly eggs and larvae) or in the nests of birds and rodents. A few species live in the open and attack chrysomelid beetles and caterpillars; the larvae of a very considerable number live in the galleries of bark beetles (mostly scolytids) and feed on the larvae.

(100) PENNSYLVANIA HISTER BEETLE: *Saprinus pennsylvanicus* (Saprininae).
Fig. 658.

Range: East of the Rocky Mts. **Adult:** Bright green, sometimes with bronze tinge. Apical half of elytra coarsely punctate; pygidium densely and coarsely punctate; propygidium densely and finely punctate. **Length:** .2". Beetles in this genus occur in carrion, dung, and rodent burrows.

(101) DESERT HISTER BEETLE: *Saprinus mayhewi.* **Fig. 659.**

Range: Southern California, in the Mojave Desert. **Adult:** Blackish brown —almost black—with light tinge of very dark bronze. Small punctures on sides of pronotum, erect golden brown marginal hairs. Elytra striated and minutely punctured. **Length:** .3". *S. lungens* and *oregonensis* are very com-

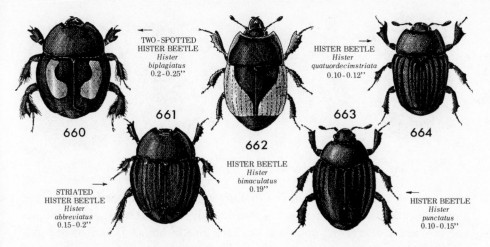

TWO-SPOTTED
HISTER BEETLE
*Hister
biplagiatus*
0.2-0.25"

STRIATED
HISTER BEETLE
*Hister
abbreviatus*
0.15-0.2"

HISTER BEETLE
*Hister
bimaculatus*
0.19"

HISTER BEETLE
*Hister
quatuordecimstriata*
0.10-0.12"

HISTER BEETLE
*Hister
punctatus*
0.10-0.15"

660   661   662   663   664

mon on dung and carrion in the Pacific Northwest, lack the pronotal fringe of hairs; in the latter species the sutural and fourth striae of the elytra often arch together near the base; they prey on fly eggs and larvae found in these media.

(102) TWO-SPOTTED HISTER BEETLE: *Hister biplagiatus* (Histerinae). **Fig. 660.**

Range: Central and eastern states. **Adult:** Black, with red markings, and fringe of short hairs on margin of pronotum. **Length:** .2–.25".

(103) STRIATED HISTER BEETLE: *Hister abbreviatus.* **Fig. 661.**

Range: New York to Florida, west to Oklahoma, Utah and the Pacific Northwest. **Adult:** Shiny black; elytra with deep striations—four complete, others short and overlapping. Pygidium densely and finely punctate; propygidium sparsely and coarsely punctate. **Length:** .15–.2". It lives in dung and carrion, feeds on maggots, and is sometimes found in fungi. *H. bimaculatus* (Fig. 662) was introduced into Hawaii from Germany for control of the horn fly. *H.* (*Dendrophilus*) *punctatus* (Fig. 663)—black, from .1" to .15" long—is occasionally found in flour mills in the central and eastern states. *H.* (*Carcinops*) *quatuordecimstriata* (Fig. 664)—black, elytra with fourteen "stripes" composed of tiny dots, from .1" to .12" long—is found in decaying organic matter in the central states.

(104) BARK HISTER BEETLE: *Platysoma depressum.* **Fig. 665.**

Range: Widely distributed from the Atlantic seaboard to the Pacific Northwest. **Adult:** Flattened, shiny black, with small punctures on the side of the pronotum, three complete striae on the dorsal surface of elytra. **Length:** .1–.12". It lives under the bark of dead and dying trees. A similar species, *P. lecontei,* is abundant under the bark of oak and elm. *P. punctigerum*—found under the bark of pines in the Pacific Northwest and California—is about .2" long, shiny black, with reddish brown legs, pronotum as

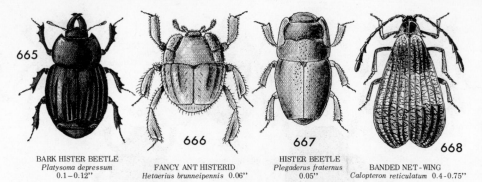

665

666

667

668

BARK HISTER BEETLE
Platysoma depressum
0.1–0.12"

FANCY ANT HISTERID
Hetaerius brunneipennis 0.06"

HISTER BEETLE
Plegaderus fraternus
0.05"

BANDED NET-WING
Calopteron reticulatum 0.4-0.75"

wide as elytra; it enters the holes of *Dendroctonus* bark beetles, lays eggs along the sides of egg galleries, has two broods, hibernates as adult; the larvae are active searchers, and feed on bark beetle larvae found in the galleries.

(105) FANCY ANT HISTERID: *Hetaerius brunneipennis* (Hetaerinae). **Fig. 666.**

Range: North central states, eastward. **Adult:** Shiny, reddish brown, with fine yellow hairs along elytral striae. Pronotum with two lobes set off from the disc by oblique sulci or furrows, and divided by transverse sulci. **Length:** .06". This genus is chiefly western; the members live in ants' nests, eat dead or dying ants and other insects, and are fed and licked by the ants.

(106) HISTER BEETLE: *Plegaderus fraternus* (Abraeinae). **Fig. 667.**

Range: Arizona, California, Nevada, Washington, Oregon, British Columbia. **Adult:** Shiny, dark reddish black; appendages paler. Pronotum divided by shallow grooves, finely punctate; elytra more coarsely punctate. **Length:** About .05". Found under the bark of hemlock and poplar. Some species of this small genus are found in the galleries of bark beetles in pine trees; they are predaceous.

Net-winged Beetles (Lycidae)

This is a small family of brightly colored flattened beetles, easily distinguished by their netlike wings (fan-shaped when folded), large pronotum which often covers most of the head, large eleven-segmented antennae, and bulging eyes. The larvae are yellow to brown, elongate, spindle-shaped, with nine segments (sometimes ten), and short, stout antennae; ocelli are usually absent, or one pair at most present. The adults are predaceous and live on plants; the larvae live under bark. There is some question as to whether the latter are carnivorous or lignivorous (wood-eating).

(107) BANDED NET-WING: *Calopteron reticulatum*. **Fig. 668.**

Range: Eastern U.S. **Adult:** Orange-yellow; elytra with black crossband and apexes, pronotum with black median longitudinal stripe. **Length:** .4–

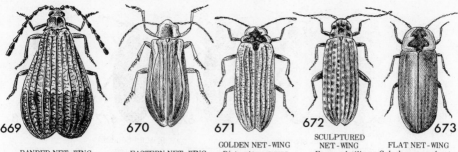

669	670	671	672	673
BANDED NET-WING	EASTERN NET-WING	GOLDEN NET-WING	SCULPTURED NET-WING	FLAT NET-WING
Calopteron terminale 0.4-0.75"	*Lycus lateralis* 0.3–0.4"	*Dictyopterus aurora* 0.3-0.45"	*Eros sculptilis* 0.3-0.5"	*Calochromus perfacetus* 0.25-0.4"

.75". *C. terminale* (Fig. 669) is similar, except that the black band is missing and the apical portion is purplish black; it ranges westward to California.

(108) EASTERN NET-WING: *Lycus (Lycostomus) lateralis*. **Fig. 670.**

Range: Eastern U.S. **Adult:** Black, with apex and sides of pronotum and elytral margin (halfway) yellow. **Length.** .3–.4".

(109) GOLDEN NET-WING: *Dictyopterus (Eros) aurora*. **Fig. 671.**

Range: Throughout most of North America, in the West from Arizona to Alaska. **Adult:** Black, with sides of pronotum orange-red. **Length:** .3–.45". The **Sculptured Net-wing,** *Eros (Platycis) sculptilis* (Fig. 672), occurs in the southern states and northward into Indiana.

(110) FLAT NET-WING: *Calochromus perfacetus*. **Fig. 673.**

Range: Eastern U.S. **Adult:** Black; pronotum with sides, and sometimes entire disc, brownish yellow. **Length:** .25–.4". It is found on linden flowers.

Fireflies (Lampyridae)

The lampyrid beetles or fireflies constitute a small family unique for their flashing light, said to be induced by an enzyme, luciferase, reacting with a substance called luciferin. (A few species are not luminous.) The color of the light and its periodicity vary with the species. Organs producing the light are located on the underside of the posterior abdominal segments; more segments are luminous in the male and the light is brighter. Luminosity persists even after death. Females are usually brachypterous, or entirely without wings; most of them fly very little or not at all, and are larva-like in form. The flashings are believed to be associated with mating.

Some larvae have luminous organs on the eighth abdominal segment, and are seen to glow inside the egg. The larvae and apterous females are referred to as "glowworms." Aside from their flashings and luminous organs—the segments are yellowish or greenish when not lighted—the adults may be distinguished by the head, which is more or less completely covered by a broad pronotum with thin flangelike margins. They are flattened, ovoid,

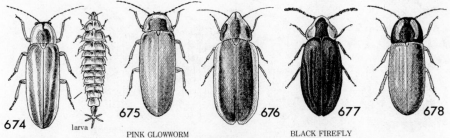

674   larva

PENNSYLVANIA FIREFLY
Photuris pennsylvanicus   0.45-0.6"

PINK GLOWWORM
Microphotus angustus   ANGULATE FIREFLY
0.4-0.6"   *Pyractomena angulata* 0.3-0.6"

BLACK FIREFLY
Lucidota atra
0.3-0.5"

BLACK FIREFLY
Ellychnia corrusca 0.35-0.55"

relatively soft, vary from brown or black to greenish, yellowish, or reddish. The antennae are serrate and eleven-segmented, the mandibles large and curved, the tarsi five-segmented. Males have eight and females have seven visible abdominal segments; the elytra have a distinct epipleural fold.

The larvae are flattened, usually spindle-shaped, with ten-segmented abdomen; the head is small, usually retracted, with curved mandibles, and prominent three-segmented antennae. Both larvae and adults are predaceous, feed mostly on snails, slugs, and insect larvae. They inject a highly toxic digestive substance into their prey that paralyzes them quickly, and suck out the liquefied body contents. The adults of some species appear not to feed. Most known species have a life cycle of two years, hibernate as larvae in chambers formed of soil, on or under the surface, and pupate in cells made in the soil. Most species are nocturnal. The larvae of some Asiatic species are aquatic and important predators of snails that act as intermediate hosts of organisms causing disease in man.

(111) PENNSYLVANIA FIREFLY: *Photuris pennsylvanicus* (Photurinae). **Fig. 674.**

Range: Eastern U.S. to Kansas and Texas. Very common. **Adult:** Elytra dark brown, with yellowish margins and oblique stripes, punctate; epipleuron incomplete toward base. Head and pronotum yellowish; disc of latter reddish, with black median stripe, punctate. They fly from early evening to midnight during July and August, emit a bright green light. Spherical eggs are laid singly or in groups in damp soil, around grass or moss, and hatch in about four weeks. **Length:** .45–.6". **Larva:** It feeds at night in the grass on snails, slugs, earthworms, cutworms. Their light is not visible unless they are turned over. An attempt to establish the species in Seattle, Washington, was unsuccessful.

(112) PINK GLOWWORM: *Microphotus angustus* (Lampyrinae). **Fig. 675.**

Range: California, Oregon, Colorado. **Adult:** Male winged, with elytra and pronotum grayish brown, eyes black, body pinkish. Female larviform, flattened, pinkish, with only vestigial elytra having parallel lines; antennae nine-segmented. Occurs in dry grass of foothills from June to August, appears in some localities only three or four nights of the year, is easily detected by the bright light emitted; females glow continuously. **Length:** .4–.6".

(113) ANGULATE FIREFLY: *Pyractomena angulata.* **Fig. 676.**

Range: Eastern third of North America. **Adult:** Blackish brown; elytra margined with yellow, finely granulate; pronotum yellowish, with darkened areas, small but distinct median longitudinal carina. **Length:** .3–.6". Light organs well developed, those of female on sides of abdomen.

(114) BLACK FIREFLY: *Lucidota atra.* **Fig. 677.**

Range: Eastern North America. **Adult:** Black; sides of pronotum yellowish, with reddish or orange marking between margin and median area; elytra granulate, with four indistinct ridges fading out toward apex. **Length:** .3–.5". It is found in trees. Light organs not well developed. *Ellychnia* (*Lucidota*) *corrusca* (Fig. 678), a closely related species, black, with brownish pubescence, wider yellowish markings on pronotum, more pronounced ridges on elytra, from .35" to .55" long, is widely distributed from coast to coast.

(115) PALE FIREFLY: *Photinus scintillans.* **Fig. 679.**

Range: Eastern U.S. from Massachusetts to Kansas and Texas. Very common. **Adult:** Dark brown, with suture and outer margin of elytra pale yellow; pronotum reddish, with yellowish margin, blackish spot in center. Elytra finely granulate, antennae with long pubescence. Males fly only at twilight, emit a yellowish light; females are brachypterous and do not fly. Light organs on underside of sixth abdominal segment in the female, on entire sixth and seventh segments of the male. **Length:** .2–.3". *P. pyralis* (Fig. 680), common east of the Rocky Mts., is similar but larger—from .4" to .55" long—with elytra finely wrinkled; both sexes are long-winged, but females seldom fly.

Soldier Beetles (Cantharidae)

The soldier beetles, or leather-wings, resemble the closely related fireflies but have no light organs and the head is much in evidence. They have leather-like (pliable) elytra, are usually dark, with orange, yellow, or red patterns, and covered with fine hairs. The antennae are long, filiform, eleven-segmented, and the eyes are lateral, bulging. The mandibles are large and curved, the maxillary palpi four-segmented and the labial palpi three-segmented. Only seven abdominal segments (sternites) are visible. The adults are often seen on flowers such as goldenrod, milkweed, hydrangea. Some feed on nectar and pollen, others are omnivorous, but the majority

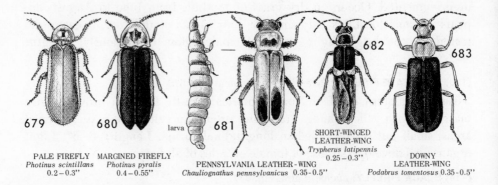

679   680   larva   681

682

683

PALE FIREFLY   MARGINED FIREFLY
Photinus scintillans   *Photinus pyralis*
0.2–0.3"   0.4–0.55"

PENNSYLVANIA LEATHER-WING
Chauliognathus pennsylvanicus 0.35–0.5"

SHORT-WINGED
LEATHER-WING
Trypherus latipennis
0.25–0.3"

DOWNY
LEATHER-WING
Podabrus tomentosus 0.35–0.5"

are believed to be predaceous in both the adult and the larval stages. Adults of many species of *Podabrus* and *Cantharis* are known to feed on grasshopper eggs, aphids, and the larvae of Lepidoptera and Coleoptera. Eggs are laid in masses in the soil or under other cover.

The larva is somewhat flattened, elongate, nearly uniform in width, with numerous fine hairs giving it a velvety appearance. The coloring is usually deep brown, purplish, or black. The abdomen is ten-segmented, the head moderate in size, with one pair of ocelli, three-segmented antennae and maxillary palpi. The known species have a brief "prelarval" stage following hatching, during which they live on yolk stored in the intestines. There are one or two broods; winter is passed in the late larval stage, and pupation occurs in cells made in the soil.

(116) PENNSYLVANIA LEATHER-WING: *Chauliognathus pennsylvanicus.* **Fig. 681.**

Range: East of the Rocky Mts., south to Arizona. **Adult:** Elytra and pronotum dull orange-yellow, with black markings; head, underside of body, and legs black. **Length:** .35–.5". **Food:** Locust eggs, cucumber beetles and other species of *Diabrotica* (many are destroyed in the fall when they become inactive). [See (368).] *C. marginatus* is similar but slightly smaller, with head mostly yellow, black longitudinal median band on pronotum, the black area variable but covering much of the elytra; it is very common in the alfalfa fields of Kansas, and is said to invade the tunnels of the corn earworm. *Chauliognathus* spp. are commonly found in tunnels of the European corn borer.

(117) SHORT-WINGED LEATHER-WING: *Trypherus latipennis.* **Fig. 682.**

Range: Connecticut, Pennsylvania, Virginia, Georgia, Indiana. **Adult:** Elytra short as in rove beetles; wings long, covering the abdomen. Blackish above, yellowish on underside, tips of elytra and margins of pronotum yellow. Very large, oval, bulging eyes in the male. **Length:** .25–.3". Often found on catnip.

(118) DOWNY LEATHER-WING: *Podabrus tomentosus* (Cantharinae). **Fig. 683.**

Range: Most of the U.S., including California. **Adult:** Elytra black, with whitish pubescence (appears grayish blue); head, base of antennae, prothorax, and legs reddish yellow. Pale yellow oblong eggs are laid in masses in or on the soil. **Length:** .35–.5″. **Larva:** Pink, with velvety covering of fine hairs, two dark longitudinal lines on dorsal surface of thoracic segments, from .6″ to .8″ full grown. It lives in the soil, where it also pupates. **Food** (adults): All kinds of aphids. Common in fields, gardens, and orchards.

(119) TWO-LINED LEATHER-WING: *Cantharis bilineatus.* **Fig. 684.**

Range: Eastern half of the U.S. **Adult:** Occiput (back of head), all but basal segments of antennae, two broad bands on pronotum, elytra, and legs black; otherwise reddish yellow. Pronotum and elytra sparsely punctate. **Length:** .25–.3″. *C. divisa* is very similar, from .25″ to .4″ long, oblique markings on pronotum variable; it is common on aphid-infested grasses, flowers, vegetables, and fruit trees from California to Alaska, and considered very beneficial.

Soft-winged Flower Beetles (Melyridae)

This is a family of moderate size—some 450 species are known in North America—characterized by a distinct clypeus or prolongation of the lower part of the head, prominent front coxae, and antennae with eleven (ten visible) segments inserted in front above the base of the mandibles. They are usually clothed with erect hairs and have a velvety appearance. The tarsi are five-segmented, the claws often with a fleshy lobe between. They have a wide range of food preference, feed on pollen, carrion, dead and live insects. *Endeodes* spp.—small wingless, predaceous beetles, yellow, or orange and black—occur along the Pacific coast from British Columbia to Baja California, are found in the cracks in rocks at or near the high-tide mark, sometimes under debris on the sandy beaches. Many species of *Collops* are valuable predators of insect pests or their eggs.

(120) FOUR-SPOTTED COLLOPS: *Collops quadrimaculatus.* **Fig. 685.**

Range: East of the Rocky Mts. **Adult:** Pronotum and elytra reddish yellow, latter with blue or bluish markings; tarsi, tibiae, and apex of antennae dark brownish black; head and abdomen black. **Length:** .15–.25″. It feeds on leafhoppers and eggs of the chinch bug (does not attack nymphs or adults of the latter).

(121) TWO-LINED COLLOPS: *Collops vittatus.* **Fig. 686.**

Range: Throughout most of the U.S. and Canada. **Adult:** Black; pronotum, sides and suture of elytra reddish yellow or dull orange, longitudinal stripes (vittae) of elytra dark bluish; pronotum sometimes with median

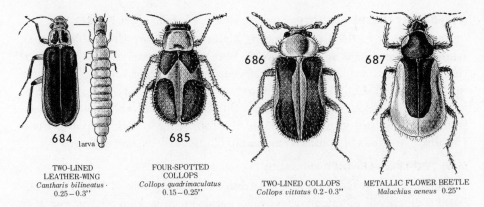

TWO-LINED
LEATHER-WING
Cantharis bilineatus ·
0.25–0.3"

FOUR-SPOTTED
COLLOPS
Collops quadrimaculatus
0.15–0.25"

TWO-LINED COLLOPS
Collops vittatus 0.2–0.3"

METALLIC FLOWER BEETLE
Malachius aeneus 0.25"

black spot. Pubescent, densely punctate. **Length:** .2–.3". It feeds extensively on live and dead larvae and pupae of the alfalfa caterpillar. *C. bipunctatus* —about the same size—has pale red pronotum, with two oblique black markings, bluish black elytra; it occurs from Kansas to California, Idaho, Oregon, feeds on the corn leaf aphid, larvae of the alfalfa weevil and eggs of the Say's stink bug; it is said to be one of the most effective natural enemies of the latter. *Collops* spp. are prevalent in cotton fields of the Southwest.

(122) METALLIC FLOWER BEETLE: *Malachius aeneus.* **Fig. 687.**

Range: East of the Rocky Mts., Pacific Northwest; an introduced species. **Adult:** Anterior angles of pronotum, much of the elytra, front of head orange or reddish brown; remainder dark green. **Length:** About .25". It is said to cause considerable unrecognized damage to developing wheat in the Midwest.

(123) PALE FLOWER BEETLE: *Attalus scincetus.* **Fig. 688.**

Range: Eastern half of the U.S. **Adult:** Yellowish, with back of head (occiput), most of pronotum, scutellum, and suture at base of elytra blackish. Elytra punctate. **Length:** .1–.12". Found on flowers of dogwood, *Viburnum,* wild rose. *A. morulus smithi,* somewhat smaller, black (except for anterior margin of head, mouthparts, basal antennal segments, and legs, which are brownish yellow), occurs in the Pacific Northwest. *A. circumscriptus* (Fig. 689) ranges eastward from Arizona, also occurs in south central states.

Checkered Beetles (Cleridae)

The checkered beetles comprise a small family distinguished by the brightly colored patterns of orange, red, or blue, and by the hairy body, the membranous lobes on the tarsal segments (the first and fourth segments often greatly reduced), and the enlarged palpi with the last segment di-

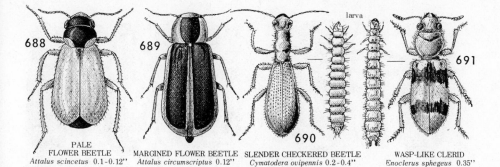

688

689

larva

691

690

PALE
FLOWER BEETLE
Attalus scincetus 0.1-0.12"

MARGINED FLOWER BEETLE
Attalus circumscriptus 0.12"

SLENDER CHECKERED BEETLE
Cymatodera ovipennis 0.2-0.4"

WASP-LIKE CLERID
Enoclerus sphegeus 0.35"

lated. The antennae have eight to eleven segments, the last three greatly enlarged to form a club. The larvae resemble those of the carabids; they are cylindrical or somewhat flattened, hairy, distinctly segmented, usually brown, yellow, red, or pink. The tenth abdominal segment is often modified as a five-lobed organ of locomotion.

Adults and larvae of a great majority of clerids are predaceous; the larvae attack immature stages of the host; the adults usually prey on the mature forms. A few are scavengers. *Cymatodera* spp. (Fig. 690) prey on gall wasps (Cynipidae); *Trichodes* spp. prey on (or parasitize) the larvae of wasps and bees; species of *Aulicus* attack the egg pods of grasshoppers. A majority of species are predaceous on scolytids and other wood-inhabiting beetles, and are believed to be important in the suppression of these forest pests. Eggs are laid at the entrance to the galleries of bark beetles or in crevices of the bark, the life cycle and number of broods being correlated with that of the host. Some pupate in the galleries, others in the soil, with or without a cell or cocoon; they hibernate as larvae, pupae, or adults.

(124) WASP-LIKE CLERID: *Enoclerus sphegeus* (Clerinae). **Fig. 691.**

Range: California to British Columbia; Idaho, Montana, North Dakota, to New Mexico. **Adult:** Pronotum and elytra dark brown or black, latter with two whitish or yellow bands; face yellow, abdomen red. **Length:** .35". **Larva:** Attacks larvae and pupae of bark beetles in their galleries under the bark. Migrates to base of tree when full grown, hibernates in cell formed in the soil and lined with foamy oral secretion. **Food:** Bark beetles of the genus *Dendroctonus. E. cupressi*—black with blue sheen and triangular orange spots on elytra, about .3" long—attacks the cypress bark beetle in California. *E. nigripes* (Fig. 692)—dull red, apical two-thirds of elytra black with whitish bands, from .2" to .3" long—attacks bark beetles in pine, spruce, juniper, and borers in hardwoods in the eastern states.

(125) RED-BLUE CHECKERED BEETLE: *Trichodes nutalli.* **Fig. 693.**

Range: East of the Rocky Mts.; Idaho, British Columbia. **Adult:** Dark blue, purplish, or greenish blue, elytra with reddish yellow markings. Pro-

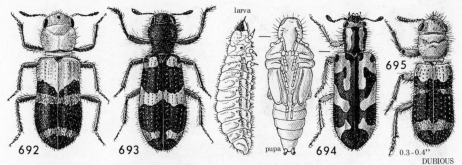

larva

pupa

695

692 693 694 0.3-0.4"

CHECKERED BEETLE
Enoclerus nigripes 0.2-0.3"

RED-BLUE CHECKERED BEETLE
Trichodes nutalli 0.3-0.45"

ORNATE CHECKERED BEETLE
Trichodes ornatus 0.5"

DUBIOUS
CHECKERED BEETLE
Thanasimus dubius

notum and elytra sparsely punctate. **Length:** .3–.45″. *T. ornatus* (Fig. 694) is very similar—a little larger, metallic blue with three irregular yellow bands across elytra—occurs in most of the western states; the larvae are yellow and dark brown, about .5″ long. Adults of both species feed on pollen and living insects; the larvae prey on solitary bees and vespid wasps.

(126) DUBIOUS CHECKERED BEETLE: *Thanasimus dubius.* **Fig. 695.**

Range: Eastern half of the U.S. **Adult:** Head, pronotum, basal area of elytra reddish brown; remainder of latter black with irregular bands of whitish pubescence. **Length:** .3–.4″. Adults feed on pollen, larvae prey on bark beetles in spruce, pine, and elm trees. *T. undulatus* (Fig. 696) is a western species, black with grayish markings, abdomen and legs reddish orange, from .25″ to .3″ long; adults and larvae prey on the fir engraver and other bark beetles.

(127) HAIRY CHECKERED BEETLE: *Phyllobaenus unifasciata* (Phyllobaeninae).

Range: Throughout most of the U.S. and Canada. **Adult:** Bluish black, silvery white band of dense hairs across middle of elytra. **Length:** .15–.2″. *P. scaber*—about the same size, brownish gray, elytra with roughened hairs, fine dense punctures—ranges from California to Washington, is said to feed on the woolly apple aphid. *P. humeralis* (Fig. 697), widely distributed throughout North America, is the same size, elytra bluish black to blue-green, each with reddish orange patch on the shoulder, legs and antennae reddish, elytra coarsely punctate, with inconspicuous short erect hairs. *Phyllobaenus* spp. are often found in tunnels of the European corn borer.

(128) RED-SHOULDERED HAM BEETLE: *Necrobia ruficollis* (Korynetinae). **Fig. 698.**

Bone or ham beetles are assigned to a separate family, the Corynetidae, in some classifications. They are usually found in spoiled meat products, fish, and cheese, sometimes in cured ham and bacon. They are believed to feed on fly maggots and other scavengers found in these places rather than the meat itself. **Range:** *N. ruficollis* is cosmopolitan in distribution. **Adult:** Front

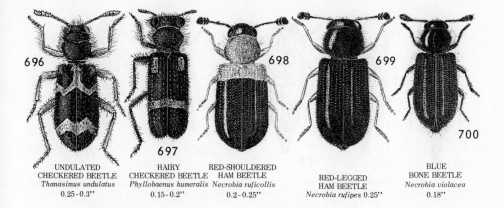

696 UNDULATED CHECKERED BEETLE
Thanasimus undulatus
0.25-0.3"

697 HAIRY CHECKERED BEETLE
Phyllobaenus humeralis
0.15-0.2"

RED-SHOULDERED HAM BEETLE
Necrobia ruficollis
0.2-0.25"

698

699 RED-LEGGED HAM BEETLE
Necrobia rufipes 0.25"

700 BLUE BONE BEETLE
Necrobia violacea
0.18"

of head, apical three-fourths of elytra metallic blue; pronotum, base of elytra, legs brownish red. Antennae, mesosternum, metasternum, abdomen dark brown. **Length:** .2–.25″. It preys on the cheese skipper and dermestid beetles.

(129) RED-LEGGED HAM BEETLE: *Necrobia rufipes.* **Fig. 699.**

Range: Cosmopolitan. **Adult:** Metallic blue or green, legs and basal segments of antennae red. Elytra with rows of punctures, dense fine punctures between. **Length:** .25″. **Larva:** Leaves fatty portions of the ham for the more fibrous parts or a nearby beam when full grown, gnawing its way; pupates in paper-like cocoon made of globules of an oral secretion. It is predaceous on the cheese skipper and dermestid larvae, also reputed to be a pest of grain and other foods, and silk.

(130) BLUE BONE BEETLE: *Necrobia violacea.* **Fig. 700.**

Range: Cosmopolitan. **Adult:** Metallic dark blue or dark green; antennae and legs very dark brown. Densely and minutely punctate, elytra with rows of coarse punctures. **Length:** .18″. It is common on skin and bones of dead fish and other animals, and is predaceous on dermestid larvae.

Tumbling Flower Beetles (Mordellidae)

The small, wedge-shaped tumbling flower beetles are easily recognized; their silky bodies are compressed laterally and arched, the abdomen is tapered sharply, not entirely covered by the elytra, and ends in a pointed process or anal style. They are common on flowers, and jump or literally tumble frantically when disturbed. The antennae are short, eleven-segmented. The hind legs are long, with enlarged femora, adapted for jumping; the tibiae have large apical spurs, the hind tarsi have four segments, the others have five. The adults are found in dead or partly dead trees or bore into the pith of plants; the assertion that some species are predaceous on the

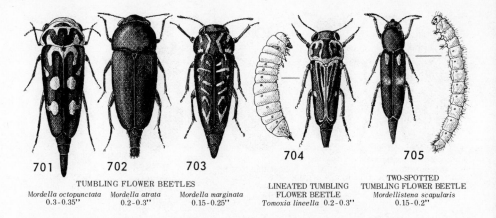

701   702   703   704   705

TUMBLING FLOWER BEETLES
Mordella octopunctata   Mordella atrata   Mordella marginata
0.3-0.35"   0.2-0.3"   0.15-0.25"

LINEATED TUMBLING
FLOWER BEETLE
Tomoxia lineella 0.2-0.3"

TWO-SPOTTED
TUMBLING FLOWER BEETLE
Mordellistena scapularis
0.15-0.2"

larvae of stem- and wood-boring insects has been questioned. Fewer than 200 species are recorded in North America.

(131) EIGHT-SPOTTED TUMBLING FLOWER BEETLE: *Mordella* (*Glipa*) *octopunctata*. **Fig. 701.**

Range: New York, Pennsylvania, Virginia, the central and southern states, to Kansas and Texas. **Adult:** Black or dark grayish, elytra marked with eight spots of yellowish pubescence; pronotum marked with yellowish pubescence as shown; finely punctate. **Length:** .3–.35". It is common on flowers of many species. *M. atrata* (Fig. 702) is black, with brownish pubescence, head and pronotum finely punctate, elytra with network of lines and moderately punctate; from .2" to .3" long; it occurs from Canada to Florida, in all but the western states, is very common on many species of flowers. *M. albosuturalis* closely resembles *atrata*, takes its place in the West; it is black, with fine reddish brown pubescence, differs in having the base and suture of elytra with silvery pubescence, the antennae more tapering to the tip, and the second segment of the maxillary palpi in the male much enlarged; from .2" to .25" long.

(132) TUMBLING FLOWER BEETLE: *Mordella marginata*. **Fig. 703.**

Range: Eastern U.S. and Canada, west to Arizona, Colorado, Manitoba. **Adult:** Black, with grayish pubescence; markings of silvery pubescence as shown (these are variable, sometimes indistinct); head minutely punctate, pronotum and elytra finely punctate, the latter also with network of lines. **Length:** .15–.25". It is abundant on many species of flowers. *Tomoxia lineella* (Fig. 704) is black to reddish brown; markings of silvery pubescence as shown; head finely punctate, pronotum coarsely punctate, elytra with rather coarse deep punctures; from .2" to .3" long; it occurs in the central and eastern states, New Brunswick; abundant in some localities, found on trees such as elm, linden, ash, birch, and hickory from June to July.

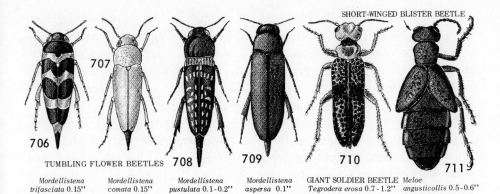

SHORT-WINGED BLISTER BEETLE

707

706

TUMBLING FLOWER BEETLES 708   709   710

711

Mordellistena
trifasciata 0.15"

Mordellistena
comata 0.15"

Mordellistena
pustulata 0.1-0.2"

Mordellistena
aspersa 0.1"

GIANT SOLDIER BEETLE
Tegrodera erosa 0.7 - 1.2"

Meloe
angusticollis 0.5-0.6"

(133) TWO-SPOTTED TUMBLING FLOWER BEETLE: *Mordellistena scapularis.*
Fig. 705.

Range: Nova Scotia to Maryland, west to Manitoba, Illinois. **Adult:** Black, with two reddish or reddish yellow spots at base of elytra; anal style reddish yellow. **Length:** .15–.2". It is common on various flowers from May to August. *M. trifasciata* (Fig. 706) is black, with two transverse yellowish bands on elytra and one across basal margin of pronotum; head yellow; about .15" long; it is widespread east of the Rocky Mts., found on ragweed and *Aralia*. *M. comata* (Fig. 707) has reddish head and pronotum, elytra with yellow pubescence, occurs from Florida to southeastern California. *M. pustulata* (Fig. 708) is black, with rows of small spots across the elytra formed by silvery pubescence; it occurs throughout the U.S. and Canada. *M. aspersa* (Fig. 709) is similar in shape, uniformly black or dark brown, with brownish gray pubescence, about .1" long, widely distributed throughout the U.S. and Canada; it is easily recognized by its small size and dark color; the larvae have been found boring into galls.

Blister Beetles (Meloidae)

The blister beetles constitute a small family, so named because the crushed insect, owing to the presence of a vesicant called cantharidin, will cause blisters on the human skin. This substance has been variously used as a counterirritant, diuretic, and aphrodisiac. Potentially dangerous, its use as a drug is largely discredited today. Blister beetles may be distinguished by the soft, elongated body, narrow neck and deflected head (often larger than the thorax), long legs, the hind tarsi with four segments, the others with five. The antennae usually have eleven segments, rarely eight or nine. The wings are well developed in most cases but are reduced or absent in a few species. The elytra are sometimes short, not always joined in the middle when folded, the posterior segments of the abdomen often being exposed. Colors are usually black, blue, purplish, green, or brownish, with metallic iridescence. Adult blister beetles are phytophagous and include some serious crop pests. The larvae of a majority of known species are predaceous on the egg pods of grasshoppers; slightly fewer species are parasitic in the cells of bees of the families Megachilidae and Andrenidae. The economic status of the family as a whole is thus debatable.

The cylindrical eggs are laid in batches in shallow burrows in the soil near the breeding grounds of the host grasshoppers or nests of the bees; other species parasitic on bees deposit their eggs in the host galleries, while some place them on leaves, stems, or blossoms of plants visited by the bees. Generally, the young larvae that hatch in the soil search for their hosts; those on foliage attach themselves to visiting bees and gain access to their nests in this way—a relationship that is referred to as phoresy. In the latter case egg production is exceedingly high compared to the others.

The larvae undergo a complex development called hypermetamorphosis and have several distinct forms. The first-instar larvae are very active and are commonly called *triungulins*. They are broadest at the thorax and head, with well-developed legs and antennae, and usually have a pair of cerci; the coloring is white, yellow, or orange at first but quickly changes to black. In the second instar they are more robust, with reduced head and legs, and resemble *carabid* larvae. In the third and fourth instars they become gorged; the abdomen is greatly enlarged, the legs are further reduced, and they resemble *scarabaeid* larvae. In the fifth or *coarctate* stage the larvae of the egg predators are markedly different from the preceding instars—legs and mouthparts are rudimentary, and body segmentation indistinct. The sixth instars are greatly reduced in size, with legs larger though still functionless, and resemble *scolytid* larvae. The coarctate or "pseudopupal" stage is essentially a resting stage and one in which hibernation and diapause usually take place. Those that develop in bees' nests cover themselves with exuviae left from the fourth and fifth molts, and take no food after the fourth molt. The length of the life cycle varies greatly among different species, requires from 35 to 50 days, and up to three years at the extremes; the majority have a single generation in a year, correlated with the host's cycle.

(134) GIANT SOLDIER BEETLE: *Tegrodera erosa* (Meloinae). **Fig. 710.**

Range: Arizona, New Mexico, southern California. **Adult:** Head and pronotum red, remainder of body, legs, antennae black; elytra reticulated (having a network of lines), yellow, with black margins, suture, and median band. **Length:** .7–1.2″. **Food:** Sagebrush and other desert plants, also alfalfa. *T. latecincta* is similar, with head and prothorax darker, black markings of elytra more pronounced. The two species are confined to the Southwest.

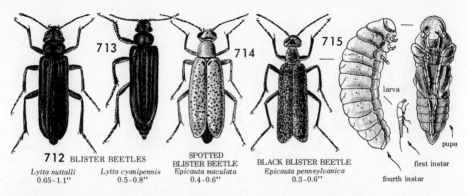

712 BLISTER BEETLES

Lytta nuttalli
0.65-1.1"

Lytta cyanipennis
0.5-0.8"

SPOTTED
BLISTER BEETLE
Epicauta maculata
0.4-0.6"

BLACK BLISTER BEETLE
Epicauta pennsylvanica
0.3-0.6"

larva

pupa

first instar

fourth instar

(135) SHORT-WINGED BLISTER BEETLE: *Meloe angusticollis.* **Fig. 711.**

Species in this genus are collectively known as oil beetles because they exude a yellow oily secretion from leg joints when disturbed; they feign death by remaining upright or falling on the side. **Range:** *M. angusticollis* occurs across Canada and northern U.S. **Adult:** Head and pronotum dark blue; abdomen and elytra violaceous, the latter short and overlapping, with surface wrinkled. The wings are missing. **Length:** .5–.6". It is found on low herbaceous plants in the spring and fall, and frequently feeds on potatoes. *M. impressus* is similar—about .6" to 1" long—shiny, bluish black, the elytra conspicuously punctate. The larvae of these species hatch in the soil and are parasitic on wild bees; they climb up the plants and hitchhike to the nests of their hosts.

(136) NUTTALL BLISTER BEETLE: *Lytta nuttalli.* **Fig. 712.**

Range: Rocky Mts. region: Montana, Wyoming, the Prairie Provinces, Idaho, Colorado, New Mexico. **Adult:** Metallic green or purplish; margins and often most of elytra violet; elytra with sparse short pubescence, antennae black. **Length:** .65–1.1". **Food:** Adults feed on legumes and are very destructive; the larvae are predaceous on grasshopper egg pods. The **Green Blister Beetle,** *L. cyanipennis* (Fig. 713), is similar, somewhat smaller, has the same habits, extends into Utah, Oregon, Idaho, Washington, British Columbia, California. *L. immaculata* is black, with ash gray or yellowish pubescence, about 1" long, appears in June, often in large swarms; it is a serious pest of potatoes, tomatoes, beets, and cabbage in the Midwest, westward to Colorado and New Mexico. These species closely resemble the European *L. vesicatoria*—the **Spanishfly**—a beautiful iridescent green species common in southwestern Europe which feeds chiefly on privet, lilac, and ash. The **Oil Beetle,** *Megetra cancellata* (Color plate 7g), is very common in Arizona, New Mexico, and western Texas, where it is found crawling on the sandy ground just before sunset in late summer and early fall.

(137) SPOTTED BLISTER BEETLE: *Epicauta maculata.* **Fig. 714.**

Range: Iowa, Colorado, northward to North Dakota and Montana. The most common and widely distributed species found in the grasshopper egg

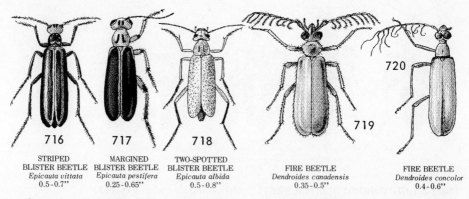

| STRIPED BLISTER BEETLE *Epicauta vittata* 0.5-0.7" | MARGINED BLISTER BEETLE *Epicauta pestifera* 0.25-0.65" | TWO-SPOTTED BLISTER BEETLE *Epicauta albida* 0.5-0.8" | FIRE BEETLE *Dendroides canadensis* 0.35-0.5" | FIRE BEETLE *Dendroides concolor* 0.4-0.6" |

pod surveys.* **Adult:** Black, with dense grayish pubescence excepting spots where the black shows. **Length:** .4–.6". **Food:** Adults are most destructive to beets, potatoes, beans, alfalfa, clover. One larva is capable of destroying a single egg pod (up to 30 eggs) of the migratory and clear-winged grasshoppers, nearly half in the case of the differential and two-striped grasshopper (each pod contains from 75 to 100 eggs). The consequence of this association is that populations of *Epicauta* fluctuate with those of the grasshoppers.

(138) BLACK BLISTER BEETLE: *Epicauta pennsylvanica.* Fig. 715.

Sometimes called the black aster bug. **Range:** Widespread over most of the U.S., very common. **Adult:** All black with sparse pubescence, finely punctate. **Length:** .3–.6". **Food:** Flowers and foliage of many cultivated ornamental plants, especially asters; also clematis, chrysanthemum, dahlia, dianthus, calendula, gladiolus, phlox, zinnia; common on goldenrod. It is destructive to beets, potatoes, tomatoes; a common species encountered in Iowa and westward in the grasshopper egg pod surveys [see (137) above].

(139) STRIPED BLISTER BEETLE: *Epicauta vittata.* Fig. 716.

Range: Central and northeastern U.S., eastern Canada. **Adult:** Yellow, with black stripes, head and pronotum striped dark brown or black; body underneath and legs black. **Length:** .5–.7". Habits similar to (138).

(140) MARGINED BLISTER BEETLE: *Epicauta pestifera.* Fig. 717.

Range: Same as (139). **Adult:** Elytra black, margined with ash gray pubescence; head and pronotum with dense gray pubescence, and two black spots on each. **Length:** .25–.65". Habits similar to (138). The **Two-spotted Blister Beetle,** *E. albida* (Fig. 718), is grayish or yellowish, pronotum with two dark longitudinal lines, from .5" to .8" long; it is a serious pest of sugar beets and garden crops west of the Missouri River.

---

* See J. R. Parker and Claude Wakeland, *Grasshopper Egg Pods Destroyed by Larvae of Bee Flies, Blister Beetles, and Ground Beetles,* Tech. Bul. No. 1165, U.S. Dept. Agr., Washington, D.C. (1957).

(141) CLEMATIS BLISTER BEETLE: *Epicauta cinerea.* **Color plate 7f.**

Also referred to as the gray blister beetle. **Range:** Same as (138). **Adult:** Black, with dense coat of gray hairs. **Length:** .4–.7″. Habits similar to (138). The **Caragana Blister Beetle,** *E. subglabra,* is similar, black, with short, sparse dark brown pubescence, from .25″ to .5″ long, ranges westward to Idaho and British Columbia. The **Ash-gray Blister Beetle,** *E. fabricii,* is also similar—from .3″ to .6″ long—black or dark brown, clothed with dense grayish hairs; the second most common species found in the grasshopper egg pod surveys [see (137) above]. *E. marginata* is black, elytra with pale yellowish margin.

(142) FIRE BEETLE: *Dendroides canadensis* (Pyrochroidae). **Fig. 719.**

The fire beetles, or fire-colored beetles, comprise a small family of flattened, soft-bodied beetles with deflected head, constricted between the eyes, and contrasting colors of black and red or yellow. The adults are found on shrubs, flowers, dead and decaying deciduous or coniferous trees; the predaceous larvae are found under the bark. **Range:** *D. canadensis* is widespread throughout the U.S. and Canada. **Adult:** Head, antennae, and elytra black, with reddish tinge; pronotum reddish yellow. The branches of the antennae are much longer in the male than in the female. **Length:** .35–.5″. *D. concolor* (Fig. 720), an eastern species, is slightly smaller, all brownish yellow. *Neopyrochroa flabellata* (Fig. 721) is dark brown or black, with head, basal segments of antennae, thorax, and legs yellowish, from .5″ to .6″ long; it is widely distributed, found on foliage in woods, occasionally under loose bark. *N. femoralis* (Fig. 722) is slightly smaller, with head, thorax, and femora yellowish red; it is widely distributed.

Ant-like Flower Beetles (Anthicidae)

The ant-like flower beetles are small in size, with head sharply deflexed, mandibles small and curved, pronotum ovoid and often sharply extended over the head in a prominent horn. They have large trochanters, slender femora, hind tarsi with four segments, middle and front tarsi with five segments. The larvae are whitish, elongated, with nine abdominal segments, a few long setae on most segments, three-segmented antennae, and one pair of ocelli. The adults are found on flowers and foliage, under stones or in the litter; the larvae are usually found in decaying vegetable matter.

(143) ANT-LIKE FLOWER BEETLE: *Notoxus constrictus.* **Fig. 723.**

Range: California and Oregon. **Adult:** Pale yellowish brown; elytra with broad black median band and black apexes. **Length:** .12–.15″. It feeds on damaged fruit. The larvae do not develop in dried fruit. It is sometimes injurious to eggplant. *N. monodon* is brownish yellow, with black spot near

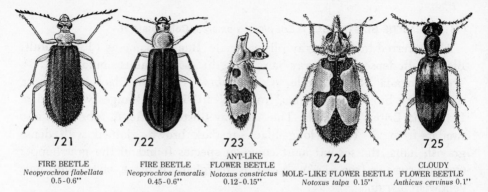

721
FIRE BEETLE
Neopyrochroa flabellata
0.5-0.6"

722
FIRE BEETLE
Neopyrochroa femoralis
0.45-0.6"

723
ANT-LIKE
FLOWER BEETLE
Notoxus constrictus
0.12-0.15"

724
MOLE-LIKE FLOWER BEETLE
Notoxus talpa 0.15"

725
CLOUDY
FLOWER BEETLE
Anthicus cervinus 0.1"

base of each elytron, submedian black band across elytra, apexes sometimes black, long erect hairs and dense pubescence; it occurs throughout much of the U.S., is reported to feed on pupae of the fruit-tree leaf roller and to bore into apple twigs.

(144) MOLE-LIKE FLOWER BEETLE: *Notoxus talpa.* **Fig. 724.**

Range: East of the Rocky Mts. **Adult:** Pronotum, antennae, legs reddish brown; elytra black, with pinkish markings. **Length:** About .15". It is found in oak and hazel, near lakes and marshes.

(145) CLOUDY FLOWER BEETLE: *Anthicus cervinus.* **Fig. 725.**

Range: Throughout North America. **Adult:** Reddish brown, with sparse pubescence, elytral markings blackish; antennae and legs dull yellowish. **Length:** About .1". It is found in litter, under stones in sandy places, in beach drift, and is attracted to lights.

(146) SHINY FLOWER BEETLE: *Anthicus floralis.* **Fig. 726.**

Commonly referred to as the "narrow-necked grain beetle." **Range:** Widely distributed pest of stored grain. **Adult:** Shiny, reddish brown; head and apical two-thirds of elytra black. Pronotum has a pair of small bumps near the front margin. **Length:** .1–.12".

(147) CONSTRICTED FLOWERBUG: *Tomoderus constrictus.* **Fig. 727.**

Range: Central and eastern U.S. **Adult:** Shiny, dark reddish brown to blackish; elytra darker toward apexes; sparsely pubescent. **Length:** About .1".

Click Beetles, Wireworms (Elateridae)

The click beetles—the larvae are called wireworms—comprise a fairly large family, with more than 800 species in North America. They are easily distinguished from any other group by the elongated, somewhat flattened body, the large freely moving prothorax, usually with pointed posterior angles, and their amazing ability to flip upright with a clicking sound when overturned. This feat is performed with a finger-like process on the pro-

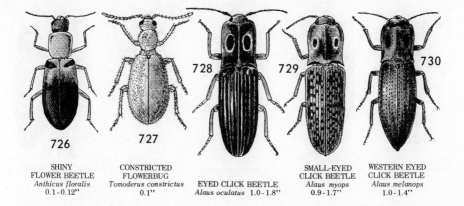

726
SHINY
FLOWER BEETLE
Anthicus floralis
0.1-0.12"

727
CONSTRICTED
FLOWERBUG
Tomoderus constrictus
0.1"

728 729 730

EYED CLICK BEETLE
Alaus oculatus 1.0-1.8"

SMALL-EYED
CLICK BEETLE
Alaus myops
0.9-1.7"

WESTERN EYED
CLICK BEETLE
Alaus melanops
1.0-1.4"

sternum which fits into a socket in the mesosternum. The prothorax and mesothorax are loosely joined, enabling the beetle to bend; the process is pulled out (not quite to the end) by bending the body, and quickly forced back into the socket by straightening the body, causing the basal part of elytra and prothorax to strike against the supporting surface and flip the beetle into the air. The action is repeated several times if necessary until the beetle lands upright.

Click beetles are minute to medium in size, usually brown, gray, or black, with hairy and sometimes scaly vestiture. The antennae are eleven-segmented, inserted close to the eyes. The legs are short, sometimes partly retractile, tarsi five-segmented, wings usually well developed. The larvae are elongate, cylindrical or somewhat depressed, with a tough, shiny, smooth cuticle, and usually reddish brown in color. The abdomen has nine segments, with a tenth reduced and only visible ventrally. The developmental period is usually long—from one to four years or more. They may be distinguished from the false wireworm (darkling beetle) by their shorter legs and less-pronounced segmentation (see p. 417).

Adults are found on the foliage of trees and other plants, in decayed wood, or on the ground. The larvae are found in the soil, under the bark of dead trees, in rotten wood and in moss. The majority of species are subterranean in the larval stage, feed on roots, tubers, stems, and seeds of a wide variety of plants; many are serious crop pests, especially in irrigated lands of the West, and some are predaceous. The tropical "fire beetles"—species of *Pyrophorus*—glow with a brilliant greenish light from two round spots at the basal corners of the pronotum, and a reddish light from the underside of the abdomen; the larvae are also luminous. The Cucubano, *P. luminosa*, of the West Indies is important as a predator of scarabaeid grubs attacking sugar cane. The larva of *Chalcolepidius silbermanni*—a nonluminous species found in the West Indies, Mexico, Central America and northern South America—is also predaceous.

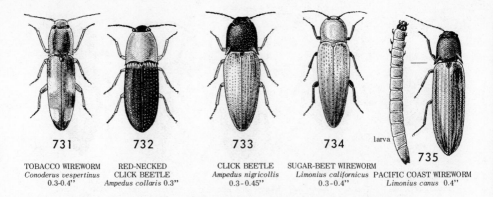

731   732   733   734   larva   735

TOBACCO WIREWORM   RED-NECKED   CLICK BEETLE   SUGAR-BEET WIREWORM
Conoderus vespertinus   CLICK BEETLE   *Ampedus nigricollis*   *Limonius californicus*   PACIFIC COAST WIREWORM
0.3-0.4"   *Ampedus collaris* 0.3"   0.3-0.45"   0.3-0.4"   *Limonius canus* 0.4"

(148) EYED CLICK BEETLE: *Alaus oculatus* (Pyrophorinae). **Fig. 728.**

Also called the big-eyed click beetle or eyed elater. **Range:** East of the Rocky Mts. **Adult:** Shiny black, with a sprinkling of silvery white scales over the whole dorsum. Large oval eyespots circled with gray scales on pronotum. Elytra distinctly striated. It occurs around rotten stumps and logs. **Length:** 1–1.8". **Larva:** Cylindrical, smooth, yellowish, about 2" to 2.5" long. It is found in dead and decaying wood. **Food** (larva): Wood-boring beetles. The **Small-eyed Click Beetle** or **Blind Elater,** *A. myops* (Fig. 729), is reddish brown to black, with sparse gray pubescence; it has smaller eyespots than *oculatus,* and the elytra are finely striated; the western form is darker, with more whitish scales, smaller eyespots.

(149) WESTERN EYED CLICK BEETLE: *Alaus melanops.* **Fig. 730.**

Range: New Mexico, to southern California and British Columbia. **Adult:** Dull black, without white scales on pronotum and around eyespots as in *A. oculatus* (148). **Length:** 1–1.4". **Larva:** Yellowish, with head and thoracic segments dark brown.

(150) TOBACCO WIREWORM: *Conoderus (Monocrepidius) vespertinus.* **Fig. 731.**

Range: Eastern and southern U.S. **Adult:** Variable, usually yellowish, with brown or nearly black markings. **Length:** .3–.4". The larvae are very injurious to tobacco, cotton, potatoes, corn, beans, and other truck crops. *C. amplicollis,* the **Gulf Wireworm,** and *C. falli,* the **Southern Potato Wireworm,** are injurious to many cultivated plants in the southern states. The latter is unusual among click beetles in having a short life cycle; it has at least two generations each year.

(151) RED-NECKED CLICK BEETLE: *Ampedus collaris* (Elaterinae). **Fig. 732.**

Range: Eastern North America. **Adult:** Shiny black, prothorax bright red, antennae and legs reddish brown. **Length:** .3". It is found on foliage, beneath stones and logs. *A. nigricollis* (Fig. 733) is similar—from .3" to .45" long—head and pronotum black, elytra dull yellowish, sparse yellow pu-

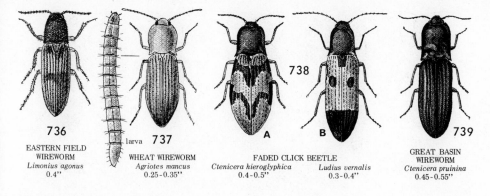

736
EASTERN FIELD
WIREWORM
Limonius agonus
0.4"

larva 737
WHEAT WIREWORM
Agriotes mancus
0.25-0.35"

738

A
FADED CLICK BEETLE
Ctenicera hieroglyphica
0.4-0.5"

B
Ludius vernalis
0.3-0.4"

739
GREAT BASIN
WIREWORM
Ctenicera pruinina
0.45-0.55"

bescence, legs reddish, antennae reddish brown; it is found under bark and in decayed wood.

(152) SUGAR-BEET WIREWORM: *Limonius californicus.* **Fig. 734.**

Range: Western North America. **Adult:** Black, elytra reddish brown, antennae and femora dark brown; dense yellowish or white pubescence; punctate. White elliptical eggs are laid in the soil. **Length:** .3–.4". **Larva:** Cylindrical, shiny, yellowish brown. It requires three years to develop; the entire life cycle takes four years. **Food** (larva): Sugar beets, beans; also potato, corn, various weeds, chrysanthemums, asters, and other ornamental flowers.

(153) PACIFIC COAST WIREWORM: *Limonius canus.* **Fig. 735.**

Range: California to British Columbia, southwestern Alberta. **Adult:** Male black, with elytra reddish brown; female reddish, with dark central spot on head and pronotum; dense white or pale yellowish pubescence; punctate. **Length:** About .4". Habits similar to (152). Related species, serious pests of agricultural crops, are: *L. infuscatus,* the **Western Field Wireworm;** *L. subauratus,* the **Columbia Basin Wireworm;** and *L. agonus,* the **Eastern Field Wireworm** (Fig. 736).

(154) WHEAT WIREWORM: *Agriotes mancus.* **Fig. 737.**

Range: Widely distributed in eastern and central North America. **Adult:** Yellowish brown, with sparse yellow pubescence, head and pronotum often dark brown; head very convex, with broad chisel-like mandibles; pronotum and elytra deeply punctate, the latter striated. **Length:** .25–.35". The larva is very destructive to wheat, other grains, and potatoes; it may be distinguished by the anal segment, cylindrical form, bright yellow color.

(155) FADED CLICK BEETLE: *Ctenicera* (*Ludius*) *hieroglyphica.* **Fig. 738A.**

Range: Central and eastern North America, British Columbia. **Adult:** Head and pronotum black; elytra pale reddish yellow, with dark brown markings; antennae and legs reddish; yellowish pubescence, denser and paler on pronotum. **Length:** .4–.5". It is found on trees and shrubs. *Ludius*

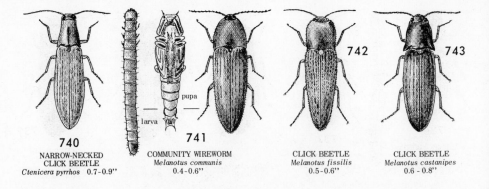

| 740 | 741 | 742 | 743 |

740
NARROW-NECKED
CLICK BEETLE
Ctenicera pyrrhos 0.7-0.9''

741
COMMUNITY WIREWORM
Melanotus communis
0.4-0.6''

CLICK BEETLE
Melanotus fissilis
0.5-0.6''

CLICK BEETLE
Melanotus castanipes
0.6 - 0.8''

vernalis (Fig. 738B) is black, basal two-thirds of elytra pale brownish with black spots, from .3'' to .4'' long; distribution scattered throughout North America.

(156) GREAT BASIN WIREWORM: *Ctenicera pruinina.* **Fig. 739.**

Range: The Great Basin and Pacific Northwest. **Adult:** All black, somewhat flattened; elytra striated and punctate. **Length:** .45–.55''. The larvae are injurious to potatoes and various crop plants. The **Dry-land Wireworm,** *C. glauca,* is similar, black, with dense white pubescence; pronotum convex, arched, wider than long, sides curved; from .25'' to .35'' long; adults feed on various prairie plants. The **Puget Sound Wireworm,** *C. aeripennis aeripennis,* is black, with elytra dark metallic green, bluish bronze or coppery, from .4'' to .6'' long, not as slender as *pruinina;* the elytra of the **Prairie Wireworm,** *C. a. destructor,* have a faint bluish or greenish tinge, are less wrinkled and have deeper punctures; they are destructive to grains and field crops in the West.

(157) NARROW-NECKED CLICK BEETLE: *Ctenicera pyrrhos.* **Fig. 740.**

Range: Eastern North America. **Adult:** Dark reddish brown, with pale yellow pubescence; elytra punctate and deeply striated. **Length:** .7–.9''. It is found on hickory, walnut, and other trees.

(158) COMMUNITY WIREWORM: *Melanotus communis* (Melanotinae). **Fig. 741.**

Range: Eastern half of North America. **Adult:** Reddish brown, with sparse pubescence; finely punctate. The third antennal segment is rarely much shorter than the fourth. **Length:** .4–.6''. The larvae are injurious to potatoes and other crop plants; they may be distinguished from wheat wireworms (154) by the reddish brown anal segment. *M. fissilis* (Fig. 742), widespread over most of North America—from .5'' to .6'' long—is brown to blackish, with sparse yellowish pubescence, difficult to separate from *communis;* the third antennal segment is nearly twice as long as second, the second and third together as long as the fourth, the pronotum coarsely punctate. *M. castanipes* (Fig. 743), also widely distributed, is from .6″ to .8″ long, dark reddish brown, with sparse yellowish pubescence; the third antennal segment is a little longer than the second, the second and third together about three-fourths as long as the fourth. These species are commonly found under loose bark, in the ground litter, or under dry manure in pastures.

Flatheaded or Metallic Wood Borers (Buprestidae)

This is a fairly large family—somewhat fewer than 700 species known in the U.S.—named for the beautiful metallic luster of the adult beetles and the broad flat enlargement of the larval thoracic segments, giving them a "flatheaded" appearance. The adults are depressed, hard-bodied, often brightly colored, usually with a transverse suture on the metasternum near the coxal plates. The head is deflected, with mandibles small and curved, eyes lateral, large and oval, antennae eleven-segmented, not set close to the eyes or mandibles. The abdomen has five visible sternites. Eggs are laid on the bark of twigs or branches of trees and are sometimes quite large.

The larvae are white or yellowish white, slender, deeply notched, tapered toward the posterior end, with ten abdominal segments ending in two fleshy lobes or sharp forceps;* the enlarged thoracic segments have horny plates on the top and bottom, and are without legs. The head is small, retracted somewhat, without ocelli, with mandibles stout, spoon-shaped, and toothed. They have been likened to a horseshoe nail in appearance, are usually found in a curved position. The larvae mine under the bark of dead, dying, or declining coniferous and deciduous trees and bushes. The adults—strong fliers, and noisy in flight—feed on leaves, excepting some species of *Chrysobothris* and *Agrilus* which feed on fungi.

(159) HAIRY BUPRESTID: *Acmaeodera pulchella* (Acmaeoderinae). **Fig. 744.**

Range: Widely distributed over eastern North America. **Adult:** Brown or bronzy black, with orange and yellow markings, brownish pubescence; punctate. **Length:** .2–.45″. Adults are common on flowers of New Jersey tea (*Ceanothus*) and other bushes. *A. connexa*—from .3″ to .5″ long—has numerous yellow curved markings and spots on elytra; the larvae mine injured oaks from Oregon to Colorado, Texas, and California.

(160) FLATHEADED CONE BORER: *Chrysophana placida.* **Fig. 745.**

Range: British Columbia to Nebraska, Colorado, New Mexico, California. **Adult:** Smooth, without punctures or hairs. Three color phases: striped, with longitudinal reddish stripe down each elytron (the most common); all green, sometimes shading to bluish green; dark bronze to black (rare).

---

* The forceps are used to grasp fecal pellets and pack them in the back of the tunnel.

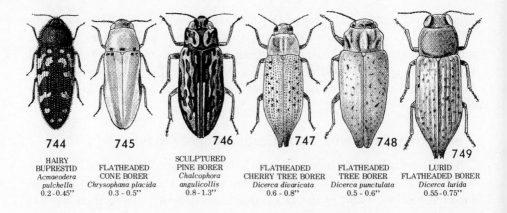

744
HAIRY
BUPRESTID
Acmaeodera
pulchella
0.2-0.45"

745
FLATHEADED
CONE BORER
Chrysophana placida
0.3 - 0.5"

746
SCULPTURED
PINE BORER
Chalcophora
angulicollis
0.8-1.3"

747
FLATHEADED
CHERRY TREE BORER
Dicerca divaricata
0.6 - 0.8"

748
FLATHEADED
TREE BORER
Dicerca punctulata
0.5 - 0.6"

749
LURID
FLATHEADED BORER
Dicerca lurida
0.55-0.75"

Length: .3–.5". The larva lives in cones of knobcone and ponderosa pines, also wood of various pines, hemlock, fir; it sometimes attacks buildings around the doors or windows.

(161) SCULPTURED PINE BORER: *Chalcophora angulicollis* (Buprestinae). **Fig. 746.**

This is one of the largest of our flatheaded borers, and the only western species of the genus. **Range:** California to British Columbia and Idaho, to South Dakota and Texas. **Adult:** Dark brown to black, upper surface marked with irregular sculptured areas; iridescent bronze luster, especially on underside. They take off suddenly in noisy flight when alarmed. **Length:** .8–1.3". The larvae feed on dead pines, fir, Douglas fir. The **Virginia Pine Borer,** *C. virginiensis,* an eastern species, is very similar—from .8" to 1.2" long—black with rough sculptures, slightly bronzed.

(162) FLATHEADED CHERRY TREE BORER: *Dicerca divaricata.* **Fig. 747.**

Range: Throughout North America. **Adult:** Brown or grayish, with brassy tinge; coarsely punctate. **Length:** .6–.8". Infests cherry, birch, alder, apple, peach, pear, and other deciduous trees. *D. punctulata* (Fig. 748) is from .5" to .6" long, grayish brown, with bronze sheen, occurs in pitch pine in the eastern and central states. *D. tenebrosa,* bronze or gray, elytra with scattered shiny black ridges and bumps—from .6" to .75" long—breeds in sickly, dying, or dead pines, northern white cedar, true fir, Douglas fir, Engelmann spruce from Alaska to California, Colorado, Minnesota, Wisconsin, and Maine.

(163) LURID FLATHEADED BORER: *Dicerca lurida.* **Fig. 749.**

Range: Eastern North America. **Adult:** Shiny, dark brown, with coppery and bronze sheen, grayish pubescence; pronotum finely, densely punctate; elytra finely striated and punctate. **Length:** .55–.75". It is found on hickory and alder.

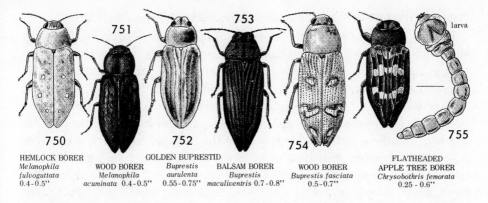

751

753

larva

750
HEMLOCK BORER
Melanophila
fulvoguttata
0.4-0.5"

WOOD BORER
Melanophila
acuminata 0.4-0.5"

752
GOLDEN BUPRESTID
Buprestis
aurulenta
0.55-0.75"

BALSAM BORER
Buprestis
maculiventris 0.7-0.8"

754

WOOD BORER
Buprestis fasciata
0.5-0.7"

755
FLATHEADED
APPLE TREE BORER
Chrysobothris femorata
0.25 - 0.6"

(164) FLATHEADED FIR BORER: *Melanophila drummondi.*

Range: Colorado, Utah, Arizona, New Mexico. **Adult:** Metallic bronze or black, with iridescent sheen, sometimes with light golden spots on elytra. **Length:** .4–.5". It attacks dying or recently felled fir, hemlock, western larch and spruce. The **Hemlock Borer,** *M. fulvoguttata* (Fig. 750), an eastern species, is almost identical, attacks hemlock and spruce. The **California Flatheaded Borer,** *M. californica,* is similar, from .3" to .4" long; the elytra have only a few short hairs, sometimes have one to three spots; it attacks pines from California to Washington and Idaho. *M. acuminata* (Fig. 751)—from .4" to .5" long, dull black—is found on pines, spruces, fir, western red cedar, Monterey cypress; it occurs throughout North America, in Europe, and Siberia. This species, and the **Charcoal Beetle,** *M. consputa,* are attracted to the smoke of forest fires, often gather around the firefighters, biting them and trying desperately to lay their eggs on the smoldering trees; *consputa* is also black, attacks badly scorched and weakened pine, spruce, firs, and other conifers, and hardwoods as well.

(165) GOLDEN BUPRESTID: *Buprestis aurulenta.* **Fig. 752, Color plate 7e.**

Range: Pacific coast and Rocky Mt. regions. **Adult:** Flattened, elytra striated, every other stria suppressed; iridescent green or bluish, suture and margins of elytra coppery. **Length:** .55–.75". The most beautiful and one of most destructive of the mountain species. They are attracted to the pitchy wood of pines, spruces, firs, Douglas fir, and other conifers. In dry wood the larvae take ten or more years to complete development, ten to twenty-six years in some houses according to records. *B. maculiventris* (Fig. 753), an eastern species—from .7" to .8" long—is black, with greenish bronze sheen, yellow lateral spots on abdomen, elytra punctate and striated; it occurs in balsam and spruce. *B. fasciata* (Fig. 754)—from .5" to .7" long— is metallic green-blue, elytra with three oblong markings near the apex in the male and two in the female, margined with black in the latter; it occurs in maple and poplar.

(166) FLATHEADED APPLE TREE BORER: *Chrysobothris femorata.* **Fig. 755.**

Range: Throughout North America. **Adult:** Dark brown with brassy sheen, indistinct gray spots and bands, coarsely punctate, with variable sculpturing. Eggs are laid in cracks or openings in the bark. **Length:** .25–.6″. **Larva:** Pale yellow, about 1″ long full grown; head deeply retracted, last abdominal segment with two lobes. It enters the bark where the egg was deposited, feeds on sapwood of young trees, and on bark of older trees. It burrows into the wood to a depth of 1″ in the fall, and constructs a chamber where it hibernates. Pupation takes place the following spring. They are not able to complete development in vigorous trees, hence do them little harm. **Food:** Apple and other deciduous fruit trees, shade and forest trees. The **Pacific Flatheaded Borer,** *C. mali,* is very similar, may be distinguished by a short lobe on the front margin of the prosternum and by the absence of a groove on the pronotum; it infests *Ceanothus,* alder, apple, walnut, and many other deciduous trees; it is the worst single beetle pest of walnut in California. The **Red-pine Flatheaded Borer,** *C. orno,* attacks living red and jack pines in Michigan, also occurs in Maine, Tennessee, Virginia, and North Carolina; eggs are laid singly under bark scales on southwest side of tree, the larva excavates a cell in the bark, causing yellowish pitch mass on red pine and white one on jack pine, enters the xylem, where it hibernates and pupates; two broods in Michigan, life cycle extends to two years.

(167) TWO-LINED CHESTNUT BORER: *Agrilus bilineatus* (Agrilinae). **Fig. 756.**

Range: Rocky Mt. region and eastward. **Adult:** Black, with greenish or bluish tinge, marked with yellowish brown or bronze stripes; elytra granulate. Eggs are laid in small groups in bark crevices. **Length:** .2–.35″. **Larva:** It bores into cambium, hibernates as full-grown larva in cell made in sapwood. One brood. It attacks injured chestnut, oak, and beech, girdling twigs spirally, causing them to die back. *A. angelicus* girdles twigs of oak similarly in California.

(168) BRONZE BIRCH BORER: *Agrilus anxius.* **Fig. 757.**

Range: Throughout the northern states and southern Canada, also Virginia, New Mexico, Arizona. **Adult:** Greenish black, with bronze tinge. Eggs are laid in groups in bark crevices. **Length:** .25–.5″. **Larva:** White, about .5″ long, with fine hairs on prothorax and spines on abdomen; anal segment with reddish brown, sharp forceps. It cuts a zigzag tunnel through the cambium, leaving much frass, causing limbs to die back, eventually killing the tree. Hibernates as full-grown larva in cell made at end of tunnel, pupates in the spring. One brood. **Food:** Birch, poplar, aspen. The **Bronze Poplar Borer,** *A. liragus,* is very similar; the larva has fewer prothoracic hairs, more spines, shorter anal forceps, and less darkened area; it girdles twigs of various poplars. Vigorous or dead trees are unsuited to the de-

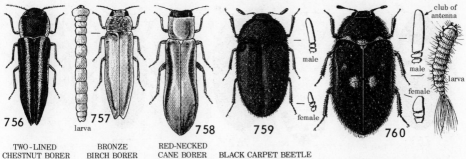

| 756 | 757 larva | 758 | 759 | male female | 760 | club of antenna male female larva |

TWO-LINED
CHESTNUT BORER
Agrilus bilineatus
0.2 - 0.35"

BRONZE
BIRCH BORER
Agrilus anxius
0.25 - 0.5"

RED-NECKED
CANE BORER
Agrilus ruficollis
0.2 - 0.3"

BLACK CARPET BEETLE
Attagenus megatoma
0.15-0.2"

TWO-SPOTTED DERMESTID
Attagenus pellio 0.2-0.25"

velopment of both species. The **Sinuate Pear Tree Borer**, *A. sinuatus*, is reddish purple, about .3" long; the larvae make winding mines under the bark, through the cambium layer, of pear trees.

(169) RED-NECKED CANE BORER: *Agrilus ruficollis*. **Fig. 758.**

Range: Eastern North America. **Adult:** Dull bluish black, with slight metallic luster; head and pronotum coppery red or brassy, elytra finely granulate. Eggs are laid in bark of cane, usually near base of leaf. **Length:** .2–.3". **Larva:** Slender, white, about .5" long full grown. It burrows upward in the sapwood and around the cane, often girdling it several times, and causing a gall or swelling. [See (344) below.] Hibernates as grub inside pith of the cane; completes growth and pupates in the spring. Adults emerge in May and June. One brood. **Food:** Raspberry, blackberry, dewberry. The **Rose Stem Borer**, *A. rubicola* (also called rose stem girdler or bronze cane borer), is similar, bronze-green; it prefers roses but also attacks raspberry in Utah and east of the Mississippi River.

Dermestid Beetles (Dermestidae)

The dermestid beetles comprise a small family, sometimes referred to as skin beetles, carpet beetles, or buffalo bugs. They are minute to small in size, generally oval and convex in shape, with a dense coating of hairs or scales. The head is small, deflexed, retractile as far as the eyes, with one median ocellus, and short, clubbed antennae (with five to eleven segments) that fit into a groove; the legs are short, retractile. The abdomen in most cases has only five visible sternites. The larvae are elongate or ovate, usually brown, covered with long hairs; they have nine or ten abdominal segments (the last only partial), up to six pairs of ocelli. They usually feed on dead animals and plant materials such as skin, furs, woolens, museum specimens (including insect collections), stored grains and other foodstuffs. Many adults are found on flowers, and a few species are known to be predaceous on the egg cases of mantids and egg masses of lepidopterous insects.

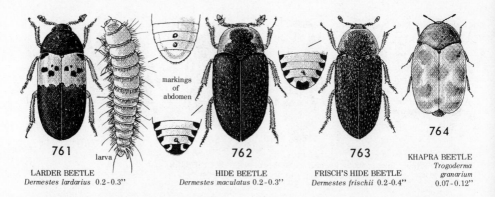

markings
of
abdomen

761
larva

LARDER BEETLE
Dermestes lardarius 0.2-0.3"

762
HIDE BEETLE
Dermestes maculatus 0.2-0.3"

763
FRISCH'S HIDE BEETLE
Dermestes frischii 0.2-0.4"

764
KHAPRA BEETLE
*Trogoderma
granarium*
0.07-0.12"

(170) BLACK CARPET BEETLE: *Attagenus megatoma* (*piceus*) (Attagenini). **Fig. 759.**

Range: Cosmopolitan. **Adult:** Head and pronotum black; elytra black or dark reddish brown, with short hairs, finely punctate; legs and antennae dark yellow. It feeds on pollen, lives only two to four weeks; a female lays up to 100 eggs. **Length:** .15–.2". **Larva:** Elongate, about .4" long, reddish or golden brown, with short scalelike hairs and long tuft of hairs at tip of abdomen. Life cycle requires from one to three years; the larval period is prolonged under unfavorable conditions. **Food:** Carpets, woolens, silks, felts, skins, feathers, milk powder, grain, cereals, seeds, insect collections. Most destructive of the carpet beetles. *A. pellio* (Fig. 760) is dark reddish brown, elytra with whitish markings, from .2" to .25" long; adults visit flowers, the larvae live in birds' nests; it occurs in Europe, Asia, and Africa as well as North America.

(171) LARDER BEETLE: *Dermestes lardarius* (Dermestini). **Fig. 761.**

Range: Cosmopolitan. **Adult:** Dark brown, with wide yellow band across base of elytra, dense coating of hairs; finely punctate. Eggs are laid near animal products. **Length:** .2–.3". **Larva:** Brown, with dense coating of hairs. Life cycle completed in 40 to 50 days normally. **Food:** Cured ham and bacon, cheese, dried fish, dog biscuits, stored tobacco, museum specimens. The **Hide Beetle,** *D. maculatus* (Fig. 762), is about the same size, black, elytra with black pubescence, scattered gray or reddish hairs, pronotum with wide lateral margins of dense gray pubescence; museums maintain cultures for cleaning bones. *D. frischii* (Fig. 763) is blackish, with gray pubescence, from .2" to .4" long; it is widely distributed, feeds on bones, carcasses, carrion, dried skins, and dead insects; F. W. Hope reported it found on Egyptian mummies in 1834.

(172) KHAPRA BEETLE: *Trogoderma granarium* (Anthrenini). **Fig. 764.**

Range: Now eradicated from the U.S. **Adult:** Pale red to brown or black; females much larger than males. **Length:** .07–.12". **Larva:** Creamy white to yellowish brown, with long hairs; second antennal segment rela-

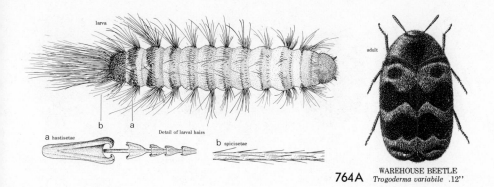

larva

adult

b   a

Detail of larval hairs

a hastisetae

b spicisetae

764A   WAREHOUSE BEETLE
       Trogoderma variabile .12"

tively short, basal segment nearly surrounded by setae. The larvae can survive for years without food. Life cycle normally requires about six weeks in warm weather. **Food:** Dried vegetable or animal matter of all kinds. A serious pest of stored grain, first found in California in 1953, apparently coming from India. The **Larger Cabinet Beetle,** *T. versicolor* (Fig. 765), is black, with reddish brown mottling and pattern of narrow, curving bands formed by gray and light brown scalelike hairs; about .12" long; the larva—same length as adult—is reddish brown, with short yellowish brown hairs and tuft on tip of abdomen, thrives on wholly vegetable diet. *T. glabrum* infests stored wheat, corn, cereal products, and cigarettes. *Trogoderma* spp. infest nests of leafcutter bees (*Megachile*). **The Warehouse Beetle,** *Trogoderma variabile* (*parabile*) (Fig. 764A), is a major pest of stored foods in many parts of the world. It is now found generally throughout the United States, with the exception of a few eastern and southern states. The adult is brownish black, about .12" long. The larva, which is about .25" long, has sharp, pointed setae (hairs) that are irritating to sensitive persons who may contact or swallow them in food.

(173) CARPET BEETLE: *Anthrenus scrophulariae.* **Fig. 766.**

Range: Cosmopolitan. **Adult:** Black, with white, brown, and yellowish scales. It sometimes appears on windows of infested buildings, is abundant on wild flowers such as wild carrot, buckwheat, *Ceanothus*. White eggs are laid on food of larvae. **Length:** .12–.15". **Larva:** Yellowish brown, with long brown hairs. **Food:** Carpets, woolens; feathers, horn, hair and other dried animal products; museum specimens and insect collections.

(174) VARIED CARPET BEETLE: *Anthrenus verbasci.* **Fig. 767.**

Range: Cosmopolitan. **Adult:** Black, with zigzag bands across elytra formed by white scales bordered with yellow; antennae eleven-segmented, eyes emarginate. **Length:** .1–.12". Adults frequent flowers, the larvae feed on dried vegetable and animal products and are often found in granaries

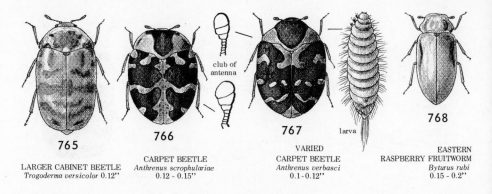

club of
antenna

765

LARGER CABINET BEETLE
Trogoderma versicolor 0.12"

766

CARPET BEETLE
Anthrenus scrophulariae
0.12 - 0.15"

767

VARIED
CARPET BEETLE
Anthrenus verbasci
0.1 - 0.12"

larva

768

EASTERN
RASPBERRY FRUITWORM
Byturus rubi
0.15 - 0.2"

and flour mills. *A. coloratus* occurs in Asia, Europe, and Africa, has been found in scattered locations in the U.S.; it is similar to *verbasci* in color, but smaller, with nine-segmented antennae, and the eyes are not emarginate.

(175) EASTERN RASPBERRY FRUITWORM: *Byturus rubi* (Byturidae). **Fig. 768.**

This genus—commonly called the fruitworm beetles—is included with the Dermestidae in some classifications; it differs in having the second and third tarsal joints with a membranous lobe underneath, the front coxal cavities closed behind, and tarsal claws toothed. **Range:** *B. rubi* occurs in northern parts of the eastern states and southern Canada. **Adult:** Light brown, with pale yellowish pubescence; head and underside dark brown, with whitish hairs; numerous coarse punctures. **Length:** .15–.2". **Larva:** Slender, about .3" long; white, with brown area on upper side of each segment, prolegs at tip of abdomen. It pupates in the soil, hibernates as adult in the soil, emerges in April. **Food:** Adult and grub stages feed on buds, blossoms, tender leaves of raspberry, blackberry, loganberry. *B. bakeri,* the **Western Raspberry Fruitworm,** is very similar; it is a serious pest of raspberries and loganberries in the Pacific Northwest.

(176) AMERICAN PILL BEETLE: *Byrrhus americanus* (Byrrhidae). **Fig. 769.**

Previously included with the Dermestidae. These are very convex, oval beetles, with deflexed head, usually shiny black, with eleven-segmented clubbed antennae, short legs; front coxal cavities open behind. When disturbed the beetles retract their antennae and legs, assume the form of a ball, thus suggesting the name, pill beetles. They live under stones or logs, in moss, near lakes or streams, among the roots of grasses, grains, seedlings in nurseries, and weeds; they are phytophagous in the grub and adult stages. *B. americanus* occurs throughout most of the northern states and Canada, excepting the Far West. **Adult:** Black, with dense grayish pubescence, gray and black striped; elytra finely striated, sparsely punctate; pronotum densely punctate. **Length:** .3–.35". *Amphicyrta dentipes* (Fig. 770), a western species, is shiny brown or black; the larva is the same

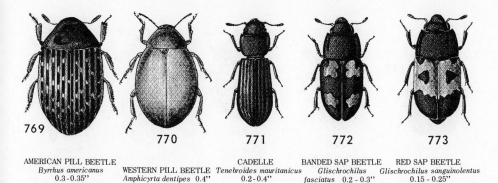

AMERICAN PILL BEETLE
Byrrhus americanus
0.3-0.35"

WESTERN PILL BEETLE
Amphicyrta dentipes 0.4"

CADELLE
Tenebroides mauritanicus
0.2-0.4"

BANDED SAP BEETLE
Glischrochilus fasciatus 0.2 - 0.3"

RED SAP BEETLE
Glischrochilus sanguinolentus
0.15 - 0.25"

color, about .4" long; it emerges from the ground to feed on grasses, clover, grains, weeds.

(177) CADELLE: *Tenebroides mauritanicus* (Ostomidae). **Fig. 771.**

The ostomatid beetles, also called grain and bark-gnawing beetles, may be recognized by the typical body shape. They occur mostly under bark in the galleries of wood-boring insects, and are predaceous, or feed on fungi. Some species, such as our example, are found in granaries. **Range:** *T. mauritanicus* is cosmopolitan, having been carried to all parts of the world. **Adult:** Flattened, head and thorax distinctly separated; shiny black to brown; head and pronotum densely and coarsely punctate, elytra deeply striated. Eggs are laid near food supply, hidden in cracks and other places. **Length:** .2–.4". This is the largest insect, next to the mealworm, attacking stored grains. **Larva:** Elongate, deeply segmented; grayish white, yellow, or pinkish, with squarely cut black head, black spots on thoracic segments, a pair of dark hooklike anal appendages. Larva and adults bore into wood between shipments of grain. **Food:** Grains, cereals, nuts. A serious pest in flour mills, granaries, warehouses, ships, and stores.

Sap Beetles (Nitidulidae)

This is a small family of minute to small beetles, broad, shiny (whence the family name), somewhat flattened, with large head and eyes; the antennae are short and have eleven segments, the last three always clubbed. The elytra are often shortened, leaving some segments or tip of abdomen exposed, and the pronotum is usually wider than long. The tarsi are five-segmented (the fourth segment small), the tibiae dilated, procoxal cavities transverse, metacoxae grooved. The larvae are elongate, usually white, with several groups of short setae and spines, three or four pairs of ocelli (sometimes reduced to pigment spots), nine abdominal segments (sometimes a small tenth segment modified as a proleg), often with a pair of short pointed appendages at the tip. They pupate in the soil and hibernate as adults.

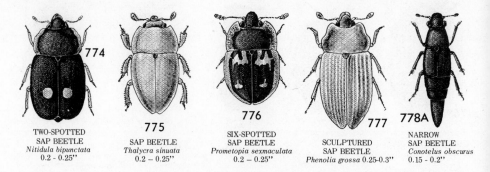

TWO-SPOTTED
SAP BEETLE
Nitidula bipunctata
0.2 - 0.25"

775
SAP BEETLE
Thalycra sinuata
0.2 – 0.25"

776
SIX-SPOTTED
SAP BEETLE
Prometopia sexmaculata
0.2 – 0.25"

777
SCULPTURED
SAP BEETLE
Phenolia grossa 0.25-0.3"

778A
NARROW
SAP BEETLE
Conotelus obscurus
0.15 - 0.2"

Adults and larvae are mostly saprophagous, feeding on decaying plant materials, sap oozing from plants, and fungi; they are found in dried or fermenting fruit, garbage, under dead bark, in the galleries of wood-boring insects and nests of ants. A few are carnivorous; the genus *Cypocephalus* is believed to be entirely predaceous, mostly on scale insects.

(178) BANDED SAP BEETLE: *Glischrochilus fasciatus* (Cryptarchinae). **Fig. 772.**

Range: Widely distributed over North America. **Adult:** Shiny black, with two yellowish or reddish bands, varying in size, on each elytron; finely punctate. **Length:** .2–.3". It is found under the bark of dying trees and often in garbage cans, feeds on sap, fungi, and decaying vegetable matter. The **Red Sap Beetle,** *G. sanguinolentus* (Fig. 773), is shiny black, from .15" to .25" long, with red elytra and black spot on middle of each elytron, tips black; last abdominal segment partly exposed in the female, covered in the male. *G. quadrisignatus* is reddish brown, with two pale yellow spots on each elytron, causes damage by entering tips of cobs of corn and cracks and injuries in fruits and tomatoes (it is also known to enter sound apples), has one brood in northern part of range. *Glischrochilus* spp. are sometimes called picnic beetles because of their intrusions into picnic grounds and barbecues; they are often a problem around fruit stands, can be dealt with effectively in small areas by luring with fermenting fruits and trapping.

(179) TWO-SPOTTED SAP BEETLE: *Nitidula bipunctata* (Nitidulinae). **Fig. 774.**

Range: Throughout North America, also Europe. **Adult:** Black, with reddish spot on each elytron, and fine pubescence. **Length:** .2–.25". Commonly found on bones and skin of animal carcasses.

(180) SAP BEETLE: *Thalycra sinuata.* **Fig. 775.**

Range: West of the Rocky Mts. **Adult:** Brown or dark reddish brown; pronotum evenly punctate, with lateral band of short setae; elytral punctures not deep and regular, with setae arising from punctures evident laterally and basally only. Distinct spines on outer margin of middle and hind tibiae. **Length:** .2–.25". **Larva:** Pale yellowish white, from .2" to .3" long, with numerous fine tubercles and a pair of short pointed processes on the ninth abdominal segment; single small ocellus at base of each antenna. **Food:** Unknown, as in the case of most species of this genus; some are known to feed on fungi. Adults usually fly at dusk and are attracted to traps baited with fermenting malt or molasses, in areas with coniferous trees.

(181) SIX-SPOTTED SAP BEETLE: *Prometopia sexmaculata*. **Fig. 776.**

Range: East of the Rocky Mts. **Adult:** Black, with pale reddish brown margins and markings; coarsely punctate, with intervening fine punctures on pronotum. **Length:** .2–.25″. It feeds on sap.

(182) SCULPTURED SAP BEETLE: *Phenolia grossa*. **Fig. 777.**

Range: Widely distributed east of the Rocky Mts. **Adult:** Black, with indistinct reddish markings; longitudinal raised ribs (costae) finely punctate, each puncture with short hair, three rows of faint punctures in between. **Length:** .25–.3″. It is found under bark, and feeds on fungi.

(183) NARROW SAP BEETLE: *Conotelus obscurus* (Carpophilinae). **Fig. 778A.**

Range: East of the Rocky Mts. **Adult:** Black, with brownish yellow legs and antennae (except club, which is blackish), irregular punctures. **Length:** .15–.2″. It is found on flowers of hollyhock, dandelion, dogwood, morning glory, sweet potato, *Hibiscus*, bindweed. The **Fruit Bud Beetle,** *C. mexicanus,* is similar, black, about .1″ long, attacks flowers of roses, peach, blackberry, citrus, and cotton in Mexico, Arizona, and California. *C. stenoides* (Fig. 778B) is brown to black, about .15″ long; elytra about as wide as long, with fine striae and coarse punctures, abdomen acutely margined and sparsely punctate, pygidium truncate and not emarginate in the male; it is abundant on sweet and field corn in the South.

(184) DRIED-FRUIT BEETLE: *Carpophilus hemipterus*. **Fig. 779B.**

Range: Cosmopolitan; carried in foodstuffs to all parts of the world, occurs in California, Arizona, New Mexico, Gulf states and northward. **Adult:** Black, with dull brownish yellow band on elytra. Eggs are laid in sour or decaying fruit. (It can be trapped easily in buckets of fermenting fruits.) **Length:** .1–.2″. **Larva:** Slender, tapered at both ends, about .25″ long; white or yellowish, with a few long hairs, head and tip of abdomen amber-brown; two large brown tubercles at extreme posterior end, two smaller ones just in front of them. Pupates in the soil, hibernates in adult stage. As larvae they are very active; the short-winged adults fly up to two miles or more, may live from six months to a year. The latter are present at all times in tropical and subtropical regions. **Food:** Fermenting fruits, also those in ripe or dried state; mainly dates and figs in California. The **Corn Sap Beetle,** *C. dimidiatus,* is very similar, varies from black with reddish tinge to brownish yellow, each elytron usually with orange spot; it occurs from Florida to

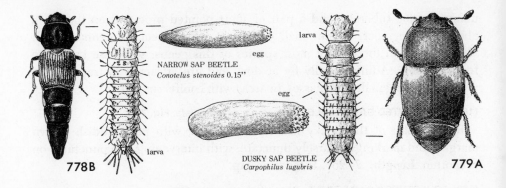

NARROW SAP BEETLE
Conotelus stenoides 0.15"

egg

larva

DUSKY SAP BEETLE
Carpophilus lugubris

larva

egg

larva

778B

779A

Quebec, Texas to Kansas and California, Washington; it is now considered a primary pest of corn in the field in the South, and has replaced *C. hemipterus* as the dominant species infesting dates in the Coachella Valley in California. The **Cactus Sap Beetle**, *C. pallipennis* (Fig. 781), is black and reddish brown or yellowish brown, common on flowers of cactus and yucca. The **Dusky Sap Beetle**, *C. lugubris* (Fig. 779A), is dark reddish brown, damages sweet corn severely in some areas, where it migrates from rotting vegetables and sap to corn in the tassel stage; *C. humeralis* is found on corn in the field and in storage, as well as on decaying fruits and vegetables; both are common along the Atlantic seaboard.

Cucujid or Flat Bark Beetles (Cucujidae)

The cucujids comprise a small group, are moderate in size, greatly flattened, with elytra usually covering all of the abdomen. The antennae are eleven-segmented, often with two- to four-segmented club. The larvae are usually elongate, cylindrical or somewhat depressed; the head is exserted, as wide or wider than pronotum, with prominent three-segmented antennae and six pairs of ocelli; the abdomen is nine- or ten-segmented, tapered, with a few prominent setae, and sometimes with cerci-like anal processes. They live under the bark of trees, under stones and debris, in decaying plant materials, in granaries, warehouses, and households. Some are beneficial, such as the scarlet *Cucujus clavipes*, one of the largest in the family (it is about half an inch long), which attacks bark beetles in their galleries; others are pests of stored foodstuffs.

(185) FLAT BARK BEETLE: *Telephanus velox.* **Fig. 780.**

Range: East of the Rocky Mts. **Adult:** Yellowish brown, head and markings darker brown; coarsely punctate and densely pubescent. **Length:** .15–.2″. It is found under bark, stones, and debris, runs rapidly when disturbed.

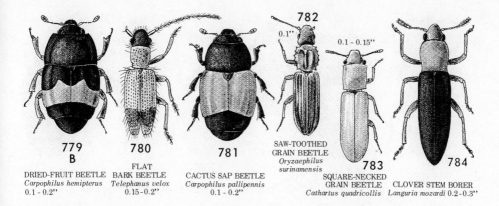

779
B

DRIED-FRUIT BEETLE
Carpophilus hemipterus
0.1 - 0.2"

780

FLAT
BARK BEETLE
Telephanus velox
0.15-0.2"

781

CACTUS SAP BEETLE
Carpophilus pallipennis
0.1 - 0.2"

782
0.1"

SAW-TOOTHED
GRAIN BEETLE
Oryzaephilus surinamensis

783
0.1 - 0.15"

SQUARE-NECKED
GRAIN BEETLE
Cathartus quadricollis

784

CLOVER STEM BORER
Languria mozardi 0.2-0.3"

(186) SAW-TOOTHED GRAIN BEETLE: *Oryzaephilus surinamensis* (Silvaninae). **Fig. 782.**

Range: Cosmopolitan. **Adult:** Very flat, dark brown to reddish brown. Pronotum with six teeth on each of the two outer edges, three ridges on top. It seldom flies, is often seen running rapidly over food. Eggs are laid on or near its food. **Length:** .1". **Larva:** Elongate, tapered, deeply segmented; brown, about .12" long. It crawls about actively, feeding here and there. Pupates under food materials, which it cements together. Hibernates in adult stage in unheated buildings or outdoors. Four to six or more broods. **Food:** Grain, cereals, dried fruits, nuts, seeds, sugar, candy, tobacco, dried meats. Adults are sometimes found outdoors under the bark of trees. The **Merchant Grain Beetle,** *O. mercator*—a household pest of cereals, not of stored grain, and also widespread in the U.S. and Canada—is very similar and sometimes mistaken for *surinamensis;* they may be separated by the length of the temple or region directly behind the eye, which is much less than half the vertical diameter of the eye in *mercator,* and more than half in *surinamensis.*

(187) SQUARE-NECKED GRAIN BEETLE: *Cathartus quadricollis.* **Fig. 783.**

Range: Cosomopolitan. **Adult:** Reddish brown, faintly punctate. **Length:** .1–.15". It is found in stored grains and cereals and abundant in the South, where it infests seed pods outdoors and stored corn.

(188) CLOVER STEM BORER: *Languria mozardi* (Languriidae). **Fig. 784.**

The languriid beetles, sometimes called lizard beetles or slender plant beetles, were formerly classed with the Erotylidae, our next group. They may be distinguished by the shape. They are black or deep blue, reddish, and orange; the antennae are eleven-segmented, with four- or five-segmented club. Five sternites are visible on the abdomen. The larvae are elongate, cylindrical, usually whitish; the abdomen is ten-segmented, with

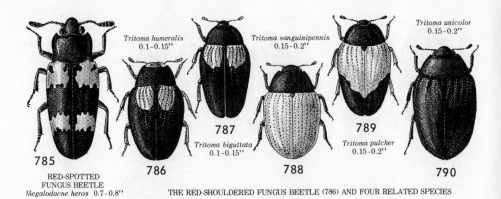

Tritoma humeralis
0.1-0.15"

Tritoma sanguinipennis
0.15-0.2"

Tritoma unicolor
0.15-0.2"

787

789

Tritoma biguttata
0.1-0.15"

Tritoma pulcher
0.15-0.2"

785

786

788

790

RED-SPOTTED
FUNGUS BEETLE
Megalodacne heros 0.7-0.8"

THE RED-SHOULDERED FUNGUS BEETLE (786) AND FOUR RELATED SPECIES

a few setae on each segment; the head is exserted, with one pair of ocelli, and three-segmented antennae. Many of them are stem borers; at least one species—our example—is a pest of economic importance. Adults feed on pollen and leaves of host plants. **Range:** *L. mozardi* is widespread east of the Rocky Mts. **Adult:** Shiny, blue-black; head, pronotum, and underside red, the last two or three abdominal segments, tarsi, and apical half of femora black. **Length:** .2–.3". **Larva:** Cylindrical, smooth along sides, yellowish, about .5" long, with curved hooklike processes at end of abdomen. It bores into stems of legumes, causing them to swell and crack and break off. **Food:** Red and sweet clover, alfalfa, wild lettuce. A subspecies, *L. m. occidentalis*, breeds in yellow sweet clover and alfalfa in the Southwest.

Pleasing Fungus Beetles (Erotylidae)

These are minute to fairly large beetles, usually oval and black, with reddish or yellowish markings. Distinctive features are the antennae—eleven-segmented, inserted in front of or between the eyes and ending abruptly in a large three- or four-segmented club—and the tarsi—five-segmented, with basal segments wide, padlike and pubescent, the fourth segment very small. The larvae are smooth or hairy, flattened or somewhat cylindrical, white to brownish, usually with short or long cerci-like anal appendages. They feed on fungi, and are found in decaying wood or debris, and under the bark of logs or dead trees.

(189) RED-SPOTTED FUNGUS BEETLE: *Megalodacne heros*. **Fig. 785.**

Range: Eastern U.S. **Adult:** Shiny, black, with red markings. Overwinters as adult. **Length:** .7–.8". It is found under the bark of logs and dead trees, and on fleshy fungi growing on trees or stumps. *M. fasciata* is very similar in color and markings, but smaller—from .4" to .6" long. Both species are gregarious and often found in large colonies.

(190) RED-SHOULDERED FUNGUS BEETLE: *Tritoma humeralis.* **Fig. 786.**

Range: Throughout eastern U.S., also Europe. **Adult:** Shiny, black; elytra with reddish yellow humeral spots, finely punctate. **Length:** .1–.15″. *T. biguttata* (Fig. 787) is very similar in color and markings, is more slender, with pronotum more rounded at the apexes. In *T. sanguinipennis* (Fig. 788) the head and pronotum are black, the elytra all red, the length from .15″ to .2″. *T. pulcher* (Fig. 789) resembles the latter closely except that a large apical area of the elytron is black, and it is slightly smaller. *T. unicolor* (Fig. 790) is all black, from .15″ to .2″ long.

(191) BANDED FUNGUS BEETLE: *Triplax festiva.* **Fig. 791.**

Range: Widely distributed in the eastern states, especially the South. **Adult:** Shiny black, with pronotum, scutellum, and transverse median band on elytra reddish yellow; finely punctate. **Length:** .2–.25″. *T. flavicollis* is similar, slightly smaller, with elytra all black, head and pronotum reddish yellow; body beneath black, legs reddish yellow. *T. thoracica* (Fig. 792) closely resembles the latter, except body beneath is reddish yellow, pronotum more rounded at apexes. Both are very similar to the western species, *T. californicus,* and have been reported from British Columbia.

Lady Beetles (Coccinellidae)

The common British name for these beetles, "ladybird," is still used to some extent in America; "ladybug" is probably the most familiar name. The lady beetles or coccinellids are easily distinguished by their shape, and the three-segmented tarsi. The red and black spotted kinds are familiar to everyone. The antennae are eight- to eleven-segmented, usually short, retractile, with a three- to six-segmented club. They hibernate as adults, singly, and in small or large aggregations. The eggs are oval, usually laid on end singly or in clusters, glued to leaves or bark. Their reproductive capacity is often high, with many generations a year in warmer climates.

The larvae are usually spindle-shaped, wrinkled, often with numerous setae, spines, or fleshy processes, and have three or four pairs of ocelli; the antennae are minute, with three to five segments. Some species secrete a wax, and are easily mistaken for the mealybugs on which they feed. They are very active, and feed exposed on plants, usually pupate with tip of abdomen glued to a leaf or the bark where feeding stopped, or in small groups assembled on twigs. Many larvae have paired thin spots on the first eight abdominal segments from which blood oozes when they curl up under attack. Some adults (as in the genus *Epilachna*) force blood from their leg joints when attacked; the fluid is sticky and traps their enemies, such as ants. Both adults and larvae are predaceous (with some exceptions,

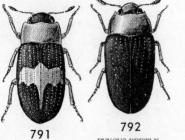

791
BANDED FUNGUS BEETLE
Triplax festiva 0.2-0.25"

792
FUNGUS BEETLE
Triplax thoracica
0.2-0.25"

793
LATERAL LADY BEETLE
Hyperaspis lateralis 0.1"

794
UNDULATED LADY BEETLE
Hyperaspis undulata 0.1"

notably in the genus *Epilachna*), and the family as a whole is highly beneficial. They prey on many different kinds of soft-bodied insects, more especially aphids, mealybugs, and scale insects. The immature forms of many larger insects and phytophagous mites are also attacked. Some 400 species of lady beetles are known in North America.

The convergent lady beetle (204)—probably the most widespread and familiar species—is interesting for its migratory habit in the far western states. Migration is usually thought of as a way of overcoming low temperatures and lack of food—for animals that do not hibernate. Western aggregations of convergent lady beetles seem to take the hard way—by migrating from the relatively warm valleys to the cold mountains, where they hibernate under a blanket of leaves and snow. These phenomena—not as yet clearly understood—are believed to be related to cyclical factors such as the length of day and gonadal or other physiological changes within the animal, as well as to temperature and nutritional requirements.

(192) LATERAL LADY BEETLE: *Hyperaspis lateralis* (Coccinellinae; Hyperaspini).
Fig. 793.

Range: Mexico, New Mexico, Arizona to Montana, Colorado, the Pacific Northwest. **Adult:** Shiny, black; front of head and front margin of pronotum and elytra, pair of median and apical spots red, orange, or yellow. **Diameter:** .1". **Larva:** Covered with long waxy filaments. It feeds mainly on mealybugs.

(193) UNDULATED LADY BEETLE: *Hyperaspis undulata.* **Fig. 794.**

Range: Throughout most of the U.S.; Canada. **Adult:** Shiny, black; face and sides of pronotum yellow in male, black in female. Three pairs of yellow marginal spots, one pair of yellow median dorsal spots on elytra. **Length:** .1". It feeds mainly on unarmored scales, is found on mesquite in Arizona. *H. fimbriolata* (Fig. 795), an eastern species, is all black, except for narrow yellow margin on elytra which may be broken near apex to form a separate spot; front and sides of head and pronotum yellow in male. The left elytra of other North American species of *Hyperaspis* are pictured in Fig. 796.

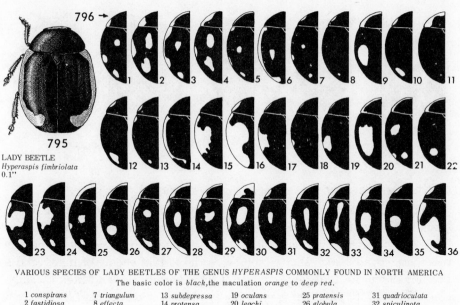

796 →

795

LADY BEETLE
Hyperaspis fimbriolata
0.1"

VARIOUS SPECIES OF LADY BEETLES OF THE GENUS *HYPERASPIS* COMMONLY FOUND IN NORTH AMERICA
The basic color is *black*, the maculation *orange* to *deep red*.

1 *conspirans*	7 *triangulum*	13 *subdepressa*	19 *oculans*	25 *pratensis*	31 *quadrioculata*
2 *fastidiosa*	8 *effecta*	14 *protensa*	20 *leachi*	26 *globula*	32 *spiculinota*
3 *bensonica*	9 *tuckeri*	15 *taeniata*	21 *proba*	27 *punctata*	33 *recurvans*
4 *gemma*	10 *elliptica*	16 *v.perpallida*	22 *oculifera*	28 *pludicola*	34 *rotunda*
5 *octavia*	11 *uniformis*	17 *v.significans*	23 *excelsa*	29 *octonotata*	35 *binotata*
6 *filiola*	12 *postica*	18 *pleuralis*	24 *taedata*	30 *fastidiosa*	36 *bolteri* V.=variety

(194) GUTTATE SCYMNUS: *Scymnus guttulatus* (Scymnini). **Fig. 797.**

Range: California; often abundant. **Adult:** Black, with brownish bands on elytra, one or two fused spots; head and front of pronotum brownish. **Length:** .05–.08". **Larva:** Covered with long waxy filaments. It feeds on mealybugs.

(195) TWO-SPOTTED SCYMNUS: *Scymnus binaevatus.* **Fig. 798.**

Range: Southern California. Introduced from South Africa in 1921. **Adult:** Black, with two reddish spots near apex of elytra. **Length:** .04". It feeds on mealybugs, especially the citrophilus mealybug.

(196) AMERICAN MINUTE LADY BEETLE: *Scymnus americanus.* **Fig. 799.**

Range: Mostly eastern North America. **Adult:** Black, elytra with narrow reddish tip, long gray pubescence; pronotum reddish along front and side margins or entirely black. **Length:** .07–.1". It feeds on aphids, is abundant in grass and low plants. *S. terminatus* (Fig. 800) is slightly smaller, black, with head, narrow lateral margins of pronotum, broad apical areas of elytra reddish. In *S. puncticollis* (Fig. 801) the pronotum is more rounded, lateral margins of pronotum and tip of elytra sometimes reddish; it has a white pubescence. The left elytra of other North American species of *Scymnus* are pictured in Fig. 802.

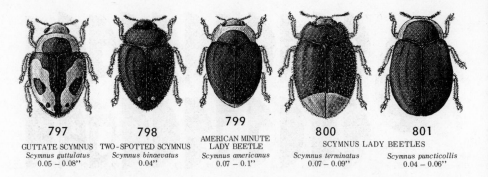

797	**798**	**799**	**800**	**801**
GUTTATE SCYMNUS	TWO-SPOTTED SCYMNUS	AMERICAN MINUTE LADY BEETLE	\ SCYMNUS LADY BEETLES /	
Scymnus guttulatus	*Scymnus binaevatus*	*Scymnus americanus*	*Scymnus terminatus*	*Scymnus puncticollis*
0.05 – 0.08"	0.04"	0.07 – 0.1"	0.07 – 0.09"	0.04 – 0.06"

(197) RED MITE DESTROYER: *Stethorus picipes.* **Fig. 803.**

Range: California, northward to British Columbia; Idaho. **Adult:** Shiny, black, with fine white pubescence; finely punctate. **Length:** .03–.04". **Larva:** Same length, dark gray or black, with numerous hairs rising from tiny tubercles, covered with shorter spines. An important enemy of mites, and a potential agent for biological control of the **Avocado Brown Mite** (*Oligonychus punicae*). *S. punctum* is very similar, black; antennae and legs, except the femora, yellow; it is widely spread over North America.

(198) MEALYBUG DESTROYER: *Cryptolaemus montrouzieri.* **Fig. 804.**

Range: California; introduced from Australia in 1892. **Adult:** Shiny, black; head, pronotum, tips of elytra, and abdomen reddish. Oval yellow eggs are laid singly in the egg sac of mealybugs or among the hosts. **Length:** .15". **Larva:** Yellow, from .3" to .4" long, covered with long, white waxy filaments; it is sometimes confused with the mealybugs. Unlike other lady beetles, they often descend on tree trunks to pupate in the litter. They are propagated in insectaries in Ventura County, California, and the adults released periodically in citrus groves for control of mealybugs. Insectary workers refer to them as "Crypts."

(199) VEDALIA: *Rodolia (Vedalia) cardinalis* (Noviini). **Fig. 805.**

Range: California, Florida; introduced from Australia in 1888. **Adult:** Red, with irregular black markings, red predominating in the female, black in the male; a dense grayish pubescence often obscures the colors and pattern. Oval red eggs are laid singly or in small clusters on egg sac of host. **Length:** .1–.15". **Larva:** Pinkish, with black markings; from .2" to .3" long full grown. It enters an egg sac to feed, also preys on young immature forms of the host. Pupates in last larval skin on branches or foliage of trees. It is capable of exerting complete control over the cottony cushion scale. [See Homoptera (43).] A closely related species, *R.* (*Novius*) *koebelei* (Fig. 806), was imported at the same time; it is slightly smaller, bright red, with head, pronotum, margins and two spots on elytra black (spots at the margins in the male, on the dorsum in the female).

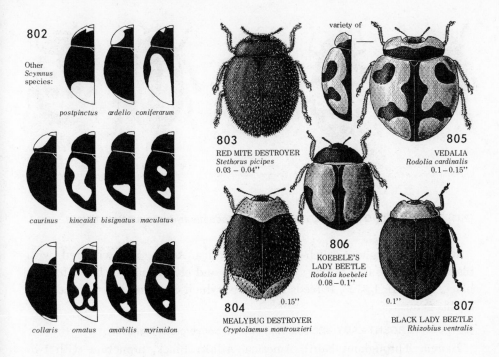

802

Other
Scymnus
species:

postpinctus ardelio coniferarum

caurinus kincaidi bisignatus maculatus

collaris ornatus amabilis myrimidon

variety of

803
RED MITE DESTROYER
Stethorus picipes
0.03 – 0.04"

805
VEDALIA
Rodolia cardinalis
0.1 – 0.15"

806
KOEBELE'S
LADY BEETLE
Rodolia koebelei
0.08 – 0.1"

804   0.15"
MEALYBUG DESTROYER
Cryptolaemus montrouzieri

0.1"   807
BLACK LADY BEETLE
Rhizobius ventralis

(200) BLACK LADY BEETLE: *Rhizobius ventralis* (Coccidulini). **Fig. 807.**

Range: California; introduced from Australia in 1892. **Adult:** Shiny (often velvety), black, with reddish abdomen. Clusters of several eggs are laid under scales. **Length:** .1". **Larva:** Brown or black, about .25" long; numerous tubercles, with long simple spines. The young larva feeds on eggs, later feeds on the young immature forms of scales settled on the foliage. (Adults feed on the latter also.) Pupates on trunk or branches. **Host:** Soft scales and mealybugs; a valuable predator of the black scale. *Lindorus lophanthae* (Fig. 808), a closely related species, is reddish brown, elytra shiny black, sometimes with black spot on pronotum, about .08" long; it feeds on mealybugs, soft and armored scales.

(201) SPOTTED LADY BEETLE: *Coleomegilla fuscilabris* (Coccinellini). **Fig. 809.**

Range: Throughout much of North America. Very common. **Adult:** Bright red to pink with black spots on elytra and pronotum (the two spots on the latter may be fused); minutely punctate. **Length:** .2–.3". It feeds on aphids. *C. floridana* (Fig. 810), an eastern species, is very similar in form and color.

(202) STRIPED LADY BEETLE: *Paranaemia* (*Ceratomegilla*) *vittigera*. **Fig. 811.**

Range: California, Washington, Oregon, Idaho, Arizona, Colorado. **Adult:** Black; pronotum reddish, with two black spots; elytra reddish, with three black stripes. **Length:** .15–.2". It feeds on aphids.

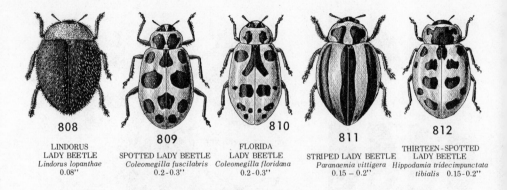

808	809	810	811	812
LINDORUS LADY BEETLE *Lindorus lopanthae* 0.08"	SPOTTED LADY BEETLE *Coleomegilla fuscilabris* 0.2-0.3"	FLORIDA LADY BEETLE *Coleomegilla floridana* 0.2-0.3"	STRIPED LADY BEETLE *Paranaemia vittigera* 0.15 – 0.2"	THIRTEEN-SPOTTED LADY BEETLE *Hippodamia tridecimpunctata tibialis* 0.15-0.2"

(203) THIRTEEN-SPOTTED LADY BEETLE: *Hippodamia tridecimpunctata tibialis.* **Fig. 812.**

Range: Throughout much of North America. **Adult:** Yellowish red, with black spots (variable in size) on pronotum and elytra; base of head, femora black. **Length:** .15–.2". It feeds on aphids, which comprise most of the food of this genus.

(204) CONVERGENT LADY BEETLE: *Hippodamia convergens.* **Fig. 813.**

Range: Throughout North America. **Adult:** Black, pronotum with two converging white stripes and white margins; elytra orange or red, with twelve black spots (not always distinct or present), scutellum black (looks like an elytral spot superficially). Spindle-shaped orange-yellow eggs are laid on end in compact clusters of 10 to 50, on leaf or bark. **Length:** .15–.25". **Larva:** Elongate, flat, from .3" to .4" long, velvety black, with orange spots. Pupates on leaf or bark. Hibernates as adult under stones, bark, or litter. In the western states, particularly along the Sierra Nevada Mts. and Coast Range of California, great swarms of them fly to the mountain canyons in late fall to hibernate, and congregate in huge aggregations under the leaves and snow, returning to the broad valleys below in early spring. The **Confused Convergent Lady Beetle,** *H. glacialis glacialis* (Fig. 814), is very similar in appearance, with the two apical spots much larger, the four subapical spots as it were fused into two larger ones, sometimes with a small spot near the humerus; it occurs east of the Rocky Mts. [For a subspecies of *H. glacialis,* see (206) below.]

(205) FIVE-SPOTTED LADY BEETLE: *Hippodamia quinquesignata.* **Fig. 815.**

Range: Western states, Rocky Mts. region; Washington, Oregon, British Columbia. **Adult:** Black, with white margins and converging white marks on pronotum. Elytra red, usually with five black markings as shown; these are variable, spots may fuse to make three bands or may be missing. **Length:** .15–.25".

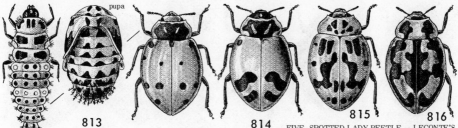

pupa

813

larva

CONVERGENT LADY BEETLE
Hippodamia convergens 0.15-0.25"

814

CONFUSED CONVERGENT
LADY BEETLE
Hippodamia glacialis glacialis
0.2-0.25"

815

FIVE-SPOTTED LADY BEETLE
Hippodamia quinquesignata
0.15 – 0.25"

816

LECONTE'S
LADY BEETLE
*Hippodamia glacialis
lecontei* 0.2 – 0.25"

(206) LECONTE'S LADY BEETLE: *Hippodamia glacialis lecontei.* **Fig. 816.**

Range: Western states. **Adult:** Similar to (205), sometimes mistaken for it. Also very variable. The three-lobed black spot over the scutellum is sometimes broken into three spots. **Length:** .2–.25".

(207) PARENTHESIS LADY BEETLE: *Hippodamia parenthesis.* **Fig. 817.**

Range: Throughout North America. Common. **Adult:** Head mostly black; pronotum black, with white markings; elytra yellowish red, with black markings; underside and legs black. The "parenthesis" is sometimes broken or fused into one blotch. **Length:** .15–.2". *H. americana* (Fig. 818A) is similar, with the black markings more extensive. The left elytra of other North American species of *Hippodamia* are illustrated and named in Fig. 818B.

(208) TWO-SPOTTED LADY BEETLE: *Adalia bipunctata.* **Fig. 819.**

Range: Throughout North America, Europe. **Adult:** Head black, with two yellow spots; pronotum black, with pale yellowish margins; elytra reddish, with two black spots. **Length:** .15–.2". A very common and familiar species. *A. frigida* (Fig. 820) and a subspecies, *A. f. humeralis* (Fig. 821), both western in distribution, are the same in size and form as *bipunctata,* shiny black, with red spots and markings.

(209) THREE-BANDED LADY BEETLE: *Coccinella trifasciata perplexa.* **Fig. 823.**

Range: Northern states, Canada, very common in the Pacific Northwest; *C. t. trifasciata* (Fig. 822) occurs in northern Europe and Asia. **Adult:** Head pale, with black band across the base in the male, black band with two pale spots in the female; pronotum black, with white marginal markings; elytra reddish or orange, with three black transverse bands. **Length:** .2". *C. transversoguttata* (Fig. 824) occurs from Newfoundland to Virginia, Texas, Arizona, California, Alaska, across southern Canada, is very similar to *trifasciata.*

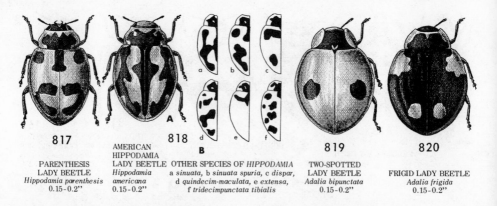

817	AMERICAN HIPPODAMIA 818	OTHER SPECIES OF *HIPPODAMIA*	819	820
PARENTHESIS LADY BEETLE *Hippodamia parenthesis* 0.15-0.2"	LADY BEETLE *Hippodamia americana* 0.15-0.2"	a *sinuata*, b *sinuata spuria*, c *dispar*, d *quindecim-maculata*, e *extensa*, f *tridecimpunctata tibialis*	TWO-SPOTTED LADY BEETLE *Adalia bipunctata* 0.15-0.2"	FRIGID LADY BEETLE *Adalia frigida* 0.15-0.2"

(210) NINE-SPOTTED LADY BEETLE: *Coccinella novemnotata.* **Fig. 825.**

Range: Quebec to Georgia, Texas, British Columbia, California; very common in the Pacific Northwest, especially east of the Cascade Mts., and in central California. **Adult:** Head and pronotum black, with pale yellowish marginal markings; elytra yellowish red or orange, with nine black spots (four on each elytron plus scutellar spot). Bright orange eggs are laid on end in small, tight clusters on underside of leaves. **Length:** .2–.3". The fourth-instar larva glues the posterior end to a leaf to pupate. Two broods in California, where they are heavy feeders on aphids in alfalfa. Adults hibernate (and summer generations estivate) in large aggregations in vegetation near feeding sites. *C. johnsoni* closely resembles the western form of *novemnotata;* the spots are usually smaller and more numerous; it occurs near the coast from Alaska to southern California.

(211) CALIFORNIA LADY BEETLE: *Coccinella californica.* **Fig. 826.**

Range: California to British Columbia. **Adult:** Head black, with white marginal markings; elytra reddish, without markings other than small scutellar spot and narrow dark brown sutural margins. **Length:** .25". Like the convergent lady beetle, they often congregate in the mountains to hibernate, and are heavy feeders on aphids.

(212) FIFTEEN-SPOTTED LADY BEETLE: *Anatis quindecimpunctata.* **Fig. 827.**

Range: Eastern North America. **Adult:** Head black, with two yellow markings; pronotum black and whitish; elytra yellowish to dark reddish brown, with fifteen black spots (seven on each elytron plus scutellar spot); spots obscured in dark specimens, humeral and scutellar spots sometimes spread out, forming band. **Length:** .25–.35". **Food:** Apple aphid, rosy apple aphid, and others.

(213) ASH-GRAY LADY BEETLE: *Olla abdominalis* (Synonychini). **Fig. 828.**

Range: Throughout much of North America; abundant in the western states (one of most important aphid feeders in California). **Adult:** Ash

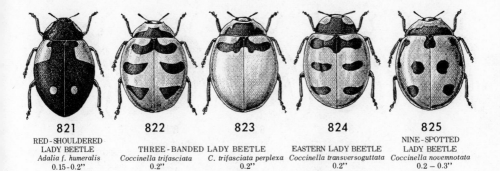

821	822	823	824	825
RED-SHOULDERED LADY BEETLE	THREE-BANDED LADY BEETLE		EASTERN LADY BEETLE	NINE-SPOTTED LADY BEETLE
Adalia f. humeralis 0.15–0.2"	*Coccinella trifasciata* 0.2"	*C. trifasciata perplexa* 0.2"	*Coccinella transversoguttata* 0.2"	*Coccinella novemnotata* 0.2 – 0.3"

gray to pale yellow, with black spots and markings as shown. **Length:** .15–.25". **Larva:** Black, forepart of head pale yellow; yellowish or orange spot on mesothorax and metathorax, lateral and dorsal spots of same color on first and fourth abdominal segments; from .6" to .8" long full grown. **Food:** Hop, rose, melon, walnut, and cabbage aphids; it has a decided preference for the walnut aphid. *O. a. plagiata* (Fig. 829) is a black form, with large red spot on each elytron.

(214) SPOTLESS LADY BEETLE: *Cycloneda sanguinea.* **Fig. 830.**

Range: South America to Arizona, east to the Rocky Mts. **Adult:** Head black and white (two white spots in the female); pronotum black and white; elytra reddish yellow, with no markings. **Length:** .15–.2". The **Western Blood-red Lady Beetle,** *C. munda* (Fig. 831) is very similar, the elytra pale to very bright red.

(215) TWICE-STABBED LADY BEETLE: *Chilocorus stigma* (Chilocorini). **Fig. 832.**

Range: Throughout much of North America. **Adult:** Shiny black, elytra with two red spots, minute dense punctures; abdomen red. Cylindrical, orange eggs are laid singly or in groups in bark crevices. **Length:** .15–.2". **Larva:** Black, with median, yellow transverse band, numerous spines. **Food:** Soft and armored scales. *C. tricyclus* and *hexacyclus,* which normally occur farther west and north, are not superficially distinguishable from *C. stigma* (some crossing or hybridization occurs naturally where the species come together). The natives *C. orbus* and *fraternus,* and the Asiatic *C. similis*— all occurring in California—are also very similar and separable only by special techniques. *C. similis* was imported many years ago and colonized in Georgia and California; it was thought to have died out, but recent evidence seems to show that it does in fact exist in California at least, under the guise of *C. orbus.*

(216) MEXICAN BEAN BEETLE: *Epilachna varivestis* (Epilachninae). **Fig. 833.**

Range: Most states east of the Mississippi; Texas to Arizona, Utah, Colorado, Nebraska. **Adult:** Yellowish brown to coppery, with sixteen small black

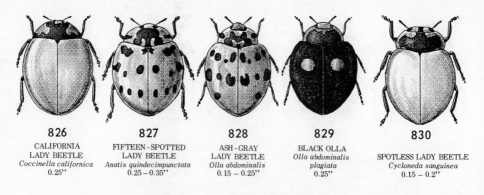

826	827	828	829	830
CALIFORNIA LADY BEETLE	FIFTEEN-SPOTTED LADY BEETLE	ASH-GRAY LADY BEETLE	BLACK OLLA	
Coccinella californica	*Anatis quindecimpunctata*	*Olla abdominalis*	*Olla abdominalis plagiata*	SPOTLESS LADY BEETLE
0.25″	0.25 – 0.35″	0.15 – 0.25″	0.25″	*Cycloneda sanguinea* 0.15 – 0.2″

spots on elytra; it darkens after overwintering, the spots become less distinct. Pronotum with fine dense punctures, elytra with fine and coarse punctures. Orange-yellow eggs are laid in large clusters on underside of leaf. **Length:** .25″. **Larva:** Orange, about .3″ long, covered with long branched spines. It feeds on underside of leaf, scrapes off lower tissues only, leaving leaf with lacelike appearance; adults often cut through the leaf. Hibernates as adult in debris of field or nearby woods. Two or three broods. **Food:** Edible beans are preferred: snap (green or string), kidney, pinto, navy, lima; it is a fairly common pest of soybeans, can breed on cowpeas, beggarweed or beggartick.

(217) SQUASH BEETLE: *Epilachna borealis.* **Fig. 834.**

Range: Eastern U.S. **Adult:** Yellow, with seven large black spots on each elytron, four small black spots on pronotum. **Length:** .25–.3″. It feeds on leaves of squash, pumpkin, and other cucurbitaceous plants.

Spider Beetles (Ptinidae)

The spider beetles may be recognized by the greatly convex elytra, small deflexed but visible head, long legs; many of them resemble spiders superficially. They are small to minute, brownish, often with patches of white scales; they have long filamentous antennae (usually eleven-segmented) inserted close together on the frons. The prosternum is very short with procoxal cavities open behind. The trochanters are very long, attached to the bases of the femora. The larvae are scarabaeoid (C-shaped white grubs) with short legs, usually five-segmented; the abdomen is ten-segmented (the ninth and tenth segments very small), often with many long setae. Some species pupate in cocoons. They are scavengers, and feed on dried animal and vegetable matter. Many of them breed indoors and are pests of stored foods and woolens; a few are inquilines in the nests of ants.

(218) HUMPBACK OR STOREHOUSE BEETLE: *Gibbium psylloides* (Ptininae). **Fig. 835.**

Range: Cosmopolitan. **Adult:** Brownish, smooth, hairless, without punc-

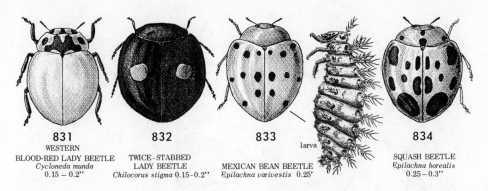

831
WESTERN
BLOOD-RED LADY BEETLE
Cycloneda munda
0.15 – 0.2''

832
TWICE-STABBED
LADY BEETLE
Chilocorus stigma 0.15-0.2''

833
MEXICAN BEAN BEETLE
Epilachna varivestis 0.25'

larva

834
SQUASH BEETLE
Epilachna borealis
0.25 – 0.3''

tures; ventral side, legs, antennae densely clothed with short, coarse yellow hairs. Elytra united along the whole length, strongly inflated. **Length:** .1–.12''. It is found in houses, warehouses, feeds on various foods, animal products, and woolen fabrics.

(219) SPIDER BEETLE: *Mezium affine*. Fig. 836.

Range: Northeastern U.S. and southeastern Canada, Europe, North Africa. **Adult:** Shiny, dark reddish brown to nearly black; strongly convex and inflated posteriorly, with dense coat of golden hairs and scales. **Length:** .1''. It is found in houses, granaries, warehouses; feeds on decaying animal and vegetable matter, dead insects, and textiles to some extent.

(220) AMERICAN SPIDER BEETLE: *Mezium americanum*. Fig. 837.

Range: Cosmopolitan. **Adult:** Very similar to *M. affine* (219). Dull yellow; elytra black, usually without hairs except for few stout, long, erect ones on each side of elytral suture near base. **Length:** .05–.1''. It occurs in dwellings, warehouses, mills, feeds on dried animal products, sometimes on tobacco seed, cayenne pepper, opium, grain.

(221) WHITE-MARKED SPIDER BEETLE: *Ptinus fur*. Fig. 838.

Range: Cosmopolitan. **Adult:** Pale reddish brown; head with dense coating of brownish yellow hairs, elytra with erect hairs at greater intervals than strial punctures, setae arising from punctures, coarse white hairs forming two transverse bands. **Length:** .08–.15''. It is found in houses, granaries, warehouses, museums.

(222) BROWN SPIDER BEETLE: *Ptinus clavipes*. Fig. 839.

Range: Nearly cosmopolitan; found in Europe and Africa; frequently seen in eastern states, Midwest, New Mexico, Arizona, California, southward into Guatemala. **Adult:** Very similar to *P. fur* (221). Pale to dark brown; elytra entirely without scales, hairs not as stout, distinctly longer than in *fur*. **Length:** .1–.15''. Mainly a scavenger, it is sometimes found feeding on rat feces, occasionally damages books, feathers, skin, dried mushrooms, dried fruit, roots, drugs, sugar, cocoa, stored seeds.

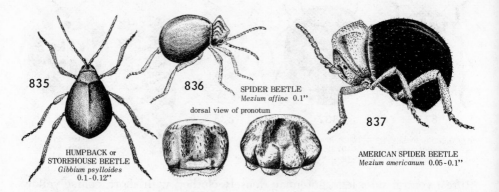

835

HUMPBACK or
STOREHOUSE BEETLE
Gibbium psylloides
0.1-0.12"

836

SPIDER BEETLE
Mezium affine 0.1"

dorsal view of pronotum

837

AMERICAN SPIDER BEETLE
Mezium americanum 0.05-0.1"

(223) HAIRY SPIDER BEETLE: *Ptinus villiger.* **Fig. 840.**

Range: Northern states, Canada; common in the Prairie Provinces and adjoining states, Europe. **Adult:** Reddish brown to blackish, with brownish yellow hairs and setae; white hairs on elytra form two irregular transverse bands. They appear in warehouses in the spring, lay their eggs in flour, feed, or other cereal products. **Length:** .1–.15". **Larva:** It feeds inside incompletely formed envelope made from mucous threads and fecal pellets, forms strong cocoon with silk spun from anus during third instar.

(224) CALIFORNIA SPIDER BEETLE: *Ptinus californicus.* **Fig. 841.**

Range: From the Middle Sierra Mts. of California northward to British Columbia; Nevada, Utah. **Adult:** Blackish brown; head, pronotum, antennae, legs, dark red. Head behind antennae densely clothed with whitish scales, pronotum with sparse short hairs; elytra with short blackish hairs between striae and conspicuous patches of white scales. **Length:** .15–.2". One of our largest species.

(225) GANDOLPHE'S SPIDER BEETLE: *Ptinus gandolphei.* **Fig. 842.**

Range: Mostly California. **Adult:** Almost identical with *P. fur* (221) in form. Elytra moderately clothed with coarse dirty-yellow hairs. **Length:** .12–.15". It feeds on cotton seed and mixed feeds.

(226) AGNATE SPIDER BEETLE: *Ptinus agnatus.* **Fig. 843.**

Range: Southern California. **Adult:** Similar in form to *P. fur* (221). Pale red except elytra, which are all blackish in the female, paler near suture in the male. Elytral punctures closely placed, with minute hairs, rows of fine bristles between striae; whitish patches formed by scalelike hairs. **Length:** .08–.12".

(227) GOLDEN SPIDER BEETLE: *Niptus hololeucus.* **Fig. 844.**

Range: Throughout North America, Canada; Europe, Asia. **Adult:** Reddish brown, densely clothed with brownish yellow scalelike hairs and

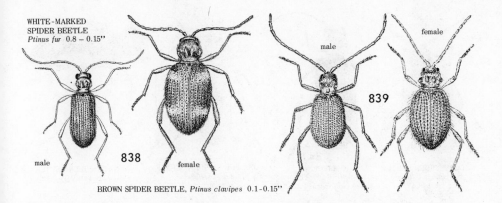

WHITE-MARKED SPIDER BEETLE
Ptinus fur 0.8 – 0.15"

male
838
female

839
male
female

BROWN SPIDER BEETLE, *Ptinus clavipes* 0.1-0.15"

bristles. **Length:** .15–.18". It feeds on miscellaneous seeds and seed products, dry plants, cotton, silk. *N. ventriculus* (Fig. 845) is somewhat smaller, occurs from Texas westward and south to Guatemala.

(228) GLOBULAR SPIDER BEETLE: *Trigonogenius globulum.* **Fig. 846.**

Range: Western North America, South America; frequently found in California and Oregon. **Adult:** Brown, densely clothed above and beneath with elongate pale yellowish brown scales; marble-like and darker on elytra, with brown spot near base on each side of suture. **Length:** .1". It feeds on grains, cereals, mixed feeds, beans, cottonseed meal, animal fertilizers, red pepper.

Deathwatch and Drugstore Beetles (Anobiidae)

The anobiids may be recognized by the short contractile legs and the greatly deflexed head and prothorax which are bent downward, nearly at right angles to the body, giving them a humped appearance when viewed from the side (Fig. 849). The antennae, with eleven segments (the last one elongated, more especially in the male), are inserted at the anterior margin of the eyes. The tarsi have five segments, decreasing in length from numbers one to four. The elytra cover the body, with a small triangular scutellum usually visible. The larvae are scarabaeoid (C-shaped fat white grubs), with or without many hairs, the abdominal segments often with infolded thickening of the margins (called plicae). In general they live on dead or dried animal and vegetable matter. Some species are destructive to a great variety of stored products; others construct galleries under the bark of dead trees. Certain species infest furniture, timbers of buildings, and other wood products. Several species that infest wood are called deathwatch beetles because of an old superstition that the strange tapping noise, made by striking their heads against the sides of the burrow, portended death. (Some species of booklice also have this tapping habit.)

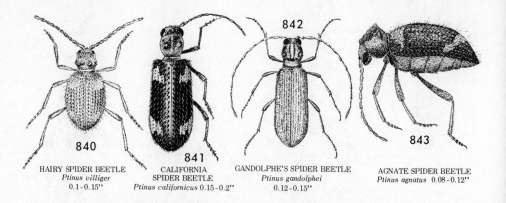

HAIRY SPIDER BEETLE
Ptinus villiger
0.1-0.15"

840

841

CALIFORNIA
SPIDER BEETLE
Ptinus californicus 0.15-0.2"

842

GANDOLPHE'S SPIDER BEETLE
Ptinus gandolphei
0.12-0.15"

AGNATE SPIDER BEETLE
Ptinus agnatus 0.08-0.12"

843

(229) DEATHWATCH BEETLE: *Xestobium rufovillosum.* **Fig. 847.**

Range: Throughout much of North America; introduced from Europe. **Adult:** Dark brown, with scattered areas of black or yellow hairs; antennae with unequal segments. **Length:** .25–.3". It causes extensive damage to wood in buildings and wood products.

(230) DEATHWATCH BEETLE: *Eucrada humeralis.* **Fig. 848.**

Range: Eastern U.S. and Canada. **Adult:** Black, marked with reddish yellow, elytra striated and punctate. Antennae, inserted near inner margin of eyes, are pectinate in the male, serrate in the female. **Length:** .15–.25". *E. robusta*—about .3" long—is black with brownish yellow pile, head and pronotum granulate, the latter convex, compressed at the base, elytra with coarse strial punctures, ranges into Oregon and British Columbia.

(231) CIGARETTE BEETLE: *Lasioderma serricorne.* **Fig. 849.**

Sometimes called towbug. **Range:** Cosmopolitan; found in all temperate, subtropical, and tropical regions. **Adult:** Reddish yellow or brownish red; elytra not striated; antennae serrate and uniform in width. Oval whitish eggs are laid on its food. **Length:** About .1". **Larva:** Curved, yellowish white, with light brown head; very hairy, about .15" long full grown. Pupates in silken cocoon. Three to six broods. **Food:** Tobacco, especially cigarettes; upholstery, seeds, black and red pepper, various dried plant products and drugs. Remains of this and the next species (232) were found in a vase in the tomb of the Egyptian king Tutankhamen after the crypt had been sealed for more than 3,000 years.

(232) DRUGSTORE BEETLE: *Stegobium paniceum.* **Fig. 850.**

Range: Cosmopolitan. **Adult:** Uniform light brown; elytra striated, covered with fine pubescence. Antennae with the third to eighth segments smaller than the first and second, and the last three segments much larger than the others. Eggs are laid on almost any dry organic material. **Length:** .1". **Larva:** White, with few hairs. Pupates in silken cocoon. Life cycle

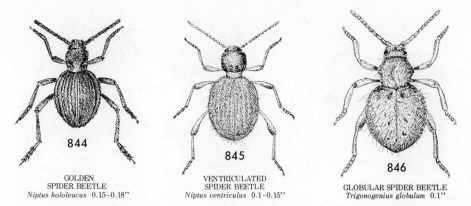

844

845

846

GOLDEN
SPIDER BEETLE
Niptus hololeucus 0.15-0.18"

VENTRICULATED
SPIDER BEETLE
Niptus ventriculus 0.1-0.15"

GLOBULAR SPIDER BEETLE
Trigonogenius globulum 0.1"

normally completed in less than two months. **Food:** Great variety of stored foods; seeds, almost any drug found in pharmacies. It is frequently found in granaries.

Darkling Beetles (Tenebrionidae)

This is a very large family, with more than 1,300 species in the U.S. They are small to large, slow-moving, black or brownish beetles, found mostly on the ground. The antennae are eleven-segmented, often clubbed, and arise from under the side of the head. The legs are stout, often long and smooth, the front and middle tarsi with five segments, the hind pair with four; the procoxal cavities are closed. The elytra are usually striated, may be ridged, and are often fused at the suture; the wings are usually absent or vestigial, seldom developed for flying. Some have the peculiar habit of standing on their heads. Many adults—flour beetles, mealworms, and others—emit foul-smelling gaseous secretions which have been identified chemically as quinones.

The larvae are cylindrical, white, yellowish, or brown, with a hard covering, distinctly segmented, pointed posteriorly, with two pointed terminal appendages and a short retractile organ. They are sometimes called false wireworms (Fig. 876) because of their resemblance to the larvae of click beetles (Fig. 741); they are more distinctly segmented, have longer legs and antennae than wireworms. Darkling beetles usually feed on fungi and decaying vegetable matter, but some are pests of stored products or feed on living plants. They are predominantly western in distribution, and generally nocturnal in habit.

(233) SHOULDERED DARKLING BEETLE: *Cnemodinus testaceus* (Tentyriinae). **Fig. 851.**

Range: Southwestern U.S. **Adult:** Brownish yellow, pubescent; scutellum distinct; scattered long yellow hairs. Mandibles grooved externally. **Length:** .25–.3".

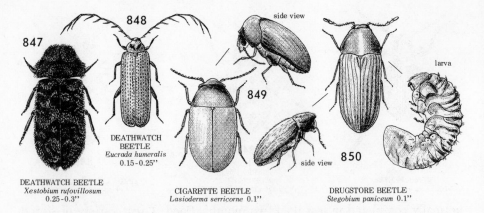

847

848

side view

larva

DEATHWATCH
BEETLE
Eucrada humeralis
0.15-0.25"

849

850

side view

DEATHWATCH BEETLE
Xestobium rufovillosum
0.25-0.3"

CIGARETTE BEETLE
Lasioderma serricorne 0.1"

DRUGSTORE BEETLE
Stegobium paniceum 0.1"

(234) ROUND DARKLING BEETLE: *Trimytis pruinosa.* **Fig. 852.**

Range: Southwestern states, Colorado. **Adult:** Black; body short, wingless. Mandibles not grooved externally. **Length:** .2–.25".

(235) HAIRY CRANIOTUS: *Craniotus pubescens.* **Fig. 853.**

Range: Arizona, southern and lower California. **Adult:** Black, pubescent. Antennae filiform, segments eight to ten enlarged, eleventh segment attached as apical process. **Length:** .4–.5".

(236) IRONCLAD BEETLE: *Phloeodes pustulosus* (Asidinae). **Fig. 854.**

Range: California; very common throughout the state. **Adult:** Depressed, very rough and hard on top; dull grayish black, with bases and tips of elytra often whitish. Pronotum with pair of short oblique black lines, elytra with three similar pairs and two pairs of small black dots at apexes. **Length:** .6–.9". It is found under the bark of dead trees, logs, or stumps, and feeds on fungi and decayed wood.

(237) SMALL IRONCLAD BEETLE: *Phellopsis obcordata.* **Fig. 855.**

Range: Eastern U.S. and Canada. **Adult:** Depressed, very rough and hard; dull brown; elytra with several rows of punctures, two long ridges. **Length:** .3–.6". It feeds on fungi. *P. porcata* is very similar, about the same size, dark brown, each elytron with three broad, interrupted ridges, the inner and outer ones ending in a large tubercle near the apex, with an additional tubercle at the apex; it occurs under bark and in the fungi of fallen trees in the Pacific Northwest and Alaska. Some classifications place the genus in a separate family, Zopheridae, the precoxal cavities being open, rather than closed as in other tenebrionids.

(238) SCULPTURED DARKLING BEETLE: *Pelecyphorus aegrotus.* **Fig. 856.**

Range: Southern and lower California. **Adult:** Black, margins of pronotum undulated, sharply pointed at corners; elytra sharply sculptured as shown. **Length:** .8–.9". *P. densicollis* is similar, black, said to be "so common in the

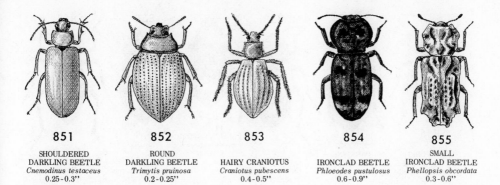

851	852	853	854	855
SHOULDERED DARKLING BEETLE	ROUND DARKLING BEETLE	HAIRY CRANIOTUS	IRONCLAD BEETLE	SMALL IRONCLAD BEETLE
Cnemodinus testaceus 0.25-0.3"	*Trimytis pruinosa* 0.2-0.25"	*Craniotus pubescens* 0.4-0.5"	*Phloeodes pustulosus* 0.6-0.9"	*Phellopsis obcordata* 0.3-0.6"

Yakima Valley [Washington] on occasion as to necessitate their being shoveled out of irrigation ditches."

(239) PUFFED DARKLING BEETLE: *Astrotus contortus.* **Fig. 857.**

Range: Texas. **Adult:** Dark brown, with scattered grayish scalelike hairs. **Length:** .4–.5". *A. regularis* is similar, less roughly sculptured, elytra with margin regular, bowlike.

(240) FALSE WIREWORM: *Eleodes suturalis* (Tenebrioninae). **Fig. 858.**

Many species of *Eleodes* occur west of the Mississippi River to the Pacific coast, from Mexico to Canada. They are found in great numbers in semiarid and desert regions in spring and early summer when migrations take place. Adults are usually found under rocks and logs during the day. *E. suturalis* is one of the common species in the drier wheat-growing areas [see (242) below]. **Adult:** Black, sometimes with red stripe down middle of back; pronotum depressed; elytra fused, and so it cannot fly. Adults feed on many different plants; many live as long as three years. Eggs are laid in the soil in groups of 10 to 60. **Length:** .9–1". **Larva:** Brown or yellowish (many species are black); resembles wireworm (see introduction to this family above). It feeds on germinating seeds, roots, underground parts of stems of young plants. It pupates in earthen cell, hibernates as larva or adult. One brood and partial second. **Food:** Wheat preferred; grasses, alfalfa, oats, millet, kafir, cotton, beans, sugar beets, other field and garden crops.

(241) PLAINS FALSE WIREWORM: *Eleodes opacus.* **Fig. 859.**

Range: See (240) above and (242) below. **Adult:** Black, with very fine grayish pubescence, five ridges on each elytron. **Length:** .4–.6". *E. humeralis* is black, with granular surface, somewhat opaque, from .55" to .8" long, occurs in the Pacific coast states, British Columbia, Idaho, Colorado, Utah, and Nevada. *E. dentipes* (Fig. 861) is smooth, pitch black, from .7" to .9" long, without wings, with elytra fused to body; it is very common in southern California, stands on its head when disturbed. *E. gigantea* is very similar but much larger—from 1.2" to 1.4" long.

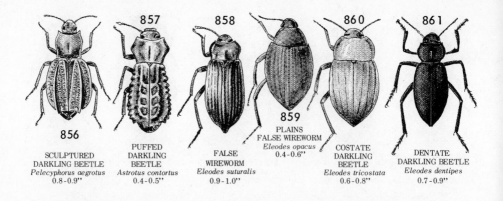

857
858
860
861

859
PLAINS
FALSE WIREWORM
Eleodes opacus
0.4-0.6"

856

SCULPTURED
DARKLING BEETLE
Pelecyphorus aegrotus
0.8-0.9"

PUFFED
DARKLING
BEETLE
Astrotus contortus
0.4-0.5"

FALSE
WIREWORM
Eleodes suturalis
0.9-1.0"

COSTATE
DARKLING
BEETLE
Eleodes tricostata
0.6-0.8"

DENTATE
DARKLING BEETLE
Eleodes dentipes
0.7-0.9"

(242) COSTATE DARKLING BEETLE: *Eleodes tricostata.* **Fig. 860.**

Range: Western half of North America, excluding Pacific Northwest; with *E. suturalis* (240) and *E. opacus* (241) it is one of the most common and widespread false wireworms in the dry wheat-growing areas. **Adult:** Black, with grayish tinge, elytra with three longitudinal rounded ridges or costae; elytra fused, hence it is flightless. **Length:** .6–.8". A related species found on grains, *Embaphion muricatum,* is unique in having the sides of pronotum and abdomen extended in a thin (sometimes bluish) plate turned upward at an angle; it is black or brownish, from .6" to .75" long, the elytra fused to the back and with minute hair-bearing tubercles; it resembles the six-spotted sap beetle (181) slightly, with the head appearing in a slot in the pronotum.

(243) MAYHEW'S DARKLING BEETLE: *Trogloderus costatus mayhewi.*

Range: Southern California. **Adult:** Shiny and black, with broad pronotum roughly sculptured, lateral margins evenly annulated, and strong black seta in each annule. Elytra each with four prominent ridges, spaces between more or less opaque with slight traces of shallow punctures. **Length:** .3–.5". *T. costatus tuberculatus* (Fig. 862) may be distinguished from *mayhewi* by the sculpturing on the pronotum and the punctures between the ridges on the elytra.

(244) COASTAL DARKLING BEETLE: *Coniontis viatica.* **Fig. 863.**

Range: California. **Adult:** Black; underside of front tibiae covered with spines. **Length:** .5–.6". *C. nemoralis borealis* is a coastal species found in Oregon. Adults of *Coniontis* are usually found under stones, logs, debris. *C. setosa*—from .3" to .45" long—black, head and pronotum with sparse fine punctures, elytra with dense pubescence and fine punctures, is common in Idaho, eastern Oregon and Washington.

(245) DARKLING GROUND BEETLE: *Blapstinus rufipes.* **Fig. 864.**

Range: California, Arizona, Lower California. **Adult:** Bluish black, with

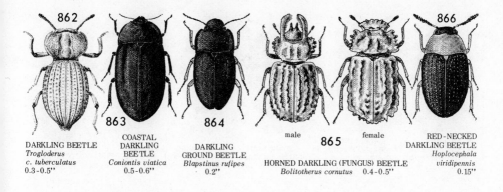

DARKLING BEETLE
Trogloderus
c. tuberculatus
0.3-0.5"

COASTAL
DARKLING
BEETLE
Coniontis viatica
0.5-0.6"

DARKLING
GROUND BEETLE
Blapstinus rufipes
0.2"

male 865 female
HORNED DARKLING (FUNGUS) BEETLE
Bolitotherus cornutus 0.4-0.5"

RED-NECKED
DARKLING BEETLE
Hoplocephala
viridipennis
0.15"

reddish legs. **Length:** .2". The adults girdle young hop and tomato plants just above the surface of the soil. Adults of *Blapstinus* spp. girdle other plants such as peppers, cucumbers, strawberries, grape. They attack young grape plants at wounds, often chew a girdle two to three inches wide around the vine. *B. oregonensis* is dull black, with dark red legs, pronotum rather densely punctate, from .2" to .25" long, common in the Pacific Northwest. *B. gregalis* occurs in the Pacific Northwest, Colorado, and California, is slightly smaller than *oregonensis* and somewhat shiny. *B. metallicus* occurs over much of the U.S. and Canada, is shiny, black, with brassy sheen, pronotum about one-third wider than long, deeply and broadly emarginate along the apex, with dense coarse punctures, elytra with deep coarse punctures in rows.

(246) HORNED DARKLING (FUNGUS) BEETLE: *Bolitotherus cornutus.* **Fig. 865.**

Range: Connecticut, Indiana, Florida. **Adult:** Black or dark brown, dorsum very rough; elytra with margins flattened and serrate, pronotum with horns in male, tubercles in female. **Length:** .4-.5". Found on fleshy fungi growing on logs, stumps, trunks of dead trees. They mimic the material in which they live, remain motionless when disturbed, and are difficult to see.

(247) RED-NECKED DARKLING BEETLE: *Hoplocephala viridipennis.* **Fig. 866.**

Range: Eastern U.S., the Midwest. **Adult:** Head, pronotum, legs, abdomen red; elytra metallic blue or green, striated, punctate. **Length:** .15". It is found under dead bark and on fleshy fungi. *H. bicornis* (Fig. 867A) is very similar, dorsum metallic bluish green with pronotum sometimes brownish; male with two short horns on lower part of head (clypeus) and two longer ones higher up.

(248) MACULATED DARKLING BEETLE: *Diaperis maculata.* **Fig. 867B.**

Range: Indiana, southward to Mexico, Guatemala. **Adult:** Black; head and elytra marked with orange-red as shown; pronotum with sparse fine

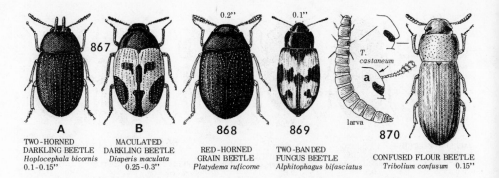

| A | B | 868 | 869 | larva | 870 |

A — TWO-HORNED DARKLING BEETLE
Hoplocephala bicornis
0.1-0.15"

B — MACULATED DARKLING BEETLE
Diaperis maculata
0.25-0.3"

868 — RED-HORNED GRAIN BEETLE
Platydema ruficorne

869 — TWO-BANDED FUNGUS BEETLE
Alphitophagus bifasciatus

870 — CONFUSED FLOUR BEETLE
Tribolium confusum 0.15"

T. castaneum

punctures, elytra with rows of shallow fine punctures. **Length:** .25–.3". They are gregarious, found under bark, especially that of elm, and in fungi.

(249) RED-HORNED GRAIN BEETLE: *Platydema ruficorne.* **Fig. 868.**

Range: Eastern half of U.S. **Adult:** Velvety black with purplish tinge, antennae reddish yellow, legs dark reddish brown; elytra striated and punctate. **Length:** About .2". It is found under the bark of trees and on fleshy fungi; abundant in shelled corn in Missouri, Illinois, Iowa, especially attracted to damp and moldy grain. *P. oregonensis* is very similar, same size, black, elytra with coarse strial punctures, tibiae and tarsi clothed in yellow hairs, common in the Pacific Northwest.

(250) TWO-BANDED FUNGUS BEETLE: *Alphitophagus bifasciatus.* **Fig. 869.**

Range: Cosmopolitan. **Adult:** Reddish brown, with two broad black bands across the elytra. **Length:** .1". It is found under bark, feeds on fungi and molds; a scavenger in refuse grain and grain products and other decaying vegetable matter, it is found around mills and warehouses, in holds of ships having wet or damaged grain.

(251) CONFUSED FLOUR BEETLE: *Tribolium confusum.* **Fig. 870.**

Range: Cosmopolitan. **Adult:** Flattened, shiny reddish brown; head and pronotum minutely punctured; elytra striated and sparsely punctate. They are very active, move quickly when alarmed, normally live about a year but often much longer. Small white eggs are laid loosely in flour or other food, covered with a sticky secretion. **Length:** .15". **Larva:** Slender, cylindrical; white, tinged with yellow, about .2" long, with two short pointed anal appendages. Pupa naked, first yellow, later changing to brown. Four or five broods annually in Kansas; all stages are present throughout the year in heated places. **Food:** Primarily flour; cereals. The most abundant and injurious insect pest of flour in the U.S.* The **Red Flour Beetle,** *T. cas-*

---

* The adults give off an offensive substance which "deleteriously affects the viscous and elastic properties of a dough made from infested flour," but "there is no evidence that injury would result from the ingestion of confused flour beetles as they may accidentally occur in cooked cereals." Louis M. Roth, *Ent. Soc. Amer. Ann.*, 34:397 (1943).

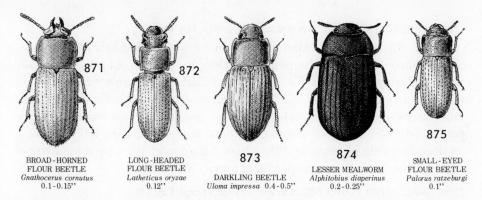

BROAD-HORNED
FLOUR BEETLE
Gnathocerus cornutus
0.1-0.15"

LONG-HEADED
FLOUR BEETLE
Latheticus oryzae
0.12"

873

DARKLING BEETLE
Uloma impressa 0.4-0.5"

874

LESSER MEALWORM
Alphitobius diaperinus
0.2-0.25"

875

SMALL-EYED
FLOUR BEETLE
Palorus ratzeburgi
0.1"

taneum, is very similar, may be distinguished by the three apical antennal segments which are enlarged abruptly and of equal size, by the absence of the notched expansion of the head above the eyes as found in *confusum,* and by the separation of the eyes (Fig. 870a): in *castaneum* they are much closer together. The two species are generally found together, but *castaneum* is more common in the South, where it also infests stored in-shell peanuts. The **Black Flour Beetle,** *T. audax,* may be distinguished by its color and larger size (about .2" long), slightly arched sides of pronotum near the base; it is found in grain products and nests of leafcutter bees (*Megachile*) but seldom in injurious numbers.*

(252) BROAD-HORNED FLOUR BEETLE: *Gnathocerus cornutus.* Fig. 871.

Range: Cosmopolitan; common over the U.S. excepting the Great Plains. **Adult:** Shiny, reddish brown; mandibles of male armed with pair of broad stout horns. **Length:** .1–.15". Its preferred food is flour and meal, but it also attacks a variety of grains.

(253) LONG-HEADED FLOUR BEETLE: *Latheticus oryzae.* Fig. 872.

Range: Widespread in southern and midwestern states. **Adult:** Flattened, pale yellowish brown, with distinctively shaped antennae and minute canthus, or raised process, behind each eye. **Length:** .12". It is common in rice and flour mills, infests grain and grain products.

(254) DARKLING BEETLE: *Uloma impressa.* Fig. 873.

Range: Eastern U.S. to Florida, Indiana. **Adult:** Shiny, chestnut brown, legs reddish brown; front of head with deep impression, pronotum with deep, curved lateral impression and sparse fine punctures; elytral striae deep, with coarse punctures. **Length:** .4–.5". It is found under bark, especially that of oak and beech.

---

* This species was previously identified as *T. madens* but "is shown to be distinct from the true *T. madens* from the Old World"; *T. audax* is smaller, less convex, and more elongate. Both species are found in stored products and are also associated with bees. See: D. G. H. Halstead, "A New Species of *Tribolium* from North America Previously Confused with *Tribolium madens* (Charp.) (Coleoptera: Tenebrionidae)," *Jour. Stored Prod. Res.,* 4:295–304 (1969).

(255) LESSER MEALWORM: *Alphitobius diaperinus.* **Fig. 874.**

Range: Cosmopolitan. **Adult:** Black or very dark reddish brown; pronotum finely and sparsely punctate. **Length:** .2–.25″. **Larva:** Yellowish brown; closely resembles the yellow and dark mealworms (257). **Food:** Grain and cereals in poor condition preferred; commonly found in flour-mill basements, in damp or musty flour or grain. They are found in many poultry houses, sometimes in great numbers; adults have been found feeding on dead carcasses of birds, and they have been reported as capable of transmitting avian leucosis, which produces tumors in broilers.*

(256) SMALL-EYED FLOUR BEETLE: *Palorus ratzeburgi.* **Fig. 875.**

Range: Cosmopolitan. **Adult:** Flattened, shiny reddish brown. **Length:** .1″; smallest of the "flour beetles." It is found in grain and milled products. The **Depressed Flour Beetle,** *P. subdepressus,* is very similar, slightly larger, may be distinguished by the strongly reflexed head, concealing the front of the eyes; it is cosmopolitan but less common in the U.S. than *ratzeburgi,* with the possible exception of the Great Plains.

(257) YELLOW MEALWORM: *Tenebrio molitor.* **Fig. 876.**

Range: Cosmopolitan; abundant only in the northern states. **Adult:** Shiny, dark brown or black; elytra striated. Beanlike white eggs, covered with sticky secretion, are laid in flour or meal in May or June. **Length:** .5–.65″; largest of the insects infesting cereal products. **Larva:** Slender, smooth, cylindrical; yellowish, shading to yellowish brown toward both ends and at margins of the segments; about 1″ long full grown. Found only in dark, damp places. Hibernates as larva, pupates in the spring. **Food:** Grain, cereals. Not normally a serious pest. The larvae are reared as food for captive animals. The **Dark Mealworm,** *T. obscurus,* is very similar, may be distinguished by the dull pitchy black coloring; the larvae are also darker. The two species are often found together; *obscurus* pupates and emerges earlier in the spring.

(258) DARKLING BEETLE: *Anaedus brunneus.* **Fig. 877.**

Range: Eastern U.S. to Florida, Indiana. **Adult:** Dark reddish brown, sparsely clothed with long yellowish hairs; antennae and legs paler; head and pronotum with sparse coarse punctures; elytra with deep coarse punctures irregularly spaced. **Length:** .2″. It is found under bark, logs, and stones in sandy areas.

(259) DARKLING BEETLE: *Meracantha contracta.* **Fig. 878.**

Range: Eastern Canada, Connecticut, Indiana, Texas. **Adult:** Very convex;

---

* J. L. Lancaster, Jr., and Joseph S. Simco, *Biology of the Lesser Mealworm: A Suspected Reservoir of Avian Leucosis,* Report Series 159 (1967), Agr. Exp. Sta., Univ. of Arkansas, Fayetteville; also, D. F. Bray et al., *Control of the Darkling Beetle in Delmarva Broiler Houses,* Pub. E. 17, Agr. Exp. Sta., Univ. of Delaware, Newark (1968).

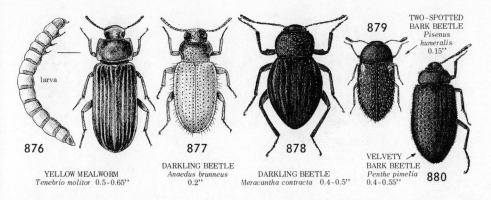

TWO-SPOTTED
BARK BEETLE
879
*Pisenus
humeralis*
0.15"

larva

876   877   878

YELLOW MEALWORM
Tenebrio molitor 0.5-0.65"

DARKLING BEETLE
Anaedus brunneus
0.2"

DARKLING BEETLE
Meracantha contracta 0.4-0.5"

VELVETY
BARK BEETLE
Penthe pimelia
0.4-0.55"   880

black, with bronze sheen; elytra, antennae, and legs deep black; pronotum with sparse coarse punctures; elytra with shallow striae, finely punctate. **Length:** .4–.5". Solitary; found under logs and bark, especially that of oak.

Melandryid Bark Beetles (Melandryidae)

Also called false darkling beetles, these are small to fairly large brown or black beetles, with hard bodies, large palpi, eleven-segmented filiform antennae, front and middle tarsi five-segmented, hind tarsi four-segmented. Adults and larvae are usually found under the bark of deciduous and coniferous trees, in dry wood and fungi; the larvae of a few species are phytophagous, others carnivorous.

(260) TWO-SPOTTED BARK BEETLE: *Pisenus humeralis.* **Fig. 879.**

Range: Eastern half of the U.S. **Adult:** Somewhat convex, shiny; dark brown to black, with sparse yellowish pubescence; pronotum densely punctate; elytra coarsely and sparsely punctate, usually with reddish spot on each shoulder. **Length:** .15". It is found on fleshy fungi.

(261) VELVETY BARK BEETLE: *Penthe pimelia.* **Fig. 880.**

Range: Eastern half of North America. **Adult:** Flattened, velvety black; pronotum densely and coarsely punctate; elytra with rows of deep punctures. **Length:** .4–.55". With *P. obliquata*—distinguished by a dense covering of yellowish orange hairs on the scutellum—it is found under the bark of dead or dying trees, or in dry fungi attached to logs and stumps, especially those of beech.

(262) BLAZED TREE BORER: *Serropalpus substriatus (barbatus).* **Fig. 881.**

Range: Widely distributed over forested areas of North America. **Adult:** Slender, reddish brown. Head and pronotum densely punctate, elytra feebly striated. It emerges in June or July through perfectly round hole cut in bark. Eggs are laid in dying or dead trees or living trees from which bark has been peeled. **Length:** .5–.75". **Larva:** Long, slim, white. It mines

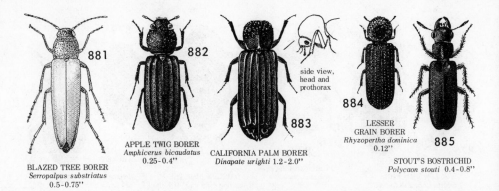

881

882

side view,
head and
prothorax

883

884

APPLE TWIG BORER
Amphicerus bicaudatus
0.25-0.4"

CALIFORNIA PALM BORER
Dinapate wrighti 1.2-2.0"

LESSER
GRAIN BORER
Rhyzopertha dominica
0.12"

885

BLAZED TREE BORER
Serropalpus substriatus
0.5-0.75"

STOUT'S BOSTRICHID
Polycaon stouti 0.4-0.8"

the sapwood, making oval tunnels in which much frass collects, requires two years to complete development. **Host:** Red fir, California incense cedar, lodgepole pine, ponderosa pine, redwood, Port Oxford cedar, Engelmann spruce, Douglas fir, and other conifers.

False Powder-post Beetles (Bostrichidae)

The false powder-post beetles are minute to very large beetles, distinguished by the large pronotum with many tubercles, the eight- to ten-segmented straight antennae with three- or four-segmented club, and the five-segmented tarsi with first segment very short. The head is small, deflexed, usually not visible from above. The larvae are stout, C-shaped, usually without hairs, with large thorax, ten-segmented abdomen with three or less folds on each segment. They are wood borers, breed in dead trees, but sometimes invade seasoned wood; the adults often bore into twigs and branches of live trees. A few species are pests of stored grains in warm climates.

(263) APPLE TWIG BORER: *Amphicerus bicaudatus (hamatus)*. **Fig. 882.**

Range: Widely distributed east of the Rocky Mts. **Adult:** Cylindrical, reddish brown to brownish black, pronotum tuberculate apically, coarsely punctate basally; elytra coarsely punctate, apexes declivous, each with a single tubercle or spine in the male. **Length:** .25–.4". The larvae are found in diseased or dying trees; the adults bore into twigs of live apple and other deciduous fruit trees, pecan, ash, and hickory, but do not oviposit here. The **Western Twig Borer,** *A. cornutus,* is similar, dark brown or black, from .4" to .5" long, with front of pronotum tuberculate, elytra declivous at apexes, each with pair of posterior horns or tubercles; the larvae breed in dead oak, mesquite, and other hardwoods; the adults bore into twigs of live fruit and nut trees such as orange, pear, fig, apricot, almond, grape. The **California Palm Borer,** *Dinapate wrighti* (Fig. 883), a closely related species, is the largest in the family—from 1.2" to more than 2" long—dark

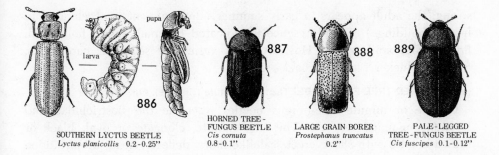

SOUTHERN LYCTUS BEETLE
Lyctus planicollis 0.2-0.25"

887

HORNED TREE-
FUNGUS BEETLE
Cis cornuta
0.8-0.1"

888 LARGE GRAIN BORER
Prostephanus truncatus
0.2"

889 PALE-LEGGED
TREE-FUNGUS BEETLE
Cis fuscipes 0.1-0.12"

brown or black, front of pronotum tuberculate, elytra distinctly striated and punctate, declivous at apexes, with two pairs of posterior horns.

(264) LESSER GRAIN BORER: *Rhyzopertha dominica* (Dinoderinae). **Fig. 884.**

Often called the Australian wheat weevil. **Range:** Cosmopolitan; tropical in origin, spread by commerce, now widespread throughout the U.S. **Adult:** Shiny, dark brown or black, antennae with three-segmented serrate club. Eggs are laid singly or in clusters in loose grain. **Length:** .12". **Larva:** Whitish grub, about .1" long, with brown head, it crawls actively about the grain, feeds on flour produced by borings of the beetles, bores into damaged grain to pupate. Cycle from egg to adult takes about a month in warm weather. This is one of the worst pests of stored grains [see (442) below]. The **Large Grain Borer**, *Prostephanus truncatus* (Fig. 888), is similar, may be distinguished by its larger size—it is nearly .2" long—and the smooth polished surface; not as widely distributed, sometimes found in corn in the South.

(265) STOUT'S BOSTRICHID: *Polycaon stouti* (Psoinae). **Fig. 885.**

Range: Arizona, California, Pacific Northwest. **Adult:** All black, head and pronotum coarsely punctate, mandibles very prominent; elytra finely punctate. **Length:** .4–.8". It breeds in dead eucalyptus, sycamore, oak, maple, California laurel, madroño, manzanita, fruit and other trees. Adults invade live trees occasionally, also cured hardwoods.

(266) SOUTHERN LYCTUS BEETLE: *Lyctus planicollis* (Lyctidae). **Fig. 886.**

The Lyctidae, or powder-post beetles, in contrast to the Bostrichidae (263), have a partially depressed body, large exposed head, two-segmented antennal club. *L. planicollis* is widespread in the South and western states. **Adult:** Pitchy black, with sparse short yellowish white pubescence; pronotum with dense coarse punctures; elytra striated, with double rows of punctures in between. Eggs are laid in pores of wood. **Length:** .2–.25". **Larva:** A white grub, about .3" long, resembling the bostrichid larva. It burrows in all directions through white wood and sapwood of hardwoods, reducing the fiber to a flourlike powder. Hibernates as larva in wood, pupates in spring; the adult appears in early summer (development is accelerated in heated buildings). Their presence is indicated by small round holes with fine powder spilling out. **Host:** A great variety of seasoned hardwoods, especially hickory, ash, and oak.

(267) HORNED TREE-FUNGUS BEETLE: *Cis cornuta* (Cisidae). **Fig. 887.**

Cisidae or minute tree-fungus beetles resemble the Bostrichidae and Lyctidae in many respects. They are minute, elongate or oval, black or brown beetles with short, erect scalelike hairs, deflexed head, often hidden from view, large pronotum, eight- to eleven-segmented antennae with three-segmented club, four-segmented tarsi. They live gregariously under bark, in fungi, and in rotten wood, are sometimes found in galleries of scolytid beetles. *C. cornuta* occurs in the eastern states. **Adult:** Dark reddish brown, with stout, erect yellowish hairs. **Length:** .08–.1″.

(268) PALE-LEGGED TREE-FUNGUS BEETLE: *Cis fuscipes.* **Fig. 889.**

Range: Eastern U.S. **Adult:** Reddish brown to black, legs and antennae paler; dense covering of stiff erect hairs. **Length:** .1–.12″. *C. versicolor* is very similar, slightly smaller, reddish brown, with stout erect bristles; pronotum with fine punctures, finely margined, sides minutely notched, elytra shiny; it occurs in British Columbia and Oregon; males have never been found.

Scarabs (Scarabaeidae)

This is a very large family with more than 30,000 known species around the world. They are small to very large, robust beetles with prominent pronotum and head, often beautifully colored and having a metallic luster. A distinctive feature is the antennal club composed of three to seven leaflike segments or lamellae, giving rise to the name "lamellicorn beetles," which is sometimes applied to the group. The lamellae are spread apart and compacted in serving their sensory function. The total antennal segments vary from seven to eleven. The maxillary palpi are four-segmented, the labial palpi are three-segmented, both pairs slender. The front tibiae are broad, toothed, modified for digging, and have one apical spur; the middle and hind pair are more slender, spined, usually with two apical spurs, sometimes with one. The tarsi are five-segmented, the front pair sometimes missing; the claws are variable but usually equal, or they may be absent. Five abdominal sternites are visible. Most scarabs have well-developed wings, are good fliers and often attracted to lights.

The larvae are C-shaped, fleshy, very "wrinkled," thicker at the posterior end, with well-developed thoracic legs; they are white or yellowish, usually with numerous setae or hairs. The abdomen is nine- or ten-segmented, usually with three annuli or rings per segment. The head is exserted, lightly

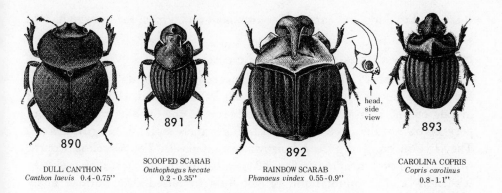

DULL CANTHON
Canthon laevis 0.4-0.75"

SCOOPED SCARAB
Onthophagus hecate
0.2 - 0.35"

RAINBOW SCARAB
Phanaeus vindex 0.55-0.9"

head,
side
view

CAROLINA COPRIS
Copris carolinus
0.8-1.1"

890

891

892

893

pigmented, with short three-segmented antennae and no ocelli. The pupae are free, enclosed in cells formed of materials available to the larvae. Adults and larvae are generally nocturnal; they feed on dung, carrion, decayed vegetable matter, and living plants, and include many serious agricultural pests. The dung beetles have attracted much attention because of their interesting habits, and are celebrated in mythology and in the writings of the great French entomologist-naturalist J. Henri Fabre. As scavengers they perform an important function in reducing the media in which many flies develop. They are said to remove 80 percent of the cattle droppings in parts of Texas. At least one species, *Onthophagus incensus,* has been used in biological control; it was imported into Hawaii from the West Indies to combat the horn fly. Some dung beetles ingest the eggs of nematodes in the feces of their intermediate hosts such as bears, pigs, and ruminants; man can be infected by drinking water containing nematode larvae released from decaying beetles. Smaller beetles like *Canthon* do not pass the parasites.

(269) DULL CANTHON: *Canthon laevis (pilularius)* (Scarabaeinae). **Fig. 890.**

One of the most common and best known of the "tumblebugs." **Range:** Generally distributed over most of the U.S. **Adult:** Dull black, with coppery, bluish, or greenish tinge; pronotum and elytra densely granulate. **Length:** .4–.75". The female deposits her eggs singly in balls of dung which are buried; the male helps her form the ball, and together they roll it some distance (apparently to compact it) before burying.*

(270) SCOOPED SCARAB: *Onthophagus hecate* (Onthophagini). **Fig. 891.**

Range: Throughout North America. **Adult:** Black, with purplish tinge, scattered short grayish hairs; pronotum granulate, with scooplike anterior projection in the male; elytra finely striated and granulate. **Length:** .2–.35".

---

* A closely related species (*Scarabaeus sacer*) had religious significance for the Egyptians, its life cycle being associated with rebirth. They wore mummified beetles strung around their bodies and a single figure with spread wings clasped over the heart. Many of the beetles were preserved in tombs. Warren E. Cox, *The Book of Pottery and Porcelain* (New York: Crown, 1953), p. 277.

It is found in dung and carrion. See comment on *O. incensus* in introduction above.

(271) RAINBOW SCARAB: *Phanaeus vindex* (Coprini). **Fig. 892.**

One of the most beautiful beetles when removed from its habitat. **Range:** East of the Rocky Mts. **Adult:** Flattened; head bronzed, clypeus extended in long curved horn in the male and tubercle in the female; pronotum coppery, elytra metallic green with bluish tinge, finely striated. **Length:** .55–.9″. They live in dung, dig deep burrows near deposits. The **Carolina Copris,** *Copris (Pinotus) carolinus* (Fig. 893), is shiny black, vertex of male extended in a short blunt horn; pronotum with raised disc, surface finely and sparsely punctate; each elytron with eight shallow striae, feebly punctate; from .8″ to 1.1″ long; it lives in dung, is strongly attracted to lights.

(272) MACULATED DUNG BEETLE: *Aphodius distinctus* (Aphodiinae, Aphodiini).
Fig. 894.

Range: Throughout North America. **Adult:** Shiny, head and pronotum black, latter sparsely punctate in the male and densely punctate in the female; elytra yellowish, with black spots, finely striated. **Length:** .2–.25″. The **Slender Dung Beetle,** *Ataenius cognatus* (Eupariini) (Fig. 895), is reddish black; legs, sides of clypeus, and narrow front margin of pronotum reddish brown, clypeus with numerous fine wrinkles; pronotum with coarse and fine punctures, elytral striae punctate; about .2″ long; it is widely distributed, hibernates in large aggregations under dry cow manure, is active in flight on warm sunny days in winter and spring. Mass flights of both species occur in spring, and great numbers are attracted to lights.

(273) FANCY DUNG BEETLE: *Bolbocerosoma farctum* (Geotrupinae, Bolboceratini).
Fig. 896.

Range: East of the Rocky Mts. **Adult:** Yellowish red, elytra with wide black band along suture and apexes; row of prominent tubercles along apical margin of pronotum in the male; antennal club large and rounded, eyes divided. **Length:** .3–.5″. *Eucanthus lazarus* is light to dark brown, with brownish antennal club, large median swelling on pronotum, large deep strial punctures on elytra, from .25″ to .45″ long; occurs in southern Ontario and eastern U.S. to South Dakota and Arizona, generally in clay soils of areas sparsely treed. Strong fliers and often found at lights.

(274) GLOSSY PILLBUG: *Geotrupes splendidus* (Geotrupini). **Fig. 897.**

Range: Eastern U.S. **Adult:** Brilliant metallic green, bronze, or purplish; pronotum with varying punctures, elytra deeply striated and punctured; antennal club smaller than in (273), composed of leaflike plates. **Length:** .5–.7″. It constructs brood cells of cow dung; in the Southeast egg laying takes place in January, pupation in May.

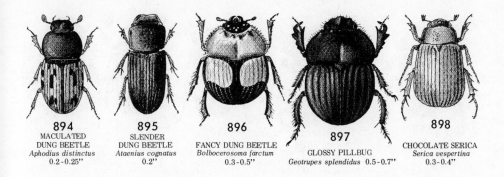

894
MACULATED
DUNG BEETLE
Aphodius distinctus
0.2-0.25"

895
SLENDER
DUNG BEETLE
Ataenius cognatus
0.2"

896
FANCY DUNG BEETLE
Bolbocerosoma farctum
0.3-0.5"

897
GLOSSY PILLBUG
Geotrupes splendidus 0.5-0.7"

898
CHOCOLATE SERICA
Serica vespertina
0.3-0.4"

(275) CHOCOLATE SERICA: *Serica vespertina* (Melolonthinae, Sericini). **Fig. 898.**

Range: Eastern North America. **Adult:** Smooth, shiny, yellowish brown to dark brown, pronotum densely punctate; elytra deeply striated, each stria with two rows of punctures. **Length:** .3–.4". The **Western Serica,** S. *fimbriata*—orange or velvety brown, elytra finely striated, about .5" long—is found on avocado and other fruit trees and on blackberry in California. The **Asiatic Garden Beetle,** *Maladera* (*Autoserica*) *castanea,* is similar in habits to the Japanese beetle (285)—has taken its place in some areas in the East—but it flies only at night and is attracted to lights; it is more rounded, velvety, chestnut brown, with short yellow hairs on underside, from .3" to .45" long; the larva has a light brown head and grayish stripe down the back, is about .75" long, moves quickly when disturbed.

(276) MAY BEETLE: *Phyllophaga rugosa* (Melolonthini). **Fig. 899.**

This large genus, known as May beetles, includes well over 100 species in North America. The larvae ("white grubs") feed on roots, and are very destructive to corn, timothy, potatoes, pasture grass, nursery plantings; adults feed on leaves of trees, especially oak, elm, hickory, beech, birch, maple, poplar, willow, locust, hackberry, walnut, and pine. Damage is severe in the North Central states and adjoining provinces; injury in the South is mostly from grubs feeding on roots of seedlings. The most injurious species have a three-year cycle which accounts for the three-year cycle of outbreaks. Broods have been designated A, B, and C; adults of A appeared in 1965, 1968, and 1971. The most severe damage occurs the year after heavy flights of the beetles—thus, in 1966, 1969, etc. Brood B is not important, Brood C may cause damage in the same areas in 1968, 1971, etc. Eggs are laid in the soil in balls of earth held together by a sticky secretion, and hatch in three to four weeks. Grubs feed the first summer mostly on decaying vegetation; they burrow deeper to hibernate for the winter, resume feeding the second year, when they cause the most damage. They hibernate again, resume feeding the following spring until June, when they form earthen cells and pupate. Adults emerge from the pupae in a few weeks

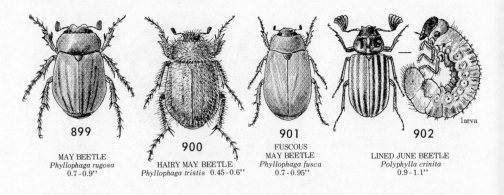

899
MAY BEETLE
Phyllophaga rugosa
0.7-0.9''

900
HAIRY MAY BEETLE
Phyllophaga tristis 0.45-0.6''

901
FUSCOUS
MAY BEETLE
Phyllophaga fusca
0.7-0.95''

902
LINED JUNE BEETLE
Polyphylla crinita
0.9-1.1''

larva

but remain in cells over winter, emerge from the ground the following spring to lay eggs, and the cycle is repeated. The larvae may be distinguished by the double row of spines on the underside of the last abdominal segment.* *P. rugosa* belongs to Brood C, is abundant in central area of Corn Belt. **Range:** Eastern half of the U.S., Montana, Oregon, Alberta, Manitoba, Ontario. **Adult:** Shiny, bright reddish brown or black; antennae ten-segmented. Clypeus (forepart of head, below the front) emarginate or notched, border reflexed or bent back. Claws curved, with median tooth; lower spur of hind tibia of male unarticulated or fixed. **Length:** .7–.9''. Adults are active from April to July.

(277) HAIRY MAY BEETLE: *Phyllophaga tristis.* **Fig. 900.**

Range: Eastern half of the U.S., Colorado, New Mexico. **Adult:** Brownish yellow, sometimes reddish; elytra clothed with short semierect hairs, pronotum with longer erect hairs at base of elytra; antennae ten-segmented. Clypeus concave. Claws slightly curved, with median tooth; lower spur of hind tibia in male articulated (free or movable). **Length:** .45–.6''. It is injurious to crops and trees. Adults are active from early March to late June. *P. futilis* is the same size, reddish brown to yellowish brown; clypeus very feebly emarginate; claws arched, with strong acute median tooth; lower spur on hind tibia unarticulated; it is very common in the eastern half of the U.S. and in Ontario, attacks many different kinds of trees; the males are strongly attracted to lights.

(278) FUSCOUS MAY BEETLE: *Phyllophaga fusca.* **Fig. 901.**

Range: Eastern U.S. and Canada; Idaho, British Columbia. Common in some states and in southwestern Ontario. **Adult:** Shiny, brown to pitchy black; antennae ten-segmented. Clypeus slightly emarginate, border moderately reflexed. Claws curved, with median tooth; lower spur of hind tibia in male fixed. **Length:** .7–.95''. Destructive to crops and trees. Adults appear

---

* Tachinid flies parasitize the grubs and frequently the adults; the latter are also attacked by pyrgotid flies [see Diptera (55)]. The maggots of robber flies and horse flies are predaceous on the grubs.

from March to July. *P. crinita* is reddish brown, head and pronotum with moderately long erect hairs, antennal club much larger than stem, claws variable in the sexes, lower spur of hind tibia in male movable; it is from .5″ to .65″ long, occurs from Mississippi to Texas, where it is one of the most abundant species. The **Prairie May Beetle,** *P. crassissima,* is also reddish brown, with antennal club a little longer than stem, clypeus very feebly emarginate and margin moderately reflexed, lower spur of hind tibia in male fixed; it is from .6″ to .85″ long, often destructive to crops in the central states. *P. lanceolata*—known as the "wheat white grub"—is broad, grayish, from .6″ to .7″ long, unable to fly; it sometimes severely damages wheat in the Great Plains.

(279) LINED JUNE BEETLE: *Polyphylla crinita.* **Fig. 902.**

Range: Pacific coast, from California to British Columbia. **Adult:** Brown, with stripes of silver scales on elytra, as shown; fairly long, erect hairs, few scales on head and pronotum; antennal club many-segmented. **Length:** .9–1.1″. The larvae feed on roots of ornamental shrubs and trees, and young fruit trees. *P. hammondi* is western, grayish brown, with three white stripes on pronotum and five on elytra (counting the wide sutural stripe as one), about 1″ long. The **Ten-lined June Beetle,** *P. decemlineata,* is eastern and western in distribution.

(280) EUROPEAN CHAFER: *Amphimallon majalis.* **Fig. 903.**

Range: First reported from New York in 1940, now occurs in Massachusetts, Connecticut, New Jersey, Pennsylvania, West Virginia, Ohio, Ontario. **Adult:** Light chocolate brown or tan, body hairy but without scales; lamellate antennal club three-segmented, small in the female, long in the male. Adults first appear in June, are most abundant in early July. They emerge from the soil at dusk, swarm into trees and bushes, making a loud buzzing sound as they do so; they take shelter in the soil or undergrowth during the day. **Length:** .5″. **Food:** Adults feed little—on the leaves of various trees—or not at all; the larvae feed on grasses, are most destructive as third instars in the spring and fall. A severe pest of lawns and pasture grasses. The **Northern Masked Chafer,** *Cyclocephala borealis* (Fig. 904), is widely distributed, pale yellowish to brownish, with sparse fine hairs, from .4″ to .5″ long; adults feed on pigweed, are strongly attracted to lights, the grubs feed on grasses; the **Southern Masked Chafer,** *C. immaculata,* takes its place in the southern states.

(281) NARROW FLOWER SCARAB: *Dichelonyx elongata* (Macrodactylini). **Fig. 905.**

Range: Eastern North America. **Adult:** Shiny, head and pronotum brown or pitch black, the latter punctate and pubescent (sparsely so in the male and densely so in the female); elytra brownish yellow, with greenish or purplish hue. **Length:** .3–.4″. It is found on oak, willow, and other trees, on

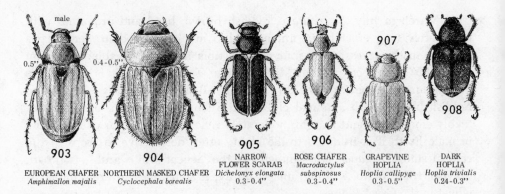

male

0.5"   0.4-0.5"

907

908

903

904

905
NARROW
FLOWER SCARAB
Dichelonyx elongata
0.3-0.4"

906
ROSE CHAFER
Macrodactylus
subspinosus
0.3-0.4"

GRAPEVINE
HOPLIA
Hoplia callipyge
0.3-0.5"

DARK
HOPLIA
Hoplia trivialis
0.24-0.3"

EUROPEAN CHAFER
Amphimallon majalis

NORTHERN MASKED CHAFER
Cyclocephala borealis

flowers of wild plum, wild rose and other plants. *D. crotchi*—the same size, reddish brown, with shiny green elytra—chews scallops in the needles of ponderosa, Jeffrey, and lodgepole pines in the West.

(282) ROSE CHAFER: *Macrodactylus subspinosus.* **Fig. 906.**

Range: Throughout the U.S. and Canada. **Adult:** Reddish brown, elytra with dense yellowish scalelike hairs, underside of body black; long-legged, female more robust than the male. Eggs are laid about 6" deep in the soil, in groups of 6 to 25, each in a separate pocket. **Length:** .3–.4". **Larva:** Feeds on roots of grasses and other plants during the summer. Hibernates in the soil in later larval stage, pupates the following May; adults appear in June and early July in great swarms. **Food** (adult): Rose, peony, iris, and other flowers; grape, apple, peach, cherry, plum, berries, vegetables; they eat flowers, leaves, and fruit. They are poisonous to young chickens, are known to contain cantharidin, the blistering agent found in blister beetles.

(283) GRAPEVINE HOPLIA: *Hoplia callipyge* (Hoplini). **Fig. 907.**

Range: California. **Adult:** Reddish brown; head and pronotum darker, the latter with iridescent green to gold center; pygidium and rest of abdomen covered with silvery, shiny scales. The color and size vary greatly, even in the same area. It leaves a round exit hole in the ground when it emerges from the pupa in April; it flies to the flowers to feed, and returns to lay eggs in the soil. **Length:** .3–.5". **Larva:** Typical scarab grub: whitish, C-shaped, with bulbous posterior. It feeds on roots of alfalfa, lawns, strawberries, and other plants during summer, hibernates in late stage. Pupates in the soil in spring. One brood. **Food** (adult): White flowers such as rose, lilies, orange blossoms; they feed in groups on forms or developing bunches, and young leaves of grape, often causing serious damage. *H. trivialis* (Fig. 908), an eastern species, is very similar, from .24" to .3" long, dark reddish brown to black, with silvery scales.

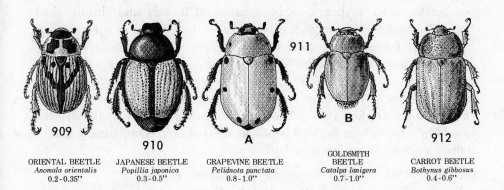

ORIENTAL BEETLE	JAPANESE BEETLE	GRAPEVINE BEETLE	GOLDSMITH BEETLE	CARROT BEETLE
Anomala orientalis	*Popillia japonica*	*Pelidnota punctata*	*Catalpa lanigera*	*Bothynus gibbosus*
0.2-0.35''	0.3-0.5''	0.8-1.0''	0.7-1.0''	0.4-0.6''

(284) ORIENTAL BEETLE: *Anomala orientalis* (Rutelinae; Anomalini). **Fig. 909.**

Range: Connecticut, New York, New Jersey; Hawaii, Japan. **Adult:** Straw-colored to yellowish brown or black, with black markings as shown. Round white eggs are laid singly in the soil, about 6″ deep. **Length:** .2–.35″. **Larva:** Distinguished from native white grubs by transverse anal slit. It feeds on roots of grasses during the summer, 5″ to 12″ below the surface, comes up near the surface to pupate the following June. Adults emerge in June to August. **Food:** More damage is caused by the larvae, to turf and nursery plantings. The adults feed on flowers. The **Pine Chafer,** A. *oblivia,* is very similar, about .4″ long; the adult chews a notch near the base of a pine needle, causing it to fall over; it infests red, jack, and Scotch pines.

(285) JAPANESE BEETLE: *Popillia japonica.* **Fig. 910.**

Range: Atlantic seaboard from North Carolina to Vermont, westward to Ohio, West Virginia; local colonies have developed farther west, north and south. **Adult:** Bright metallic green; elytra coppery brown, body underneath covered with grayish hairs; two tufts of white hairs on tip of abdomen, five along each side. The female may be distinguished from the male by the rounded front tibial spurs (pointed in the male). They fly only in the daytime [see Asiatic garden beetle (275) above]. Elliptical creamy white eggs are laid in the soil, one to four at a time. The female returns above ground to feed between ovipositions. **Length:** .3–.5″. **Larva:** Grayish white, with dark brown head, about 1″ long full grown; may be distinguished from other white grubs by V-shaped arrangement of two rows of spines on underside of the last abdominal segment. It feeds on roots of plants, causes serious damage to grasses, lawns, vegetables, nursery stock. Three instars. Hibernates as full-grown larva in the soil, pupates in the spring. Adults emerge in July. One brood. The complete life cycle sometimes requires two years. **Food** (adult): Foliage and ripening fruit of grape, apple, cherry, plum, to a less extent peach, quince, nectarines, various berries; corn, soybeans, many ornamental flowers and shrubs, shade trees. They often skeletonize leaves, and feed in a mass on ripening fruit, leaves, and flowers. They feed on some plants that are toxic to them— geranium, castor bean, and bottlebrush-buckeye; the flowers are more toxic than the leaves. Many predators attack the grubs; the most effective parasites are the imported *Tiphia vernalis* and *popilliavora*. [See Hymenoptera (63); also see Diptera (97).] They are also susceptible to "milky disease," caused by a bacterium, *Bacillus popilliae;* it is cultured commercially and the spore preparations used as a control agent.

(286) GRAPEVINE BEETLE: *Pelidnota punctata* (Rutelini). **Fig. 911A.**

Range: Eastern U.S., Indiana. **Adult:** Convex; dull reddish brown or brownish yellow; underside, top of head, and scutellum black, tinged with green; small round black dots on pronotum and elytra as shown; sparse fine punctures irregularly spaced. **Length:** .8–1″. **Food** (adult): Wild and cultivated grapes. The larvae are found in decaying roots and stumps of trees. The **Goldsmith Beetle,** *Catalpa lanigera* (Fig. 911B), is similar in color, habits, and distribution.

(287) CYPRESS BEETLE: *Plusiotis gloriosa.* **Color plate 7i.**

Six species of *Plusiotis* occur in the U.S., all in the Southwest. These handsome beetles are especially interesting because of their sharply restricted range and food plants. **Range:** *P. gloriosa* is restricted to certain zones in the mountains of Arizona, at the lower elevations. **Adult:** Convex, shiny, brilliant green, with silver stripes; edges of pronotum and elytra, scutellum and tarsi match the stripes. On juniper these distinctive beetles are almost impossible to see unless they move. When it is dry they feed at night and spend the day underground; when the humidity is high enough they feed during the day also. **Length:** About 1″. **Food** (adult): Juniper. *P. beyeri* (Color plate 7h) is about 1.3″ long, convex, shiny, yellowish green, with purplish legs, elytra with very fine shallow punctures in rows; it feeds on oaks in the lower elevations of the mountains in Arizona. *P. lecontei* feeds exclusively on pines at higher elevations.

(288) CARROT BEETLE: *Bothynus* (*Ligyrus*) *gibbosus* (Dynastinae, Oryctini). **Fig. 912.**

Range: Throughout most of North America. **Adult:** Reddish brown to black; elytra striated and punctate. Eggs are laid in the soil in early spring, hatch in one to three weeks. **Length:** .4–.6″. **Larva:** White, with bluish tinge; head and spiracles brown, 1.2″ long full grown. It feeds on roots of grasses, cereals, pigweed. Hibernates as adult in the soil. One brood. **Food** (adult): Feeds above and below the surface of the ground on carrot, celery, parsnip, potatoes, beets, corn, cotton, dahlias, elm, oak.

(289) SUGARCANE BEETLE: *Euetheola rugiceps.* **Fig. 913.**

Range: Oklahoma to Kentucky, southward. **Adult:** Jet black; pronotum and elytra with minute punctures; elytra with faint striae. Legs strong, with coarse spines. Mating takes place in the ground in spring. Eggs are laid in earthen cells in groups of three or four. **Length:** About .5″. **Larva:** Dirty white, darker near the posterior end; head brick red, legs pale brown; about 1.25″ long full grown. It feeds on decaying vegetable matter. Hibernates in adult stage. One brood. **Food** (adult): Corn, sugar cane, rice. It causes severe damage to young corn by chewing into the outer wall of the stalk just below the surface.

(290) RHINOCEROS BEETLE: *Dynastes tityus* (Dynastini). **Fig. 914.**

Range: Southern U.S., from Arizona eastward. **Adult:** Yellowish gray, with brown to black spots; rarely, all reddish brown. Male with three horns on pronotum—two short ones and one long one in the middle, the latter curving to meet a long horn on the head; female with small tubercle on the head. **Length:** 1.5–2″. This genus contains the largest beetles, some tropical species being several inches long.

(291) GREEN JUNE BEETLE: *Cotinis nitida* (Gymnetini). **Fig. 915.**

Range: Southern states, northward to Long Island and southern Illinois. **Adult:** Dull velvety green on top, margins of pronotum and elytra brownish yellow; shiny green and orange-yellow on underside. Grayish spherical or oval eggs are laid in the soil in a ball of earth. It is attracted to humus, decaying plants, and manure. **Length:** .8–.9″. **Larva:** Creamy white; head dark brown; about 2″ long full grown. It comes to the surface after a rain, has the habit of crawling on its back. Hibernates as larvae deep in the soil; completes development in the spring, pupates in earthen cell. Adults emerge in June and July. One brood. **Food:** The larva is a serious pest in tobacco plant beds; the tunnels as well as feeding cause damage by drying out the roots. It also attacks lawns, field crops such as corn, oats, sorghum, alfalfa; strawberries, vegetables, and ornamentals. The **Fig Beetle**, *C. texana,* is very similar, occurs from Texas to southern California.

(292) BUMBLE FLOWER BEETLE: *Euphoria inda* (Cetoniini). **Fig. 916.**

Range: Throughout much of North America. **Adult:** Elytra yellowish brown, mottled with black; head and pronotum black, slightly bronzed, densely pubescent. Underside dark, with yellowish hairs. Its buzzing sound when flying accounts for the name. **Length:** .5–.6″. The larvae breed in decaying vegetation, dung, and rotten wood; adults feed on fruits and corn.

(293) YELLOW-SCALED VALGUS: *Valgus seticollis* (Valgini). **Fig. 917.**

Range: Eastern half of the U.S. **Adult:** Black, underside coated with yellow scales; elytra and exposed tip of abdomen densely granulate, pro-

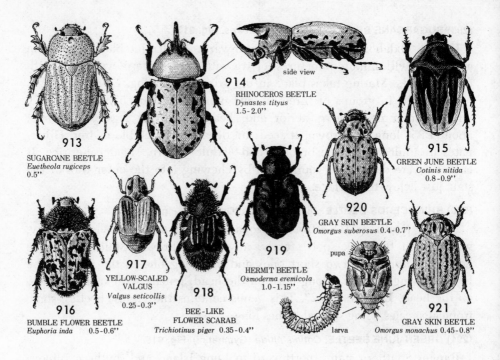

913 SUGARCANE BEETLE
Euetheola rugiceps
0.5''

914 RHINOCEROS BEETLE
Dynastes tityus
1.5-2.0''
side view

915 GREEN JUNE BEETLE
Cotinis nitida
0.8-0.9''

916 BUMBLE FLOWER BEETLE
Euphoria inda   0.5-0.6''

917 YELLOW-SCALED VALGUS
Valgus seticollis
0.25-0.3''

918 BEE-LIKE FLOWER SCARAB
Trichiotinus piger 0.35-0.4''

919 HERMIT BEETLE
Osmoderma eremicola
1.0-1.15''

920 GRAY SKIN BEETLE
Omorgus suberosus 0.4-0.7''

pupa

larva

921 GRAY SKIN BEETLE
Omorgus monachus 0.45-0.8''

notum sparsely punctate. **Length:** .25–.3''. Adults are found on flowers in spring and summer.

(294) BEE-LIKE FLOWER SCARAB: *Trichiotinus piger* (Trichiini). **Fig. 918.**

Range: Eastern half of the U.S. **Adult:** Head and pronotum black, with bronze tinge, clothed in dense, erect yellowish hairs; elytra reddish brown to black, with sparse pubescence and white markings as shown. Underside densely clothed in long white or yellowish hairs. **Length:** .35–.4''. It frequents wild flowers.

(295) HERMIT BEETLE: *Osmoderma eremicola.* **Fig. 919.**

Range: Eastern half of the U.S. **Adult:** Shiny, dark reddish brown, sparsely punctate. **Length:** 1–1.15''. It breeds in rotten wood; adults fly about wooded areas and are attracted to lights.

(296) GRAY SKIN BEETLE: *Omorgus* (*Trox*) *suberosus** (Troginae). **Fig. 920.**

Some classifications place this group in a separate family, Trogidae (from the Greek meaning "gnaw"); they are called skin beetles—a term that is sometimes applied to the Dermestidae as well. Skin beetles feed on carrion, skin, feathers, and dung. Some species are carnivorous in part at least. The larvae are white to straw yellow, with top of head and prothoracic shield

* Genus *Omorgus* restored: Charles W. Baker. See Bibliography: Technical.

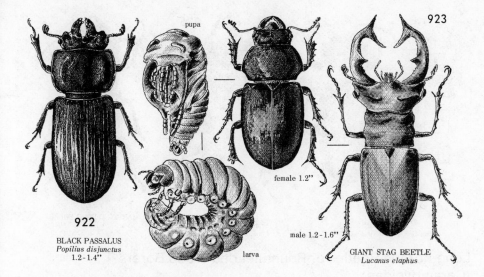

pupa

female 1.2"

male 1.2-1.6"

922

BLACK PASSALUS
Popilius disjunctus
1.2-1.4"

larva

GIANT STAG BEETLE
Lucanus elaphus

orange, red, or brown to black; they pupate in smooth-walled pupal cell in the soil. **Range:** *O. suberosus* ranges over most of the U.S. and Canada. **Adult:** Dull grayish brown, with small tubercles and patches of matted hairs. **Length:** .4–.7". Adults are found in cow dung, carrion, and feathers, and are attracted to lights. The larvae feed on grasshopper eggs and probably dead and decaying grasshoppers around the egg pods. *O. monachus* (Fig. 921) is about the same size, with tubercles larger and more rounded; it is also found on carrion and feathers.

(297) BLACK PASSALUS OR **BESS BEETLE:** *Popilius disjunctus* (Passalidae). **Fig. 922.**

Also called peg beetle, horned passalus, or horn beetle. **Range:** Eastern U.S. **Adult:** Depressed, shiny, black. Pronotum smooth, with deeply impressed median line; elytra with deep striae, finely punctate. **Length:** 1.2–1.4". They live in decayed logs and stumps in loosely organized colonies; adults attend the larvae and feed them premasticated food. They are said to communicate by stridulating.

(298) GIANT STAG BEETLE: *Lucanus elaphus* (Lucanidae). **Fig. 923.**

Range: North Carolina, Virginia, westward to Illinois and Indiana; Oklahoma. **Adult:** Shiny, reddish brown; antennae and legs blackish. **Length** (male): 1.2–1.6" excluding mandibles; add another half-inch or so for the mandibles. The female is about 1.2" long, rarely found. They live under decaying logs or stumps, lay their eggs in crevices of the bark. Adults are reported to feed on honeydew or plant exudations, and the larvae on juices of partially decayed wood; they are attracted to lights.

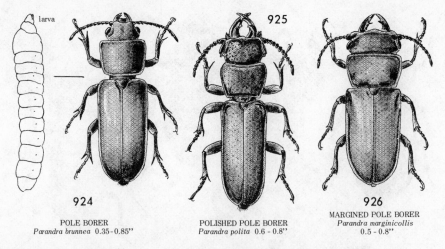

larva

925

924

POLE BORER
Parandra brunnea 0.35-0.85"

POLISHED POLE BORER
Parandra polita 0.6 - 0.8"

926

MARGINED POLE BORER
Parandra marginicollis
0.5 - 0.8"

Long-horned Beetles or Roundheaded Wood Borers (Cerambycidae)

This is a very large family, with some 20,000 species described world-wide, more than 1,200 of them in North America. They are generally large, slender beetles, smooth, hairy, or scaled, mostly brown but often brightly colored. The antennae are normally very long (longer in the male), frequently much longer than the body, with eleven segments but sometimes as many as twenty-five or more. The mandibles are curved, usually stout, pointed, or may be very long and toothed; the maxillary and labial palps are four- and three-segmented respectively. The femora are usually large, tarsi five-segmented, the fourth segment indistinct. The elytra usually cover the abdomen but are sometimes short, exposing part of it; a few are apterous. The abdomen has five or six sternites. The larvae are generally white, elongated, somewhat cylindrical, with head rounded on the sides, antennae inconspicuous and retractile, three-segmented; they have one to five pairs of ocelli, wide thorax with small legs or none. The abdomen is nine-segmented, telescopic, with lateral folds on most of the segments, one or two spines on the dorsal side of the ninth segment and three anal lobes.

Adults commonly visit flowers, and feed on pollen, flowers, leaves, bark, and wood; this food is seldom the plant on which they oviposit. Eggs are laid in wood, under bark, or in the bark crevices. The larvae are wood borers, mine live, dead, and dying trees, often cause considerable damage to forest, shade, and fruit trees, or shrubs. Larval development generally requires from one to three years, but in some cases it may be as short as two or three months. The length of this period is influenced to some extent by temperature and the type and age of wood. The long-horned beetles have numerous enemies. Many birds include the adults as part of their diet, and woodpeckers are very fond of the larvae. Lizards lie in wait on

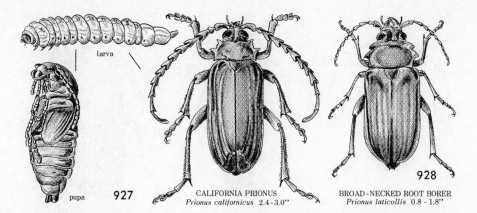

larva

pupa **927**

CALIFORNIA PRIONUS
Prionus californicus 2.4-3.0"

928

BROAD-NECKED ROOT BORER
Prionus laticollis 0.8 - 1.8"

fallen trees or logs and capture many of the beetles as they fly in to mate and oviposit. The most important insect predators are other beetles, especially the clerids, cucujids, and ostomatids. The ambush bug and wheel bug prey on the flower visitors. They are attacked by a host of parasites, chiefly braconid and ichneumonid wasps.

(299) POLE BORER: *Parandra brunnea* (Parandrinae). **Fig. 924.**

Range: Eastern half of the U.S. to Colorado. **Adult:** Shiny, yellowish brown to reddish brown, head darker; pronotum margined, coarsely and densely to finely and sparsely punctate; elytra margined, densely to sparsely punctate. The antennae in this genus are shorter than in most other longhorn beetles; the fourth tibial joint is distinct. **Length:** .35–.85". The larvae are gregarious, attack soft maple and other shade trees, any structural timber, poles, or ties in contact with the ground. The larval period lasts from three to four years. *P. polita* (Fig. 925) is smooth, shiny, pale brown, sparsely punctate, from .6" to .8" long, occurs with *brunnea* in southern part of range, extends west to Texas. *P. marginicollis* (Fig. 926) is smooth, shiny, reddish brown, finely and sparsely punctate, from .5" to .8" long, occurs in southern California west of the Colorado Desert.

(300) CALIFORNIA PRIONUS: *Prionus californicus* (Prioninae, Prionini). **Fig. 927.**

Range: Pacific coast, California to Alaska, eastward to the Rocky Mts. **Adult:** Shiny, uniformly dark reddish brown; pronotum polished, with sparse minute to large punctures, lateral margins each with three prominent sharp teeth; elytra evenly punctate. Antennae 12-segmented. It is nocturnal, attracted to lights, flies with a loud humming sound. **Length:** 2.4–3". **Larva:** A huge white grub, from 2.4" to 3" long. Pupates in cell at end of long frass-lined tunnel in the soil. **Food:** The larva feeds on roots of oak, alder, poplar, madroño, apple, peach, cherry, and other fruit trees, also lives in dead or decomposing wood of deciduous and coniferous trees. The larvae of this and closely related species have been used as food by Indians.

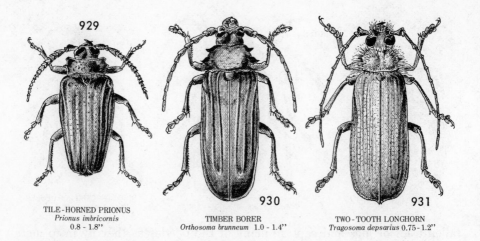

929

TILE-HORNED PRIONUS
Prionus imbricornis
0.8 - 1.8"

930

TIMBER BORER
Orthosoma brunneum 1.0 - 1.4"

931

TWO-TOOTH LONGHORN
Tragosoma depsarius 0.75 - 1.2"

(301) BROAD-NECKED ROOT BORER: *Prionus laticollis.* **Fig. 928.**

Range: Eastern half of North America. **Adult:** Shiny, brownish black, with reddish cast; pronotum sparsely punctate, with three blunt lateral teeth on a side, the basal one not prominent; elytra with irregular punctures, three indistinct longitudinal ridges on each elytron. Antennae 12-segmented. **Length:** .8–1.8". The larva attacks roots of trees and shrubs, is injurious to fruit trees and grapevines.

(302) TILE-HORNED PRIONUS: *Prionus imbricornis.* **Fig. 929.**

Range: Central and eastern states, Ontario. **Adult:** Shiny, dark reddish brown; pronotum flattened at lateral margins, with prominent teeth, dense fine punctures; elytra each with three longitudinal ridges. Antennae very thick in the male, with 18 to 20 overlapping segments, more slender and shorter in the female, with 16 to 18 segments. The female deposits 100 to 200 eggs around the base of trees. **Length:** .8–1.8". The larva feeds on bark first, then the roots; it is injurious to grape, pear, especially oak and chestnut. The larval period lasts three years or more. The larvae of *Derobrachus brevicollis* feed on the roots and stolons of Bahia grass in the Atlantic Coastal Plain; the adult is dark brown, about 1.5" long; eggs are laid singly in the soil; the enormous yellowish grub attains a length of 4"!

(303) TIMBER BORER: *Orthosoma brunneum.* **Fig. 930.**

Range: Eastern half of the U.S., eastern Canada. **Adult:** Shiny, light brown to brown; head and pronotum wrinkled and coarsely punctate, lateral margin of the latter armed with three sharp teeth; elytra unevenly but distinctly punctate, with three costae. Antennae extend to apical third of elytra in the male, to basal third or middle in the female. "Like other prionids, the adult flies late in summer after the flowering of the chestnut" (Craighead). It is attracted to lights. Eggs are inserted in timbers that

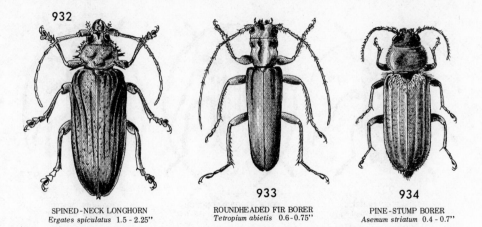

932

933

934

SPINED-NECK LONGHORN
Ergates spiculatus 1.5 - 2.25"

ROUNDHEADED FIR BORER
Tetropium abietis 0.6 - 0.75"

PINE-STUMP BORER
Asemum striatum 0.4 - 0.7"

have been dead several years. **Length:** 1–1.4". **Larva:** It makes an extensive mine, tightly packed with frass. They feed in large numbers in moist heart-wood, leave little of the original wood. The larval period lasts two to three years. Pupation takes place in a cell deep in the wood, requires about one month to complete. **Food:** Cross ties, telephone poles, all structural timbers in contact with the ground.

(304) SPINED-NECK LONGHORN: *Ergates spiculatus* (Ergatini). **Fig. 932.**

Also called spiny wood-borer, ponderosa pine borer, pine sawyer. **Range:** California, Arizona, Pacific Northwest, Montana. **Adult:** Dark brown; head and pronotum somewhat darker than elytra; coarsely punctate; pronotum as wide as base of elytra and with four impressed pits in the male, much narrower than base of elytra and with three raised calluses in the female, lateral margins armed with a few large or many small teeth or spines; elytra without contrasting pale blotches between the costae. It is often attracted to lights. Eggs are laid in crevices of bark of dead trees or stumps. **Length:** 1.5–2.25". One of the largest beetles on the Pacific coast. **Larva:** Creamy white, with reddish head armed with four spines just above the mandibles; about 2.5" long full grown. Excavates large tunnels in sapwood, then heart-wood, packing them with wood fibers. **Food:** Douglas fir and pines. Very destructive to killed or felled coniferous trees, logs, poles. A subspecies, *E. s. neomexicanus*, occurs from Mexico north to Wyoming and South Dakota; the elytra are reddish brown, with pale blotches between the costae.

(305) TWO-TOOTH LONGHORN: *Tragosoma depsarius* (Tragosomini). **Fig. 931.**

Range: Coniferous forests of the Northern Hemisphere, more common above 5,000 feet; western, north central, and northeastern states; Canada. **Adult:** Shiny, brown. Lateral margins of pronotum hairy, with single sharp tooth in the middle; elytra with raised lines, coarse shallow punctures. **Length:** .75–1.2". **Larva:** Resembles that of *Ergates* (304), differs from

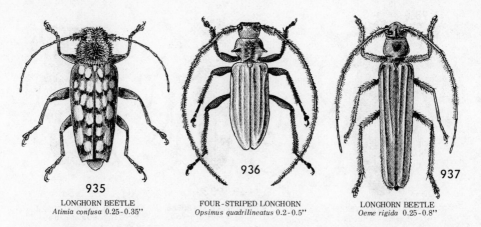

935
LONGHORN BEETLE
Atimia confusa 0.25-0.35"

936
FOUR-STRIPED LONGHORN
Opsimus quadrilineatus 0.2-0.5"

937
LONGHORN BEETLE
Oeme rigida 0.25-0.8"

that of *Prionus* (300) in having cerci. It feeds in the sapwood of decaying logs. **Food:** Pines.

(306) ROUNDHEADED FIR BORER: *Tetropium abietis* (Aseminae, Asemini). **Fig. 933.**

Range: Mountains of western U.S. and Canada. **Adult:** Brown, with fine, short pale yellow pubescence; granulate-punctate. **Length:** .6–.75". The larvae are commonly found working under the bark of felled true firs, are believed to cause the death of standing trees. Most injurious of the roundheaded wood-boring species in western coniferous trees. *T. cinnamopterum* is reddish brown, occurs in the northeastern states and eastern Canada, and in the Pacific Northwest; the larvae feed for a time under the bark of dead spruce, fir, and pine in British Columbia, later bore into the sapwood, causing much damage to logs intended for lumber.

(307) PINE-STUMP BORER: *Asemum striatum* (*atrum*). **Fig. 934.**

Range: Coniferous belt of North America, Europe, Asia; throughout the U.S. and Canada. **Adult:** Opaque; dull brownish or blackish, with dense fine pubescence, elytra often brownish yellow; pronotum wrinkled, densely punctate; elytral striae indistinct; subapical antennal segments small. Eggs are laid in crevices of bark at base of recently dead or fallen trees, telephone poles, stumps still solid and moist. **Length:** .4–.7". **Larva:** Somewhat depressed, with sparse fine brownish hairs, about .9" long full grown; prothorax with fine asperites, abdomen with pair of conical anal spines. It feeds under bark at first, in the sapwood later. Hibernates as larva. Pupates in spring, adults emerge in May and June. **Food:** Pine, larch, fir.

(308) LONGHORN BEETLE: *Atimia confusa* (Atimini). **Fig. 935.**

Range: Eastern Canada and U.S., to Kansas and Texas. **Adult:** Brown; elytra, legs, antennae often reddish; clothed with dense recumbent grayish hairs; pronotum with close, coarse punctures; design on elytra formed by denuded areas. **Length:** .25–.35". Eggs are laid in newly cut trees in the

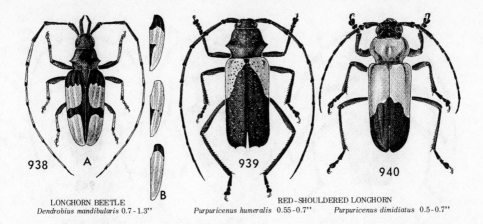

938 A

B

939

940

LONGHORN BEETLE
Dendrobius mandibularis 0.7-1.3"

RED-SHOULDERED LONGHORN
Purpuricenus humeralis 0.55-0.7"   *Purpuricenus dimidiatus* 0.5-0.7"

spring and fall. The larva mines under bark, pupates in cell constructed in the sapwood. Some hibernate as adult, others as larva. **Food:** Cedars, juniper, cypress. In *A. c. dorsalis*, a subspecies, the sides of the pronotum are nearly parallel, denuded areas of elytra are longitudinal, with broad band of dense pubescence at the sides; from .25" to .5" long; it occurs in the Pacific coast states, British Columbia, Baja California on juniper, cypress, cedar, redwood, incense cedar. *A. c. maritima*, dark reddish brown, with less distinct markings, occurs only along the coast of California on Monterey cypress.

(309) FOUR-STRIPED LONGHORN: *Opsimus quadrilineatus* (Cerambycinae, Opsimini). **Fig. 936.**

Range: Pacific coast from California to Alaska. **Adult:** Brown to bluegray, with fine pale yellow pubescence, elytra with dense fine punctures. Antennae about as long as body in the female, much longer in the male. **Length:** .2–.5". **Larva:** It bores into sapwood and heartwood, tunneling with the grain. Hibernates as adult in pupal cell in vertical tunnel. Life cycle normally requires one year. **Food:** Sitka spruce, Douglas fir, true fir, shore pine, Monterey cypress. Sound, recently dead, or dying wood is attacked, sometimes seasoned wood of rustic homes.

(310) LONGHORN BEETLE: *Dendrobius mandibularis* (Trachyderini). **Fig. 938A.**

Range: Mexico to California, Arizona, New Mexico, Texas. **Adult:** Shiny, smooth; elytra yellow, with broad basal band, median band, suture and lateral margin black. Antennae flattened, extending beyond elytral apexes by two segments in the female, by five in the male. **Length:** .7–1.3". The larvae feed on dry dead branches. **Food:** Male paloverde, citrus. In *D. m. reductus* (Fig. 938B), a subspecies, the black markings are greatly reduced or lacking; it feeds on willow in southeastern California, Arizona, and Baja California.

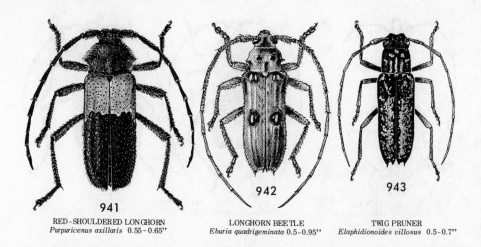

941
RED-SHOULDERED LONGHORN
Purpuricenus axillaris 0.55-0.65"

942
LONGHORN BEETLE
Eburia quadrigeminata 0.5-0.95"

943
TWIG PRUNER
Elaphidionoides villosus 0.5-0.7"

(311) LONGHORN BEETLE: *Oeme rigida* (Melthiini). **Fig. 937.**

Range: Atlantic seaboard to Missouri and Texas. **Adult:** Yellowish brown to reddish brown; top of pronotum densely punctate; elytra with fairly uniform pubescence, and three costae each. **Length:** .25–.8". The larvae mine under the bark of dead and dying cedar and juniper. There may be two or an overlapping of broods in the South.

(312) RED-SHOULDERED LONGHORN: *Purpuricenus humeralis* (Purpuricenini).
Fig. 939.

Range: Central and eastern states, Canada. **Adult:** Dull black with large red or orange triangular humeral areas. Pronotum with single median tooth on lateral margins, five tubercles on dorsal surface, punctate; elytra deeply punctate. **Length:** .55–.7". The larvae feed in dead branches of oak, hickory, birch, and other trees; the adults frequent various flowers. *P. axillaris* (Fig. 941), an eastern species, and *P. dimidiatus* (Fig. 940), found in California, are similarly marked.

(313) LONGHORN BEETLE: *Eburia quadrigeminata* (Hesperophanini). **Fig. 942.**

Range: Southern Canada and the Atlantic seaboard to Kansas and Texas. **Adult:** Yellowish brown; elytra with dense, uneven pale pubescence; pronotum coarsely punctate, with pair of black tubercles on top and a single pair of short lateral spines; elytra each with two pairs of white spots usually bordered with black, and a pair of apical spines. **Length:** .5–.95". The larvae bore contorted mines in the heartwood of solid, dry wood, packing them with frass. **Food:** Oak, hickory, ash, chestnut, locust, maple, beech, elm, cherry.

(314) TWIG PRUNER: *Elaphidionoides* (*Elaphidion*) *villosus* (Elaphidionini). **Fig. 943.**

Range: Eastern U.S. **Adult:** Brown, clothed with irregular patches of

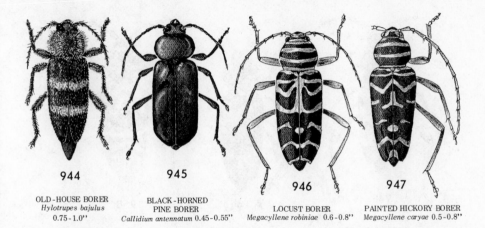

944
OLD-HOUSE BORER
Hylotrupes bajulus
0.75-1.0"

945
BLACK-HORNED
PINE BORER
Callidium antennatum 0.45-0.55"

946
LOCUST BORER
Megacyllene robiniae 0.6-0.8"

947
PAINTED HICKORY BORER
Megacyllene caryae 0.5-0.8"

grayish yellow pubescence; legs and antennae reddish brown; coarsely punctate. Antennae stout, extending over one segment beyond apexes of elytra in the male, more slender and not reaching apexes in the female. **Length:** .5–.7". Eggs are laid on small twigs of living hardwood trees. The larva mines center of twig downward and later cuts outward, causing twig to break off.

(315) OLD-HOUSE BORER: *Hylotrupes bajulus* (Callidiini). Fig. 944.

Range: Cosmopolitan, European in origin; it occurs here only east of the Mississippi River, and in Texas. **Adult:** Flattened; brownish black, sides of pronotum with dense gray pubescence, elytra with two grayish bands of white pubescence. Head with pit in front between lower part of eyes. Antennae extend a little beyond base of elytra in the male. Adults appear in June or July, live from 8 to 16 days. Whitish spindle-like eggs are laid in fan-shaped clusters, in rows or layers in crevices of wood. **Length:** .75–1". **Larva:** Whitish, deeply segmented, with broad thorax, three pairs of ocelli; about 1.25" long full grown. Developmental period from egg to adult requires from three to five years and longer. **Food:** Seasoned wood, preferably pine and spruce. They infest sound wood in newer buildings (in Europe they occur in old buildings). They work in the sapwood; signs of infestation are rasping or ticking sounds of larvae, blistering of wood, boring dust, exit holes of adults.

(316) BLACK-HORNED PINE BORER: *Callidium antennatum hesperum*. Fig. 945.

Range: Throughout North America. **Adult:** Flattened, opaque; brilliant bluish black or purple; finely punctate; elytra with faint costae. **Length:** .45–.55". **Larva:** Yellowish white, legless, about .75" long full grown. It works under bark, extends mine into sapwood. **Food:** Logs and limbs of dead ponderosa pine, sugar pine, Douglas fir, spruce, pine, and other conifers; attacks wood in lumberyards, seasoned wood in rustic structures.

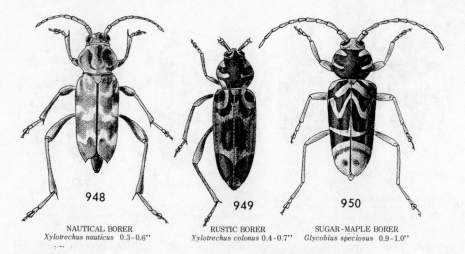

948

949

950

NAUTICAL BORER
Xylotrechus nauticus 0.3-0.6"

RUSTIC BORER
Xylotrechus colonus 0.4-0.7"

SUGAR-MAPLE BORER
Glycobius speciosus 0.9-1.0"

(317) LOCUST BORER: *Megacyllene robiniae* (Clytini). **Fig. 946.**

Range: Eastern and southern U.S. **Adult:** Velvety black, with golden yellow pubescence, forming crossbars; antennae dark brown, extending beyond middle of elytra in the male, falling short of it in the female. **Length:** .6–.8". Adults appear in the fall, feed on pollen of goldenrod. Eggs are laid in the bark. Hibernates as larva, pupates the following August. The larvae bore in the sapwood of black locust. The **Painted Hickory Borer,** *M. caryae* (Fig. 947), is almost identical in general appearance, slightly longer, the elytra somewhat more tapered, middle of second segment of hind tarsi bare rather than densely hairy, antennae extending beyond apexes of elytra in the male and to the middle of the elytra in the female; it breeds in hickory, black walnut, butternut, mulberry, Osage orange.

(318) NAUTICAL BORER: *Xylotrechus nauticus.* **Fig. 948.**

Range: California and Washington, east to Montana. **Adult:** Grayish brown to black, elytra with zigzag white bands; antennae short. **Length:** .3–.6". **Larva:** Creamy white, with amber head, about .7" long full grown. It bores into heartwood of dead, damaged, and weakened trees. **Food:** Oak, madroño, walnut, eucalyptus. The **Rustic Borer,** *X. colonus* (Fig. 949), is similar, slightly larger, dark brown or black, pronotum with four dim white or yellowish spots, elytra with yellowish crossbands similar to *nauticus;* it occurs in eastern Canada and U.S., breeds in oak, maple, beech, and other deciduous hardwoods.

(319) SUGAR-MAPLE BORER: *Glycobius speciosus.* **Fig. 950.**

Range: Northern U.S. and eastern Canada. **Adult:** Black; head, pronotum and elytra marked with dense yellowish pubescence as shown. Antennae extend to middle of elytra in the male. **Length:** .9–1". Adults appear in July. Hibernation takes place in the larval stage; two years are required to

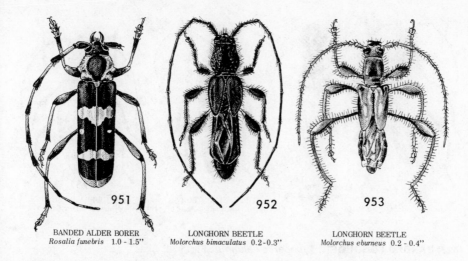

951

952

953

BANDED ALDER BORER
Rosalia funebris 1.0 - 1.5"

LONGHORN BEETLE
Molorchus bimaculatus 0.2 - 0.3"

LONGHORN BEETLE
Molorchus eburneus 0.2 - 0.4"

complete the life cycle. The larvae mine under the bark of hard maples, work into the heartwood; they are very destructive to open stands.

(320) BANDED ALDER BORER: *Rosalia funebris* (Rosaliini). **Fig. 951.**

Also called California laurel borer. **Range:** Alaska to California and New Mexico. **Adult:** Black and pale blue as shown. **Length:** 1–1.5". The larvae bore into dead California laurel, Oregon ash, willow, alder.

(321) LONGHORN BEETLE: *Molorchus bimaculatus* (Molorchini). **Fig. 952.**

Range: Atlantic states, Connecticut, Indiana, Ohio. **Adult:** Dark brown, with lighter spot at base of each elytron; head and pronotum coarsely punctate; elytra short, barely extending to middle of abdomen, coarsely punctate; antennae extend beyond abdomen in the male, beyond elytra in the female. **Length:** .2–.3". Adults are abundant on flowers of red haw in May and June; they breed in dead wood of hickory, maple, ash, dogwood, sweet gum, chestnut, oak, grape, Juneberry, witch hazel. *M. semiustus* occurs in Florida, *M. eburneus* (Fig. 953) in California.

(322) MARGINATED FLOWER LONGHORN: *Leptura emarginata* (Lepturinae, Lepturini). **Fig. 954.**

Range: Eastern Canada and U.S. to Texas. **Adult:** Velvety black, pronotum sparsely punctate; elytra reddish brown, with apexes black, finely punctate. Antennae of male extend beyond middle of elytra. **Length:** 1–1.2". The larvae feed on decaying hardwoods. Also an eastern species, *L. lineola* (Fig. 955) is black, elytra brownish yellow, with narrow black sutural and median stripes, narrow stripe or spots along margin, from .3" to .4" long; the adults visit wild flowers and breed in birch. *L. brevicornis* is dull black, finely punctate, about .6" long, breeds in yellow pine and other conifers in California, Oregon, Washington, Nevada.

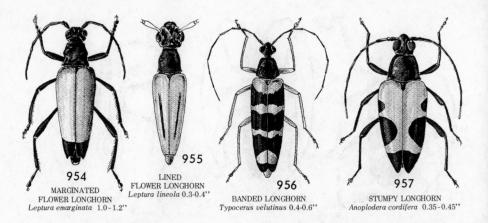

954
MARGINATED
FLOWER LONGHORN
Leptura emarginata 1.0-1.2''

955
LINED
FLOWER LONGHORN
Leptura lineola 0.3-0.4''

956
BANDED LONGHORN
Typocerus velutinus 0.4-0.6''

957
STUMPY LONGHORN
Anoplodera cordifera 0.35-0.45''

(323) BANDED LONGHORN: *Typocerus velutinus*. Fig. 956.

Range: Massachusetts to Florida, west to Indiana, Wisconsin. **Adult:** Black; pronotum with yellowish pubescence; elytra reddish brown, with transverse yellow bands as shown. **Length:** .4–.6''. Adults are abundant on flowers, breed in decayed wood of conifers, birch and other hardwoods.

(324) STUMPY LONGHORN: *Anoplodera cordifera*. Fig. 957.

Range: Eastern U.S. **Adult:** Black, elytra with golden yellow pubescence as shown; pronotum finely punctate. It is common on flowers from May to July. **Length:** .35–.45''. A. *vittata* is about the same length, much narrower, black, elytra each with wide, red longitudinal median stripe which is sometimes reduced to dim basal spots or lacking; it is abundant on flowers in spring and summer, breeds in conifers and hardwoods throughout much of North America.

(325) STRIPED TOXOTUS: *Toxotus vittiger*. Fig. 958.

Range: Eastern Canada and U.S. to Minnesota. **Adult:** Black; elytra reddish or light brownish, with black stripes as shown, densely and finely punctate. **Length:** .4–.8''. The adults are found on flowers, more especially viburnum and hydrangea.

(326) TWO-LINED FLOWER LONGHORN: *Acmaeops bivittatus*. Fig. 959.

Range: Eastern U.S. **Adult:** Brownish yellow, with black markings as shown. Pronotal spots sometimes missing; elytra coarsely punctate, often all bluish black. **Length:** .2–.35''. Adults frequent various wild flowers, especially those of wild crane's bill (*Geranium*).

(327) FLOWER LONGHORN: *Strangalina (Strangalia) soror*. Fig. 960.

Range: High Sierras of California and Nevada. **Adult:** Black and pale yellowish brown; elytra pale, with black median band and tip. **Length:** .45–.7''. Adults frequent flowers of false hellebore or corn lily, and breed in

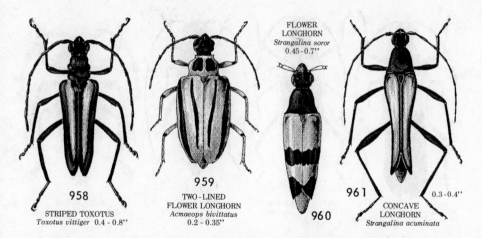

FLOWER
LONGHORN
Strangalina soror
0.45 - 0.7"

958
STRIPED TOXOTUS
Toxotus vittiger 0.4 - 0.8"

959
TWO - LINED
FLOWER LONGHORN
Acmaeops bivittatus
0.2 - 0.35"

960

961
CONCAVE
LONGHORN
Strangalina acuminata
0.3 - 0.4"

dead branches of yellow pine. The **Concave Longhorn,** S. *acuminata* (Fig. 961), is black, elytra dull yellow or brownish, with narrow black margins and suture, sometimes all black, sides curved concavely, narrower at apex than at shoulders, finely punctate, from .3" to .4" long; adults frequent viburnum, and breed in dead ironwood, alder, viburnum in eastern U.S. and Canada.

(328) ELDER BORER: *Desmocerus palliatus* (Desmocerini). **Fig. 962.**

Range: Central U.S. and Canada. **Adult:** Bright metallic blue, basal third of elytra golden yellow. Elytra densely, coarsely punctate, each with three fine ridges. **Length:** .65–.9". The larvae bore into living elderberry; the adults are found on flowers.

(329) GOLDEN ELDER BORER: *Desmocerus auripennis*. **Fig. 963.**

Range: California, Nevada. **Adult:** Black, with bluish sheen; elytra all golden in the male, middle area bluish or greenish in the female; coarsely punctate; posterior margin of pronotum toothed. **Length:** .9–1.1". The larvae require two years for development; they mine trunks and limbs of blue elderberry. In *D. californicus* the elytra are margined with orange in both sexes.

(330) LION BEETLE: *Ulochaetes leoninus* (Necydalini). **Fig. 964.**

Range: Oregon, Nevada, California, in the higher altitudes. **Adult:** "Looks more like a bumble bee than a beetle." Black, pronotum covered with dense yellow pubescence. Elytra very short, tipped with yellow; membranous hindwings almost entirely exposed, extending beyond end of abdomen, which is usually held upward. **Length:** .8–1". The larvae bore into roots of pines, Douglas fir, true firs, spruce.

(331) HAIRY BORER: *Ipochus fasciatus* (Lamiinae, Dorcadiini). **Fig. 965.**

Range: Southern California. **Adult:** Dark mahogany brown, entire body

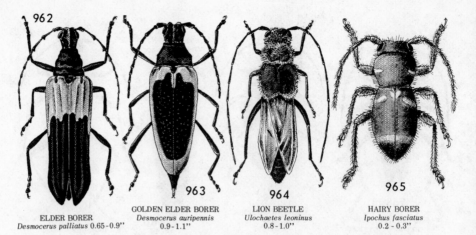

962

963

964

965

ELDER BORER
Desmocerus palliatus 0.65-0.9"

GOLDEN ELDER BORER
Desmocerus auripennis
0.9-1.1"

LION BEETLE
Ulochaetes leoninus
0.8-1.0"

HAIRY BORER
Ipochus fasciatus
0.2-0.3"

sparsely clothed in long whitish hairs; elytra with two irregular bands of white scales, suture not plainly evident. **Length:** .2–.3". The larvae mine the small twigs of Persian walnut.

(332) SPOTTED TREE BORER: *Synaphaeta guexi* (Mesosini). **Fig. 966.**

Range: California, British Columbia. **Adult:** Bluish gray, with minute black and orange markings, two zigzag bands across elytra. Adults emerge in March and April, feed on new growth. **Length:** .6–1". **Larva:** Creamy white, head amber; more than 1" long full grown. **Food:** Walnut, fruit and other trees. The larvae mine dead and dying trees.

(333) LONG-NECKED LONGHORN: *Dorcaschema wildii* (Dorcaschematini). **Fig. 967.**

Range: Eastern U.S. **Adult:** Reddish brown, with dense grayish pubescence; elytra with numerous small bare spots and larger one on margin near the middle. **Length:** .6–.9". The larvae are found in mulberry, Osage orange. The **Twig Girdler,** *Oncideres cingulata* (Onciderini), is rough-surfaced, grayish brown or yellowish brown, with ash gray pubescence forming a broad median band across elytra, from .35" to .6" long; the female deposits eggs in the tips of twigs, which she then girdles lower down with a deep V cut; grubs overwinter in severed portions of twigs and tunnel them in the spring; it is a pest of pecan in the South, works other hardwoods throughout the East. The work of the **Oak Twig Pruner,** *Hypermallus villosus,* is often confused with that of the twig girdler; the *larva* severs the twig from the *inside,* leaving the bark intact. The severed end of a *girdled* twig is jagged; that of a *pruned* twig is smooth and shows the plugged tunnel.

(334) LIVING-HICKORY BORER: *Goes pulcher* (Monochamini). **Fig. 968.**

Range: Eastern U.S. **Adult:** Brownish, with grayish yellow pubescence; elytra with brown band not complete to suture. **Length:** .8–1". The larvae live in hickory and oak. The **Oak Sapling Borer,** *G. tesselatus,* is similar.

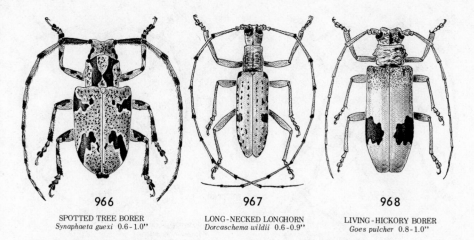

966
SPOTTED TREE BORER
Synaphaeta guexi 0.6-1.0"

967
LONG-NECKED LONGHORN
Dorcaschema wildii 0.6-0.9"

968
LIVING-HICKORY BORER
Goes pulcher 0.8-1.0"

(335) SPOTTED PINE SAWYER: *Monochamus maculosus.* **Fig. 969.**

The larvae of this genus, called "sawyers" because their chewing sounds can be heard, are very destructive to dying, recently killed and felled trees before they can be harvested or milled. Adults feed on needles of conifers or gnaw the bark from young twigs. Females chew pits in the bark and lay from one to six eggs in each. The whitish grubs are footless, from 1" to 1.75" long; they feed from four to eight weeks between the bark and wood, filling the space with wood fragments and loosening the bark. They then bore into the heartwood and turn outward, thus forming a U-shaped burrow ending near the point of entry. Early borings are pushed out and form piles of sawdust; later they are packed in the gallery except in the rounded cell formed later for pupating. Adults emerge by cutting holes in the thin layer of wood and in the bark. Males have very long antennae and front legs. The life cycle is usually completed in one year, sometimes requires two. **Range:** *M. maculosus* occurs throughout the western states. **Adult:** Drab brown, dense pubescence forming grayish spots as shown. Pronotum with prominent lateral tooth on each side. **Length:** .5–1". The larvae work pines and Douglas fir.

(336) WHITE-SPOTTED SAWYER: *Monochamus scutellatus.* **Fig. 970.**

Range: Eastern North America. **Adult:** Black, mottled with gray and brown pubescence; scutellum with white pubescence. Pronotum with a tooth in the middle on each side. **Length:** .6–1.1". The larvae feed on pine. The **Oregon Fir Sawyer,** *M. oregonensis,* is the western counterpart of *scutellatus;* it is black, with gray patches, from .5" to 1.25" long; spine on each side of pronotum, male antennae twice as long as body; the white larvae—from 1" to 1.75" long—bore large holes in scorched, injured, or recently felled Douglas fir, true fir, and lodgepole pine in western forests. The **Northeastern Sawyer,** *M. notatus* (Fig. 971), is dark yellowish brown, with gray and

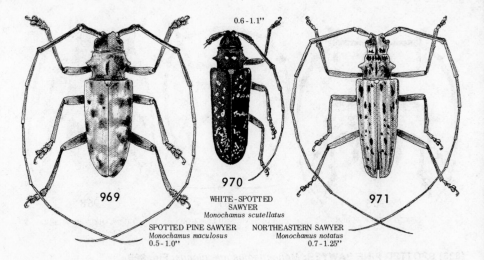

0.6 - 1.1"

970
WHITE - SPOTTED
SAWYER
Monochamus scutellatus

969

971

SPOTTED PINE SAWYER
Monochamus maculosus
0.5 - 1.0"

NORTHEASTERN SAWYER
Monochamus notatus
0.7 - 1.25"

white mottling and black spots, tooth on each side of pronotum, from .7" to 1.25"; male antennae more than twice as long as body; the larvae feed on pine.

(337) SOUTHERN PINE SAWYER: *Monochamus titillator.* **Fig. 972.**

Range: Eastern North America. **Adult:** Brown, mottled with grayish and black pubescence; spine on each side of pronotum, male antennae much longer than body; elytra coarsely punctate, with sharp spine at the apexes. **Length:** .8–1.2". The larvae live in pine.

(338) COTTONWOOD BORER: *Plectrodera scalator.* **Fig. 973.**

Range: District of Columbia to Montana, Indiana, Louisiana. **Adult:** Shiny black, with markings of dense white pubescence forming design as shown; other areas with sparse fine punctures; antennae extend three segments beyond apexes of elytra in the male, just beyond them in the female. **Length:** 1–1.5". Eggs are laid in August and September. Hibernates as larva. Two to three years required to complete development. The larvae develop in cottonwood, poplar, aspen, willow; they attack a tree mostly at the base or below the ground line, sometimes completely girdling it.

(339) PONDEROSA-PINE BARK BORER: *Acanthocinus princeps (spectabilis)*

(Acanthocinini). **Fig. 974.**

Range: New Mexico to California, northward to British Columbia. **Adult:** Speckled gray, with black zigzag bands across elytra as shown; minute black punctures on body and legs; antennae several times longer than body, pronotum with spine on each side, abdomen ending in long ovipositor in the female. **Length:** .75–.9". The larvae mine in dying or dead yellow and other pines, and in stumps. They are found in ponderosa pine killed by the western pine beetle, and are sometimes mistakenly blamed for death of the

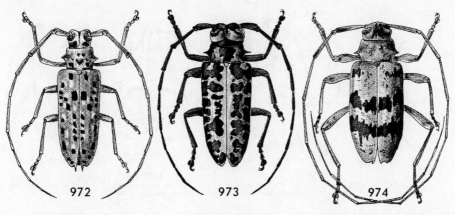

SOUTHERN PINE SAWYER
Monochamus titillator 0.8-1.2"

COTTONWOOD BORER
Plectrodera scalator 1.0-1.5"

PONDEROSA-PINE BARK BORER
Acanthocinus princeps 0.75-0.9"

972   973   974

trees. Keen states that they are "more beneficial than otherwise, in that they rob the bark beetles of their food."

(340) ELM BORER: *Saperda tridentata* (Saperdini). **Fig. 975.**

Range: Midwestern and northeastern U.S., Canada. **Adult:** Brownish black, with grayish and orange pubescence; pronotum and elytra with orange stripes longitudinally and laterally, joined, and with round black spots. Hibernates as larva; adults appear in May, are active until August. The life cycle requires one year ordinarily. **Length:** .35–.6". The larvae breed in twigs and branches of elms, work from the top down, gradually killing the tree; they attack trees of low vitality, primarily street trees.

(341) POPLAR BORER: *Saperda calcarata.* **Fig. 976.**

Range: Throughout most of North America. **Adult:** Reddish brown, with dense grayish pubescence, irregular yellowish patches, small black dots; elytra coarsely punctate, acute apically in the male, rounded in the female, in both cases ending in a spine. Long hornlike ovipositor extends from end of abdomen in the female. Hibernates in larval stage; adults appear in July and August. Two to three years are required to complete the life cycle. **Length:** .8–1.2". The larvae work in the trunk and large limbs of felled and weakened poplar, aspen, cottonwood, willow.

(342) ROUNDHEADED APPLE TREE BORER: *Saperda candida.* **Fig. 977.**

Range: Eastern half of the U.S. and Canada, westward to Texas, Nebraska. **Adult:** Light brown, with two wide, white stripes; coarsely and densely punctate; underside and head white. It emerges from pupal cell in the wood through a circular hole in the bark. Adults appear from June to August; they feed on the bark of twigs, middle rib or stem of leaves, but do little harm. Eggs are laid in slightly curved slit in the bark, singly or in

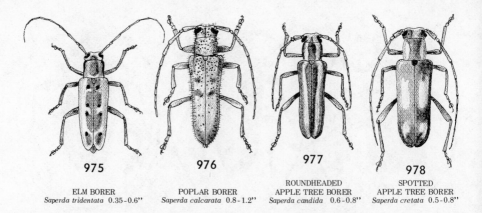

975	976	977	978
ELM BORER	POPLAR BORER	ROUNDHEADED APPLE TREE BORER	SPOTTED APPLE TREE BORER
Saperda tridentata 0.35-0.6"	*Saperda calcarata* 0.8-1.2"	*Saperda candida* 0.6-0.8"	*Saperda cretata* 0.5-0.8"

groups, usually just above the ground line. They attack three- to ten-year-old trees. **Length:** .6–.8". **Larva:** Whitish, with brown head and black mandibles, footless; nearly 1.5" long full grown. Hibernates in larval stage, requires two to four years to complete its life cycle. It first attacks the inner bark, spends the first winter in a burrow near the ground, next spring extends burrow into wood and up the trunk. Several larvae may girdle a tree the first season and kill it. **Food:** Apple chiefly; also pear, quince, possibly other trees. The most destructive borer attacking apple in the East. The **Spotted Apple Tree Borer,** S. *cretata* (Fig. 978), is similar, from .5" to .8" long, brown, with white stripe along each side of pronotum, a large white spot at sides and near center of elytra, a smaller spot near the apexes; it attacks the larger limbs (not the trunk), lays eggs in pairs; the larvae mine in opposite directions.

(343) LINDEN BORER: *Saperda vestita.* Fig. 979.

Range: Northeastern U.S. and southern Canada. **Adult:** Black, with grayish or yellowish green pubescence, usually three bare spots on each elytron; densely and coarsely punctate. Adults appear in midsummer, feed on green bark, larger veins and stems of leaves. Eggs are laid in groups of two or three, in notches cut in the bark, near ground level. **Length:** .5–.8". The larvae mine under the bark near ground line, work into wood and downward into surface roots, occasionally attack the lower limbs. **Food:** Linden (basswood).

(344) RASPBERRY CANE BORER: *Oberea bimaculata* (Phytoeciini). Fig. 980.

Range: Throughout the U.S. and Canada. **Adult:** Shiny black; pronotum bright yellow or orange, with two black spots, deep coarse punctures; elytra with coarse punctures in rows. Adults appear in June. The female cuts two rows of punctures around stem of new shoot a half inch or so apart, six to eight inches from the tip, places eggs between. **Length:** .3–.5". **Larva:** Whitish, legless, about .75" long full grown. It bores into pith,

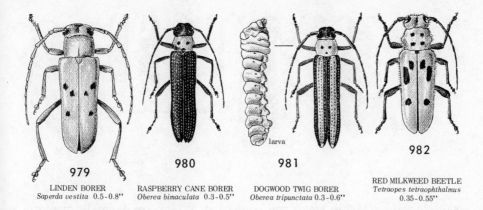

979
LINDEN BORER
Saperda vestita 0.5-0.8"

980
RASPBERRY CANE BORER
Oberea bimaculata 0.3-0.5"

larva

981
DOGWOOD TWIG BORER
Oberea tripunctata 0.3-0.6"

982
RED MILKWEED BEETLE
Tetraopes tetraophthalmus
0.35-0.55"

works its way down toward base of cane. [See (169) and p. 632.] Hibernates as larva. Two years are required to complete development. **Food:** Raspberry primarily; also rose, blackberry, asters. The **Dogwood Twig Borer,** *O. tripunctata* (Fig. 981), is similar, from .3" to .6" long, head usually dark brown, pronotum yellow, with three black spots, elytra black, with wide yellowish longitudinal stripe on each; the female girdles the shoot as described above; the larvae tunnel twigs of dogwood, elm, fruit trees, viburnum and other shrubs. The **Azalea Stem Borer,** *O. myops,* is yellowish, with two small black spots on pronotum; about .5" long; it attacks tips of azalea, rhododendron, mountain laurel in the manner of *bimaculata.*

(345) RED MILKWEED BEETLE: *Tetraopes tetrophthalmus* (Tetraopini). **Fig. 982.**

Range: Eastern North America. **Adult:** Red, pronotum with four black spots, elytra with three large black spots as shown; humerus, scutellum, body beneath black. **Length:** .35–.55". Breeds in milkweed. *T. femoratus* is very similar, with four small black spots on pronotum, two pairs of black spots on elytra (the oblong pair of *tetrophthalmus* missing), from .3" to .6" long; it is widely distributed throughout the U.S., common in the Southwest.

Leaf Beetles (Chrysomelidae)

This is one of the largest families in the order, with approximately 20,000 species described, close to 1,400 in North America. They are minute to small beetles of varying shapes—elongated or oval, flattened or convex—often brightly colored, with iridescent and metallic hues. The antennae are short and simple, as a rule; they are eleven-segmented (a few are ten-segmented), inserted in front of the eyes. The mandibles are short, inconspicuous. Legs are short, with hind femora often enlarged for jumping, tibial spurs usually lacking; tarsi are five-segmented, the third segment bilobed in a majority of

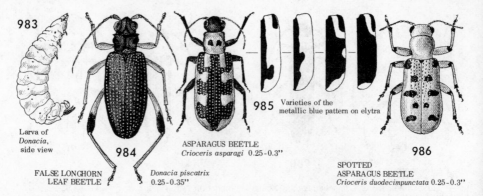

983

Larva of
Donacia,
side view

984

FALSE LONGHORN
LEAF BEETLE

Donacia piscatrix
0.25-0.35"

ASPARAGUS BEETLE
Crioceris asparagi 0.25-0.3"

985 Varieties of the
metallic blue pattern on elytra

986

SPOTTED
ASPARAGUS BEETLE
Crioceris duodecimpunctata 0.25-0.3"

species, the fourth very small and concealed. Wings are generally well developed, seldom wanting, elytra covering the body. The abdomen has five visible sternites. The females are mostly oviparous. Eggs are laid on plants, in the tissues, or in the soil.

The larvae are greatly varied in form, have eight abdominal segments visible dorsally, the ninth and tenth tubelike. The head is small, exserted, with one to six pairs of ocelli or none; antennae with one, two, or three segments. Those feeding on plants are somewhat caraboid (resemble carabid larvae), often wrinkled, with spines or tubercles (Fig. 9C); they move about on plants, mine leaves or stems, or live in galls. Underground forms are usually scarabaeoid (resemble scarabaeid larvae), feed in or on plant roots. The grublike larvae of the subfamily Donaciinae (Fig. 983) (containing some 60 species) are truly aquatic. Leaf miners occur in the subfamilies Galerucinae and Hispinae. The larvae of Cryptocephalinae and Chlamisinae are casebearers. Both adults and larvae of the chrysomelids feed on plants; many species are serious agricultural pests.

(346) FALSE LONGHORN LEAF BEETLE: *Donacia piscatrix* (Donaciinae). **Fig. 984.**

Range: Eastern North America. **Adult:** Shiny, bronze, with green or brownish yellow sheen; antennae and legs reddish yellow. Pronotum finely punctate with two rounded tubercles on the sides; elytra with coarse strial punctures. The female crawls down a flower peduncle of the water lily to deposit her eggs below the water surface. **Length:** .25–.35". **Larva:** Grublike, tapered sharply at anterior end, with a pair of forklike processes (spiracles) on the eighth abdominal segment (Fig. 983) which are inserted in the plant stem to capture air; it feeds on submerged stems, petioles, and roots. It spins a cocoon on a submerged plant from silk produced by glands in the mouth; the adult emerges ten months later. **Food:** Yellow water lily. See (365).

(347) ASPARAGUS BEETLE: *Crioceris asparagi* (Criocerinae). **Fig. 985.**

Range: Throughout much of North America; Europe. **Adult:** Smooth, shiny; head metallic blue; pronotum red, with two blue spots; elytra bluish black, each with three large yellowish spots and reddish outer margins. It gnaws out buds from tender tips, causing them to scar and turn brown. Eggs are laid on end, in rows. Hibernates as adult under loose bark or other shelter, in stems of old plants. **Length:** .25–.3″. **Larva:** Plump, wrinkled, bare; dull gray, with black head and legs; about .35″ long full grown. It excretes a black fluid that stains the plant, feeds on tips of plant for two weeks or less, then pupates in the soil. Two to five broods. **Food:** Asparagus.

(348) SPOTTED ASPARAGUS BEETLE: *Crioceris duodecimpunctata.* **Fig. 986.**

Range: Throughout much of North America; Europe. **Adult:** Smooth, shiny; pronotum reddish brown, elytra with twelve black spots. It appears later in the spring than *asparagi;* greenish eggs are glued on their sides to leaves of plant. **Length:** .25–.3″. **Larva:** Orange, otherwise similar to (347). It feeds on developing berries. Life history similar to C. *asparagi* (see above); two broods. **Food:** Asparagus. The larvae are not very destructive, but the adults are. The latter injure plants in same way that the asparagus beetle does; they are more numerous and harmful in some sections. *Syneta albida* (Orsodacninae), light gray or yellowish, about .25″ long, with long slender antennae, is a pest of pear and cherry from northern California to British Columbia, also feeds on apple, plum, prune, small fruits, some nut trees; the beetles eat the skin and flesh of immature fruit, leaving scars and deformities, also feed on stems, causing fruit to drop; eggs are dropped on the ground, the larvae feed on roots, hibernate in the soil and pupate the following spring.

(349) THREE-LINED POTATO BEETLE: *Lema trilineata.* **Fig. 987.**

This is the familiar "old-fashioned potatobug." **Range:** Central and eastern U.S., Texas, California. **Adult:** Reddish yellow; pronotum with two small black spots (not always); elytra with three black stripes, and rows of punctures, coarser at basal half. Sometimes mistaken for the smaller striped cucumber beetle (370), it is readily distinguished by the pronotum, which is constricted at the middle. Hibernates as adult. Yellow to brownish yellow eggs are laid singly or in jumbled clusters, some on end, some flat, on either side of leaf. **Length:** .25–.3″. **Larva:** Grayish yellow, easily distinguished from any other potato pest by the covering of a wet mass of its own excrement. They are gregarious, first skeletonize the leaves, later devour all but the midvein. Pupates in oval white cell in the soil. Two broods. **Food:** Potato, other plants of the nightshade or potato family. See (362).

(350) CEREAL LEAF BEETLE: *Oulema (Lema) melanopus.* **Fig. 988.**

Range: Michigan, Indiana, Ohio, Illinois, Ontario; a native of Turkey and other Mediterranean countries, Europe, first found in Michigan in 1962

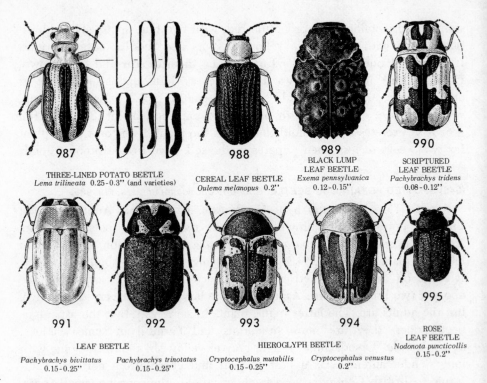

987

THREE-LINED POTATO BEETLE
Lema trilineata 0.25-0.3'' (and varieties)

988

CEREAL LEAF BEETLE
Oulema melanopus 0.2''

989
BLACK LUMP
LEAF BEETLE
Exema pennsylvanica
0.12-0.15''

990
SCRIPTURED
LEAF BEETLE
Pachybrachys tridens
0.08-0.12''

991

LEAF BEETLE

Pachybrachys bivittatus
0.15-0.25''

992

Pachybrachys trinotatus
0.15-0.25''

993

HIEROGLYPH BEETLE
Cryptocephalus mutabilis
0.15-0.25''

994

Cryptocephalus venustus
0.2''

995
ROSE
LEAF BEETLE
Nodonota puncticollis
0.15-0.2''

and recently reported from Kentucky, Pennsylvania, New York, and West Virginia. **Adult:** Elytra and head metallic bluish black, pronotum light reddish orange; coxa and tarsus black, femur and tibia light reddish orange. **Length:** .2''. **Larva:** Yellowish, with brownish black legs; coloring obscured by covering of fecal matter; slightly longer than adult. Thoracic segments with three, five, and five sclerites respectively surrounding base of each leg; three tubercles, each with seta, partially surrounding spiracle at side of each abdominal segment. **Food:** Small grains: chiefly oats, barley, rye, wheat; grasses. Oats are preferred. They chew long strips from leaves (adults usually eat through the tissue), leaving plants white-tipped; yellowish white patches appear in fields heavily infested.

(351) BLACK LUMP LEAF BEETLE: *Exema pennsylvanica* (Chlamisinae). **Fig. 989.**

Range: Eastern and southern U.S. and Canada. **Adult:** Black; head mostly yellowish, pronotum and elytra with yellowish markings, latter coarsely and densely punctate, with scattered tubercles. **Length:** .12–.15''.

(352) SCRIPTURED LEAF BEETLE: *Pachybrachys tridens* (Cryptocephalinae, Pachybrachini). **Fig. 990.**

Range: Eastern U.S. **Adult:** Yellow, with black markings, pattern variable; the black Y on pronotum is sometimes fused with a rectangular spot, with smaller round spots on either side; underside black, antennae and legs pale yellow. Pronotum coarsely punctate, elytra with coarse strial punctures. **Length:** .08–.12". It feeds on hickory, elm, willow, poison ivy. *P. obsoletus* is similar, with markings more diffuse, from .1" to .12" long; it occurs from New Hampshire to Florida, and in British Columbia.

(353) LEAF BEETLE: *Pachybrachys bivittatus.* **Fig. 991.**

Range: Throughout much of North America. **Adult:** Yellow; pronotum with reddish suffusion, elytra each with black or brown median stripe and one to three lateral spots; underside and legs reddish. **Length:** .15–.25". It feeds on willows. *P. othonus,* an eastern species, is similar, occurs on elm and ash. *P. trinotatus* (Fig. 992) ranges from Florida to Connecticut and Indiana, is black, with red triangular spot on head and two red basal spots on pronotum, occurs on wild or false indigos and *Ceanothus.*

(354) HIEROGLYPH BEETLE: *Cryptocephalus mutabilis* (Cryptocephalini). **Fig. 993.**

Range: Eastern North America, to Texas, Kansas, Nebraska, Iowa, Minnesota. **Adult:** Shiny, reddish brown; pronotum orange to red or black, with narrow yellow margin (except along the base), sometimes with two elongate, yellowish, oblique basal spots; elytra creamy yellow to very light orange, with orange to red or black spots and more or less distinct strial punctures (pronotal and elytral spots black in the male). The head is retracted under the pronotum when at rest, a characteristic of this large genus. **Length:** .15–.25". It feeds on a variety of plants, including peanuts. *C. sanguinicollis* is shiny, elytra dark brown to black, pronotum orange to bright reddish, sometimes suffused with black; elytra each with eight rows of punctures; it normally feeds on willow and sagebrush in California, Nevada, Utah, and the Pacific Northwest, but is also found on blackberries, strawberries, prunes, roses, and wild licorice. A subspecies, *C. s. nigerrimus,* found in California, is dark brown to black throughout.

(355) HIEROGLYPH BEETLE: *Cryptocephalus venustus.* **Fig. 994.**

Range: Widely distributed east of the Rocky Mts. **Adult:** Head and pronotum orange to dark reddish brown, latter with yellow margin and two light, oblique basal spots, finely punctate; elytra creamy yellow to light orange, each with two broad black or brown oblique stripes and nine rows of fine strial punctures. **Length:** .2". It frequents fleabane and other flowers in open meadows. Four subspecies besides *venustus* are recognized; the elytra are nearly all black or yellow in the extremes.

(356) ROSE LEAF BEETLE: *Nodonota puncticollis* (Eumolpinae, Iphimeini). **Fig. 995.**

Range: Most of the U.S. as far west as Montana and Arizona. **Adult:** Metallic greenish or bluish; legs and base of antennae reddish yellow; elytra with irregular coarse punctures. They appear in May and June, bore into buds and partly expanded flowers, also feed on tender shoots and foliage;

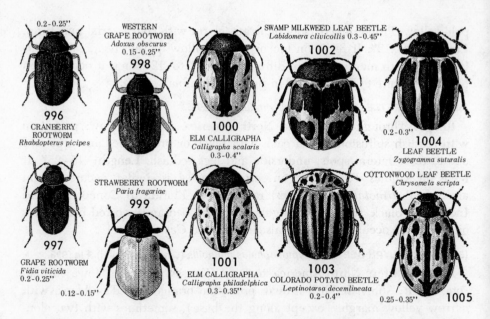

0.2-0.25"

996
CRANBERRY
ROOTWORM
Rhabdopterus picipes

WESTERN
GRAPE ROOTWORM
Adoxus obscurus
0.15-0.25"
998

1000
ELM CALLIGRAPHA
Calligrapha scalaris
0.3-0.4"

SWAMP MILKWEED LEAF BEETLE
Labidomera clivicollis 0.3-0.45"
1002

0.2-0.3" 1004
LEAF BEETLE
Zygogramma suturalis

STRAWBERRY ROOTWORM
Paria fragariae
999

997
GRAPE ROOTWORM
Fidia viticida
0.2-0.25"

0.12-0.15"

1001
ELM CALLIGRAPHA
Calligrapha philadelphica
0.3-0.35"

1003
COLORADO POTATO BEETLE
Leptinotarsa decemlineata
0.2-0.4"

COTTONWOOD LEAF BEETLE
Chrysomela scripta

0.25-0.35" 1005

they frequently swarm over the flowers and riddle them with small holes. **Length:** .15–.2". **Food**(adult): Chiefly rose; also iris, peony, blackberry, raspberry, strawberry, clover, pear, peach, plum. This is a serious pest of roses. The larvae feed on the roots of various plants in wastelands.

(357) CRANBERRY ROOTWORM: *Rhabdopterus picipes* (Colaspini). **Fig. 996.**

Range: Massachusetts to South Carolina, west to the Dakotas. **Adult:** Shiny, dark brown to black, elytra sometimes with margins bronzed or greenish; antennae, tarsi and tibiae reddish yellow; elytra with coarse punctures. **Length:** .2–.25". **Food:** Cranberry, blueberry. The larvae feed on small roots, adults chew holes in the leaves. The closely related **Grape Colaspis,** *Colaspis flavida,* is pale brown, about .15" long, breeds on roots of lespedeza, red clover, and grass; the larvae are very destructive to germinating seeds and seedlings of cultivated rice when this crop follows lespedeza and other legumes; farther north they are common on corn roots after the seedling stage, but the insect is not now a significant pest of grape or strawberries.

(358) GRAPE ROOTWORM: *Fidia viticida* (Leptotini). **Fig. 997.**

Range: East of the Rocky Mts. **Adult:** Reddish brown, with grayish yellow pubescence; pronotum with dense fine punctures, elytra with coarse punctures in rows. **Length:** .2–.25". Adults feed on upper surface of grape leaves, eating a series of chainlike holes scattered over the leaf. Their damage is unimportant compared to the larvae, which devour small roots and burrow into larger ones. The closely related **Western Grape Rootworm,** *Adoxus* (*Bromus*) *obscurus* (Adoxini) (Fig. 998), is brown or black, with short gray pubescence, pronotum with fine punctures, elytra with fine punctures in rows, from .15″ to .25″ long; it is widely distributed, attacks grape in California in the manner of the preceding species.

(359) STRAWBERRY ROOTWORM: *Paria fragariae* (Typophorini). **Fig. 999.**

Range: Throughout North America. **Adult:** Highly variable: pronotum black or yellow, smooth or punctate, elytra all black or yellowish, or yellowish with two to four black spots which may be fused to form stripes; fine strial punctures. They feed on foliage at night, causing small holes in the leaves, and are found at base of plant during the day. Hibernates as adult in trash. Eggs are laid around crowns of plants in the spring. **Length:** .12–.15″. **Larva:** White, with brown head; it feeds on roots, and may be distinguished from root *weevils* by the presence of legs. Damage to plants is caused by the larvae rather than the adults. One and partial second brood. See (410).

(360) ELM CALLIGRAPHA: *Calligrapha scalaris* (Chrysomelinae, Doryphorini). **Fig. 1000.**

Range: East of the Rocky Mts. **Adult:** A beautiful beetle—shiny, head and thorax dark metallic green, elytra pinkish cream with green markings as shown, moderately punctate. Eggs are laid on end attached to leaf, limb, or trunk, usually in clusters. **Length:** .3–.4″. **Larva:** Humped, sluglike, yellowish or cream-colored, about .4″ long full grown; head with black mark on each side, collar of six spots behind the head; abdomen with black line down the middle and gray oblong spots on each side of line, eight shiny black spots on the sides. It chews holes in leaves from underside, pupates in the soil. Hibernates as adult under loose bark. One brood in north central states. **Food:** Elm, hazel, linden. *C. philadelphica* (Fig. 1001) is dark metallic green, elytra pale yellowish, with black spots and stripes as shown, from .3″ to .35″ long; it occurs on dogwood, linden, elm in the Atlantic states, west to Nebraska.

(361) SWAMP MILKWEED LEAF BEETLE: *Labidomera clivicollis*. **Fig. 1002.**

Range: Eastern seaboard west to Texas and Nebraska. **Adult:** Head and pronotum bluish black or greenish black, elytra reddish yellow with bluish or greenish markings as shown; finely, sparsely punctate. **Length:** .3–.45″. It is commonly found on swamp and other milkweeds.

(362) COLORADO POTATO BEETLE: *Leptinotarsa decemlineata*. **Fig. 1003.**

Range: Throughout North America. **Adult:** Yellowish brown, with variable number of black spots on pronotum and five black stripes on each elytron (counting the sutural stripe as two); row of punctures on each side of stripes. Yellow or orange eggs are laid on end, grouped in rows, usually on underside of leaves. **Length:** .2–.4″. **Larva:** Cherry red, with black head and legs in early stage, changing to yellowish red or orange, with two rows of black spots on each side of body, about .6″ long full grown. Hibernates as adult in the soil. One to three broods. Both larvae and adults attack the leaves. **Food:** Chiefly potatoes; also tomatoes, eggplant, peppers, tobacco, buffalo bur, ground cherry and other solanaceous weeds (the original host). See (349).

(363) LEAF BEETLE: *Zygogramma suturalis* (Zygogrammini). **Fig. 1004.**

Range: East of the Rocky Mts. **Adult:** Brown, with metallic bronze or greenish sheen; elytra yellow, with broad brown median and sutural stripes; coarsely punctate. **Length:** .2–.3″. It is found on ragweed and goldenrod. The **Sunflower Beetle**, Z. (*Zygospila*) *exclamationis*, devours the foliage of sunflowers, occurs from Arizona and New Mexico to Montana and Manitoba.

(364) COTTONWOOD LEAF BEETLE: *Chrysomela scripta* (Chrysomelini). **Fig. 1005.**

Range: Throughout North America. **Adult:** Dull greenish yellow or reddish yellow, with black spots; elytra coarsely punctate. Yellowish or reddish eggs are laid in clusters on underside of leaves. **Length:** .25–.35″. **Larva:** Black. It skeletonizes the leaves (adults leave only the midrib and main veins). Five broods in the West. **Food:** Willow, cottonwood, poplar. Trees are often defoliated. The **Willow Leaf Beetle**, *C. interrupta* (Fig. 1006), is slightly smaller, dark metallic green and yellow, each elytron with seven black spots which may merge to form transverse bands; it ranges from California to Alaska and eastward, feeds mostly on willow. The **Poplar Leaf Beetle**, *C. lapponica* (Fig. 1007), is reddish, with large black spots, some fused to form transverse bands; it is injurious to willow and poplar in the Northwest. The **Aspen Leaf Beetle**, *C. crotchi*—black, with light brown elytra, from .25″ to .3″ long—frequently defoliates trembling aspen in western Canada; its eggs are often sucked dry by syrphid fly larvae, which also attack the early instar larvae of *crotchi*. The closely related *Chrysolina gamelata* and *hyperici* were imported from Australia in 1943 to control the Klamath weed (St.-John's-wort, or goatweed) in northern California; *gamelata* is about .25″ long, usually some shade of dark blue, blue-green, or bronze; the larva is gray-green, mottled, about .35″ long.*

(365) WATERLILY LEAF BEETLE: *Pyrrhalta* (*Galerucella*) *nymphaeae* (Galerucinae, Oidini). **Fig. 1008.**

Range: Throughout North America. **Adult:** Brownish yellow, with fine pubescence; head and elytra black, pronotum with impressed median line and three black spots; elytra coarsely and densely punctate. Eggs are laid in small clusters on upper surface of leaf. **Length:** About .2″. **Larva:** Un-

* See: A. H. Murphy et al., *Improving Klamath Weed Ranges,* Cir. 437, California Agr. Exp. Sta., Davis (1954).

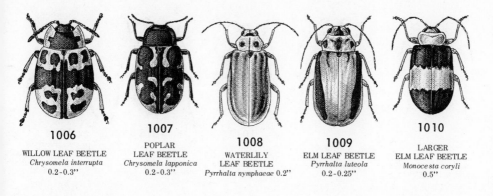

1006	1007	1008	1009	1010
WILLOW LEAF BEETLE	POPLAR LEAF BEETLE	WATERLILY LEAF BEETLE	ELM LEAF BEETLE	LARGER ELM LEAF BEETLE
Chrysomela interrupta 0.2–0.3"	*Chrysomela lapponica* 0.2–0.3"	*Pyrrhalta nymphaeae* 0.2"	*Pyrrhalta luteola* 0.2–0.25"	*Monocesta coryli* 0.5"

like the false longhorn leaf beetle (346), it has prolegs, lacks spines on the eighth abdominal segment; it can't swim, is not truly aquatic. **Food:** Yellow pond lily, white or pink water lily. All stages are found on leaf surfaces. The **Cherry Leaf Beetle,** *P. cavicollis,* occurs east of the Rocky Mts., is similar in size and shape, orange or reddish brown, with fine pubescence, elytra coarsely but not densely punctate. The **Gray Willow Leaf Beetle,** *P. decora decora,* is also eastern in range, the same size and shape (sides, or pronotum, are less angular, more curved), yellowish to dark brown with fine silky pubescence; its western counterpart is the **Pacific Willow Leaf Beetle,** *P. d. carbo.*

(366) ELM LEAF BEETLE: *Pyrrhalta (Galerucella) luteola.** Fig. 1009.

Range: Throughout North America. **Adult:** Yellowish to olive green, with fine pubescence; three black spots on pronotum, two lateral stripes on eiytra, latter finely and evenly punctate. Orange-yellow, spindle-shaped eggs are laid on end in groups on underside of leaves during May and June. **Length:** .2–.25". **Larva:** Black when newly hatched, turns yellow, with two black stripes down the back, black head, legs, tubercles; about .5" long full grown. They feed gregariously on underside of leaves, skeletonizing them. Pupates in bark crevice, hibernates as adult on or near the trees or in buildings. One or two broods, sometimes a partial third. **Food:** Elm. A chalcid wasp, *Tetrasticus brevistigma,* frequently kills a large percentage of the pupae in the northeastern states; in moist seasons a fungus often kills many pupae and adults.

(367) LARGER ELM LEAF BEETLE: *Monocesta coryli.* Fig. 1010.

Range: Mississippi to Florida, north to Pennsylvania, west to Kansas. **Adult:** Yellowish, elytra with broad greenish blue basal and apical bands. Yellow egg masses are glued to underside of leaves in June and July. **Length:** About .5". **Larva:** Metallic reddish brown. They feed gregariously for three to six weeks. Three instars. Hibernates as mature larva in cell

* The name *luteola* (Müller) = *xanthomelaena* (Schrank).

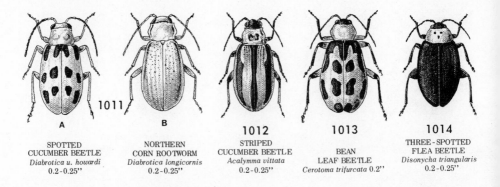

A	B	1012	1013	1014
SPOTTED CUCUMBER BEETLE *Diabrotica u. howardi* 0.2-0.25"	NORTHERN CORN ROOTWORM *Diabrotica longicornis* 0.2-0.25"	STRIPED CUCUMBER BEETLE *Acalymma vittata* 0.2-0.25"	BEAN LEAF BEETLE *Cerotoma trifurcata* 0.2"	THREE-SPOTTED FLEA BEETLE *Disonycha triangularis* 0.2-0.25"

formed in the soil. Pupates in the spring. One brood. **Food:** Elm, hawthorn, hazelnut, red birch, pecan.

(368) SPOTTED CUCUMBER BEETLE (SOUTHERN CORN ROOTWORM):
Diabrotica undecimpunctata howardi. **Fig. 1011A.**

Range: Most of the U.S. east of the Rocky Mts., southern Canada, Mexico. **Adult:** Greenish yellow, with black head, and eleven black elytral spots (counting the common basal spot as one); antennae dark brown excepting three basal segments which are yellow; elytra finely and irregularly punctate. In the North they hibernate in adult stage in any shelter near plants; many migrate north and south with the seasons, and may travel 500 miles in three or four days. Yellow oval eggs are laid in the soil around plants. **Length:** .2–.25".* **Larva:** Yellowish white, with brown head and brownish patch on top of last body segment, about .5" long full grown. It feeds on roots and stems for two to four weeks, then constructs a cell in the soil and pupates. One or two broods in the North, three or more in the South. **Food:** Corn, potatoes, cucurbits, peanuts, many other field and garden crops, fruits, weeds. Adults eat the leaves. The **Western Spotted Cucumber Beetle,** *D. u. undecimpunctata,* is slightly smaller, with body, legs, and antennae entirely black; it occurs west of the Rocky Mts., damages citrus as well as many other plants. The **Banded Cucumber Beetle,** *D. balteata,* is yellowish green, with three black lateral stripes, occurs in the southern states from coast to coast, is most abundant in the fall. The most important parasite of cucumber beetles is *Celatoria diabrotica,* a widely distributed tachinid fly; the maggot consumes most of the body contents, leaving only a brittle shell.

(369) NORTHERN CORN ROOTWORM: *Diabrotica longicornis.* **Fig. 1011B.**

Range: Mostly north central states, scattered east and south. **Adult:** Greenish to yellowish green; antennae, tibiae, and sometimes head and

* See: R. E. Roselle and G. T. Weekman, *Corn Rootworm Control,* E.C. 66–1596, Ext. Serv., Univ. of Nebraska, Lincoln (1966); H. L. Hansen and C. K. Dorsey, *Southern Corn Rootworm Control in West Virginia,* Cir. 102, Agr. Exp. Sta., West Virginia Univ., Morgantown (1957).

pronotum brown; elytra with irregular, close coarse punctures. Very active, it will tumble from flowers and corn silk when disturbed. Eggs are deposited around roots of corn in September and October, and hatch the following spring. **Length:** .2–.25″. **Larva:** Wrinkled, threadlike, white with yellowish brown head. It burrows and feeds in roots of corn until July, when it constructs a cell in the soil and pupates; adults emerge in July and August. Hibernates in egg stage. **Food:** Corn, grasses. One of the most important pests of corn in the Upper Mississippi Valley. The **Western Corn Rootworm,** *D. virgifera,* is yellowish green, with black stripe on the side of each elytron and along the suture, similar to the striped cucumber beetle (370), or wholly black except for margins and apexes; it occurs in the Plains states, east of the Rocky Mts., causes severe injury to corn.

(370) STRIPED CUCUMBER BEETLE: *Acalymma (Diabrotica) vittata.* **Fig. 1012.**

Range: Eastern states and Canada, west to Colorado, New Mexico, and Mexico. **Adult:** Pale yellow to orange-yellow; head and scutellum black, pronotum smooth; elytra with three black stripes as shown, striations and double rows of punctures. Hibernates in adult stage, under litter in wooded areas; emerges in April, feeds on pollen and weeds until cucurbits appear. Orange-yellow eggs are laid at base of plants. **Length:** .2–.25″. **Larva:** Slender; white, brownish at the ends; about .35″ long full grown. The last abdominal segment is more flattened than in the larvae of the rootworms (368) and (369). It feeds for two to six weeks on roots, then pupates in the soil. One brood in northern part of range. **Food:** The larvae feed on roots of cucurbits only. Adults feed on beans, peas, corn, blossoms of many cultivated and wild plants, and transmit bacterial wilt and cucumber mosaic from one plant to another. The **Western Striped Cucumber Beetle,** *A. trivittata,* is very similar, except that the first antennal segment is pale yellow (antennae all black in *vittata*); abundant in southern California, ranges into neighboring states. See (349).

(371) BEAN LEAF BEETLE: *Cerotoma trifurcata* (Monoleptini). **Fig. 1013.**

Range: Eastern states, Canada, west to New Mexico. **Adult:** Shiny, reddish to yellowish; head black, elytra marked with black as shown; specimens vary, some with only the scutellar triangle black, others with the spots running together. **Length:** About .2″. A pest of beans, it also attacks cowpeas, bush clover, hog peanuts, soybeans, and many other legumes.

(372) THREE-SPOTTED FLEA BEETLE: *Disonycha triangularis* (Alticini). **Fig. 1014.**

Range: Widely distributed throughout the U.S. **Adult:** Shiny, black or bluish black: pronotum yellow, with three black spots arranged in a triangle, finely punctate; elytra bluish black, with dense fine punctures. **Length:** .2–.25″. It is commonly found on beets, chickweed, spinach, and

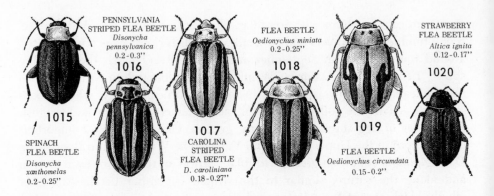

PENNSYLVANIA
STRIPED FLEA BEETLE
*Disonycha
pennsylvanica*
0.2-0.3''
1016

FLEA BEETLE
Oedionychus miniata
0.2-0.25''
1018

STRAWBERRY
FLEA BEETLE
Altica ignita
0.12-0.17''
1020

1015

SPINACH
FLEA BEETLE
*Disonycha
xanthomelas*
0.2-0.25''

1017
CAROLINA
STRIPED
FLEA BEETLE
D. caroliniana
0.18-0.27''

1019

FLEA BEETLE
Oedionychus circumdata
0.15-0.2''

related plants. The **Spinach Flea Beetle,** *D. xanthomelas* (Fig. 1015), is very similar, about the same in size and range; pronotum yellow, without spots, elytra bluish black, both almost without punctures; it is found on chickweed, beets, spinach, pigweed, amaranth.

(373) PENNSYLVANIA STRIPED FLEA BEETLE: *Disonycha pennsylvanica.* **Fig. 1016.**

Range: Widely distributed throughout the U.S. **Adult:** Shiny, black; pronotum pale yellow, with three black spots which are sometimes fused into one band; elytra yellow, with five black stripes as shown, finely punctate. **Length:** .2–.3''. It feeds on arrowhead, dock, knotweed, willow. The **Carolina Striped Flea Beetle,** *D. caroliniana* (Fig. 1017), is very similar, slightly smaller, yellow, with two black spots on pronotum, narrower black stripes on elytra.

(374) FLEA BEETLE: *Oedionychus miniata.* **Fig. 1018.**

Range: Atlantic states to Florida, Texas, Indiana. **Adult:** Brownish yellow, pronotum with broad brown spot, elytra with black stripes as shown, body underneath reddish brown. The stripes are sometimes lacking. **Length:** .2–.25''. It is found on many kinds of small shrubs and plants.

(375) FLEA BEETLE: *Oedionychus circumdata.* **Fig. 1019.**

Range: Texas, east to Massachusetts. **Adult:** Brownish yellow, with black markings as shown; these are variable, with stripes sometimes fused or lacking. Margin of pronotum flattened, elytra coarsely punctate. **Length:** .15–.2''. It is found on low plants such as plantain and verbena, also on beech and walnut.

(376) STRAWBERRY FLEA BEETLE: *Altica ignita.* **Fig. 1020.**

Range: Throughout much of the U.S. **Adult:** Shiny, greenish metallic bronze, sometimes bluish, underside and legs bluish black; deep groove across pronotum near base, elytra sparsely punctate near base. Eggs are laid on strawberry foliage. **Length:** .12–.17''. **Larva:** Yellowish to dark olive-green, hairy, about .2'' long full grown. Hibernates in adult stage. One or

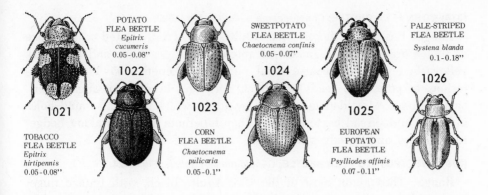

TOBACCO
FLEA BEETLE
*Epitrix
hirtipennis*
0.05-0.08"

1021

POTATO
FLEA BEETLE
*Epitrix
cucumeris*
0.05-0.08"

1022

1023

CORN
FLEA BEETLE
*Chaetocnema
pulicaria*
0.05-0.1"

SWEETPOTATO
FLEA BEETLE
Chaetocnema confinis
0.05-0.07"

1024

1025

EUROPEAN
POTATO
FLEA BEETLE
Psylliodes affinis
0.07-0.11"

PALE-STRIPED
FLEA BEETLE
Systena blanda
0.1-0.18"

1026

two broods. Both adults and larvae feed on leaves, flowers, and fruit. **Food:** Strawberry, evening primrose. The **Grape Flea Beetle,** *A. chalybea,* is similar, a little wider and more rounded, from .17" to .2" long, shiny, dark metallic blue (sometimes greenish or coppery); it overwinters as adult, feeds on buds of grape in most of the eastern states and in Ontario; eggs are laid on secondary shoots, developing larvae feed on leaves; injury to buds resembles that of climbing cutworms. The **Blueberry Flea Beetle,** *A. sylvia,* is at times a serious pest of low-bush blueberries in Maine and eastern Canada; the larvae feed on leaves and blossoms, pupate on the ground; adults feed on leaves and fruit during July and August, when overwintering eggs are laid. *A. carduorum* was released in Canada for control of Canada thistle, a native of Eurasia and its preferred host plant.

(377) POTATO FLEA BEETLE: *Epitrix cucumeris.* **Fig. 1022.**

Range: Mid-Nebraska to Maine, North Carolina, South Carolina. **Adult:** Shiny black, antennae and legs reddish; pronotum finely punctate, elytra with coarse, shallow strial punctures. **Length:** .05–.08". This is a serious pest of potatoes and related plants; it also feeds on a great variety of other plants. Hibernates in adult stage. One or two broods. The **Tuber Flea Beetle,** *E. tuberis,* is difficult to differentiate, ranges from western Nebraska and Colorado to the Pacific Northwest, Arizona, California; it is also very injurious to potatoes. The **Western Potato Flea Beetle,** *E. subcrinita,* shiny bronze or black, occurs west of the Great Plains, including Arizona, California. The **Tobacco Flea Beetle,** *E. hirtipennis* (Fig. 1021), brownish black, occurring in the southern, central, and western states, and the **Eggplant Flea Beetle,** *E. fuscula,* very similar to *cucumeris,* are also injurious to potatoes. Adults chew small round holes in the leaves, causing them to wilt and die; *cucumeris* sometimes defoliates the plants. The larvae attack underground parts of the host plants; those of *tuberis* are destructive to tubers, eating out long black tunnels.

(378) CORN FLEA BEETLE: *Chaetocnema pulicaria.* **Fig. 1023.**

Range: East of the Rocky Mts. **Adult:** Black, faintly tinted with bronze or bluish green; basal segments of antennae orange, tibiae and tarsi dark yellowish; elytra with coarse strial punctures. **Length:** .05–.1″. Most injurious to corn, also feeds on millet, sorghum, broomcorn, sugar beets, oats, cabbage. Most damage is done by adults which eat small round holes in leaves and spread bacterial wilt. They hibernate as adults along hedgerows, fences, and the edges of woods.

(379) SWEETPOTATO FLEA BEETLE: *Chaetocnema confinis.* **Fig. 1024.**

  Range: Throughout most of the U.S. **Adult:** Black, with bronze tinge; legs reddish yellow; elytra with deep, coarse strial punctures. **Length:** .05–.07″. Hibernates as adult, attacks young sweet potato plants shortly after coming out of hibernation. Eggs are laid mostly around bindweed in June; the larvae feed on roots. A new generation of adults appears in July and August, feeds on bindweed and morning glory; sweet potato is attacked only by the hibernating adults. Other plants attacked include corn, wheat, oats, rye and other grasses, sugar beet, raspberry; long narrow grooves are eaten out of upper surface of leaves along the veins. The closely related **European Potato Flea Beetle,** *Psylliodes affinis* (Fig. 1025)—light tan, with black head, from .07″ to .11″ long—is now established in North America, was first reported from New York in 1968 on bitter nightshade, *Lycium,* potato, tobacco, tomato, hops. The native **Hop Flea Beetle,** *P. punctulata,* resembles *affinis* closely, is the same size, black, with bronze or greenish tinge; it is found on hops, rhubarb, strawberry, and various weeds.

(380) PALE-STRIPED FLEA BEETLE: *Systena blanda.* **Fig. 1026.**

Range: New York, southern states to California, north to Utah, Colorado, Idaho. **Adult:** Shiny, pale yellow to brown or blackish, head reddish; pronotum with black lateral margins, sparse fine punctures; elytra usually darker, with pale broad median stripe, dense fine punctures. **Length:** .1–.18″. Adults feed on foliage of various weeds, truck crops, potatoes, grapevines, pear; they occasionally damage young avocado trees in California.

(381) WESTERN BLACK FLEA BEETLE: *Phyllotreta pusilla.* **Fig. 1027.**

Range: Texas to California, north to the Dakotas and Montana. **Adult:** Shiny, black to dark olive-green. **Length:** .06–.08″. It often occurs in swarms, which are injurious to cabbage, cauliflower, turnip, radish, horseradish and other crucifers, corn, sugar beet. Hibernates as adult in litter. Two or more broods. The **Horseradish Flea Beetle,** *P. armoraciae,* is shiny black, elytra dull yellowish, with narrow black margins and sutural stripe, from .1″ to .17″ long; it occurs in the eastern half of the U.S., attacks horseradish and other crucifers; the right elytron of this and some other species of *Phyllotreta* are shown in Fig. 1028.

1027

0.06 - 0.08"

WESTERN BLACK FLEA BEETLE
Phyllotreta pusilla

1028

c

a b

d e f

g h i

WESTERN STRIPED FLEA BEETLE
Phyllotreta ramosa (a),

and other species of
Phyllotreta:
b. *vittata*   f. *libecki*
c. *armoraciae*   g. *oblonga*
d. *discedens*   h. *utana*
e. *zimmermanni* i. *bipustulata*

1029

BASSWOOD LEAF MINER
Baliosus ruber 0.2-0.3"

1030

ARGUS TORTOISE BEETLE
Chelymorpha cassidea
0.3-0.45"

(382) WESTERN STRIPED FLEA BEETLE: *Phyllotreta ramosa.* **Fig. 1028a.**

Range: California, Nevada. **Adult:** Shiny black, with brassy sheen, white or yellow curved stripe on each elytron. **Length:** .08". It is often destructive to crucifers in California. *P. cruciferae,* an Asiatic species, is now a pest of crucifers across southern Canada and northern U.S. The **Striped Flea Beetle,** *P. striolata,* also attacks crucifers, is black with yellowish curving stripe, from .06" to .1" long.

(383) BASSWOOD LEAF MINER: *Baliosus ruber* (Hispinae, Chalepini). **Fig. 1029.**

Range: Eastern half of North America. **Adult:** Flattened, reddish yellow; reddish underneath, legs reddish yellow; elytra deeply sculptured, antennae short and pointing forward. **Length:** .2–.3". Found on basswood, soft maple, locust, oak.

(384) ARGUS TORTOISE BEETLE: *Chelymorpha cassidea* (Cassidinae, Stolaini).
Fig. 1030.

Also called milkweed tortoise beetle. **Range:** Throughout North America. **Adult:** Bright red, pronotum with six black spots, elytra with thirteen similar spots; deeply punctate. It resembles some spotted lady beetles superficially. Hibernates in adult stage. Eggs are laid in clusters of 15 to 30, each attached to leaf by a pedicel. **Length:** .3–.45". **Larva:** Oval, flattened, about .4" long full grown; branched spines around margin and a very long pair posteriorly, called the "anal fecal fork." The fork is turned over the back and used to carry excrement and debris; for this trait they are referred to as "trash carriers" or "peddlers." **Food:** Adults and larvae chew small holes in the leaves of sweet potato, morning glory, milkweed, plantain, cabbage, corn, raspberry.

(385) GOLDEN TORTOISE BEETLE: *Metriona bicolor* (Cassidini). **Fig. 1031.**

Sometimes called goldbug. **Range:** Eastern U.S., including Texas. **Adult:** Brilliant greenish gold; elytra each with conspicuous small circular depression near base, shallow strial punctures; margins of pronotum and elytra flattened, extending beyond sides of body, hiding the head and most of the body. **Length:** .2–.25″. **Larva:** Oval, flattened, brown, with branched marginal spines and anal fecal fork [see (384) above]; they are also "trash carriers." The larvae and adults riddle leaves with small holes. Pupates on leaf, hibernates as adult. **Food:** Sweet potato, eggplant, and related plants of the morning glory family. The **Lineated Tortoise Beetle,** *M. bivittata* (Fig. 1032), is similar in shape and habits, more elongate, about the same size; pale yellow, pronotum with reddish brown triangular spot at base; elytra each with two black wide longitudinal stripes that meet at the apex, and a black sutural stripe; it occurs in the eastern states, westward to Arizona.

(386) BLACK-LEGGED TORTOISE BEETLE: *Jonthonota nigripes.* **Fig. 1033.**

Range: Eastern states, westward to Texas. **Adult:** Dull red, with three indistinct spots on each elytron; black underneath, legs black. **Length:** .2–.3″. It feeds on sweet potato and other plants of the morning glory family. The closely related **Beet Tortoise Beetle,** *Cassida nebulosa* (Fig. 1034), is pale green, with black markings; the larva is a "trash carrier" [see (384) above]. *Gratiana pallidula*—green, .2″ long—feeds only on eggplant and purple nightshade in southern Texas, where it is sometimes a severe pest; the larva—another "trash carrier"—is green, with 16 pairs of branched appendages along the sides, and a pair of long, dark brown fecal forks.

(387) MOTTLED TORTOISE BEETLE: *Deloyala (Chirida) guttata.* **Fig. 1035A.**

Range: Most of North America. **Adult:** Yellow; pronotum usually with large black spot and two small pale ones within; elytra black, with irregular yellow spots and pale margins. **Length:** .2–.25″. It feeds on plants of the morning glory family. The larva (Fig. 1035B) is a "trash carrier."

Seed Beetles (Bruchidae)

Also called pea and bean weevils, or bean weevils, this family was known as Lariidae, is sometimes named Mylabridae. Seed beetles are oval, have a prominent, somewhat triangular pronotum, small head, wide exposed pygidium, and the habit of feigning death. They are minute to small beetles, usually colored black or brown, sometimes reddish or yellowish, often with mottled patterns and fine hairs or scales. The antennae are eleven-segmented and arise in front of the eyes. The tarsi are five-segmented—the first joint elongated, the fourth very small—and have claws with basal hooks. The larvae undergo "hypermetamorphosis." In the first, active stage, they are somewhat caraboid, with well-developed legs, and spined or toothed thorax; after entering the host seed and completing the first molt, most species become more like caterpillars, except that the legs become

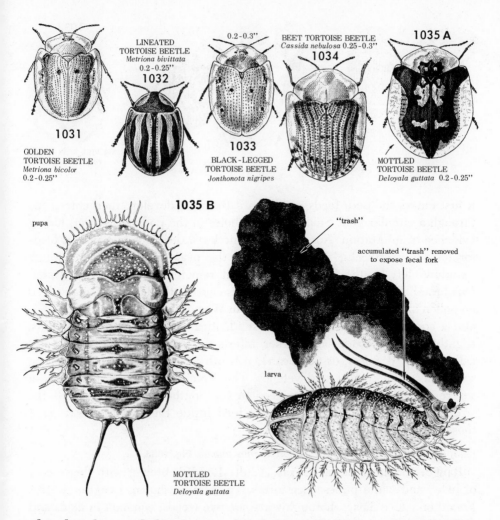

LINEATED
TORTOISE BEETLE
Metriona bivittata
0.2-0.25"
1032

0.2-0.3"

BEET TORTOISE BEETLE
Cassida nebulosa 0.25-0.3"
1034

1035 A

1031

GOLDEN
TORTOISE BEETLE
Metriona bicolor
0.2-0.25"

1033

BLACK-LEGGED
TORTOISE BEETLE
Jonthonota nigripes

MOTTLED
TORTOISE BEETLE
Deloyala guttata 0.2-0.25"

1035 B

pupa

"trash"

accumulated "trash" removed
to expose fecal fork

larva

MOTTLED
TORTOISE BEETLE
Deloyala guttata

reduced or lost, and they become blind. Most of them are found in the seeds of leguminous plants.

(388) PEA WEEVIL: *Bruchus (Mylabris) pisorum.* **Fig. 1036.**

Range: Cosmopolitan; North America, also Europe, Africa, Japan, Australia. **Adult:** Black, with dense coating of reddish brown, yellowish, or whitish hairs forming pattern as shown; pronotum with coarse punctures; elytra with fine strial punctures. Hibernates in adult stage, under any shelter outdoors; they remain in the pea in storage. Oval, orange eggs are glued singly or in pairs to the outside of a pod; unlike the bean weevil (390) they can't breed continuously on dry seed. **Length:** About .2". **Larva:** Crescent-shaped; white or cream-colored, with head and mouthparts brown; about .25" long full grown, when it fills the pea. Spines are lost and the short legs are reduced after entering the pea and completing the first molt;

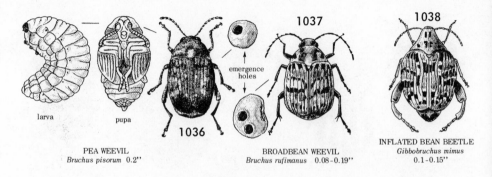

1037

1038

larva   pupa

emergence
holes

1036

PEA WEEVIL
Bruchus pisorum 0.2"

BROADBEAN WEEVIL
Bruchus rufimanus 0.08-0.19"

INFLATED BEAN BEETLE
Gibbobruchus mimus
0.1-0.15"

it first enters the pod, feeds inside until the peas develop, then enters one through a circular hole. Small entrance holes in the pods grow over, hiding evidence of infestation. Normally only one larva develops in a pea; it feeds for four to six weeks, then pupates in the pea. Adult emerges through "window" left by the larva; in storage it remains in the pea all winter. One brood. **Food:** Cultivated peas. Adults cause little damage; females feed on pollen of pea blossoms. The **Broadbean Weevil,** *B. rufimanus* (Fig. 1037), also a cosmopolitan species, is almost identical in appearance and habits, slightly smaller, broader and darker, infests the European broad bean (also called horse, Windsor, or tick bean); it is often a serious pest in California, sometimes attacks peas and vetches. The **Vetch Bruchid,** *B. brachialis*—a severe pest of vetch grown for seed in the South—is similar but half the size, stubby, almost black; the larvae feed inside the growing seeds (not those in storage).

(389) INFLATED BEAN BEETLE: *Gibbobruchus mimus.* **Fig. 1038.**

Range: Indiana, Kansas, Texas. **Adult:** Reddish brown, with dense coat of black and white pubescence forming pattern as shown. **Length:** .1–.15". Found on oxeye daisy, thoroughworts, joe-pye weeds; common in fields and along roadsides.

(390) BEAN WEEVIL: *Acanthoscelides obtectus.* **Fig. 1039.**

Range: Cosmopolitan. **Adult:** Flattened, black, with velvety gray or brown pubescence; elytra marked with irregular brownish transverse bands as shown, having punctate striae, with intervals densely punctate; legs reddish. The white eggs are laid on beans. **Length:** .1–.15". Life history and habits similar to those of pea weevil (388); it can, however, breed continuously in stored beans, and many larvae are normally found in a single bean. **Food:** Most varieties of beans (rarely lima beans), cowpeas; peas and lentils in storage. The **Cowpea Weevil,** *Callosobruchus maculatus* (Fig. 1040), is more elongate, from .11" to .18" long, with more of abdomen exposed, reddish brown, marked with black and gray, two black median-marginal spots on elytra, two similar ones on pygidium; it prefers cowpeas

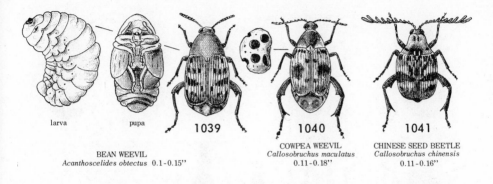

larva   pupa

1039   **1040**   **1041**

BEAN WEEVIL
Acanthoscelides obtectus 0.1-0.15"

COWPEA WEEVIL
Callosobruchus maculatus
0.11-0.18"

CHINESE SEED BEETLE
Callosobruchus chinensis
0.11-0.16"

but will attack beans and peas in storage. The **Chinese Seed Beetle,** *C. chinensis* (Fig. 1041), is about the same size, cosmopolitan in distribution.

Fungus Weevils (Anthribidae)

The members of this family closely resemble the snout weevils (Curculionidae). Distinctive features are the deflexed head, with short broad beak, and the elongated pronotum, with transverse ridge at or near the base; also, the scalelike hairs and variegated color patterns of brown, black, and gray. The elytra cover the abdomen—five abdominal segments are visible on the underside—and the antennae are eleven-segmented, usually with a three-segmented apical club. The larvae are curved, usually elongate-cylindrical in shape, and hairy; they feed on fungi, dead wood and vegetation, and seeds. The adults are usually found on the same plants; some are believed to feed on pollen.

(391) COFFEE BEAN WEEVIL: *Araecerus fasciculatus.* **Fig. 1042.**

Range: Almost cosmopolitan; abundant in the southern states. **Adult:** Color varies from dull to dark brownish, with dense, mottled light and dark brown pubescence; elytra with alternate rows of light and dark spots; legs partly reddish brown. It is common in cornfields of the South, where eggs are laid in soft kernels of corn. **Length:** .1–.2". **Larva:** Creamy white; curved, hairy; dorsal folds of the abdominal segments with short, parallel longitudinal wrinkles. **Food:** Fruit, coffeeberries, cornstalks, corn, many different kinds of seeds and seed pods. It is a pest of a wide variety of dry stored vegetable products, often found in warehouses and shipments.

(392) FUNGUS WEEVIL: *Toxonotus* (*Anthribus*) *cornutus.* **Fig. 1043.**

Range: Ontario to Florida, west to Texas and Missouri. **Adult:** Brownish, with dense covering of hairs; pronotum with five tufts of hair; elytra marked with spots of darker hairs and wide band across suture near the base; elytral punctures coarse, indistinct, in rows. **Length:** .15–.25". It is found on dead branches of oak and on woody fungi.

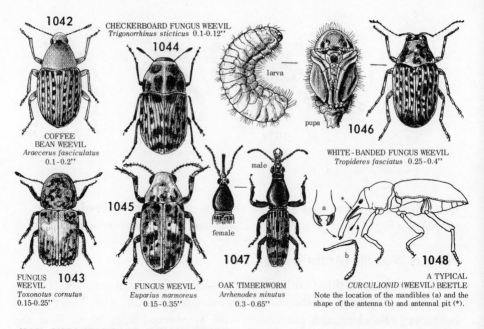

1042

CHECKERBOARD FUNGUS WEEVIL
Trigonorrhinus sticticus 0.1-0.12"

1044

larva

pupa

1046

COFFEE
BEAN WEEVIL
Araecerus fasciculatus
0.1-0.2"

male

WHITE-BANDED FUNGUS WEEVIL
Tropideres fasciatus 0.25-0.4"

1045

female

FUNGUS 1043
WEEVIL
Toxonotus cornutus
0.15-0.25"

FUNGUS WEEVIL
Euparius marmoreus
0.15-0.35"

1047

OAK TIMBERWORM
Arrhenodes minutus
0.3-0.65"

a

b

1048

A TYPICAL
CURCULIONID (WEEVIL) BEETLE
Note the location of the mandibles (a) and the
shape of the antenna (b) and antennal pit (*).

(393) CHECKERBOARD FUNGUS WEEVIL: *Trigonorrhinus (Brachytarsus) sticticus.* **Fig. 1044.**

Range: Quebec and New England to Florida, west to Iowa. **Adult:** Brown, with dense dark brown and grayish yellow pubescence forming pattern as shown; antennae and legs pale reddish; pronotum with dense fine punctures; elytra finely striated, with faint punctures. **Length:** .1–.12". It is known to breed in fungus (smut) on corn and wheat; adults are common on various flowering plants and the blossoms of buttonbush, under logs in the winter.

(394) FUNGUS WEEVIL: *Euparius marmoreus.* **Fig. 1045.**

Range: Ontario and New England to Florida, west to Texas and Iowa. **Adult:** Sooty brown, with pale brown and grayish yellow pubescence forming spots as shown; pronotum with dense coarse punctures, elytra with rows of deep coarse punctures. **Length:** .15–.35". Adults frequent flowers of goldenrod and thoroughwort. Hibernates as adult in rotten beech and maple stumps. Adults and larvae are found mainly in woody fungi and oak logs and stumps.

(395) WHITE-BANDED FUNGUS WEEVIL: *Tropideres (Eurymycter) fasciatus.* **Fig. 1046.**

Range: Widely distributed over the U.S., southern Canada. **Adult:** Dark brown to blackish, with white pubescence forming spot on beak and white lateral band on elytra near apex; pronotum with roughly sculptured surface, elytra with short wrinkles. **Length:** .25–.4". Found on dead twigs and fungi growing on dead trees such as beech.

(396) OAK TIMBERWORM: *Arrhenodes minutus* (Brentidae). **Fig. 1047.**

The brentid beetles—also called straight-snouted or primitive weevils—may be recognized by the long, narrow body and long beak projected straightforward. The ten- or eleven-segmented antennae are inserted in front of the eyes, near the beak; labrum, maxillary and labial palpi are absent. The femora are greatly enlarged. The beak of the female is much longer and more slender than that of the male, and is used for boring holes deep in the wood of trees to accommodate her eggs; the male stands by during the drilling and egg laying and assists his mate in withdrawing the beak when it gets stuck in the wood. There are only six known species of brentids in North America. *A. minutus* is widely distributed in eastern North America. **Adult:** Dark reddish brown to black, elytra with variable yellowish spots, deep striae. **Length:** .3–.65″. The larvae feed on wood and fungi, adults feed on fungi, tree exudations, and other insects; both are found under the bark of dying or fallen oak, beech, maple, and poplar.

Snout Beetles or Weevils (Curculionidae)

This is believed to be the largest family of insects, the number of known species throughout the world being generally estimated at close to 40,000. The most noticeable feature of the group is the prolongation of the head in the form of a snout, which is usually long and slender, curved downward, with the antennae (generally elbowed and clubbed) inserted in the sides. They are minute to large beetles (Fig. 1048), commonly oval, elongated, cylindrical, or somewhat flattened, often covered with a dense coating of scales. Mouthparts are located at the end of the snout, with the mandibles pincer-like, and toothed; the labrum is usually absent. Generally three-segmented, the palpi are short and often concealed. The antennae are nine- to eleven-segmented, usually geniculate (bent abruptly), with a three-segmented club; they are often inserted in grooves, one on each side of the beak. The legs may be short or long; the tarsi are five-segmented, simple or padlike, the fourth segment often very small. Wings may be well developed, reduced, or lacking, with the elytra usually covering the abdomen, but sometimes exposing the pygidium. Five abdominal sternites are visible. Eggs are normally placed in plant tissues after the female makes a hole with her snout.

The larvae are robust, C-shaped, smooth, or wrinkled, with few hairs, and usually without legs. The abdomen is nine- or ten-segmented; the antennae are generally very small, with one segment. One or more pairs of ocelli or pigment spots are usually present but inconspicuous. They pupate in cells made in the soil, in cocoons made of plant fibers or silk spun from anal glands. Adults and larvae feed on plants in a great variety of ways—they burrow in the roots, stems, and seeds, feed on flowers and fruit, mine

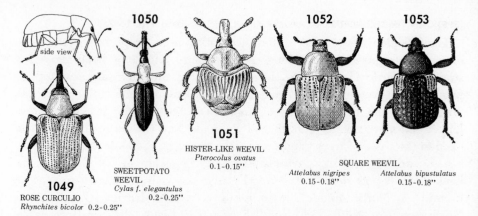

1050

1052   **1053**

side view

1051

HISTER-LIKE WEEVIL
Pterocolus ovatus
0.1-0.15"

SWEETPOTATO
WEEVIL
Cylas f. elegantulus
0.2-0.25"

SQUARE WEEVIL
Attelabus nigripes   *Attelabus bipustulatus*
0.15-0.18"   0.15-0.18"

1049

ROSE CURCULIO
Rhynchites bicolor 0.2-0.25"

and roll leaves, live under the bark of dead, dying, and vigorous trees. The majority live in plant tissues, and many are serious agricultural pests. They exhibit a high degree of host specificity; some species have thus become useful in biological control of noxious weeds. Scattered species are aquatic; a few burrow in submerged stems and roots of aquatic plants without themselves coming in contact with the water.

(397) ROSE CURCULIO: *Rhynchites bicolor* (Rhynchitinae; Rhynchitini). **Fig. 1049.**

Range: Widely distributed throughout North America, and Europe; most common in the northern states and adjacent Canada. **Adult:** Head (behind the eyes), pronotum, elytra bright red; snout, legs, underside black. Pronotum finely punctate, elytra with fine strial punctures and coarse ones between. Snout curved, shorter and stouter in the female. Adults appear in June, eat holes in buds, flower stems, and leaves. Eggs are laid in buds, later in rose hips (fruit). **Length:** .2–.25". **Larva:** White, legless; it feeds on seeds. Pupates and hibernates in the soil. One brood. **Food:** Wild and cultivated roses, peonies.

(398) SWEETPOTATO WEEVIL: *Cylas formicarius elegantulus* (Cyladinae). **Fig. 1050.**

Range: South Carolina to Florida, west to Texas. **Adult:** Antlike in form, smooth, shiny; head, snout, and elytra bluish black, pronotum and legs bright orange red; elytra with faint strial punctures. Last antennal segment cylindrical, longer than the others combined in the male, oval and shorter in the female. It can fly long distances. Minute white eggs are laid in punctures made in vines near the ground or in roots, and in stored potatoes. **Length:** .2–.25". **Larva:** Cream-colored, with head pale brown; about .4" long full grown. It burrows in stems or roots, feeds for two or three weeks, pupates in a burrow. Six to eight broods. It causes severe damage to sweet potatoes. Egg punctures, usually close together, feeding and exit holes are evidence of infestation. **Food:** Sweet potatoes, in field or storage; morning glory.

(399) HISTER-LIKE WEEVIL: *Pterocolus ovatus* (Pterocolinae). **Fig. 1051.**

Range: New England to Florida, west to Iowa and Missouri. **Adult:** Flattened; resembles hister beetle somewhat. Indigo blue; snout, antennae, legs darker; elytra short, with wide shallow striae, irregularly and coarsely punctate; three abdominal segments exposed. **Length:** .1–.15″. It is found on grape, peach, plum, oak.

(400) SQUARE WEEVIL: *Attelabus nigripes* (Attelabinae). **Fig. 1052.**

Range: Eastern North America, to Kansas and Colorado. **Adult:** Bright red, except front of head and underside; latter dark red, remainder black. Pronotum with fine punctures, elytra with rows of coarse punctures. **Length:** .15–.18″. It occurs on leaves of hickory and especially sumac. *A. bipustulatus* (Fig. 1053) is very similar in distribution, in shape and size; black, with bluish tinge, large red spot on each shoulder; elytra with small indistinct punctures; it is found on hickory, walnut, oak, especially red, post, and laurel oaks. Species of this genus are leaf rollers.

(401) SNOUT BEETLE: *Apion porcatum* (Apioninae). **Fig. 1054.**

Range: Massachusetts, Virginia, to Texas. **Adult:** Robust, shiny black; pronotum coarsely punctate; elytra with deep strial punctures, strongly convex intervals, sparse pubescence. **Length:** .1″. *A. impunctistriatum* (Fig. 1055) is slender, slightly smaller, black, with very slender clubbed antennae; elytra shorter, exposing more of abdomen; front femora of male with large tubercle; it ranges from Ontario to Texas. These species and others of this large genus occur on many legumes, flowering plants, and trees. *A. longirostre*—the **Hollyhock Weevil**—is a widespread pest of ornamental hollyhocks. *A. fuscirostre* was recently introduced and established in California for biological control of Scotch broom (*Cytisus scoparius*), an introduced noxious perennial weed which is unpalatable to cattle, deters coniferous seedlings, and burns with such intense heat that it constitutes a fire hazard; the beetle feeds on broom foliage, buds, flowers, and pods in which it oviposits; the larva eats the seeds, develops only in the pods of this species; it hibernates as adult in cracks of plant stems, feeds only slightly on some other leguminous weeds. The seed weevil *A. ulicis* was introduced into Oregon and California (also Hawaii, Australia, and New Zealand) for control of gorse (*Ulex europaeus*), which is used as a hedge and ornamental but has become a pest on the range.

(402) SNOUT BEETLE: *Tachygonus lecontei* (Tachygoninae). **Fig. 1056.**

Range: New Jersey to Florida, west to Texas. **Adult:** Reddish brown, with dense coating of white and brownish yellow flattened hairs and black and white erect bristles; pronotum with tuft of black hairs each side of the middle; elytra with four smaller tufts of white hairs, and rows of very coarse punctures. **Length:** .08–.1″. It is found on leaves of young oak trees.

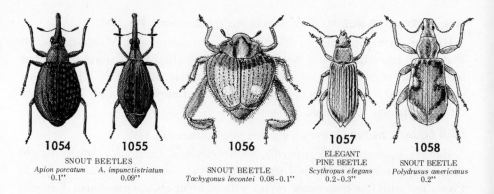

1054
SNOUT BEETLES
Apion porcatum
0.1"

1055
A. impunctistriatum
0.09"

1056
SNOUT BEETLE
Tachygonus lecontei 0.08-0.1"

1057
ELEGANT
PINE BEETLE
Scythropus elegans
0.2-0.3"

1058
SNOUT BEETLE
Polydrusus americanus
0.2"

(403) SNOUT BEETLE: *Polydrusus americanus* (Otiorhynchinae, Polydrusini). **Fig. 1058.**

Range: Canada, northeastern U.S., to Kansas. **Adult:** Shiny brown; dense coating of gray and orange-brown scales forming stripes and pattern as shown; elytra with strial punctures, intervals flat, with row of short, partly erect hairs. **Length:** About .2". It is found on beech, various shrubs and plants.

(404) ELEGANT PINE BEETLE: *Scythropus elegans*. **Fig. 1057.**

Range: Northern Rocky Mt. region, other western states. **Adult:** Variable; metallic blue-green, gold, brass, or bronze, sometimes with lighter stripes along margins of elytra. **Length:** .2–.3". It bites out chunks of needles, leaving edges saw-toothed, feeds on lodgepole pine and other pines.

(405) SNOUT BEETLE: *Aphrastus taeniatus*. **Fig. 1059.**

Range: New England to Missouri. **Adult:** Convex, with dense coating of pale gray scales; head and elytral stripes of pale brown scales; antennae reddish brown; elytra with closely placed punctures in rows, intervals somewhat convex, with row of flattened minute white hairs. Head and beak together shorter than thorax. **Length:** .2–.35". Adults are found on ragweed, pokeweed, papaw, white birch, sassafras, hazel, alder; the larvae are abundant in roots of various grasses.

(406) IMBRICATED SNOUT BEETLE: *Epicaerus imbricatus* (Epicaerini). **Fig. 1060.**

Range: Most of the U.S., west to Colorado, Utah, New Mexico, Arizona. **Adult:** Dense coating of brownish gray scales; elytra with paler zigzag bands, and large, round deep punctures almost as wide as intervals. **Length:** .3–.45". **Food:** Many kinds of plants; blackberry, raspberry, gooseberry, apple, cherry, cucurbits, many field and garden crops are injured.

(407) FULLER ROSE BEETLE: *Pantomorus godmani.** **Fig. 1061.**

Range: Throughout much of North America; very common in California. **Adult:** Pale brown, with sparse covering of scales; elytra each with oblique

* Proposed change of name: Fuller's Rose Weevil, *Pantomorus cervinus*.

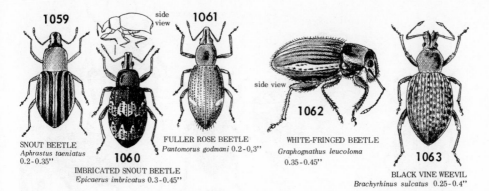

1059

side view

1061

side view

1062

SNOUT BEETLE
Aphrastus taeniatus
0.2-0.35"

1060

IMBRICATED SNOUT BEETLE
Epicaerus imbricatus 0.3-0.45"

FULLER ROSE BEETLE
Pantomorus godmani 0.2-0.3"

WHITE-FRINGED BEETLE
Graphognathus leucoloma
0.35-0.45"

1063

BLACK VINE WEEVIL
Brachyrhinus sulcatus 0.25-0.4"

white band near margin. Short, broad snout; wingless. Eggs are laid in batches around base of small plants or under bark of trees. **Length:** .2–.3". **Larva:** White, including head (unlike other weevils); only the mandibles are dark brown. It feeds on roots, pupates in earthen cell. Hibernates mostly in larval stage, adults emerging in July; the latter sometimes hibernate. **Food:** Almost any kind of plant; the larvae are very injurious to rose and other ornamentals, and strawberries; adults feed on many kinds of fruit trees.

(408) WHITE-FRINGED BEETLE: *Graphognathus leucoloma.* **Fig. 1062.**

Range: Southeastern states; Louisiana, Arkansas to New Jersey. A native of South America. **Adult:** Black, with dense coating of dark gray and grayish brown scales; pronotum and elytra with fringe and stripes of white hairs; elytra with coarse strial punctures, intervals slightly convex, with scattered fine granules. All adults are females, reproduce by parthenogenesis. They feed on foliage, causing little damage. Eggs are laid in the soil around plants, in clusters of 15 to 20 eggs each, covered with a sticky secretion that hardens. They usually hatch in about 17 days, but some remain over winter and hatch in the spring. **Length:** .35–.45". **Larva:** Yellowish white, legless; about .5" long full grown. It feeds on roots, and is very destructive. Eleven months are required to complete development, sometimes two years or more. Usually hibernates as larva, sometimes as egg. One brood. **Food:** Almost any kind of plant; especially potatoes, tobacco, corn, cotton, peanuts, soybeans, velvet beans, ornamental flowers, wild aster, cocklebur, goldenrod. Three species of *Graphognathus* occur over this range, are very similar in appearance and habits.

(409) BLACK VINE WEEVIL: *Brachyrhinus (Otiorhynchus) sulcatus* (Brachyrhinini). **Fig. 1063.**

Range: California, Pacific Northwest, other parts of the country. **Adult:** Black, with patches of yellow scales; elytra with coarse strial punctures. Adults are nearly all females and cannot fly. It feeds at night, chews scal-

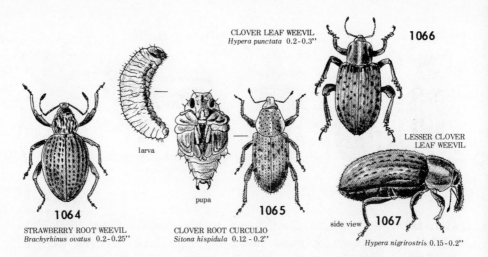

CLOVER LEAF WEEVIL
Hypera punctata 0.2-0.3"

1066

larva

pupa

LESSER CLOVER
LEAF WEEVIL

1064

STRAWBERRY ROOT WEEVIL
Brachyrhinus ovatus 0.2-0.25"

1065

CLOVER ROOT CURCULIO
Sitona hispidula 0.12 - 0.2"

side view 1067

Hypera nigrirostris 0.15-0.2"

lops in the edges of leaves without serious damage; found around the base of plants during daytime. Eggs are laid in the soil around plants. **Length:** .25–.4". **Larva:** White or pinkish, with brown head; about .4" long full grown. It feeds on roots, causes more damage than the adult. Hibernates as partly grown larva in earthen cell, resumes feeding in the spring. One brood. **Food:** Strawberry, blackberry, blueberry, cranberry; yew, azalea, and many other ornamentals. Several closely related species also attack strawberries in California and elsewhere: *B. cribricollis, B. meridionalis, B. rugosostriatus;* they are light brown to black, shiny, without scales, with coarse strial punctures, similar in habits to *sulcatus. B. rugosostriatus*—known as the **Rough Strawberry Root Weevil**—is common on fruit trees in Ohio and often tries to enter houses to hibernate. The **Obscure Root Weevil,** *Sciopithes obscurus,* is a pest of strawberries and cane fruits in Oregon, causes "ragging" of leaves; it is brownish, with white lines, about .25" long.

(410) STRAWBERRY ROOT WEEVIL: *Brachyrhinus ovatus.* **Fig. 1064.**

Also known as strawberry crown girdler. **Range:** Throughout most of North America; sometimes found in California, a serious pest in the Pacific Northwest. **Adult:** Shiny black, with sparse yellowish pubescence; antennae and legs reddish brown; elytra with coarse, deep strial punctures. It is wingless (the elytra are fused); males have never been found. They feed at night on leaves and berries, can crawl rapidly; many are seen in the fall and late summer on the sides of houses trying to get in. It hibernates as adult or larva. **Length:** .2–.25". **Food:** Strawberry, raspberry, grape, apple, peach, young pine and spruce in nurseries and plantations. Most of the damage is done by the root-feeding larvae, which are legless [see (359) above]. *Brachyrhinus* spp. are quite susceptible to the fungus *Beauveria bassiana.*

(411) CLOVER ROOT CURCULIO: *Sitona hispidula* (Curculioninae, Sitonini) **Fig. 1065.**

Range: Throughout most of North America. **Adult:** Shiny black, with dense coat of coppery and grayish scales, the latter forming three stripes on pronotum; elytra with faint strial punctures, row of stiff grayish setae between; snout wider than in the alfalfa weevil (413). Eggs are laid on the underside of leaves. **Length:** .12–.2″. The larvae feed on roots, the adults on leaves. Hibernates as adult. Two broods. **Food:** Clover, alfalfa, bluegrass, wild grasses. The **Sweetclover Weevil,** S. *cylindricollis,* is widely distributed, grayish, with dark and yellow patches and black stripes, from .12″ to .25″ long; the larvae feed on the roots of alfalfa and various clovers; adults prefer the foliage of sweet clover, often try to enter houses in late summer. The **Pea Leaf Weevil,** S. *lineata,* is confined to the Pacific Northwest and California, where it is a pest of peas, vetch, red clover, and alfalfa primarily, and of strawberries occasionally.

(412) CLOVER LEAF WEEVIL: *Hypera punctata* (Hyperini). **Fig. 1066.**

Range: Northern U.S., southern Canada; European or Siberian in origin. **Adult:** Black, with dense coating of brownish or yellowish brown scales, numerous bristles; scales, longitudinal stripes, and underside whitish, grayish, or yellowish; pronotum finely punctate, elytra with fine strial punctures. It emerges in May to July, feeds on leaves at night. Eggs are laid in September or October, hatch quickly except for a few that overwinter; they are inserted in punctures made in plant stems and sealed with a fecal plug. **Length:** .2–.3″. **Larva:** Dark green, head and anal segment dark brown; white longitudinal line down the back, dark brown or black lines along the sides below the spiracles; from .3″ to .55″ long full grown. It climbs up the plant at night to feed on leaves. Hibernates as immature larva in the soil. Pupates in cocoon in the soil or litter the following summer. One brood. **Food:** Clover, alfalfa; adults also eat timothy and Jerusalem artichoke. The **Lesser Clover Leaf Weevil,** *H. nigrirostris* (Fig. 1067), is smaller, from .15″ to .2″ long, reddish brown to black, with greenish, yellowish, or gray scales and stiff hairs; eggs are inserted in leaf sheaths; pale yellowish green, striped larvae feed on tender leaves and flower buds. The **Clover Head Weevil,** *H. meles,* is similar in size and coloring to *nigrirostris* —blackish, with gray, reddish, or greenish scales. The minute **Clover Seed Weevil,** *Miccotrogus picirostris,* an introduced species—now abundant in the Pacific Northwest, Minnesota, and across Canada—is blackish or grayish, with whitish scales; snout, tibiae and tarsi reddish apically; prothorax wider than long, narrower than elytra; it overwinters as adult in litter.

(413) ALFALFA WEEVIL: *Hypera postica.* **Fig. 1068.**

Range: Throughout the U.S., first found in Utah in 1904. **Adult:** Brown, with darker, wide stripe down the back, along the suture; often turns uni-

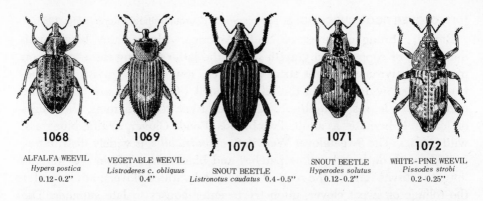

1068
ALFALFA WEEVIL
Hypera postica
0.12-0.2''

1069
VEGETABLE WEEVIL
Listroderes c. obliquus
0.4''

1070
SNOUT BEETLE
Listronotus caudatus 0.4-0.5''

1071
SNOUT BEETLE
Hyperodes solutus
0.12-0.2''

1072
WHITE-PINE WEEVIL
Pissodes strobi
0.2-0.25''

formly darker brown or almost black as it ages. It feeds on leaves and buds. Lemon yellow eggs are laid in clusters in stems of plants. **Length:** .12–.2''. **Larva:** Yellowish at first, changing to pale green, with white median-dorsal stripe and thinner white line along the side, shiny black or dark brown head, from .25'' to .4'' long full grown. It is very active when handled, first appears in April, feeds within plant tips as they open, then on lower foliage, skeletonizing leaves, which turn whitish. Pupates in cocoon on plant or in curl of fallen leaf. Three or four instars. Hibernates in egg stage (the eastern form) or as adult (the western form), the latter in crown of plant or other shelter. One brood. **Food:** Chiefly alfalfa; bur clover, yellow sweet clover, vetch and other clovers rarely. An imported parasite, the ichneumon wasp *Bathyplectes curculionis,* often destroys up to 90 percent of the larvae.

(414) VEGETABLE WEEVIL: *Listroderes costirostris obliquus.* **Fig. 1069.**

Range: Most of the southern states, parts of California. **Adult:** Dull grayish brown with pale gray V near apex of elytra. Only females are known. When disturbed it feigns death, falls over on its back, retracts antennae and legs, remains motionless. It commonly lives nearly two years. Parthenogenetic eggs are laid in September, in buds or other parts of the plant. **Length:** About .4''. **Larva:** Slender, convex, about .5'' long full grown. It feeds on foliage and in roots, pupates in earthen cell. One brood. Adults cause most of the damage, cut off young plants near the ground line, as do cutworms, or feed on buds. **Food:** Various vegetables (potatoes are severely damaged), weeds.

(415) SNOUT BEETLE: *Listronotus caudatus.* **Fig. 1070.**

Range: Eastern North America. **Adult:** Black, with dense coat of brownish yellow scales; pronotum with yellowish stripes, finely punctate; elytra with fine strial punctures. **Length:** .4–.5''. It is found on smartweed and arrowhead. The **Carrot Weevil,** *L. oregonensis,* is copper-colored, about .15'' long, oviposits on stems; the larva tunnels the roots of carrots and parsley, pupates in the soil; hibernates as adult, has two or three broods.

(416) SNOUT BEETLE: *Hyperodes solutus.* **Fig. 1071.**

Range: East of the Rocky Mts. **Adult:** Reddish brown to black, with dense covering of rounded brownish yellow scales; antennae and legs reddish brown; elytral spots black, striae fine with few punctures. **Length:** .12–.2″. It occurs on arrowhead, arrow arum, and other plants found in swamps, shallow water, along the banks of rivers and ditches.

(417) WHITE-PINE WEEVIL: *Pissodes strobi* (Pissodini). **Fig. 1072.**

Range: Eastern Canada, New England, Pennsylvania, south to Mississippi, west to Minnesota. **Adult:** Light to dark brown, pronotum with small spots of whitish scales, scutellum covered with white scales; elytra marked with white scales as shown, fine strial punctures. Eggs are deposited in small pits made in the leader bark. **Length:** .2–.25″. Adults and larvae feed on inner bark, killing the leader; the larvae converge in a "feeding ring" at base of stems, encircling them. Hibernates as adult. Glistening drop of resin indicates adult feeding or egg laying. **Food:** White pine preferred; jack, red, and Scotch pines and Norway spruce are also attacked in plantations. The **Northern Pine Weevil,** *P. approximatus,* is similar, rusty brown, excavates egg pits in the bark of logs, stumps, and weakened standing pines, including young trees in plantations; it works the lower branches rather than leaders as in the case of *strobi,* ranges from North Carolina to Quebec, west to Manitoba. The **Lodgepole Terminal Weevil,** *P. terminalis,* kills the tops of young pines in California and Alberta. The **Yosemite Bark Weevil,** *P. schwarzi* (*yosemite*), and the **Monterey Pine Weevil,** *P. radiatae* (Fig. 1073), work the base of saplings.*

(418) BLACK ELM BARK WEEVIL: *Magdalis barbita* (Magdalini). **Fig. 1074.**

Range: Eastern Canada, New England, to Texas, North and South Dakota. **Adult:** Black; pronotum with sharp spines at anterior angles, coarsely punctate; elytra with coarse strial punctures; snout slender and longer in the female (about the length of head and pronotum). **Length:** .15–.25″. The larvae tunnel in the cambium and bark, adults feed on the leaves. **Food:** Dead and weakened elm, hickory, oak, and walnut. The **Red Elm Bark Beetle,** *M. armicollis,* has similar range and habits; the female is pale reddish brown, the male all black or with reddish tinge; beak of male stout, shorter than thorax, finely punctate, that of female slender, longer than thorax, coarsely punctate; pronotum densely punctate, sides curved; elytral striae deep, with coarse punctures; the larvae burrow under the bark of elms, adults usually attack upper branches. Both of these weevils can transmit the fungus causing Dutch elm disease, but neither is considered important as a

---

* Most native species of *Pissodes* can be identified by knowing the species of host tree and breeding sites. See S. G. Smith and B. A. Sugden, "Host Trees and Breeding Sites of Native North American *Pissodes* Bark Beetles, with a Note on Synonymy," *Ent. Soc. Amer. Ann.,* 62(1):146–148 (1969).

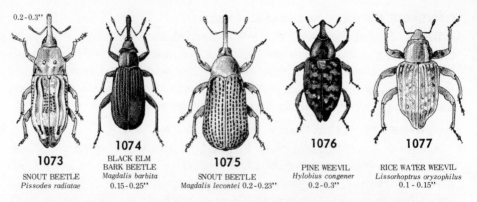

0.2-0.3"

1073
SNOUT BEETLE
Pissodes radiatae

1074
BLACK ELM
BARK BEETLE
Magdalis barbita
0.15-0.25"

1075
SNOUT BEETLE
Magdalis lecontei 0.2-0.23"

1076
PINE WEEVIL
Hylobius congener
0.2-0.3"

1077
RICE WATER WEEVIL
Lissorhoptrus oryzophilus
0.1 - 0.15"

vector [see (447) and (448) for scolytid vectors]. *M. lecontei* (Fig. 1075) is dark blue, from .2″ to .23″ long, frequents recently killed or dying ponderosa pine along the Pacific coast. *M. performata* is bluish black, with coarse sculpturing, from .2″ to .25″ long; it deposits eggs at the base of pine needles near tips of recently killed branches; the larvae tunnel in the pith, adults feed on new shoots of young trees, attacking soft tissues near the base of needles; all pines are attacked from Florida to Ontario. The **Bronze Apple Tree Weevil,** *M. aenescens,* metallic bronzy black, lays its eggs in punctures made in injured or dead branches of apple, cherry, prune, hawthorn, occurs in the western states and British Columbia.

(419) PINE WEEVIL: *Hylobius congener* (Hylobiini). **Fig. 1076.**

Range: Massachusetts to Alaska. **Adult:** Dark reddish brown to black, elytra with three or four narrow irregular cross-stripes formed by white scales. Scutellum smooth; inner edge of fore tibiae with fringe of white hairs in male. Eggs are deposited in cavities made in the bark of logs or stumps. **Length:** .2–.3″. The larvae tunnel in bark, hibernate in prepupal stage; adults feed on inner bark of logs and slash, remove irregular patches of bark. **Food:** Red, white, and Scotch pines. The **Pine Root Collar Weevil,** *H. radicis,* is very similar, a pest of northern pine plantations; it feeds on bark and sapwood near soil surface, prefers young Scotch, red, and jack pines, also attacks white and Austrian pines; evidence of infestation is blackened pitch oozing from bark where roots and trunk join at soil surface; the larvae live a year, hibernate under bark; adults may live three or four years, hibernate in litter. The **Pales Weevil,** *H. pales,* is dark reddish brown, elytra speckled with white or yellowish scales, forming faint oblique crossbands near middle; it inserts eggs in bark of freshly cut pine logs and stumps, feeds on bark at base of pine saplings; it ranges from eastern Canada to Florida, west to the Lake Superior region.

(420) RICE WATER WEEVIL: *Lissorhoptrus oryzophilus* (Hydronomini). **Fig. 1077.**

Also known as the rice-root maggot. It is aquatic in the adult and larval stages. **Range:** Eastern U.S. and Canada, west to Iowa and Texas. **Adult:** Grayish brown, with dense coating of olive-gray scales; antennae and tarsi reddish brown; elytral striae with small close punctures. Snout wide, as long as thorax; basal half of antennal club smooth and shiny. Hibernates as adult in matted grasses in vicinity of lakes, ponds, and rice fields. It migrates to rice field in the spring, feeds on newly emerged plants, making longitudinal slits on upper surface of leaves. It swims equally well on or below surface of water, mates as often below as above, is believed to store bubbles of air under the elytra for breathing under water. **Length:** .1–.15″. **Larva:** Milky white, legless, about .5″ long full grown. It feeds on roots of rice plants. Hooklike spiracles are inserted in plant for trapped air. **Food:** Rice, arrowhead, bulrushes, water lily. This is one of the most serious pests of cultivated rice.

(421) BOLL WEEVIL: *Anthonomus grandis* (Anthonomini). **Fig. 1078.**

"Most costly insect in the history of American agriculture." It is said to have "crossed the Rio Grande at Brownsville [Texas] in 1892." **Range:** All the cotton-growing states from Texas to the East Coast, Mexico; now occurs in Arizona and California (where it was first reported in 1965).* **Adult:** Tan to dark gray or brown, with pale yellowish scalelike hairs; pronotum densely and coarsely punctate, elytra with deep striae and close punctures; snout of female slender and shiny, half as long as body. Adults feed on bolls and leaves, hibernate in woods, along ditches, in trash and litter. Eggs are laid singly (sometimes two or three) on a square or boll. **Length:** .15–.3″. **Larva:** White, with brown head and mouthparts, about .5″ long full grown. It feeds inside a square and pupates here. Seven broods or less. Both egg punctures and feeding of adults are damaging; egg punctures cause squares to flare, turn yellow, and fall to the ground. **Food:** Cotton.

(422) PEPPER WEEVIL: *Anthonomus eugenii*. **Fig. 1079.**

Range: California to Florida. **Adult:** Black, sparsely covered with gray or yellowish scalelike hairs; snout smooth, half as long as body in the female. It feeds on foliage, blossom buds, and tender pods; hibernates in litter. Eggs are laid in punctures made in buds or immature pods. **Length:** .1–.12″. **Larva:** Grayish white, with pale brown head; about .25″ long full grown. It feeds on buds or among seeds inside pod; pupates in bud or pod. Five to eight broods. **Food:** Peppers, eggplant, potato, and other plants of the nightshade family. The **Plum Gouger,** *A. scutellaris*† (Fig. 1080), is dark reddish brown; beak, head, pronotum, and narrow sutural stripe clothed in

---

* Owing to certain variations, the name "boll weevil complex" (*Anthonomus grandis* complex) has been proposed for forms occurring in the Southwest and in northwest Mexico.

† Genus changed to *Coccotorus*.

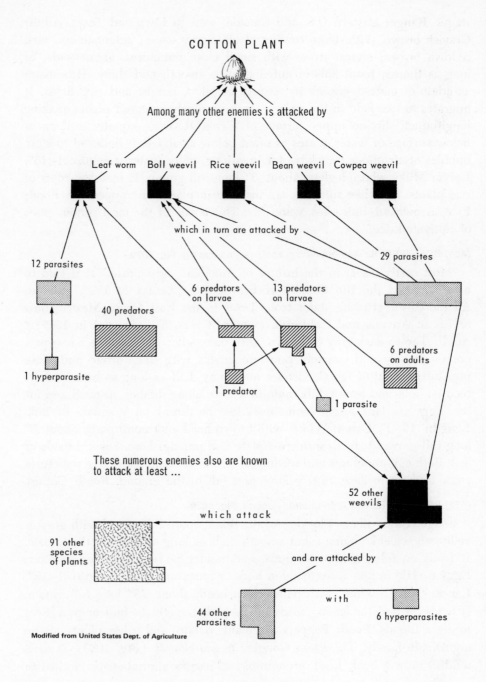

COTTON PLANT

Among many other enemies is attacked by

Leaf worm   Boll weevil   Rice weevil   Bean weevil   Cowpea weevil

which in turn are attacked by

12 parasites

29 parasites

6 predators
on larvae

13 predators
on larvae

40 predators

6 predators
on adults

1 hyperparasite

1 predator

1 parasite

These numerous enemies also are known
to attack at least ...

52 other
weevils

which attack

91 other
species
of plants

and are attacked by

44 other
parasites

with

6 hyperparasites

Modified from United States Dept. of Agriculture

Diagram showing interrelationship of some beneficial and injurious
insects associated with the cotton boll weevil

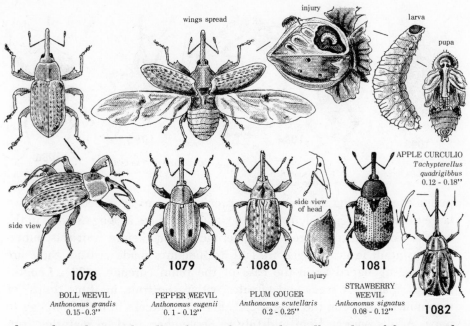

wings spread

injury

larva

pupa

side view of head

APPLE CURCULIO
Tachypterellus quadrigibbus
0.12 - 0.18"

side view

1078

1079

1080

injury

1081

1082

BOLL WEEVIL
Anthonomus grandis
0.15-0.3"

PEPPER WEEVIL
Anthonomus eugenii
0.1 - 0.12"

PLUM GOUGER
Anthonomus scutellaris
0.2 - 0.25"

STRAWBERRY WEEVIL
Anthonomus signatus
0.08 - 0.12"

dense, long, brownish yellow hairs; elytra with small patches of fine grayish hairs, striae finely punctate, intervals wide and flat; from .2″ to .25″ long; the larva eats through the flesh of wild and cultivated plums, pupates near the pit, also attacks apples; it occurs from the eastern states to Kansas and Texas. (Proposed name for plum gouger: *Coccotorus suctellaris*.)

(423) STRAWBERRY WEEVIL: *Anthonomus signatus*. **Fig. 1081.**

Range: Eastern half of the U.S., to Texas, Colorado. **Adult:** Black, with white pubescence, scutellum and broad area in the middle without hairs; elytra chestnut brown, with two black spots, deeply striated, with large punctures. Snout turned under, about half as long as body. It girdles stems, causing buds to fall off or hang from stems. Eggs are laid in punctures made in buds. **Length:** .08–.12″. The larva feeds inside the bud. Adults emerge in midsummer, feed on pollen and blossoms for a short time, then hibernate. One brood. **Food:** Strawberry, raspberry, blackberry, dewberry, redbud. The **Cranberry Weevil,** *A. musculus,* is a serious pest of cranberry and blueberry in the northeastern states; it is dark reddish brown, about .12″ long, with slightly curved snout about one-third to one-half as long as body; the slender larva is yellowish white with brown head; overwintering adults appear early in spring, feed on buds and leaves; the larvae develop in buds.

(424) APPLE CURCULIO: *Tachypterellus quadrigibbus*. **Fig. 1082.**

Range: East of the Rocky Mts. **Adult:** Dark red; snout, antennae, and legs paler; pronotum and adjacent half of elytra thinly coated with grayish

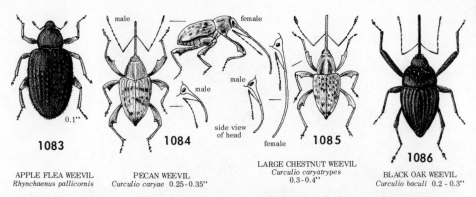

	male	female			

1083

male

male

side view of head

1084

female

1085

1086

LARGE CHESTNUT WEEVIL
Curculio caryatrypes
0.3-0.4"

APPLE FLEA WEEVIL
Rhynchaenus pallicornis

PECAN WEEVIL
Curculio caryae 0.25-0.35"

BLACK OAK WEEVIL
Curculio baculi 0.2 - 0.3"

pubescence, forming three stripes on pronotum; coarsely punctate, with two large tubercles on each elytron. Snout slender, curved, as long as body in the female. Feeds on buds, fruit spurs, terminal twigs before fruit sets; hibernates as adult in litter. Eggs are laid in punctures made in fruit; these are not crescent-shaped as in the case of the plum curculio (436). **Length:** .12–.18". **Larva:** It feeds and pupates within the fruit, in "June drops" or mummified apples on the tree, spends five to six weeks in fruit from egg to pupa. Adults emerge from mid-July to September, feed on maturing fruit until they hibernate. One brood. **Food:** Apple, quince, pear, hawthorn, wild crab, haw, shadbush. The **Apple Flea Weevil,** *Rhynchaenus pallicornis* (Fig. 1083)—shiny black, antennae red with black tip, elytra with punctured striae, about .1" long—injures new growth, the larva mines leaves; it hibernates as adult, ranges from Nova Scotia and New England to Oregon and Texas.

(425) PECAN WEEVIL: *Curculio caryae* (Curculionini). **Fig. 1084.**

Range: Widely distributed throughout the U.S., causes greatest damage in southern part of range: from Oklahoma and Texas to Georgia, North and South Carolina. **Adult:** Brownish or grayish, with scattered yellow hairs, reddish brown underneath scales; elytral striae punctured, intervals flat, roughly punctured. Snout slender, curved, much longer than body in the female, half as long in the male. The female drills holes through shells of nuts, deposits two to four or more eggs in separate pockets (with a single puncture). **Length:** .25–.35". **Larva:** Full grown in a month, eats hole in shell and emerges from September to January, enters soil and forms earthen cell, remains here for a year or two before pupating, which takes place in the fall; the adult emerges in a short time but remains in the soil until the following summer. Life cycle requires from two to three years. One brood. **Food:** Pecan, hickory. Females are better fliers than males and tend to fly to higher parts of trees; males more often crawl to the trees upon emerging. The weevils are easily jarred loose early in the morning (when they are sluggish) and can be caught in a sheet. [See Lepidoptera (260).]

(426) LARGE CHESTNUT WEEVIL: *Curculio caryatrypes (proboscideus).* **Fig. 1085.**

Range: Eastern half of North America. **Adult:** Dark brown, with dense coating of yellowish hairs, reddish black spots; snout slender, curved, much longer than body in the female. **Length:** .3–4″. The female drills holes in nuts to deposit her eggs. The larva feeds on chestnut until it falls, leaves the nut to hibernate in the ground, pupates here the following July. The **Small Chestnut Weevil,** *C. sayi,* is black, with coating of brownish scale-like hairs, irregularly marked with yellow scales, from .2″ to .3″ long.

(427) BLACK OAK WEEVIL: *Curculio baculi.* **Fig. 1086.**

Range: Eastern half of the U.S. **Adult:** Reddish brown to black, with thin coating of grayish scales, elytra with patches of brownish scales, faint strial punctures; snout slender, curved, three-fifths as long as body in the female, less than half as long as body in the male. **Length:** .2–.3″. It infests acorns of white and black or red oaks. *C. rectus* (Fig. 1087) ranges westward to Arizona, is brownish, with coating of pale brown scales, elytra with yellow or orange spots; from .3″ to .35″ long; snout longer than body and curved near tip in the female (shorter, curved throughout in the male); it also breeds in acorns of white and black or red oaks. In Missouri, *baculi* shows little preference for any of these oaks, while *rectus* shows a definite preference for the black or red oaks.

(428) HAZELNUT WEEVIL: *Curculio neocorylus (obtusus).* **Fig. 1089.**

Range: Canada to Texas. **Adult:** Brown to black, with coating of gray to yellowish scalelike hairs, pronotum and elytra marked with paler scales, male with long yellowish hairs on last abdominal segment; snout about two-thirds as long as body in the female, shorter than this in the male. **Length:** .25–.3″. It infests hazelnuts. The **Filbert Weevil,** *C. occidentalis* (Fig. 1088), is mottled brownish yellow, from .2″ to .3″ long, infests cultivated filbert nuts (which are closely related to the common wild hazelnut) along the Pacific coast.

(429) MULLEIN WEEVIL: *Gymnetron tetrum* (Mecinini). **Fig. 1090.**

Range: Michigan and Ohio, eastward to New York, West Virginia, and Georgia. **Adult:** Broad, somewhat depressed; black, with dense, flattened yellowish gray pubescence; pronotum finely punctate; elytra with deep striae, coarse punctures. Antennae five-jointed between club and ring joints. Femora stout, toothed. **Length:** .1–.18″. It is found on mullein, usually within a capsule. Hibernates in rosette of lower leaves, often in small aggregations. The seed weevil *G. antirrhini* attacks toadflax in eastern Canada; attempts to establish it in the western provinces for control of this noxious weed were unsuccessful.

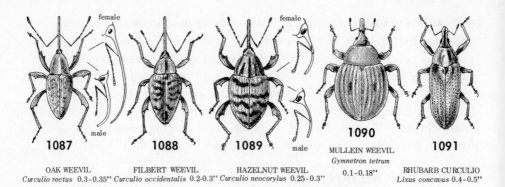

female — female

1090

1087 male   **1088**   **1089** male   MULLEIN WEEVIL   **1091**

Gymnetron tetrum

OAK WEEVIL   FILBERT WEEVIL   HAZELNUT WEEVIL   0.1-0.18"   RHUBARB CURCULIO

Curculio rectus 0.3-0.35" *Curculio occidentalis* 0.2-0.3" *Curculio neocorylus* 0.25-0.3"   *Lixus concavus* 0.4-0.5"

(430) RHUBARB CURCULIO: *Lixus concavus* (Cleonini). **Fig. 1091.**

Range: Throughout much of the U.S. **Adult:** Black, with sparse grayish pubescence; specimens are often covered with orange or reddish pollen when taken; antennae and tarsi reddish brown; elytra with fine strial punctures. Hibernates as adult. **Length:** .4–.5". Adults puncture stems of rhubarb; the larvae are found in stems of dock, sunflower, and thistle. The **Knotweed Weevil,** *L. parcus,* is similar but smaller, black, slender, with thin white pubescence, from .2" to .3" long; the larvae produce galls on the lower branches of knotweed in California. *L. musculus* also produces galls on knotweed, occurs in Colorado and Texas, eastward. The closely related *Microlarinus lypriformis,* a stem-boring weevil, and *M. lareyniei,* a seed weevil, were imported in a biological control program against the puncture vine in California; the seed weevil has also been released in Washington and Texas.

(431) POTATO STALK BORER: *Trichobaris trinotata* (Barini). **Fig. 1092.**

Range: Eastern U.S., Canada, west to Colorado. **Adult:** Blackish, covered with dense coat of flattened, gray scalelike hairs which create a frosted appearance; scutellum and spot at base of pronotum on each side are bare; elytra with fine striae, intervals with three rows of scalelike hairs. Eggs are deposited singly in deep stem or petiole cavities. **Length:** .12–.2". **Larva:** Yellowish white, with brownish head; about .3" long. It bores up and down the stem, causing plant above to wilt. Pupates in stem; the adult matures in a week, delays emergence until the following spring. One brood. **Food:** Potato, eggplant, and related weeds such as ground cherry, cocklebur, jimson weed, horse nettle. The **Tobacco Stalk Borer,** *T. mucorea,* is almost identical, a trifle larger, breeds in jimson weed, attacks potato, tomato, and eggplant in Arizona, New Mexico, California.

(432) PINE REPRODUCTION WEEVIL: *Cylindrocopturus eatoni* (Zygopini). **Fig. 1093.**

Range: California; also believed to be in Oregon. **Adult:** Black, densely clothed with grayish scales. It punctures needles of young trees, later feeds

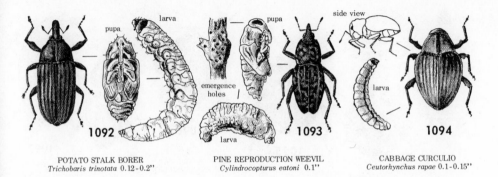

POTATO STALK BORER	PINE REPRODUCTION WEEVIL	CABBAGE CURCULIO
Trichobaris trinotata 0.12-0.2"	*Cylindrocopturus eatoni* 0.1"	*Ceutorhynchus rapae* 0.1-0.15"

on new stems and twigs. Drops of pitch form where feeding and egg punctures are made in the bark. Very active; a good flier but often hops when disturbed, and mistaken for a leafhopper. Minute pear-shaped white eggs are laid singly in niches excavated in bark of main stem and twigs. **Length:** About .1". **Larva:** Cream-colored, about .15" long full grown. It constructs tunnel in outer layers of wood, feeds on cambium, causes tree to turn brown, die from the top down. Hibernates in cell constructed deeper in the wood. Pupates in spring. One brood. **Food:** Young ponderosa and Jeffrey pines. The insect is very destructive to plantations and new growth in burned areas.

(433) CABBAGE CURCULIO: *Ceutorhynchus rapae* (Ceutorhynchini). **Fig. 1094.**

Range: Throughout most of North America; Europe. **Adult:** Black, clothed with yellowish or grayish hairlike scales above, larger whitish scales underneath; pronotum with coarse punctures; elytra with fine striae, punctate. Snout slender, slightly longer than head and pronotum, antennae inserted near middle. Gray oval eggs are inserted in stems of plants. **Length:** .1–.15". **Larva:** It feeds within the stalks and leaf stems and on the edge of leaves of seedling plants. Pupates in earthen cell, hibernates as adult. Several broods. **Food:** Cabbage, cauliflower, horseradish, turnip, mustard, peppergrass. The **Cabbage Seedpod Weevil,** *C. assimilis,* is a serious pest of cabbage, turnip, and rutabaga seed crops in Washington (where much of our cabbage seed is grown); it is from .07" to .1" long, grayish black, covered with gray scales, beak longer than head and pronotum; eggs are inserted in seed pods, the larva—lemon yellow to white with light brown head, about .2" long full grown—feeds inside the pod.

(434) GRAPE CURCULIO: *Craponius inaequalis.* **Fig. 1095.**

Range: New England to Florida, west to the Mississippi Valley. **Adult:** Dark brown; elytra marked with lines formed by whitish and brownish scalelike hairs, antennae and legs reddish brown; elytral striae punctate, intervals raised, broken. It feeds on leaves, causing characteristic short

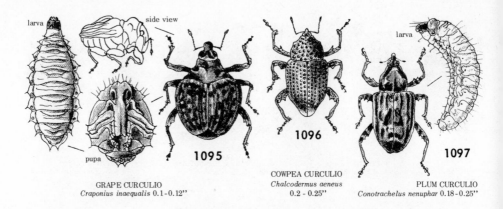

larva

side view

larva

pupa

1095

1096

1097

GRAPE CURCULIO
Craponius inaequalis 0.1-0.12"

COWPEA CURCULIO
Chalcodermus aeneus
0.2 - 0.25"

PLUM CURCULIO
Conotrachelus nenuphar 0.18-0.25"

curved slits, usually in groups. Eggs are laid in punctures made in berries during July and August. **Length:** .1–.12". **Larva:** It feeds on flesh and seeds inside the berry for about three weeks, then pupates in earthen cell; adult emerges in three or four weeks. Hibernates as adult in trash. **Food:** Wild and cultivated grapes.

(435) COWPEA CURCULIO: *Chalcodermus aeneus* (Cryptorhynchini). **Fig. 1096.**

Range: Maryland to Florida, west to Oklahoma and Texas. **Adult:** Black, sometimes with bronze sheen; elytra with coarse strial punctures. Snout nearly straight, longer than pronotum. It feeds on pods and peas. **Length:** .2–.25". **Larva:** Whitish, with yellowish head, about .35" long full grown. It feeds within the green seed, and when grown cuts a hole in the pod and drops to the ground to pupate. Hibernates as adult in rubbish. **Food:** Cowpeas, and closely related legumes, stems of cotton (where it is sometimes mistaken for the boll weevil), strawberries, and other crops.

(436) PLUM CURCULIO: *Conotrachelus nenuphar.* **Fig. 1097.**

Range: East of the Rocky Mts. **Adult:** Dark brown and black, marked with yellowish and whitish pubescence; elytra with conspicuous crests or humps. Beak stout, curved, slightly longer than head and pronotum. Hibernates in adult stage under litter; emerges when peaches begin to bloom in the South, when in "shuck-split stage" in the North. It makes round feeding punctures in skin of fruit. The female cuts crescent-shaped flap in skin where a whitish, elliptical egg is inserted. **Length:** .18–.25". **Larva:** Yellowish white, with small brown head, about .4" long full grown. It feeds inside the fruit for about two weeks, penetrating to the pit, leaves fruit to pupate in the soil. The adult emerges in 30 to 35 days, feeds for a while before hibernating. One brood in the North, two in the South. **Food:** Plum, peach, apple, pear, cherry, and blueberry. Feeding and egg punctures permit entry of fungi, result in "brown rot"; stone fruits ordinarily drop, apples and pears usually remain on trees but are deformed. The **Walnut Curculio,** *C. jug-*

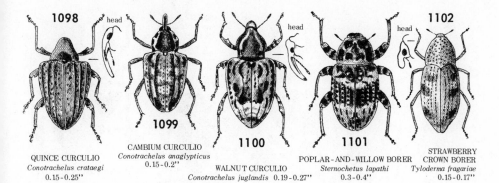

1098

head

1099

CAMBIUM CURCULIO
Conotrachelus anaglypticus
0.15-0.2''

QUINCE CURCULIO
Conotrachelus crataegi
0.15-0.25''

head

1100

WALNUT CURCULIO
Conotrachelus juglandis 0.19-0.27''

head

1101

POPLAR-AND-WILLOW BORER
Sternochetus lapathi
0.3-0.4''

1102

head

STRAWBERRY
CROWN BORER
Tyloderma fragariae
0.15-0.17''

landis (Fig. 1100), is very similar in shape, color, and size; the crests or humps on the elytra are less abrupt, the median ones less prominent than in *nenuphar;* it attacks walnut, butternut, hickory; the larvae feed in the green fruit, causing premature drop.

(437) QUINCE CURCULIO: *Conotrachelus crataegi.* **Fig. 1098.**

Range: Iowa to the East Coast, southward. **Adult:** Black, with dense coating of yellowish and grayish scales; the latter form a double line on each side of the pronotum, meeting at the center and extending back across the shoulder and basal third of the elytra; body underneath with thin coating of grayish yellow scales. Elytra each with four conspicuous carinae or ridges, double rows of coarse punctures between. Snout longer than head and pronotum, deeply striated, with ridges and coarse punctures. **Length:** .15-.25''. It feeds on quince, hawthorn, and related plants. The **Cambium Curculio,** *C. anaglypticus* (Fig. 1099) feeds on the cambium and inner bark of many fruit, forest, and shade trees, has also been found attacking peaches and cotton bolls; it works around the edge of tree wounds, is injurious to columbine, whose roots and crowns it prefers; it is broad, moderately depressed, dark reddish brown to black, with fine yellowish hairs forming two narrow lines on each side of pronotum, and a wide oblique humeral stripe on elytra; latter also with faint whitish band behind the middle; from .15'' to .2'' long; it is abundant from New England to Florida, west to Iowa and Texas.

(438) POPLAR-AND-WILLOW BORER: *Sternochetus (Cryptorhynchus) lapathi.*
Fig. 1101.

Range: Southern Canada, most of the U.S. except Alaska and the southern states from South Carolina to Arizona; also occurs in Europe. **Adult:** Black, with dense coating of black and whitish scales, scattered tufts of black bristles; antennae and tarsi reddish brown; elytra striated, with large square punctures. Beak coarsely and densely punctate, as long as head and

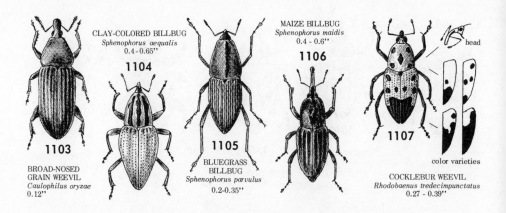

BROAD-NOSED
GRAIN WEEVIL
Caulophilus oryzae
0.12"

1103

CLAY-COLORED BILLBUG
Sphenophorus aequalis
0.4 - 0.65"

1104

BLUEGRASS
BILLBUG
Sphenophorus parvulus
0.2-0.35"

1105

MAIZE BILLBUG
Sphenophorus maidis
0.4 - 0.6"

1106

1107

head

color varieties

COCKLEBUR WEEVIL
Rhodobaenus tredecimpunctatus
0.27 - 0.39"

pronotum together. It cuts a small hole in the bark to feed or to oviposit (generally one egg to a hole). It usually feeds on green smooth-barked shoots of current year, and oviposits in stems more than one year old. **Length:** .3–.4". **Larva:** Starts enlarging egg niche as soon as it emerges. Fine brown wood chips indicate borings. Mines average about one to one and one-half inches long in the bark and wood respectively. Six instars. Life cycle may extend up to three years. Hibernates as egg, larva, pupa, or adult. Most adults emerge in the fall. **Food:** Poplar, willow, alder, birch. The insect is a pest of ornamental and plantation trees.

(439) STRAWBERRY CROWN BORER: *Tyloderma fragariae.* **Fig. 1102.**

Range: Throughout the U.S., except the Rocky Mt. states and higher altitudes. **Adult:** Black; elytra and legs reddish brown. Elytra with six darker spots, rows of widely separated punctures, coarse toward base. Snout longer than head. It is not able to fly. The adult feeds on foliage, hibernates in litter. Glistening white, elliptical eggs are laid from May to August, in punctures made in crown and base of leaf stalks. **Length:** .15–.17". **Larva:** White or light orange, with brownish or black head and mouthparts; about .2" long full grown. It feeds for four to eight weeks, then pupates within the crown; the adult emerges in late summer, feeds for a while before hibernating. One brood. The larvae cause serious damage to strawberry plants. [See Lepidoptera (246) and (288).]

(440) BROAD-NOSED GRAIN WEEVIL: *Caulophilus oryzae* (Cossonini). **Fig. 1103.**

Range: Florida, Georgia, South Carolina. **Adult:** Dark reddish brown. Pronotum finely punctate; elytra striated, with deep coarse punctures. It breeds in soft, damaged grains, is a pest of stored grains. Strong fliers, they fly to cornfields and infest grain before it hardens. Eggs are usually laid in broken portions of grain. The cycle from egg to adult requires about one month; adults live about five months. **Length:** .12".

(441) CLAY-COLORED BILLBUG: *Sphenophorus (Calendra) aequalis* (Sphenophorini). **Fig. 1104.**

Range: East of the Rocky Mts. **Adult:** Black, with shiny bluish gray crust. Pronotum with three dark elevated stripes; elytra finely striated, often with dark longitudinal stripes. It hibernates as adult. One brood. **Length:** .4–.65". It breeds in bulrushes or sedges, also attacks young corn plants. The **Bluegrass Billbug,** S. *parvulus* (Fig. 1105), is similar but much smaller, from .2" to .35" long, black, with gray crust, antennae reddish brown excepting club, elytra with fine striae and coarse punctures; it breeds in Kentucky bluegrass, also attacks corn, timothy, and wheat. The **Maize Billbug,** S. *maidis* (Fig. 1106), an eastern species, is similar but without the gray crust, black or dark reddish, from .4" to .6" long, pronotum with slightly elevated median stripe, elytra with fine strial punctures; it often damages corn in southern states. The **Southern Corn Billbug,** S. *callosus,* is grayish brown to black, about .4" long; the adult is attracted to nut grass but causes much damage to seedling corn in the South; it is usually found at base of plant, head downward, ranges from Florida to Maine, west to Wisconsin and Arizona. The **Cocklebur Weevil,** *Rhodobaenus tredecimpunctatus* (Fig. 1107), is red, with variable black markings, underside and beak black, from .27" to .39" long; it is found on a great variety of plants in the western states.

(442) GRANARY WEEVIL: *Sitophilus (Calendra) granarius.* **Fig. 1108.**

Range: Cosmopolitan; prefers temperate climate, more frequently found in northern states and Canada. **Adult:** Moderately shiny, chestnut brown or blackish; pronotum with *elongated* punctures [contrast with *oryzae* (443)]; elytra with deep striae and punctate. Snout slender, two-thirds as long as pronotum, finely and sparsely punctate. It is wingless, feeds on kernels of grain, lives about seven to eight months. Eggs are laid in holes bored into the grain berry, and sealed over with a gelatinous secretion. **Length:** .12–.18". **Larva:** It feeds and pupates inside the kernel. Cycle from egg to adult requires four weeks in warm weather. The head capsule serves to distinguish the larva of this and the next species from those of other stored-food pests. **Food:** All kinds of stored grains. Most of the insect damage to stored grains is caused by four species: this one, the rice weevil (443), lesser grain borer (264), and Angoumois grain moth [see Lepidoptera (286)].

(443) RICE WEEVIL: *Sitophilus (Calendra) oryzae.* **Fig. 1109.**

Also called the black weevil. **Range:** Cosmopolitan; most abundant in warm regions. **Adult:** Reddish brown to black, elytra usually with four reddish or yellowish spots. Pronotum with dense, coarse *round* punctures [contrast with *granarius* (442)]; elytra with deep striae, coarse punctures. Snout slender, three-fourths as long as pronotum, with four rows of coarse punc-

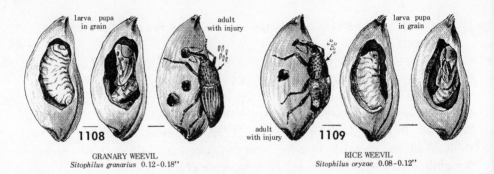

larva pupa
in grain

adult
with injury

1108

GRANARY WEEVIL
Sitophilus granarius 0.12-0.18"

larva pupa
in grain

adult
with injury

1109

RICE WEEVIL
Sitophilus oryzae 0.08-0.12"

tures. An adult lives four or five months. Early stages as described for *granarius*. **Length:** .08–.12". Like the broad-nosed grain weevil (440), they are good fliers, and fly to grain fields, where new infestations are started. **Food:** All kinds of grains. This is one of the worst pests of stored grains [see (442) above].

Bark Beetles (Scolytidae)

This is a fairly large family of minute to small beetles, characterized by their stout, compact form and short snout (the family name, derived from Greek, means "cut short"), and by the antennae. They are cylindrical in shape, usually black or brown and covered with fine hairs, short bristles, or stout setae. The abdomen has five visible sternites. The antennae, which are short and inserted in front of the eyes, have a variable number of segments (often 11 or 12), with the first segment enlarged and the last three or four forming a compact club. The mandibles are small and hidden, labrum absent, and palpi three-segmented. Usually the wings are well developed, the legs are short, with toothed tibiae and five-segmented tarsi (the first segment as a rule minute).

Many scolytids stridulate by rubbing specialized parts located on the edge of the elytra and abdominal tergite, or on the head and prothorax.* Usually it is either the male or female (seldom both) which is so equipped—evidently the sex opposite that which starts the entrance tunnel to the gallery; the sound is its password. The species of *Dendroctonus* represented here have the elytra-tergite type of mechanism in the males, as does *Hylurgopinus rufipes* (448); the *Ips* species have the head-prothorax type in the females (usually the vertex is rubbed against the underside of the anterior edge of the pronotum). Only the *Ips* associated with pine have the stridulatory organ; those found on spruce do not.

The larvae resemble those of the curculionids or weevils—they are white,

* See Barbara A. Barr, "Sound Production in Scolytidae (Coleoptera) with Emphasis on the Genus *Ips*," *Canad. Ent.*, 105(6):636–672 (1969).

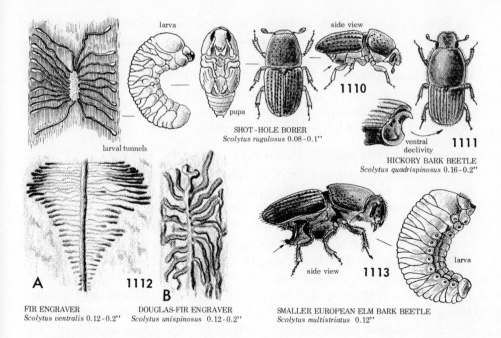

larva

side view

1110

pupa

SHOT-HOLE BORER
Scolytus rugulosus 0.08-0.1"

larval tunnels

ventral
declivity **1111**

HICKORY BARK BEETLE
Scolytus quadrispinosus 0.16-0.2"

side view **1113**

larva

A **1112** B

FIR ENGRAVER
Scolytus ventralis 0.12-0.2"

DOUGLAS-FIR ENGRAVER
Scolytus unispinosus 0.12-0.2"

SMALLER EUROPEAN ELM BARK BEETLE
Scolytus multistriatus 0.12"

curved, legless, with prominent head—have nine or ten abdominal segments, live internally in plants. Most of them attack trees, and are most often injurious to conifers; a few live in seeds, stems, or roots of herbaceous plants. Some species attack only dead or dying trees, others prefer vigorous ones; they invade sapwood and sometimes the heartwood, and the layer between wood and bark or the bark itself.

The principal enemies of bark beetles appear to be clerid or checkered beetles. Recent investigations by Lindquist and others show that certain (*Iponemus*) mites parasitize the eggs of *Ips* beetles and may also be significant in their control. The mites develop in the egg niches of the host, and newly emerged adult mites ride on the backs (in the elytral declivity) of their adult hosts as they fly from the first host tree to another.*

(444) SHOT-HOLE BORER: *Scolytus rugulosus* (Scolytinae). **Fig. 1110.**

This genus is easily recognized by the ventral declivity, or concavity, on the underside of the abdomen at the posterior end. *S. rugulosus* is also called fruit-tree bark beetle. **Range:** Most of North America; an introduced species. **Adult:** Dark brown or black; antennae, tibiae and tarsi, and tips of elytra reddish brown; pronotum densely punctate, elytra striated and densely punctate. Adults drill holes (as might be made by small shot) in the bark and wood when entering and leaving. The female constructs an egg gallery

* See Evert E. Lindquist, "Review of Holarctic Tarsonemid Mites (Acarina: Prostigmata) Parasitizing Eggs of Ipine Bark Beetles," *Mem. Ent. Soc. of Canada,* No. 60 (1969), Ottawa.

about 2″ long under bark, parallel *with* the grain, deposits eggs in niches made in the sides. **Length:** .08–.1″. **Larva:** White, with reddish head, legless, about .1″ long full grown. It feeds on sapwood for about 30 days, then pupates. Second-brood larvae hibernate. Two or three broods. Both adults and larvae feed between bark and sapwood, making a centipede-like grooved configuration, prefer injured or dying trees. **Food:** Apple, quince, stone fruits, mountain ash, Juneberry, hawthorn.

(445) HICKORY BARK BEETLE: *Scolytus quadrispinosus.* **Fig. 1111.**

Range: Throughout eastern U.S. **Adult:** Shiny dark brown or black, head of male and pronotum fringed with long hairs; pronotum finely punctate; elytral striae coarsely punctate, intervals finely punctate. Ventral declivity deeply concave, with four lateral spines in the male. Limbs of declining trees (preferably) and vigorous ones are attacked. Eggs are laid in niches along sides of vertical galleries (running *with* the grain). **Length:** .16–.2″. **Larva:** Each one constructs its own tunnel at right angles to gallery. Hibernates as mature larva in cell at end of tunnel, pupates the following spring or summer. One brood. **Food:** Hickory (especially shagbark and pignut), butternut, pecan. The most destructive insect attacking hickory.

(446) FIR ENGRAVER: *Scolytus ventralis.* **Fig. 1112A.**

Range: All of the western states, British Columbia. **Adult:** Shiny black; punctate, elytra striated. Ventral declivity without prominent spines. It attacks the main trunk of the tree, the female excavating a "nuptial chamber" *across* the grain, then a horizontal egg gallery, one in each direction, cutting deep in the bark and sapwood. The male assists in removal of the dust, remains during the egg-laying period of two to three weeks. Eggs are laid in niches made in sides of the gallery. **Length:** .12–.2″ **Larva:** It feeds separately in individual mine excavated at right angles to the egg gallery (*with* the grain). A brown fungus stain is always found in feeding area. Hibernates as larva. Pupates the following summer in cell at end of mine; adults emerge in ten to fourteen days. Generally one brood, with partial second in lower altitudes; developmental period two years in higher altitudes. **Food:** True fir. This insect has been very destructive to white fir, can breed where only part of the cambium has been killed; it attacks windfalls and green logs. The **Douglas-fir Engraver,** *S. unispinosus* (Fig. 1112B), is similar, has a long spine projecting from the middle of the ventral declivity; it attacks injured, dying, or recently killed Douglas fir in the Pacific coast states and British Columbia; egg galleries follow the grain of the wood, larval cavities turn with the grain.

(447) SMALLER EUROPEAN ELM BARK BEETLE: *Scolytus multistriatus.* **Fig. 1113.**

Range: Southern Ontario and most of the states east of the Mississippi

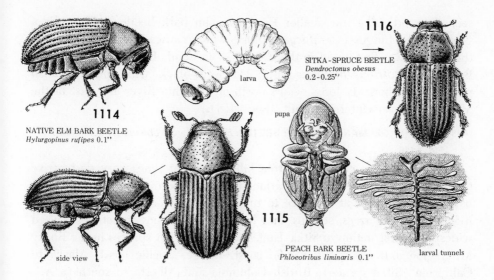

1116

SITKA-SPRUCE BEETLE
Dendroctonus obesus
0.2-0.25"

larva

pupa

1114

NATIVE ELM BARK BEETLE
Hylurgopinus rufipes 0.1"

1115

side view

PEACH BARK BEETLE
Phloeotribus liminaris 0.1"

larval tunnels

River; Missouri to Oklahoma and Louisiana, west to Colorado, Nevada, California, the Pacific Northwest. **Adult:** Shiny, dark reddish brown; punctate, elytra striated. Ventral declivity with blunt spine in the middle. Adults emerge in spring, feed on crotches of living elm twigs, later bore through bark of recently cut, dead, or dying elm material. Entrance and exit holes resemble those of the shot-hole borer (444). **Length:** About .12". Egg laying and larval habits similar to other species of *Scolytus* described above. Egg galleries in inner bark are *parallel* with grain, larval mines cut *across* grain. [Contrast with (448) below.] Hibernates as larva; adults emerge in June and July in northern parts of range. One and partial second brood. The beetles carry on their bodies the fungi causing Dutch elm disease, spread it to healthy trees when feeding on the twig crotches or when forming galleries in trees too vigorous for development of the larvae. **Food:** Elm. A braconid wasp, *Dendrosoter protuberans*, which parasitizes the larval stage in Europe, has recently been introduced into Ohio and Missouri. Nematodes are recorded as sterilizing a large percentage of adult beetles in England.

(448) NATIVE ELM BARK BEETLE: *Hylurgopinus rufipes* (Hylesininae). **Fig. 1114.**

Range: New Brunswick, south to North Carolina, Mississippi, west to Manitoba, Kansas, Minnesota, Colorado, Arkansas, Oklahoma. **Adult:** Brown to black; less shiny than preceding species, and without the ventral declivity and spine; head and pronotum densely punctate, elytra with rows of coarse punctures. **Length:** About .1". Egg laying and larval habits similar to those of *Scolytus* above. In contrast to the other elm bark beetle (447), egg galleries extend *across* the grain and the larvae mine *with* the grain. Hibernates as larva or adult. Most adults and the main population overwinter in tunnels in the bark of healthy trees, emerge for feeding in May. One and partial second brood. Like the smaller European elm bark beetle (447), it is a vector of the fungus (*Ceratocystis ulmi*) which causes Dutch elm disease; the malady was first noted in the East in 1930. **Food:** Elm. The **Peach Bark Beetle,** *Phloeotribus liminaris* (Fig. 1115), is black, with sparse whitish pubescence, about .1″ long, occurs north of the Ohio River and east of the Mississippi, overwinters as adult. [See (418).]

(449) SITKA (ENGELMANN) SPRUCE BEETLE: *Dendroctonus obesus* (*engelmanni*).
Fig. 1116.

Members of this genus—the "tree killers"—are at times very destructive to spruce and pine. Evidence of attack is reddish boring dust caught in the bark, or pitch tubes at entrance to tunnels. They breed under the bark of living and dying trees, stumps, and logs. *D. obesus* has sometimes caused widespread destruction to Sitka, Engelmann, and other spruces in the Rocky ·Mt. region and along the Pacific coast. **Range:** Pacific coast, Alaska to California (Sitka spruce); British Columbia and Alberta to southern Arizona and New Mexico, from Oregon to the Black Hills of South Dakota (Engelmann spruce). **Adult:** Dark reddish brown to black, sparsely clothed with long hairs; pronotum with irregular coarse punctures; elytral striae moderately impressed, punctures coarse in the female, both more obscure in the male. Adults excavate short, straight longitudinal egg galleries, wider than the beetles, in the inner bark; eggs are laid in masses in elongate cavities alternating from side to side. **Length:** .2–.25″. **Larva:** The larvae at first bore en masse transversely from the egg gallery, later make separate mines. Pupal cells are constructed in the inner bark. In the range of Engelmann spruce, they require two years for development, hibernate as half- or full-grown larvae; new adults emerge from August to October, migrate to base of trunk and the root collar, where they hibernate until the following June and July. In the range of Sitka spruce, there is one generation which hibernates as adult, and a partial generation that overwinters as larva. **Food:** Engelmann and Sitka spruce; also other spruces, sometimes lodgepole pine. The most important natural enemies of the beetles are woodpeckers. *D. obesus*, previously known as the **Sitka Spruce Beetle;** *D. engelmanni,* the **Engelmann Spruce Beetle;** *D. piceaperda,* the **Eastern Spruce Beetle;** *D. borealis,* the **Alaska Spruce Beetle;** and *D. rufipennis,* the **Red-winged Pine Beetle,** are now considered to be one species, *D. obesus.**

(450) WESTERN PINE BEETLE: *Dendroctonus brevicomis* (*barberi*). Fig. 1117.

Range: California to British Columbia; Idaho, Montana, to Arizona and New Mexico. **Adult:** Light brown to brown or nearly black; pronotum with a narrow transverse elevation near the front margin in the female, and a transverse depression similarly located in the male; faint elytral striae, with

* According to Wood (1969), *D. rufipennis* now takes precedence.

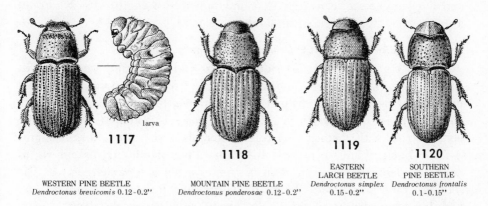

larva

1117

1118

1119

1120

WESTERN PINE BEETLE
Dendroctonus brevicomis 0.12-0.2''

MOUNTAIN PINE BEETLE
Dendroctonus ponderosae 0.12-0.2''

EASTERN
LARCH BEETLE
Dendroctonus simplex
0.15-0.2''

SOUTHERN
PINE BEETLE
Dendroctonus frontalis
0.1-0.15''

very small punctures. **Length:** .12–.2″. It attacks the trunk from the base to two-thirds of the height. The work of these beetles may be distinguished from that of other bark beetles in the same range by the winding egg galleries crisscrossing each other, forming irregular markings on inner surface of bark and surface of sapwood. The larvae feed on the inner bark. Adults hibernate in the galleries. One to four broods. **Food:** Ponderosa pine (western yellow pine), Coulter pine. It normally breeds in overmature and weakened trees, and windfalls, but may attack vigorous trees. The most important natural enemies of the beetles are woodpeckers, clerid and ostomatid beetles. *D. barberi,* known as the **Southwestern Pine Beetle,** is now considered a variation of *D. brevicomis.*

(451) MOUNTAIN PINE BEETLE: *Dendroctonus ponderosae (monticolae).* **Fig. 1118.**

Range: Mountain regions from California to New Mexico, British Columbia; Idaho, Montana, Wyoming, South Dakota. **Adult:** Black; elytra finely punctate and striated. **Length:** .12–.2″. It constructs perpendicular egg galleries through the inner bark, usually straight but sometimes winding. Only one male is found in the gallery with a female. Eggs are deposited singly in niches on alternate sides of the gallery; the larvae tunnel at right angles to the gallery, pupate in cells at end of their tunnels. Hibernates in egg, larval, or adult stage, emerging in groups from June to September. One brood. **Food:** Various pines such as lodgepole, western white, sugar, ponderosa; sometimes fir, spruce, hemlock. It attacks small to large living trees; reddish pitch tubes and boring dust indicate point of entrance. *D. ponderosae,* previously known as the **Black Hills Beetle,** and *D. monticolae,* the **Mountain Pine Beetle,** are now considered to be one species, *D. ponderosae.*

(452) EASTERN LARCH BEETLE: *Dendroctonus simplex.* **Fig. 1119.**

Range: Eastern Canada, Maine to Maryland and West Virginia, west to Minnesota and Alaska. **Adult:** Reddish brown to black; pronotum coarsely and finely punctate, elytra striated and punctate. **Length:** .15–.2″. **Food:** Tamarack (eastern or American larch).

(453) SOUTHERN PINE BEETLE: *Dendroctonus frontalis.* **Fig. 1120.**

Range: Pennsylvania to Florida, west to Oklahoma and Texas. **Adult:** Light brown to reddish brown or nearly black; front of head with prominent tubercle on each side of a distinct median groove; pronotum with transverse elevation near front margin in the female, coarsely and sparsely punctate; elytral striae with distinct punctures, intervals with raised lines. **Length:** .1–.15″. **Food:** Pine, spruce. The **Arizona Pine Beetle,** *D. arizonicus,* is now considered to be a variation of *frontalis,* not a separate species. The southern pine beetle, and the **Black Turpentine Beetle,** *D. terebrans,* are very injurious to pine in the South; *terebrans* is about .25″ long, usually works the lower six feet of tree, causing large pitch tubes; the much smaller *frontalis* attacks the full length of the tree, and the tubes are much smaller or absent. They are distinguished from *Ips* beetles by the manner in which the trees are worked [see (456) below], as well as the rounded posterior end.

(454) RED TURPENTINE BEETLE: *Dendroctonus valens.* **Fig. 1121.**

Range: Throughout northeastern and western U.S. and southern Canada. **Adult:** Dark red, never black except underside of body, which is light red to black; pronotum densely, irregularly punctate; elytral striae indistinct, with obscure punctures. It excavates irregular longitudinal egg galleries, and the larvae mine between bark and wood of healthy, injured, or dying trees, freshly cut logs and stumps. **Length:** .25–.4″. Largest species in the genus. **Food:** Pines, sometimes spruce, larch, or fir. The **Douglas-fir Beetle,** *D. pseudotsugae,* is reddish brown, about .2″ long, a serious pest of Douglas fir in western forests; egg galleries are straight or slightly curved, mostly in the inner bark but slightly scoring the sapwood, with larval mines branching off at right angles to either side; it usually hibernates as adult; one brood.

(455) APPLEWOOD-STAINER: *Monarthrum mali* (Ipinae). **Fig. 1122.**

Members of this genus belong to a group known as ambrosia beetles. They bore small round tunnels (nuptial chambers) directly into the sapwood; secondary tunnels branch off from this in several directions, with "larval cradles" in turn branching off from these. The enlarged forecoxal cavity of *M. mali* females are used as repositories for the ambrosial fungi that are brought into the galleries; this serves to inoculate the host trees, or to put it another way, to seed the fungus gardens which provide their food. **Range:** Eastern North America. **Adult:** Brownish to black; antennae and legs yellowish; elytra with yellowish pubescence, fine strial punctures, two small teeth at apex. **Length:** .1–.12″. Besides apple, it attacks logs or weakened trees of

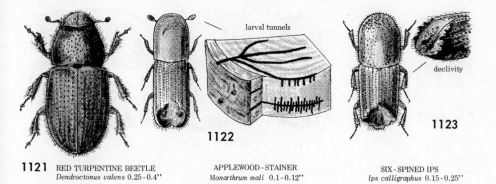

1121 RED TURPENTINE BEETLE
Dendroctonus valens 0.25-0.4"

APPLEWOOD-STAINER
Monarthrum mali 0.1-0.12"

SIX-SPINED IPS
Ips calligraphus 0.15-0.25"

1122

larval tunnels

declivity

1123

oak and other hardwoods; the holes become surrounded by a dark brown or black stain caused by the fungi. Another ambrosia beetle, *Xyleborus dispar,* is sometimes damaging to newly planted apple, cherry, or plum in the Pacific Northwest; it is dark reddish brown to black, with apex of elytra rounded (not excavated); the female is about .12" long, the wingless male much smaller; it tunnels into the hardwood, sometimes girdling the tree; the larvae do not excavate, but rather remain in the parental gallery, where they feed on the growing fungus. Larvae of the tiny **Black Twig Borer,** *Xylosandrus compactus,* are also ambrosia feeders; the adult female is black to dark brownish, about .05" long, excavates egg galleries in twigs of orchid, avocado, coffee, cacao, and many other plants; a native of Asia and a serious pest, it now occurs in Hawaii and Florida.

(456) SIX-SPINED IPS (ENGRAVER): *Ips calligraphus.* **Fig. 1123.**

The beetles in this genus are called engravers, and normally feed on the cambium layer of weakened, dying, or recently felled pines, but sometimes attack vigorous trees. In contrast to *Dendroctonus,* they work in the trunks and larger branches with thin bark, and frequently attack young trees. The egg galleries radiating from the central nuptial chamber form distinctive patterns, often grooving the sapwood when the cambium layer is thin. Unlike *Dendroctonus,* the egg galleries are kept clean, boring dust is much in evidence, and pitch tubes are rare. Eggs are laid in niches along sides of the gallery; the larvae work away from it individually. The larvae are more tapering, thicker at the anterior end than those of *Dendroctonus.* They usually hibernate as adults, may have two to five broods. A distinctive feature of the adult is the pronounced concavity or declivity of the posterior end, with two to six pairs of spines along the edge of the elytra. **Range:** *I. calligraphus* (*ponderosae*) occurs in eastern Canada and throughout most of the U.S., south to Honduras. **Adult:** Black, antennae and legs dark brown; pronotum with fine to coarse punctures, elytra with coarse strial punctures. The male has a larger median tubercle in front of head than the female,

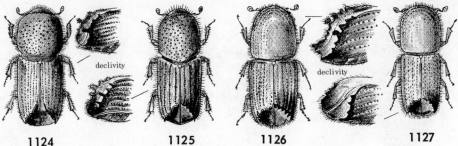

1124	1125	1126	1127
MONTEREY PINE ENGRAVER *Ips mexicanus* 0.15-0.2"	CALIFORNIA PINE ENGRAVER *Ips plastographus* 0.15-0.2"	CALIFORNIA FIVE-SPINED IPS *Ips confusus* 0.17-0.2"	PINE ENGRAVER *Ips pini* 0.15-0.2"

coarser punctures, third declivital spine with notch like a crochet hook. **Length:** .15–.25". It mines under the green bark of stumps, logs, injured and dying pines.

(457) CALIFORNIA FIVE-SPINED IPS (ENGRAVER): *Ips confusus.* **Fig. 1126.**

Range: Western U.S., south into Mexico. **Adult:** Black, underside dark brown, antennae and legs lighter brown; front of head with coarse granules in the female, large median tubercle in male; pronotum with dense medium punctures, elytra with coarse strial punctures. Declivital spines coarser in the male. Egg galleries usually comprise three to five straight tunnels radiating from the central entrance chamber; typically each gallery consists of three tunnels forming an inverted **Y. Length:** .17–.2". Hibernates as adult under the bark of a dead tree. Two to five broods; the summer generations develop in felled logs, frequently attack young growth nearby. **Food:** Ponderosa, sugar, western white, Coulter, digger, and Monterey pines; other pines less frequently.

(458) PINE ENGRAVER: *Ips pini (oregonis).* **Fig. 1127.**

Range: Throughout the forested areas of North America, south to Tennessee and Georgia on the east, and northern Mexico on the west. **Adult:** Black; antennae and tarsi brown. Front of head densely and coarsely punctate above, granulate below, usually with median tubercle in the male; pronotum with fine and coarse (lateral) punctures, elytra with fairly deep, coarse strial punctures. Four pairs of declivital spines on posterior margins of elytra; the third spine is the largest in the male, about equal to the second and fourth in the female. **Length:** .15–.2". **Food:** Pine, spruce, larch. The **California Pine Engraver,** *I. plastographus* (Fig. 1125), attacks trunks and branches of felled lodgepole, Monterey, and bishop pines in the Coastal Range of California and in the Sierra Nevada Mts.; it is reddish brown, from .15" to .2" long, with four pairs of declivital spines; the egg galleries resemble those of *I. confusus* (457); it is usually associated with the red turpentine beetle (454) and the Monterey pine engraver (below) in the killing of living and injured trees, has three to five broods. The **Monterey Pine Engraver**, *I. mexicanus* (*radiatae*) (Fig. 1124), attacks living, injured, dying, and recently felled Monterey, lodgepole, knobcone, bishop, and whitebark pines from British Columbia to California, Idaho, and Wyoming; it is shiny dark brown, from .15″ to .2″ long, with one long and two shorter pairs of declivital spines; the egg galleries are sharply curved, with three or four larval mines emanating from each egg niche.

21

HYMENOPTERA: Bees, Ants, Wasps, Sawflies, Horntails

The third largest order of insects in number of species described to date, the Hymenoptera comprise the most significant group in the regulation of insect populations (see Introduction, p. 18). The order is named for the membranous or transparent wings of the adults; they have two pairs, excepting some that are wingless. The hindwing, which is smaller than the forewing, has a row of tiny hooks on the front margin which clasp the forewing, holding them together almost as one; some can be easily mistaken for flies (Diptera), which have a single pair of wings. The veins are relatively few, the forewing being more completely veined than the hindwing. The adults are generally moderate to small or minute in size, some egg parasites (Trichogrammatidae) being among the smallest insects known; a few, however, such as the tarantula hawk and the *Megarhyssa* wasp, are very large. Mouthparts vary from chewing to chewing-lapping types.

The Hymenoptera are commonly bisexual, that is, they reproduce by mating, as most insects do. Males, however, are normally produced from unfertilized eggs (parthenogenesis) and females from fertilized eggs—the sperm being retained or released from the spermatheca when the female lays her eggs. Thus, the sex of her progeny rests with the gravid female at the moment of oviposition—an instinctive response, in part to certain external stimuli. Some species are unisexual and not known to mate, and only females appear. Metamorphosis is complete. The order is divided into two main suborders: the Symphyta (Chalastogastra) and the Apocrita (Clistogastra).

The Symphyta include the horntails, sawflies, and stem sawflies—our examples (1) to (18). The abdomen of the adults in this group is sessile or closely seated, attached over nearly the full width of the thorax. The sawflies are robust, the horntails are long and cylindrical; head, thorax, and abdomen are about the same width. The ovipositor of the female sawfly consists of two sawlike blades enclosed in two outer plates or guides—an adaptation for cutting and inserting eggs into plant tissues; that of the horntails is long and sheathed, for inserting eggs in bark and wood. The trochanters in either case are two-jointed, and the hindwings have three basal cells. The larvae of sawflies are similar to caterpillars (Lepidoptera); they are free-living, feeding on plants, sometimes mining leaves, forming galls, spinning webs, or rolling leaves. They have a single pair of ocelli, and usually well-developed legs and prolegs; the latter number from six to eight pairs and lack crochets. (In the Lepidoptera, there are rarely more than five pairs of prolegs, one pair always being at the end of the abdomen, and they have crochets.) Sawfly larvae are often gregarious, feeding in compact colonies which sometimes methodically strip the branches of their leaves.

The Apocrita—our examples from (19) on—include the solitary and parasitic wasps, social wasps, ants, and bees. The abdomen of the adult is petiolate, or stalked, with one or two (in some ants) segments sharply constricted, forming a slender petiole or "waist" at point of attachment to the thorax. Most of the adults in this group feed either extensively or principally on nectar and pollen of flowers, or on honeydew and plant exudates. Many female parasites require supplemental protein occasionally for continued egg production and imbibe the blood that oozes from the oviposition wound, or they may puncture the host expressly for feeding. Where the host is concealed in a cocoon, in the host plant (stem, pod, or folded leaf), or in scale armor, the ovipositor serves as a mold around which a fluid is poured and allowed to solidify, forming a tube; the blood rises by capillary action.* In the bees, certain wasps, and ants, the ovipositor is modified to form a stinger purely for defensive or offensive purposes; these are the only insects that have such a stinger. The larvae of the Apocrita are not free-living; they are parasitic or live on a supply of food stored or fed by the mother or female workers. They are legless, the head is exposed, with mouthparts and antennae greatly reduced and ocelli absent.

Ants, bees, and the larger wasps—yellow jackets, hornets, mud daubers and the like—are familiar to everyone. The solitary wasps build nests in

---

* Flanders observed a chalcid (*Spintherus* sp.)—it parasitizes the stem-enclosed eggs of the alfalfa weevil—sucking the *egg* of the host dry by constructing a feeding tube. She located the host eggs from the fecal plug with which the weevil seals the stem puncture. See Stanley E. Flanders, "The Parasitic Hymenoptera: Specialists in Population Regulation," *Canad. Ent.*, 94(11):1138 (1962).

burrows or of mud on exposed surfaces, and provision them with paralyzed insects and spiders which serve as food for the developing young. The social wasps, such as hornets and yellow jackets, feed the larvae continuously with partially masticated insects. The activities of the parasitic wasps are much less obvious—mostly because they are generally so small; their impact on the insect community, however, is very profound. The ovipositor of the female parasite is not normally used for stinging other than the host insect (some large ones have been known to sting collectors), and is used for depositing her eggs in or on the host, singly or several to a host (most often the larva), according to the species of wasp. The larvae complete their development in or on a single host, which succumbs or is unable to complete its developmental cycle.

As a group, ants occupy a unique place among insects. They are the most numerous of all terrestrial animals. W. M. Wheeler attributed the success of ants to their variability, great numbers, and wide geographical distribution; to their "remarkable longevity," the "abandonment of certain overspecialized modes of life," and plasticity in relationships with other animals, including man. Worker ants live from four to seven years, queens live from thirteen to fifteen years; in contrast, worker bees live only a few weeks or months, while queens live only a few years. Termites live longer than bees or wasps but are not so widespread. Some colonies of ants outlast a generation of men; colonies of wasps and bumble bees, which are relatively scarce, are "merely annual growths." Ants have not restricted their diet, do not require elaborate combs; many ants do not depend on a single fertilized queen as the reproductive unit of the colony. They live in all manner of terrestrial environments, are not restricted like bees and termites.

Ants have remarkably few enemies, despite the many uninvited guests and hangers-on (some harmful, others neutral or mutualistic) that are tolerated in their nests. Wheeler concluded that "as a group ants are eminently beneficial, that many species deserve our protection." They hasten the breakdown of organic matter and aerate the soil; many species are carnivorous and important predators. Some species are a nuisance in the home, others damage crops, but their greatest harm comes from protecting aphids, mealybugs, and other insects for their honeydew excretions; predators are driven off or killed and parasites are often prevented from ovipositing in their hosts.

SYMPHYTA

Sawflies (Tenthredinidae)

The Tenthredinidae is the largest family of sawflies. Adults have nine- to sixteen-segmented antennae, and front tibiae with two spurs (in the horntails they have only one). Many feed externally on leaves; some mine them.

(1) CHERRY FRUIT SAWFLY: *Hoplocampa cookei.* **Fig. 1128.**

Range: California, Oregon, Washington. **Adult:** Black, with part of head, antennae, and legs yellow or reddish brown. Shiny white, kidney-shaped eggs are laid singly on developing blossoms. **Length:** .2″ or less. **Larva:** Crescent-shaped, with seven pairs of prolegs; clear white, with brown head, about .2″ long full grown. The young larva enters the fruit, eats its way into the kernel, leaves the fruit through a hole in the side (its trademark). Hibernates in silken cocoon in the soil. One brood. **Food:** Wild and culti-vated cherry, plum, occasionally prune in Oregon. The fruit turns yellow, drops prematurely. The **European Apple Sawfly,** *H. testudinea,* and the **Pear Sawfly,** *H. brevis,* the only members of this genus found on pear and apple, are minor pests of these fruits in the eastern states. *H. testudinea*— about .25″ long—is blackish, with large black spot on top and front of head; *brevis*—about .2″ long—is brownish, with head entirely brown and wings slightly yellowish; adults frequent the blossoms and larvae mine the fruit.

(2) ELM LEAF MINER: *Fenusa ulmi.* **Fig. 1129.**

Range: Eastern U.S. and Canada; Europe. **Adult:** Black, except tibiae and hind tarsi, which are brownish; prominent furrow on front of head; wings almost transparent, gray-brown along the veins. **Length:** .15″. The larvae mine leaves of European and American elms. The **European Alder Leaf Miner,** *F. dohrnii,* is about the same size, black, except tibiae and tarsi, which are dark brown; without distinct furrow on head; it ranges across the northern part of Europe and North America. The **Birch Leaf Miner,** *F. pusilla* (Fig. 521C), is black, about .07″ long, lays its eggs in unfolding young leaves of birch; two or more broods; European in origin, it was discovered in Connecticut in 1923, now occurs in eastern Canada and from Maine to Delaware, west to Minnesota, Oregon, Washington.

(3) LARCH SAWFLY: *Pristiphora erichsonii.* **Fig. 1130.**

Range: Across Canada, the northern states as far west as Montana. A native of Europe, it was first reported in Massachusetts in 1880; also occurs in Russia and Japan. **Adult:** Body and antennae of female black; abdomen with broad orange band and tapered toward the rear. The male is smaller, with yellowish antennae and broad orange abdominal band; ab-domen rounded at the tip and not tapered. Less than 2 percent are males, reproduction is by parthenogenesis, without mating. Translucent eggs are laid in rows under bark of terminal twigs; egg slits cause twig to curl slightly. **Length:** About .4″. **Larva:** Gray-green, with shiny jet black head, about .75″ long full grown; seven pairs of prolegs: on segments 2 to 7 and

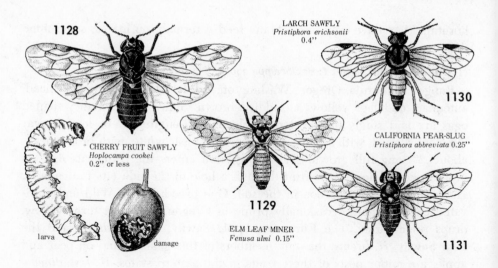

1128

LARCH SAWFLY
Pristiphora erichsonii
0.4''

1130

CHERRY FRUIT SAWFLY
Hoplocampa cookei
0.2'' or less

CALIFORNIA PEAR-SLUG
Pristiphora abbreviata 0.25''

larva

damage

1129

ELM LEAF MINER
Fenusa ulmi 0.15''

1131

10. They feed in groups on foliage in the upper crowns, stripping branches. Hibernates as larva in tough, papery, brown cocoon in forest litter; some larvae spend two winters in cocoon before pupating. One brood. **Food:** Tamarack (eastern larch), western larch, any species of *Larix*. Serious outbreaks have occurred in eastern Canada and in Minnesota. [For parasites, see Diptera (97).] The larvae are susceptible to fungus diseases and microsporidian infections; shrews, where present, destroy many cocoons.

(4) CALIFORNIA PEAR-SLUG: *Pristiphora abbreviata*. **Fig. 1131.**

Range: California, Oregon, Washington; also New York and Connecticut. **Adult:** Black, with yellow markings on prothorax. White eggs are inserted in leaf tissue. **Length:** .25''. **Larva:** Slimy, broadest at anterior end; bright green, matching leaf closely. It feeds on foliage only, rests along the edge of a circular or oval hole, usually cut out near the margin of a leaf, and is difficult to see. It drops to the ground to pupate. Hibernates as pupa in a thin brown cocoon in the soil. One brood. **Food:** Pear.

(5) PEAR-SLUG: *Caliroa cerasi*. **Fig. 1132.**

Range: Throughout North America; also occurs in Europe. **Adult:** Shiny black, with smoky wings. White, oval eggs are inserted under epidermis of leaves. First-brood females are parthenogenetic, producing large second brood. **Length:** .3''. **Larva:** Sluglike, front end wide, tapered posteriorly; olive green, covered with a slimy secretion; about .5'' full grown. It feeds on upper surface of leaf, which is skeletonized, leaving characteristic network of veins. Hibernates as pupa in the soil. Two broods in the western states. **Food:** Pear and cherry preferred; also plum, quince, hawthorn, buttonbush, Juneberry, mountain ash.

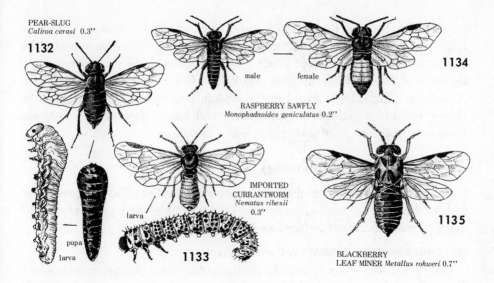

PEAR-SLUG
Caliroa cerasi 0.3"
1132

male   female

RASPBERRY SAWFLY
Monophadnoides geniculatus 0.2"

1134

IMPORTED
CURRANTWORM
Nematus ribesii
0.3"

larva

pupa

larva   **1133**

1135

BLACKBERRY
LEAF MINER *Metallus rohweri* 0.7"

(6) IMPORTED CURRANTWORM: *Nematus ribesii.* **Fig. 1133.**

Range: Throughout the U.S. and southern Canada. A native of Europe. **Adult:** Body brownish black, with pale yellow markings. Elongated, flattened white eggs are laid end to end along veins on underside of leaves. **Length:** About .3". **Larva:** Early stage has black head, numerous spots; uniform green when mature, about .8" long, with six pairs of prolegs. It raises head and posterior end of body when disturbed, feeds on edges of leaves, sometimes stripping bushes. Hibernates as larva or pupa in a capsule-like cocoon on the ground. One to three broods. **Food:** Currant, gooseberry.

(7) RASPBERRY SAWFLY: *Monophadnoides geniculatus (rubi).* **Fig. 1134.**

Range: Widespread throughout North America. **Adult:** Black, with yellowish and reddish markings. It appears in spring at blossoming time. Eggs are inserted in leaf tissues. **Length:** About .2". **Larva:** Pale green, with white spiny tubercles; about .6" long full grown. It feeds on underside of leaves, making numerous small holes. **Food:** Raspberry, blackberry, loganberry. The **Blackberry Leaf Miner,** *Metallus rohweri* (Fig. 1135), mines the leaves of blackberry and other brambles, is occasionally abundant in the Northeast; it hibernates as larva, emerging in the spring.

(8) DOCK SAWFLY: *Ametastegia glabrata.*

Range: Throughout Canada and the northern states. A native of Europe. **Adult:** Bluish black, with reddish legs. Eggs are laid in leaves. **Length:** .4-.5". **Larva:** Bright green, with numerous prominent white spots (tubercles), seven pairs of prolegs; about .6" long full grown. Hibernates as larva in stem of host plants, dried alfalfa, mustard, or sometimes in an apple. Four broods. **Food:** Dock, wild buckwheat, sheep sorrel, and related plants. It occasionally attacks apples in nearby orchards; round holes bored in fruit resemble those made by the codling moth; sometimes causes injury in orchards of Washington and British Columbia. The **Violet Sawfly,** A. *pallipes* (Fig. 1136), found in the eastern part of the U.S. and Canada, closely resembles *glabrata* in appearance and habits; later-stage larvae eat holes at or near margin of leaves of violets and pansies.

Conifer Sawflies (Diprionidae)

Adults of the conifer sawflies have very stout bodies. The antennae have at least thirteen segments, are serrate in the females and pectinate in the males. Some of our most serious forest pests are in this group.

(9) EUROPEAN SPRUCE SAWFLY: *Diprion hercyniae.*

Range: Eastern Canada and the northeastern states, west to northwestern Ontario and Minnesota. **Adult:** Black; it emerges through a smooth, circular hole cut in the end of the cocoon. It is sluggish except in warm weather, when it flies swiftly. Males are generally scarce. Pale green eggs are inserted in slits made in the needles. **Length:** About .3″. **Larva:** Green, like the foliage, with five narrow longitudinal white stripes appearing in the fourth instar; the stripes become more distinct in the fifth instar but disappear in the sixth and final stage when it drops to the ground to pupate; about .75″ long full grown. The cocoon is golden brown, capsule-like, about .4″ long. The larvae feed in groups, stripping the needles, which they chew off from the tip backwards. They prefer foliage from the previous year or older and, like the *Neodiprion* sawflies, are upper-crown feeders. Hibernates as larva in cocoon in the forest litter; some hibernate three winters. One or two broods. **Food:** White, black, red, and Norway spruce. The first serious outbreak occurred in the Gaspé (Quebec) in 1930.

(10) EUROPEAN PINE SAWFLY: *Neodiprion sertifer.* **Fig. 1137.**

Range: First discovered in New Jersey in 1925, now extends to southern Ontario and west to Wisconsin; occurs in Japan, Korea, Siberia, Europe. **Adult:** Blackish, with translucent wings and dark venation typical of the group. Males have conspicuous branched antennae, characteristic of the group. Normally, there are two to three times as many females as males; males are produced by parthenogenesis, females from fertilized eggs. They mate soon after emerging from cocoons; eggs are deposited in rows of slits made in needles of the season's current growth and are hermetically sealed within the needles. **Length:** About .4″. **Larva:** Head black, body dirty gray-green to mottled yellowish, with lighter dorsal stripe, on either side of which is a dark green or black stripe bordered on both sides with white; about .9″ long full grown. They feed in groups, starting at the tips, con-

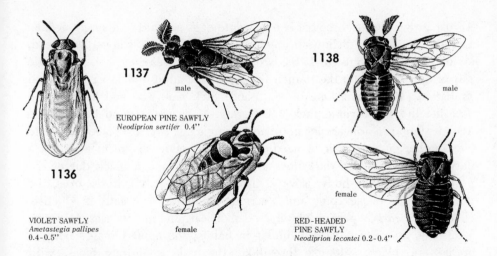

1137

male

EUROPEAN PINE SAWFLY
Neodiprion sertifer 0.4"

1138

male

female

1136

VIOLET SAWFLY
Ametastegia pallipes
0.4-0.5"

female

RED-HEADED
PINE SAWFLY
Neodiprion lecontei 0.2-0.4"

sume the entire needles and strip the branches; when disturbed they rear their heads. Pupation takes place in cocoon on the ground—the pupa is first golden brown, later black—and hibernation in the egg stage. **Food:** Two-needled (hard) pines, also spruce close by; jack, red, and Scotch pines in plantations. The principal parasites are *Dahlbominus fuscipennis* (Eulophidae), *Drino bohemicus* (Tachinidae), and the ichneumonid *Pleolophus basizonus* (which also parasitizes *N. swainei*). The larvae are also very susceptible to a polyhedrosis virus, and shrews (*Sorex, Blarina*) destroy many cocoons.

(11) RED-HEADED PINE SAWFLY: *Neodiprion lecontei.* **Fig. 1138.**

Range: Eastern Canada, states east of the Mississippi River; Texas, Louisiana, Arkansas, Missouri, Minnesota. **Adult:** Head and thorax of female pale reddish brown, abdomen black, legs reddish brown; antennae saw-toothed. The male is smaller, all black; antennae feather-like. The female is monogamous. **Length:** .2–.4". **Larva:** Yellowish, with six longitudinal rows of black dots, pale reddish brown head, about 1" long full grown. Second instars strip edges of needles, causing them to dry up in characteristic reddish, strawlike masses. Habits in general similar to (10) above, except that hibernation takes place as mature larva in the cocoon; in emerging, the adult cuts a smooth circular hole, as do other diprionids. One brood in the North, four or five in the Southeast. **Food:** Hard pines, such as jack, shortleaf, slash, red, loblolly, Virginia, pitch, longleaf; other native and exotic species. Severe outbreaks occurred in the Lake states in the 1940's. The most important parasite is *Closterocerus cinctipennis* (Eulophidae), which attacks the eggs.

(12) JACK PINE SAWFLY: *Neodiprion pratti banksianae.*

Range: Eastern Canada, to Manitoba; Michigan, Wisconsin, Minnesota.

Adult: Brown. From three to six oval white eggs are laid in a single needle. Length: .6″. Larva: Yellowish green, with black stripes, shiny black head, about 1″ long full grown. It feeds on old needles, leaving a characteristic tuft of new growth on the branch tip. Habits of adult and larva in general similar to (10) above. One brood. Food: Jack pine; red and Scotch pine occasionally. The **Swaine Jack-pine Sawfly**, *N. swainei*, found in the same area, is similar in appearance and habits; single egg slits are placed opposite one another on each pair of needles; the pale yellowish green larva has a dark brown head, two dark lines along the back, and a black dot at the posterior end of the body, is very susceptible to a polyhedrosis virus; the pest is specific to jack pine and a serious defoliator, especially in Quebec. *N. taedae linearis*, a pest of loblolly and shortleaf pine in Arkansas, Missouri, Louisiana, and Texas, is also similar in habits; the adult female's body is orange and black, antennae threadlike; the male is entirely black, with feathery antennae; older larvae are dull green, with black stripes. The **Lodgepole Sawfly**, *N. burkei* (Fig. 1139), has caused serious damage to lodgepole pine in Montana and Oregon, is also found in Idaho, Wyoming, Alberta; the male is black and the female brownish, about .35″ long; the larva is greenish or grayish, with lighter lateral stripes and brown head, about 1″ long full grown. The **Hemlock Sawfly**, *N. tsugae*, occasionally defoliates extensive areas of western hemlock in Oregon and northward into Alaska. The **White-pine Sawfly**, *N. pinetum*, most often attacks white pine but occurs on pitch and shortleaf pine in the East; the larva is yellowish white, with black head and four longitudinal rows of rectangular black spots.

(13) WEB-SPINNING SAWFLY: *Cephalcia californica* (Pamphiliidae).

The web-spinning sawflies differ from other sawflies in that the larvae have no prolegs and remain attached to the food plant (conifers) by means of a silk webbing. The cylindrical eggs are laid singly or in groups on the foliage. The larvae sever the needles near the base and usually drag them back to their shelters to eat. They drop to the ground to overwinter and pupate; many spend two or more winters in the soil before pupating and emerging. **Range:** *C. californica* ranges from British Columbia to Montana, Wyoming, and California. **Adult:** Robust, black except for most of appendages and the ninth abdominal sternum (in the male), which are yellowish orange; antennae 23- to 29-segmented; wings cloudy, with black veins. **Length** (forewing): .3–.4″. **Larva:** Tan, about .12″ long full grown; five (male) and six (female) instars. **Host:** Lodgepole and ponderosa pines. *C. marginata* is found on red, white, and jack pines in eastern U.S. and Canada.

(14) ELM SAWFLY: *Cimbex americana* (Cimbicidae). **Fig. 1140.**

The cimbicid sawflies are characterized by the capitate antennae; *C.*

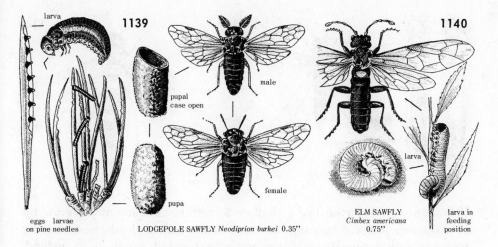

larva

1139

male

pupal
case open

female

pupa

eggs   larvae
on pine needles

LODGEPOLE SAWFLY *Neodiprion burkei* 0.35"

1140

larva

ELM SAWFLY
Cimbex americana
0.75"

larva in
feeding
position

americana is the largest common species. **Range:** Eastern United States and Canada, west to Colorado and British Columbia. **Adult:** Steel blue, with three or four oval spots on each side of body; wings smoky. Adults girdle the bark on twigs, killing many of them near the tops of trees. Oval, transparent flattened eggs are inserted in the leaves. **Length:** .75". **Larva:** Naked, wrinkled, pale yellowish, with black middorsal stripe and eight pairs of prolegs; from 1" to 2" full grown. It feeds on leaves, lies coiled up at rest. Hibernates as larva in a cocoon on the ground, pupates in the spring. **Food:** Elm and willow; also poplar, alder, maple, and other broad-leaved trees. *C. a. pacifica* has similar habits, attacks willow in Oregon and Washington, is brownish red, about 1" long. *C. rubida* attacks willows in California and Nevada, is rusty red-brown, with black markings; the wings are metallic blue and smoky brown.

Horntails (Siricidae)

The horntails, or wood wasps, have long cylindrical bodies with a horny spikelike projection at the posterior end. The front tibiae have one spur (in the sawflies they have two). The larvae, which also have the anal spines or "horntails," bore into plant stems, trunks, and limbs of dead and dying trees.

(15) WESTERN HORNTAIL: *Sirex areolatus.* **Fig. 1141.**

Range: British Columbia, California, Arizona, New Mexico, Colorado. **Adult:** Metallic blue-black; legs entirely black or blue-black; female with wing completely clouded, 20- to 23-segmented antennae; male with wings only slightly clouded at apexes, third to seventh abdominal segments yellowish orange. Eggs are deposited in dead trees by means of long, sheathed ovipositor as long as or longer than forewing. **Length:** 1–1.4". **Larva:**

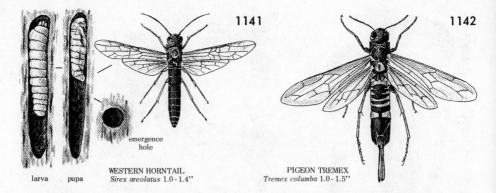

emergence
hole

larva   pupa

WESTERN HORNTAIL
Sirex areolatus 1.0 - 1.4"

PIGEON TREMEX
Tremex columba 1.0 - 1.5"

Yellowish white; shaped like a shallow **S**, with small spine at posterior tip of abdomen. It bores into sapwood and heartwood of dead trees, forming perfectly round tunnels, pushing partially digested wood fibers behind. One to two seasons are required to complete development. **Food:** Redwood, cypress, cedars; occasionally pine and other conifers. Unseasoned and frequently seasoned lumber are attacked.

(16) BLUE HORNTAIL: *Sirex cyaneus*. Color plate 7j.

Range: Throughout North America. **Adult:** Metallic blue-black; female with legs reddish brown except coxae and trochanters, wings transparent except clouded apical margin; male with front and middle legs reddish brown beyond trochanters, basal segments of antennae reddish brown or black, thorax metallic blue or greenish, wings slightly yellowish, with faint smoky margins. **Length:** .8". **Food:** Pine, fir, spruce, Douglas fir. *S. juvencus californicus* (=*S. obesus*) is very similar, the male with coxae and basal segments of abdomen metallic greenish black, remainder of legs and abdomen reddish brown; female with legs bluish black, forewing clouded across the middle and on apical margin; from 1" to 1.2" long; it attacks various pines, Monterey cypress, Douglas fir, from British Columbia to California, and in Nevada, Arizona, New Mexico. The eastern form of *juvencus* occurs chiefly in weakened and newly dead balsam fir in eastern Canada, requires two years to complete its life cycle; a fungus deposited in the wood at the time of oviposition is essential to development beyond the first-instar larval stage.

(17) PIGEON TREMEX: *Tremex columba*. Fig. 1142.

Range: Throughout North America. **Adult:** A large black or brownish horntail, with head and thorax sometimes reddish and black; abdomen marked with yellow bands; antennae yellow at base and black at tip; wings smoky yellow. Short, blunt yellow ovipositor in the female. Eggs are drilled into bark and wood. **Length:** 1–1.5". **Larva:** Creamy white, about 2" long full grown. It constructs a round tunnel in the sapwood and heartwood. Two years are normally required to complete its life cycle. **Food:** Various deciduous hardwoods, such as maple, beech, oak, and elm. Dead and dying trees, and logs, are attacked. See (27).

(18) WHEAT STEM SAWFLY: *Cephus cinctus* (Cephidae). **Fig. 1143.**

Also known as the western grass stem sawfly. **Range:** West of the Mississippi River; Manitoba to Alberta. A native species, originally found on hollow-stemmed grasses. **Adult:** Shiny black, abdomen with dorsal, transverse yellow bands; femora, stigma and costa mostly yellow. Eggs are thrust into plant tissues on upper part of stems. **Length:** About .5″. **Larva:** Slender, S-shaped, slightly enlarged at thorax; whitish, with brown head capsule and mandibles, posterior end armed with stiff bristles. It feeds within the stem, boring through the joints, reaches lower part of plant by late summer; cuts a V-shaped groove around inside of stem at base, forms sealed chamber, where it hibernates in a cocoon. Pupates here in the spring. Weakened plant stems break off at the groove. One brood. **Food:** Wheat preferred; barley, spring rye, timothy, wild grasses. The most important parasite of this sawfly is the braconid wasp *Bracon cephi* (24). The **European Wheat Stem Sawfly,** *C. pygmaeus,* is similar to *cinctus*—about the same size, with yellow abdominal bands, black femora, dark brown costa and stigma; the **Black Grain Stem Sawfly,** *C. tabidus,* is without yellow bands, males have horseshoe-shaped depressions on underside of last two abdominal segments; both are European in origin, and presently confined to the East. [See (29) and p. 526.]

APOCRITA

Cynipids or Gall Wasps (Cynipidae)

Also called gallflies (not to be confused with the gall midges), cynipids are small to minute wasps with short 11- to 16-segmented antennae. The larvae cause deformities called galls, which are growths of plant tissues and serve as food and living quarters during larval and pupal development. They are found mostly on oak but also occur on rose and some members of the thistle family; the plants do not appear to be harmed by these abnormal growths. There are some 200 species of western gall wasps alone, found on various oaks, each producing a characteristic gall which aids in identification of the insect. Some are attached to twigs, others to leaves or stems. They are variously shaped—globular, pear-shaped, starlike, conical, spiny, and so on. Some oak galls are rich in tannins and in the Old World have been used for making some of the best "permanent" writing inks. The Aleppo gall, produced on several species of oak in southeastern Europe, is 65 percent tannic acid; a Chinese gall, produced on sumac, has an even

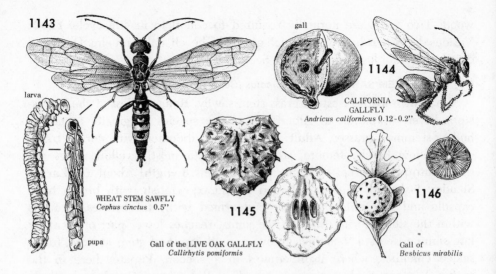

1143

larva

WHEAT STEM SAWFLY
Cephus cinctus 0.5"

pupa

gall

1144

CALIFORNIA
GALLFLY
Andricus californicus 0.12-0.2"

1145

Gall of the LIVE OAK GALLFLY
Callirhytis pomiformis

1146

Gall of
Besbicus mirabilis

higher tannin content and is also used for making ink. Two small sub-families of parasitic wasps are included in the family.

(19) CALIFORNIA GALLFLY: *Andricus californicus.* **Fig. 1144.**

Range: California, Oregon, Washington. **Adult:** Brownish or reddish brown. It appears from October (in the North) to February (in the South). Eggs are deposited on leaves of the host plant. **Length:** .12–.2". **Larva:** Minute, legless, creamy white; it causes galls on twigs of oak. The galls—largest and most common in the West—are formed on twigs in the spring. They are spherical or kidney-shaped, from 1.5" to 4" in size, and inhabited by several or many larvae. Green at first (and often called "green apples" or "oak apples"), they turn whitish, tan, or dark later. Large galls are inhabited by other insects, including the filbertworm, a pest of walnuts and filberts [see Lepidoptera (276)]. The oak galls serve as a reservoir for orchard infestations. **Host:** Valley, blue, scrub, and leather oaks. The **Live Oak Gallfly,** *Callirhytis (Andricus) pomiformis,* produces a gall (Fig. 1145) on coast and interior live oak in California. *Besbicus (Cynips) mirabilis* produces a beautiful smooth, spherical, green or yellow thin-walled gall (Fig. 1146) with red spots, on upper side of the leaves of Garry oak [see (20) below for another occupant of this gall]; it occurs in northern California, Oregon, Washington, and British Columbia. The **Mossy-rose-gall Wasp,** *Diplolepis rosae,* causes a large globular growth of cells surrounded by mosslike filaments on the stems of rose in the Midwest. *D. polita* produces a globular red and green gall, from .2" to .4" in diameter, with long sharp spines; the galls occur in clusters on the leaves and stems of wild rose in Colorado and westward. *Antron (Dryophanta) douglasii* produces starlike pink or coral-red galls, usually with eight points; they are from

.3″ to .5″ in diameter, occur on underside of the leaves of valley and blue oak in central California. *Neuroterus saltatorius* produces minute round or ovate jumping galls, resembling a seed; they occur on underside of the leaves of various oaks in California, Arizona, and Texas; the galls drop to the ground in great numbers when ripe and move, seemingly all at once, as the larvae inside throw themselves violently.

(20) PACIFIC GALL WASP: *Synergus pacificus.*

This species is an inquiline or guest of other oak gall cynipids. **Range:** Pacific coast states and British Columbia. **Adult:** Black, with antennae, most of legs, part of head yellowish brown (head mostly yellow in the male); antennae 14-segmented in the female, 15-segmented in the male; wings clear, pubescent, with yellow veins. Eggs are laid singly in soft immature galls of other wasps, and hatch in about four days. **Length:** .08″. **Larva:** Minute, transparent, cream-colored; stout, tapered posteriorly, body segments well defined, from .05″ to .1″ full grown. When it hatches it burrows a little farther into the gall but stays close to the outside wall; it forms a subgall, often causing deformities on the host gall, as do the other guests. In galls of *Besbicus mirabilis* (Fig. 1146) the larval stage lasts about two months, and the pupal stage about two weeks. The galls fall to the ground in the fall, and the adults emerge next summer, from May to August. **Host:** Galls of *Heteroecus* (*Andricus*) *pacificus* (Fig. 1147) on canyon live oak, *Besbicus mirabilis* and others on Garry oak. The hosts on Garry oak have a sexual and agamic generation; only the galls of the latter are available when *S. pacificus* adults emerge.

Braconids (Braconidae)

The parasitic wasps fall mainly into two superfamilies: the Ichneumonoidea and the Chalcidoidea. The dominant families among the former are Braconidae and Ichneumonidae; both include species that attack a large majority of the economically important plant pests, and many have been used in biological control.

The braconids are nearly all primary parasites and are almost entirely beneficial. Very few are secondary parasites, and only a few parasitize predatory species; the latter are mainly in the genus *Perilitus* and attack lady beetles. The braconids are parasitic mostly on the larvae of Lepidoptera and Coleoptera, but attack many Diptera and Homoptera (notably aphids), fewer Hymenoptera and Hemiptera. *Dendrosoter* spp. parasitize bark beetles (*Ips*) and stem-boring weevils (*Cylindrocopturus*); *D. protuberans* has been imported recently from France for control of the smaller European elm bark beetle, a scolytid vector of Dutch elm disease.

Adults of the braconids are generally smaller than those of the ichneu-

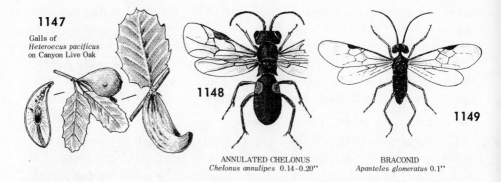

1147

Galls of
Heteroecus pacificus
on Canyon Live Oak

1148

ANNULATED CHELONUS
Chelonus annulipes 0.14-0.20"

BRACONID
Apanteles glomeratus 0.1"

1149

mons, seldom as much as one-half inch long. They feed mostly on honey-dew and plant exudations, but females often supplement this with body fluids oozing from the oviposition wound. Some puncture the host expressly for feeding and the fluids may constitute a principal part of the diet; where the host is hidden—as in a stem, pod, or leaf roll—a feeding tube is constructed. Egg production is often high; a gravid female of *Apanteles glomeratus,* a parasite of cabbageworms, may contain up to 2,000 eggs. The life cycle of many species is short, allowing several generations or broods in a year.

The braconid larvae may feed externally or internally on the host; commonly they are external parasites where the host is hidden, and internal parasites where the host feeds exposed. Some species are solitary, others are gregarious, with a few or many individuals feeding on a single host. However, superparasitism (more eggs laid in a single host than can complete development) occurs in some solitary species. Apparently the females cannot always tell whether the host is already parasitized. The excess is usually eliminated by cannibalism, and the winner takes all. Females in the case of gregarious internal parasites usually deposit all the eggs in a single host with one insertion of the ovipositor. *Apanteles militaris,* a parasite of the armyworm, is said to oviposit up to 72 eggs with one thrust of the ovipositor, in a second or less. Where feeding is external, the female usually paralyzes the host permanently before ovipositing; where feeding is internal this is seldom the case. The endoparasites usually hibernate as first-instar larvae in the living hosts; others overwinter as mature larvae or pupae in the host cocoons. (Only the female of one species is known to hibernate as an adult; the male dies after mating in the fall.) In emerging from their hosts, the mature larvae usually spin cocoons separately or in a mass on or near the hosts and pupate; *Meteorus* spp. hang their cocoons by a slender thread on a leaf or twig. Some species complete feeding after emerging from the host, before spinning cocoons; *Ascogaster quadridentata,* a parasite of the codling moth, and *Chelonus annulipes*

(Fig. 1148), an imported parasite of the European corn borer, have this habit, as do all species of *Macrocentrus*. Among the genera *Ascogaster, Chelonus,* and *Phanerotoma,* eggs are deposited in the host egg, with development taking place in the larval stage of the host.

Males are commonly produced from unfertilized eggs, and females from fertilized eggs. Some species are unisexual and males seldom if ever appear; in others mating appears to be necessary to produce any progeny. In a large number of species males outnumber females; in only a few do the females greatly outnumber the males. *Macrocentrus gifuensis,* an imported parasite of the European corn borer, reproduces by polyembryony. One egg is placed in the host larva with each insertion of the ovipositor, and ten or more larvae develop from this single egg; 40 or more, with an average of 21, emerge from a single host as the result of one or several eggs laid. The female's reproductive capacity of 200 to 300 eggs is thus multiplied significantly.

(21) BRACONID: *Apanteles glomeratus.* **Fig. 1149.**

Range: Throughout North America. Imported from England, established in 1884. **Adult:** Black, with yellowish and pale reddish markings. The female deposits from 15 to 30 eggs in a single host larva. **Length:** About .1″. **Larva:** Gregarious; mature larvae emerge from the host larva, spin oval, yellowish cocoons in an irregular mass on the leaf, around the host. (The latter may survive but is unable to pupate.) Hibernates usually as early instar larva in living host, sometimes as mature larva in the cocoon. Several broods. **Hosts:** Southern cabbageworm, imported cabbageworm, other butterflies and moths. Related species, all solitary parasites, imported for biological control, are: *A. solitarius* against the satin moth, *A. melanoscelus* (Fig. 1150) against the gypsy moth, and *A. lacteicolor* against the brown-tail moth; all are important factors in the complex of parasites operating against these forest and shade tree pests.

(22) BRACONID: *Apanteles congregatus.* **Fig. 1151.**

Range: Throughout North America. **Adult:** Black, with legs partially yellow. The female oviposits many eggs in a single host. **Length:** .1″. **Larva:** Gregarious; mature larvae emerge from the host larva, spin oval, white cocoons separately at point of emergence. The cocoons stand on end, commonly cover more than half of the exposed area of their large hornworm hosts. The adult emerges through a circular hole cut in the tip of the cocoon. The host is not killed immediately but is unable to complete its development. **Hosts:** Tomato hornworm, catalpa sphinx, and other sphinx moths. *A. harrisinae,* a gregarious internal parasite of the western grape leaf skeletonizer, is also a native species and important in natural control of its host. *A. carpatus* is a parasite of clothes moths (*Tinea, Tineola*).

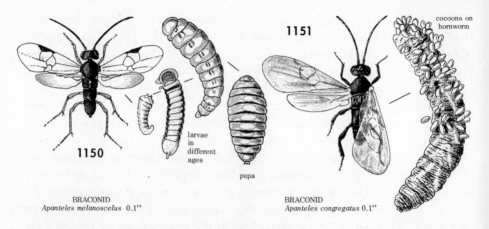

1151

1150

larvae
in
different
ages

pupa

cocoons on
hornworm

BRACONID
Apanteles melanoscelus 0.1"

BRACONID
Apanteles congregatus 0.1"

(23) BRACONID: *Apanteles medicaginis.* **Fig. 1152.**

Range: California, Arizona, Nevada, Utah, Idaho, east of the Rocky Mts. to Kansas. **Adult:** Black; maxillary palpi yellowish except near base; antennae considerably longer than the body in the male, somewhat shorter than the body in the female; legs reddish yellow and black; forewing with dark brown patch on costal margin. Eggs are inserted in body cavity of first- to third-instar larvae of host. **Length:** .1–.15". **Larva:** First stage translucent, distinctly segmented; third and final stage creamy white, opaque, distinctly segmented, with numerous small spines; about .18" long; head small, with well-developed mouthparts. It is a solitary parasite. Pupates outside the host in a yellowish (sometimes white) cocoon, which is often found on upper leaves of alfalfa plants. Overwinters as first-instar larva in the host. Several broods. **Host:** Alfalfa caterpillar. This "is the most prevalent parasite of the alfalfa caterpillar and is an important factor in its control." The eastern species, *A. flaviconchae,* is very similar but is gregarious; it attacks the alfalfa caterpillar and larva of the clouded or common sulphur, but favors the latter, whose main food plant is red clover. Both caterpillars feed to a lesser extent on white clover.

(24) BRACONID: *Bracon hebetor.*

Range: Throughout North America; found wherever grains, flour, and other dried foods are stored. **Adult:** Black, with yellowish markings. The female deposits several eggs on the host larva after paralyzing it by stinging several times. **Length:** About .1". **Larva:** Grublike; it sucks body juices of host through minute lacerations made with mandibles, moves about as it feeds. Spins cocoon close to host remains. Under favorable conditions, the complete life cycle requires less than two weeks. **Hosts:** Mediterranean flour moth, Indian meal moth, tobacco moth, almond moth, sugarbeet crown borer (*Hulstia undulatella*), occasionally the greater wax moth. *B. cephi* is an important parasite of the wheat stem sawfly (18), has habits

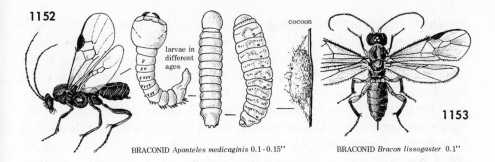

1152

larvae in different ages

cocoon

1153

BRACONID *Apanteles medicaginis* 0.1-0.15''   BRACONID *Bracon lissogaster* 0.1''

similar to *hebetor,* hibernates as larva in uncut stems, has two broods; *B. lissogaster* (Fig. 1153) parasitizes the sawfly in grasses but seldom in wheat. *B. mellitor,* a robust, rusty red-brown wasp, is a parasite of several Lepidoptera and Coleoptera, including the boll weevil and other curculionid pests. *B. kirkpatricki*—imported but possibly not established—has many hosts but prefers the pink bollworm and boll weevil. *Atanycolus* (*Bracon*) *charus* is a parasite of the flatheaded apple tree borer. The closely related *Opius melleus* deposits its eggs in the eggs of the apple maggot and other species of *Rhagoletis.*

(25) BRACONID: *Macrocentrus ancylivorus.* **Fig. 1154.**

Range: Massachusetts to Virginia, west to Texas; Oregon, California. Between 1929 and 1935 it was propagated and colonized on a large scale in all the peach-growing areas of the eastern and North Central states and Ontario in a program sponsored by the USDA for biological control of the Oriental fruit moth. **Adult:** Amber to yellowish; slender and delicate, about the size and shape of a mosquito. Crepuscular or nocturnal in habit, it attacks very young to nearly mature larvae. **Larva:** Develops at rate determined by host, matures when it cocoons; emerges (only one) from host larva, forms own closely woven, mahogany-brown cocoon within that of host. Normally completes life cycle in about a month. Hibernates as young larva within hibernating host larva. **Hosts:** Oriental fruit moth, strawberry leaf roller, ragweed borer and others of similar habits; filbertworm of Oregon and California. *M. delicatus* is the second most abundant parasite of the Oriental fruit moth; *M. instabilis* is also common in peach orchards. *M. gifuensis* is a parasite of the European corn borer (see introduction, p. 522).

(26) BRACONID: *Lysiphlebus* (*Aphidius*) *testaceipes.* **Fig. 152C.**

Range: Widespread in the U.S. and Canada. **Adult:** Abdomen brownish or dusky, excepting part of the second segment which is yellowish; head and thorax black, legs yellowish brown. Eggs are deposited one in each aphid. **Length:** .08''. **Larva:** It feeds inside the aphid, eventually consuming the entire body contents, leaving only a "mummy." Spins a cocoon inside the mummy after cutting a hole in the bottom and securing it to the surface with silk. Circular holes in the tops of mummies denote emergence of the adult parasites. **Host:** Cotton or melon aphid, corn leaf aphid, black citrus aphid, and others. Because of its wide host range this species is generally considered to be the most effective parasite of aphids. In 1907 it was used against the greenbug in Kansas and Oklahoma in the first large-scale attempt at biological control with a native parasite. It often oviposits in the spirea aphid but is unable to complete its development in this host. *Aphidius smithi*—imported from India—is an important parasite of the pea aphid and now established in California, Washington, Idaho, Colorado, Kansas, Wisconsin, and Hawaii. *A. matricariae*—almost cosmopolitan in distribution—is the most important parasite of the green peach aphid.

Ichneumons (Ichneumonidae)

This large family is often referred to as ichneumon-flies. They are generally small to medium in size, but vary in the extreme from minute forms to the large *Megarhyssa* and *Rhyssa* wasps. A few species are without wings; in some the female is wingless and the males with or without wings. This variation among individuals of the same sex is believed to be due to a difference in the amount of food available. While the family is extremely important in the natural control of insect pests there has not been the same success with biological control in this group as with some others. Like the braconids, the ichneumons are mostly primary parasites, although there are many hyperparasitic species that attack braconid and ichneumonid larvae in exposed cocoons. The majority are endoparasites and attack the larvae of Lepidoptera, but not a few attack the pupal stage, and many are internal or external parasites of Hymenoptera and Coleoptera. They are important in the natural control of sawflies and many wood- and stem-boring insects. Those attacking sawflies are generally solitary parasites. Some species prey on spider eggs; others destroy the eggs and larvae of bees and wasps.

Most ichneumons oviposit in the host stage where they complete their development, but many attack the larva and emerge from the pupa. *Collyria calcitrator,* an introduced parasite of the wheat stem sawfly, oviposits in the egg stage but develops (and overwinters) in the larval stage of the host. Hibernation among the ichneumons commonly takes place in the cocoon as a mature larva, but all species of several genera hibernate as adult females. Jumping cocoons are an interesting occurrence in some species of *Bathyplectes;* one is said to jump an inch by the sudden straightening of the body of the larva within the cocoon. The number of broods varies considerably throughout the family—anywhere from one to ten—

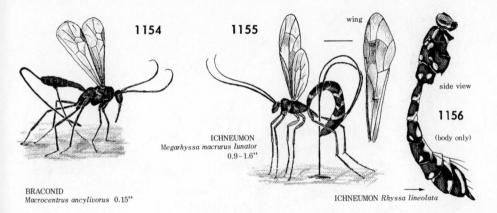

1154

1155

wing

side view

1156

(body only)

ICHNEUMON
Megarhyssa macrurus lunator
0.9 - 1.6"

BRACONID
Macrocentrus ancylivorus 0.15"

ICHNEUMON *Rhyssa lineolata*

but is most often from one to three. Egg production is generally not high though quite variable—50 or less in some species but frequently from 400 to 600.

Many adult females feed largely on the body juices of their hosts and attack them expressly for feeding; *Itoplectis conquisitor* females kill significant numbers of the gypsy moth pupae by stinging them and feeding on the juices, without ovipositing. Unlike braconids and chalcids, ichneumons are not known to construct feeding tubes. Where the parasite feeds externally on a concealed host, the latter is usually permanently paralyzed; the fluid injected for this purpose has a preservative quality and is potent enough to kill the host larva in some cases. Where superparasitism by a solitary species occurs, the excess individuals are eliminated by cannibalism, the first to hatch usually being the survivor. Males are commonly produced from unfertilized eggs—by parthenogenesis—and the females from fertilized eggs. The female usually deposits unfertilized eggs in the smaller, less desirable hosts, and the resulting progeny are males. There are normally slightly fewer males than females. A few species are unisexual and produce females for an indefinite period or number of generations.

(27) ICHNEUMON: *Megarhyssa macrurus lunator.* **Fig. 1155.**

Range: Eastern Canada and the U.S.; Texas, north to South Dakota. **Adult:** Brown, with yellow head and markings on thorax; yellow V-shaped markings bordered with black along sides of abdomen; legs yellow and brown. Wings partially transparent, with brownish markings. Male smaller, with less extensive brown markings on legs. Eggs are laid in tunnel of host in dead or dying hardwoods; long, sheathed ovipositor is looped over abdomen when drilling through bark and depositing eggs. It probably utilizes cracks and crevices in the wood, possibly oviposition holes of host, to get through to tunnel. **Length:** .9–1.6" (body); the female ovipositor extends 3" or more beyond this. **Larva:** White, grublike; pupates in tough cocoon

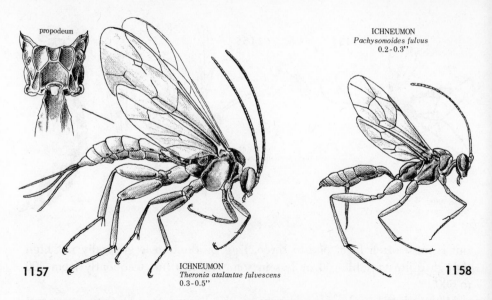

propodeum

ICHNEUMON
Pachysomoides fulvus
0.2-0.3''

1157

ICHNEUMON
Theronia atalantae fulvescens
0.3-0.5''

1158

in tunnel. **Host:** Pigeon tremex. *M. m. macrurus* has more extensive brown markings of the wings, is found in the extreme south, from Florida to Texas and Mexico. The closely related genus *Rhyssa* is a world-wide group parasitic on borers in conifers. They are not quite as large as *Megarhyssa,* are usually black, with white face, large white spots and other white markings along the sides. *R. lineolata* (Fig. 1156), with extensive white markings, and one of the most striking, ranges across North America. Species of *Rhyssa* and *Ibalia* have been introduced from Europe into New Zealand for control of horntails (*Sirex*) infesting pines.

(28) ICHNEUMON: *Theronia atalantae fulvescens.* **Fig. 1157.**

Range: Transcontinental, in the U.S. and Canada. Subspecies occur in Europe, Japan, Korea. **Adult:** Brownish yellow, with some indefinite yellowish markings; often with brown markings on head, thorax, legs, and abdomen; wings tinged with brownish yellow. It has large sharp claws—each has a fluid-filled pocket—with which it clings to a net or its captor; this has suggested the possibility of their use as poison fangs. It "makes a conspicuous pale streak" as it flies through the forest. **Length** (front wing): .3–.5''. It is believed to hibernate as an adult female. **Host:** Lepidoptera of medium or larger size. It is an important parasite of forest pests, especially the pine butterfly, but is sometimes a hyperparasite of ichneumonids attacking the same host.

(29) ICHNEUMON: *Scambus tecumseh.** **Fig. 1159.**

Range: Transcontinental, in the U.S. and Canada. One of the most common species, abundant in forested areas. **Adult:** Black; male with tro-

* According to Townes, "This species may not prove to be distinct from *buolianae* of Europe."

chanters and front and middle coxae white, front femur and tibia reddish, middle femur and tibia reddish above to white underneath, antennae 21- to 22-segmented; female with legs mostly red, hind tibia and basal segments of hind tarsus whitish, antennae 21- to 24-segmented. **Length** (front wing): .1–.2″ (male); .2–.35″ (female). **Host:** Spruce budworm, eye-spotted bud moth, western tent caterpillar, European pine shoot moth, and other Lepidoptera. *S. detritus* (Fig. 1160) is also transcontinental in distribution, fairly abundant from May to August; it is black, legs largely reddish, wings tinged with yellowish, forewing from .15″ to .3″ long; it is a parasite of the wheat stem sawfly and the European wheat stem sawfly.

(30) ICHNEUMON: *Horogenes punctorius.* **Fig. 1161.**

Range: Ontario and Massachusetts to Delaware, west to Ohio and Michigan. Imported from Europe and the Orient. **Adult:** Black, with reddish legs. The female normally mates only once and oviposits in the host larva on the plant or in the tunnel. More than one egg may be deposited in a single host, but only one larva will develop; the others become encysted and die. Females outnumber males. **Length:** .5″. **Larva:** Dirty white; tapered toward both ends. A solitary internal parasite, it consumes all but head capsule and skin of the host, spins closely woven cocoon incorporating host remains in loosely woven outer threads. Hibernates as first-instar larva in host larva. One or more generations, depending on the weather and host cycle. **Host:** European corn borer. One of most valuable of the parasites introduced during large-scale program for biological control of the European corn borer, from 1921 to 1949. [See (41) below.]

(31) ICHNEUMON: *Pachysomoides (Polistiphaga) fulvus.* **Fig. 1158.**

Range: Throughout the U.S. and southern Canada; Mexico, Cuba. **Adult:** Brownish yellow, with whitish and dark brown markings; front and middle coxae whitish, especially in the male; specimens west of the Rocky Mts. are normally darker, with less extensive white markings. Adults are commonly seen resting on the walls and windows of buildings. Eggs are laid on mature larvae in the nests of the host just before they spin their cocoons. **Length** (front wing): .2–.3″. **Larva:** Feeds on larva of host, before or after it spins a cocoon, or on the pupa. Spins own cocoon inside that of host. It is gregarious, with two to six parasites per host. **Host:** *Polistes* wasps. Late in the season, when the *Polistes* nests are abandoned, one can usually find cocoons in them containing the parasites.

(32) ICHNEUMON: *Cryptanura banchiformis.* **Fig. 1162.**

Range: Massachusetts, south to Florida, west to Texas, Kansas, and Michigan. **Adult:** Black, with extensive white markings on head, thorax, and abdomen; male with front and middle coxae and trochanters all white, hind tarsi white; female with legs mostly brownish yellow. According to

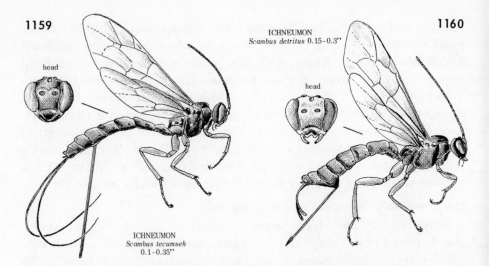

ICHNEUMON
Scambus detritus 0.15-0.3"

head

head

ICHNEUMON
Scambus tecumseh
0.1-0.35"

Townes, "The sting of this species is moderately severe and of long duration. One sting received by [him] was still painful on the night of the third day." They fly rapidly—from late spring to early fall—the male with a dancing flight and a little faster than the female. They are found in heavy undergrowth in or on the edge of deciduous woods. **Length** (front wing): .4–.45".

(33) ICHNEUMON: *Exochus nigripalpis tectulum.* **Fig. 1163.**

Range: Transcontinental, in Canada and the U.S., except California. **Adult:** Black; process between antennae with broad white border; legs brownish yellow, male with front and middle coxae yellowish underneath; female with legs more uniformly brownish yellow. They fly throughout the summer in forested areas. **Length** (front wing): .15–.25". **Host:** Commonly the spruce budworm. *E. n. subobscurus* occurs in California; the process between the antennae has a narrow white border or is all black; male with front and middle coxae dark reddish brown underneath; it has been reared from the orange tortrix, apple skinworm (*Argyrotaenia franciscana*), and other leaf rollers. This species does not give off the pungent odor characteristic of this genus, at least not at all times.

(34) ICHNEUMON: *Chorinaeus excessorius.* **Fig. 1164.**

Range: Widely distributed—common in eastern half of the U.S. and Canada, less common west of the 100th meridian. **Adult:** Black; front of head, front and middle coxae and trochanters, apex of femora and base of tibiae pale yellow; legs otherwise brownish yellow. They fly throughout the summer, are numerous in the undergrowth of deciduous forests. **Length** (front wing): .15–.2". **Host:** Reported rearings are from the spruce budworm and strawberry leaf roller.

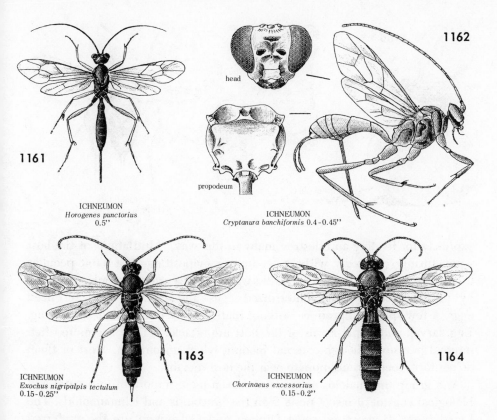

1162

head

propodeum

1161

ICHNEUMON
Horogenes punctorius
0.5"

ICHNEUMON
Cryptanura banchiformis 0.4-0.45"

1163

ICHNEUMON
Exochus nigripalpis tectulum
0.15-0.25"

1164

ICHNEUMON
Chorinaeus excessorius
0.15-0.2"

Chalcids (Chalcidoidea)

This superfamily includes many families and many thousands of species throughout the world. The name "chalcid" is commonly applied to members of the various families comprising the entire group and not simply to the family Chalcididae. Most of them are very small, often no more than a hundredth of an inch long. The great majority are parasites or hyperparasites. Notable exceptions occur in the Eurytomidae and include such well-known pests as the wheat jointworm and the wheat straw-worm, which forms galls in wheat and other grasses, and the clover seed chalcid, a pest of clover and other legume seeds; and others that are parasites as young larvae but resort to plant feeding later. The fig wasp demonstrates the interesting relationship of a single insect and plant completely dependent on each other.

Adult chalcids feed mostly on plant exudations and honeydew excreted by their insect hosts. Females commonly supplement this with body fluids oozing from oviposition wounds, and construct feeding tubes where the hosts are hidden under scale coverings or in plants. Some puncture the host

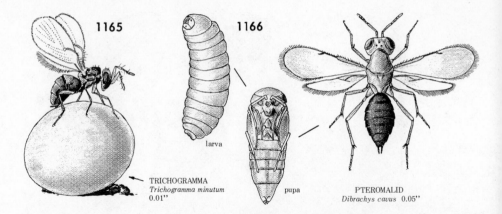

1165   1166

larva

TRICHOGRAMMA
Trichogramma minutum
0.01"

pupa

PTEROMALID
Dibrachys cavus 0.05"

expressly for feeding and destroy many in this way. "Mutilation" of the host —puncturing repeatedly without feeding or ovipositing—is a habit peculiar to a few species. Like other parasitic wasps, males are commonly produced by parthenogenesis—from unfertilized eggs—and females from fertilized eggs. A few species reproduce without mating, with only females occurring. Egg, larval, and pupal forms of the host are attacked. They are more often internal parasites, though external feeding is not uncommon. Most of them do not form cocoons, and pupate near the host remains.

As a group the chalcids appear to be even more important in natural and biological control of insect pests than the braconids and ichneumonids. The Lepidoptera, Homoptera, some Diptera and Coleoptera are the preferred hosts. The coccids—scale insects and mealybugs—are more severely attacked by parasites than any other comparable group, and largely by the chalcids, which are particularly amenable to use in biological control. Because of the ease with which coccids spread and their damage to citrus crops, much attention has been directed toward the use of their enemies as a means of suppression. According to DeBach, control of no fewer than 38 species of coccids has been achieved in this way, or about 40 percent of the total cases of successful biological control.

(35) TRICHOGRAMMA: *Trichogramma minutum* (Trichogrammatidae). **Fig. 1165.**

Range: Throughout North America. **Adult:** Minute, yellowish, with brownish abdomen and pink eyes. From one to fifty eggs are deposited in a single host egg, sometimes twenty-five at a time; the number depends on size of host. Host eggs turn black. Females normally outnumber males, the ratio depending on food available. **Length:** .01". **Larva:** The newly hatched larva moves about in host egg, thus stopping development of the embryo. Hibernates as larva in host egg. The number of broods varies from 13 to 50 or more, depending on climate. **Host:** Eggs of more than 200 species of insects are known to be parasitized, ranging from the bollworm in the South to the spruce budworm in northern forests. It is important in control of the corn earworm and southern cornstalk borer in some areas; with *T. evanescens* it has been used extensively in biological control of field and orchard crops.

(36) PTEROMALID: *Dibrachys cavus* (Pteromalidae). **Fig. 1166.**

Range: Eastern U.S. and Canada, Colorado, Utah, California, Oregon, Idaho, British Columbia; South America, Europe, Asia, Australia. According to Allen, this species is "one of the most widely distributed, abundant, and polyphagous of all hymenopterous parasites." **Adult:** Shiny, head and thorax greenish, abdomen dark brown, often appearing to be all black. It normally oviposits several to many eggs in the host pupa after paralyzing it. **Length:** About .05″. **Larva:** Several or many larvae develop in a single host. Pupates in host cocoon, hibernates here as mature larva. Several broods. **Host:** Many species of Lepidoptera, Diptera, Hymenoptera, several species of Coleoptera, and even some spiders. Muesebeck et al. list 97 hosts. It is a primary parasite of many species of Lepidoptera and phytophagous Hymenoptera but is often hyperparasitic on other parasites of the same hosts. The closely related *Nasonia* (*Marmoniella*) *vitripennis* parasitizes the pupae of various insects, particularly blow flies, flesh flies, and muscid flies.

(37) GOLDEN CHALCID: *Aphytis chrysomphali* (Eulophidae). **Fig. 1167.**

Range: New York, Florida, Louisiana, California; a native of China, but present in California for a long time, probably came in with the California red scale on nursery stock. **Adult:** Pale lemon yellow, with big green compound eyes. Eggs are inserted under scale covering, on host's body. It destroys many hosts by mutilation (see introduction, p. 532). The California strain is uniparental, does not produce males. **Length:** .04″. **Larva:** Ovoid, top globular, tapered toward posterior end. It feeds externally on the scale, under armor, where it pupates; emerges by cutting hole in cover or pushing out under the margin. One to ten or more broods, depending on climate. **Hosts:** California red scale, yellow scale, coconut scale and others. The closely related *Aphytis lingnanensis* is similar in appearance and habits, except that it is biparental and males are produced by parthenogenesis. Both are important in control of California red scale. *A. lepidosaphes*, a solitary or gregarious parasite of the purple scale, is similar in appearance; the adult female feeds on body juices of the host by constructing a feeding tube to surface of armor; it is highly specific to this scale, gives good control in many areas. All of these parasites have been imported from the Orient by California in large-scale biological control programs. *A. maculicornis*, a solitary external parasite of the olive scale (*Parlatoria oleae*), similar in habits to the above, was imported from Iran and Iraq in a large-scale program for control of the olive scale in California between 1952 and 1960.

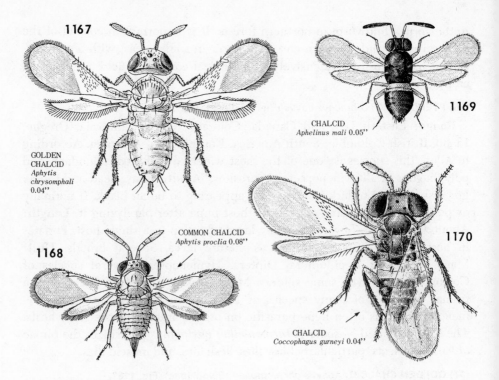

1167

GOLDEN
CHALCID
*Aphytis
chrysomphali*
0.04"

CHALCID
Aphelinus mali 0.05"

1169

COMMON CHALCID
Aphytis proclia 0.08"

1168

1170

CHALCID
Coccophagus gurneyi 0.04"

(38) COMMON CHALCID: *Aphytis proclia* (*Aphelinus diaspidis*). **Fig. 1168.**

Range: Cosmopolitan; throughout the U.S. **Adult:** Dull yellow, with dusky broken band near middle of each of first five abdominal segments. Minute yellowish eggs have long stalk at one end, are inserted under scale covering, on body of host. **Length:** .03"; wingspread .08". **Larva:** Feeds externally on body of scale, also on the eggs. **Hosts:** Walnut scale, San Jose scale, oystershell scale, and many other armored scales.

(39) CHALCID (EULOPHID): *Aphelinus mali*. **Fig. 1169.**

Range: Throughout North America. A native species, introduced from the East into the apple-growing areas of the Pacific Northwest and British Columbia for control of the woolly apple aphid. Exported to many other countries. **Length:** .05". **Adult:** Dark brown to nearly black, with base of abdomen yellow, legs and antennae partly yellow. One or more eggs are deposited in body of aphid, but only one develops into mature wasp. **Larva:** Internal parasite of immature and mature forms of host; consumes all but the skin. Body juices ooze from the aphid, gluing it to the twig; aphid shells become black and mummified. Pupates in host, the adult emerging by cutting a hole in the top of the "mummy." Hibernates as mature larva in host remains. Life cycle requires about 30 days. **Host:** Woolly apple aphid, cabbage aphid, greenbug, rose aphid, and other aphids.

(40) CHALCID (EULOPHID): *Coccophagus gurneyi.* **Fig. 1170.**

Range: Imported from Australia in 1927 for control of the citrophilus mealybug, established in California. *Hungariella (Tetracnemus) pretiosus* (*Encyrtidae*), which parasitizes the same host, was also imported at this time. **Adult:** Black, with conspicuous yellow band across base of abdomen; male entirely black, not so easily recognized. It feeds on honeydew excreted by the host. Males are hyperparasites. The female probes the host with its ovipositor, deposits an unfertilized egg (future male) if the host is already parasitized by its own or other species such as *pretiosus.* **Length:** .04″. **Larva:** Has twelve well-defined segments, very small mandibles. Embryonic development of the hyperparasite is completed in three days, about the time it takes the female to hatch, but it does not hatch until the female (or other primary parasite) consumes body fluids of the host; it feeds internally or externally on the primary parasite. Since *C. gurneyi* females predominate at low densities of the host, the hyperparasitic trait tends to conserve males and is advantageous to the species. *H. pretiosus* deposits eggs in the early nymph stage of the mealybug, can itself control the citrophilus mealybug; *C. gurneyi,* however, is dominant because of its hyperparasitic habit. Their control over the citrophilus mealybug is comparable to that of the vedalia beetle over the cottony cushion scale. The closely related *Aspidiotiphagus citrinus citrinus*—a cosmopolitan species widespread in the U.S.—is the second most important parasite of the yellow scale in California; it is yellowish black, with black compound eyes and red ocelli; in contrast to *Comperiella bifasciata* (43), its hosts include many other scales.

(41) CHALCID (EULOPHID): *Sympiesis (Eulophus) viridula.* **Fig. 1171.**

Range: New York, Pennsylvania, Ohio, Indiana, Iowa, Wyoming; Nova Scotia, Ontario, Alberta. First imported from Italy in 1930 for biological control of the European corn borer. **Adult:** Black; front of head and the thorax (except prothorax) metallic green, legs lemon yellow, wings clear; abdomen with yellowish blotches, sometimes mostly yellowish. The females are much larger than the males, outnumber them two to one. Eggs are deposited externally on host larva after paralyzing it permanently. **Length:** .06–.12″. **Larva:** An average of 28 larvae develop externally on a single host in summer colonies, three times this in winter colonies, reducing it to an empty shell in four or five days. Pupates in host tunnel, hibernates here in pupal stage. Several broods. **Host:** European corn borer, smartweed borer, and *Pyrausta penitalis.* [See (30) above, and Diptera (96).]

(42) CHALCID (ENCYRTID): *Aphycus helvolus* (Encyrtidae).

Range: First imported into California from South Africa in 1937. It occurs in nearly all citrus-growing areas of the world. **Adult:** Female orange-yellow; male dark brown, slightly smaller. The long-lived female lays an

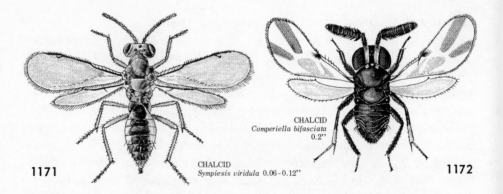

CHALCID
Comperiella bifasciata
0.2"

1171

CHALCID
Sympiesis viridula 0.06-0.12"

1172

average of 400 to 500 eggs, oviposits into the side of the black scale (which is unarmored) and feeds on body fluids issuing from the wound. The egg has a stalk which remains fixed at point where it is inserted. If a suitable host can't be found, egg production can be suspended. **Length:** .04". **Larva:** Solitary internal parasite of early nymph stage of the host. In California there are eight or more broods in "even hatch" or single-brooded areas of the black scale—in the interior regions—and two for each of the two broods of the host in the "uneven hatch" areas of the coastal regions. **Hosts:** Black scale, European fruit lecanium, and other soft scales. *A. helvolus* is the most important parasite of black scale in California, followed by *A. stanleyi* and *A. lounsburyi,* which are also orange-yellow, and similar in appearance. *Pseudaphycus malinus* has been an important factor in biological control of the Comstock mealybug on apples in Connecticut and other eastern states. *Leptomastix dactylopii* is a widespread parasite of mealybugs; it is cultured in insectaries and released periodically in southern California for control of mealybugs, more especially the citrus mealybug. *Ooencyrtus kuwanai*—together with another imported egg parasite, *Anastatus disparis* (Eupelmidae)—is an important enemy of the gypsy moth.

(43) CHALCID (ENCYRTID): *Comperiella bifasciata.* **Fig. 1172.**

Range: Citrus-growing areas. Introduced into California from the Far East. **Adult:** Brilliant metallic sheen, front and top of head creamy white, with dark, longitudinal median stripe; the wings are held semierect at rest, straighten out after death. The eggs are stalked—a characteristic of the family—but the stalk in this case apparently serves no purpose, since the whole egg is deposited in the body cavity of the host. Hosts are punctured frequently for feeding or are simply mutilated, adding greatly to their mortality. **Length:** About .2". **Larva:** Solitary internal parasite, mostly of the female scale, preferably the second instar. **Hosts:** This is the most important parasite of yellow scale; it can also develop on Florida red scale and Chinese red scale (*Aonidiella taxus*). Its host specificity is shown by the fact that it

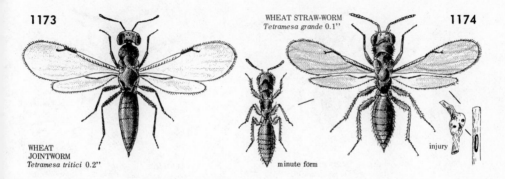

WHEAT STRAW-WORM
Tetramesa grande 0.1"

WHEAT
JOINTWORM
Tetramesa tritici 0.2"

minute form

injury

will not develop on California red scale, which is hardly distinguishable from the yellow scale. For the second most important parasite of the yellow scale, see (40) above.

(44) WHEAT JOINTWORM: *Tetramesa* (*Harmolita*) *tritici* (Eurytomidae). **Fig. 1173.**

Range: Eastern North America, Nebraska, Kansas, Utah, California, Oregon. **Adult:** Black, both sexes winged. Eggs are inserted in wheat stems above joints; each female lays upwards of 70 eggs. **Length:** .2". **Larva:** White, very small. It forms a cell in the wall of a stem, feeds on the sap. Several larvae form a hard woody gall, each in its individual cell just above the second or third joint, causing the stem to bend or break. (Hessian fly maggots are greenish, the pupae brown, and are found between leaf sheath and stem.) It remains in wheat stubble after harvest, changing to yellowish pupa within the cell. Hibernates as pupa; the adult gnaws its way out of the cell in the spring. One brood. **Host:** Wheat. Ranks next to Hessian fly as pest of wheat in states where it occurs. [See Diptera (18).] The **Wheat Straw-worm**, *T. grande* (Fig. 1174), has "alternate generations"; the first or minute form is black, about .1" long, wingless in both sexes, appears in early spring; eggs are laid in short stems and the larvae destroy developing heads. The next generation or large form, which does less damage, is fully winged, shiny black, about .15" long, and consists of females only; the greenish yellow larvae infest stems near the joints, usually in the largest and most vigorous plants. The **Rye Jointworm**, *T. secale*, occurs in eastern U.S., westward to Utah and Alberta; the larvae form galls in the stems of rye, usually in clusters just above the second and third joints, which causes stems to bend and reduces the size of kernels. It hibernates as larva, pupates and emerges in spring; the male is distinguished by its petiolate waist and plumose antennae. The closely related **Clover Seed Chalcid,** *Bruchophagus platyptera* (*gibbus*) (Fig. 1175), is black, with legs partly brown, inserts eggs in the developing seeds of alfalfa and clover, is very injurious to seed crops. [See Diptera (19).] The **Alfalfa Seed Chalcid,** *B. roddi,* often causes severe

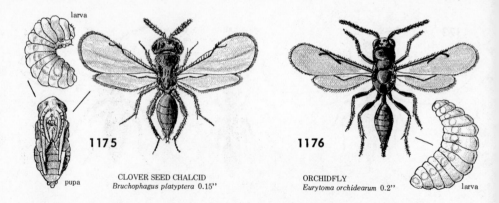

larva

pupa

1175

CLOVER SEED CHALCID
Bruchophagus platyptera 0.15"

1176

ORCHIDFLY
Eurytoma orchidearum 0.2"

larva

losses in alfalfa seed production; it has two or three broods, overwinters as a full-grown larva inside the seed. (*Tetramesa* accords with Peck's *Catalogue.*) [See (45) below.]

(45) ORCHIDFLY: *Eurytoma orchidearum.* **Fig. 1176.**

Also called cattleyafly or orchid isosoma. **Range:** Cosmopolitan; common in greenhouses. **Adult:** Black. **Length:** About .2". **Larva:** White, legless, about .25" full grown. It feeds in the bulbs, stems, leaves, and buds of host plants. **Host:** Many kinds of orchids. A serious pest in greenhouses. *E. parva* parasitizes the wheat jointworm (44), closely resembles it, though smaller (about .15" long); the female inserts her ovipositor into the host's cell in a wheat stem, deposits her eggs; the parasitic larva is a solitary external parasite of the first or second instar of the host larva, quickly consumes it and completes development by feeding on plant sap, hibernates as mature larva in the wheat stem. The wheat jointworm is also parasitized by two other chalcids: *Ditropinotus aureoviridis* (Torymidae) and *Eupelmus allynii* (Eupelmidae). *D. aureoviridis* females are dark brownish, males golden green; the larva is a solitary external parasite of the mature *Tetramesa* larva. Several eggs may be thrust through the wall of a wheat stem, but the oldest larva destroys the others. *E. allynii* adults are greenish black or purplish black, with yellow compound eyes; the female places about 50 percent of her eggs on the wall of the host cells, fastening each with a series of strands believed to be secreted by a gland adjunct to the oviducts, as in species that cement their eggs; it is common and widespread throughout the U.S. and eastern Canada.

(46) FIG WASP: *Blastophaga psenes* (Agaonidae). **Fig. 1177.**

Range: Native of the Mediterranean region of Europe, Africa, and Asia Minor; introduced into California in 1890 for cross-fertilization of a Smyrna-type fig known as the Calimyrna (or Lob Injür). **Adult:** Female shiny black, with large compound eyes, three ocelli, ten-jointed antennae, almost vein-

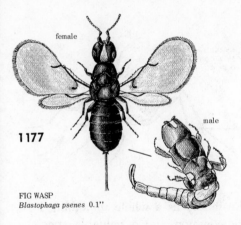

female

male

1177

FIG WASP
Blastophaga psenes 0.1"

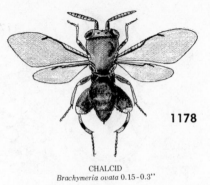

1178

CHALCID
Brachymeria ovata 0.15-0.3"

less wings; male amber-colored, wingless, with small compound eyes, no ocelli, long abdomen usually curved under body. They breed only in non-edible wild figs called caprifigs. The fig flower is contained in a receptacle connected to the exterior only through a very small orifice. The minute eggs are inserted through the hollow style into the ovaries of the flowers. The short-styled flowers of the caprifigs are shaped to receive the eggs of the wasp, the long-styled flowers of the Calimyrna figs are not; coupled with this, the caprifig flowers are staminate while the Calimyrna are pistillate only and contain no pollen. As eggs are laid, some pollen of the caprifigs rubs off on the wasps and is carried to the flowers of the Calimyrna figs which they also visit, and which are absolutely dependent on this pollen for fertilization and development of the fruit. **Length:** .1". **Larva:** Lives in seed which in this case is referred to as a gall. After completing their development the males emerge and gnaw holes in the galls containing the females; mating is accomplished through the holes. One day after fertilization the females are fully developed, enlarge the holes, emerge, and fly away to lay their eggs. Caprifigs produce three crops; only the crop called "profichi" contains pollen when the wasps (second brood) are mature and flowers of the Calimyrna figs are receptive. (The winter crop contains no pollen.) The process of transferring pollen is called "caprification"; one method used in orchards is to hang containers with caprifigs on the trees. They are supplied at intervals, the quantity depending on the size of the trees. Three to five caprifig trees supply enough wasps to pollinate 100 Calimyrna fig trees.

(47) CHALCID: *Brachymeria ovata* (Chalcididae). **Fig. 1178.**

This family contains some of the largest chalcids; they have conspicuous markings, greatly enlarged hind femora adapted for jumping, do not fold their wings at rest. They are valuable for their attacks on moths and blow

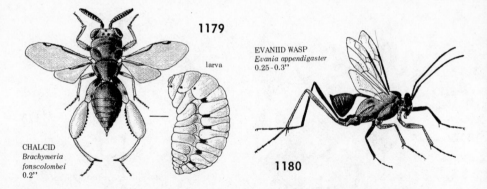

1179

larva

EVANIID WASP
Evania appendigaster
0.25-0.3"

CHALCID
Brachymeria
fonscolombei
0.2"

1180

flies, but many hyperparasites make the family as a whole something less than beneficial. *B. ovata ovata* is very common and a valuable species. **Range:** Throughout North America. **Adult:** Robust, black and yellow; femur of hind leg black, with white or yellow tip. Tegula—small sclerite at base of costa of forewing—also white or yellow. **Length:** .15–.3". **Larva:** Solitary internal parasite of pupa. **Host:** White-marked tussock moth, western tussock moth, fruit-tree leaf roller, range caterpillar, cotton leafworm, and many other Lepidoptera. Easily mistaken for the black and yellow subspecies, *B. o. abiesae*, a parasite of the California oakworm and California tent caterpillar. *B. fonscolombei* (Fig. 1179) is an important parasite of the mature larvae of flesh flies and blow flies throughout the South, hibernates in the pupal stage. *B. compsilurae*, an eastern species, is a hyperparasite, attacks the larvae of tachinid flies while they are still in the dead bodies of their hosts.

(48) EVANIID WASP: *Evania appendigaster* (Proctotrupoidea, Evaniidae). **Fig. 1180.**

This family differs from most Hymenoptera in having the abdomen attached near the top of the propodeum instead of down near the coxae. They are parasitic in the egg capsules of cockroaches and are most abundant in the tropics; some species have become widely distributed in commerce. **Range:** *E. appendigaster* is common in Arizona and in cities of the Gulf states and Atlantic states as far north as New York. **Adult:** Entirely black, clothed with a very fine pubescence. Eggs are deposited in the egg capsules before they harden. Only one egg is inserted within one of the eggs in a capsule; only one individual is able to develop in a single capsule. **Length:** .25–.3". **Larva:** The first instar has 13 distinct segments and prominent mandibles, feeds within a single egg; after the first instar other eggs are attacked, and it becomes in effect an egg predator. **Host:** Cockroaches.

(49) SCELIONID WASP: *Eumicrosoma beneficum* (Scelionidae). **Fig. 1181.**

The Scelionidae is a moderate-sized family of minute wasps that parasitize the eggs of a wide variety of other insects. Like *Trichogramma* wasps, they prefer freshly laid eggs, but unlike them, they are decidedly selective in their choice of host. Phoresy is common, some females attaching themselves to the back of the female host and remaining there until she lays her eggs. The females of *Scelio* dig into the soil to deposit their eggs in the egg pods of grasshoppers shortly after they have been laid, often remaining here several hours to complete oviposition in the egg mass with one thrust of the ovipositor. A number of species have been used effectively for biological control of insect pests in Hawaii and the Far East; releases of *Telenomus emersoni* were effective in reducing the horse fly *Tabanus dorsifer* in Texas. **Range:** *E. beneficum* occurs in Virginia, Kansas, Oklahoma, and throughout the British West Indies. **Adult:** Head and thorax shiny, black; antennae reddish yellow, with four-segmented club in the female, dark brown and without a club in the male; palpi and legs reddish yellow; entire abdomen reddish yellow in the female, apical half dark brown in the male. Eight or nine generations in a season, sometimes four or five in the eggs of a single host brood. **Length:** .03″. **Host:** Chinch bug. *Telenomus minimus* parasitizes the eggs of the armyworm.

(50) PLATYGASTERID WASP: *Platygaster zosine* (Platygasteridae). **Fig. 1182.**

This species belongs to a small family of very small black wasps that are parasitic in the larvae of midges (Cecidomyiidae). **Range:** *P. zosine* is generally found wherever the Hessian fly occurs. **Adult:** Shiny, black; antennae ten-segmented, brownish; all tibiae and tarsi with white pubescence; wings clear, extending far beyond the abdomen. Eggs are deposited singly in the host egg, development and emergence take place in the larval and pupal stages of the host. The female appears to recognize eggs that have been previously attacked. **Length:** .05″. **Host:** Hessian fly. *P. herrickii* and *P. hiemalis* are very similar but smaller, also parasitize the Hessian fly. *P. hiemalis* normally deposits several eggs in a single host egg, and by a simple form of polyembryony two individuals develop from each parasite egg.

(51) CUCKOO WASP: *Parnopes edwardsi* (Chrysididae). **Fig. 1183.**

The family name means "golden," referring to the brilliant metallic coloring in shades of green, blue, red, or purple. They are called cuckoo wasps because of their habit of laying eggs in exposed cells or nests of potter wasps and mud daubers. They are solitary external parasites of the host larvae, hibernate as larvae and pupate in cocoons in the host cells. **Range:** *P. edwardsi* occurs in the Pacific coast states, Idaho, and British Columbia. **Adult:** Brilliant green to deep blue. Like other cuckoo wasps, it has a habit of curling the body into a ball, with wings protruding to the sides, when alarmed; this is believed to be a way of feigning death for protection. They have no sting. **Length:** .25–.5″.

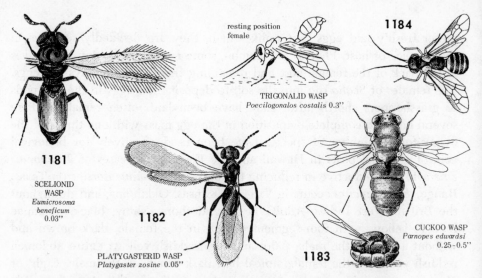

1184

resting position
female

TRIGONALID WASP
Poecilogonalos costalis 0.3"

1181

SCELIONID
WASP
*Eumicrosoma
beneficum*
0.03"

1182

PLATYGASTERID WASP
Platygaster zosine 0.05"

1183

CUCKOO WASP
Parnopes edwardsi
0.25 - 0.5"

(52) TRIGONALID WASP: *Poecilogonalos costalis* (Trigonalidae). **Fig. 1184.**

The members of this very small family of robust wasps are found in the nests of vespid wasps (as primary parasites) and in the pupae of ichneumonid and tachinid parasites (as secondary parasites). Eggs are laid in minute slits made in leaves near the margins; they never hatch here, but some manage to reach the host in a devious manner. In the case of their ichneumonid and tachinid hosts, it is by way of the caterpillars and sawfly larvae that eat the leaves (and eggs) and are themselves parasitized by suitable hosts. In the case of vespid wasps, it is believed to be by way of the body contents of caterpillars and sawfly larvae containing trigonalid larvae which are fed to the vespid young by their female attendants. The adult trigonalid wasp emerges from the cocoon or puparium of its parasite hosts, and from a cell in the nest of its vespid host. **Range:** *P. costalis* ranges from Massachusetts and New York to North Carolina and West Virginia. **Adult:** Black, clothed in pile; antennae pale reddish brown; markings on thorax, most of the legs, and on apical margins of the abdominal segments yellowish white; wings mostly clear, venation and stigma reddish brown. It is found in moist wooded areas. **Length:** .3". **Host:** *Phosphila turbulenta,* "probably secondary" (Muesebeck).

Hornets, Yellow Jackets, and Potter Wasps (Vespidae)

The vespid wasps are medium to large in size, usually black and white or black and yellow. They fold their wings fanlike along the length of the body when at rest, with the exception of the subfamily Masarinae, which have clubbed antennae and build mud nests attached to rocks or twigs, provisioning them with nectar and pollen. Most of the vespids are solitary wasps and construct cells in wood or the soil, in hollow stems and reeds, abandoned *Sceliphron* nests, and old burrows of bees, or build nests of mud on rocks and twigs. The nests in most cases are provisioned with caterpillars. Wasps in the subfamilies Vespinae, Polistinae, and Polybiinae are social. In the Zethinae they come close to social behavior but lack a caste system and division of labor; they utilize wood cavities, or several females may construct a colony of cells made from plant materials glued together, but each wasp feeds its own young.

The adult social wasps have well-developed mandibles (for capturing and chewing insects) and tongue (for sucking nectar, fruit and other juices). They build paper nests in trees, under eaves and in buildings, in the ground, hollow trees, and thickets. The paper is made from wood chewed into a pulp. Their relatively simple caste system consists of queens, workers (sterile females), drones or males whose only function is to provide sperm. The queen lays fertilized eggs to produce females, unfertilized eggs to produce males. The sterile workers can lay eggs, and they sometimes do toward the end of the season, but they produce males only. Some species are monogynous—having only one queen—others are polygynous. Old queens, drones, and workers die at summer's end and the colony breaks up. Only the young queens, now mated, survive to hibernate—under the bark of trees or in other crevices. In the spring they form new colonies, build the first combs, and start laying fertilized eggs as soon as a few cells are made. Later, workers take over the building of additional cells and care of the young while the queen lays eggs. The white, legless larvae are fed partially masticated insects daily. The adults feed on nectar, ripe fruit, bits of insects captured for the young, and secretions from the mouths of the grubs, often withholding their food until this oral tidbit is forthcoming. When mature the grub seals the cell and pupates.

(53) GIANT HORNET: *Vespa crabro germana* (Vespinae). **Color plate 7m.**

Range: Southern Massachusetts and New York to North Carolina; scattered colonies in South Carolina, Georgia, Tennessee, Kentucky, Ohio, and Indiana; introduced from Europe. **Adult:** Brown and yellow. It usually nests in hollow trees, stumps or logs, or in buildings. **Length:** .7–.9″.

(54) YELLOW JACKET: *Vespula* (*Vespula*) *pennsylvanica.* **Color plate 7l.**

The name "yellow jacket" is commonly applied loosely to all black and yellow hornets. Yellow jackets belong to the subgenus *Vespula* and usually nest in the ground or in hollow logs and stumps close to the ground. Hornets belong to the subgenus *Dolichovespula* and usually build aerial nests.*
Range: *V. pennsylvanica* occurs in western North America, mostly west

* The subgenera may be distinguished by the space separating the lower margin of the compound eye and the base of the mandible—it is never longer than one-half of the penultimate (second from last) antennal segment in *Vespula,* and always longer in *Dolichovespula.*

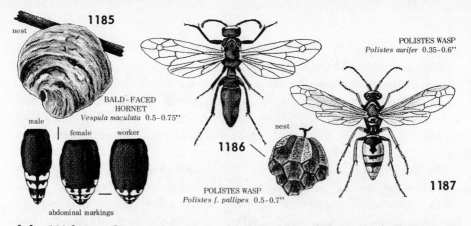

1185

nest

BALD-FACED
HORNET
Vespula maculata 0.5-0.75"

male

female   worker

abdominal markings

1186

POLISTES WASP
Polistes f. pallipes 0.5-0.7"

nest

POLISTES WASP
Polistes aurifer 0.35-0.6"

1187

of the 100th meridian, and is very common. **Adult:** Black and yellow; basal segment of antennae yellow underneath; compound eyes close to mandibles, with upper three-fourths of the eye encircled by yellowish band. It nests in the ground. **Length:** .5–.7". *V. maculifrons* is very similar, black and yellow, with basal segment of antennae mostly or entirely black; it occurs in the eastern half of North America, nests in the ground or in decayed stumps and hollow logs. *V. vulgaris*—a northern species—is very similar to the above two, black and yellow, may be distinguished by the broad, dark longitudinal stripe (sometimes anchor-shaped) on the middle of the clypeus; it is one of the most common yellow jackets, transcontinental in distribution. *V. squamosa*—a southeastern species—and *V. sulphurea*—found in Oregon and California—are similar to the above three, may be distinguished by the two broad, slightly curved, yellowish longitudinal stripes on the mesonotum.

(55) BALD-FACED HORNET: *Vespula (Dolichovespula) maculata.* **Fig. 1185.**

Range: Throughout the U.S. and Canada; one of the most common and widespread wasps in North America. **Adult:** Black, with white markings on tip of abdomen; face and thorax with white or pale yellowish markings. The colony builds a large globular paper nest suspended from the limb of a tree or bush, or an overhang. **Length:** .5–.75". **Food:** The larvae are fed all kinds of insects, many of which the workers catch on the wing. *V. arenaria*—also transcontinental in range—is black and yellow, the latter extending over entire length of abdomen, the first two segments of which are deeply notched with black; it builds a globular nest in thickets above ground or under eaves. *V. adulterina* is without a worker caste and the larvae live as guests in the nests of *arenaria*.

(56) POLISTES WASP: *Polistes fuscatus pallipes* (Polistinae).* **Fig. 1186.**

Polistes wasps resemble hornets in color pattern; the abdomen is more slender, attached to the thorax by a short petiole. In the hornets and yellow

* The Polybiinae are very similar in habits to the *Polistes* wasps; their nests in our range are generally smaller, consisting of a single paper comb of hexagonal cells suspended by short pedicel. They range from the Far West and South into South America.

POTTER WASP
Eumenes fraterna
0.75"

nest

MASON WASP
Monobia quadridens
0.8"

jackets, the clypeus is broadly truncate (and slightly notched) at the apex, and the hindwing is without a lobe at the anal angle; in the Polistinae, the clypeus is pointed at the apex and the hindwing has an anal lobe. *Polistes* wasps are generally more peaceful and less vicious than hornets and yellow jackets. Their paper nests consist of circular combs of cells suspended by stem attached to a limb, eaves, or other structure; the ends of the cells are open and the heads of the larvae exposed to view. There are numerous subspecies of *fuscatus*, their range extending throughout the U.S. and Canada. **Range:** *P. f. pallipes*—sometimes referred to as the "dark paper wasp"—occurs in eastern Canada and the northeastern states, south to West Virginia. **Adult:** Largely black, with narrow yellowish markings; propodeum smooth or with fine transverse striae. **Length:** .5–.7". *P. annularis*—the "large paper wasp"—is from .8" to 1" long, with coarse transverse striae on propodeum. *P. exclamans*—known as the "zebra paper wasp"—has blackish, reddish, and yellowish markings on abdomen, reddish mesonotum, and yellowish band behind the ocelli; a valuable predator of the tobacco hornworm, it is encouraged in tobacco-growing areas by being provided shelter boxes for nesting; it is said to be troublesome to citrus workers in Arizona, who are frequently stung by them. *P. rubiginosus* is practically all orange in color. *P. aurifer* (Fig. 1187)—a black and yellow species, from .35" to .6" long—is very common in the West, ranging from Colorado to the Pacific, and from British Columbia to Baja California; it is a subspecies of *fuscatus*. See (31).

(57) POTTER WASP: *Eumenes fraterna* (Eumeninae). **Fig. 1188.**

Range: Eastern U.S., west to Texas, Oklahoma, Kansas, Nebraska, Minnesota; Ontario. **Adult:** Black, with yellow markings. It constructs a rounded vaselike nest of clay on twigs of trees and bushes. As in the case of other vespid mud wasps, the egg is placed in the cell, attached to the roof or wall, *before* the cell is stored with insects (sphecid mud wasps place the egg *on* the prey). When filled with larvae, the cell is sealed. **Length:** .75". **Food:** The nest is provisioned with paralyzed cankerworms and other small caterpillars for the developing larva.

(58) MASON WASP: *Monobia quadridens.* **Fig. 1189.**

Range: Eastern and southern U.S., west to New Mexico, Kansas, Illinois, Wisconsin. **Adult:** Black, with yellow markings. It usually nests in wood, partitioning the burrow into a series of cells, using mud; it takes over abandoned carpenter bee tunnels and those of ground bees, also old *Sceliphron* nests and other cavities. The cells are separated by mud partitions, with little silk being used. **Length:** .8″. **Food:** Nests are provisioned with large paralyzed cutworms.

Spider Wasps (Pompilidae = Psammocharidae)

This fairly large family contains many species of common wasps—small to large in size—that build nests in the ground and provision them with spiders. Some are parasitic in the nests of other spider wasps, a few nest in mud cells, and others construct cells in existing holes in wood. A number of species, such as *Ceropales maculata,* are "social parasites"; the female surreptitiously inserts an egg in the prey captured by another pompilid and left momentarily unguarded before being placed in the nest. The larva of *Ceropales* hatches first and devours the host's egg and lives off its stores.

Pompilids may be distinguished from any other wasps by the oblique transverse groove (suture) on the side of the mesothorax (or mesopleuron) dividing it into upper and lower halves. The majority of spider wasps are black, with blackish wings; many are colored with metallic blue, orange, red, or white. They are long-legged, very active, and often seen on flowers or running rapidly on the ground or low plants. The antennae of the males have 13 segments (except in *Pepsis formosa* they have 12) and are usually straight; those of the females have 12 segments and are often curled. Some of the largest and showiest species are in the genus *Pepsis;* these occur mostly in Mexico and the Southwest and provision their nests with mygalomorph spiders—the so-called tarantulas.

(59) TARANTULA HAWK: *Pepsis mildei.* **Color plate 7k.**

Range: Texas to Kansas, west to California and northern Mexico. **Adult:** Metallic blue, antennae reddish, wings fiery red except the outer margin and base, which are dusky. The female slowly approaches, then stings the much larger tarantula, which offers no resistance and acts as if hypnotized. The paralyzed spider is dragged to a previously prepared burrow, where it is implanted with an egg and serves as food for the developing larva. **Length:** .8–1.2″. **Host:** Tarantulas. *P. formosa formosa* (Fig. 1190) ranges from northwestern Mexico to Kansas, west to Nevada and Arizona; its wings are mostly orange. *P. pattoni* occurs in western Mexico, southern California, and southwestern Arizona; its wings are mostly blackish.

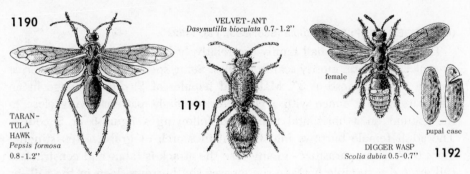

1190

VELVET-ANT
Dasymutilla bioculata 0.7 - 1.2''

1191

TARAN-
TULA
HAWK
Pepsis formosa
0.8 - 1.2''

female

pupal case

DIGGER WASP
Scolia dubia 0.5 - 0.7''   1192

Velvet-ants (Mutillidae)

The velvet-ants are so named for the resemblance of the wingless females to ants. They lack the dorsal node or projection on the petiole characteristic of ants, and have a heavy coating of brightly colored hairs. The males are larger than the females, usually have wings, and are sometimes more than an inch long. The antennae of the females are curled, 12-segmented; those of the males are straight, 13-segmented. They are common in the warmer regions. Most of them are solitary external parasites; they attack the larvae or pupae of bees (including the honey bee), vespid and sphecid wasps, and are therefore injurious. A few species attack ants. The females use stingers for defense against adult bees and wasps when invading their nests; they run rapidly and can inflict a painful sting to humans if molested.

(60) SACKEN'S VELVET-ANT: *Dasymutilla (Mutilla) sackenii.* **Color plate 7n.**

Range: Oregon to California, Arizona, Nevada. **Adult:** Black, with dense coating of white hairs. **Length:** About .5''. **Host:** A sphecid wasp, *Bembix occidentalis beutenmulleri.* The **Gray Velvet-ant** or **Thistle Down Mutillid,** *D. gloriosa,* is very similar, black, with long white hairs; from .5'' to .65'' long; it occurs in Utah, Nevada, and Texas, west to California.

(61) VELVET-ANT: *Dasymutilla bioculata.* **Fig. 1191.**

Range: Manitoba, south to Louisiana, west to British Columbia and New Mexico. **Adult:** Black, with dorsum of thorax and abdomen reddish. **Length:** .7–1.2''. The size of this species shows a remarkable variation according to the size of its hosts; the larvae feeding on small prey develop into small specimens; those feeding on large prey become proportionately larger. **Hosts:** The sphecid wasps *Bembix pruinosa* and *Microbembex monodonta.* The **Red Velvet-ant,** *D. occidentalis* (Color plate 7o)—sometimes called the "cow-killer ant" in the mistaken belief that it stings cattle—is red, from .6'' to 1'' long, occurs from Connecticut to Florida, west to Missouri and Texas; it invades the nest of the bumble bee, *Bombus fraternus.* A similar species, *D. fulvohirta,* ranges from North Dakota to Texas, west to Alberta and California; it invades the nests of solitary bees, *Anthophora occidentalis* and *Megachile* spp.

(62) DIGGER WASP: *Scolia dubia* (Scoliidae). **Fig. 1192.**

The wasps in this small family are relatively large solitary external parasites of Coleoptera, mostly scarab beetles. Some species of tropical America have a wing expanse of 3″. Males and females of *S. dubia* engage in an elaborate mating dance with a characteristic flight pattern—flying close to the ground on a horizontal plane and following a figure-8 or S course. The adult female burrows into the soil in search of grubs, often stinging more than it can parasitize—in any case the attack is fatal—and constructs a cell around a suitable host. In some cases she burrows deep in the soil—as much as four feet—dragging the grub behind, before forming the cell and implanting an egg. The parasite pupates and hibernates in a cocoon in the cell. Egg production is generally low, with a single brood in temperate climates. *S. manilae* was successfully introduced from the Philippines into Hawaii for control of the Oriental beetle. **Range:** *S. dubia dubia* ranges from Massachusetts to Florida, west to Colorado and Arizona. **Adult:** Hairy, black, abdomen with bright red and yellow markings; wings black. **Length:** .5–.7″. **Host:** Green June beetle. The grubs are sometimes attacked on the ground in the early morning. *S. d. haematodes* occurs in Texas, northern Mexico, Arizona, and California. *S. d. monticola* ranges from Mexico north to western Texas, New Mexico, and Arizona.

(63) SPRING TIPHIA: *Tiphia vernalis* (Tiphiidae).

Range: Throughout the eastern U.S. Imported from Korea in 1926. **Adult:** Shiny, black. It feeds exclusively on honeydew excretions of aphids. The female burrows under the sod to reach a grub, stings it, causing temporary paralysis, then deposits an egg on the underside. Unfertilized eggs are normally placed on smaller hosts, the resulting progeny in this case being male. Adults appear in May, the cycle being well adapted to that of its host. **Length:** .75″. **Larva:** A solitary external parasite; it consumes all but the head capsule and legs of the grub. Hibernates as an *adult* in a cocoon within earthen cell of the host. One brood. **Host:** Japanese beetle. Up to 65 percent of the grubs in an area are often parasitized by this species. The **Fall Tiphia,** *T. popilliavora* (Fig. 1193), is similar in appearance and habits, except that it hibernates as a *larva,* and emerges in August when only the smaller host grubs are available, thus giving rise to a higher percentage of males and fewer progeny; the adults feed almost exclusively on the blossoms of wild carrot. *T. asericae* was imported to combat the Asiatic garden beetle. *T. intermedia,* a native, parasitizes the Japanese beetle only sporadically and then only a small percentage of the grubs. *T. femorata* and *T. morio* were imported from Europe and released in New York in 1956 and 1957 for biological control of the European chafer.

Ants (Formicidae)

Ants are perhaps the most familiar of all insects, being the most abundant and widespread; they are found in every type of habitat from the Arctic to the tropics. They are black or some shade of red, brown, or yellow. The body texture may be smooth or rough; the surface may be naked, hairy, or spined. A distinctive feature is the large head with geniculate, or elbowed, antennae (Fig. 7C); the first segment, or scape, is greatly elongated except in most of the males. The mouthparts are generally well developed, compound eyes usually small (absent in some species), and three ocelli are present except in the workers.

Ants may be distinguished from other Hymenoptera by the abdomen—with a greatly enlarged portion called the gaster, and the narrow basal segment (petiole), the latter having a characteristic dorsal projection or node. Two basal segments are narrowed in some species, the second being called the postpetiole. The gaster has one more visible segment in the male than in the female. The legs are well developed, with one-jointed trochanters and five-segmented tarsi with claws. The sexual females and males have two pairs of wings; the workers have none (unlike their counterparts among the social bees and wasps); the males retain their wings, the females do not. The newly emerged adults are called callows and lack the color and hardness of the mature forms; several days are required to reach maturity.

Many species of ants have stridulating organs consisting of a file on the postpetiole and a surface against which this is rubbed on the first segment of the gaster. Stridulation is an important means of communication among the primitive ants and the Myrmicinae; it is used for warning or assembling other members of the colony, and is believed to supplement the sense of smell (located mainly in the antennae) as a means of recognition. Stridulations of the Texas leaf-cutting ant can be heard plainly by humans at a distance of a foot from the ear. All ants possess hearing organs in their hind tibiae similar to the chordontal organs of crickets and grasshoppers. Johnson's organs—the hearing mechanism of mosquitoes located in the second antennal joint—also occur in some ants.

All ants are social and live in colonies, usually in the ground or in rotting wood. The size of the colonies varies greatly, according to the species, location, and age of the colony. There are three distinct castes (Fig. 1194): workers, females, and males. The workers of some species are polymorphic—occurring in various forms, each varying in size and structure and having specialized functions; in some species they are dimorphic—having two forms—and in the others they are monomorphic, with only one form. The workers ordinarily enlarge and repair the nest, defend the colony against attack, forage, and care for the young and the queen, but some are capable of

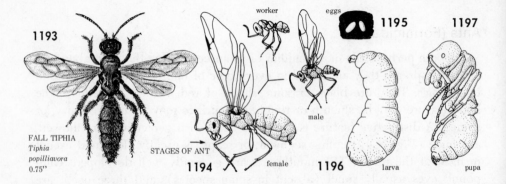

1193

FALL TIPHIA
*Tiphia
popilliavora*
0.75"

STAGES OF ANT →

1194

worker

eggs

1195

male

female

1196

1197

larva

pupa

laying eggs.* The males—intermediate in size between the workers and females—apparently have only one function, to mate with the unfertilized females. The queens are distinguished by their larger stature and huge gaster. The long-lived queens mate only once and lay eggs continuously in tropical and subtropical regions; egg laying is interrupted only during the winter in colder climates, except possibly for some species that nest in houses.

When mature, most winged or sexual forms gather in great swarms and take off on their nuptial flight; mating takes place in flight or on the ground, after which the male dies. Unless she meets with some misfortune, the mated female—now a queen—founds a new colony. After locating a suitable nesting site, she removes her wings and burrows into the ground to await the maturing of her eggs. She feeds and tends the first brood of larvae; later broods are cared for by the workers while she continues to lay eggs. This is the most common method of forming new colonies, but in some species they are formed by budding or splitting—the queen or a fertilized daughter leaves the nest with a few workers who aid her in establishing a new colony—and in others by usurping the nest of another species. The worker ants, being the most common, and also the most troublesome of the pest species, are the best known and the easiest to identify. Our descriptions apply to the workers.

The microscopic eggs (Fig. 1195) are white or pale yellow, cylindrical or oval in shape, and covered with a thin membrane. The larvae (Fig. 1196) are segmented, have no legs, and are often shaped like a crook-necked squash, with the tiny head at the smaller end; they are blind, and carried about by workers or nurses from one chamber to another as conditions of temperature and moisture change. They are fed regurgitated liquid food from the mouths of the nurses or tiny bits of partially masticated insects,

* It was thought that the progeny in this case would always be males, as is usual with most Hymenoptera when the eggs are not fertilized; it now appears that workers and females may also be produced by the egg-laying workers of at least some species.

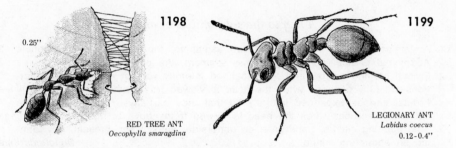

1198 **1199**

0.25"

RED TREE ANT
Oecophylla smaragdina

LEGIONARY ANT
Labidus coecus
0.12-0.4"

or the hyphae of fungi in the case of the fungus-eating ants. The pupa (Fig. 1197) resembles the adult in shape but is soft and colorless. Among the more primitive ants, the larvae pupate in parchment-like cocoons; in the higher forms the pupa is usually naked.

The more primitive subfamilies of ants—the Ponerinae and Dorylinae—are largely carnivorous and live on other insects or invertebrates; to this latter group belong the vicious legionary ants—the equivalent of the Old World driver or army ants—which devour small vertebrate animals as well. The higher groups—the Myrmicinae, Dolichoderinae, and Formicinae—are mostly herbivorous but include many species that feed on living or dead insects or other animals. According to Clausen, "their value in a predatory capacity is perhaps greatly underestimated." The well-known red tree ant *Oecophylla smaragdina* (Formicinae) of tropical Asia was the first insect used in biological control of insect pests. It builds conspicuous nests in trees, binding the edges of leaves together using a thread of silk squeezed from an ant larva (as depicted in Fig. 1198). These ants were collected and released in citrus groves in southern China centuries ago (they attack many caterpillars and destroy or drive off beetles and bugs); in Yemen they have been used in date groves to control another species of ant causing damage.

After its accidental entry into Hawaii, *Pheidole megacephala* (Myrmicinae) was reported to "hold the housefly down to negligible numbers"; it was also credited with destroying three-fourths of the larvae of the Mediterranean fruit fly after they mature and reach the ground. "One of the most efficient groups of predaceous ants" is said to be in the genus *Solenopsis* (Myrmicinae). There is considerable evidence to show that certain species of fire ants (*Solenopsis*) do more good than harm, and that reports of their damage have been exaggerated. Ants are presently propagated and released for control of forest insect pests in Germany. Ants in general like the nectar of flowers and other sweet exudates from plants as well as from homopterous insects. In visiting flowers they aid in pollination.

A key for the separation of the most important subfamilies of ants is given below. The structures referred to here and in the descriptions are illustrated in Fig. 7C. (A footnote at the bottom of p. 93 explains how to use a key.)

A key to the subfamilies of ants*

1. Abdominal pedicel consisting of two segments, the petiole and postpetiole 2
   Abdominal pedicel consisting of one segment, the petiole 3
2. Frontal carinae narrow and not expanded laterally so that the antennal insertions are fully exposed when the head is viewed from above .... *Dorylinae*
   Frontal carinae expanded laterally so that they partially or wholly cover the antennal insertions when the head is viewed from above .... *Myrmicinae*
3. Anal opening slitlike, transverse, on underside; hairs, when present, not forming an encircling fringe .... *Dolichoderinae*
   Anal opening distinctly circular, at the tip of the abdomen, and usually surrounded by a fringe of hairs .... *Formicinae*

(64) LEGIONARY ANT: *Labidus (Eciton) coecus* (Dorylinae). **Fig. 1199.**

Range: Widely distributed from Oklahoma and Arkansas to Texas and Louisiana, south to Argentina; a native species. **Adult:** Bright reddish brown or chestnut; pedicel two-segmented; body smooth and shiny, except posterior half of thorax; antennae very short, 12-segmented, with short scape; eyes absent or rudimentary, resembling ocelli; tarsal claws with tooth between base and apex. Nests are temporary, in decayed stumps or logs, under stones or other objects; colonies are very large. They travel under the protection of covered roads which the blind workers construct rapidly of leaves, keying them together in a convex form so they won't collapse. Workers are polymorphic. **Length** (workers): .12–.4″. **Food:** Many injurious insects, including the secondary screw-worm. They bite and sting viciously, can kill chickens and small pets.

(65) BIG-HEADED ANT:† *Pheidole bicarinata vinelandica* (Myrmicinae). **Fig. 1200.**

Range: New York to Nebraska, south to Florida and Arizona; a native species. **Adult:** Yellowish or light to dark brown; head very large, with median furrow, posterior half smooth and shiny; pedicel two-segmented; antennae twelve-segmented, with short scape, three-segmented club. Nests in rotting wood, in open spaces under objects; occurs in the deserts and the mountains, and along the beaches. Workers are dimorphic. **Length** (workers): .06–.12″. **Food:** Small insects, seeds, and honeydew in its natural environment; meat, grease, bread indoors. An intermediate host of tapeworm of wild and domestic fowl.

(66) RED (TEXAS) HARVESTER: *Pogonomyrmex barbatus.*

Range: Lower altitudes of Oklahoma, Colorado, Kansas, Utah, Texas, New Mexico, Arizona, California, Mexico. This genus is largely western. **Adult:** Worker reddish brown, female slightly darker than worker and

* Adapted from W. S. Creighton, *Ants of North America*, Harvard University Museum of Comparative Zoology, Bulletin 104, 1950.

† An extraordinarily large head in proportion to the body is characteristic of this genus. The name "big-headed ant" is more properly applied to *Pheidole megacephala*— a species occurring in Florida—in accordance with *The Common Names of Insects.*

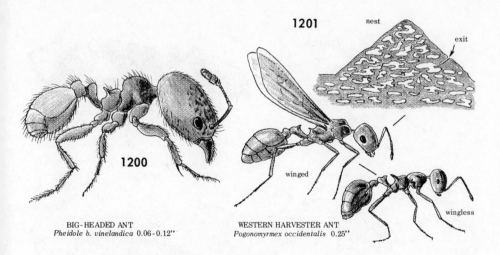

1201

nest

exit

winged

wingless

1200

BIG-HEADED ANT
Pheidole b. vinelandica 0.06-0.12"

WESTERN HARVESTER ANT
Pogonomyrmex occidentalis 0.25"

male. Swarming may take place any time from April to October. Tiny milk-white, capsule-like eggs are laid in clusters in chambers of the colony. The nest is a series of subterranean tunnels, about .25" wide, and chambers which are flat-bottomed, with dome-shaped ceilings, ranging from .25" wide and .5" long to 10" wide and 1 foot long. Grain and seeds are stored in the chambers. Circles ranging from 3 to 25 feet in diameter are cleared of vegetation in forming the nest, with an entrance in the center. Barren pathways 1 to 4 feet wide, sometimes 200 feet long, lead away from the circle. The workers are monomorphic. **Length** (workers): .25–.5" (largest of the harvester ants). **Larva:** Cream-colored, about .25" long full grown. The naked pupa is pale cream, with legs and antennae folded on the underside. **Food:** Seeds, grains. The greatest damage results from barren areas, totaling up to one-quarter of an acre per 20 acres. Adults can inflict a painful bite or sting. Several subspecies and variations occur throughout the area. The closely related **Western Harvester Ant,** *P. occidentalis* (Fig. 1201), occurs in higher altitudes of the same states, has similar habits to *barbatus* except that a mound is formed at the entrance to the nest; workers are about .25" long and can sting. The **California Harvester Ant,** *P. californicus,* occurs from Texas to California and Nevada; workers are from .2" to .3" long, reddish brown (like the queens), the males are black and red; their nests are built in sand and fine gravel, with a fan-shaped crater on one side of the entrance; it bites and stings fiercely, harvests seeds but does not cut down vegetation. The **Florida Harvester Ant,** *P. badius,* likewise harvests seeds and does not cut down vegetation.

(67) PHARAOH ANT: *Monomorium pharaonis.* **Fig. 1202.**

Range: Throughout North America. Tropical in origin, it has been carried to all parts of the world. **Adult:** Yellowish or light brown to reddish, sparsely

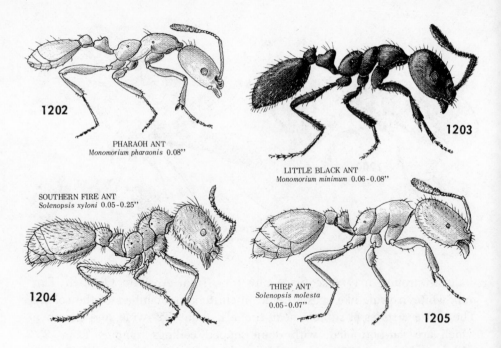

1202

PHARAOH ANT
Monomorium pharaonis 0.08"

1203

LITTLE BLACK ANT
Monomorium minimum 0.06-0.08"

SOUTHERN FIRE ANT
Solenopsis xyloni 0.05-0.25"

1204

THIEF ANT
Solenopsis molesta
0.05-0.07"

1205

pubescent; head, thorax, petiole, and postpetiole punctate; eyes small, segments of antennal club increase in size apically; clypeal carinae feebly defined. Commonly nests in the soil, but is also a "house ant," will nest in crevices of walls and woodwork, where it probably breeds continuously. Workers are monomorphic. **Length** (workers): .08". **Food:** Dead and live insects in its natural environment; sweets, greasy and protein foods indoors; it is practically omnivorous. A stubborn pest.

(68) LITTLE BLACK ANT: *Monomorium minimum.* **Fig. 1203.**

Range: Throughout North America; a native species, one of the most common "house ants." **Adult:** Slender, smooth, typically shiny black, sometimes dark brown; sparsely pubescent; antennae twelve-segmented, with well-defined three-segmented club. It commonly constructs small craters in the soil and rotted wood, invades houses. Workers are monomorphic. **Length** (workers): .06–.08". **Food:** Sweets, meats, vegetables and other dead food indoors; other insects outdoors.

(69) SOUTHERN FIRE ANT: *Solenopsis xyloni.* **Fig. 1204.**

Range: California to South Carolina and Florida, most common in the Gulf states; a native species. **Adult:** Yellowish to reddish, gaster dark; dense coating of hairs. Head moderate in size, eyes well developed; antennae ten-segmented, with two-segmented club; frontal carinae far apart, partly concealing antennal inserts; clypeus with two carinae, two to five teeth on anterior margin; mandibles not sharply incurved, with three well-defined teeth. Commonly nests in the soil, making irregular mounds of loose soil; colonies are often large. Workers are polymorphic. **Length** (workers): .05–.25″. **Food:** Live and dead insects; practically omnivorous, fond of honeydew and dead foods of high protein content.

(70) THIEF ANT: *Solenopsis molesta.* **Fig. 1205.**

Range: Throughout North America; a native species. **Adult:** Mostly smooth, shiny, yellowish to bronze, with rather dense hairs. Similar to *xyloni* (69) except that head is sparsely punctate, with minute eyes, antennal club relatively large. Commonly nests in the soil or wood but invades houses; nests independently or in nests of other ants. Nuptial flight occurs in late July or early fall. Workers are monomorphic. **Length** (workers): .05–.07″; one of our smallest ants. **Food:** Larvae of other insects; practically omnivorous, prefers proteins and fats as house pest, fond of honeydew, sometimes destroys seeds of grains. It is named for its habit of robbing the food and brood of other ants.

(71) FIRE ANT: *Solenopsis geminata.* **Fig. 1206.**

Range: South Carolina and Florida to Texas, south to Costa Rica; a native species. **Adult:** Color highly variable; yellowish or reddish to blackish. Similar to *xyloni* (69), may be distinguished by the very large head and sharply incurved mandibles, usually without teeth. It commonly nests in the ground around clumps of vegetation, where it leaves large mounds of dirt; also nests in rotting wood or under objects. Workers are polymorphic. **Length** (workers): .1–.25″. **Food:** Other insects, particularly the larvae of flies; strongly predaceous, considered one of the most important predatory species of ants.

(72) IMPORTED FIRE ANT: *Solenopsis saevissima richteri.* **Fig. 1207.**

Range: South Carolina to Florida, westward to Arkansas and Texas. First reported in 1930, it is believed to have entered Alabama in ship cargoes before 1920. **Adult:** Highly variable in color; reddish to reddish brown, except much of the gaster, which is blackish, sometimes with broad yellowish red band across the base of the gaster. Similar to *xyloni* (69) and *geminata* (71), except that head is moderate in size, mandibles with four well-defined teeth and not curved inward sharply. Nests are marked by mounds from a few inches to 3 feet in height (if supported by stump or plants), from 1 to 5 feet in diameter, with an average of 25 to 40 mounds per acre in open areas not cultivated. Colonies average 25,000 workers, with only a few sexual forms until early spring before the mating flight, when they may reach a few thousand. The young queen excavates a small cavity in the ground, lays about 100 eggs which hatch in about 10 days. Workers are polymorphic. **Length** (workers): .12–.25″. **Larva:** Develops in about

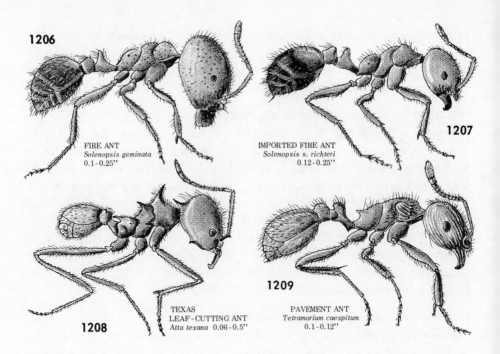

1206

FIRE ANT
Solenopsis geminata
0.1-0.25"

IMPORTED FIRE ANT
Solenopsis s. richteri
0.12-0.25"

1207

TEXAS
LEAF-CUTTING ANT
Atta texana 0.06-0.5"

1208

1209

PAVEMENT ANT
Tetramorium caespitum
0.1-0.12"

9 days; the pupal stage lasts about 10 days. **Food:** House fly larvae, boll weevil grubs, cutworms, and many other pests. Occasionally consumes seeds or other crops, but "damage to plants in general is rare"; sometimes "young plants are seriously damaged by girdling." They do not damage livestock. Their stings (which cause a burning sensation and tiny pustules) and the mounds which interfere with the operation of machinery are causes of annoyance.*

(73) TEXAS LEAF-CUTTING ANT: *Atta texana*. Fig. 1208.

Range: Texas, Louisiana, in well-drained loamy or sandy soils. **Adult:** Light to dark brown, body with dense minute punctures; antennae eleven-segmented, without distinct club; a spine on each frontal lobe, a large and small one on each occipital lobe, a small one on prothorax; legs very long, pedicel two-segmented. It constructs a large nest 10 to 20 feet below the surface of the ground, consisting of numerous chambers close together, connected by tunnels. On the floor of the chambers, which are sometimes nearly 2 feet high and wide, and more than 3 feet long, are fungus gardens which provide food for the developing larvae. Outward evidence of the nests is large mounds of dirt piled up at the entrance to the craters and galleries leading to the chambers. The workers are the most highly polymorphic

---

* It is now confirmed that two species are actually involved here: *S. richteri,* commonly referred to as the "dark form," and *S. invicta,* the "red form," which was introduced 20 years later and is gradually replacing *richteri.* (David R. Smith, Agr. Res. Serv., in *Cooperative Economic Insect Report,* U.S. Dept. Agr., 22(9): 103, Mar. 3, 1972.)

of our species. One caste brings in the leaves gathered by the foragers, grinds and macerates the leaves, and builds up the garden; another weeds the garden, while soldiers guard the nest. The new queen brings a pellet of fungus with her on the nuptial flight, plants and manures it with her own excrement as she lays eggs and tends the larvae.* **Length** (workers): .06–.5″; one of our largest species. **Food:** Leaves are cut from wild and cultivated plants, and seedlings, often causing considerable damage; houses are sometimes invaded for seeds and cereals.

(74) PAVEMENT ANT: *Tetramorium caespitum.* **Fig. 1209.**

Range: The U.S., most abundant along the Atlantic seaboard; occurs in Europe and Asia. **Adult:** Light to dark brown or blackish, hairy; antennae twelve-segmented, with three-segmented club; epinotum with pair of short spines or tubercles; head and prothorax with fine paralleled grooves; femora and tibiae large. Common in lawns; nests under stones and at edge of pavement, in sandy or rocky places, also invades houses and nests in crevices of woodwork and masonry. Workers are monomorphic. **Length** (workers): .1–.12″. **Food:** Live and dead insects; practically omnivorous. Sometimes a pest of gardens, eating roots of young plants and planted seeds. More of a nuisance in households and greenhouses, fond of honeydew, meats and grease.

(75) ARGENTINE ANT: *Iridomyrmex humilis* (Dolichoderinae). **Fig. 1210.**

Range: Most of southern states, including California. An introduced species, first noted in New Orleans in 1890. Very aggressive, it kills or drives out native species. **Adult:** Soft-bodied, slender, light to dark brown, with sparse hairs; antennae long, with long scape and not clubbed; eyes well developed; one-segmented erect pedicel; anal opening on underside, transverse, slit-shaped, without fringe hairs. The workers have no sting and only a feeble bite. They emit a strong musty odor when crushed [see (76) and footnote]. Nests in large colonies in the soil, under stones, pavement, logs, sidewalks, also in houses, often with numerous queens in a single colony. Mating is assumed to take place in the nest, since nuptial flights have not been observed; males are taken at lights, never females. Several colonies sometimes unite for the winter, when they are inactive, and disperse again in the spring. Entire life cycle from egg to adult takes about 78 days. Workers are monomorphic. **Length** (workers): .04–.12″. **Food:** All kinds. They invade houses, bees' nests, are very fond of honeydew secretions of aphids, mealybugs and scales, which they distribute about

* The fungus cultured by similar (attine) ants has been found to be an effective decomposer of cellulose (in the leaves) and a rich source of carbohydrates. Recent investigations indicate that salivary as well as fecal materials are applied to the gardens by the ants attending them, with the possible involvement of antibiotic and growth-stimulating substances. M. M. Martin and R. M. Carman, *Science*, Oct. 27, 1967, p. 531.

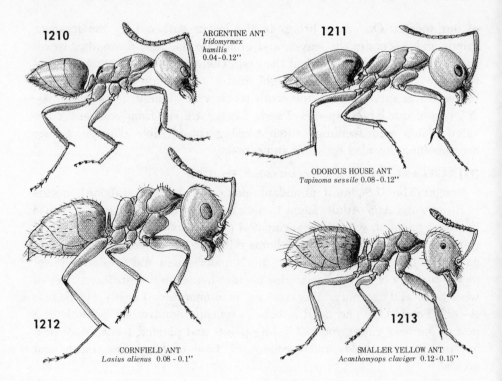

1210 ARGENTINE ANT
Iridomyrmex humilis 0.04-0.12"

1211 ODOROUS HOUSE ANT
Tapinoma sessile 0.08-0.12"

1212 CORNFIELD ANT
Lasius alienus 0.08 - 0.1"

1213 SMALLER YELLOW ANT
Acanthomyops claviger 0.12-0.15"

gardens and orchards and protect from their natural enemies; protection sometimes includes a "tent" or "shed" to cover the aphids. Their habit of crawling everywhere suggests the possibility of their spreading disease organisms.

(76) ODOROUS HOUSE ANT: *Tapinoma sessile.* **Fig. 1211.**

Range: Widely distributed over North America, very common in the Far West. **Adult:** Soft-bodied, dark reddish brown to black. One-segmented pedicel hidden by overhanging gaster. Similar to the Argentine ant (75), but broader and darker, slower in movements; transverse ventral anal opening with erect hair on each side. Generally nests outdoors, under stones and logs, in trees, stumps, and bird nests, occasionally in foundations of buildings. Gives off distinctive musty odor—it has been described as "rotten coconut"—when crushed.* Workers are monomorphic. **Length** (workers): .08–.12". **Food:** All kinds, especially sweets and honeydew; it often invades houses. A native species, it has been driven away from many places by the Argentine ant.

* Defensive secretions from anal glands of *T. nigerrimum* have been identified chemically as terpenoids and ketones, the latter being the cause of this odor. Similar substances have been found in other dolichoderine ants.

(77) CORNFIELD ANT: *Lasius (niger) alienus* (Formicinae). **Fig. 1212.**

Range: Widespread throughout most of North America. **Adult:** Light to dark brown, robust, soft-bodied; one-segmented pedicel, with long pointed node; eyes well developed; antennae twelve-segmented, without club; maxillary palpi long, six-segmented; anal opening at end of abdomen, circular, surrounded by fringe of hairs. When crushed, they emit a strong odor of formic acid.* It has no sting. Nuptial flights are mostly during August and September. Workers are monomorphic. **Length** (workers): .08–.1″. **Food:** Live and dead insects; they gather the nectar of flowers, are very fond of honeydew, and invade homes for sweets and meats. They transport strawberry root aphids to the crowns and roots of strawberry plants, construct tunnels and chambers to accommodate them; damage results from drying out of the undermined plants as well as from aphids feeding on the roots. The ants care for the eggs of the corn root aphid, carry the young to the roots of smartweed (and other weeds), transfer later generations to the roots of corn; in this remarkable association the aphids have become dependent on the ants to complete their complicated life cycle involving alternate host plants. [See Homoptera (23) and (29).]

(78) SMALLER YELLOW ANT: *Acanthomyops (Lasius) claviger.* **Fig. 1213.**

Range: Ontario to Florida, west to Washington, Utah, New Mexico. Very common, especially in the eastern and central states. **Adult:** Stout, smooth and shiny, quite hairy; pale yellow to yellowish red; maxillary palpi short and three-segmented, eyes small, antennae twelve-segmented (inserted close to edge of clypeus), pedicel one-segmented with long pointed node, anal opening as in (77) above. It has no sting and smells strongly of citronella when disturbed.† They nest in the soil, exposed or under stones or in rotting wood, in rather large colonies. Nuptial flights occur mostly in September. Workers are monomorphic. **Length** (workers): .12–.15″. **Food:** Mostly honeydew of subterranean aphids and mealybugs. They stay underground during the day, forage at night. The **Larger Yellow Ant,** A. *interjectus* (Fig. 1214), is similar in appearance and habits, may be distinguished by its larger size (from .15″ to .18″ long), longer scape, fewer and more scattered hairs; their nests in open areas have large mounds; nuptial

---

* Formicine ants spray formic acid from anal glands with a mechanism similar to that of whipscorpions. They bend the abdomen under and forward to spray their enemies, usually as they bite them.

† A. *claviger* secretes citronellal in mandibular glands connected to a large reservoir in the head, uses it along with formic acid for defense; citronellal hastens penetration of formic acid through the integument of the attacker. See: M. S. Chada et al., "Defense Mechanisms of Arthropods—VII: Citronellal and Citral in the Mandibular Gland Secretions of the Ant *Acanthomyops claviger* (Roger)," *Jour. Ins. Physiol.,* 8:175–179 (1962).

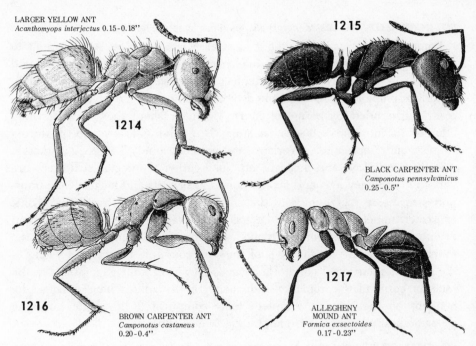

LARGER YELLOW ANT
Acanthomyops interjectus 0.15-0.18"

1214

1215

BLACK CARPENTER ANT
Camponotus pennsylvanicus
0.25-0.5"

1216

BROWN CARPENTER ANT
Camponotus castaneus
0.20-0.4"

1217

ALLEGHENY
MOUND ANT
Formica exsectoides
0.17-0.23"

flights occur at nighttime, with mating taking place on the ground near the nest. The northern form of the **Honey Ant,** *Myrmecocystus mexicanus hortideorum* (Fig. 1220), is pale yellow, from .2" to .4" long, nocturnal in habit, feeds on honeydew of aphids, coccids, and oak galls; foraging workers gather honeydew and transport it in their crops back to the nest, where it is transferred to "repletes"—workers whose sole function is to store the food in greatly distended crops (which cause their gasters to swell to enormous spheres) as they hang by their claws side by side from the vaulted ceilings of the nest chambers; droplets of the sugary food are disgorged when solicited by the other ants;* as many as 300 repletes, which are prized by Indians as food, may be found in the several chambers of a large nest; a nest is marked by the large circular entrance in the center of a cone-shaped crater of small pebbles; the ant occurs in Colorado, New Mexico, southern California, and Mexico.

* Wheeler wrote: "Those who, in anthropomorphic mood, are wont to extol the fervid industry and extraordinary feats of muscular endurance in ants, should not overlook the beatific patience and self-sacrifice displayed by the replete *Myrmecocystus* as it hangs from the rafters of its nest, month in, month out—for years, perhaps—a reservoir of temperament as well as liquid sweetness." Occasionally one falls to the floor and is put back in place by the other workers. The steadfastness of purpose alluded to here by William Morton Wheeler can hardly be questioned. But some doubt concerning the reputed industry of ants has been expressed by George Wheeler, who is investigating their behavior. He observed that ants actually "spend a great deal of their time just loafing," and the worker ant in "preening"; they only *appear* to be engaged in endless activity.

(79) BLACK CARPENTER ANT: *Camponotus pennsylvanicus (herculeanus).* **Fig. 1215.**

Range: Ontario and Quebec to Florida, west to North Dakota and Texas. **Adult:** Black (thorax and petiole sometimes light to dark brown), gaster with long pale yellow to grayish hairs; pedicel one-segmented, antennae twelve-segmented (not clubbed), scape enlarged apically; eyes well developed, front edge of clypeus slightly lobed at center. Anal opening as in (77) above. It nests in wood: trunks of trees, logs, building timbers, telephone and power poles. Large galleries usually start where the wood has decayed, may be extended below the ground. The ants do not eat the wood; they will bite but not sting humans. Workers are polymorphous. **Length** (workers): .25–.5″; one of our largest ants. **Food:** Live and dead insects, honeydew, juice of ripe fruit. It is fond of sweets, often invades houses. The **Red Carpenter Ant,** *C. ferrugineus,* occurs from New York to Georgia, west to Nebraska and Kansas, has similar habits. The **Florida Carpenter Ant,** *C. abdominalis floridanus,* is the most common species of *Camponotus* in Florida, also occurs in Georgia, Alabama, North and South Carolina; it infests beehives and wood in buildings. The **Brown Carpenter Ant,** *C. castaneus* (Fig. 1216) is abundant in the southeastern states, ranges west to Iowa and Texas.

(80) ALLEGHENY MOUND ANT: *Formica exsectoides.* **Fig. 1217.**

Range: Throughout eastern U.S. and Canada. **Adult:** Abdomen and legs blackish brown, head and thorax rust red. Nests are marked by huge conical mounds, sometimes measuring nearly 3 feet high and 6 feet in diameter, and may be very numerous in an area. Young queens are unable to found a colony without the help of workers, which migrate with one or more queens. If this fails, workers of *F. fusca* are seized as pupae and used later to care for her brood. New colonies are often founded by fission or extension of and breaking off from existing colonies. **Length** (workers): .17–.23″. **Food:** Insects, honeydew.

(81) SILKY ANT: *Formica fusca.* **Fig. 1218.**

Range: Throughout North America and the entire north temperate zone. **Adult:** The numerous subspecies and varieties vary from reddish brown to all reddish, black, or brownish black. Common in varied habitats: fields, forests, hilly and mountainous areas. As a rule they build "masonry domes" —small compacted mounds of dirt. **Length** (workers): .12–.2″. **Food:** Insects, honeydew. It is known to feed on larvae of Swaine's jack pine sawfly and the eastern tent caterpillar. The **Western Thatching Ant,** *F. obscuripes,* which builds rather large mounds of coarse twigs and grass, also preys on tent caterpillars. Many *fusca* workers are enslaved by *F. sanguinea* and by species of *Polygerus,* the so-called amazons. *F. sanguinea* workers are dark

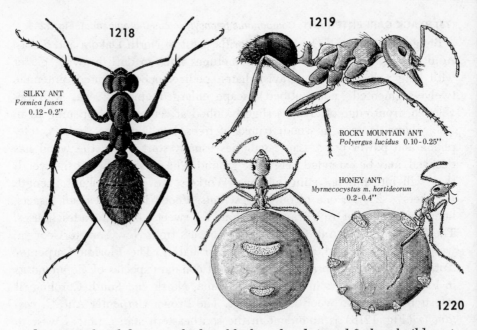

1218

SILKY ANT
Formica fusca
0.12-0.2"

1219

ROCKY MOUNTAIN ANT
Polyergus lucidus 0.10-0.25"

HONEY ANT
Myrmecocystus m. hortideorum
0.2-0.4"

1220

red, except the abdomen, which is black and red-tipped;* they build nests with small mounds or under logs and stones; young queens raid the nests of *fusca* for worker pupae, being unable to start a colony going themselves. The **Rocky Mountain Ant** or **Shining Amazon**, *Polyergus lucidus* (Fig. 1219), is brilliant red, from .1" to .25" long; the queen invades a weak *fusca* colony, kills its queen, and takes over the nest. To maintain the colony periodic raids are made on other colonies of *fusca;* workers in the nest are killed and the pupae carried off. Colonies thus become a mixture of both species. *F. fusca* adults are often parasitized by mermithid nematodes; the long, coiled worm is visible in the abdomen before it emerges.

Mud Daubers, Sand Wasps, and Cicada Killers (Sphecidae)

This is a large family of solitary wasps that build their nests in the soil, in hollow stems, in cracks or crevices, or fashion them of mud on exposed surfaces, and provision them with paralyzed insects or spiders. The mud nests of sphecid wasps may be distinguished from those of vespid wasps by the manner in which the egg is placed; the sphecids deposit the egg on the prey, the vespids attach it to the roof or wall of the cell (before provisioning it). In temperate climates hibernation takes place as a cocooned larva in the cell, with pupation occurring the following summer. Host preference is often quite specific. Some species, as in the genus *Bembix*, store

* According to W. M. Wheeler the North American forms are subspecies of the typical European *sanguinea;* they are all characterized by a notching of the anterior border of the clypeus.

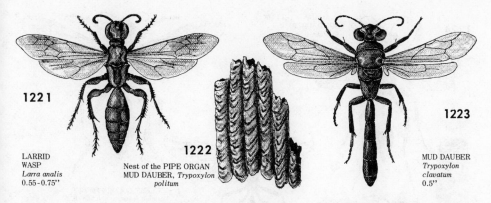

1221

LARRID
WASP
Larra analis
0.55-0.75"

1222

Nest of the PIPE ORGAN
MUD DAUBER, *Trypoxylon
politum*

1223

MUD DAUBER
*Trypoxylon
clavatum*
0.5"

their nests almost entirely with flies (Diptera), while the common hosts of *Sphex* and *Larra* are Orthoptera. In contrast to bees, the hairs of sphecid wasps are simple and undivided, and the hind tarsi and tibiae are never modified for carrying pollen.

(82) LARRID WASP: *Larra analis.* **Fig. 1221.**

Range: Eastern states, north to New York, south to Florida, Louisiana, west to Texas. **Adult:** Black, opaque; much of head, thorax, and legs with dense silvery pile; wings dark smoky brown, with coppery reflections; abdominal segments with apical bands of silvery pile. It feeds mostly on nectar, but imbibes host's blood oozing from puncture made in foreleg. The wasp enters the burrow of a cricket, routs it into the open, stings and temporarily paralyzes it. An egg is deposited in a rasp made just behind one of the hind legs. **Length:** .55–.75". **Larva:** Legless, with 13 distinct segments, large stout mandibles, .8" long full grown; it feeds externally. The host dies in its burrow in about eight days, just before the fourth molt of the parasite larva; the latter completes feeding in the fifth instar, pupates in a cocoon near the host remains. The cycle from egg to adult requires about seven weeks. Four broods. **Host:** Northern mole cricket; the parasite is important in its control. Species of *Larra* have been introduced into Puerto Rico and Hawaii for biological control of mole crickets.

(83) PIPE ORGAN MUD DAUBER: *Trypoxylon politum.*

Wasps in this genus are all black or black with red markings, and build mud nests of long parallel tubes on exposed surfaces, or use open plant stems or twigs, cracks in masonry or walls, abandoned nests of other wasps, old burrows of other insects. The nests are partitioned off with mud and each cell provisioned with spiders and implanted with an egg. The male of some species guards the nest while the female forages. *T. politum* is the largest species in the genus. **Range:** Massachusetts to Florida, west to Kansas and Texas. **Adult:** Shiny, black; hind tibiae white; wings dark, violaceous. Hind trochanters of the male unarmed; underside of first ab-

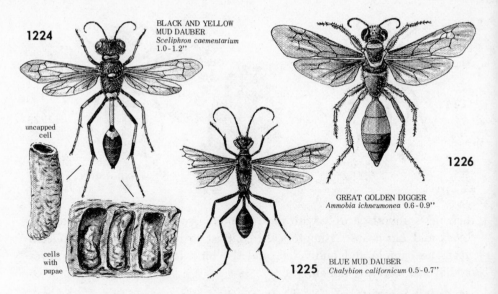

1224

BLACK AND YELLOW
MUD DAUBER
Sceliphron caementarium
1.0-1.2"

uncapped
cell

1226

GREAT GOLDEN DIGGER
Ammobia ichneumonea 0.6-0.9"

cells
with
pupae

1225

BLUE MUD DAUBER
Chalybion californicum 0.5-0.7"

dominal segment with hooked process. Its nest is a series of parallel mud tubes of varying length (like a pipe organ), with several or many tubes in a row (Fig. 1222). **Length:** .7". **Food:** Various spiders. *T. clavatum* (Fig. 1223) is about .5" long, with white pubescence, hind tarsi partly pale, hind trochanters of male armed, occurs east of the Rocky Mts. (excepting New England) and in Ontario; it constructs mud cells in deserted nests of *Sceliphron caementarium* (84), and in old beetle burrows, provisions them mostly with hunting spiders. *T. striatum* attacks orb weavers mostly, occurs in eastern U.S. and Canada, nests in holes in posts and in abandoned burrows of mining bees.

(84) BLACK AND YELLOW MUD DAUBER: *Sceliphron caementarium.* **Fig. 1224.**

Range: Entire U.S. and southern Canada. **Adult:** Thread-waisted, black or brown, with yellow markings, partially yellow legs. One of the most common mud daubers, often seen around wet places digging up balls of mud for its nest. This may be a single cell but is more commonly several side by side. The nests are placed on the underside of rocks, logs, boards, and other places. The female wasp paralyzes spiders, packs them into the cell with her head. When full, one egg is laid inside and the cell sealed. **Length:** 1–1.2". **Larva:** Pale yellowish, about .8" long full grown, with prominent oval callus on each side of thorax. Pupates in cocoon inside cell. Hibernates in cocoon. Two broods. **Host:** Spiders.

(85) BLUE MUD DAUBER: *Chalybion californicum.* **Fig. 1225.**

Range: Canada and the entire U.S. **Adults:** Metallic blue to blackish, with blue wings. It does not make a nest of its own—it is dependent on *Sceliphron caementarium* (84), taking over the latter's nests. It opens a cell by moistening the clay with water, empties it of spiders and egg deposited by the other wasp, and deposits its own paralyzed spiders; an egg is laid on the first spider and the cell sealed over when full. **Length**: .5–.7″. **Host**: Mostly black widow spiders. The wasp is probably the most effective predator of this spider.

(86) SAND WASP: *Prionyx (Sphex) atratus.*

Range: Entire U.S. and southern Canada. **Adult**: Black, with black pubescence; wings, legs dark brownish; abdomen dark brown, varied with black at apex, petiole black. The female digs an inclined tunnel, about 6″ deep, with one cell at the end; she provisions it with a paralyzed grasshopper and places one egg on the host. **Length**: .6–.7″. The **Great Golden Digger,** *Ammobia (Sphex) ichneumonea* (Fig. 1226)—abdomen in part and legs are reddish—has same range as *atratus,* digs a burrow with several cells radiating from a common chamber, stores one or more paralyzed grasshoppers and deposits a single egg in each compartment.

(87) SAND WASP: *Sphex (Ammophila) urnarius.* **Fig. 1227.**

Range: Eastern provinces and states, west to Texas. **Adult**: Thread-waisted; black, with sparse pubescence; second segment of petiole and the segment following pale red; wings clear, with apexes dusky; face of the male much narrower than that of the female, pubescence on head and thorax dense, that of head silvery. The female excavates a short burrow, carefully removing the dirt away from the entrance. The cell is provisioned with a caterpillar, an egg is laid on the host, and the burrow is closed with well-tamped dirt. **Length**: .75–.85″ (female); .7–.75″ (male). This is the celebrated "tool user" that has attracted much attention for its habit of using a pebble or other small object (gripped in the mandibles) to tamp the dirt sealing off the burrow. According to Evans, the individuals of this species and S. *junceus* "use a tool" only occasionally, most of the time simply using the head to tamp. S. *alberti* and S. *xanthopterus*—western and southwestern species respectively—are believed to "use a tool" consistently.

(88) GIANT CICADA KILLER: *Sphecius speciosus.* **Fig. 1228.**

Range: East of the Rocky Mts.; widespread and common. **Adult**: Black, with yellow markings. It appears in middle or late summer. It digs a tunnel about 6″ deep, makes a right-angle turn and tunnels for another 6″ or more, ends with one or more globular cells. The tunnel is on an incline, and may have branches; favored sites are along the roadside or other bank. After stinging a large cicada, the female wasp drags it up a tree (if the capture is made low down or on the ground), straddles it, and takes off toward the burrow, partly gliding. If a tree is not available, the prey is dragged on the ground. One or two cicadas are placed in a cell, with one

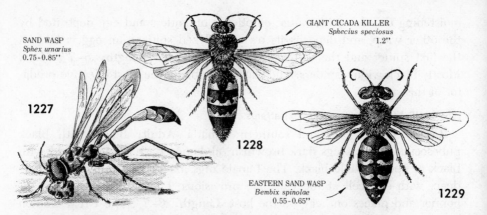

SAND WASP
Sphex urnarius
0.75-0.85"

1227

GIANT CICADA KILLER
Sphecius speciosus
1.2"

1228

EASTERN SAND WASP
Bembix spinolae
0.55-0.65"

1229

egg. **Length:** 1.2" or larger. **Larva:** Hatches in two or three days. Pupates in cocoon in cell and hibernates here. One brood. They damage lawns to some extent and sting if molested, and are thus considered as pests.

(89) EASTERN SAND WASP: *Bembix (Epibembex) spinolae.* **Fig. 1229.**

Range: Eastern states, west to South Dakota, Colorado, Arizona, Texas; southern Canada. **Adult:** Stout; black, with short white pubescence; clypeus, spot above and on sides of prothorax, interrupted apical bands on abdominal segments, tibiae (mostly), and tarsi light or greenish yellow; wings clear, with pale brown venation. The female digs a simple inclined burrow about 6" long, terminating in an oval cell about 4" below the surface; she stings and paralyzes a fly, places it in the cell, and lays an egg upright alongside it. Provisioning is resumed just before the egg hatches. **Length:** .55–.65". **Larva:** Its food is supplied "progressively," that is, at intervals during the feeding period. With each delivery of food the mother opens and closes the entrance to the burrow. The larva is completely developed in five days and spins a cocoon, and the mother starts building a new nest nearby. **Host:** Various flies. *B. hinei* nests along the beaches of Louisiana and Texas, is black, with spots on side of thorax, lateral spots on the first five abdominal segments, and lower part of legs yellow, from .6" to .75" long; it constructs a nest of several cells with a common tunnel, provisions them one time only, mostly with horse flies; it can complete two nests in a day.

(90) WESTERN SAND WASP: *Bembix (Epibembex) comata.**

Range: Pacific coast from California to British Columbia; Alberta, Nevada, New Mexico; common on the sand dunes along the seashore. **Adult:** Very similar to *B. spinolae* (89), with the abdominal bands white in males, the pubescence longer and denser and the apical segment usually spotted in the females. The female digs an inclined burrow 8" to 10" long, with a

* According to Evans, this may be a subspecies of *spinolae.*

single brood chamber at the end, about 4″ below the surface (all in less than an hour). A single paralyzed fly is placed on its back in the cell and an egg attached to it in an upright position (this fly serves as a support only and is not eaten). **Length:** .5–.65″. **Larva:** As in the case of *spinolae*, provisioning is progressive. The tunnel is closed when wasp leaves, and closed from within when wasp enters the brood chamber; the female usually sleeps in the "antechamber" or takes shelter here (males dig their own burrows for this purpose). When full grown the larva pupates in a cocoon composed of particles of sand cemented together with a salivary secretion. Unlike *spinolae*, the female digs a branch from the existing tunnel for her next brood chamber, and another from the other side when feeding of the second larva is completed. Two or three broods. **Host:** A wide variety of flies.

(91) SAND WASP: *Bembix pruinosa*.

Range: Generally distributed over the U.S. and southern Canada. **Adult:** Black; head, thorax, and first abdominal segment with dense pale grayish pubescence; dorsum of first five abdominal segments with greenish white bands, that of last segment spotted in the male. The female digs a shallow tunnel parallel with the ground surface, and from the far end of this (after closing the entrance) digs an inclined burrow to a depth of about 12″; she pushes her excavated material into the first tunnel, eventually filling it. Near the bottom she constructs a long narrow branch which is the brood chamber and lays an egg in it. Provisioning does not begin until just before the egg hatches. **Length:** .75″. **Larva:** Like *spinolae*, it is supplied in the progressive manner. The mother arranges each delivery of paralyzed flies neatly in a row in the brood chamber, keeps the cell clean by pushing debris to the far end. The larva starts at one end of the row, eats its way to the other end. **Host:** Various flies.

(92) BURROWING WASP: *Philanthus ventilabris*. **Fig. 1230.**

The black and yellow wasps in this genus are among the most beautiful of the burrowing wasps and provision their nests with bees. The common European species, *P. triangulum*, is known as the bee wolf; it stings honey bees in the throat, malaxates them, and feeds on the honey from their crops before storing them in the brood cells as food for the larvae. Some species dig short oblique burrows in the level ground; others dig long winding tunnels from a bank, with short branches each ending in an oval brood chamber. **Range:** *P. ventilabris* is generally distributed throughout the U.S. and Canada. **Adult:** Black with yellowish apical bands on the abdominal segments, yellow legs; the wings have a yellowish cast. **Length:** .5″. Nests are provisioned with several kinds of small bees. *P. gibbosus*, also widely distributed over North America, preys chiefly on halictid bees; the

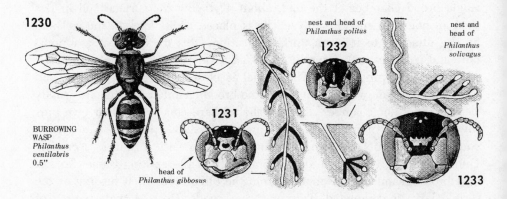

1230

BURROWING
WASP
*Philanthus
ventilabris*
0.5"

1231

head of
Philanthus gibbosus

nest and head of
Philanthus politus

1232

nest and
head of
*Philanthus
solivagus*

1233

heads and nests of this species (Fig. 1231), *P. politus* (Fig. 1232), and *P. solivagus* (Fig. 1233) are compared. The closely related **Weevil Wasp,** *Cerceris clypeata* (Fig. 1234), is dark blackish brown, from .7" to .8" long, provisions its nest with beetles—mostly curculionids, chrysomelids, and buprestids; it occurs in the Mississippi Valley and eastward.

(93) CONTORT DRYINID: *Gonatopus contortulus* (Dryinidae). **Fig. 1235.**

Range: Throughout North America. **Adult:** Shiny black, with reddish legs; the female is wingless, antlike, with slender thorax, long forelegs with large pincer-like claws used to grasp the adult host when ovipositing; two basal segments of antennae yellow, remainder reddish; the males are winged when they appear. Females are normally reproduced by parthenogenesis and without mating. Eggs are usually laid in the body of the host, or may be placed externally. **Length:** .15". **Larva:** First larval stage passed inside the host; second-instar larva assumes U shape, forces itself partly out between two anterior segments, head and posterior third of body remaining within the host. The larval sac or extended membrane is oval in shape, shiny black, has the appearance of a hernia. The mature larva frees itself, spins a white cocoon, and pupates on foliage. Hibernates as larva in cocoon. Two broods. **Host:** Leafhoppers. Adult females are predaceous and destroy many leafhoppers in this way. Parasitization by the larvae results in "parasitic castration," destroying the host's ability to reproduce.* [For another parasite of leafhoppers, see Diptera (43).]

---

* The stylopids called twisted-wing parasites (order Strepsiptera), which are closely related to beetles, also parasitize leafhoppers in a similar way. The wingless female, however, spends its entire life cycle in the host except for a brief period as first-instar larva (or triungulin) when it emerges from the host and drops to a plant to look for another leafhopper. Mating takes place on the host, the adult male being winged and free-living (as when a triungulin). Dispersal is thus largely dependent on the chance that the tiny larvae will find suitable hosts on their own—a risk which is compensated for by a high egg production.

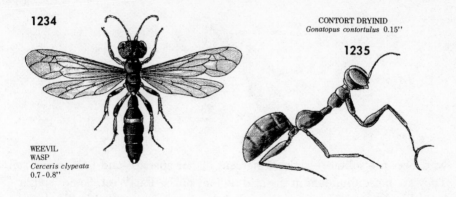

1234

WEEVIL
WASP
Cerceris clypeata
0.7-0.8"

CONTORT DRYINID
Gonatopus contortulus 0.15"

1235

Bees (Apoidea)

Bees are characterized by their dense coating of plumose or branched hairs, and the enlargement of the first or basal joint of the hind tarsi (Fig. 1236). In many bees this joint has rows of comblike hairs (on the inside)—the pollen brush—which are used for gathering pollen. Moistened pollen is packed between marginal rows of bristles on the outside of the hind tibiae—the pollen baskets—and carried back to the nest.* In some bees the pollen brush is on the underside of basal abdominal segments. Parasitic bees lack these pollen-collecting adaptations. A large crop—the honey stomach—at the anterior end of the alimentary tract is used to store nectar and carry it to the nest, where it is manipulated and stored in cells as honey. Bees are found wherever flowers grow and conditions are suitable for nesting. Flowers provide their basic food, and for many species a mating site and place for sleeping. The food of adult bees is largely nectar (which is mostly carbohydrate), with some pollen; the larval food is largely pollen (which is rich in protein), usually mixed with some nectar. "Constancy in the collection of pollen is a characteristic of bees in general" (Linsley). Many species consistently collect pollen from a single species or a group of related species; this is a characteristic of the great majority of solitary bees.

Bumble bees are dominant in the higher altitudes and are more abundant in temperate climates than in the tropics. It has been estimated that in total species bees constitute about 25 percent of the Hymenoptera; these are contained in relatively few genera, perhaps not more than one hundred. Most

* Pollen is transferred from the pollen brush to the pollen basket by rubbing the two hind legs together, scraping the grains from the tarsus of one leg, and forcing them into the basket on the tibia of the opposite leg. The scraping is done by the pollen comb (or pectin)—a row of stiff bristles located at the end of the tibia on the inside; the pollen is forced through a gap between the tibia and tarsus, into the concave pollen basket on the outside of the tibia.

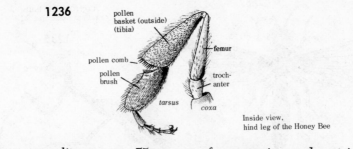

pollen
basket (outside)
(tibia)

pollen comb

pollen
brush

femur

troch-
anter

tarsus

coxa

Inside view,
hind leg of the Honey Bee

wild bees are solitary—some 75 percent of our species—and nest in the soil. They are most abundant in the arid regions of the Far West. Some "solitary" bees, like "solitary" wasps, are gregarious and construct their cells close together, but there is no division of labor which characterizes the social colony. Some nesting sites of alkali bees in Utah have been estimated to contain as many females as moderately sized apiaries.

There are three kinds of social bees: the bumble bee, honey bee, and stingless bee. They are generally grouped into one family, the Apidae. Stingless bees are strictly tropical and not found north of Mexico. Like the honey bee, they establish new colonies by swarming, but with the difference that *new* queens leave the nest to found a new colony rather than the mother queen. Colonies of the stingless bee are kept by the Indians in Yucatán; nesting sites or apiaries are constructed by stacking short pieces of hollow logs. Though stingless, swarms of these bees can annoy humans, since they are sticky and can bite slightly. Unlike the honey bee, and like wasps, bumble bees can sting repeatedly without harming themselves. The honey bee's stinger has barbs, and usually the bee (worker) is unable to remove it and free herself without tearing it from her own body, which is always fatal.* There are three distinct castes among social bees: the fertile female (queen), the infertile female or worker, and the drone or male.

The common bumble bees belong mostly to the genus *Bombus* (or used to, at any rate) and are usually black and yellow. As in the case of social wasps, colonies break up in the fall and only the young queens, now mated, remain to hibernate; all others perish. In the spring, after emerging from hibernating quarters in the soil or litter, the queen looks for a place to nest, which may be on or below the ground surface—under rocks, in a clump of grass, the abandoned nest of a mouse or bird—and makes preparations for her first brood. Egg cells are constructed—usually of pollen in the case of the first brood—and one egg is laid in each. (*B. balteatus* places all first-brood eggs in a single cell.) One or more cups are fashioned of pollen and wax and filled with honey before the young arrive.

The queen bumble bee feeds pollen to the larvae by pushing pellets

---

* In defense against her own kind—other bees and insects—the honey bee does not normally lose her stinger, and can sting repeatedly.

under the brood mass through a gap in the wax-pollen covering. She usually requires three to four weeks or more to produce the first brood of workers, which consist on an average of from eight to ten individuals. After the first brood arrives (commonly all workers), the queen confines herself to egg laying while her daughters forage and enlarge the nest, constructing brood cells, building and storing honey and pollen pots, caring for the young. The egg cells of the second brood are often built in straight parallel lines, and usually three or four eggs on an average are placed in each. Brood cells are not used again, but cocoons are used for storage of honey and pollen after undergoing alterations. In some colonies the second brood is all male and the third brood is female. Many species construct pollen-pockets under and alongside the worker broods, feeding them from these collectively until the final instar. The broods that are to become males or queens and the last-instar larvae of workers are fed a regurgitated mixture of honey and pollen. Workers of the later broods become progressively larger. The queens and males usually leave the nest and fly off to mate (some mate before leaving, or near the entrance); the males are often seen flying about the entrance at this time, waiting for the young queens to come out.

Bumble bee colonies are never large, probably average less than 200 individuals and seldom contain more than a few hundred. (The population of workers in a hive of honey bees may be 20,000 to several times this.) Bumble bee nests are frequently invaded by "cuckoo bees"—closely related species (*Psithyrus*) that resemble bumble bees but have no worker caste and whose queens lack pollen baskets on the hind legs. They make their egg cells in the center of the brood mass, laying a few eggs in each, and usually manage to get most of their young reared by the hosts. The common culprits are *P. variabilis* in the East and *P. insularis* in the West. Their presence sometimes reduces the host brood considerably. Their hum is more subdued than that of bumble bees (*Psithyrus* means "whisperer"), and they are stouter, thus able to sting harder.

Colonies of honey bees are maintained throughout the winter by their habit of storing honey and clustering. By continued movement of wings, legs, and bodies and the normal process of metabolism, enough heat is generated to keep the compact cluster of individuals alive. They draw closer together as the temperature drops and shift their position to gain access to more stores of honey. The queen remains within the cluster and by early February begins laying eggs, if the stores of honey and pollen are adequate. The small, elongated, white eggs are attached to the base of the cell (one to each) and remain upright for about three days, when they hatch. As the days grow longer and warmer, the cluster expands and drones are produced in preparation for the division of the colony. The colony expands by building new cells, regardless of crowding.

Wax combs are built up of two horizontal rows of individual cells backed up to one another. The worker broods are kept in the lower central part of the comb, and the pollen stored around them in cells of the same size. The larger cells of the drones are usually in the lower corners of the comb; honey is stored in cells of the same size toward the back of the hive. Queen cells are large thimble-shaped cells which hang down vertically from the brood comb. Beekeepers add "supers" to the upper part of the hive to give the bees more space to store honey and prevent swarming. With ready-made wax comb foundations, bees have more time to make honey and consume less stores.* Brood cells are kept open and the larvae fed daily. All get royal jelly (a secretion of the workers) for two days; workers and drones are then given honey and pollen, while queens remain on the royal jelly diet. The white grublike larva is full grown and fills its cell in about six days; the cell is then capped and the larva pupates after spinning a cocoon. Pupation requires about seven and one-half days for queens, twelve days for workers, and fourteen and one-half days for drones.

Swarming by honey bees occurs when new queens appear in the hive, prompting the mother queen to leave and half the colony to follow her. They cluster on a limb or other overhang and remain until the scouting bees find a new home. The cause of swarming is believed to be the production of certain pheromones by the queen which affect the workers.† Swarming can often be prevented by providing more room above the brood nest area into which the colony can expand. In the parent colony the new queens fight one another until only one remains alive. The survivor flies out to mate and returns as the new queen mother. Young workers first serve as nurses, later produce wax; after about three weeks of various chores in the hive they become foragers. Workers live about six weeks during the peak of activity. Honey bees are proficient pollinators because they visit only one kind of flower on a trip. The queen is fed royal jelly throughout her life, which lasts a year or more. Drones are fed honey as long as it is

* These are thin sheets of beeswax imprinted with a pattern of the bee cells; they serve as "starters," being fixed to the base of the removable wooden frames placed in hives, and on them the bees build their combs.

† Pheromones are substances secreted by the bees, the scent of which prompts certain behavior by other members of the colony. Some of these pheromones are known, others can only be conjectured. One of these substances is used to identify the hive when a new colony is established (each colony has a distinctive odor); some bees are stationed at the entrance, and fan with the wings as they bend the tip of the abdomen downward, exposing a droplet. The same scent gland is used to mark new sources of food. An alarm or sting pheromone has also been identified; when a bee stings, this substance prompts other bees in the vicinity to sting the same place. A "queen substance" assures the social unity of the colony and is believed to prevent the ovaries of the workers in the colony from developing. Bees communicate in other ways, the best known being the "dance" performed by foraging bees to tell the others in which direction to fly, how far to fly, and what kind of food they will find when they get there. Bees also recognize and are guided by colors and are receptive to the polarized light in the sky, which aids them in navigation.

plentiful but are ousted from the hive when they become a burden; their life span is about eight weeks. The queen and drone (possibly from another colony) mate in the air, then drop to the ground. To free herself, the female pulls the male genitalia from his body, thus causing his death. With the help of workers, she is able to remove the male organs from her own body.

The true honey bee, *Apis mellifera*, a native of Europe, Asia, and Africa, was brought to North America by the early settlers sometime before 1638. For many decades it was important only as a source of honey and beeswax. It has been estimated that 80 percent of our commercial crops today are pollinated by honey bees. However, reproduction of many forest trees, wild plants, and forage crops are completely dependent on native wild species. Clover and alfalfa are largely "self-sterile" and require cross-pollination for good seed production. Clovers have a heavy, sticky pollen which is not carried by the wind. Red clover is largely dependent on bumble bees for pollination; other bees do not visit these flowers readily, since it is difficult—with their shorter tongues—for them to reach through the long corolla for nectar. Leaf-cutting bees (*Megachile*) are important pollinators of alfalfa in the West. Alkali bees (*Nomia*) are mainly responsible for the high yields of alfalfa grown for seed in Washington and Utah, and are so valuable that artificial nesting sites have been devised to increase the numbers.*

The principal insect enemies of bees among the predatory species are robber flies, assassin bugs, ambush bugs, ants, and sphecid wasps. The best known among the wasps are in the genus *Philanthus* (92). The parasites of bees are largely Diptera and include thick-headed flies, humpbacked flies, flesh flies, and tachina flies; the most important of these are the thick-headed flies, which are solitary internal parasites of adult bees and oviposit on the host in flight [see Diptera (49)].

(94) SWEAT BEE: *Halictus farinosus* (Halictidae).

The Halictidae have one suture below each antennae. [See (96) and (100) below.] **Range:** *H. farinosus* ranges from California to New Mexico, northward to British Columbia, Nebraska, Montana. **Adult:** Black, with yellow legs, reddish yellow markings on abdomen; head and thorax with

---

* In the alfalfa flower the stigma (tip of pistil receiving the pollen) and the anthers (upper part of stamen containing the pollen) are enclosed by two petals under tension which trip when a slight pressure is applied. When the bee trips the flower, it unwittingly leaves pollen from a previously visited flower on the stigma, thereby effecting cross-pollination. Honey bees can get nectar without tripping the flowers, but prefer sweet clover and other sources where available; they have been forced to trip flowers in the Southwest (where other sources of pollen are scarce). *Nomia melanderi* and *Megachile rotundata*—both pollen-seeking bees—are the species used by alfalfa seed growers. See W. D. Fronk, *Increasing Alkali Bees for Pollination,* Mimeo. Cir. No. 184, Agr. Exp. Sta., Univ. of Wyoming, Laramie (1963), also reference at foot of p. 577.

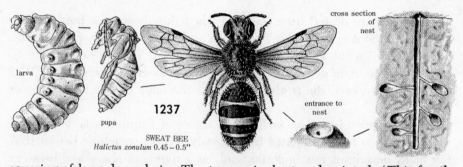

cross section
of
nest

larva

pupa

1237

entrance to
nest

SWEAT BEE
Halictus zonulum 0.45 – 0.5"

covering of long dense hairs. The tongue is short and pointed. (This family and the next two—Andrenidae and Colletidae—are called "short-tongued bees"; the remaining families are known as "long-tongued bees.") **Length:** .45". This genus shows a tendency toward social behavior, some species resembling bumble bees in habit.* Typically, a female emerges from her hibernation burrow in the spring and makes a nest consisting of a curving burrow with cells branching off to the sides; these are provisioned with pollen and nectar, and an egg laid in each. The first brood—all females—appears in four to six weeks and constructs cells branching off from the parental burrow, or builds entirely new nests. Since there are no males, they are unfertilized and produce males. The mother continues to lay eggs—on pollen balls made by her daughters. Her new brood is again female, and they now mate with the brood of males. The males die, and the mated females dig hibernation burrows for overwintering. Some species produce a third brood of males and females. They are important pollinators of agricultural crops. *H. rubicundis* and *H. confusus* are important pollinators of low-bush blueberries in eastern U.S. and Canada, are also common on buckwheat. *H. zonulum* (Fig. 1237) occurs from Michigan to New York and southeastern Canada, also in Europe and Asia. The related species *Dufourea linanthi* (Fig. 1238) is black, about .25" long, common on *Linanthus.*

(95) ALKALI BEE: *Nomia melanderi.* **Color plate 8a.**

Range: Wyoming, Colorado, Utah, Idaho, Washington, California. **Adult:** Black; abdominal segments two to four with broad light emerald-green apical bands; anterior part of mesothorax with pale yellowish brown hairs and black bristles intermixed. It flies during July and August, nests in sandy or clayey, alkali soil—for which it gets the name "alkali bee." Each female builds a single nest of 15 to 20 cells in about 30 days; the entrance to the shaft is marked by a mound of dirt. They are gregarious and often form huge aggregations of individual nests. **Length:** .5". Hibernates as naked prepupa in brood cell. One brood. In some localities a partial second brood,

---

* They sometimes nest in colonies with a common entrance, which is always guarded by a female, but construct and provision their own individual cells.

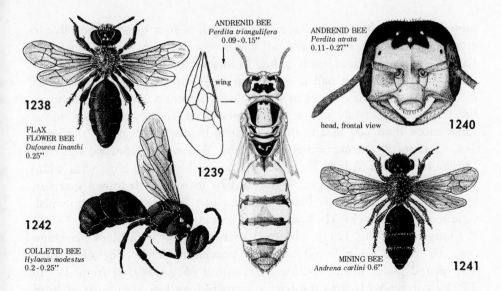

ANDRENID BEE
Perdita triangulifera
0.09-0.15"

wing

ANDRENID BEE
Perdita atrata
0.11-0.27"

head, frontal view

1240

1238

FLAX
FLOWER BEE
Dufourea linanthi
0.25"

1239

1242

COLLETID BEE
Hylaeus modestus
0.2-0.25"

MINING BEE
Andrena carlini 0.6"

1241

mostly females, may occur in late summer; scarcity of males at the time results in many unfertilized eggs being laid and a large percentage of males in the overwintering brood. It is an important pollinator of alfalfa grown for seed, but has "a rather broad host range for a wild bee," according to Ribble. *N. howardi*—now considered a separate species—is recorded from Utah, California, Arizona, Mexico, and "probably nests in alkali soil."

(96) MINING BEE: *Andrena carlini* (Andrenidae). **Fig. 1241.**

The Andrenidae may be distinguished from other families by the two sutures below each antenna. [See (94) and (100).] **Range:** *A. carlini* occurs from Nova Scotia to New Jersey, west to Minnesota, Colorado, Wyoming. **Adult:** Black, with dense covering of long hairs on head, thorax, and legs. The tongue is short, pointed. (The family is sometimes referred to as the "acute-tongued bees.") Both sexes emerge from larval cells in the spring. The female builds a nest consisting of a long branching tunnel with a cell at the end of each branch; she provisions the cells with pollen and nectar, and lays an egg in each. **Length:** .6". **Larva:** Develops rapidly and pupates, but the adult does not emerge from the pupal cell until the following spring. One brood. It is an important pollinator of blueberry in Maine and eastern Canada, as is *A. regularis*.

(97) ANDRENID BEE: *Perdita zebrata.* **Color plate 8b.**

The genus *Perdita* is a large group of small bees confined largely to the western states. Close to 750 species have thus far been described, according to Timberlake. They are remarkable in often restricting their choice of flowers to one or a few closely related species. They are gregarious, nest in the ground, usually in groups with one female to a nest. The entrance tunnels of a *P. zebrata* nest descend to a depth of four to six inches; five to eight branches are constructed at the bottom of the main tunnel and running in all directions, each ending in an oval brood cell; the cell is provisioned with a pollen ball in which an egg is inserted. (The females of *P. opuntiae*—a species found in Colorado—construct a long, smooth, mud-lined tunnel in gravelly soil, with cells placed at irregular intervals, sometimes more than 200 in all, and several individuals occupying the same nest.) **Range:** *P. zebrata* occurs in Montana, Wyoming, Nebraska, and New Mexico to Nevada. **Adult:** Head and thorax dark metallic blue, with white pubescence; abdomen yellow, segments two to four with brown margins; legs yellow, marked with black; wings clear, iridescent, with brownish venation. **Food:** Rocky Mt. bee plant (*Cleome serrulata*). **Length:** .2–.25″. *P. triangulifera* (Fig. 1239) is highly variable, yellow, with black markings, from .09″ to .15″ long, found on clover, acacia, mistletoe (*Phoradendron*), and related plants in the Southwest. *P. atrata* (Fig. 1240) is from .11″ to .27″ long, largely black in the female, brownish yellow with parts of head and thorax black in the male, found at higher elevations on blazing star (*Mentzelia*) and related plants.

(98) COLLETID BEE: *Colletes compactus* (Colletidae).

Range: Nova Scotia to Georgia, west to Wisconsin, Missouri, Colorado, Arizona. **Adult:** Black, with dense covering of long hairs on head, thorax, and legs. It has a short, slightly bilobed tongue. (In contrast to the Andrenidae above, the family is sometimes referred to as the "obtuse-tongued bees.") It constructs a burrow 18 to 28 inches deep, with two to four lateral branches 2 to 6 inches long and a cell at the end of each, provisions the cells with pollen and nectar, lays an egg in each. The burrows are marked by low mounds of dirt. **Length:** .45″. **Food:** Asters, goldenrod, sneezeweed, sunflower and closely related composites. A subspecies, *C. c. hesperius,* occurs in Oregon, Washington, and British Columbia.

(99) COLLETID BEE: *Hylaeus modestus.* Fig. 1242.

Range: Eastern Canada and the Atlantic states, Colorado, New Mexico. **Adult:** Black; side of face lemon yellow in the male, with triangular yellow patch in the female; wings clear, tinged with brown, venation reddish brown; tibiae yellow basally; tarsi reddish brown in the female, yellow in the male; apical margins of abdominal segments brownish. Very common on many flowers. It nests in the upper 3 inches or less of the hollow twigs of sumac, provisions the cells with pollen and nectar. **Length:** .2–.25″. **Larva:** It constructs a very thin transparent cocoon. The cocoons are separated by heavy dividing walls of a waxy substance. *H. affinis* is the same size, black, with yellow markings similar to those of *modestus,* occurs throughout North America.

(100) MASON BEE: *Hoplitis producta* (Megachilidae).

In this family there is one suture below each antenna [see (94) and (96) above], the labrum is longer than wide and widened at the extreme base [compare with Apidae (102)]; they are sometimes called the "thick-jawed bees." **Range:** *H. producta* occurs from Quebec to Georgia, west to Alberta, Colorado, and Texas. **Adult:** Black, with marginal bands of short white hairs along either side of abdomen. It lacks pollen baskets on the tibiae, carries pollen in "pollen brush" on underside of abdomen. It nests in the hollow stems of sumac or other pithy plants (or excavates the pith from the stem), partitions off a few cells at the bottom of the tunnel with a mixture of clay and plant material. The cells are supplied with pollen and nectar and an egg laid in each, with space left between the last cell and walled entrance. **Length:** .3″. **Larva:** Hibernates in prepupal stage in cocoon inside the cell. Several subspecies of *producta* occur farther west.

(101) LEAFCUTTING BEE: *Megachile* (*Xanthosarus*) *latimanus.* **Fig. 1243.**

Range: Nova Scotia to Georgia, west to Alberta, Wyoming, and Colorado; British Columbia. **Adult:** Black, densely clothed with pale brownish yellow pubescence, faint white apical bands on abdominal segments two to five. Pollen brush of female on underside of abdominal segments two to five pale red. It bores a tunnel in wood or the ground, or uses a cavity, in which thimble-like cells are shaped from pieces of leaves cut evenly from rose or other plant. When a cell is provisioned with pollen and nectar and an egg laid, it is sealed over with pieces of leaves cut round and slightly larger than diameter of the cell, so they fit tightly. A series of several cells may be made in a tunnel, end to end. **Length:** .5–.6″. *M. perihirta* is very similar, slightly smaller, pollen brush of female brighter red; it occurs from Alberta to Nebraska and Texas, west to British Columbia, California, and Mexico. These species and *M. dentitarsus* are the most important pollinators of alfalfa in western Canada; the pollen brush of *dentitarsus,* on segments five and six, is red on apical half only. They are more active than bumble bees and honey bees and, unlike them, consistently trip flowers, which is essential to pollination. *M. rotundata* (Fig. 1244)—black, with yellow-green abdominal bands, whitish yellow pollen brush—is Eurasian in origin, was accidentally introduced into the eastern states in the 1930's, has spread westward to Oregon and California; it has one brood, overwinters in cell as prepupa; almost specific to alfalfa, it can be managed by providing artificial nesting sites (boxes of malt straws cut in two and sealed at one end with a beeswax-paraffin mixture).[*]

---

[*] See R. J. Walstrom and P. A. Jones, *Alfalfa Leaf-cutter Bee Management for Alfalfa Pollination in South Dakota,* Bul. 544, Agr. Exp. Sta., South Dakota State Univ., Brookings (1968).

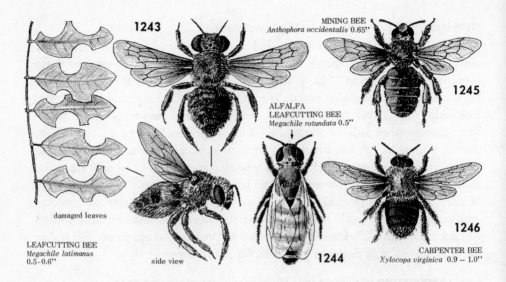

1243

MINING BEE
Anthophora occidentalis 0.65''

1245

ALFALFA
LEAFCUTTING BEE
Megachile rotundata 0.5''

damaged leaves

LEAFCUTTING BEE
Megachile latimanus
0.5-0.6''

side view

1244

1246

CARPENTER BEE
Xylocopa virginica 0.9 – 1.0''

(102) MINING BEE: *Anthophora occidentalis* (Apidae, Anthophorinae). **Fig. 1245.**

In the Apidae the labrum is wider than long and narrowed at the extreme base [compare with Megachilidae (100)]. The subfamily Anthophorinae is sometimes referred to as the "flower-loving bees." *Anthophora* is a fairly large genus of "mining bees," which usually occur in large colonies and show a decided preference for sand or clay banks. Their nests are similar to those of *Halictus* (94), but the tunnel entrances are protected by chimneys curving downward with the slope of the bank and made of mud pellets; the inside is smooth and glazed, as are the brood cells. Each cell is provisioned with a semiliquid mass of pollen and nectar in which one pearly white egg is laid; the cell is sealed tightly and is impervious to water. They pupate in the brood cells without a cocoon and hibernate as larvae in the cells. Adults visit many different kinds of flowers. **Range:** *A. occidentalis* occurs in Montana, Wyoming, Oregon, South Dakota, Kansas, Colorado, Utah, New Mexico. **Adult:** Black, clothed with a dense, short yellow pubescence; tip of abdomen dusky, wings clear with apexes dusky; male with face on each side lemon yellow, tarsi brownish yellow. **Length:** .65''. A large complex of inquilines and parasites are normally found in the nests of *occidentalis;* most important in their effect on the bees are: a large green chalcid parasite *Monodontomerus montivagus* (Torymidae), the velvet-ant *Dasymutilla fulvohirta,* the cuckoo wasp *Chrysis densa,* and blister beetles (*Hornia* spp.). Bombyliid and conopid flies are parasitic on *Anthophora* generally. *Anthophora abrupta* is similar to *occidentalis,* occurs from Massachusetts to North Carolina, west to Wisconsin, Kansas, and Texas. *A. bomboides* with several subspecies is widespread throughout North America. *A. furcata terminalis* nests in wood, much like *Xylocopa*

(104), occurs from Quebec to North Carolina, west to Alberta, Colorado, Utah, Arizona.

(103) LITTLE CARPENTER BEE: *Ceratina dupla (Xylocopinae).*

Range: Quebec to Florida, west to Wisconsin, Louisiana. **Adult:** Metallic blue. It excavates the pith from dead twigs, partitions the hollow stem with cemented chips, places pollen and nectar together with an egg in each cell. The mother waits in her own—the top—cell for emergence of her brood; the lower ones mature first, each gnawing away the ceiling of its own cell. When they reach the top all fly off with the mother but return and clean out the old nest, which is used again by one of them. **Length:** .25″. Females of the second brood hibernate in their nests. Two broods. **Host:** Sumac and other pithy plants. Adults visit many kinds of flowers in various families. *C. acantha* (Fig. 1249) is similar, occurs in California and Lower (Baja) California.

(104) CARPENTER BEE: *Xylocopa virginica* (Xylocopinae). **Fig. 1246.**

Range: Connecticut to Florida, west to Illinois, Kansas, and Texas. **Adult:** Black, resembling a bumble bee, except that the abdomen is bare and without conspicuous yellow markings. It lacks pollen baskets on the hind tibiae, a dense brush of hairs serving the purpose. In *Xylocopa* females the antennae have 12 segments, and the hind tibiae have two spurs; in males they have 13 segments and one spur. It excavates a tunnel up to 12 inches long in solid wood, partitions it into several cells separated by cemented wood chips; pollen and nectar are placed in each cell together with an egg. **Length:** .9–1″. **Larva:** Hibernates in prepupal stage in cell. One brood. Adults visit a number of flowers.

(105) MOUNTAIN CARPENTER BEE: *Xylocopa tabaniformis orpifex.* **Fig. 1247.**

Range: Lower California, California, Oregon. **Adult:** Black, male with white or yellow and black hairs on head and prothorax. Habits as in (104) above. Burrows are 5″ to 12″ long, curved upward, in sound wood—sometimes in colonies. **Length:** .5–.7″. **Larva:** Yellowish white, about .3″ long. **Host:** Douglas fir, redwood, or other sound timber; structures are sometimes weakened. *X. t. androleuca* occurs in Utah, Nevada, Arizona, and eastern California, *X. t. parkinsoniae* in Texas. *X. brasilianorum varipuncta* (Fig. 1248)—the **Valley Carpenter Bee**—is larger, iridescent blue-black (female) or yellowish (male), occurs in partially decayed oak, eucalyptus, pepper trees, and other hardwoods in northwestern Mexico, California, and Arizona.

(106) BUMBLE BEE: *Bombus fraternus* (Apinae).

Unlike the honey bee, bumble bees have one or two spurs on the hind tibia. **Range:** East of the Rocky Mts. Next to the Pennsylvanian bumble bee (109), this is the most common eastern species. **Adult:** Black; thorax with

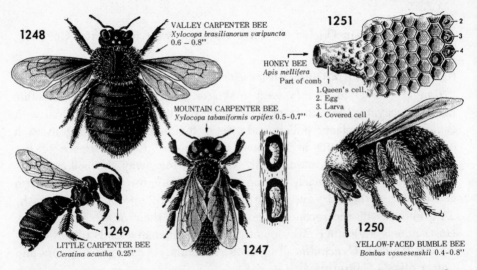

1248

VALLEY CARPENTER BEE
Xylocopa brasilianorum varipuncta
0.6 – 0.8"

1251

HONEY BEE
Apis mellifera
Part of comb 1
1. Queen's cell
2. Egg
3. Larva
4. Covered cell

MOUNTAIN CARPENTER BEE
Xylocopa tabaniformis orpifex 0.5-0.7"

1249

LITTLE CARPENTER BEE
Ceratina acantha 0.25"

1247

1250

YELLOW-FACED BUMBLE BEE
Bombus vosnesenskii 0.4-0.8"

anterior portion above and on sides yellow, broad black band between the wings; abdomen with two basal segments above yellow. Wings black-violaceous, darker toward the bases. **Length:** .85–1" (queens); .5–.9" (males); .5–.85" (workers). (Proposed genus: *Pyrobombus*.)

(107) YELLOW BUMBLE BEE: *Bombus fervidus.*

Range: Widespread throughout the U.S. and southern Canada. **Adult:** Black; most of thorax, and dorsum of first four abdominal segments, yellow. Hairs dense, coarse and medium in length. Often confused with the Pennsylvanian bumble bee (109). Queens come out of hibernation mostly in May, usually make their nests of dry grass *on* the ground; however they are very adaptable and may use a great variety of sites on and below the surface of the ground. **Length:** .6–.9" (queens); .4–.6" (males); .3–.6" (workers). A long-tongued species, very valuable as a pollinator of red clover. It is common on the prairies and in open spaces. (Proposed genus: *Mega-bombus*.)

(108) BLACK-FACED BUMBLE BEE: *Bombus californicus.*

Range: Along the Pacific coast, in the western states and provinces, east to Kansas. **Adult:** Black; head entirely black, front of thorax and usually dorsum of only the fourth abdominal segment yellow. Hairs moderately long and coarse. Queens come out of hibernation mostly in June, usually nest *on* the ground or other surface location; they are less adaptable in nesting than the yellow bumble bee (107) or the Pennsylvanian bumble bee (109). **Length:** .6–.9" (queens); .5–.65" (males); .35–.6" (workers). A long-tongued species, very valuable as a pollinator of red clover. It occurs in the foothills and wooded areas, in open spaces along streams bordered with trees. The **Yellow-faced Bumble Bee,** *B. vosnesenskii* (Fig. 1250), occurs along the Pacific coast, east to Colorado; it is black, with face, top of head, pronotum, mesonotum in front of wings, and dorsum of fourth abdominal segment yellow; hairs dense, short and fine; from .4″ to .8″ long; it is the most common bumble bee in parts of California, and sometimes mistaken for the yellow bumble bee (107) or the black-faced bumble bee. (Proposed genus: *Megabombus* and *Pyrobombus* respectively.)

(109) PENNSYLVANIAN (AMERICAN) BUMBLE BEE: *Megabombus pennsylvanicus* (*Bombus americanorum*). Color plate 8c.

Range: Throughout the U.S., southern Canada, and northern Mexico. **Adult:** Black; front of head clothed with black pile; males somewhat variable, with greater part of abdomen above yellow, apex reddish brown; females with front of abdomen yellow, the remainder black. The majority of queens come out of hibernation in May, usually nest on the ground, often in deserted mouse nests; some nest underground. **Length:** .7–.9″ (queens); .55–.9″ (males); .5–.7″ (workers). They visit a wide variety of flowers, are usually found in open grasslands. The average number of individuals in a colony is commonly less than 200. Easily irritated, the workers often pursue an intruder attempting to escape; their sting is painful to sensitive persons. Advanced nests may be detected by the presence of many males flying about the entrance. It is commonly parasitized by the closely related "cuckoo bee," *Psithyrus variabilis*, which is mostly black, with dense yellow pile on vertex, hairs on dorsum yellow posteriorly, hairs on thorax mainly yellow in the male. *Pyrobombus* (*Bombus*) *impatiens*—widely distributed throughout the East and Midwest, and in southeastern Canada—is black, with tuft of yellow hairs on vertex in the female, head mostly yellow in the male, thorax yellow, first abdominal segment yellow; its favored flowers are sunflowers, thistles, partridge pea, clovers, nettles, rose.

(110) NEVADA BUMBLE BEE: *Megabombus* (*Bombus*) *nevadensis nevadensis*.

Range: Pacific coast, the western states and provinces, east to Illinois. **Adult:** Black; head of female all black, face and occiput of male yellow; dorsum of thorax and first three abdominal segments yellow; tip of abdomen reddish in the male. Hairs dense and short. Queens come out of hibernation from mid-May to mid-June, nest under or on the ground. Only one egg is laid in a cell—in all broods. The larvae are fed individually, the first brood averaging about 12 individuals. Cells of the second and third broods are not built in straight parallel rows. It is a prolific producer of wax. **Length:** .7–.9″ (queens); .5–.7″ (males); .6–.7″ (workers). A long-tongued species, very valuable as a pollinator of red clover. It occurs in the wooded areas of the foothills, and is common on the prairies. *M. n. auricomis* occurs in the eastern states, west to Colorado, and in southern Alberta; it is found in the foothills and river valleys in the western part of its range but not on the prairies.

(111) HONEY BEE: *Apis mellifera.* **Color plates 8d, e, Figs. 1236, 1251.**

Range: World-wide. Wild swarms usually occur wherever bees are "domesticated." **Adult:** Colors variable, generally some shade of black, gray, or brown, intermixed with yellow; thorax with dense coat of fine short hairs, usually black or paler in color; abdomen with thin coat of hairs and often banded with yellow. Unlike bumble bees, the hind tibia is without spurs. **Length:** .6–.8″. There are now several races of *Apis mellifera* in use in North America. The most popular has been the Italian bee, partly because of its brighter color and the greater ease in finding the queen; it is black, with scattered yellowish hairs, three to five yellow bands on abdomen. This is a gentle bee, fairly easy to handle, most resistant of the races to European foulbrood. The Caucasian bee is mild-mannered (though a persistent stinger when aroused), black in color and banded with gray, has a longer tongue than the others. It is a heavy producer of propolis or bee glue—a brown resinous substance which serves as a wax cement; it tends to be more economical with its stores when there is no nectar flow. The Carniolan is gray and otherwise closely resembles the Caucasian; it is the most gentle of the three but tends to swarm most. Carniolans are good comb builders, generally produce whiter cappings on cells than the others.

22

DIPTERA: Flies, Mosquitoes, Gnats, Midges

This order is the fourth in size—next to the Hymenoptera—measured by number of species described. The name comes from the Greek meaning "two wings." The single pair of membranous wings which characterize this group are attached to a greatly enlarged mesothorax. A pair of knoblike appendages called "halteres," attached to the metathorax, take the place of the second pair of wings possessed by other insects; they are thought to act as balancers. Nevertheless, many of the Diptera are swift, powerful fliers; a few are wingless. The prothorax and metathorax are greatly reduced and are fused with the mesothorax, which occupies most of the dorsal surface of the thorax. The head, attached to the body by a slender neck, is usually vertical and freely movable, has two large compound eyes and three ocelli; the antennae have various shapes, and are commonly three- to sixteen-segmented. The legs are usually alike, and the tarsi five-segmented, with two claws. Coloring is generally somber, though some are brightly striped or have a metallic luster.

Metamorphosis in this group is complete. Most flies deposit eggs, but some parasitic species—certain sarcophagid, tachinid, and other flies—deposit larvae (larviposit); others—some of the gall midges—produce larvae by parthenogenesis, and these or pupal forms (more rarely) give birth to other larvae (paedogenesis). The larvae of flies, called maggots, are almost legless, often without a distinct head and with retracted mouthparts. The pupa is either free or enclosed in the last larval skin (puparium).

The Diptera include most of the insect vectors of human disease and some exceedingly disagreeable if not dangerous pests of animals and man. *Anopheles* mosquitoes transmit malaria, *Aedes aegypti* is the vector of yellow fever, and several species of *Culex* (including the common house mosquito) are implicated in encephalitis. Flies transmit bacteria that cause typhoid fever, dysentery, diarrhea, and other diseases. According to the U.S. Public Health Service, "flies constitute one of the greatest of public health hazards." Some are serious plant pests, such as the gall midges and fruit flies. On the other hand, the order includes many beneficial species—important in the natural control of insect pests, or as pollinators, and scavengers which dispose of dead animals and plant wastes. Recently it has been shown that the somewhat neglected family of Sciomyzidae—the larvae are aquatic and terrestrial and parasitic or predaceous on fresh-water snails—may be significant in the control of snails which act as intermediate hosts for flatworms (flukes) causing disease in man and other animals; the introduction of sciomyzids into Hawaii has helped indirectly to control liver flukes which parasitize cattle.*

The time-honored division of the Diptera was into two main suborders: the Orthorrhapha (including the Nematocera and Brachycera) and the Cyclorrhapha. Stone et al. recognize three suborders: the Nematocera, Brachycera (both divisions of the Orthorrhapha in the older classification), and Cyclorrhapha. The Orthorrhapha—our examples (1) to (42)—are called "straight-seamed flies," and the Cyclorrhapha—the remainder—are called "circular-seamed flies," referring to the manner in which the adult emerges from the pupa. Imms stresses the "absence of development of a puparium" as the main point of difference.†

In the Orthorrhapha the antennae are usually longer than the head and have six or more segments, the larva has a well-developed head—it may be retractile—and the mandibles move vertically; the pupa is free (not enclosed in the last larval skin) and the adult usually emerges through a longitudinal slit ("straight seam") that opens near the anterior end of the pupal case. There is no suture or lunule in the adult. In the Cyclorrhapha the antennae have less than six segments, the head of the larva is vestigial, and mouthparts consist of vertically moving hooks; the larva pupates within the last larval skin (puparium), and the adult emerges through a circular slit at the anterior end of the puparium. In most families the head has a frontal suture or slit through which a bladder-like sac—the ptilium—is pushed; with this organ the adult breaks through the end of the puparium, thus making a "circular seam." The ptilium is later withdrawn into the head and the frontal suture closed; it can be pushed out and withdrawn at will and is

* See Lloyd V. Knutson, "Snail-killing Sciomyzid Flies," in *The Cornell Plantations,* 17:59–63 (1962), Ithaca, N.Y.

† A. D. Imms, *A General Textbook of Entomology,* rev. by O. W. Richards and R. G. Davies (London: Methuen, 1964), p. 607.

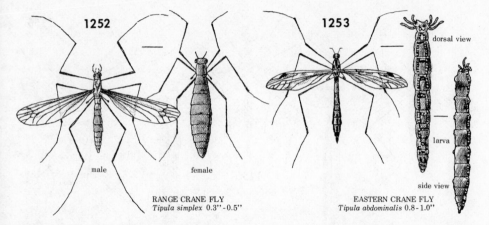

1252

1253

dorsal view

larva

male female

side view

RANGE CRANE FLY
Tipula simplex 0.3"-0.5"

EASTERN CRANE FLY
Tipula abdominalis 0.8-1.0"

used to push through the soil when pupation takes place there. All flies in this group have a frontal lunule—an oval or crescentric area above the base of the antennae and bounded by the frontal suture; the latter is visible in most of the families in this group, numbered (43) to (101) in our examples.*

NEMATOCERA

Crane Flies (Tipulidae)

The Tipulidae—called crane flies because of their long legs and slow flight—comprise the largest family of Diptera. Commonly found in wet or humid areas, often numerous near lakes and streams, they are frequently mistaken for mosquitoes (the common species are much larger). Distinctive features are the prominent V-shaped suture or fold on the thorax between the wings, and the long feathery or toothed antennae of the males (they are filiform and shorter in the females). The halteres are also prominent. When captured, the legs collapse regardless of care in handling. The larvae are found in moist soil, in wet places along bodies of water—a few are aquatic—and in decaying leaves and organic matter in wooded areas. Some species are pests of sod grasses, grains, and other cultivated plants.

(1) RANGE CRANE FLY: *Tipula simplex.* **Fig. 1252.**

Range: California. **Adult:** Female grayish brown, with small wing stubs, rather short legs. The male is winged. Elongated black eggs are laid in soil in the spring and summer. **Length:** .4–.5" (female); .3–.4" (male). **Larva:** Pale brown, .7–1" long full grown. They appear from January to March, live in small holes in the ground, emerge at night or on dark days to feed on green grasses and grains. Pupates in the soil. The pupa is somewhat flat-

---

* The frontal suture and ptilium are absent in the Pipunculidae (43), Syrphidae (44) to (48), and Phoridae (50).

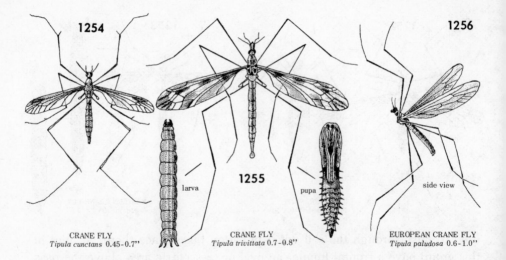

1254 **1256**

larva **1255** pupa side view

CRANE FLY
Tipula cunctans 0.45-0.7"

CRANE FLY
Tipula trivittata 0.7-0.8"

EUROPEAN CRANE FLY
Tipula paludosa 0.6-1.0"

tened, with two "horns" at the anterior end. One brood. *T. abdominalis* (Fig. 1253) ranges from Newfoundland to Wisconsin, south to Kansas and Florida. *T. cunctans* (Fig. 1254) occurs in Canada, from Manitoba to New Brunswick, south to Colorado, Pennsylvania, and Alabama; head and thorax gray, abdomen yellowish with median brown stripe, from .45" to .7" long; the larvae feed on roots in the late fall, hibernate as half-grown larvae; sometimes a serious pest of Japan clover. *T. trivittata* (Fig. 1255) occurs from Newfoundland to Iowa, south to Tennessee and South Carolina.

(2) EUROPEAN CRANE FLY: *Tipula paludosa.* **Fig. 1256.**

Also called marsh crane fly. First discovered in Canada in 1955. **Range:** Newfoundland, Nova Scotia, British Columbia, Washington; Europe. **Adult:** Clear brownish; antennae with 14 to 15 segments, the first two yellow; sides of thorax covered with grayish frosting; wings with milky tinge, brownish along margin. They normally fly in the evening, sometimes in the morning. The shiny black eggs are usually pushed into the soil. **Length:** .75–1" (female); .6–.7" (male). **Larva:** Young larva pinkish white, feeds on decayed vegetable matter in the soil; older larva grayish, with black head, from 1.4" to 1.6" long, migrates at night to feed on plants. Pupates late in May, the adult emerging in August or September. Hibernates as larva in or on the ground. The larvae are sometimes called leatherjackets. **Food:** Barley, oats, wheat, rye, other grasses; turnips, strawberries, peas, corn, tobacco. A serious pest of lawns, grains, and sod grasses. The most effective parasite in Europe is a tachinid fly, *Siphona geniculata,* now established in North America.

(3) MOTH FLY: *Psychoda alternata* (Psychodidae). **Fig. 1257.**

The Psychodidae, or moth flies, have hairy bodies and wings, resemble moths; the species commonly found in homes are variously called drain,

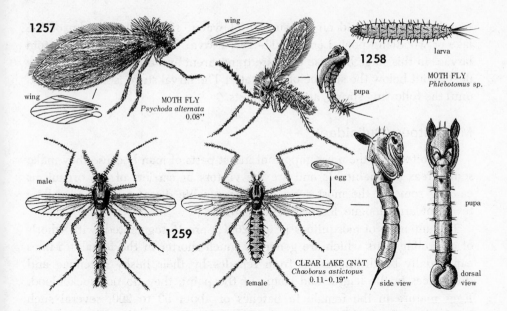

1257

wing

MOTH FLY
Psychoda alternata
0.08"

wing

wing

larva

1258

pupa

MOTH FLY
Phlebotomus sp.

male

egg

pupa

1259

CLEAR LAKE GNAT
Chaoborus astictopus
0.11-0.19"

female

side view

dorsal
view

filter, or sewage flies. The wings are held rooflike when at rest; their flight is short and jerky. *Psychoda* and *Phlebotomus* breed in moist decaying organic matter. *Phlebotomus* (Fig. 1258) (and closely related genera) are bloodsucking, attack mammals, including man, fowl, lizards, and frogs. Tropical species, called sand flies, are vectors of "sand-fly fever" and other diseases. *Psychoda alternata* is a cosmopolitan species. **Range:** Throughout North America. **Adult:** Body yellowish or light tan, wings lighter, with faint black and white mottling; dense coating of long hairs; antennae 15-segmented, terminal segments fused, the terminal being very small and button-like. Eggs are laid in masses of 30 to 100 on the surface film of sewage filter beds or in drains. **Length:** .08". **Larva:** Feeds on film deposits, decaying vegetation, microscopic plants and animals. *P. phalaenoides*—brownish, with terminal segments of 15-segmented antennae distinct—is widely distributed, often seen emerging from sinks, washbasins, and drains. Psychodids associated with filth may cause myiasis in man.

(4) CLEAR LAKE GNAT: *Chaoborus astictopus* (Chaoboridae). **Fig. 1259.**

Chaoborids, or phantom midges, are sometimes placed with the Culicidae; the adults have shorter mouthparts, do not suck blood, and the larvae have prehensile antennae. The larvae are predaceous, use their unique antennae to capture prey. *C. astictopus* is chiefly of local importance around Clear Lake, a resort area in northern California, but has attracted wide attention. **Range:** California, Oregon. **Adult:** Brown, with straw-colored hairs. They are present from May to October, are attracted to lights, and appear in such great swarms as to almost smother one, getting into the eyes, ears, nose, and mouth.* Eggs are laid on the surface of water, sink to the bottom of the lake or woodland pool. **Length:** .11–.19″. **Larva:** They are called "phantom larvae" in this stage because they are transparent and difficult to see as they float about below the surface of the water. The larval stage is not completed until the following spring, after four molts.

Mosquitoes (Culicidae)

Mosquitoes are the most important insect pests of man because they make some areas uninhabitable and are the vectors or carriers of the organisms causing some of the most serious communicable diseases, such as yellow fever, malaria, dengue, filariasis, encephalitis.†

The antennae of mosquitoes (Fig. 1260C) are 15-segmented, with whorls of encircling hairs which are generally much shorter in the females. Males can usually be distinguished from females by their bushy antennae and differences in the length and shape of the palpi; they do not suck blood. Eggs mature in the female in batches of about 50 to 200; several such batches may be laid by one female during her lifetime of a few weeks. In the bloodsucking species a blood meal is usually required for egg production. Mouthparts of the female mosquito (Figs. 1260A, B) consist of a long piercing-sucking organ composed of six stylets enclosed in a flexible sheath (the proboscis or labium). Two pairs of stylets are cutting organs with fine teeth (the mandibles and maxillae), the other two are called the *hypopharynx* and *labrumepipharynx;* the latter is channeled, and when the two are pressed together they form a tube. Blood is drawn up the tube by two pumps operated by pharyngeal muscles. A secretion of the salivary glands, which causes the itching, is injected through a smaller duct in the hypopharynx.

Male mosquitoes feed on sugar which they get from the nectar of flowers and fruit juices. (Most females also feed on nectar, and need sugar to sustain flight and normal longevity.) The larvae are aquatic and most of them free-swimming. They possess tracheal gills, but must come to the surface for air; their food consists of detritus and small organisms such as

* For some effects of the program to eradicate the gnat, see Robert L. Rudd, *Pesticides and the Living Landscape* (Madison: Univ. of Wisconsin Press, 1964).

† *Yellow fever* is an acute infectious disease caused by a virus (*Stegomyia*). Jungle yellow fever is endemic principally in monkeys and other primates in the American tropics and Africa. Urban reservoirs of infection are man and mosquitoes. *Dengue,* or "breakbone fever," is a nonfatal, acute infectious disease caused by a virus and endemic in tropical and subtropical countries. *Malaria* is an infectious disease caused by protozoan parasites (*Plasmodium* spp.). *Encephalitis,* or "inflammation of the brain," commonly called sleeping sickness, is caused by a virus. Three forms of the disease occur in the United States: eastern and western encephalitis and St. Louis encephalitis. *Filariasis* is a tropical disease caused by a microscopic nematode, *Wucheria bancrofti,* which obstructs the lymph vessels, causing immense swelling of affected parts.

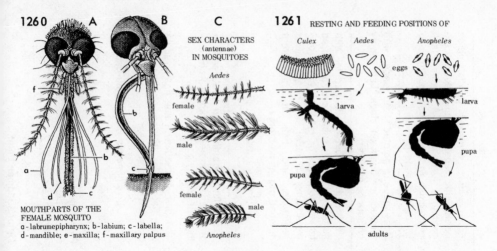

1260
A B C SEX CHARACTERS
(antennae)
IN MOSQUITOES

Aedes

female

male

female

male

Anopheles

MOUTHPARTS OF THE
FEMALE MOSQUITO
a - labrumepipharynx; b - labium; c - labella;
d - mandible; e - maxilla; f - maxillary palpus

1261 RESTING AND FEEDING POSITIONS OF

Culex　　*Aedes*　　　*Anopheles*

eggs

larva

larva

pupa

pupa

adults

flagellates, diatoms, and green algae. The developmental period lasts from four to ten or more days, during which they molt four times.

The most important genera are *Aedes, Anopheles,* and *Culex,* which can be distinguished from one another readily as shown in the table on page 590 and Fig. 1261. Most species of *Aedes* bite. *A. aegypti* is believed to be the only vector involved in the yellow fever and dengue fever epidemics which have occurred in this country; it is also capable of transmitting the three most important strains of encephalitis organisms found in the United States. Some species of *Aedes* have a single brood, but there may be several hatchings of overwintering eggs with various floodings. Large areas of salt marshes, pastures, and flood plains may be heavily seeded with eggs, causing severe infestations.

Culex pipiens, the common house mosquito, is primarily a tormentor of birds and poultry, and to a lesser extent of horses, cows, and dogs. It was considered the principal vector involved in the outbreak of St. Louis encephalitis which occurred in this country in 1933, and it can transmit filarial worms. *Culex* mosquitoes overwinter as fertilized females, breed continuously during warm weather or throughout the winter in warmer parts of the South.

The most important species of *Anopheles* is the eastern *A. quadrimaculatus,* the principal vector of malarial organisms. *A. crucians* (Fig. 1268B), which occurs in the eastern and southern coastal plains and the lower Mississippi Valley, is sometimes infected in nature but is not known to transmit the organisms; it is not entirely ruled out, however, as a possible vector. *A. freeborni* (*maculipennis*) is believed to be the chief carrier of malarial organisms in southwestern Canada and the western states; *A. albimanus* (Fig. 1268C) plays this role in the Caribbean area. Anopheline larvae are identified by the clypeal hairs (on the front, lower part of the head), the sutural hairs (higher up on the head), the lateral hairs on the fourth and

Distinction Between Various Genera of Mosquitoes (See Fig. 1261)

	Aedes	Culex	Anopheles
Breeding places	Damp soil—places periodically flooded by rainwater, tides, overflows: salt marshes, pastures, flood plains, tree holes, water containers.	More or less permanent collection of water: ditches, ponds, gutters, catch basins, eaves, troughs, cans; temporary collections of water.	Fresh water: permanent ponds, marsh pools with vegetation and debris, excepting A. atropos and A. bradleyi (salt or brackish water), A. albimanus (brackish or fresh water).
Eggs	Spindle-shaped, with hexagonal reticulations. Scattered over soil.	Capsule-like. Laid on surface of water in clusters or "rafts" of a hundred or more.	Boat-shaped, pointed both ends, with air sacs or floats on sides. Laid singly on surface of water.
Larvae	Have breathing tube with single pair of hair tufts near middle.	Have breathing tube with 4–7 pairs of ventro-lateral hair tufts or single hairs.	No breathing tube. Feeds parallel with surface of water, suspended by paired fan-shaped tufts or modified hairs (palmate hairs) on upper side of abdominal segments, and pair of transparent notched organs on thorax.
	Comb on 8th abdominal segment in form of patch or single row.	Comb scales in either single row or large patch.	
Adults	Rest with body parallel to resting surface.	Rest with body parallel to resting surface, as do Aedes.	Rest with abdomen and proboscis in straight line, pointed at angle with resting surface.
	Ornamented with white scales. Some have patches of white or yellow scales on thorax.		Abdomen clothed with fine hairs rather than flat scales. Wings with spots of dark and white scales.
	Some have proboscis banded with white rings; many have white-banded tarsi.	Unbanded proboscis and tarsi, excepting C. bahamensis, C. tarsalis and C. opisthopus.	Only A. albimanus has white-banded tarsi.
	Female palpi short; male palpi very short or about as long as proboscis.	Female palpi short, male palpi longer than proboscis.	Palpi as long as proboscis in both sexes, clubbed in males. All dark or with white rings.
	Tip of abdomen tapered (6th and 7th segments; 8th usually completely retracted); two short, slender cerci pro-	Tip of abdomen blunt.	Tip of abdomen blunt.

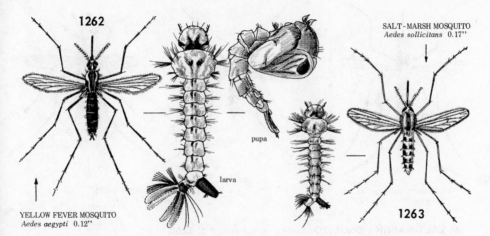

1262

SALT-MARSH MOSQUITO
Aedes sollicitans 0.17"

pupa

larva

YELLOW FEVER MOSQUITO
Aedes aegypti 0.12"

1263

fifth abdominal segments, and the palmate hairs (paired, fan-shaped tufts on the abdominal segments).

The genus *Psorophora* is closely related to *Aedes*. *P. confinnis* (*columbiae*), the **Florida Glades Mosquito** (or dark rice-field mosquito), is a severe biter, occurs throughout the Southeast but is most troublesome in the Florida Everglades and the rice-growing areas of Arkansas and Louisiana. A medium-sized, sooty black mosquito with white-ringed proboscis and tarsi, the females attack at night or in grassy and shaded areas during the day. They sometimes appear in large swarms but are not as aggressive as the salt-marsh *Aedes*. *P. ciliata*, **Gallinipper**, a large eastern species with wing spread of .6", is also a severe biter. It may be distinguished by the median longitudinal band of yellow scales on the mesonotum, with a narrow, bare polished band on each side, and by the bushy hind tibiae and tarsi. Not all bad, it breeds in rain pools and flood waters, where the larvae prey on other mosquito larvae.

(5) YELLOW FEVER MOSQUITO: *Aedes aegypti.* **Fig. 1262, Color plate 8f.**

The most thoroughly domesticated mosquito. **Adult:** Dark, mesonotum with four silvery lines, the outer pair curved to form a lyre-shaped pattern; head with narrow median lateral stripes, palpi of female short, white-tipped; abdominal segments with white bands, hind tarsi with conspicuous white bands, front and middle tarsi with narrow white bands on first two segments only; wing scales all dark. **Length** (wing): .12". **Larva:** Antennal hair single, head hairs single. Single curved row of scales on comb (eighth abdominal segment, where air tube and anal segment attach); each scale with long central spine, several shorter lateral spines. **Food** (adult female): Blood of man preferred. Besides the viruses of yellow fever and dengue fever, it transmits the nematode *Dirofilaria immitis,* the cause of heartworm in dogs, cats, and wild carnivores [see (8) below].

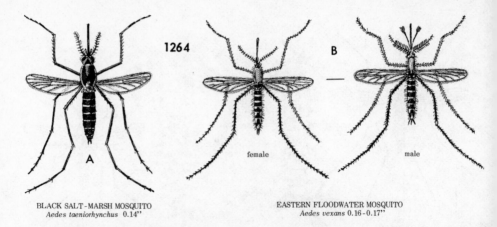

1264   B

female   male

BLACK SALT-MARSH MOSQUITO
Aedes taeniorhynchus 0.14"

EASTERN FLOODWATER MOSQUITO
Aedes vexans 0.16-0.17"

(6) SALT-MARSH MOSQUITO: *Aedes sollicitans.* **Fig. 1263.**

Range: Atlantic and Gulf coasts, from New Brunswick to Texas; also the Midwest, in areas flooded intermittently with salt water from oil or artesian wells, and in salt-water swimming pools. This is the notorious "New Jersey mosquito." **Adult:** Thorax golden brown; abdomen ringed with white or yellow bands, and having median white line the entire length; wide white rings on tarsi and proboscis, wings speckled with white and brown. **Length** (wings): .17". **Larva:** Head hairs single, antennal tuft multiple. Comb scales in patch, thornlike, fringed on each side toward base. **Food** (adult female): Blood of man and domestic animals. Fierce biters and strong fliers, they commonly migrate in large swarms 40 to 50 miles from any marsh, and attack in full sunlight. The **California Salt-marsh Mosquito,** *A. squamiger,* a large brownish gray species with ringed tarsi, breeds in salt marshes and tide pools along the Pacific coast; swarms often invade towns and cities. *A. nigromaculis* is fairly large, blackish, with yellowish mesonotum and white rings on tarsi, palpus all dark, proboscis sometimes with median band or patch of pale scales; a fierce biter, it breeds in standing irrigation water and rain pools, is an important pest in California and other western states.

(7) BLACK SALT-MARSH MOSQUITO: *Aedes taeniorhynchus.* **Fig. 1264A.**

Range: Atlantic and Gulf coasts from Maine to Mexico; all southeastern states except Tennessee; Pacific coast from southern California to Peru. **Adult:** Black, abdomen ringed with broad white bands; lacks longitudinal stripe of *sollicitans* (6) and *mitchellae* (a similar but uncommon southeastern species). Sharply contrasting white rings on proboscis and tarsi; lacks central ring on first tarsal segment of *sollicitans.* **Length** (wings): .14". **Larva:** Hairs on upper and lower part of head are single. Comb scale in a patch, each rounded at lower end, fringed on top with spinules. **Food** (adult female): Blood of man and domestic animals. Hard biters, they are migratory, dispersing many miles from breeding places, often in great

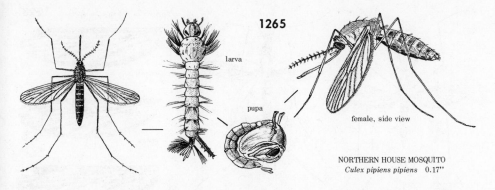

1265

larva

pupa

female, side view

NORTHERN HOUSE MOSQUITO
Culex pipiens pipiens 0.17"

swarms. They are the most abundant and troublesome mosquito along two-thirds of the Florida coasts. *A. dorsalis* is widely distributed in northern and western United States and Canada, breeds in brackish and fresh water, may be recognized by the white speckling of the proboscis, palpi, and legs, and the large area of pale scales on the head; it is an important pest of man and animals, the dominant species in many places. *A. vexans* (Fig. 1264B), a floodwater breeder in the East, has narrow white rings on tarsal segments, broad white bands on abdomen, unmarked proboscis and thorax. The **Floodwater Mosquito**, *A. sticticus,* occurs in the northern and western states and Canada under similar circumstances.

(8) NORTHERN HOUSE MOSQUITO: *Culex pipiens pipiens.* **Fig. 1265.**

Range: Northern states, Tennessee, North Carolina; northern half of South Carolina, Georgia, Alabama, Mississippi. **Adult:** Wings, proboscis, and tarsi dark; thorax pale brown or grayish; abdominal tergites (dorsal sclerites) have broad white basal bands continuous with lateral spots, their posterior borders fairly straight (not evenly rounded as in *C. p. quinque-fasciatus*). **Length** (wing): .17". **Larva:** Head hairs multiple, equal in size. Lateral hairs on third to fifth abdominal segments double; comb scales in patch, each rounded and fringed. **Food** (adult female): Blood of fowl, birds, domestic animals, man. It is regarded as the most abundant night-flying house mosquito of the northern states. The **Southern House Mosquito,** *C. p. quinquefasciatus,* is the most common mosquito in the Southeast, ranges west to Missouri, Kansas, Utah, and California. Like *Aedes aegypti* (5) and *Anopheles punctipennis* (9), *C. pipiens* is a vector of *Dirofilaria immitis,* a nematode causing heartworm in dogs [see also Siphonaptera (3)]; as noted above, it is also a vector of the virus causing St. Louis encephalitis and of a nematode causing filariasis in man. *C. tarsalis*—the principal vector of western equine encephalitis—is medium-sized, pale brown, with broad white ring near middle of proboscis, golden brown scales on mesonotum; abdominal segments with broad yellowish white basal bands dorsally, tarsi with white basal and apical bands.

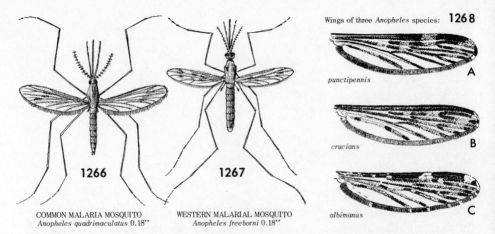

Wings of three *Anopheles* species: **1268**

punctipennis   A

crucians   B

albimanus   C

1266   **1267**

COMMON MALARIA MOSQUITO
Anopheles quadrimaculatus 0.18''

WESTERN MALARIAL MOSQUITO
Anopheles freeborni 0.18''

(9) COMMON MALARIA MOSQUITO: *Anopheles quadrimaculatus.* **Fig. 1266.**

Range: South of the Great Lakes: North Dakota to Quebec, Florida, Texas, and Mexico. **Adult:** Fairly large, dark brown, with four darker spots near center of wing; proboscis, palpi, thorax, and tarsi have dark scales, top of head has patch of pale scales. The angle at which the body is held at rest is not as pronounced as in other anophelines, may not be characteristic when heavy with blood. **Length** (wing): About .18''. **Larva:** Antennae have large, well-branched tuft near the middle. Inner clypeals usually simple, outer ones densely branched, posterior ones very small and forked at tip. Sutural hairs with 8 to 10 branches along shaft; palmate hairs well developed on third to seventh segments. **Food** (adult female): Blood of man and domestic animals; horses and cows are more attractive than sheep, goats, pigs, or dogs. The attractiveness of individuals varies. They disperse at dusk, are active again at daybreak, with occasional forays at night. The percentage of infected *quadrimaculatus* in malarious areas in this country is comparatively low. *A. freeborni* (*maculipennis*) (Fig. 1267), the western malarial mosquito, is medium-sized, blackish; palpi all black, wings with black spots and bronze patch at apex. *A. punctipennis* (Fig. 1268A) ranges from Canada to Mexico and from the Atlantic to the Pacific, may be recognized by the unbanded palpi and white spot on costal margin of the wing; it is common in fall and early spring, prefers margins of flowing streams for breeding; the females are active and bite mostly at night but are seldom found indoors. *A. punctipennis* is an intermediate host of *Dirofilaria immitis*, a cause of dog heartworm, but is not an important carrier [see (5) and (8) above]; it is not believed to be significant in the transmission of malaria.

Biting Midges (Ceratopogonidae = Heleidae)

The biting midges are small to minute mosquito-like flies that breed in aquatic and semiaquatic environments. They include some very annoying

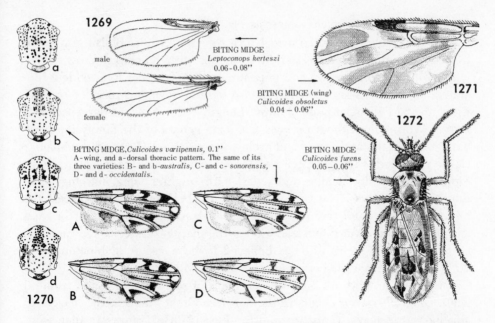

1269

male

BITING MIDGE
Leptoconops kerteszi
0.06–0.08"

female

BITING MIDGE (wing)
Culicoides obsoletus
0.04 – 0.06"

1271

1272

BITING MIDGE, *Culicoides variipennis*, 0.1"
A - wing, and a - dorsal thoracic pattern. The same of its
three varieties: B- and b-*australis*, C- and c- *sonorensis*,
D- and d- *occidentalis*.

BITING MIDGE
Culicoides furens
0.05–0.06"

A   C

B   D

1270

pests, variously called no-see-ums, punkies, sand flies, moose flies, and gnats. Some are vectors of organisms causing disease in horses, cattle, sheep, birds, and man; certain species in tropical countries transmit filarial worms to humans. A few take blood from wing veins of the Neuroptera, Lepidoptera, and Odonata; others attack turtles and lizards. A blood meal is required for egg production; both males and females often visit flowers for nectar. Most species lay their eggs in a gelatinous clump or string in the water or at its edge. The larvae develop in damp or wet places, in mud, at the edge of streams, lakes, and ponds, in salt marshes and water-holding plants; they feed on decaying organic matter and dead insects, or on newly hatched larvae of other flies and insects in the same media.

(10) BITING MIDGE: *Leptoconops torrens.*

Referred to locally as the valley black gnat. **Range:** California to New Mexico, Texas, and Colorado. **Adult:** Shiny black; venation very faint, female antenna 14-segmented. **Length:** .06–.08". Common along streams, a fierce biter. It requires two years to complete its life cycle. *L. (Holoconops) kerteszi* (Fig. 1269) occurs from Washington to Nebraska, south to Utah and California (where it is called the Bodega black gnat, since the larvae were first found in the humus-laden sand just above tide level at the mouths of the streams emptying into Bodega Bay);* it is also found in Egypt and Tunisia; the antenna is 13-segmented; the orange-colored larva is found around the edge of tidal marshes or saline lakes.

* Leslie M. Smith and Homer Lowe, "The Black Gnats of California," *Hilgardia,* 18(3):157–183 (1948), Univ. of California, Berkeley.

(11) BITING MIDGE: *Culicoides variipennis.* **Fig. 1270.**

Range: Widely distributed over North America. **Adult:** Bluish gray, with frosted appearance, brown markings; thorax with numerous brown dots; wings with irregular brown and gray streaks, legs banded. Eyes widely separated; antennae 15-segmented, the third to tenth segments rounded or oval. Eggs are laid in gelatinous clumps. **Length:** About .1″. A large species. **Larva:** Wormlike, without prolegs. (In some genera of the family the body has strong bristles and spines and prolegs are present.) **Host** (adult female): Cattle preferred. Five subspecies of *C. variipennis* are recognized; western forms are generally smaller and paler, with brighter wing markings than those in the East. *C. v. variipennis* (Fig. 1270Aa) is distributed over the forested northern and eastern regions of North America, breeds in the mud and pools around water tanks and along the margin of creeks and marshes. *C. v. australis* (Fig. 1270Bb) is believed to be a vector of bluetongue, a viral disease of sheep, occurs in the lower Mississippi Valley and Gulf Coastal Plain around salt marshes and pools of salt water pumped from oil wells. *C. v. albertensis* occurs in the northern Great Plains at the margin of alkaline cattle pools. *C. v. sonorensis* (Fig. 1270Cc) ranges farther west, breeds in mud at the margin of stock ponds and slow streams and in sewage effluent. *C. v. occidentalis* (Fig. 1270Dd) occurs at the edge of alkaline pools in the Pacific coastal region.

(12) BITING MIDGE: *Culicoides obsoletus.* **Fig. 1271.**

Range: Alaska to Quebec, south to northern California, Oklahoma, Tennessee, and Virginia; Europe. **Adult:** Shiny, brown; wings with two light spots on costal margin, other markings diffused, hairs confined to apical margins. Eyes close together; antennae as in (11). **Length:** .04–.06″. *C. furens* (Fig. 1272) is a salt marsh pest, occurs along the Gulf and Atlantic coasts to Massachusetts and Cuba. Species of *Culicoides* have a tendency to bite humans around the head, especially at the edge of the hat band and collar.

Midges (Chironomidae = Tendipedidae)

The midges are nonbiting and generally harmless, except for the occasional annoyance of large swarms. They are minute to small, delicate flies (sometimes mistaken for mosquitoes) with long, slender bodies and narrow, short wings (Fig. 1273A). They often "dance" in large swarms over the water, inciting the fish to jump. Midges may be distinguished from biting midges by the posteriorly flattened head, the reduced mouthparts (without mandibles), the long plumose antennae of the male (Fig. 1273B) (the flagellum—the portion beyond the two basal segments—is 11- to 14-segmented in the male, 5- to 14-segmented in the female, nearly always 13-

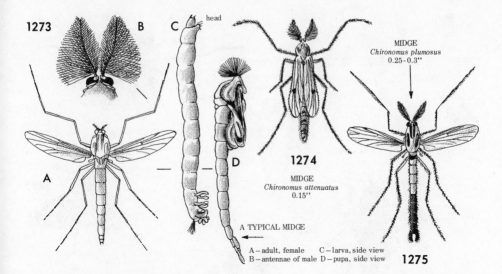

1273

B C head

A

D

MIDGE
Chironomus plumosus
0.25-0.3"

1274

MIDGE
Chironomus attenuatus
0.15"

A TYPICAL MIDGE

A—adult, female   C—larva, side view
B—antennae of male D—pupa, side view   **1275**

segmented in both sexes of the biting midges); the thorax often projecting over the head, with a median furrow or keel posteriorly (but without the V-suture), tibiae usually with double comb. As in the Ceratopogonidae, the eggs are usually laid in gelatinous strings or in gelatinous masses.

The larvae are mainly aquatic (Fig. 1273C), mostly in fresh water, and constitute an important source of food for fish; they inhabit environments as diverse as torrential streams and the anaerobic ooze of deep lakes. A substantial number are predaceous. Some build silk-lined tubes from sand, mud, and debris, and feed from within the case; in cultivated rice fields they sometimes damage seedlings. They are slender, wormlike, with 12-segmented body; the head capsule is not retractile; prolegs are usually present on the prothorax and last abdominal segment. Some pupae (Fig. 1273D) are free-living. The genus *Chironomus* (*Tendipes*) contains the larger and better-known species; their red larvae are called bloodworms because (unlike most other insects) the blood contains hemoglobin.

(13) MIDGE: *Chironomus attenuatus.* **Fig. 1274.**

Range: British Columbia to Hudson Bay and Quebec, south to California and Florida. This species (synonymous with *C. decorus*) is the most widespread and abundant of the larger midges. **Adult:** Pale green to fawn or light brown, with antennae, mouthparts, thoracic markings, and abdominal bands dark brown; wing venation pale brown. It breeds in lakes, ponds, rivers, streams, also sewage oxidation ponds, irrigated rice fields. Males swarm in early evening. Egg masses are deposited on the hind femora of the female and thrown into the water. **Length:** .22"; wing: .15". **Larva:** Inhabits silty ooze at the bottom, feeds on plankton from within a silken tube.

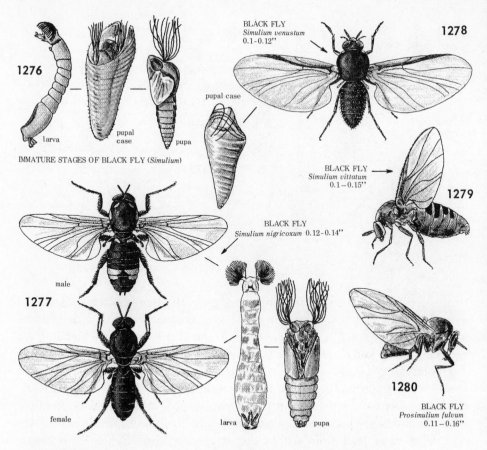

1276

larva | pupal case | pupa

IMMATURE STAGES OF BLACK FLY (*Simulium*)

pupal case

BLACK FLY
Simulium venustum
0.1–0.12″

1278

BLACK FLY
Simulium vittatum
0.1–0.15″

1279

BLACK FLY
Simulium nigricoxum 0.12–0.14″

male

1277

female | larva | pupa

BLACK FLY
Prosimulium fulvum
0.11–0.16″

1280

(14) MIDGE: *Chironomus plumosus.* **Fig. 1275.**

Range: British Columbia to Quebec, south to California and North Carolina; Greenland, Bermuda, Europe. **Adult:** Rather stout; straw-colored or light brown, often tinged with green; antennae, mouthparts, thoracic markings, abdominal bands light to dark brown; abdomen green, without markings in some western specimens; anterior wing veins light brown. **Length** (wing): .25–.3″. Largest species in the family; the body of some males is more than .5″ long.

Black Flies (Simuliidae)

This family—also referred to as buffalo gnats and turkey gnats—is of considerable economic and medical importance. Adult females of most species are bloodsucking, attack domestic and other warm-blooded animals, birds, and man; like mosquitoes, both sexes feed on nectar. Some transmit diseases of birds; others in the tropics carry a filarial parasite (*Onchocerca volvulus*) causing onchocerciasis, one form of which may cause blindness. Common species transmit blood parasites to turkeys and ducks, sometimes causing serious loss. They are gray-brown to black with broad wings, stout bodies, head pointed down, without ocelli, thorax arched, giving a humpbacked appearance. Immature stages are aquatic and found attached to stones, sticks, and vegetation in flowing streams. The smooth whitish yellow eggs (later turning almost black) are found in masses on vegetation or rocks or scattered loosely on the bottom of streams.

The larvae (Fig. 1276) attach themselves by means of a posterior sucker, and point the head downstream; most of them have a pair of mouth brushes (cephalic fans) to strain food particles from the water. The body is soft, pouchlike, enlarges at the posterior third, has anal gills near the tip. On the underside of the first thoracic segment is a proleg with a circlet of hooks; it moves about with the aid of this prothoracic leg and by undulating the posterior sucker, or by letting out a silken thread. Cocoons are pocket- or boot-shaped, with thoracic respiratory filaments (tracheal gills) of the pupa protruding (Fig. 1276). The adult emerges in an air bubble if the cocoon is still submerged. Larvae and cocoons often cover rocks in mosslike masses, and sometimes cover the submerged surface of irrigation canals and flumes, seriously impeding the flow of water.

(15) BLACK FLY: *Simulium venustum.* **Fig. 1278.**

Range: Alaska to Greenland, south to Texas, Mississippi, and South Carolina. **Adult:** Shiny, black; female with 11-segmented antennae, basal segments reddish brown, yellowish pubescence on thorax, legs mostly yellowish; male darker, legs marked with yellow and white. **Length** (wing): .1–.12″. They fly from May to October, are most numerous in June and early July. Immature forms are found in small streams (permanent and temporary), rivers, drainage and irrigation ditches. The pupa has six respiratory filaments protruding from the closely woven pocket-like cocoon. Hibernates in larval stage. Three or four broods. **Host** (adult female): Horses, cattle, sheep, man, deer, birds. It is an intermediate host of a protozoan blood parasite of ducks, *Leucocytozoon anatis;* mortality from the disease is often high among ducklings. *S. jenningsi* (*nigroparvum*) occurs from Manitoba to Maine, south to Texas and Florida; it is glossy black, mesonotum gray pollinose in the female; palpi longer than antennae, latter 11-segmented, halteres pale yellow; legs yellowish in the female, brownish to black in the male; each of two respiratory tufts of the pupa has ten filaments; it feeds on turkeys, is an intermediate host of *Leucocytozoon smithi,* which causes a blood disease in turkeys. *S. nigricoxum* (Fig. 1277) occurs in the Far North—in Alaska and Canada.

(16) BLACK FLY: *Simulium vittatum.* **Fig. 1279.**

Range: From the Arctic to Mexico and from the Atlantic to the Pacific; Europe, Greenland. One of the most common and widely distributed species. **Adult:** Black; female with opaque, gray pollinose coating, sparse pale pubescence, 11-segmented antennae, the two basal segments brownish, five dark brown to black longitudinal stripes on mesonotum, basal half of tarsi yellow; male much darker, more slender. **Length** (wing): .1–.15″. It breeds in almost any flowing water, often in great numbers. Hibernates in egg or larval stage. The pupa has sixteen respiratory filaments. One to five broods, depending on the latitude. **Host** (adult female): Chiefly horses and cattle; they swarm about man but don't normally bite.

(17) BLACK FLY: *Prosimulium mixtum.*

Range: Ontario to Labrador, south to New Brunswick, Nova Scotia, Quebec, New York; possibly Wisconsin and Minnesota. **Adult:** A dark species, member of the so-called *hirtipes* complex; basal segments of antennae only slightly paler than the rest; thorax with yellow pile; sides light brown, mottled, or all dark brown; abdomen dark brown. It emerges in early spring or summer. Eggs are scattered in the water and sink to the bottom. **Length** (wing): .1–.15″. Immature stages found in cold, small temporary streams. Hibernates in egg or larval stage. Usually one brood. **Host** (adult female): Man, horses, deer. This and closely related species are fierce biters, often very abundant. *P. mixtum* is difficult to distinguish from other members of the complex: *P. fuscum*—a larger dark species, from .12″ to .16″ long—and *P. fontanum*, a yellowish species (found only in Ontario and Quebec). The pupae in this complex have sixteen respiratory filaments. *P. fulvum* (Fig. 1280)—a bright yellow species with simple claws, from .11″ to .16″ long—ranges widely from Alaska southward to California and Colorado.

Gall Midges (Cecidomyiidae = Itonididae)

The gall midges comprise a large family of delicate, slender flies with broad wings and long legs. The coxae are not long, the tibiae are without spurs, and the wings usually have three (five at most) long veins and no cross-veins. Probably not more than two-thirds of the gall midges cause galls on plants. Many live on decaying organic matter, others live as inquilines or guests in the deformities caused by other insects, or are predaceous. Paedogenesis—the parthenogenetic reproduction of larvae which give birth to similar larvae—occurs in a few genera. A number of species are of considerable economic importance, but the great majority are not. The Hessian fly is one of the most destructive pests of wheat.

(18) HESSIAN FLY: *Mayetiola (Phytophaga) destructor.* **Fig. 1281.**

Range: Wheat-growing areas of North America. It is believed to have come in with the straw bedding of Hessian troops during the Revolutionary War. **Adult:** Sooty black, abdomen of female reddish orange. Reddish, elongated eggs are laid in the grooves on upper side of leaf. **Length:** .12″.

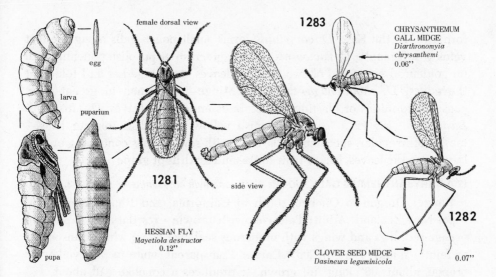

egg

larva

puparium

pupa

female dorsal view

1281

side view

HESSIAN FLY
Mayetiola destructor
0.12"

1283

CHRYSANTHEMUM
GALL MIDGE
*Diarthronomyia
chrysanthemi*
0.06"

1282

CLOVER SEED MIDGE
Dasineura leguminicola
0.07"

Larva: Reddish when newly emerged, turns white or greenish white; shiny, legless, a little less than .2" long full grown. It works its way down to lower part of plant, behind the leaf sheath, usually below the ground, sucks plant juices after rasping the straw; the weakened plant falls to the ground. Hibernates as full-grown maggot in brown puparium (the so-called flaxseed stage) hidden in sheath of newly sown wheat or stubble. Normally two broods—spring (March) and fall (October); there may be three to five broods under the most favorable conditions; generally a single generation in California. **Food:** Wheat, barley, rye, some wild grasses. Late fall planting (following the so-called fly-free date), after adults have emerged and eggs are laid, reduces their number. [See Hymenoptera (44).]

(19) CLOVER SEED MIDGE: *Dasineura leguminicola.* **Fig. 1282.**

Range: British Columbia to Nova Scotia, south to Oregon, New York, Virginia. **Adult:** Thorax black, abdomen bright red; wings dusky, fringed with hairs. The male is more slender and fragile than the female, with shorter abdomen; the female has a long ovipositor. Oval, pale yellow eggs, which turn orange later, are laid singly or in clusters on hairs of calyx or inside leaf sheath near head. **Length:** .07". **Larva:** Pale yellow with bright orange spot, changing to creamy white and finally pink; about .1" full grown. The young larva wriggles to top of flower and inside unopened petal to the ovary, where it sucks sap and prevents seed from forming; injured heads resemble those infested by the clover head caterpillar [see Lepidoptera (270)]. Hibernates as larva in white silken cocoon on the ground or crawls under debris without forming a cocoon, pupates in the spring. Three broods in the Northwest. **Food:** Mostly red clover. These midges, if abundant, can destroy an entire seed crop but do not impair its value as forage. [See Hymenoptera (44).] Many other species of *Dasineura* cause plant deformities. In the East, *D. communis* forms small pouch galls along the leaf veins of soft maple; *D. flavicornis* forms a greenish or purplish pouchlike gall on goldenrod composed of two or more leaves stuck together and folded at the edges; *D. rhodophaga*—the **Rose Midge**—is common in greenhouses, causes deformities or blasting of rose buds and young leaves. *D. vitis*, with *Lasioptera vitis* (21), causes greenish yellow to reddish "tomato galls"— masses of irregular, succulent galls, each with several cells containing orange larvae—on the leaves, leaf or fruit stalks, and tendrils of grape.

(20) CHRYSANTHEMUM GALL MIDGE: *Diarthronomyia chrysanthemi.* **Fig. 1283.**

Range: Oregon to Quebec, south to California, and Rhode Island; Europe, New Zealand. **Adult:** Pale yellowish brown; eyes dark red, abdomen orange-red, legs and wings with numerous scalelike hairs; wings transparent or faintly smoky. **Length:** .06″. **Larva:** Transparent white to pale yellow or orange, about .06″ long full grown. It produces a conelike gall about .12″ high on leaf, leaf petiole, stem, or bud, causing the shoot to become deformed. Hibernates as larva in the gall. Two broods. **Host:** Chrysanthemums, outdoors or in greenhouses.

(21) SORGHUM MIDGE: *Contarinia sorghicola.* **Fig. 1284.**

Range: Kansas to Virginia and southward to New Mexico, Texas, and Florida; Hawaii, Australia, India, Sudan. **Adult:** Orange-red; head and palpi yellow, antennae brown (as long as body in the male, half as long in the female); black spot on mesonotum, legs brown, wings transparent grayish, female with long ovipositor. Tiny white eggs are laid in the spikelets or seed husks. **Length:** .08″. **Larva:** Oval; pink, changing to orange. Sucks juice from developing seed, causing it to dry up; 8 to 10 larvae may mature in one seed. The pupa is dark orange. Hibernates as larva in light brown cocoon in spikelet, pupates in spring; some do not emerge until the second or third spring, many do not form a cocoon and die in the winter. Thirteen broods in southern Texas. **Food:** Sweet sorghums, sudan grass, broomcorns; most serious pest of grain sorghums, causes great loss to seed crops of all three. A chalcid wasp, *Eupelmus popa* (Eupelmidae), attacks the larvae and pupae. The **Pear Midge,** *C. pyrivora,* causes deformed fruit with dark blotches, and their premature drop; eggs are laid among the blossoms, and the tiny whitish or orange-colored larvae feed inside the young fruit. Numerous species of *Contarinia* cause various kinds of plant deformities, among them the eastern *C. spiraeina,* which causes "cabbage" bud galls on spirea; *C. johnsoni,* the **Grape Blossom Midge,** which causes grape buds to swell and turn yellow-green or reddish; and *C. negundifolia,* which causes midrib or vein swellings on leaves of box elder. The larvae of *C. washingtonensis* tunnel the cone scales of Douglas fir; those of *C. coloradensis* form budlike apical swellings on the needles of ponderosa pine. The closely related *Lasioptera cylindrigal-*

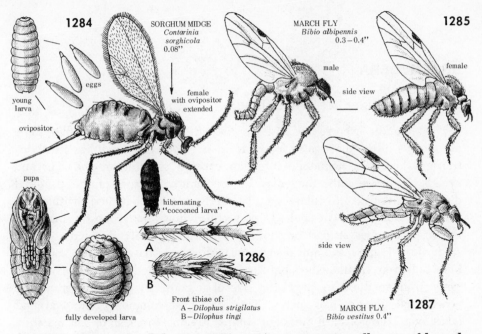

1284 SORGHUM MIDGE
Contarinia
sorghicola
0.08"

eggs

young
larva

ovipositor

female
with ovipositor
extended

pupa

hibernating
"cocooned larva"

A

B

1286

Front tibiae of:
A—*Dilophus strigilatus*
B—*Dilophus tingi*

fully developed larva

MARCH FLY **1285**
Bibio albipennis
0.3–0.4"

male   female

side view

side view

MARCH FLY **1287**
Bibio vestitus 0.4"

lae (Fig. 520) and *solidaginis* (Fig. 519) cause stem galls on goldenrod. [See Diptera (60), and Lepidoptera (285) for other gallmakers on goldenrod.] *L. vitis* causes galls on grape [see (19) above].

(22) MARCH FLY: *Bibio albipennis* (Bibionidae). **Fig. 1285.**

March flies appear in large numbers early in the spring. They are small to moderate in size, usually quite hairy and dark in color. They have a distinctive head—longer and narrower in the female, almost covered by the eyes in the male. The latter's enormous eyes are divided into two parts, the lower third having smaller facets. Ocelli are present. The front tibiae usually have large apical spurs; in the genus *Dilophus* (Fig. 1286) they have two or three widely separated rows of spurs that look like claws. The female digs a burrow one to four inches deep, deposits an egg mass of 200 to 300 eggs, dies soon after. The larvae are from .25" to 1" long, 12-segmented, yellowish, with darker spiracles and shiny brown head. They are primarily scavengers, live on decaying organic matter among plant roots. Some have been reported as causing crop damage; they sometimes enter potatoes already damaged by wireworms. Adults are attracted to flowers and are at times important pollinators. **Range:** *B. albipennis* occurs throughout North America. **Adult:** Shiny, black; body with dense pale golden to gray pile; wings clear, whitish, veins and stigma yellowish brown; inner spur of front tibiae short (about one-third as long as the outer one). **Length** (wing): .3–.4". *B. vestitus* (Fig. 1287) is similar, widely distributed from Nova Scotia to California. *Dilophus strigilatus* occurs from Alaska to California and Utah, *D. tingi* in California (see above).

BRACHYCERA

Deer and Horse Flies (Tabanidae)

The tabanids, or horse flies and deer flies, are also known as gad flies; some are vicious biters that cause injury to man. Species of *Chrysops* in Africa transmit loiasis, a nematode infection causing swelling of the skin and eyelids; with *Tabanus*, they may be mechanical vectors of the bacterial diseases anthrax and tularemia. Tabanids are stout-bodied, with huge heads (mostly eyes in males), powerful wings, body mostly black or brown, sometimes striped or spotted. The antennae are distinctive—short, stout, three-jointed, the apical segment with a large basal division and from four to seven smaller divisions tapered and pointed apically. The flies range from medium to large—the size of a house fly to a wingspread of nearly two and one-half inches. They are most commonly found around lakes, bogs, the northern muskeg swamps, streams, brackish or salt marshes, desert springs, beaches, irrigated land, and ponds. The eggs of most species are laid in masses on vegetation over or near the water.

The larvae are elongate, whitish, those of *Tabanus* often marked with rings of black; they usually mature in damp or wet soil and litter, though some are found in running water, others in dry situations; they are often predaceous and cannibalistic. The adult females of most species are presumed to take a blood meal, but some are nonbiting, occasionally with mandibles reduced or absent. As with other biting flies, sugar is also necessary for the female, and both sexes feed on nectar or other exudations. Most species attack the larger mammals, but others feed on monkeys, crocodiles, lizards, turtles. None has been recorded as feeding on amphibians or on birds.

(23) BLACK HORSE FLY: *Tabanus atratus.* **Fig. 1288.**

Range: Washington to Quebec, south to New Mexico and Florida. **Adult:** Jet black; thorax with whitish or yellowish pubescence, abdomen often with bluish white tinge; wings black to smoky brown, unspotted. As in other species of *Tabanus*, ocelli are absent, the third antennal segment has five divisions, the hind tibiae are without apical spurs. Eggs are laid on end in masses, on leaves or stems of aquatic plants and trees overhanging the water. **Length:** .8–1.1″. **Larva:** Very striking in appearance—whitish, banded with black, pointed at both ends, elevated ring on each body segment, about 2″ long full grown. It buries itself in the mud, lives on insects or other small animals. Hibernates as larva, sometimes for two winters, pupates in the spring. **Host** (adult female): Wild and domestic mammals: cattle,

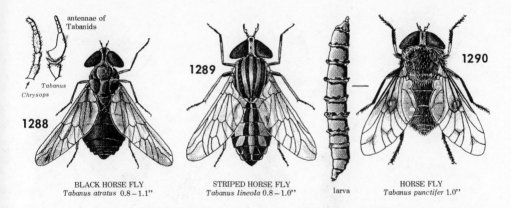

antennae of
Tabanids

Tabanus
Chrysops

1288

1289

1290

BLACK HORSE FLY
Tabanus atratus 0.8 – 1.1"

STRIPED HORSE FLY
Tabanus lineola 0.8 – 1.0"

larva

HORSE FLY
Tabanus punctifer 1.0"

horses, mules, hogs, and other animals. The flies alight on the backs, necks, and heads of their hosts, slit the skin with their knifelike mouthparts, and suck blood for several minutes, leave with blood oozing from the wound; the loss of blood from these flies is debilitating to some animals. The **Striped Horse Fly,** *T. lineola* (Fig. 1289), ranges over most of North America; the eyes have three green stripes on a purple background; thorax distinctly striped; abdomen dark slate gray to light brown, the median stripe narrow or quite wide, chalky white; dorsolateral stripes wide and distinct to almost obsolete, reddish yellow. *T. punctifer* (Fig. 1290) occurs from British Columbia to Kansas, south to California and Texas, is about 1" long, with whitish or golden pile on thorax, and usually a small dark spot on the wings. *T. sulcifrons*—the most harmful horse fly attacking cattle in southern Illinois—is about 1" long, reddish brown, with row of whitish or yellowish triangles down the middle of the abdomen.*

(24) DEER FLY: *Chrysops callidus.* **Fig. 1291.**

Range: British Columbia to Maine, south to Texas and Florida. **Adult:** Black, with yellowish markings; callus below the eyes yellow; abdomen mainly yellowish, black V-shaped mark on second segment reaching the anterior margin of the segment; wings marked with dense smoky black. Like other species of *Chrysops,* it has brilliant green or golden eyes with zigzag stripes; ocelli are present, the antennae more slender than in *Tabanus,* the third segment with five divisions, hind tibiae with two apical spurs [compare with (23) above]. Black shiny eggs are laid in masses of 100 or more on vegetation near the water. It is found in open woodlands near standing water, flies during June and July. **Length:** .4–.5". **Larva:** Yellowish white or greenish, abdomen with brown annuli, about .5" long. **Host** (adult female): Wild and domestic mammals, man. A pest of cattle and horses. *C. excitans* ranges from Alaska to Labrador, south to

* See Willin N. Bruce and George C. Decker, *Control of Horse Flies on Cattle,* Biol. Notes No. 24, Illinois Nat. Hist. Survey, Urbana (1951).

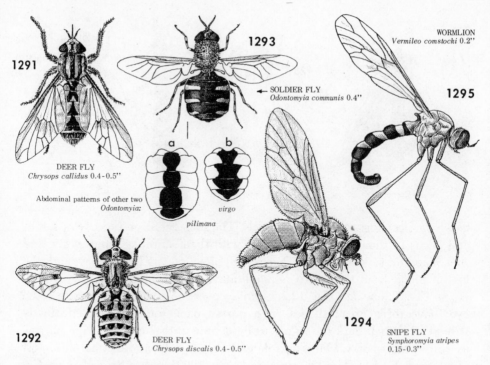

1291

DEER FLY
Chrysops callidus 0.4-0.5"

Abdominal patterns of other two
Odontomyia:

a

b

pilimana

virgo

1293

← SOLDIER FLY
Odontomyia communis 0.4"

WORMLION
Vermileo comstocki 0.2"

1295

1292

DEER FLY
Chrysops discalis 0.4-0.5"

1294

SNIPE FLY
Symphoromyia atripes
0.15-0.3"

California and New Jersey; the first and second abdominal segments are largely yellow, with a thick black V-shaped mark on the middle of the second segment; the male is almost wholly black, with gray median abdominal triangles; it is found in damp woods during June and early July. *C. discalis* (Fig. 1292), a western species, is a serious pest of cattle near lakes, causing as much as 50 percent loss in milk production during the peak of the deer fly season;* it is also a vector of bacteria causing "deer fly fever," a type of tularemia infecting man (reservoirs of infection are believed to be its favored host, rabbits.)† *C. niger* is numerous in mid-May along the Atlantic Coastal Plain.‡

(25) SOLDIER FLY: *Odontomyia communis* (Stratiomyidae). **Fig. 1293.**

The soldier flies are small to moderate in size, often brightly colored, and characterized by a soft pubescence and absence of bristles. They are found mainly in wooded or forested areas or in grassy fields near water. The

* See Marion Bacon, *A Study of the Arthropods of Medical and Veterinary Importance in the Columbia River Basin,* Tech. Bul. 11, Agr. Exp. Sta., State Col. of Washington, Pullman (1953).

† The bacteria (*Pasteurella tularensis*) causing tularemia are also transmitted by certain ticks and possibly fleas.

‡ Red or black helium-filled weather balloons, floating four to six feet above the ground, proved very attractive to these flies in the Georgia coastal area; covered with a sticky substance, the balloons are an effective trap with possibly some practical value in recreational areas. See E. L. Snoddy, "Trapping Deer Flies with Colored Weather Balloons," *Jour. Georgia Ent. Soc.,* 5(4):207 (Oct. 1970).

adults frequent flowers, and the larvae are terrestrial or aquatic. The aquatic larvae develop in sluggish streams or in ponds, feed on algae, microorganisms, and decaying vegetation. They are elongated, usually spindle-shaped and somewhat flattened; the caudal or last segment has a breathing tube, the slit opening fringed with conspicuous hairs, and the second to last segment in some species (*Odontomyia* and *Hedriodiscus*) has two to four strong, curved hooks on the underside. **Range:** *O. communis* ranges from Washington to Michigan, south to California and Texas. **Adult:** Quite variable; head yellow, top with transverse black band in the female, face of male black; dorsum of thorax black, sides yellow or so marked; legs yellow; abdomen yellow, with black bands, usually confined to basal half of segments in the female and to median third in the male, the black sometimes predominating in either case. **Length:** .4″.

(26) SNIPE FLY: *Symphoromyia atripes* (Rhagionidae). **Fig. 1294.**

The Rhagionidae, or snipe flies, small to medium in size, are sluggish, robust flies with short antennae, long legs, abdomen often pointed. Some resemble the yellow-banded wasps and bees. The bloodsucking habit is widespread in the family; species of *Symphoromyia* and *Suragina* attack mammals, including man, and are vicious biters. The larvae of snipe flies are all terrestrial, so far as known, excepting the genus *Atherix,* and are commonly found in rotten wood or decaying vegetation. Adults and larvae are predaceous. Species of *Vermileo,* called "wormlions," dig sand pits much like those of the antlions to trap other insects; one of them, *V. comstocki* (Fig. 1295), lives in the High Sierras. Certain species of *Atherix* and *Rhagio* frequent flowers and honeydew. The females of *Atherix variegata,* which occurs throughout Canada and the United States, congregate in a dense mass, suspended from a limb or rock overhanging the water. Here they deposit their eggs and die; other females are attracted to the cluster and add their eggs. The larvae drop to the water as they hatch and undergo development there. *S. atripes* inflicts painful bites on man. **Range:** Alaska to Alberta, south to California and Colorado. **Adult:** Black with reddish legs, hind tibia with one spur. **Length:** .15–.3″. *S. hirta*— about the same size, black with yellow markings—ranges from Alberta to Pennsylvania, south to New Mexico and Alabama. *Rhagio dimidiatus* (Fig. 1296) is a common western species; the hind tibia has two terminal spurs, anal cell of wing is open.

Stiletto Flies (Therevidae)

The stiletto flies are small to medium in size, slender, with varied colors and tapering bodies resembling some robber flies. They are not very active, making short but swift flights, and are seen on blossoms of shrubs and on

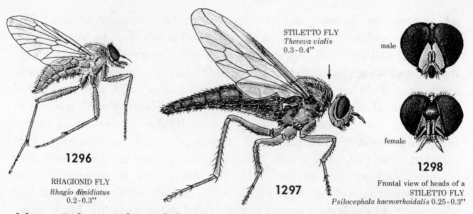

STILETTO FLY
Thereva vialis
0.3-0.4"

male

female

1296

RHAGIONID FLY
Rhagio dimidiatus
0.2-0.3"

1297

1298

Frontal view of heads of a
STILETTO FLY
Psilocephala haemorrhoidalis 0.25-0.3"

foliage. It has not been definitely established that they are predatory; the fleshy proboscis is not adapted for this habit. The curious larvae are long (1" or more) and slender, appear to have twenty segments but in reality have six or less; they live in the soil, in decaying vegetation and rotten wood, and so far as known are predaceous and often cannibalistic.

(27) STILETTO FLY: *Thereva vialis.* **Fig. 1297.**

Range: California, Oregon, British Columbia, and Manitoba. **Adult:** Black, with yellowish pile; front of head yellowish brown pollinose; antennae with white pile in the male, black with the first two segments gray pollinose in the female; mesonotum dark gray pollinose, with two pale longitudinal stripes, scutellum with two pairs of black marginal bristles; wings clear or whitish; stem of halteres dark gray, knob yellowish in the female and white in the male. **Length:** .3-.4". *T. candidata,* an eastern species, is very similar.

(28) STILETTO FLY: *Psilocephala haemorrhoidalis.* **Fig. 1298.**

Range: Michigan to New York, south to Alabama and Texas. **Adult:** Black, with white pile; antennae black, the first two and base of third segments with black bristles; thorax black, with bluish gray pollinose, median black stripe with a faint white stripe on each side; scutellum with two pairs of black bristles; base of tibiae and basal segments of tarsi yellow; abdomen black, with gray pollinose; wings clear, stigma and veins brownish; stem of halteres yellowish, knob blackish. **Length:** .25-.3". *P. frontalis* is very similar, occurs farther north, from Wyoming to Quebec, south to Louisiana, New York, New Jersey. *P. aldrichi* (Fig. 1299) occurs from British Columbia to Manitoba, south to California and Colorado, and in New Jersey.

(29) WINDOWPANE FLY: *Scenopinus (Omphrale) fenestralis* (Scenopinidae). **Fig. 1300.**

The habits of the scenopinids are not well known, but the larvae are believed to be all predators upon various insects. **Range:** *S. fenestralis* is wide-

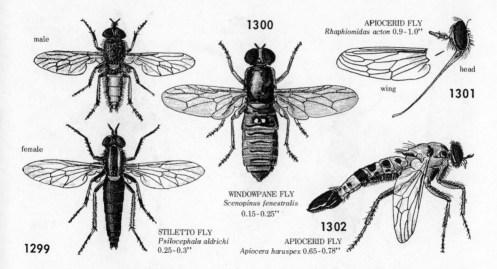

male

female

1300

APIOCERID FLY
Rhaphiomidas acton 0.9-1.0"

head

wing

1301

WINDOWPANE FLY
Scenopinus fenestralis
0.15-0.25"

1302

STILETTO FLY
Psilocephala aldrichi
0.25-0.3"

APIOCERID FLY
Apiocera haruspex 0.65-0.78"

1299

spread over North America and most of the world through commerce. **Adult:** Black; antennae brownish black; thorax clothed with short recumbent pile; knobs of halteres brownish above, white below; legs yellowish red, tarsi brownish; abdomen brownish black. Eyes narrowly separated in the male; front finely wrinkled, with distinct median furrow in the female. They are frequently seen around the windows of mills and warehouses. **Length:** .15–.25". **Larva:** Threadlike, white, about .75" long. They are common in accumulations of flour and grain dust. **Food:** Stored food pests, household pests, bark beetles, powder-post beetles, other insects. *S. glabrifrons* is similar, but slightly smaller, with front smooth and polished, eyes distinctly separated in the male, halteres all white; it is also widespread, similar in habits.

(30) APIOCERID FLY: *Rhaphiomidas acton* (Apioceridae). **Fig. 1301.**

The apiocerids are largely restricted to the arid and semiarid regions of the West. The striking adults make a loud noise in flight, and have the habit of hovering. Little is known about the immature stages. A species closely related to the widespread *Apiocera haruspex* (Fig. 1302) is known to lay eggs in loose sandy soil at the base of desert plants. **Range:** *R. acton* occurs in California. **Adult:** Robust, dark brown or black, clothed with dense yellow pile; abdomen orange, first segment all black, second to fifth segments with median basal dark spots, getting progressively larger toward the apex; abdominal pile shorter, more sparse in the female. **Length:** .9–1".

(31) MYDAS FLY: *Mydas clavatus* (Mydidae). **Fig. 1303.**

This species belongs to a small family commonly called mydas flies. They are elongated, moderate to very large in size, resemble wasps and the next group, the robber flies. Some are among the largest Diptera. *M. heros* of

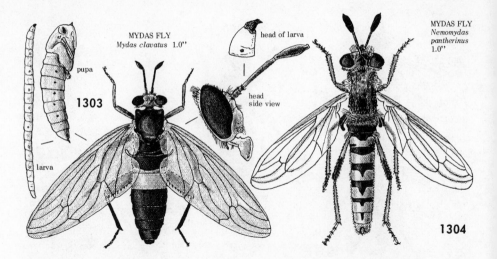

MYDAS FLY
Mydas clavatus 1.0"

pupa

1303

larva

head of larva

head
side view

MYDAS FLY
Nemomydas
pantherinus
1.0"

1304

Brazil is more than 2" long, with a wingspread of 4". Adults and larvae are said to be predaceous. They may be distinguished from other flies by the four-jointed, clubbed antennae and the peculiar wing venation, with the fourth vein curved forward and ending at or before the apex. **Range:** *M. clavatus* ranges throughout North America. **Adult:** Black, abdomen marked with red (second segment red on dorsum only), body without bristles; the large hind legs are black, with red and yellow markings. **Length:** About 1". **Larva:** Smooth, ivory white, 1.6" long full grown. It is found in decaying wood, preys on the larvae of Coleoptera found in the same medium. The pupa is reddish brown and bright red, 1" long, with two horns at the apex. *M. maculiventris* is said to be "an important larval predator of phytophagous scarab beetle larvae in Florida." *Nemomydas pantherinus* (Fig. 1304) is found in California and the Pacific Northwest.

Robber Flies (Asilidae)

The asilids, or robber flies, are large, powerful flies, usually with robust, hairy body and long slender abdomen. They are generally gray, but some are bright black and yellow, and can be mistaken for bees or wasps by their buzzing and droning. Sun-loving, they are found in dry, open fields and pastures, in sandy areas, more especially at the edge of woods, and on the range.* The adults capture all kinds of flying insects on the wing, and are sometimes destructive to bees. They have a small fovea between the eyes, three ocelli, and a stout beak with which they pierce their victims and drain them dry. Many have a beardlike brush of hairs (the mystax) above the mouth. Eggs are laid in the soil or on plants, or scattered over the ground.

* Robber flies are reported to play an important role in destroying grasshoppers on rangeland. See R. J. Lavigne and R. E. Pfadt, *Parasites and Predators of Wyoming Rangeland Grasshoppers*, Sci. Monog. 3, Agr. Exp. Sta., Univ. of Wyoming, Laramie (1966).

The larvae live in the soil or in wood and prey mostly on white grubs or pupae of Coleoptera; they are also reported to feed on grasshopper egg pods. They are cylindrical or slightly flattened, distinctly segmented, with the underside of the thoracic segments having two long hairs on each side. More than a year is generally required to complete the life cycle. Robber flies as a whole are considered beneficial, and probably rank with the dragonflies as predators. Some species of *Cyrtopogon* (Fig. 1305A) are said to perform a mating dance.

(32) ROBBER FLY: *Efferia (Erax) pogonias (barbatus).*

Range: Widely distributed throughout North America. **Adult:** Black; light grayish brown pollinose; mystax grayish, with black; abdomen of female white pollinose, third to sixth segments each with a pair of black spots; abdomen of male with segments one and two and posterior margins of three to five white pollinose, six and seven silvery; tibiae orange excepting black tips. They are common in late summer, make a high-pitched buzzing sound. **Length:** .5–.8″. **Larva:** Slender, shiny, white. Hibernates as larva in the soil. The adults feed on a wide variety of small insects—halictid bees, winged ants, flies, butterflies, moths, bugs. The larvae of *E. interrupta* (Fig. 1305B)—a species widespread in North America—are valuable predators of white grubs, and the adults are predaceous on adult horse flies, the larvae of which also prey on white grubs. *Efferia* spp. prey on rangeland grasshoppers, which are also the principal food of *Stenopogon* spp.

(33) ROBBER FLY: *Promachus fitchii.*

Sometimes called the "Nebraska bee-killer." **Range:** Nebraska to Connecticut, south to Texas and Florida. **Adult:** Brownish yellow; terminal segments of abdomen with silver hairs. Eyes emerald green in live specimens, reddish in dead specimens. **Length:** .8–1.2″. The adults prey mostly on Hymenoptera, the larvae feed on May beetles. *P. bastardii,* sometimes called the "false Nebraska bee-killer," is similar, about the same size, dark brown, occurs in much of the same area; adults prey mostly on Hymenoptera. *P. vertebratus* (Fig. 1306) is a Midwest species, has yellowish thorax, gray and black abdominal bands.

(34) ROBBER FLY: *Laphria flavicollis.* **Fig. 1307.**

The genus *Laphria* includes some large, colorful species resembling bumble bees. **Range:** *L. flavicollis* ranges over most of the U.S. and southern Canada. **Adult:** Black; mystax mostly yellow, scutellum with yellow hairs, abdomen with black pile. **Length:** .5–.75″. Adults prey on beetles, leafhoppers, honey bees. *L. sacrator* (Fig. 1308) is stout-bodied, black, with yellowish hairs over all but the terminal segments of the abdomen, resembles a bee; it ranges from Wisconsin to Quebec and Nova Scotia, south to Connecticut.

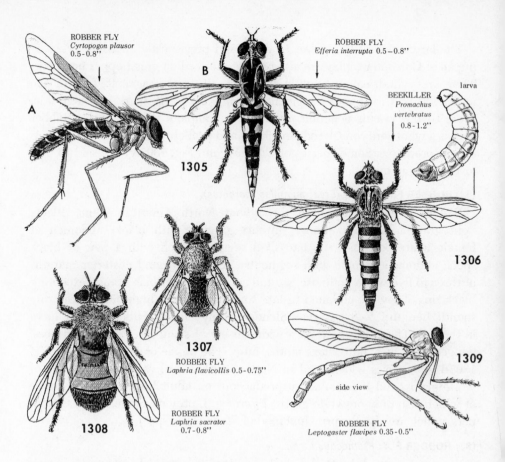

ROBBER FLY
Cyrtopogon plausor
0.5-0.8"

A

B

ROBBER FLY
Efferia interrupta 0.5-0.8"

larva

BEEKILLER
*Promachus
vertebratus*
0.8-1.2"

1305

1306

1307

ROBBER FLY
Laphria flavicollis 0.5-0.75"

1309

side view

1308

ROBBER FLY
Laphria sacrator
0.7-0.8"

ROBBER FLY
Leptogaster flavipes 0.35-0.5"

(35) ROBBER FLY: *Leptogaster flavipes.* **Fig. 1309.**

Range: Nebraska and Minnesota, east to Maine, south to Kansas and Georgia. **Adult:** Slender, fragile; black, thorax above dull reddish; abdomen dark, with extensive pale markings. **Length:** .35–.5". Inconspicuous but very common on grasses and ferns at the edge of woods. Adults prey on leafhoppers and small flies.

Bee Flies (Bombyliidae)

The bombyliids, or bee flies, are stout-bodied, usually black, covered with a dense pile. They have slender legs, short three-segmented antennae, and rather long proboscis. Nectar- and pollen-feeders, they hover over flowers and are easily mistaken for bees. Eggs are laid in the soil near the host. Upon hatching, the larvae immediately set out in search of their hosts. Most of the larvae so far as known are internal or external parasites of the larvae of other soil-inhabiting insects or predaceous on the egg pods of grasshoppers. The active first-instar larva is a planidium type—wormlike, 12-segmented, the head strongly sclerotized; a pair of lateral spines on each thoracic segment and a pair of long caudal spines aid in locomotion. In the inactive second instar the spines are lost; the third instar becomes smooth, grublike, crescent-shaped, and tapered at both ends. The pupae are the free type, resemble those of the Asilidae, Nemestrinidae, and closely related families. The life cycle in temperate climates usually covers a period of one year. Bee flies are most common in sandy areas and in the Southwest.

(36) BEE FLY: *Anthrax analis.* **Color plate 8g.**

Range: Montana to Ontario, south to Arizona and Florida. **Adult:** Bright velvety black; basal two-thirds of wing black, the remainder transparent. Eggs are deposited close to the entrance to burrows of tiger beetles. **Length:** About .3″. **Larva:** Long slender mandibles; body deeply segmented, pointed at the posterior end. It attaches itself to underside of host larva's thorax. Hibernates in this position in host tunnel. When the host pupates the following spring, the parasite bores into the pupa to complete its development. The full-grown larva, about .7″ long, then undergoes pupation itself. The pupa—with four curved hooks at the anterior end and stiff bristles on the body—works its way through the soil to the surface, precisely at which time the adult is ready to emerge. **Host:** Tiger beetles. *A. tigrinus* (Fig. 1310) is purplish black with much of the wing blackened, about .6″ long, ranges from California to Kansas, Ontario, Pennsylvania, south to Florida. The closely related *Systoechus vulgaris* is widely distributed, and like *Anastoechus,* is very hairy on the face and abdomen; the larvae are predators of grasshopper egg pods. *Systoechus* spp., *Anastoechus barbatus,* and *Aphoebantus* spp. (along with blister beetle larvae and ground beetles) are important factors in natural control of grasshopper on the prairies. *Poecilanthrax californicus* (Fig. 1311) ranges from California and Arizona to Colorado and Idaho.

(37) BEE FLY: *Bombylius major.* **Fig. 1312.**

Range: British Columbia to Maine, California, and Georgia. **Adult:** Black; face and front gray pollinose, with long black and brown hairs; thorax clothed in yellowish pile, with whitish tuft in front of wings, large brown tuft on sides below; abdomen clothed in yellowish pile, with black tuft on side of third segment. Proboscis black, nearly as long as body. Eggs are placed in the entrance tunnels of solitary bees when females of the latter are absent. **Length:** .3–.5″. **Larva:** Parasitic on the larvae of bees. **Host:** Solitary bees: *Andrena, Colletes, Halictus. B. lancifer* (Fig. 1313) is more subdued in color, with reddish legs, front of wings brownish, shading to gray, about .4″ long, ranges from California and Oregon to New Mexico.

(38) TANGLEWING FLY: *Neorhynchocephalus sackenii* (Nemestrinidae). **Fig. 1314.**

The tanglewing (or tangle-veined) flies are a small group, usually found in meadows hovering over flowers or darting about as they make a loud

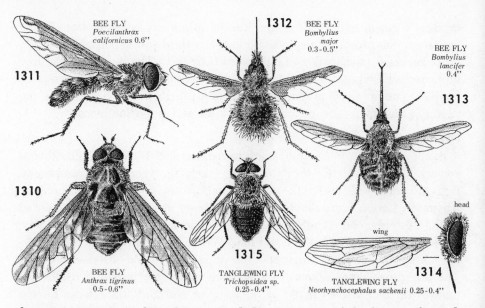

BEE FLY
Poecilanthrax californicus 0.6"

1311

1312 BEE FLY
Bombylius major
0.3-0.5"

BEE FLY
Bombylius lancifer
0.4"

1313

1310

head

wing

1314

1315

BEE FLY
Anthrax tigrinus
0.5-0.6"

TANGLEWING FLY
Trichopsidea sp.
0.25-0.4"

TANGLEWING FLY
Neorhynchocephalus sackenii 0.25-0.4"

buzzing noise. Some have been recorded as parasites of the larvae of wood-boring beetles, and at least two species (represented here) are known to be effective parasites of grasshoppers. **Range:** *N. sackenii* occurs from Washington to Iowa, Michigan, south to California and Georgia; Mexico. **Adult:** Black, with dense yellowish pile; abdomen wholly black, not distinctly banded or marked. Eggs are laid in cracks in fence posts and other upright objects—in large numbers, as many as 1,000 in 15 minutes. (The tiny maggots are scattered by the wind.) **Length:** .25-.4". **Larva:** Bores into abdomen of grasshopper, forms a breathing tube attached to the host's tracheal system. Emerges from host to overwinter in the ground, pupates the following spring. One brood. **Host:** The migratory grasshopper and others. *Trichopsidea clausa* (Fig. 1315) is similar, occupies about the same range, attacks *Metator pardalinus* and other grasshoppers in the West. They may be distinguished by the proboscis, which is long and stiletto-like in *N. sackenii*, and short, truncate, and concealed by dense pile in *T. clausa*. "Grasshopper populations have been noted to undergo drastic reduction after severe attack" by these parasites.

Small-headed Flies (Acroceridae)

The Acroceridae (formerly Cyrtidae), or small-headed flies, are small to medium-sized, with inflated abdomen, large humped thorax, and very small head composed mostly of eyes. Eggs are deposited in large numbers near the hosts. The larvae are solitary internal parasites of spiders. They undergo a hypermetamorphosis. The active first instar (a planidium type) is distinctly segmented, armed with spiny plates, two long apical bristles, and a caudal

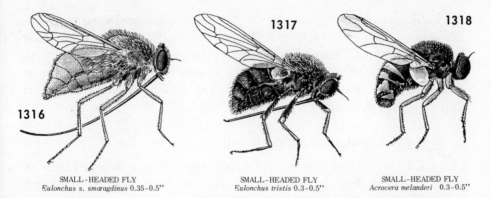

SMALL-HEADED FLY
Eulonchus s. smaragdinus 0.35-0.5"

1317

SMALL-HEADED FLY
Eulonchus tristis 0.3-0.5"

1318

SMALL-HEADED FLY
Acrocera melanderi 0.3-0.5"

disc. It is propelled by a wriggling motion or by jumping, which is accomplished by "standing" on the caudal disc, bending the body, and "springing" with the apical bristles. On entering the host it changes rapidly to a sluggish, indistinctly segmented maggot without bristles. Pupation takes place outside the host.

(39) SMALL-HEADED FLY: *Eulonchus smaragdinus smaragdinus.* **Fig. 1316.**

Range: Northern California to Mexico (Baja California). **Adult:** Metallic green to blue (sometimes with purplish reflections or abdomen all purple), with long, dense golden yellow or yellowish white pile; legs bright yellow, with coxae metallic green or blue, tarsi light brown; antennae blunt in the male, terminal segment tapered and sharp-pointed in the female; long curved proboscis (usually extending beyond apex of abdomen when at rest). Eyes contiguous above and below the antennae, small area below (the frontal triangle) smooth and polished. **Length:** .35–.5". **Host:** Not known. Schlinger "presumes that certain avicularoid spiders will be found to be their hosts." The subspecies *E. s. pilosus* occurs in southern California—it is a little larger, with white pile, extremely dense on thorax—as does *E. halli,* which may be distinguished by the shorter proboscis (less than the wing length) and dark brown legs. *E. tristis* (Fig. 1317) has black legs, is confined to the San Francisco Bay area. *Acrocera melanderi* (Fig. 1318) is blackish, abdomen with yellowish markings, antennae sharply pointed, tibial spurs lacking; it ranges from California and Washington to Maine and Georgia, has been reared from a species of *Hololena* by Schlinger.

(40) SMALL-HEADED FLY: *Ogcodes adaptatus.* **Fig. 1319.**

Range: Western North America, from Alaska to California. **Adult:** Black, with greenish white pile; antennae dark brown, tibiae dark brown basally; thorax and abdominal pile with slight golden tinge in the female. They are found in moist grassy areas, from April to October in the lowlands of California. The active adult period is short at high mountain elevations; large numbers ("literally thousands") of flies have been observed during late June and early July laying eggs on the dead twigs of sagebrush, some twigs being found to contain 15,000 to 20,000 black parasite eggs per linear foot. **Length:** About .2″. **Larva:** First instar (planidium) 12-segmented, with various types of setae. They contact the spiders on the plants at night or drop to the ground to find them. **Host:** Wolf spiders (*Pardosa*), crab spiders (*Xysticus, Philodromus*).

Long-legged Flies (Dolichopodidae)

The Dolichopodidae—known as the long-legged flies—is a large family of small, rather slender, metallic green or blue flies with short proboscis and long legs. They are common in lightly shaded areas near water and are most numerous in cold regions or higher altitudes. Both adults and larvae are predaceous. The latter are mostly aquatic (excepting a few species that mine the stems of grasses and sedges or live under the bark of trees) and are found in a great variety of habitats—in the mud and water, on moist and dry beaches, and among intertidal rocks. They pupate in cocoons made of sand, mud, and various fragments cemented together; two long thoracic breathing tubes project from the pupa through the cocoon wall. Some adult males have attracted attention for their interesting mating rituals.

(41) LONG-LEGGED FLY: *Dolichopus plumipes.*

Range: Alaska to the Northwest Territory, south to California, Arkansas, and Michigan; Mexico, Greenland, Europe. **Adult:** Green, sometimes with bronze reflections; face yellow—shorter, wider, and more gray in the female; first two segments of antennae yellow, the third black; sides of thorax and abdomen white pollinose, front coxae yellow, middle and hind coxae black with yellow tips, femora and tibiae yellow, tarsi black; halteres yellow, wings grayish, sometimes tinged with brown. **Length:** .15–.2″. The larvae are found in the soil in a variety of situations from wet to dry. The courtship behavior of the male *Dolichopus* is often elaborate and centers in a display of the legs before the female. The posture and movements vary with the species. The middle tibia and tarsus of the male of *D. plumipes* are adorned with a feather-like fringe; he flies up to the resting female, hovers overhead, dangling the fringes in front of her. *D. pugil* (Fig. 1320) ranges from Ontario to Nova Scotia, south to New Jersey.

(42) DANCE FLY: *Rhamphomyia scolopacea* (Empididae).

The dance flies comprise a fairly large family, and are named for their mating swarms and dances. The adults of both sexes are usually predaceous on small insects, and many are frequent flower visitors. The chivalrous male in the genera *Rhamphomyia* (Fig. 1321), *Empis*, and *Hilara* captures a prey and carries it about until mating, when it is offered to the female; some species first envelop the offering in a frothy web. This is no idle gesture,

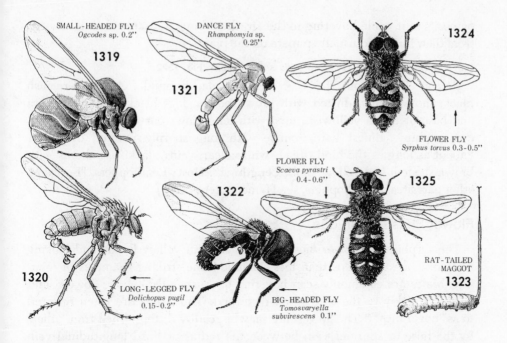

SMALL-HEADED FLY
Ogcodes sp. 0.2''
1319

DANCE FLY
Rhamphomyia sp.
0.25''

1321

1324

FLOWER FLY
Syrphus torvus 0.3-0.5''

FLOWER FLY
Scaeva pyrastri
0.4-0.6''

1322

1325

RAT-TAILED
MAGGOT
1323

1320

LONG-LEGGED FLY
Dolichopus pugil
0.15-0.2''

BIG-HEADED FLY
*Tomosvaryella
subvirescens* 0.1''

since only the male in this case is a hunter and indeed may not feed itself. Dance flies are found in moist places, along the water's edge and seashores, and are abundant in spring and early summer in temperate climates. Both males and females gather in swarms but apparently leave the crowds individually to mate. The larvae live in the soil, decaying wood and vegetation, under the bark of trees, or are aquatic; they are believed to be mostly if not entirely predaceous. They resemble the dolichopodid larvae except that the last body segment is rounded in the terrestrial forms, or has tapered processes in the aquatic forms; the latter also have well-developed footlike pseudopods on most of the body segments. **Range:** *R. scolopacea* occurs in Pennsylvania and states along the Atlantic seaboard, from Rhode Island to Maryland. **Adult:** Grayish, with silvery reflections; head black, antennae reddish brown, proboscis yellowish; abdomen silvery, with brighter reflections than on thorax. **Length:** .25''.

CYCLORRHAPHA

Big-headed Flies (Pipunculidae)

The big-headed flies (or big-eyed flies) are parasites of Homoptera and important in the natural control of leafhoppers. The head is usually larger than the thorax and composed mostly of compound eyes. Like the syrphids, the pipunculids are capable of swift flight, hover for long periods, and are frequently found on flowers. The female grabs a leafhopper from the plant, oviposits in it while hovering in the air, then drops it. Mature larvae emerge from their hosts and usually pupate on the plants.

(43) BIG-HEADED FLY: *Tomosvaryella subvirescens.* **Fig. 1322.**

Range: Throughout the U.S. and Canada; Bermuda. **Adult:** Greenish black; thorax shiny, dusted with brown pollen; legs black, apex of femora and base of tibiae yellowish, tarsi with long claws curved at the tips; abdomen shiny, almost bare, female with long sharply pointed ovipositor (about as long as the hind tibia); wings clear, veins blackish, halteres with brownish stem and yellow knob. **Length:** .1". **Host:** Leafhoppers. [For some other parasites of leafhoppers, see Hymenoptera (93).]

Flower Flies (Syrphidae)

The syrphids, or flower flies—often black with yellow bands and looking more like bees or wasps than flies—include some striking examples of mimicry. Males are commonly seen hovering or "suspended" above flowers, and darting quickly to the side when disturbed. Thus, they are often referred to as "hover flies."* The flower flies may be readily distinguished from others by the false or spurious vein between the radius or third longitudinal vein and the media or fourth vein. The oval, chalky white eggs have a reticulated pattern, are laid horizontally and singly (in the case of aphidophagous species) or in masses on plants near the food of the larvae. Some species lay their eggs among those of host beetles and the larvae suck out the contents of the host's eggs, leaving only the collapsed chorions stuck to the leaf.

Syrphid larvae are more varied in form and habit than the adults; they live chiefly on decaying organic matter or are predaceous on aphids and other soft-bodied Homoptera or Thysanoptera and on the immature forms of other insects. A few attack bulbs and other parts of plants. Aphid-feeders —they are usually pointed at the head—grasp the prey with their jaws, raise it into the air, and drain the body contents. The typical larva is sluglike, colored green, brown, gray, or with a mottling of the same. Those that breed in putrid or stagnant water and moist excrement have a long tail-like anal breathing tube with spiracle at the end, and are called "rat-tailed maggots" or "mousies" (Fig. 1323). Others living in bees' or wasps' nests have short tails; those living among plant roots have pointed tails. The puparium is teardrop- or pear-shaped in the subfamily Syrphinae (our first three examples) and barrel-shaped in the Milesiinae. The Syrphinae are almost as important in the control of aphids as are the lady beetles. The

* By way of contrast, bees weave back and forth and bob up and down when visiting flowers, and their antennae are longer and elbowed; at rest they fold their wings with the tips overlapping, while syrphid flies hold theirs half-spread. (The fact that the latter have one pair of wings and bees two is not always apparent at a glance.) Syrphid flies do not sting; as pollinators they probably rank next to the bees.

principal enemies of syrphid flies are ichneumon wasps. Strangely enough, many syrphids are trapped by the sticky pollen of the common milkweed when visiting these flowers.

(44) FLOWER FLY: *Syrphus ribesii* (Syrphinae).

Range: Alaska to New Brunswick, south to Central America and North Carolina. A native of Europe, also found in Asia. **Adult:** Blackish, with pale yellow bands on abdomen, the first broken and the third constricted at the middle; face yellowish, eyes usually bare. **Length:** .3–.5″. The larvae feed on various aphids. *S. torvus* (Fig. 1324) is similar, widespread throughout the U.S. and Canada, and in Eurasia.

(45) FLOWER FLY: *Scaeva pyrastri.* **Fig. 1325.**

Range: Alaska to Alberta, south to California, New Mexico, and Arkansas; Eurasia. Most abundant of the western species. **Adult:** Shiny black, with three sets of broken yellow bands on abdomen; face yellow, eyes and antennae reddish brown, eyes densely haired; wings very clear. **Length:** .4–.6″. **Larva:** Pale green, with white longitudinal stripes, pointed at the anterior end, .6″ long full grown. Pupates on plants or ground, under leaves. Hibernates in puparium. Several broods. **Food** (larva): Various aphids. One larva is said to "destroy about a thousand aphids in the two or three weeks of [its] larval existence."

(46) FLOWER FLY: *Didea fasciata.* **Fig. 1326.**

Range: British Columbia to Nova Scotia, south to Oregon, New Mexico, and North Carolina; Europe and Asia. **Adult:** Black, abdomen with yellow bands; black stripe on face, cheeks yellow or light brown, scutellum yellowish brown; eyes with scattered moderately long hairs; wings slightly clouded, third vein rather deeply bent. **Length:** About .5″. **Food** (larva): Aphids and other soft-bodied insects. In *D. alneti* the abdominal bands are greenish or bluish, somewhat narrower, the third wing vein less deeply bent; it ranges from Alaska to Labrador, south to Nova Scotia and Colorado. *Baccha clavata* [see Pictured Key to the Insect Orders: (*) Diptera]—widespread in North and South America, and known as the four-spotted aphis fly—is a slender wasplike fly with two spots on each of two enlarged, bulbous posterior abdominal segments; the larva—dull green with a red dorsal line —is a common predator of the spirea aphid and other aphids in Florida and California.

(47) DRONE FLY: *Eristalis tenax* (Milesiinae). **Fig. 1327.**

Range: Alaska to Labrador, south to California and Florida, cosmopolitan; Europe. **Adult:** Brownish to black, with moderate pile; scutellum yellowish; eyes not spotted, with pile forming a vertical stripe. It "closely mimics the honey bee in color, size, and actions," is most often found on

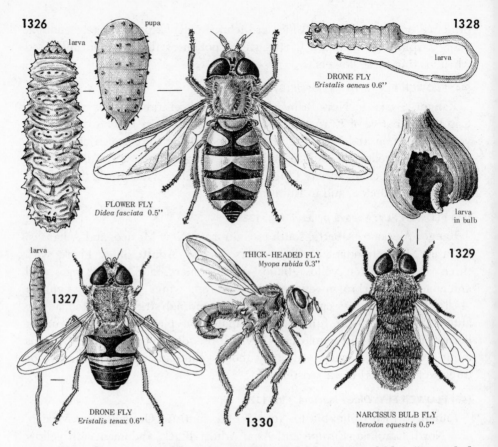

1326

larva

pupa

FLOWER FLY
Didea fasciata 0.5"

1328

larva

DRONE FLY
Eristalis aeneus 0.6"

larva
in bulb

larva

1327

DRONE FLY
Eristalis tenax 0.6"

THICK-HEADED FLY
Myopa rubida 0.3"

1330

1329

NARCISSUS BULB FLY
Merodon equestris 0.5"

composite flowers. **Length:** .6". **Larva:** Cylindrical, grublike "rat-tailed maggot," with long filamentous breathing tube. It lives in putrid, stagnant water and in moist excrement; commonly occurs in open latrines. *E. aeneus* (Fig. 1328) is widespread throughout North America, Europe, Eurasia, and Africa.

(48) NARCISSUS BULB FLY: *Merodon* (*Lampetia*) *equestris.* **Fig. 1329.**

Range: British Columbia to California, Wisconsin to New Brunswick and Georgia. Not a native; occurs in Europe. **Adult:** Resembles a small bumble bee. Black, with bands of long, colored hairs, in combinations of black, yellow, orange, buff. It appears in late spring or early summer, and is seen zigzagging along rows of plants. Eggs are laid singly on leaves near base of plant or in the soil. **Length:** .5". **Larva:** Hemispherical, unsegmented; dirty tan, with wrinkled skin; short, black breathing tube at the hind end. It migrates from leaf to bulb, eats a large cavity in the center; normally one larva to a bulb. Sunken brown area in the white root ring, at edge of basal plate, indicates infestation. Hibernates as larva in the bulb, moves near surface of the soil in spring to pupate. Puparium large, brown, with

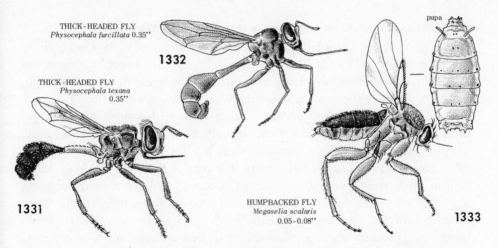

THICK-HEADED FLY
Physocephala furcillata 0.35"

1332

THICK-HEADED FLY
Physocephala texana
0.35"

pupa

1331

HUMPBACKED FLY
Megaselia scalaris
0.05-0.08"

1333

two hornlike breathing tubes near the front. Some larvae hibernate a second winter. One brood. **Food:** Narcissus preferably; hyacinth, lilies, tulips, amaryllis and other bulbs. The **Lesser Bulb Fly**, *Eumerus tuberculatus*, is blackish green, almost devoid of hairs, with white crescent markings on sides of abdomen, about .3" long; the yellowish gray larva is about .5" long and believed to feed on the basal rot rather than healthy tissues; it occurs across Canada and northern U.S. The **Onion Bulb Fly**, *E. strigatus*, is similar in appearance and distribution, was accidentally introduced from Europe with *E. narcissi*.

(49) THICK-HEADED FLY: *Physocephala burgessi* (Conopidae).

The thick-headed flies are generally slender, petiolate flies resembling wasps. Some species, as in the genus *Myopa* (Fig. 1330), are stout. They are usually dark in color, sometimes with yellow bands on the abdomen. Adults are flower-loving like the syrphids; the larvae are solitary internal parasites of adult Hymenoptera. "Pincers" at the tip of the female fly's abdomen are used to grasp the prey (bees and wasps) during oviposition, which takes place in flight and with no apparent resistance from the victims. **Range:** *P. burgessi* ranges from British Columbia to Alberta, southward to California and Texas, is commonly found at the edge of forested areas. **Adult:** Dark red, face and lower half of front yellow; proboscis short, nearly black, with red tip; mesonotum with wide median stripe, end of abdomen yellowish, front coxae shiny black; knobs of halteres yellow, front of wings blackish, posterior margin grayish. **Length:** .35". **Host:** Bumble bees. *P. texana* (Fig. 1331) is widespread from Canada to Mexico, usually found "in open brush lands"; its preferred hosts are said to be *Bembix* wasps. *P. furcillata* (Fig. 1332) occurs from eastern Canada and New England to Alberta, California, and the Canal Zone, at high and low elevations, where it parasitizes bees.

(50) HUMPBACKED FLY: *Megaselia aletiae* (Phoridae). **Fig. 1334.**

The phorids, or humpbacked flies, are minute to small in size, usually black, brown, or yellowish. The head is small, with prominent bristles on the front, three-segmented antennae, and rather large eyes; the legs are large, hind femora generally stout, tibiae often with bristles near the tip. The wings are clear or brownish (reduced or lacking in some females), and unusual in that the heavy costal vein extends to only about the middle of the front margin, and generally has a fringe of short bristles. Except for two or three other short, stout veins near the costa, the rest of the wing has only three or four other weak veins. Humpbacked flies are common around decaying organic matter—vegetable and animal—and are sometimes found in the nests of ants, termites, and bees. They fly with quick jerky movements. Some larvae are parasitic, others are scavengers or live in fleshy and woody fungi; they have been known to occur in human feces, probably through the use of unwashed containers such as bottles. **Range:** *M. aletiae* is widely distributed throughout the U.S., from California to New York and Connecticut, south to Alabama and Florida. **Adult:** Yellowish, front dark above, antennae brown; eyes black, angular, prominent; thorax reddish, abdomen usually darker, sometimes nearly black, banded with yellow; hind femora with brownish tips. Eggs are deposited over host caterpillar in haphazard manner. **Length:** .05–.08″. The larva is parasitic, enters host body cavity through the anus. **Host:** Elder shoot borer (Noctuidae). *M. scalaris* (Fig. 1333) ranges from Indiana to Massachusetts, south to Texas, Florida, and Cuba, also occurs in subtropical and tropical regions of the world; in a case on record a patient passed larvae, pupae, and adults of this species over a period of time (the larvae apparently being able to mature and pupate and adults emerge in the alimentary tract if not to repeat the cycle).

(51) CARROT RUST FLY: *Psila rosae* (Psilidae). **Fig. 1335.**

The flies in this family are small, slender, and shiny, with long antennae; they are generally seen resting on foliage in the shade. The larvae live in the roots and stems of plants or under the bark of trees, which they enter by way of wounds. The most injurious species is the carrot rust fly. **Range:** *P. rosae* is widely distributed north of the 40th parallel: from British Columbia and Alberta to Utah and Colorado, eastward to Newfoundland, Wisconsin, Illinois, Maryland; Europe. **Adult:** Dark green to shiny blue-black, with yellow head and black eyes, sparse coat of yellow hairs on body, yellow legs. Eggs are deposited around the base of plants. **Length:** .15″. **Larva:** Yellowish to dark brown, about .3″ long full grown. It works its way down to the roots, attacks crowns or root tips, causing leaves to turn yellow and plant to wilt; tunnels in roots are rusty red. Pupates in brown puparium about .2″ long, near surface of soil. Two and partial third broods. **Food:** Carrots, celery, parsnips, especially where grown in muck soil.

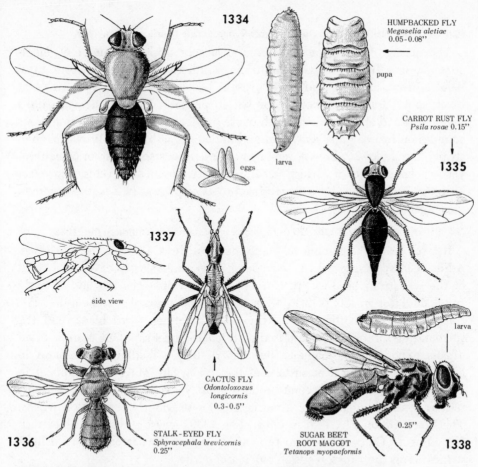

1334

HUMPBACKED FLY
Megaselia aletiae
0.05-0.08"

pupa

eggs

larva

CARROT RUST FLY
Psila rosae 0.15"

1335

1337

side view

larva

CACTUS FLY
Odontoloxozus longicornis
0.3-0.5"

1336

STALK-EYED FLY
Sphyracephala brevicornis
0.25"

SUGAR BEET
ROOT MAGGOT
Tetanops myopaeformis

0.25"

1338

(52) STALK-EYED FLY: *Sphyracephala brevicornis* (Diopsidae). **Fig. 1336.**

Range: Minnesota to Quebec, south to Colorado and North Carolina. **Adult:** Black; head reddish, top brown, with peculiar eyes—located at the end of two fixed stalks; thorax with lateral spines short, black, longer and reddish posteriorly; tarsi reddish (the front femora and tibiae in the male appear to be "raptorial," are used to grasp the female's wings during mating, and apparently are not for seizing prey); wings with transverse bands near apex brown; halteres white. They are found on skunk cabbage and other plants along streams or near water, often in great numbers, and are attracted to nectar and other sweet exudates. Pear-shaped eggs are laid singly, in small groups, on sphagnum moss or the mud beneath. **Length:** .25". **Larva:** White, smooth, shiny; cylindrical, 12-segmented, tapered anteriorly, posterior end blunt, ending in a pair of spiracular stalks; about .2" long full grown; head telescopes into prothorax, mouth hooks brown. They are semiaquatic. **Food** (larva): Decaying organic matter. They hibernate as adults in crevices among rocks and roots along the banks of streams, or near water, usually in large aggregations. Two broods are possible.

(53) CACTUS (NERIID) FLY: *Odontoloxozus longicornis* (Neriidae). **Fig. 1337.**

The neriids comprise a small family, largely tropical. **Range:** *O. longicornis* ranges from Texas to California, south to Mexico and Costa Rica. **Adult:** Brownish; face and lower part of head yellow; each eye with two black spots along front border, brown stripe behind; antennae with third segment broad apically, arista with white pubescence; thorax with numerous spinules on brown dots; scutellum and abdomen with broad, brown median stripe; legs yellowish brown, femora with numerous spines; stem of halteres yellow, knobs brown; wings clear, grayish, costa and veins brownish. **Length:** .3–.5″. It has been reared from roots of papaya in Texas, is common in cactus areas.

(54) SUGAR BEET ROOT MAGGOT: *Tetanops myopaeformis* (Otitidae). **Fig. 1338.**

Range: British Columbia to Manitoba, south to California, New Mexico. **Adult:** Glossy black, with smoky patch on costal margin of wing a third of the way from base to tip. Curved, glossy white eggs are laid in early June, singly or in lots of 2 to 40 in the ground near plants. **Length:** .25″. **Larva:** It feeds on the taproot; the point of attack turns black, and the surrounding soil becomes saturated with leaking beet sap. One plant is often attacked by 10 to 30 maggots; one can destroy a seedling. Hibernates in larval stage; pupates near surface of the soil in late May, adults appear in about two weeks. Small second brood in most years. **Food:** One of the most serious pests of sugar beets; a native species previously feeding on weeds. The **Black Onion Fly,** *Tritoxa flexa* (Fig. 1339)—western as well as eastern according to Cole—is black to brownish black, with narrow, curving light stripes across the wings, feeds on onion and garlic.

(55) PYRGOTA FLY: *Pyrgota undata* (Pyrgotidae). **Fig. 1340.**

Pyrgotid adults are mostly nocturnal, and the larvae are parasites of scarabaeid beetles. They are medium to large flies with long wings often partly clouded and forming beautiful patterns. Ocelli are usually absent. The tip of the abdomen is curved downward, the ovipositor is conical and sharply pointed. *P. undata* is widespread east of the Rocky Mts. **Range:** Manitoba to Quebec, south to Texas and Florida. **Adult:** Body and legs brownish; wings, excepting posterior margins, same color. The female lights on a feeding beetle, causing it to take flight, inserts its ovipositor in the beetle's back through a tender part of the integument exposed only in flight; the scarabs resist the attack violently. The fly usually attacks *female* beetles and finds a happy hunting ground under lights where many of the beetles congregate. **Length:** .4″. **Larva:** White, very robust, tapered anteriorly. It kills a beetle in 10 to 14 days, becomes a scavenger, consuming the entire contents of the host. Pupates in host remains, then hibernates; the adult emerges the following spring. One brood. **Host:** May beetles. *Sphecomyiella*

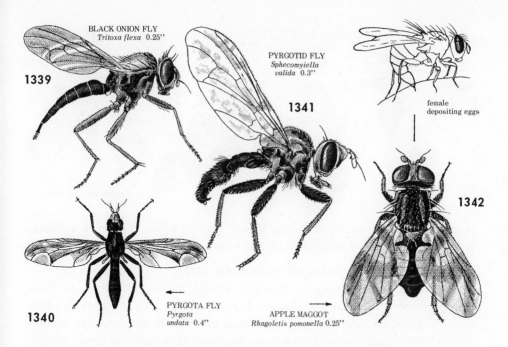

BLACK ONION FLY
Tritoxa flexa 0.25"

1339

PYRGOTID FLY
*Sphecomyiella
valida* 0.3"

1341

female
depositing eggs

1342

1340

PYRGOTA FLY
*Pyrgota
undata* 0.4"

APPLE MAGGOT
Rhagoletis pomonella 0.25"

(*Pyrgota*) *valida* (Fig. 1341) is almost as common, about .3" long, dull black, wings mottled with black; it ranges westward to Arizona.

Fruit Flies (Tephritidae)

The tephritids, or fruit flies, are small to medium in size with varied coloring and beautiful patterns on the body and wings. They include many flies that are among the worst pests of cultivated fruits. Females oviposit in healthy plant tissues, can be recognized by their habit of raising and lowering the wings slowly as they rest or walk about on the fruit; the larvae tunnel through the fruit. Some species form stem and root galls, and a few mine leaves.

(56) APPLE MAGGOT: *Rhagoletis pomonella.* **Fig. 1342.**

Range: North Dakota to Nova Scotia, south to Texas and Florida. **Adult:** Body shiny, black, with yellow markings; legs yellowish; wings crossed with black bands. It feeds by "scrubbing" the surface of leaves and fruit. The female punctures the skin with her ovipositor and inserts an egg. **Length:** .25". **Larva:** Creamy white or yellowish, about .25" long full grown. Sometimes called the "railroadworm" for its extensive tunneling of the pulp. There may be no sign of the maggot's presence except slight depressions in the skin or dark lines which indicate tunnels. Hibernates in brown puparium in the soil; adults emerge in July, some remain in the puparium two or more winters. **Food:** Apple, plum, haws, blueberries. The **Blueberry Maggot,**

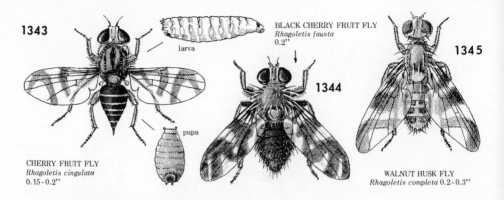

1343

BLACK CHERRY FRUIT FLY
Rhagoletis fausta
0.2''

larva

1344

1345

pupa

CHERRY FRUIT FLY
Rhagoletis cingulata
0.15-0.2''

WALNUT HUSK FLY
Rhagoletis completa 0.2-0.3''

R. *mendax*, occurs in Nova Scotia, Maine, and New Jersey; the fly is black, with white abdominal bands and black bands across the wings.

(57) CHERRY FRUIT FLY: *Rhagoletis cingulata*. **Fig. 1343.**

Also called cherry maggot. **Range:** Middle states, Michigan to New Hampshire, south to Florida. **Adult:** Shiny, black; abdomen with four white crossbands, tibiae and tarsi yellow; wings clear, with black bands. Eggs—yellow, with small pedicel at the end—are inserted in half-ripe fruit, usually one but sometimes more in a single cherry. **Length:** .15–.2''. **Larva:** White, about .25'' long full grown. The young maggot feeds near the pit, later in the flesh. (The fruit appears normal until it begins to shrivel.) Hibernates in brown puparium in the ground. One brood. **Food:** Cherry, pear, plum, wild cherry. The **Western Cherry Fruit Fly,** *R. indifferens,* is found in the Pacific Northwest and California; its native host is bitter cherry. The **Black Cherry Fruit Fly,** *R. fausta* (Fig. 1344), is all black, with wing bands narrower, ranges from British Columbia to Quebec, south to California and New York, prefers sour to sweet cherries. [See Hymenoptera (24).]

(58) WALNUT HUSK FLY: *Rhagoletis completa*. **Fig. 1345.**

Range: Minnesota to California and Texas. **Adult:** Brownish, with yellow-ish markings, blue eyes; yellowish white spot on thorax with lateral stripe same color, darker crossbands on abdomen, black bands on wings. It emerges in California from July to September. Pearly white eggs have fine reticulation, are inserted in stem region of husk, 12 to 15 at a time. **Length:** .2–.3''. **Larva:** Yellow, with black mouth hooks, about .4'' long full grown. It leaves the husk to hibernate in a straw-colored puparium in the soil, looks like a wheat kernel at this stage. One brood. **Food:** English walnuts, black walnuts, peach. The principal injury to walnuts is staining of the shells. The **Pepper Maggot,** *Zonosemata electa,* is a serious pest of hot cherry peppers (never bell or sweet peppers) in Delaware, commonly occurs on horse nettle; it ranges from Ontario to Florida, west to Indiana, Oklahoma, and Texas. The **Currant Fruit Fly,** *Epochra canadensis,* is a serious pest of currants and

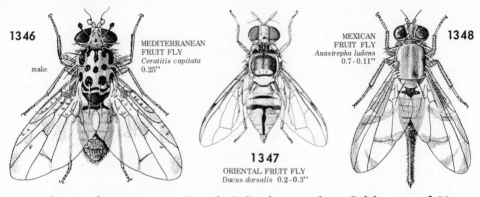

1346

MEDITERRANEAN
FRUIT FLY
Ceratitis capitata
0.25"

male

MEXICAN
FRUIT FLY
Anastrepha ludens
0.7-0.11"

1348

1347

ORIENTAL FRUIT FLY
Dacus dorsalis 0.2-0.3"

gooseberries from Maine to British Columbia, south to California and New York. The **Sunflower Maggot,** *Strauzia longipennis,* mines the pith of sunflowers.

(59) MEDITERRANEAN FRUIT FLY: *Ceratitis capitata.* **Fig. 1346.**

Also called medfly. **Range:** Outbreaks have occurred in Florida, but it is "not now known to occur on the continent north of Nicaragua." It is found in tropical North America, South America, Hawaii, the Mediterranean region, Australia. **Adult:** Thorax shiny black, with yellowish white markings; abdomen yellowish with two silvery crossbands; wings marked with yellow, brown, black. It has sponging mouthparts, feeds on fluids only; eggs are inserted in the fruit, two to ten at a time. **Length:** About .25". **Larva:** It feeds for ten days to six weeks inside the fruit, pupates in the soil. In cooler regions it hibernates as pupa or adult, in warmer regions production is continuous. One to twelve broods. **Food:** Many wild and cultivated fruits, especially citrus, the common deciduous fruits, and coffee. The **Olive Fruit Fly,** *Dacus oleae,* is a serious pest of olives in the Mediterranean area. The **Oriental Fruit Fly,** *D. dorsalis* (Fig. 1347), is an important pest of mangoes in Hawaii and a potential pest of avocados on the Mainland. The **Melon Fly,** *D. cucurbitae,* is a pest of cucurbits in Hawaii. Substantial or partial control of all but *D. oleae* has been achieved with parasites; control by release of sterilized males is being tried. The **Mexican Fruit Fly,** *Anastrepha ludens* (Fig. 1348), occurs in Central America, Mexico, and southern Texas, attacks citrus and mangoes chiefly; it is conspicuous—considerably larger than a house fly, brightly colored with attractive wing pattern, the female's ovipositor about as long as its thorax and abdomen combined; in Mexico it is known as the orangeworm.* [See Homoptera (38), Citrus Whitefly.]

* Potentially destructive to citrus and avocados, it has been confined for the most part within Mexico by means of rigid quarantine programs. A capture of the fly was made recently in San Diego, California (*The San Diego Union,* April 1, 1970). It "has no cold-resistant overwintering stage that undergoes diapause or hibernation," but "adult flies periodically extend their normal range by dispersal from northeast Mexico into the lower Rio Grande Valley in Texas." Studies indicate that the flies could probably maintain themselves along the southern coast of California and the coast of the Gulf of Mexico, and in Florida.

A. *suspensa,* normally found in Cuba and other Caribbean islands, is occasionally found in Florida. The **West Indian Fruit Fly,** A. *mombinpraeoptans,* occurs in Cuba and Puerto Rico, occasionally in Florida and Texas.

(60) TEPHRITID FLY: *Eurosta solidaginis.* **Fig. 1349.**

Range: Northwest Territory to Maine, south to British Columbia, Kansas, Ohio, New Jersey. **Adult:** A comparatively large, heavy species, with reddish brown body; scutellum dark, with two bristles; ovipositor heavy, conical in shape; wings brownish, with fine network of lines, large clear triangle extending partly into discal cell. A variation called *fascipennis* differs in having a clear oblique band near the apex of the wing, a narrower and longer spot at apex of the third longitudinal vein along the border, and a smaller spot at apex of the fourth vein. A variation called *subfasciata* has an oblique interrupted stripe near the triangle at apex of the first longitudinal vein, a transverse clear spot between the third and fourth veins. **Length:** About .35″. The larva causes a symmetrical globular gall about 1.25″ in diameter on the main stem of goldenrod, and is found in the center of the pithy gall (Fig. 518). [For other gallmakers on goldenrod see Diptera (21) and Lepidoptera (285).]

(61) SEAWEED FLY: *Coelopa frigida* (Coelopidae). **Fig. 1350.**

Range: Northern Manitoba to Labrador, south to Rhode Island and islands of the North Atlantic; Europe, China. **Adult:** Black; antennae, palpi, and proboscis reddish; last three abdominal segments bordered with yellowish red above (less widely in the female), the last four so bordered on the sides; legs reddish yellow, femora and tibiae spiny, apical half of tarsi with yellow brush. **Length:** .12–.25″. In fall and winter huge swarms of the flies are seen around kelp washed up on the seacoast; the decaying seaweed is sometimes "a quivering mass of maggots." *C. nebularum* is very similar, the posterior part of abdomen much more bristly in the male, legs invariably blackish in the female; it occurs from the Bering Sea south to Alaska and Vancouver Island. *C. vanduzeei* occurs along the coast of California, north to Alaska.

(62) CHEESE SKIPPER: *Piophila casei* (Piophilidae). **Fig. 1351.**

Range: Cosmopolitan, probably came from Europe. **Adult:** Shiny black or dark bronze; face, mouthparts, antennae yellow. Slender white eggs are pointed, slightly curved, laid singly or in clusters on food and in cracks. **Length:** .15″. **Larva:** White or yellowish, legless, about .3″ long full grown. They "skip" or jump by bending almost double and straightening out quickly. **Food:** Animal fats, especially cheese and smoked meats; pork is preferred to beef, and cream cheese to other kinds. They are rare in houses unless food materials are left exposed.

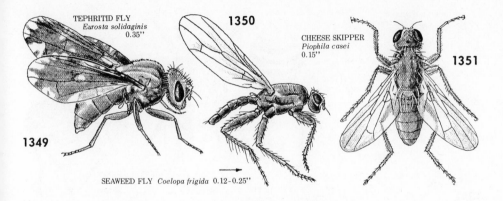

TEPHRITID FLY
Eurosta solidaginis
0.35"

1350

CHEESE SKIPPER
Piophila casei
0.15"

1351

1349

SEAWEED FLY *Coelopa frigida* 0.12-0.25"

(63) RICE LEAF MINER: *Hydrellia griseola* (Ephydridae). **Fig. 1352.**

The ephydrids, commonly known as shore flies, are found around water and serve as an important source of food for waterfowl. They are aquatic or semiaquatic. Abundant around salt and alkaline lakes in the West—where they are called brine flies—the puparia were gathered by aboriginal Indians for food. All known species of *Hydrellia* are leaf and stem miners. *H. griseola* is a "common leaf-mining pest of irrigated cereals." **Range:** Widespread in the U.S., Canada, and Europe. **Adult:** Olivaceous brown; front of head varying from white to yellow and golden brown; palpi club-shaped, mostly yellow; antennae dark, arista with five or six branching hairs; thorax and abdomen shading to metallic blue-green laterally; femora metallic blue-green, tibiae dark gray; tarsi black, inside of first segment with dense golden setae; wings clear. **Length:** .07–.1" (body); .1–.15" (wing). **Larva:** Yellowish to greenish, tapered at both ends, 13-segmented, with numerous setae, about .1" long. It mines one or more leaves, pupates in leaf. Several broods. **Food:** Rice, sometimes barley, oats, wheat, onions; wild aquatic and semiaquatic plants. Adults feed on insects that have fallen on the water.

(64) VINEGAR FLY: *Drosophila melanogaster* (Drosophilidae). **Fig. 1353.**

The drosophilids, or vinegar flies (also known as pomace flies), are found mostly around rotting fruit, feeding primarily on microorganisms associated with the decay. Some feed on other decaying organic matter or on plant exudations; a few are leaf miners, parasites, or predators. They may cause myiasis if swallowed; eggs and larvae in the cracks of tomatoes are a problem in processing factories. **Range:** *D. melanogaster* is cosmopolitan. **Adult:** Yellowish, with dark crossbands on abdomen; the feathered arista is characteristic of the family. A female lays up to 2,000 pearly white eggs, each with a pair of "wings" or respiratory "horns" near the anterior end; the eggs of all known *Drosophila* have one or more of these horns, the tips of which extend above the surface of the moist media in which the eggs develop.

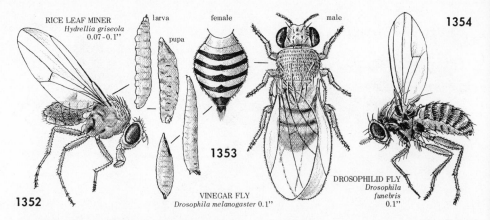

RICE LEAF MINER
Hydrellia griseola
0.07-0.1"

larva

pupa

female

male

1354

1353

1352

VINEGAR FLY
Drosophila melanogaster 0.1"

DROSOPHILID FLY
Drosophila
funebris
0.1"

Length: .1". **Larva:** Cream-colored or transparent, about .2" long full grown. Because it is very prolific, easy to rear, and has a short life cycle (thus many generations in a short time), this fly has provided laboratory subjects for studies on which much of our knowledge of heredity is based. Experiments with radiation-induced mutations in these flies led to the discovery of the sterile-male technique for control of insects.* *D. funebris* (Fig. 1354) is also cosmopolitan, occurs in Europe and over the entire U.S.

Chloropid Flies (Chloropidae)

The chloropid flies are very small, are numerous in grass and low herbage, and have a great variety of habits. Some visit flowers; others swarm about the face and eyes and rasp the eye membranes with their mouthparts. Many feed on decaying organic matter; others—such as the frit flies—are destructive to grasses; a few are gallmakers; others still are predaceous. *Thaumatomyia glabra* is an important factor in the control of the sugar beet root aphid (*Pemphigus populivenae*); *Pseudogaurax signatus*—found in the southern part of the country—is predaceous on the egg sacs of the black widow spider. The eye gnats (*Hippelates*) are common in warm, dry regions, where they swarm about the head with an annoying persistence—darting at the eyes, mouth, nose, or wounds of man and other animals. Females get protein (necessary for egg production) from the exposed animal mucus, and sometimes transmit acute bacterial conjunctivitis or "pinkeye"; in tropical regions they are implicated in yaws and trachoma (an eye disease caused by a virus).

* The idea of inundating an area with sterilized male insects to eradicate them was conceived by E. F. Knipling of the U.S. Department of Agriculture's Research Division. The natural male population at the time of these massive releases must be sufficiently low to enable the sterile males to "crowd them out," as it were, in competing for the females. The result of course is that the eggs laid by the females will be nonviable. The technique has been used successfully against the screw-worm and certain fruit flies, and appears to work in the case of the codling moth in British Columbia; it is being tried against the pink bollworm and other pests.

(65) EYE GNAT: *Hippelates pusio.* **Fig. 1355.**

Range: Washington to North Dakota, Pennsylvania, south to California, Florida, and Mexico. **Adult:** Dark gray to black. **Length:** About .05″. It breeds in excrement or in soils having a high content of organic matter. The life cycle is completed in three to four weeks. Males are attracted to flowers; females prefer animals, including man. *H. collusor* (which may be a subspecies of *pusio*) is common in California, Nevada, Arizona, and northern Mexico, and most active in spring and fall; it has been implicated in an epidemic of pinkeye in a school in southern California. In the West Indies and Brazil *H. flavipes* is said to be "responsible for the majority of cases of yaws" (a spirochetal infection causing skin lesions).

(66) FRIT FLY: *Oscinella frit.* **Fig. 1356.**

The name comes from *frits,* as wheat damaged by the fly is called in Sweden. **Range:** Canada, the U.S. (excepting Florida), Mexico, and Europe. **Adult:** Black, with yellow halteres; some forms have yellow markings on the legs. White finely ridged eggs are laid on leaves, sheaths, and heads or panicles of grains and grasses. **Length:** .04–.08″. **Larva:** Yellow, with curved mouth hooks, about .1″ long full grown. Spring and fall broods live in the panicles or in stems. Pupates in brown puparium under leaf sheath, hibernates in winter grasses or grains. Three or four broods. **Food:** Wheat, oats, rye, barley, corn, timothy, sedge, lawn grasses, wild grasses. The larvae feed in stems of grasses at the highest node, causing "silver top."

(67) HOLLY LEAF MINER: *Phytomyza ilicis* (Agromyzidae).

Most agromyzids are leaf miners, and a species usually feeds on only one genus of plant.* **Range:** *P. ilicis* is found in the Northeast and from British Columbia to California; it was introduced from Europe. **Adult:** Grayish black; frons dark below, more yellowish above. Eggs are deposited in punctures made with ovipositor in lower surface of new foliage; punctures are made in upper surface for feeding before egg laying. **Length** (wing): .11–.12″. **Larva:** Pale yellow, about .15″ long full grown. It makes a winding tunnel between margin and midrib, gradually widening the mine as it develops (Fig. 1357A). Hibernates in puparium inside the leaf, adults emerge in the spring. One brood in the North. **Food** (host plants): English holly, American or Christmas holly, more often the latter. Mined tissues turn yellowish green. The **Native Holly Leaf Miner,** *P. ilicicola,* occurs in Ontario and from Massachusetts to Alabama and Texas; it is smaller than *ilicis*—wing about .08″ long, with frons dark in front, paler brownish gray

* According to Spencer, they can often be identified from the leaf mine and knowing the plant. Most species mine the upper layer of cells immediately below the epidermis; the mines are usually linear, serpentine, or form a blotch, and may be near the margin, in the center, or at the midrib. Some species feed in the center layer of cells and their mines are easily overlooked; others mine both layers of cells and the mines are transparent. Species vary also in the way the larvae deposit their fecal pellets in the mines.

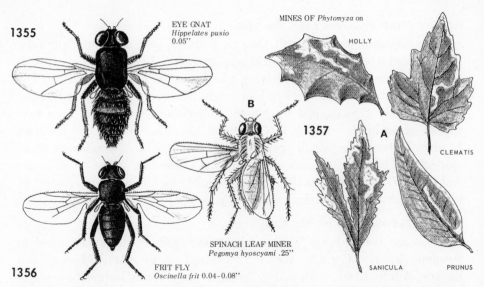

1355

EYE GNAT
Hippelates pusio
0.05"

MINES OF *Phytomyza* on

HOLLY

B

1357

A

CLEMATIS

SPINACH LEAF MINER
Pegomya hyoscyami .25"

1356

FRIT FLY
Oscinella frit 0.04-0.08"

SANICULA

PRUNUS

above; some of its eggs are laid in *feeding* punctures. The **Columbine Leaf Miner,** *P. aquilegivora* (*minuscula*), is widespread; the adult is shiny black, lays its eggs on the underside of columbine and aster leaves; the whitish maggot makes a winding mine in the leaf, pupates in a ball-like puparium on the leaf; there are four broods, hibernation occurs in pupal stage in the soil. The **Chrysanthemum Leaf Miner,** *P. syngenesiae* (*atricornis*), is widespread, European in origin; the mesonotum is gray, frons yellow, wing about .1" long. The **Murdock Leaf Miner,** *Phytobia* (*Amauromyza*) *maculosa,* is widespread in the U.S., occurs in Hawaii, Central and South America; the adult is black, except the halteres, which are white above; the larva makes a blotch or blister-like mine in chrysanthemum leaves.

Anthomyiid Flies (Anthomyiidae)

Anthomyiids closely resemble the flies in the next family (Muscidae), and are sometimes included with them. The best-known members are probably the root maggots, but many attack other parts of plants, as in the case of the **Raspberry Cane Maggot,** *Pegomya rubivora,* and the **Spinach Leaf Miner,** *P. hyoscyami* (Fig. 1357B). Females of *rubivora* lay their eggs in leaf axils on new canes of raspberry, blackberry, rose, and other brambles, the larvae girdle the tips and mine down into the new shoots; *hyoscyami* larvae mine out blotches in leaves of spinach and beets. Some species of this family feed on dung; many are predaceous in the adult stage, often attacking other flies, including those of the same family. The larvae of some species are predators of grasshopper egg capsules; some are aquatic or semiaquatic and partly predaceous. The small, slender, gray **Kelp Fly,** *Fucellia rufitibia* (with reddish tibiae), gathers in great swarms about fresh and

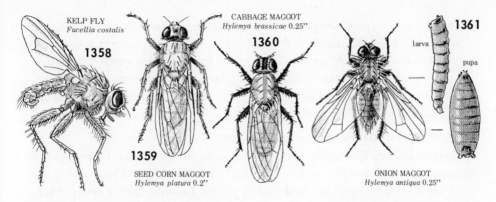

drying piles of kelp along the Pacific coast from British Columbia to California. *F. costalis* (Fig. 1358) is larger, quick in flight, and part of the time predaceous on sand flies along the beaches of southern California.

(68) SEED CORN MAGGOT: *Hylemya platura (cilicrura)*. **Fig. 1359.**

Also called the bean maggot. **Range:** Cosmopolitan; Alaska to Greenland, south to California and Florida; Europe. **Adult:** Grayish brown; resembles the house fly but smaller and more slender; hind femur with continuous series of short bristles on underside from base almost to apex. Eggs are laid in moist soil containing decaying plant materials. **Length:** .2". **Larva:** Yellowish white, .25" long full grown. It is attracted to sprouting seeds, being in the soil when the seeds sprout, feeds between the cotyledons or halves of the seed. (Those that hatch after seeds sprout bore into stems beneath the soil surface.) Hibernates as free maggot around roots of clover and in manure, or as maggot in brown puparium in infested field. Two or three broods in the North, four or five in the South. **Food:** Beans, corn, peas, and melons, to lesser extent potato seed pieces, onions, turnips, radishes; sometimes egg capsules of locusts.

(69) CABBAGE MAGGOT: *Hylemya brassicae*. **Fig. 1360.**

Sometimes called the radish maggot. **Range:** Alaska to California and Colorado, Manitoba to Newfoundland, south to Illinois and North Carolina; Europe. An introduced species. **Adult:** Ashy gray, with black stripes on thorax, black bristles over body; resembles the house fly but smaller. Hind femur with clump of short bristles on underside near base. White, finely reticulated eggs are laid on plants at the ground line. **Length:** .25". **Larva:** White, about .25" long full grown; blunt at rear end, pointed at front end. It tunnels into roots or fleshy parts, eats off small roots, causing plant to wilt; feeds three or four weeks before pupating. Usually hibernates as pupa in brown puparium in the soil. Two or three broods. **Food:** Cabbage, cauliflower, broccoli, Brussels sprouts, radish, turnip. Injury is seldom serious in southern part of range. Its principal enemies are rove beetles,

Aleochara spp. [see Coleoptera (96)]. *Hylemya seneciella* was introduced into the Pacific coast states from France to control the noxious weed tansy ragwort, supplementing an earlier import—the cinnabar moth *Tyria jacobaeae;* eggs are laid in the buds, the maggots (unlike the orange and black-striped moth larvae) feed singly on seeds within the heads and excrete a frothy substance that cements the florets.

(70) ONION MAGGOT: *Hylemya antiqua.* **Fig. 1361.**

Range: British Columbia to Quebec, south to Kansas and New Jersey; Europe. **Adult:** Slender, pale gray, with large wings. White, elongated eggs are laid in soil crevices near onion plants. **Length:** .25″. **Larva:** White, blunt at hind end and pointed at front end, about .3″ long full grown. It crawls down the plant behind the leaf sheath or tunnels down into the bulb; the mine causes the plant to turn yellow and wilt. Normally pupates in the soil near the plant, hibernates as larva or pupa in puparium. Two or three broods. **Food:** Onions; other plants are rarely attacked. Cull onions left on the ground or in piles are the most important source of infestation.

House Flies, Stable Flies, and Their Allies (Muscidae)

The muscids comprise a rather large family of flies small to medium in size, some shade of gray or brown, yellowish or black, with plain wings. As adults many visit flowers and are important pollinators; others are predaceous. The larvae feed on dung and excrement, dead or decaying animal matter, thus aiding in its decomposition; some are harmful, feeding on the flesh or blood of living animals. Because of their association with dead foods and filth, adults may cause myiasis of man and of domestic animals, or transmit disease-causing organisms. This family includes the notorious **Tsetse Fly** (*Glossina* spp.) of Africa, the sole carrier of the trypanosomes causing African sleeping sickness in man. Most muscid flies hibernate as adult; many try to enter houses in the fall.

(71) HOUSE FLY: *Musca domestica.* **Fig. 1362.**

Range: Cosmopolitan; one of the most widely distributed insects and the one most frequently associated with man. It is now believed to have been introduced from the Eastern Hemisphere. **Adult:** Thorax gray, with four darker longitudinal stripes; abdomen gray or yellowish, with darker median line and irregular pale yellowish spot at anterior lateral margins; arista plumose. The wings fold straight back at rest. Oval, white eggs are laid in batches of 75 to 100 on moist animal manure, excrement, and garbage, and hatch in 12 to 24 hours; each female lays five or six batches. **Length:** About .25″. **Larva:** Anterior end pointed, posterior end blunt, with two spiracles (useful in identification). It burrows deep into breeding material, using two mouth hooks. There are two or more broods per month in warm weather.

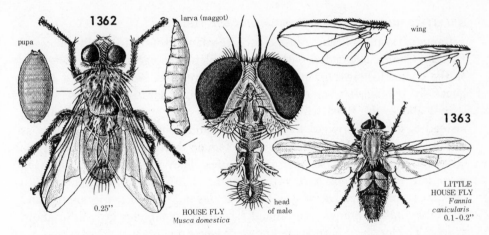

1362

pupa

larva (maggot)

wing

1363

0.25"

HOUSE FLY
Musca domestica

head
of male

LITTLE
HOUSE FLY
Fannia
canicularis
0.1-0.2"

Food: Adults are attracted to moist feces and decaying organic matter, but seek a variety of liquid foods (mouthparts are equipped for sponging as seen in Fig. 11D). Straw-colored spots left on surfaces are regurgitated liquid from the crop; dark spots are fecal matter. By this means or other contact with food they can transmit typhoid, dysentery, diarrhea, cholera, pinworm, hookworm, tapeworms. The house fly is a greater threat to human health than any other species of insect. They are susceptible to a fungus infection which may reach epizootic proportions; discharged fungus spores cause dead flies to stick to ceilings and windows.

(72) FACE FLY: *Musca autumnalis.*

Range: Northern half of the U.S. and across Canada, south to northern California, Kansas, Tennessee, and Georgia; an introduced species, first recorded in North America in 1952 and spreading rapidly. It prefers moist, shaded pastures, is scarce on dry, treeless pastures and the open plains. **Adult:** Difficult to separate from the house fly (71), having four stripes on the thorax and same wing venation; it is slightly larger, however, and darker; in *autumnalis* the eyes of the male almost meet (not so in *domestica*), stripe around eyes of the female silvery (it has a golden sheen in *domestica*), thorax slate gray. It breeds in cow dung, which *domestica* will rarely do, since it prefers a moist medium (dung dries quickly). **Length:** .3". **Food:** Blood oozing from wounds on cattle inflicted by blood-sucking flies; this is a "facultative bloodsucker"—attracted to animal blood and skin exudations but not able to puncture the skin. A pest of cattle, it is commonly seen around the eyes and nostrils of the animals (or resting on nearby rocks and fence posts); they leave the cows when the latter enter the barns. It has been said that "the disappearance of leprosy from Europe may have depended, at least in part, on the control of *M. autumnalis* brought about by the general improvement in sanitary standards."*

* L. S. West. See Bibliography: Technical.

(73) LITTLE HOUSE FLY: *Fannia canicularis.** **Fig. 1363.**

Range: Cosmopolitan; Alaska to Greenland, south to California and Florida. Closely associated with man, completely replaces *Musca domestica* as the common house fly in many subarctic regions. **Adult:** Similar to the house fly but smaller, more slender; dull gray, thorax with three darker stripes above; the third wing vein continues straight to the apical margin instead of curving upward to meet the second vein; arista not plumose as in the house fly. Abdomen darker in the female than in the male; that of the latter has larger yellowish area on the sides. They are often seen hovering in midair, darting back and forth. **Length:** .1–.2″. **Larva:** Flattened, with prominent lateral spines (Fig. 1364A). It develops in decaying animal and vegetable matter, human and animal excrement, completes growth in about a week. The larvae of *Fannia* are known to cause myiasis in man.

(74) LATRINE FLY: *Fannia scalaris.* **Fig. 1364B.**

Range: Cosmopolitan; Gulf of Mexico to Canada. It is found in all faunal areas of the world, except truly tropical and arctic regions. **Adult:** Head of male with 9 to 12 frontal bristles; thorax dark grayish, with four brownish stripes; legs black, wings clear, halteres yellow; abdomen gray pollinose, with median black stripe. It is attracted to lights, carcasses, and aphid-infested plants, breeds in the media indicated below. It is uncommon in houses. **Length:** .2–.3″. **Larva:** Segments with lateral spines, those on abdomen having slender forked projections on basal half of each, those on thoracic segments bare. It is found in fungi, wasps' and bees' nests, birds' nests, rodent burrows, dead land snails, dung, manure, cesspools, garbage, excrement of humans and other animals. "It has been implicated in cases of intestinal, aural, and vesicular myiasis." *F. trianguligera* and *F. incisurata* are almost identical with *scalaris,* slightly smaller on the average. *F. incisurata* is almost cosmopolitan, widespread from Canada to Mexico, but rare in the western half of the continent; *trianguligera* occurs in New Mexico, Texas, and Arizona. Females of *scalaris* are distinguished by the distinct stripes on the thorax and strong basal bristle on the middle of the femora; males of *incisurata* have 6 to 9 frontal bristles, and those of *trianguligera* have 7 to 10 bristles on the front of the head.

(75) COASTAL FLY: *Fannia femoralis.* **Fig. 1365.**

Range: Illinois to West Virginia and Georgia, west to Mexico, California, mostly near the coast. **Adult:** Shiny, black; head of male with 10 to 13 frontal bristles; thorax brownish black, without stripes; abdomen bluish gray, with median stripe. **Length:** .15–.2″. It breeds in carcasses, human excreta, cow manure, chicken manure. Common in poultry manure around

---

* Some authors place *Fannia* and *Ophyra* (76) with the preceding family, Anthomyiidae.

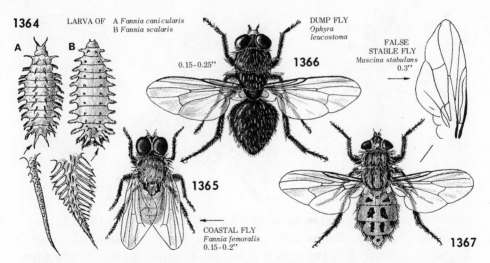

1364 LARVA OF A *Fannia canicularis* B *Fannia scalaris*

A B

0.15-0.25"

1366 DUMP FLY *Ophyra leucostoma*

FALSE STABLE FLY *Muscina stabulans* 0.3"

1365

COASTAL FLY *Fannia femoralis* 0.15-0.2"

1367

chicken ranches. Life cycle from egg to adult requires from 10 to 21 days. *F. pusio* is almost identical, the females being virtually inseparable; males of *pusio* may be distinguished by the row of long, fine bristles (lacking in *femoralis*) on the underside of the hind tibiae, and the shorter preapical bristles on the hind femora; it occurs along the coast from New York and New Jersey to Florida, west to Texas, Arizona, California, Mexico; also Hawaii, Guam, and Cuba.

(76) DUMP FLY: *Ophyra leucostoma.* **Fig. 1366.**

Also called the black garbage fly. **Range:** British Columbia to Newfoundland, south to Arizona and Florida; world-wide. At times abundant in urban communities, the principal species found around garbage disposal areas. **Adult:** Black, a little larger and stouter than the little house fly (73). It does not normally enter houses, but in the Pacific Northwest it may be numerous around restaurants and in some cases replaces the house fly as the dominant species. **Length:** .15–.25". **Larva:** Develops in mixed garbage, fowl excrement, decaying vegetation, animal matter. **Food:** Animal manure and vegetable matter, other fly larvae in the same medium.

(77) FALSE STABLE FLY: *Muscina stabulans.* **Fig. 1367.**

Also known as the nonbiting stable fly. **Range:** Alaska to Newfoundland, south to Arizona and Georgia. **Adult:** Often mistaken for the house fly, in having the four dark longitudinal stripes on the thorax; it is larger than *Musca domestica* and has a pale spot on top of the posterior end of the thorax; the third wing vein is bent slightly upward near the apical margin but does not meet the second vein. Eggs are laid in manure, human excrement, decayed vegetable matter, scattered garbage. It frequently enters houses and lays eggs in raw and cooked meats, feeds on milk and other liquid foods. **Length:** About .3". **Larva:** Becomes carnivorous as it nears

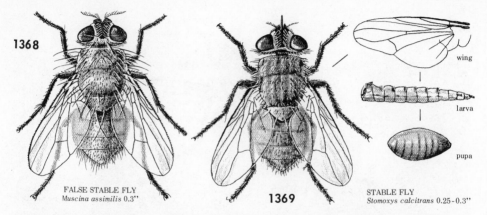

1368

FALSE STABLE FLY
Muscina assimilis 0.3"

1369

STABLE FLY
Stomoxys calcitrans 0.25-0.3"

wing

larva

pupa

maturity and destroys other fly larvae. Larval period is from 15 to 25 days' duration. Normally hibernates in pupal stage, sometimes as larva. Several broods in summer. It is believed that human intestinal myiasis results from eating food containing eggs deposited by the flies. *M. assimilis* (Fig. 1368) is similar but darker, often occurs with *stabulans* though less abundant.

(78) STABLE FLY: *Stomoxys calcitrans*. **Fig. 1369.**

Range: Cosmopolitan; Alaska, British Columbia to Nova Scotia, throughout the U.S., south to South America. **Adult:** Resembles the house fly but is distinguished from all other "domestic" flies by its long, sharp awl-like proboscis; the shape and venation of the wing is also distinctive, and it is held at an angle rather than flat when at rest. Eggs are laid on old straw stacks, piles of fermenting hay, weeds, grass, seaweeds. **Length:** .25–.3". **Larva:** Feeds on fermenting vegetation, never on feces or garbage. The life cycle from egg to adult is longer than that of house fly, averages from 21 to 25 days. **Food** (adult): Blood of domestic animals and man; it visits flowers for sugar (nectar), as do all bloodsucking flies. A fierce biter. Not known to be a vector of human disease, but does transmit surra (a trypanosomal disease) and infectious anemia (a virus disease) among horses, and causes myiasis of man and of domestic animals.

(79) HORN FLY: *Haematobia irritans*. **Fig. 1370.**

Range: British Columbia to Quebec, south to California and Florida. **Adult:** Resembles the house fly but is more slender and half the size. It has a conspicuous pointed proboscis similar to that of the stable fly (78), black and yellow palpi nearly same length as the proboscis; ground color black, cheeks with yellowish hairs, body with brownish dust and dark median stripe; wings are spread at a 60-degree angle when resting and folded when feeding. Eggs are laid on fresh cow dung. **Length:** .15". **Larva:** Ventral side of segments 6 to 12 with raised portion and fleshy spinelike processes; from .25" to .3" long full grown. It enters the manure upon hatching, pupates here or in the ground beneath. Life cycle from egg to adult requires about two

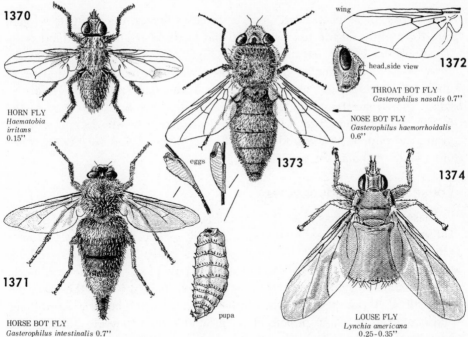

1370

HORN FLY
*Haematobia
irritans*
0.15"

wing

1372

head, side view

THROAT BOT FLY
Gasterophilus nasalis 0.7"

NOSE BOT FLY
Gasterophilus haemorrhoidalis
0.6"

eggs

1373

1374

1371

HORSE BOT FLY
Gasterophilus intestinalis 0.7"

pupa

LOUSE FLY
Lynchia americana
0.25-0.35"

weeks. **Food:** Blood of cattle mostly, of sheep to some extent, and horses. Strictly a range fly. The flies often rest at base of horns during cool weather but attack the back of the animal, out of reach of the tail and swing of the head (they rarely bite man); the number of flies on a single animal is frequently in the thousands. Cattle lose weight from loss of blood and fretting. The puparia are often parasitized by pteromalid wasps in cooler regions where populations of the flies are consequently smaller.

(80) HORSE BOT FLY: *Gasterophilus intestinalis* (Gasterophilidae). **Fig. 1371.**

Range: Throughout Canada and the U.S.; cosmopolitan. **Adult:** Brownish, very hairy, resembling a bumble bee; faint smoky markings on wings. It takes no food, lives from three days to three weeks. Eggs are glued to hairs, usually on front legs of the horse; they hatch only after a sudden rise in temperature, usually induced by licking of the tongue. **Length:** About .7". **Larva** (or "bot"): Yellowish white or pinkish, from .5" to .7" long full grown. It resembles a screw: posterior end blunt, anterior end pointed, ending in two mouth hooks, segments marked by circles of spines. It enters the mouth, feeds on mucous membrane of tongue for about two weeks, then passes to the stomach. It is full grown by late winter or spring; in May the mature larva releases its hold and passes to the ground in feces, where it pupates. Adults appear in June. One brood. **Host:** Horses. The **Throat Bot Fly,** *G. nasalis* (Fig. 1372), has rust red thorax, a band of black hairs around the middle of the abdomen, and clear wings; it glues eggs to base of hairs on head, jaw, or throat of horses; the maggots wriggle into the mouth, invade tissues around the teeth, later enter the duodenum. The **Nose Bot Fly**, *G. haemorrhoidalis* (Fig. 1373), considered worst of the three horse bots, is smaller than the other two, with reddish hairs on tip of abdomen; it glues eggs on lip hairs close to the skin (not around the nostrils); the larvae tunnel under epidermis of lip, later find their way to the mouth and intestinal tract, attach to wall of rectum before dropping to the ground to pupate; it is found mostly in the Midwest and Northwest. (For other kinds of "bots," see p. 651.)

Louse Flies (Hippoboscidae)

These strange insects bear little resemblance to flies superficially. They are flat and leathery, live externally on *one* host, as lice do, and like the Anoplura are bloodsucking. They are parasitic on birds and mammals, and are unusual in that they come close to the mammals in the way they are born. Reproduction is by pupiparity (integral viviparity), in which the larva develops in a uterine pouch in the mother, sustained by glandular secretions, and is mature, ready to pupate at birth; at this time they are glued to the hairs of the host.* The bird parasites have wings "just in case something happens to their transportation," Cole states, though in some cases they are short and functionless; most of the mammal parasites have no wings; some lose them when attaching to their hosts.

(81) LOUSE FLY: *Lynchia albipennis.*

Range: British Columbia to Quebec, south to Washington, Texas, Florida, Central and South America. **Adult:** Dark brown; mesonotum with deeply impressed longitudinal line and transverse median groove forming a cross; sides with grayish pile; dorsal surface of abdomen light brown, darkening laterally; wings whitish, with costal veins, and longitudinal veins basally, brown. **Length:** .2"; to tip of wing .3". They are found on shore birds (plovers and sandpipers), herons, mourning doves. *L. americana* (Fig. 1374) has about the same range, is very similar, slightly larger, occurs on our common birds of prey such as the great horned owl, screech owl, and hawks. *Pseudolynchia canariensis* is very damaging to squab pigeons and will bite man.

(82) SHEEP KED: *Melophagus ovinus.* **Fig. 1375.**

Sometimes called sheeptick. **Range:** British Columbia to Maine, south to California, Louisiana, North Carolina, Mexico, Central and South America; Europe. Not a native species. **Adult:** Ticklike; posterior edge of

* The Streblidae, which are bloodsucking external parasites of bats and found mostly in the tropical and subtropical regions in caves, and the Nycteribiidae—wingless spiderlike flies—are also pupiparous.

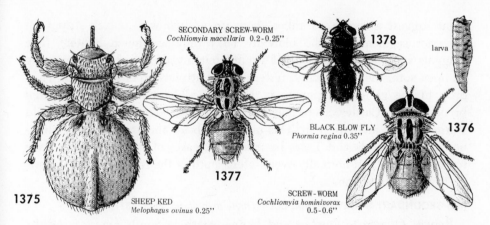

SECONDARY SCREW-WORM
Cochliomyia macellaria 0.2-0.25"

1378

larva

BLACK BLOW FLY
Phormia regina 0.35"

1376

1377

1375

SHEEP KED
Melophagus ovinus 0.25"

SCREW-WORM
Cochliomyia hominivorax
0.5-0.6"

head strongly arched and inserted in concave anterior slope of prothorax; antennae hidden in pits; palpi well developed, very long or short; legs short and thick, femora swollen, front tibia with one apical spur, middle tibiae with two spurs; wings rudimentary, reduced to knobs sometimes mistaken for halteres, which are absent. **Length:** .25". **Host:** Sheep. It may get on the sheephandler and inflict a painful bite.

Blow Flies (Calliphoridae)

The metallic blue and green or black blow flies—also called bottle flies—are named for their habit of depositing eggs on animal carcasses and meat products, causing them to "blow" or "bottle" with maggots. The larvae leave breeding media, bore into the ground to form puparia and transform to adults. They are mechanical vectors of disease organisms, like house flies, have similar mouthparts and feed in the same way. These flies have a very keen sense of smell and are capable of flying long distances. They cause animal and human myiasis but have less opportunity to contact food than house flies, since they invade houses less often. Blow flies overwinter as adults and will often try to enter houses for shelter in the fall.

(83) SCREW-WORM: *Cochliomyia (Callitroga) hominivorax.* **Figs. 1376, 1381,**
   Color plate 8h.
   Range: Southern Florida and Texas, occasionally parts of Oklahoma and New Mexico, south to South America; it may occur in New Jersey, Indiana, Minnesota, Montana, and California in summer, owing to cattle shipments. **Adult:** Dark, shiny blue-green, about twice the size of the house fly; three black longitudinal stripes on thorax, reddish yellow face. Eggs are laid near a wound in masses of 10 to 400, cemented together and to dry skin, hatch in 10 to 24 hours; the fly is especially attracted to the navel of a newborn calf. **Length:** .5–.6". **Larva:** Blunt at hind end, pointed at front end, about .7" long full grown; elevated spines encircling each seg-

ment suggest a screw; air tubes show up as two wide, black stripes on posterior segments [in the blow fly (85) they are much lighter]. The maggots start feeding at edge of wound but invade sound tissue, causing an open sore; the odor from a festering sore attracts other blow flies. They drop to the ground to pupate in brown seedlike puparia. Eight to ten broods. **Host:** Cattle, sheep, horses, mules, goats, hogs, domestic and wild warm-blooded animals, man. This insect has been almost completely eliminated from Florida and Texas by massive releases of sterilized males (see footnote, p. 630). It normally overwinters south of the frost line and spreads northward in the summer.

(84) SECONDARY SCREW-WORM: *Cochliomyia macellaria.* **Figs. 1377, 1382.**

Range: Oregon to Quebec and Maine, south to South America. **Adult:** Dark bluish green, with three black longitudinal stripes on thorax, yellow or reddish face; somewhat lighter in color than *hominivorax* (83) and *smaller* than the house fly. It lives two to six weeks on a variety of food—from garbage to nectar. Creamy white eggs are laid in masses of 40 to 250 in carcass or in open wound, in ears or eyes of living animal, in navel of newborn calf, and hatch in about four hours. **Length:** .2–.25". **Larva:** Unlike the screw-worm (83), it feeds on dead tissues only. It matures in 4 to 5 days on living animals, in 6 to 20 days on carcasses, drops to the ground to pupate. Six to fourteen broods. **Host:** Same as (83) above. Frequently involved in "blowing" of meat in shops, abattoirs, and homes.

(85) BLACK BLOW FLY: *Phormia regina.* **Fig. 1378.**

Range: Alaska to Labrador, south to Georgia and Mexico. **Adult:** Blue-green, dark blue, or greenish black. Very similar to green bottle fly (86), but a little larger and more robust, and darker. **Length:** About .35". Habits as described for green bottle fly. Abundant in the West. Responsible for most of the trouble from wool maggots in the Southwest [see (86)].

(86) GREEN BOTTLE FLY: *Phaenicia (Lucilia) sericata.* **Fig. 1379.**

Range: Canada to Mexico; cosmopolitan. **Adult:** Brilliant metallic bluish green, without markings, almost twice the size overall of the house fly. Eggs are laid in batches of about 180 in decomposing animal matter or in garbage containing a mixture of animal and vegetable matter. Strongly attracted to flesh, it attacks an animal soon after death or fresh meat a few minutes after exposure; it also deposits eggs on wounds, occasionally causing myiasis, and on soiled sheep's wool, where the larvae are called fleeceworms or sheep-wool maggots. The flies are most active on warm sunny days, are attracted to garbage, fruit, and nectar, enter houses in spring and fall. **Length:** .4–.5". **Larva:** It completes development in two to ten days, burrows into the ground to pupate. Hibernates as full-grown larva in the soil. Four to eight broods.

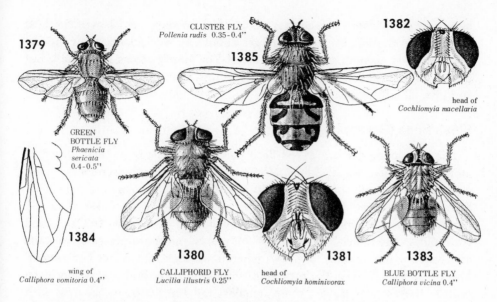

1379

1382

CLUSTER FLY
Pollenia rudis 0.35-0.4''

1385

head of
Cochliomyia macellaria

GREEN
BOTTLE FLY
*Phaenicia
sericata*
0.4-0.5''

1384

1380

1381

1383

wing of
Calliphora vomitoria 0.4''

CALLIPHORID FLY
Lucilia illustris 0.25''

head of
Cochliomyia hominivorax

BLUE BOTTLE FLY
Calliphora vicina 0.4''

(87) CALLIPHORID FLY: *Lucilia illustris (Phaenicia caesar).* **Fig. 1380.**

Range: Throughout North America; Germany. **Adult:** About the size of the house fly, but easily recognized by the bright metallic green thorax and abdomen. They often congregate in large numbers around garbage cans, from where they enter houses. They prefer to oviposit in decaying meat but will breed in human excrement, animal manure, and living flesh. **Length:** About .25''.

(88) BLUE BOTTLE FLY: *Calliphora vicina (erythrocephala).* **Fig. 1383.**

Range: Alaska to Quebec, south to Mexico and South America. **Adult:** Metallic blue abdomen, dark grayish thorax, covered with long fine hairs; large red eyes, orange cheeks, black beard. It is attracted to flowers, feces, overripe fruit, decaying vegetable matter, sores on animals, and may cause intestinal myiasis. **Length:** .4''. **Larva:** Normally develops in dead animal matter, frequently "blows" exposed meat. Cycle from egg to adult requires 15 to 20 days or more. *C. vomitoria* (Fig. 1384) is very similar in appearance and habits, can be distinguished by the black cheeks and reddish beard; it is more often found in slaughterhouses than in homes, is attracted to dead flesh but will oviposit on wounds and sores.

(89) CLUSTER FLY: *Pollenia rudis.* **Fig. 1385.**

Range: British Columbia to Nova Scotia, and throughout the U.S.; Europe and Africa. **Adult:** Resembles the house fly, but is distinctly larger and darker; thorax with thick coating of short, golden, crinkly hairs among the larger bristles; abdomen checkered with black and silver. Wings overlap at rest. They are sluggish and fly with a buzzing sound. Eggs are deposited indiscriminately in the soil. **Length:** .35–.4''. **Larva:** It enters the body of an earthworm, where it feeds for about thirteen days, and leaves the host to pupate in the soil; the adult emerges about two weeks later. Four broods. Swarms often accumulate in attics, closets, or empty rooms in the fall, but do no harm except to stain walls and curtains; many die during the winter and give off a disagreeable odor.

Flesh Flies (Sarcophagidae)

The Sarcophagidae are called flesh flies because many of them breed in animal flesh and meat. Some species breed in animal excrement, more especially dog stools, but the majority are parasitic and many are beneficial. Some feed on dead insects trapped in plants; others eat those killed by parasites. Strong fliers, the adults feed on nectar, honeydew, tree sap, and fruit juices. The males "take stations" on plants and other places of vantage and wait for the females to go by, whereupon they take off in hot pursuit. Certain species feed on the provisions in nests of bees and wasps and often destroy the eggs or larvae; some females trail the wasps as they fly with provisions to their nests. Unlike other "domestic" flies, but like the closely related Tachinidae, many females deposit living larvae rather than eggs. They may be distinguished from the muscids by the bare-tipped antennae (those of the Muscidae are feathery) and by the three black longitudinal stripes on the thorax (the muscids usually have four). They are often abundant in urban areas but do not normally enter houses. Sarcophagids do not appear to be vectors of human disease organisms but may cause myiasis.

(90) FLESH FLY: *Wohlfahrtia vigil.* **Fig. 1386.**

Range: Alaska to Nova Scotia, south to California, New Mexico, Iowa, and New Jersey; Europe. **Adult:** Black; second antennal segment and palpi yellow; thorax gray pollinose, with three broad, black longitudinal stripes; abdomen thinly gray pollinose; wings clear, tinged with yellow along costa and basally. **Length:** .45″. Living larvae are deposited on young mammals, where they bore into the skin, forming pustules; when development is completed they emerge and drop to the ground to pupate.

(91) SARCOPHAGID FLY: *Sarcophaga cooleyi.*

Range: British Columbia to Saskatchewan, south to Arizona and Kansas; Ontario, New York. **Adult:** Light gray; frontal stripe and second antennal segment brownish, back of head clothed with silvery white or slightly yellowish hairs, two rows of black cilia behind the eyes, cheeks clothed with black hairs; thorax and abdomen covered with short bristly hairs. **Length:** .4–.5″. It is a scavenger, breeding in human excrement and in garbage cans, especially the latter which contain fish remains. *S. haemorrhoidalis* (Fig. 1387) is similar, gray pollinose, from .4″ to .55″ long, almost

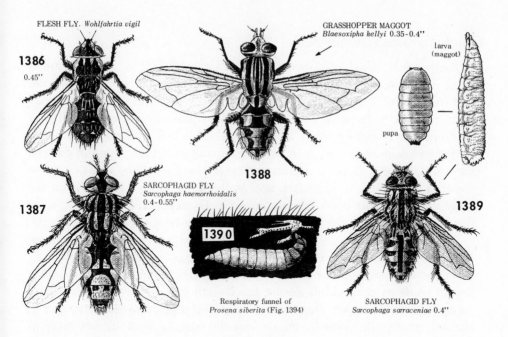

FLESH FLY. *Wohlfahrtia vigil*

1386
0.45"

GRASSHOPPER MAGGOT
Blaesoxipha kellyi 0.35-0.4"

larva
(maggot)

1388

pupa

SARCOPHAGID FLY
Sarcophaga haemorrhoidalis
0.4-0.55"

1387

1390

1389

Respiratory funnel of
Prosena siberita (Fig. 1394)

SARCOPHAGID FLY
Sarcophaga sarraceniae 0.4"

cosmopolitan in distribution, occurs from Oregon to Quebec, south to California and Florida, and in Europe; it has been reared from human excrement and the differential grasshopper, may cause myiasis in man. *S. aldrichi* is also similar, gray pollinose, from .35" to .45" long, ranges from British Columbia to southeastern Canada, south to Illinois and Connecticut; it is an important parasite of the forest tent caterpillar, larviposits on the host's cocoon, into which the maggots bore to feed on the larval contents; it also attacks the spruce budworm and gypsy moth pupae. *S. sarraceniae* (Fig. 1389) is about the same size, occurs in New York, to North and South Carolina, Alabama, and Georgia, breeds in dead insects and excrement.

(92) GRASSHOPPER MAGGOT: *Blaesoxipha (Sarcophaga) kellyi.* **Fig. 1388.**

Range: British Columbia to Ontario, south to New Mexico. **Adult:** Ash gray, slightly yellowish; top of head and frontal stripe brown, thorax with five ill-defined black stripes; abdomen yellowish gray pollinose, with three ill-defined dorsal stripes; female more grayish, with stripes faint and fifth abdominal segment red; legs black, halteres brown, wings clear. The female strikes a grasshopper in flight, larviposits under the open hindwing, causing host to fall to the ground. **Length:** .35–.4". **Host:** High plains grasshopper and others. An important factor in the natural control of grasshoppers. *B. falciformis* females larviposit in the hind femur of grasshoppers (*Melanoplus* and *Oedaleonotus*); the larva migrates to and develops in the thoracic cavity of the host.

Tachina Flies (Tachinidae)

Tachina flies, which comprise the second largest family in the order, are of great economic significance in the suppression of insect pest populations, and many have been used effectively in biological control. They are brisk flies, medium to large in size, somber black, gray, or brown with lighter markings, and characterized by a strongly bilobed metanotum. Robust, swift fliers, and frequently seen walking rapidly over the ground or foliage, they visit flowers, feed on nectar and other plant exudates and honeydew secreted by aphids, scale insects, and leafhoppers. Only a few feed on the body fluids of their hosts (as the parasitic Hymenoptera often do), since this is possible only in the few that have a piercing-type ovipositor.

Parthenogenesis is very infrequent in the group so far as known. Some female tachinids deposit minute (microtype) eggs on the foliage eaten by their hosts; these eggs are usually incubated (the larvae fully developed) before being oviposited but do not hatch until ingested by the host. More commonly they deposit larger (macrotype) eggs in or on the host, and in the soil, or deposit larvae in, on, or near the host; in some species maggots are placed on plants, where they latch on to the host as it comes by, or in the soil, where they search out their grub hosts. Both grubs and adults of May beetles are frequently parasitized.

The larvae of tachina flies are internal parasites of many different kinds of insects: Coleoptera, Hemiptera, Orthoptera, Hymenoptera (mostly sawflies), and more especially the Lepidoptera; a few parasitize sowbugs.* Most often they attack the larval or pupal stages of the host, though a large number of species attack only adults (especially among the Hemiptera and Coleoptera). Typically, the larvae are robust, with thin transparent cuticle and minute spines on each segment; those living free for a time before finding a host are armored with plates or scales covering the top and sides, with rows of spines on the underside. Species living free in the host have "spiracular hooks" at the rear end for temporary attachment to the trachea or outer wall of the host. Many tachinid larvae are unique (except for a few species of Sarcophagidae and Nemestrinidae) in obtaining oxygen directly from the air through a "respiratory funnel" (Fig. 1390) connected to a perforation in the host's skin or tracheal system. Attachment to the opening is secured by a funnel-like growth of the host tissue around the posterior end (and spiracles) of the parasitic larva; the growth is the host's defensive reaction to the irritation. Those attached directly to the outside (usually at

---

* *Melanophora roralis*, an intensely black fly, about .2″ long, is a common parasite of sowbugs in both Europe and North America. The puparium is deep orange-yellow, with the anal respiratory tubercles black; it occupies the entire body space under the chitinous remains of the sowbug.

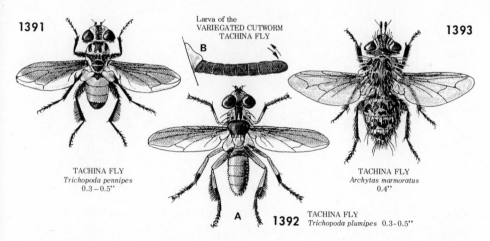

1391

Larva of the
VARIEGATED CUTWORM
TACHINA FLY
B

1393

TACHINA FLY
Trichopoda pennipes
0.3–0.5"

TACHINA FLY
Archytas marmoratus
0.4"

A   1392   TACHINA FLY
Trichopoda plumipes 0.3-0.5"

the point where the parasite enters the host) are dark and show through the host's skin.

(93) TACHINA FLY: *Trichopoda pennipes.* Fig. 1391.

Range: Throughout most of North America; introduced into Washington. **Adult:** Abdomen yellowish; scales covering halteres yellow. The female oviposits hard-shelled eggs on the adult host. **Length:** .3–.5". **Larva:** It bores through the chorion of its own egg, then through the integument of the host, attaches itself to the trachea, where a respiratory funnel forms. When full grown it leaves through or near the anus of the host before the latter dies; it pupates in the soil, excepting the winter generation which pupates inside the host. **Host:** Hemiptera. This parasite sometimes destroys 90 percent of the squash bug or stink bug population. There appear to be different "strains," one in the East that parasitizes the squash bug, one in the Southeast that attacks the southern green stink bug, and one in the West that attacks the bordered plant bug. Attempts to establish the eastern strain in California were not successful. *T. plumipes* (Fig. 1392A) occurs from Connecticut to the Carolinas, west to Kansas and south to Florida and Texas.

(94) VARIEGATED CUTWORM TACHINA FLY: *Archytas californiae.*

Range: British Columbia to Nova Scotia, and the entire U.S. **Adult:** Thorax brownish yellow pollinose; abdomen coal black; sides of face clothed with numerous fine white hairs, without bristles; scales covering halteres white. Adults feed on nectar and some honeydew. The female deposits many minute larvae on stem, leaves, and flowers of host food plants, fastening a membranous cup on the posterior end securely to the plant surface. **Length:** .4". **Larva:** The first instar—about .02" long—is a planidium-type larva, protected by a mosaic of blackish polygonal plates. The free-living maggots remain where seated by the mother, but bent over, pressed to the plant surface, and when there is a disturbance rear upward, swinging

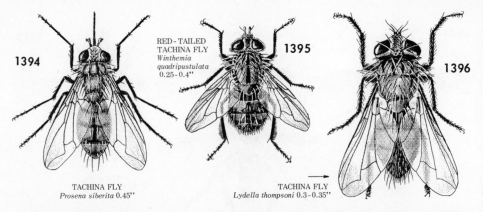

1394

1395

1396

RED-TAILED
TACHINA FLY
*Winthemia
quadripustulata*
0.25-0.4"

TACHINA FLY
Prosena siberita 0.45"

TACHINA FLY
Lydella thompsoni 0.3-0.35"

their heads in wide circles in an attempt to contact the host (Fig. 1392B). After doing so, they bore into the caterpillar, and the parasite becomes white, soft, and transparent; the large third instar almost completely fills the host pupa where it pupates. No breathing funnel develops at any stage, but air is obtained from the outside by applying the posterior spiracles to a rupture in the host. Two broods in the South. **Host:** Cutworms, armyworms. It is an important parasite of the variegated cutworm and armyworm. *A. marmoratus* (Fig. 1393) is similar, occurs from North Carolina to Florida, to Arizona, and from Kansas south to Peru.

(95) TACHINA FLY: *Prosena siberita.* **Fig. 1394.**

Range: Eastern U.S., imported from Europe. **Adult:** Black, with gray pollinose, dark crossbands. They are easily captured during the long period of copulation. The female deposits maggots on the soil (about 800 in all), and the larvae search for their hosts. **Length:** .45". **Larva:** Enters body cavity of host to feed, attaching anal spiracles to the host's trachea (see Fig. 1390). Pupates outside host in the soil. One brood. **Host:** Japanese beetle. Adults feed on flowers of umbelliferous plants.

(96) RED-TAILED TACHINA FLY: *Winthemia quadripustulata.* **Fig. 1395.**

Range: Washington to Vermont, south to California and Florida. **Adult:** Black, with gray pollinose; head with median red stripe, cheeks red, palpi yellow; thorax with five opaque black stripes, often poorly defined, and thin median stripe, scutellum red; abdomen black, sides tinged with red, often obscured by thick gray pollen, fourth segment red apically; wings grayish, transparent, tinged with brown at base. **Length:** .25–.4". The female extends her long ovipositor forward between the legs, oviposits two or more eggs quickly, usually on the last two thoracic segments where the caterpillar can't reach with its mandibles. **Larva:** It escapes from the egg through a slit in the chorion, bores into the host integument, and fixes itself at the point of entry; the respiratory funnel is distinguishable within a few hours. It feeds for a few days, leaves the host to pupate in the soil. Several broods. **Host:** More than fifty species of caterpillars are parasitized. Fifty percent of the armyworm population in an area is frequently destroyed; it is an important enemy of the corn earworm on alfalfa and vetch. A related introduced species, *Lydella thompsoni* (*grisescens*) (Fig. 1396), is a valuable parasite of the European corn borer now established throughout the Corn Belt; females deposit living larvae in or at the entrance to tunnels in the corn stalks; the larva finds the host caterpillar, bores into it or enters through a natural opening, and makes contact with a tracheal trunk, where a respiratory funnel forms; it hibernates as larva in the host, pupates in the tunnel, has two broods; a female produces up to 1,000 larvae, some of which will hatch in the uterus and destroy her (by working their way out) if they are not deposited in time. [See Hymenoptera (30) and (41).] *Lespesia archippivora* is a parasite of the bollworm, tobacco budworm, beet armyworm, cabbage looper, and salt-marsh caterpillar and potentially a biological control agent in cotton-growing areas.

(97) TACHINA FLY: *Bessa harveyi.*

Range: South of the Northwest Territory to Newfoundland, south to Washington and Ohio, also California and Arizona. **Adult:** Black; face and cheeks silvery, proboscis dark brown or blackish, thorax silvery pollinose with four black longitudinal stripes, wings transparent and grayish; abdomen shiny, black, base of segments two to four silvery pollinose. **Length:** .2″. Habits similar to *Winthemia* (96). **Host:** Larch sawfly, European spruce sawfly. It is an important factor in natural control of the larch sawfly, will parasitize spruce sawflies when its favored host is not present. Some multiple parasitism occurs with the ichneumon wasp *Mesoleius tenthredinis* (an imported parasite of the larch sawfly), in which case the tachinid is successful.* A related species, *Hyperecteina aldrichi* (Fig. 1397), introduced and established in the East, is about .25″ long, attacks adults of the Japanese beetle; hard-shelled white eggs are placed on the thorax (usually of the female), and are plainly visible against the dark green color; they hatch in two days and the maggot bores into the body cavity but does not make permanent attachment to the tracheal system; it pupates and hibernates in the host, has one brood.

(98) TACHINA FLY: *Compsilura concinnata.* **Fig. 1398.**

Range: Minnesota to Prince Edward Island, south to New Jersey; also British Columbia to California. Originally introduced into New England for control of the gypsy moth, colonies have been liberated on other hosts in many states. **Adult:** Black; face and top of head white, latter with deep black stripe; palpi orange, antennae brown; thorax whitish, with blackish

---

* The larvae of some strains of the larch sawfly are able to encapsulate eggs oviposited in them by *M. tenthredinis* and prevent their hatching.

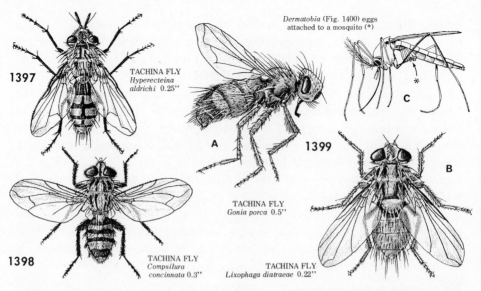

1397   TACHINA FLY
*Hyperecteina
aldrichi* 0.25"

Dermatobia (Fig. 1400) eggs
attached to a mosquito (*)

C

A   1399

B

TACHINA FLY
Gonia porca 0.5"

1398   TACHINA FLY
*Compsilura
concinnata* 0.3"

TACHINA FLY
Lixophaga diatraeae 0.22"

iridescence and four deep black stripes; wings almost glossy transparent. The female has a piercing-type ovipositor, larviposits one to four maggots in the body cavity of each host caterpillar. **Length:** .3". It hibernates as an immature larva in the host (not including the gypsy and browntail moths), pupates in the soil after emerging from the host; pupa dark brown, barrel-shaped. Three or four broods. **Host:** Gypsy moth, browntail moth, satin moth; nearly 200 species of Lepidoptera representing 20 families, and three families of sawflies are attacked. Parasitization of the satin moth is sometimes 50 to 70 percent. Another introduced species, *Townsendiellomyia nidicola,* is a valuable parasite of the brown-tail moth; fully incubated eggs are laid on the underside of host caterpillars in late summer; the maggot bores into the body cavity, migrates to the esophagus, where it hibernates, and moves back to the body cavity the following spring to feed, where a respiratory funnel forms at an opening in the host integument. *Blepharipa* (*Sturmia*) *scutellata*—also introduced and established in the East—is a valuable parasite of the gypsy moth, its only host; up to 5,000 minute hard-shelled black eggs are deposited by a female on foliage eaten by host caterpillars, but only those that reach the host's digestive tract will hatch; the mature larva emerges from the host pupa and drops to the ground, where it hibernates in the pupal stage. *Gonia porca* (Fig. 1399A)—a parasite of cutworms—occurs from British Columbia to Manitoba, south to California and Arizona; it is about .5" long, with abdomen all black except for red base, tip, and middorsal line. *Lixophaga diatraeae* (Fig. 1399B) was introduced from Cuba and established in Florida, where it helps control the sugarcane borer; it is about .22" long, the thorax gray pollinose with bluish tinge and black vittae, abdomen orange with gray pollinose patches.

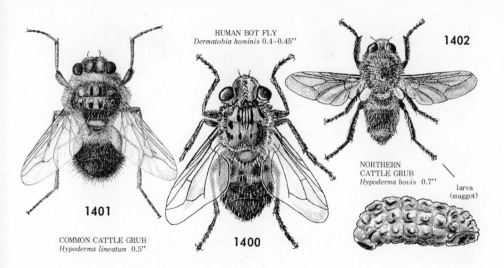

HUMAN BOT FLY
Dermatobia hominis 0.4-0.45"

1402

NORTHERN
CATTLE GRUB
Hypoderma bovis 0.7"

larva
(maggot)

1401

COMMON CATTLE GRUB
Hypoderma lineatum 0.5"

1400

Bot and Warble Flies (Oestridae)

The bot and warble flies are robust and hairy, resembling bumble bees, and swift in flight. The larvae are endoparasites of mammals and are called "bots." [See (80) for another kind of bot.] The tumorous swelling of the skin caused by bots is called a "warble." The closely related rabbit and rodent bots (Cuterebridae) include the large, blue, brown-winged **Human Bot Fly**, *Dermatobia hominis* (Fig. 1400), which parasitizes birds, man, and other mammals from Mexico to northern Argentina. It is noted for its diabolical manner of dispersal. The insidious fly seizes a mosquito, domestic fly, or tick, glues 15 to 20 eggs on the captive's body, and lets it go (see Fig. 1399C). When the latter alights on a warm-blooded animal, the eggs of the bot fly fall off, whereupon the larvae hatch and bore into the host's skin; they mature in 50 to 100 days and drop to the ground to pupate.

(99) COMMON CATTLE GRUB: *Hypoderma lineatum.* **Fig. 1401.**

Sometimes called ox bot, ox warble, or heel fly. **Range:** Throughout Canada and the U.S. **Adult:** Black, with white longitudinal stripes on thorax, tufts of whitish hairs on side of prothorax, reddish orange hairs on tip of abdomen. Oval white eggs are laid in a row on hairs of the animal, mostly about the heels and legs, never on the animal's back, where the maggots become lodged later. **Length:** .5". **Larva** (bot): It bores into the skin, and for the next six months migrates through the body: first to the gullet, thence to the diaphragm, and up along the ribs to the back, where it cuts a hole through the skin, places the posterior spiracles near the opening, and remains for about two months; the cysts cause irritation and suffering to the cattle. Hibernation takes place as a maggot in the animal's back; the mature larva turns brown or black, squeezes out through the hole, drops to the

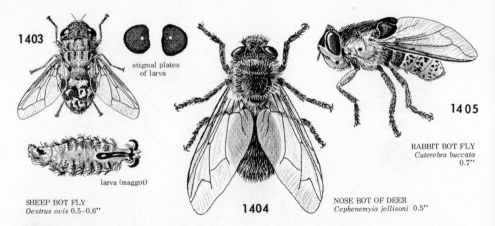

1403

stigmal plates
of larva

larva (maggot)

SHEEP BOT FLY
Oestrus ovis 0.5-0.6"

1404

1405

RABBIT BOT FLY
Cuterebra buccata
0.7"

NOSE BOT OF DEER
Cephenemyia jellisoni 0.5"

ground to pupate. Adults emerge five weeks later—from April to June. **Host:** Cattle, bison; rarely horses, goats, or man. *H. bovis* (Fig. 1402), the **Northern Cattle Grub** (also called large warble fly or bomb fly), has a similar life history, but lays its eggs singly on hairs of the hind legs near the hocks and knees, and on the belly, with much buzzing and darting about, which greatly agitates the animals; the adult is larger than that of *lineatum*— about .7" long—yellowish, with more clearly defined reddish tip on the abdomen; the maggots migrate to the animal's back by a less devious route, are never found in the gullet. In migrating through the body, cattle grubs may injure the esophagus (and interfere with swallowing), impair other organs, damage the spinal cord (causing paralysis); carcasses lose value, milk production is reduced.* *H. tarandi* infests reindeer and caribou herds throughout Alaska.

(100) SHEEP BOT FLY: *Oestrus ovis.* Fig. 1403.

Range: Throughout North America, wherever sheep are grazed. **Adult:** Head and thorax dull yellowish, with numerous tiny black nodules; abdomen shiny gray. Eggs hatch in uterus of the female fly, and tiny maggots are deposited on edge of nostrils of sheep and goats. **Length:** .5–.6". **Larva** (bot): Creamy white to yellowish, about .8" long full grown, with wide dark or brownish crossbands on upper side; lower surface spiny. It moves up the nasal passages to the sinus, where it feeds until mature, causing inflammation of the mucous membranes and a nasal discharge like that of a head cold. In hot weather larvae remain in the sinuses about a month, return to the nasal passage, and drop to the ground to pupate. In colder regions those reaching the sinuses in the fall remain here for the winter. **Host:** Sheep,

* See M. A. Khan, *Control of Cattle Grubs in the Prairie Provinces*, Pub. 1309, Canada Dept. Agr., Ottawa (1967); H. J. Stockdale, *Dairy Cattle Insects and Their Control*, Pm. 322, Coop. Ext. Serv., Iowa State Univ., Ames (1966); *Beef Cattle Insects and Their Control*, Pm. 294, Coop. Ext. Serv., Iowa State Univ., Ames (1967); *Control Insect Pests on Dairy Cattle Safely*, Leaf. 279, Ext. Serv., Pennsylvania State Univ., University Park.

goats. The nose bot of deer, *Cephenemyia jellisoni* (Fig. 1404), is said to be "the swiftest of all flies," and is credited with flying at the rate of 800 miles an hour!

(101) RABBIT BOT FLY: *Cuterebra princeps* (Cuterebridae).

Range: Washington to South Dakota, south to Arizona, New Mexico, Texas, Mexico. **Adult:** Bluish black; antennae with first two joints dark brown; thorax dull, with thin black pubescence, sides and underside with dense yellowish white pubescence; abdomen shiny steel-blue, second segment with or without tuft of white hairs on each side, sides and underside with white pollinose. **Length:** .75″. **Host:** The larva (bot) causes tumorous swelling or warble under the skin of cottontail and jack rabbits. *C. tenebrosa* is black, abdomen without pollen; it attacks pack rats and grasshopper mice from British Columbia and Oregon to South Dakota, south to California and Colorado. *C. buccata* (Fig. 1405) was the principal bot fly found on rabbits in an Illinois survey. Cases of nasal and cutaneous myiasis in man caused by *Cuterebra* larvae have been reported.

23

SIPHONAPTERA: Fleas

As indicated by the name of the order, fleas have sucking mouthparts and no wings. The laterally compressed body, long powerful legs and smooth cuticle with short spinelike hairs directed backwards enable them to move swiftly through the host's hairy or feathery covering (see Fig. 7D). Fleas are parasitic and require the blood of warm-blooded animals to reproduce. The majority of species infest burrowing mammals; some are associated with large carnivores, others with birds and bats. Only man among the primates is a host of fleas. Several species of rat fleas infest homes and buildings which harbor rats, and will bite humans as well. The Oriental rat flea and others transmit plague by biting first an infected rat and then a human; they are vectors of bacteria called *Pasteurella pestis,* the cause of plague, and may themselves die or recover from the infection. Many fleas also carry rickettsiae called *Rickettsia typhi* (*mooseri*), the cause of "murine or endemic fleaborne typhus," from infected rats to man.* Fleaborne typhus occurs in Europe, Africa, South and Central America, and our southeastern states; it is a milder form of typhus than "epidemic louseborne typhus," which is caused by *Rickettsia prowazeki* and is transmitted by the body louse (see Anoplura: Sucking Lice).†

* Rickettsiae are considered as intermediate between bacteria and viruses. Like bacteria (and unlike viruses) they can be seen under an ordinary microscope. Like viruses (and unlike bacteria) they do not grow on artificial media and can be propagated only in the host.
† Twenty-some species of fleas found in the U.S. can transmit plague organisms under laboratory conditions. Some are possible vectors of bacteria (*Pasteurella tularensis*) causing tularemia; the common vectors of this organism are ticks and a deer fly [see Diptera (24)].

Metamorphosis (Fig. 1406) in fleas is complete. The female of most species lays its eggs while on the host, but the eggs usually fall to the floor, ground, or bedding, where they hatch in a few days. The tiny whitish, legless larvae have biting and chewing mouthparts, live in the dirt or dust, and feed on organic matter. They wriggle violently when disturbed, are not normally found on the host. The life cycle in warm, moist climates averages six to seven weeks. Adult fleas can live several weeks without food and are not necessarily eliminated by a short absence from the animals harboring them. Excepting the sticktight fleas, they do not stay with one host but transfer from one individual or one species to another. Fleas do not develop and live in sand for successive generations without feeding on animals; these places are therefore not reservoirs of infestations as some people believe. Throughout the world, more than 1,000 species are known; the authority Karl Jordan believes there are actually twice this number.

(1) HUMAN FLEA: *Pulex irritans* (Pulicidae). **Fig. 1407.**

Range: Throughout the world. **Adult:** Light at first changing to dark brown. Preantennal region of head with two long bristles: one below the eye (see Oriental Rat Flea), the other at the base of the maxilla; each thoracic segment with a row of bristles alternately long and short. Genal and pronotal spines absent. It can jump 6 to 8 inches high and 12 to 15 inches horizontally. Adult life may extend to two years or more. The female lays upwards of 500 eggs on animals or floors of houses, barns, or other shelter. **Length:** .04″. **Larva:** Whitish, cylindrical, sparsely covered with hairs. Life cycle from egg to adult varies from 15 to 65 days, depending on temperature, moisture, and food supply. **Host:** Man, dogs, cats, rabbits, coyotes, deer, horses, mules, rodents, and pigs (mainly the last two).* A possible vector of plague, an intermediate host of the rodent tapeworm *Hymenolepis nana*, which is known to infest man [see (2) below]. *P. simulans*—"long misidentified as *P. irritans*"—is the common flea of coyotes.†

(2) DOG FLEA: *Ctenocephalides canis*. **Fig. 1406.**

Range: Cosmopolitan. **Adult:** It is easily differentiated from the human flea by the elongated anterior portion of the body, long labial palpi, genal ctenidium (comb) of 7 or 8 spines and pronotal ctenidium of 14 to 16 spines. The female lays about 70 eggs on hairs of the host or on the floor. **Length:** .07″ (male) to .15″ (female). **Larva:** Whitish, elongate, cylindrical, with sparse covering of hairs. Life cycle from egg to adult requires from 27

---

* The world-wide occurrence of the "so-called human flea" and its association with man (and not other primates), is believed by Holland to be through the pig. He notes (1969) that "in New Guinea pigs are kept close to and often inside the household where piglets may be nursed by native women along with their own children."

† H. T. Gier and D. J. Ameel, *Parasites and Diseases of Kansas Coyotes*, Tech. Bul. 91, Agr. Exp. Sta., Kansas State Univ., Manhattan (1959).

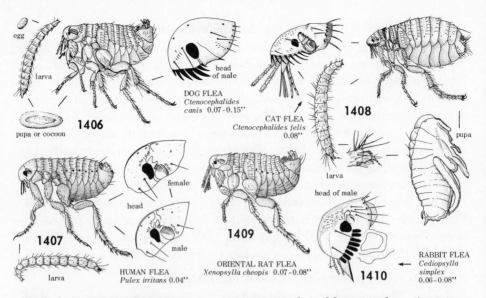

egg

larva

pupa or cocoon

1406

DOG FLEA
*Ctenocephalides
canis* 0.07-0.15"

head
of male

CAT FLEA
Ctenocephalides felis
0.08"

1408

larva

pupa

1407

larva

head

female

male

HUMAN FLEA
Pulex irritans 0.04"

1409

ORIENTAL RAT FLEA
Xenopsylla cheopis 0.07-0.08"

head of male

1410

RABBIT FLEA
*Cediopsylla
simplex*
0.06-0.08"

to 40 days. **Host:** Dogs, cats, man, rats, squirrels, rabbits, poultry. An intermediate host of the dog tapeworm *Dipylidium caninum* and the rodent tapeworm *Hymenolepis nana,* both of which occur in man, especially children. Eggs of the tapeworm are ingested by the flea larva, and the tapeworm completes its development in the adult flea. Dogs are infected by nipping the fleas and pass them on to the children when allowed to lick their faces. The dog flea may also be a vector of plague, and has been known to harbor *Rickettsia typhi,* the cause of murine or endemic typhus.

(3) CAT FLEA: *Ctenocephalides felis.* Fig. 1408.

Range: Cosmopolitan. **Adult:** Very similar to *C. canis* and often mistaken for it; they were considered the same species for a long time. It may be distinguished from other fleas by the shallow, gradually sloping head, with first two genal spines equal in length; inner side of hind femur has 7 to 10 bristles (there are 10 to 13 in the dog flea). **Length:** .08" (female); male slightly less. **Host:** Same as dog flea. With *C. canis* (2), it may be a vector of a nematode causing "heartworm" of dogs and occasionally cats [see Diptera (8)]. The cat flea can also transmit murine or endemic typhus.

(4) ORIENTAL RAT FLEA: *Xenopsylla cheopis.* Fig. 1409.

Range: Cosmopolitan. **Adult:** Like the human flea (1) and chigoe (12), it lacks the genal and pronotal ctenidia displayed by the dog and cat fleas; it has ocular bristle in front of eye (see human flea). **Length:** .07–.08". **Larva:** The larval period lasts 30 days or more. The pupal stage is comparatively long: from 25 to 34 days. Adults live as long as a year. **Host:** Rats and other rodents, man. It is the principal vector of *Pasteurella pestis,* the cause of plague; also a carrier of *Rickettsia typhi,* the cause of murine

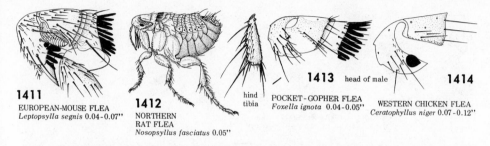

1411
EUROPEAN-MOUSE FLEA
Leptopsylla segnis 0.04-0.07"

1412
NORTHERN
RAT FLEA
Nosopsyllus fasciatus 0.05"

hind
tibia

1413 head of male
POCKET-GOPHER FLEA
Foxella ignota 0.04-0.05"

1414
WESTERN CHICKEN FLEA
Ceratophyllus niger 0.07-0.12"

or endemic fleaborne typhus. This is the flea primarily responsible for the bubonic plagues—Black Death—of the Old World during the fourteenth and seventeenth centuries.* It is an intermediate host of the rodent tapeworms *Hymenolepis diminuta* and *H. nana* [see (2) above], both of which frequently infest children. The **Rabbit Flea,** *Cediopsylla simplex* (Fig. 1410)—a true parasite of the eastern cottontail (*Sylvilagus floridanus*), also found on the bobcat, red fox, and short-tailed shrew—may be recognized by the angulate head, nearly vertical genal ctenidium of 8 spines on a side, and arrangement of bristles on the head; pronotum has single row of bristles and ctenidium of 6 or 7 spines on a side; it is from .06" to .08" long.

(5) EUROPEAN-MOUSE FLEA: *Leptopsylla* (*Ctenopsyllus*) *segnis* (Ceratophyllidae). Fig. 1411.

Range: Widely distributed in Europe and North America, especially the port cities. **Adult:** Structural features of head and pronotum, as shown in the figure, serve to distinguish this flea. **Length:** .04–.07". **Host:** Rats, mice, men. A true parasite of the European house mouse (*Mus musculus*), now common in North America. It is a "weak vector of plague," capable of transmitting murine or endemic typhus; it seldom bites man, and is considered a negligible factor in outbreaks of plague.†

(6) NORTHERN RAT FLEA: *Nosopsyllus fasciatus.* **Fig. 1412.**

Range: Cosmopolitan. **Adult:** It may be recognized by the shape of the head, well-developed eyes, pronotal ctenidium of 18 to 20 teeth, lower row of three setae in the male (two in the female) in front of the eye. **Host:** Brown or Norway rat (*Rattus norvegicus*), black or roof rat (*R. rattus*), various house rats, wood rats, mice, other rodents, man. It is not believed to be a significant factor in natural outbreaks of plague. Like the Oriental rat flea (4), it is an intermediate host of *Hymenolepis diminuta*, a common tapeworm of rats and mice which frequently infests children. The **Pocket-gopher Flea,** *Foxella ignota* (Fig. 1413)—a true parasite of the eastern

---

* A pneumonic form which spreads directly from one person to another usually occurs during epidemics. This form, resembling the Black Death of medieval Europe, caused the death of 60,000 people in Manchuria in 1911; it is believed to have stemmed from the Siberian marmot, which was hunted for furs.

† This species is sometimes placed in the family Hystrichopsyllidae.

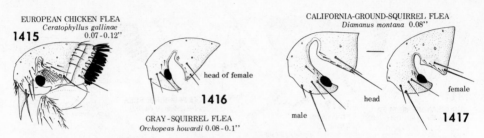

EUROPEAN CHICKEN FLEA
Ceratophyllus gallinae
1415 0.07-0.12"

head of female
1416

GRAY-SQUIRREL FLEA
Orchopeas howardi 0.08-0.1"

CALIFORNIA-GROUND-SQUIRREL FLEA
Diamanus montana 0.08"

head

male

female

1417

(*Geomys bursarius*) and northern (*Thomomys talpoides*) pocket gophers, also found on white-footed or deer mice, and moles—is distinguished by the bristles on the head and hind tibia, pronotum with row of alternately long and short bristles and ctenidium of 10 or 11 spines; nine geographic variations, or subspecies, have been named.

(7) WESTERN CHICKEN FLEA: *Ceratophyllus niger*. Fig. 1414.

Range: Western part of North America. **Adult:** A very dark species, as implied by its name; it may be distinguished from other fleas found in chicken houses by frontal outline of head with sharp in-cut, comparatively large, egg-shaped eyes, and arrangement of bristles in front of the eyes. Pronotal ctenidium consists of 26 to 28 spines. It contacts the host only when it wishes to feed. **Length:** .07" (male)–.12" (female). **Host:** Fowl, sparrows, pigeons, man. They are destructive to young poultry during warm summer days if unchecked. The closely related **European Chicken Flea**, *C. gallinae* (Fig. 1415), occurs in the New England states, may be identified by the lateral row of 4 to 6 bristles on inner surface of the hind femur.

(8) CALIFORNIA-GROUND-SQUIRREL FLEA: *Diamanus montana*. Fig. 1417.

Range: Washington to California, east to Nevada, Utah, Arizona, Colorado, and New Mexico. **Adult:** Distinguished by arrangement of bristles on the head, as shown in the figures, and by the long spine at the tip of the second joint of the hind tarsus extending beyond the third and over the fourth joints; abdominal tergites each with two rows of bristles. **Length:** .08". **Host:** California (or Beechey) ground squirrel (*Otospermophilus beecheyi*); seldom found on other rodents, usually the only flea found on this squirrel, whose burrows are "teeming with them" during July and August. A moderately efficient vector of plague.* **Frankin's-ground-squirrel Flea,** *Opisocrostis bruneri*—a large flea, from .12" to .15" long—has two rows of bristles on preantennal region of head, a postantennal row of four or five bristles, a single row of alternately stout and weak bristles on the pronotum, two or three rows of bristles on the mesonotum, and about five rows on the metanotum; it ranges from Manitoba to Alberta, Illinois and Wis-

---

* Sylvatic (wild rodent) plague is endemic in the western third of North America. Natural epidemics flare up occasionally among rodents, but only a few outbreaks have involved man.

consin to Colorado and Montana, is a true parasite of Franklin's ground squirrel (*Citellus franklinii*) and an efficient vector of plague. The **Gray-squirrel Flea**, *Orchopeas howardi* (*wickhami*) (Fig. 1416)—from .08″ to .1″ long—may be recognized by the arrangement of bristles on the head; pronotum with single row of bristles and ctenidium of nine spines on a side; it is abundant in the East, a true parasite of the gray squirrel (*Sciurus carolinensis*).

(9) AMERICAN MOUSE FLEA: *Stenoponia americana* (Hystrichopsyllidae). **Fig. 1418.**

Range: New Brunswick to Manitoba, south to Virginia, Tennessee, Alabama, west to Montana. **Adult:** Genal ctenidium composed of 13 teeth on each side; pronotal ctenidium consisting of 25 or 26 teeth on each side. **Length:** .16–.2″. A large species, abundant in the fall and winter months. **Host:** Rodents and insectivores; favored host is the white-footed (or deer) mouse (*Peromyscus*).

(10) CHIPMUNK FLEA: *Tamiophila grandis*. **Fig. 1419.**

Range: Massachusetts, New York, Michigan; Nova Scotia to Manitoba. **Adult:** It may be distinguished by the arrangement of bristles on the head and the genal ctenidium of two spines: the first much wider and shorter than the other; pronotum with two rows of bristles and a ctenidium of 10 or 11 spines on each side. In this family the eyes are absent or vestigial. **Length:** .16–.2″. A large species. **Host:** Eastern chipmunk (*Tamias striatus*), its true host; also cottontail rabbit, red squirrel, and weasel.

(11) STICKTIGHT FLEA: *Echidnophaga gallinacea* (Tungidae). **Fig. 1420.**

Range: Eastern, southern, and southwestern states; along the Pacific coast into Oregon. **Adult:** Dark brown, almost black. Distinctive features are: head angulate in front, divided by groove, with two bristles in each of pre- and post-antennal regions; eyes heavily pigmented, mandibles deeply serrated and tapering toward the end; single row of bristles on pronotum and mesonotum, no ctenidia; patch of 18 to 20 bristles on inner side of hind coxa. **Length:** .04″ (male)–.08″ (female). **Host:** Poultry, birds, dogs, cats, rabbits, ground squirrels, rats, horses, man. They attach themselves in masses on the face, wattles, and earlobes of chickens and turkeys, and remain there, excepting the males which tend to move about at night. They are very damaging to chickens if unchecked, and occasionally get under the skin of poultry handlers. A vector of murine or endemic typhus.

(12) CHIGOE: *Tunga penetrans*. **Fig. 1421.**

Also called such names as sand fleas, tique, chique, pico, bicho, and jiggers (not to be confused with chiggers—mites which cause reddish welts on the human skin). **Range:** Mexico, Central and South America, other tropical regions; southern United States sporadically. **Adult:** It may be

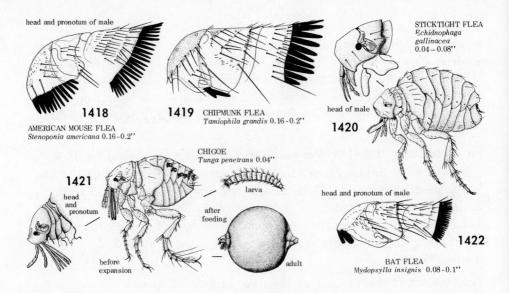

head and pronotum of male

1418

AMERICAN MOUSE FLEA
Stenoponia americana 0.16-0.2"

1419 CHIPMUNK FLEA
Tamiophila grandis 0.16-0.2"

STICKTIGHT FLEA
*Echidnophaga
gallinacea*
0.04-0.08"

head of male

1420

CHIGOE
Tunga penetrans 0.04"

1421

head
and
pronotum

larva

after
feeding

before
expansion

adult

head and pronotum of male

1422

BAT FLEA
Mydopsylla insignis 0.08-0.1"

distinguished from the sticktight flea (11) by its smaller size and the absence of a patch of bristles on the hind coxa. The female acts like any other flea until it has mated; it then bores under the skin to lay eggs, and remains there. As the eggs develop, the body of the flea becomes enormously distended, forming a sphere which may reach the size of a pea. The eggs are usually extruded from an opening in the skin, and most of them fall to the ground, where they hatch in a few days. **Length:** .04". **Larva:** Most of the tiny white, hairy larvae develop on the ground like other fleas. Some eggs remain in the wound and the larvae develop in the tissues, causing a sore similar to the lesion caused by blow flies. It normally feeds for six to twelve days, spins a cocoon, pupates, and emerges in about five days. **Host:** Poultry, birds, dogs, cats, man. Badly infested Brewer's blackbirds and English sparrows have been found in Texas. On poultry the fleas are found mostly around the eyes and comb; on dogs and cats they lodge around the ears; they attack man mainly under the toenails and between the toes, causing lesions which may result in tetanus infection or gangrene.

(13) BAT FLEA: *Myodopsylla insignis* (Ischnopsyllidae). **Fig. 1422.**

This family is distinguished from the others by the hardened "preoral plates" or flaps which resemble ctenidia and are located near the front on each side of the oral margin. *M. insignis* is the most common and widely distributed of the North American bat fleas. **Range:** Kentucky to Maine, west to the Rocky Mts. and across Canada. **Adult:** Four or five stout bristles near front of antennae, postantennal bristles in four rows. Pronotal ctenidium with 18 to 21 slender spines on each side. **Length:** .08–.1".

Appendixes

Appendixes

Geologic Eras or Periods: The Place of Insects in Time

ERA	PERIOD	EPOCH	MILLIONS OF YEARS AGO	EVENTS
Archaeozoic ("Precambrian")			1,000–2,000	Period of rock formation. No fossils have been traced to this period.
Proterozoic ("Precambrian")				Period of prolonged erosion, great volcanic activity and upthrusting of mountains. The Mesabi iron ores are formed. Fossils from this era are scarce, but many forms of primitive life are believed to have existed.
Paleozoic era or Primary period	Cambrian		600	Alternate submergence and uplift of great land masses. Large continental land masses under water. Rich in fossils of trilobites,* brachiopods (lamp shells),† snails, and sponges. Fossils from the oldest Cambrian include the Onychophora, forms of which still exist.‡
	Ordovician		500	Land low in elevation, more limestone formed than sandstone and shale. Rich in invertebrate marine life. Some fishlike vertebrates appear.
	Silurian		425	Deserts formed, salts deposited. Primitive fish increase. Beginnings of the first life on land, with the appearance of primitive plants. Scorpions—perhaps the first animals to extract oxygen from the air—and myriapods (millipedes and centipedes) appear. The primitive wingless insects are believed to have evolved during this period.

* Trilobita, a class of Arthropoda now extinct. They resemble isopods such as woodlice slightly and are probably closely related to the crustaceans.

† Brachiopoda, a small phylum of animals now few in numbers and species. They are bivalves and resemble mussels somewhat, though not closely related.

‡ A phylum intermediate between the Arthropoda and Annelida. They are caterpillar-like, have numerous feet and a pair of antennae, are carnivorous and live in rotten wood in the tropical countries.

Geologic Eras or Periods: The Place of Insects in Time (Continued)

ERA	PERIOD	EPOCH	MILLIONS OF YEARS AGO	EVENTS
	Devonian		405	Continental land masses mostly above sea, large areas later flooded and sediments deposited, sandstones formed. Among the invertebrates, trilobites, corals, starfish, and sponges are abundant. Fish dominate marine life, include sharks and armored fish. Amphibians appear; the land is forested with giant ferns. The Collembola—the most ancient of all insects —are known from fossil imprints in the Old Red Sandstone of this period. Cockroaches probably had their beginning in this period.
Carboniferous		Lower Carboniferous (Mississippian)	345	This and the preceding (Devonian) period are referred to as the Age of Sharks.
		Upper Carboniferous (Pennsylvanian)	310	Repeated submergence and rising of land, with great deposits of sediment; a period of mountain formation and volcanic activity. Rapid growth of vegetation and extensive marshes of the warm, humid Pennsylvanian period paved the way for later deposits of coal. Land animals include primitive amphibians and reptiles, spiders, snails, scorpions, and huge archaic dragonflies. Insects attain their greatest size during this period (fossils of ancient dragonflies have a wingspread of more than two feet). Cockroaches—most ancient of the winged insects—are so abundant in the strata of this period that it is referred to as the Age of Cockroaches.
	Permian		280	Marked by extreme cold and aridity, with extensive salt and gypsum deposits. Long submergence during Early Permian resulted in thick deposits of sandstone and limestone. Most of the continent is above sea level during the Upper Permian, during which the Appalachians were thrust up. The Odonata arise in this period, parallel with but not directly descended from the gigantic dragonflies of the Pennsylvanian which now

become extinct. Primitive Ephemeroptera (mayflies), Plecoptera (stoneflies), and Mecoptera (scorpionflies) appear in the Lower Permian. Modern families of the Mecoptera are known from the Baltic amber of the Lower Oligocene and fossils of the Miocene. Permian deposits are rich in the remains of early Homoptera, from which the Hemiptera (Heteroptera) are believed to be derived. True Coleoptera are known from the Upper Permian.

Mesozoic era or Secondary period		Most of the continent is above sea level, more especially the Appalachian region, which undergoes erosion. Beetles are abundant in the deposits of this era. Most of the phytophagous or plant-feeding kinds do not appear until the Cretaceous period of this era. Among the Hymenoptera, only the Symphyta (sawflies and horntails) are known from the Mesozoic; the parasites and other wasps and the bees come later with the flowering plants. This era is known as the Age of Reptiles.
Triassic	230	Reptiles dominant in the sea and on land, dinosaurs numerous. Conifers dominant plant life; cycads—intermediate between ferns and palms—appear. True flies are believed to have appeared during the Triassic. Primitive Trichoptera (caddisflies) are abundant in the Upper Triassic, where the earliest Hemiptera (Heteroptera) are also found.
Jurassic	181	Deserts of the Southwest flooded and sediments deposited. The period closes with upheaval of mountains and much volcanic activity in the West. Reptiles dominate sea, land, and air; plants include cycads, tree ferns, and conifers. Planipennia (lacewings) and phasmatids (walkingsticks) flourished during this time. The early Dermaptera (earwigs) and Psocoptera (booklice) are found in the deposits of this period. (The Mallophaga and Anoplura—chewing and sucking lice respectively—are not known as fossils.) Modern families of Hemiptera (bugs) and Ephemeroptera (mayflies) are evident in this period.

Geologic Eras or Periods: The Place of Insects in Time (Continued)

ERA	PERIOD	EPOCH	MILLIONS OF YEARS AGO	EVENTS
	Cretaceous		135	The Late Cretaceous is marked by uplift of land and receding of water, and beginnings of the Rockies. Dinosaurs dominant on land until close of the period, when primitive mammals appear. Many of the modern plants are established with flowering plants (angiosperms) gaining ascendency over the spore-forming and more primitive types, giving impetus to the development of phytophagous insects and parasitic Hymenoptera. Trees include maple, oak, birch, elm, and willow.
Cenozoic era or Tertiary and Quaternary periods	Tertiary			The present landscape begins to take shape, mammals become dominant.
				Fossils of all living families of the Hymenoptera are found in the Tertiary.
		Paleocene ⎫ Eocene ⎬ Sometimes considered as one epoch.	63 58	Submergence of the Atlantic and Gulf coastal plains and the Great Valley of California. The ancestors of modern mammals appear, including that of the horse. The earliest fossils of the Lepidoptera belong to this epoch, but their ancestry is believed to go back to the Mesozoic.
		Oligocene	36	More land above sea level. The more ancient mammals give way to carnivores (dogs and cats), beavers, mice, squirrels, and rabbits. Giant hogs and camels appear, while the horse develops. The Diplura, Thysanoptera (thrips), Isoptera (termites), Siphonaptera (fleas), Strepsiptera, and mantids are known first from the Baltic amber which accumulated in the Lower Oligocene. Remains of the Embioptera have been found in the Baltic amber and deposits of the Miocene, but their ancestors are believed to go back to the Lower Permian.

	Miocene	25	The Atlantic and Gulf coasts and a large part of the Great Valley of California are submerged, the Cascades and Coast Range thrust up. Grass increases in relation to forests. Mastodons, weasels, cats, and camels appear, and the horse continues development.
	Pliocene	13	Outline of North America assumes that of the present. The epoch closes with uplift of the Cascades, Rockies, Colorado plateau, and Appalachians, and much volcanic activity in the West. Northwestern Europe partly submerged; the Alps are thrust up, and volcanoes—including Vesuvius and Etna—are active.
Quaternary	Pleistocene	1	Known as the Great Ice Age. The ice advanced south four times, covering large parts of Europe, North and South America, and Asia. Warmer periods followed each retreat. Animals included elephants, bisons, camels, true horses, giant wolves, sabertooth cats, pigs, armadillos, sloths, and man. Man's early history is divided into two periods: **Paleolithic or Old Stone Age** Man in this early period was a hunter and cave dweller. Neanderthal man of the Middle Paleolithic made carefully shaped flake tools. Cro-Magnon man of the Upper Paleolithic is considered a race of modern man, *Homo sapiens.* Flint blades and lance points, needles and other tools made of bone, and more highly developed cave art are features of the Upper Paleolithic. **Neolithic or New Stone Age** The period following the Old Stone Age features the development of polished stone tools, pottery, carpentry, and weaving, the domestication of animals and the beginning of agriculture.

It will be seen from this brief outline of the geologic periods that insects are very ancient. "They are twice as old as the Reptiles, which are old inhabitants of the Earth, and three times as old as the Mammals: perhaps a thousand times as old as the human species."* Atomic physics has provided a method of measuring this lapse of time with a fair degree of accuracy by estimating the products of the decay of radioactive materials in the various rock layers.

The existence of certain land masses, or "asylums," which were rarely inundated after the Cambrian period, permitted the terrestrial fauna to develop. These areas are designated: Laurentia (the Laurentian Plateau, or Canadian Shield in North America), Angara Land (the Siberian Plateau), and Gondwanaland (in the Southern Hemisphere). The Canadian Shield—the earth's greatest area of exposed Archaean rock, composed largely of granite, gneiss, and schist—is rich in minerals, forests, and furs; no marine deposits later than the Cambrian are found here.

* R. Jeannel. See Bibliography: General.

Orders and Families* Represented in This Book

1. Thysanura: Bristletails
        Lepismatidae: Silverfish, Firebrats
2. Collembola: Springtails
        Sminthuridae (1)
        Hypogastruridae (3)
        Poduridae (4)
3. Orthoptera: Cockroaches, Grasshoppers, and Their Allies
        Blattidae: Cockroaches (1)
        Mantidae: Mantids (6)
        Phasmatidae: Walkingsticks (9)
        Acrididae: Grasshoppers (10)
        Tettigoniidae: Longhorn Grasshoppers and Katydids (17)
        Gryllidae: Crickets (22)
        Gryllotalpidae: Mole Crickets (25)
        Tridactylidae: Pigmy Mole Crickets (25)
        Gryllacrididae: Cave and Camel Crickets (26)
4. Dermaptera: Earwigs
        Forficulidae
        Labiduridae
5. Plecoptera: Stoneflies
        Perlidae (1)
        Pteronarcidae (2)
        Perlodidae (4)

* The arrangement, with subdivisions in the larger groups, is as follows:
Order
    Suborder
        Superfamily   -oidea (name ending)
            Family   -idae
                Subfamily   -inae
                    Tribe   -ini
The number in parentheses is the first species in that family, the main species being numbered separately in each order.

6. Embioptera: Embiids or Webspinners
    Oligotomidae
7. Isoptera: Termites
        Rhinotermitidae: Subterranean Termites (1)
        Kalotermitidae: Nonsubterranean Termites (3)
8. Psocoptera: Booklice, Psocids
        Liposcelidae: Booklice
        Caeciliidae: Psocids
9. Mallophaga: Chewing Lice
        Menoponidae: Poultry Body Lice (1)
        Philopteridae: Feather Chewing Lice (3)
        Trichodectidae: Mammal Chewing Lice (4)
10. Anoplura: Sucking Lice
        Pediculidae: Human Lice (1)
        Linognathidae: Smooth Sucking Lice (4)
        Haematopinidae: Wrinkled Sucking Lice (4)
11. Thysanoptera: Thrips
        Thripidae (1)
        Aeolothripidae (8)
12. Hemiptera (Heteroptera): True Bugs
        Corixidae: Water Boatmen (1)
        Notonectidae: Backswimmers (2)
        Nepidae: Waterscorpions (3)
        Belostomatidae: Giant Water Bugs (4)
        Gerridae: Water Striders (6)
        Veliidae: Small Water Striders (7)
        Mesoveliidae: Water Treaders (8)
        Hydrometridae: Marsh Treaders (9)
        Hebridae: Velvet Water Bugs (10)
        Gelastocoridae: Toad Bugs (11)
        Saldidae: Shore Bugs (12)
        Cimicidae: Bat, Bed, and Bird Bugs (13)
        Miridae: Plant Bugs (14)
        Nabidae: Damsel Bugs (17)
        Anthocoridae: Flower Bugs, Minute Pirate Bugs (18)
        Phymatidae: Ambush Bugs (19)
        Reduviidae: Assassin Bugs (20)
        Tingidae: Lace Bugs (27)
        Lygaeidae: Lygaeid Bugs (29)
        Coreidae: Coreid Bugs (33)
        Berytidae: Stilt Bugs (35)
        Pentatomidae: Stink Bugs (36)
        Cydnidae: Burrower Bugs (46)
        Pyrrhocoridae: Pyrrhocorid Bugs (47)
13. Homoptera: Aphids, Leafhoppers, Scale Insects, and Their Allies
        Cicadidae: Cicadas (1)
        Cercopidae: Spittlebugs (2)
        Membracidae: Treehoppers (3)
        Cicadellidae: Leafhoppers (4)

Pterophoridae: Plume Moths (277)
Cosmopterigidae: Cosmopterigid Moths (Webworms) (278)
Tortricidae: Leaf Roller Moths (279)
Yponomeutidae: Ermine Moths (Plutellid Moths) (284)
Gelechiidae: Gelechiid Moths (Leaf Miners, Borers) (285)
Gracillariidae: Leaf Blotch Miners (291)
Tineidae: Clothes Moths (292)
Coleophoridae: Casebearer Moths (294)
Lyonetiidae: Lyonetiid Moths (Leaf Skeletonizers and
Perforators) (295)
Incurvariidae: Yucca Moths (296)
20. Coleoptera: Beetles
Adephaga
Caraboidea
Cicindelidae: Tiger Beetles (1)
Carabidae: Ground Beetles (16)
Haliplidae: Crawling Water Beetles (51)
Dytiscidae: Predaceous Diving Beetles (53)
Gyrinoidea
Gyrinidae: Whirligig Beetles (57)
Polyphaga
Hydrophiloidea
Hydrophilidae: Water Scavenger Beetles (59)
Staphylinoidea
Silphidae: Carrion Beetles (65)
Leptinidae: Mammal Nest (Parasitic) Beetles (71)
Limulodidae: Ant Nest Beetles (72)
Staphylinidae: Rove Beetles (73)
Pselaphidae: Ant-loving Beetles (97)
Scaphidiidae: Shining Fungus Beetles (99)
Histeridae (Histeroidea):* Hister Beetles (100)
Cantharoidea
Lycidae: Net-winged Beetles (107)
Lampyridae: Fireflies (111)
Cantharidae: Soldier Beetles (Leather-wings) (116)
Melyridae (Cleroidea): Soft-winged Flower Beetles (120)
Cleridae (Cleroidea): Checkered Beetles (124)
Mordelloidea (Meloidea)
Mordellidae: Tumbling Flower Beetles (131)
Meloidae: Blister Beetles (134)
Pyrochroidae: Fire Beetles (142)
Anthicidae (Tenebrionoidea): Ant-like Flower Beetles (143)
Elateroidea
Elateridae: Click Beetles, Wireworms (148)
Buprestidae (Buprestoidea): Flatheaded or Metallic Wood Borers
(159)
Dascilloidea

* Superfamilies in parentheses indicate rearrangements as shown in *The Beetles of the United States* by Ross H. Arnett, Jr.

Chironomidae (= Tendipedidae): Midges (13)
Simuliidae: Black Flies (15)
Mycetophiloidea
Cecidomyiidae (= Itonididae): Gall Midges (18)
Bibionoidea
Bibionidae: March Flies (22)
Brachycera (Orthorrhapha)
Tabanoidea
Tabanidae: Deer Flies, Horse Flies (23)
Stratiomyidae: Soldier Flies (25)
Rhagionidae: Snipe Flies (26)
Asiloidea
Therevidae: Stiletto Flies (27)
Scenopinidae: Windowpane Flies (29)
Apioceridae: Apiocerid Flies (30)
Mydidae: Mydas Flies (31)
Asilidae: Robber Flies (32)
Bombyliidae: Bee Flies (36)
Nemestrinidae: Tanglewing Flies (38)
Acroceridae: Small-headed Flies (39)
Empidoidea
Dolichopodidae: Long-legged Flies (41)
Empididae: Dance Flies (42)
Cyclorrhapha: Circular-seamed Flies
Syrphoidea
Pipunculidae: Big-headed Flies (43)
Syrphidae: Flower Flies (44)
Conopidae: Thick-headed Flies (49)
Phoroidea
Phoridae: Humpbacked Flies (50)
Nothyboidea
Psilidae: Carrot Rust Fly and others (51)
Diopsidae: Stalk-eyed Flies (52)
Micropezoidea
Neriidae: Neriid Flies (53)
Tephritoidea
Otitidae: Otitid Flies (54)
Pyrgotidae: Pyrgotid Flies (55)
Tephritidae: Fruit Flies (56)
Sciomyzoidea
Coelopidae: Seaweed Flies (61)
Pallopteroidea
Piophilidae: Skipper Flies (62)
Drosophiloidea
Ephydridae: Shore Flies (63)
Drosophilidae: Vinegar Flies (64)
Chloropoidea
Chloropidae: Chloropid Flies (65)

GLOSSARY: Definition of Terms*

abdomen—third or posterior part of the insect body, attached to the thorax, usually having only nine or ten segments apparent, without legs in the adult stage.

agamic—reproducing without mating, as when males are absent.

alate—winged; any winged form; except in the case of termites, it commonly applies to the winged female only.

alimentary canal—digestive tube between the mouth and the anus.

alitrunk—part of thorax to which the wings are attached; includes the propodeum or first abdominal segment in the Hymenoptera.

anal plate (shield)—hard covering of dorsum on last segment of caterpillar.

anal proleg—fleshy unjointed leg on the last abdominal segment of certain insect larvae, used for clinging to plants and in locomotion.

anal vein—unbranched longitudinal vein (A) near **anal** or **inner margin** of wing, starting at base and often extending to the outer margin (*see* **costa; subcosta; radius; media; cubitus**).

androconia—specialized scent scales appearing on various parts of the wings of certain butterflies.

annuli (*sing.,* **annulus**)—rings encircling a segment or joint. **annulate** = ringed.

antennae (*sing.,* **antenna**)—paired sensory appendages, segmented, exceedingly varied in form—one on each side of the head—sometimes referred to as "horns" or "feelers."

anterior—front or toward the front; ahead of. Opposite of **posterior**.

anus—opening of the digestive tract through which waste products are discharged; *adj.,* **anal**—at or near the posterior end of the body.

apex (*pl.,* **apexes** or **apices**)—that part opposite or farthest from the base or point of attachment; *adj.,* **apical**—at or near the apex.

* Based mainly on *A Glossary of Entomology,* by J. R. de la Torre-Bueno (Brooklyn Entomological Society); *A Dictionary of Biology,* by M. Abercrombie, C. J. Hickman, and M. L. Johnson (Penguin Books); *Webster's New World Dictionary* (The World Pub. Co.), and *Webster's New Collegiate Dictionary* (G. & C. Merriam Co.).

aphidophagous—feeding on aphids; **aphidivorous.**

appressed—closely applied to.

apterous—wingless.

arista—long spine or bristle on the antennae of certain flies; terminal portion of the antennae in the higher Diptera.

articulate—*v.*, to connect at a joint; *adj.*, jointed or segmented.

asexual—reproducing without the union of sperm and egg; parthenogenetic.

asperate—roughened. Surface roughenings of tiny dotlike elevations are called **asperites.**

bacteria (*sing.*, **bacterium**)—single-celled microscopic organisms with a simple nucleus and no chlorophyll, which multiply by simple division. *See* **protozoa; virus.**

base—that part nearest the body or point of attachment. The base of the thorax is that part nearest the abdomen; the base of the abdomen is that part nearest the thorax. Opposite of **apex.**

bipectinate—pectinate on both sides; feather-like.

bisexual—having two distinct and separate sexes.

blasted—blotching of leaves or petals, resulting in wilting, caused by feeding of insects such as thrips.

boll—the seed pod or capsule of a plant, especially cotton or flax. *See* **square.**

bot—larva of a bot fly or warble fly. *See* **warble.**

brachypterous—having short or abbreviated wings, thus unable to fly; a trait more often found in the female insect.

bristles—stiff (and usually short, stout) hairs. True bristles articulate with the skin in a cavity or socket.

callus (*pl.*, **calluses**)—hardened lump or swelling of cuticle, one at base of wing.

cambium—*See* **sapwood.**

Cambrian—first period of the Paleozoic era (600 million years ago).

canthus—raised chitinous or hardened area sometimes dividing the eyes.

capitate—having a head. Applied to antennae *abruptly* enlarged at the tip; **clavate** antennae are *gradually* enlarged.

Carboniferous—fifth period of the Paleozoic era; in North America it is divided into Lower Carboniferous or Mississippian (345 million years ago), and Upper Carboniferous or Pennsylvanian (310 million years ago).

carina (*pl.*, **carinae**)—elevated ridge or keel.

carnivorous—flesh-eating, as opposed to **herbivorous.**

caste—specialized form of individual in a **social** colony with special functions.

cauda (*pl.*, **caudae**)—the tail; tail-like appendage at anal end. *See* **cerci.**

cerci (*sing.*, **cercus**)—Slender paired appendages, believed to be sensory in function, at the hind end of many insects; they may be modified for grasping. *Syns.*, **caudal filaments, anal forceps** or **stylets.**

cervical shield—hardened plate on the prothorax of caterpillars just behind the head; also called **prothoracic shield.**

cervix—upper part of neck. In the Diptera, that part of the **occiput** lying between the **vertex** and **neck.**

chafer—a name—from the Greek, meaning "devourer"—applied to some scarab beetles: rose chafer, pine chafer, European chafer, and others.

chalcid—a wasp belonging to the superfamily Chalcidoidea (Hymenoptera).

chitin—a constituent of the integument or cuticle of insects and other arthropods (also certain other invertebrates, and of the walls of fungi); the hardened parts of the insect covering are said to be **chitinous.**

chordontal organs—sound receptors; hearing mechanism.

chorion—the outer membrane or shell covering the insect egg.

chrysalis or **chrysalid** (*pl.*, **chrysalises** and **chrysalides**)—parchment-like covering protecting the pupal stage of butterflies.

cilia (*sing.*, **cilium**)—fringes arranged in tufts or single hairs; also applied to scattered hairs.

cirrate—applied to pectinate antennae with long, curved lateral branches which are sometimes fringed with hairs. *See* **plumose.**

claspers—a pair of clasping appendages, part of the male genitalia.

clavate—clubbed or thickened toward the tip. *See* **capitate.**

clavola—"all the insect antennae except the first and second segments." *See* **ring joints.**

clypeal hairs—in mosquito (*Anopheles*) larvae, hairs on lower part of head. *See* **sutural hairs.**

clypeus—lower part of the head below the front and to which the labrum or upper lip is attached.

coccid—a scale insect belonging to the superfamily Coccoidea (Homoptera).

cochineal—crimson dye from scale insects (*Dactylopius*), more especially *D. coccus;* the insect itself.

cocoon—protective covering of the pupa, spun from silk fibers produced by the larvae of moths and many other kinds of insects.

coition—act of mating. *Syn.*, **copulation.**

colony—community or aggregation of individuals, either unorganized or organized and social in behavior. *See* **gregarious** and **social.**

commensalism—intimate association of two dissimilar organisms in which one is benefited without harming the other; they are said to be **commensals.** *See* **symbiosis.**

compound eyes—the large lateral pair of eyes possessed by most insects; they consist of a few or many separate facets (planes) and associated cones or sensillae called ommatidia.

connate—united along the whole length, as in the elytra of some beetles.

connectives—longitudinal nerve cords or fibers connecting the **ganglia.**

copulation—joining together of the male and female in mating.

corbicula (*pl.*, **corbiculae**)—a smooth concavity with fringe of hairs, on outside of hind tibiae of certain bees, used for carrying pollen; **pollen basket.**

cordate—heart-shaped.

corm—fleshy underground stem or bulb of certain plants, such as the gladiolus.

cornicles—pair of tubes placed posteriorly, one on each side of the aphid's back, which against an attacker emit a defensive fluid (not to be confused with the anal honeydew prized by ants and other insects).

costa (*pl.*, **costae**)—raised longitudinal ridge, rounded at the crest; also, longitudinal vein (C) along anterior margin of wing (*see* **subcosta; radius; media; cubitus; anal vein**).

coxa (*pl.*, **coxae**)—first or basal segment of the leg, between the body and trochanter; it fits into the **coxal cavity** on the underside of the body.

cremaster—spiny process at the tip of the abdomen by which some pupae are suspended; it is used by some subterranean pupae in working their way to the surface before the adult emerges; apex of terminal segment of abdomen.

crepuscular—active or flying at dusk. *See* **nocturnal** and **diurnal**.

crochets—curved spines or hooks on the bottom of the prolegs of caterpillars; curved spines on the cremaster of pupae.

ctenidium (*pl.*, **ctenidia**)—short, thickened spines arranged in a row like the teeth of a comb; unlike true bristles, they are nonarticulated extensions of the integument. The stout spines on the hind margin of the pronotum of certain fleas compose the **pronotal comb** (**ctenidium**); those on the lateroventral border of the head compose the **genal comb** (**ctenidium**).

cubitus—fifth longitudinal vein (Cu) of the wing extending from the base; it usually has two branches (*see* **costa; subcosta; radius; media; anal vein**).

cuneus—small triangular area at end of hardened portion of **hemelytra**.

cuticle (**cuticula**)—noncellular outer covering of the insect body. *Syn.*, **integument**. *See* **epidermis** and **chitin**.

dactyl—toe, finger, or digit.

declivity—a declivous or downward slope.

decumbent—bending down at the tip, with the base upright. *See* **recumbent**.

deflexed—bent downward abruptly.

denticulate—having small teeth or notches.

depressed—flattened.

detritus—any fragment rubbed off; debris.

Devonian—fourth period of the Paleozoic era (400 million years ago).

diapause—period of suspended animation or development "irrespective of the environmental conditions," with metabolic processes greatly reduced; may be correlated with the seasons as during **hibernation** and **estivation**.

diatoms—microscopic unicellular plants occurring singly or in colonies, abundant among other plankton of salt and fresh water.

dimorphism—two distinct forms in the same species; sexual or seasonal variations in color and other characteristics. *See* **polymorphism**.

disc (**disk**)—central upper surface within the margins, often applied to the pronotum, and to the wing (discal area, discal cell).

discal area—central area of the wing occupied by the discal cell.

discal cell—the large cell in the median area of the wing (between the stems of R and Cu veins in the Lepidoptera), usually called "the cell."

distal—end farthest from point of attachment. Opposite of **proximal**.

diurnal—active or flying in the daytime. *See* **nocturnal** and **crepuscular**.

dorsum—generally, the upper surface. In the Coleoptera it usually refers to the mesothorax and metathorax; in the Diptera it applies to the area between the neck and scutellum, limited laterally by the dorsopleural suture separating the side and top.

ectoparasite—*see* **external parasite**.

elytra (*sing.*, **elytron**)—the forewings serving as wing covers for the hindwings and possessed by the Coleoptera. Sometimes applied to **tegmina** of the Orthoptera and **hemelytra** of the Hemiptera.

emarginate—with margin notched or section cut out.

endemic—prevalent, confined to, or continuously occurring in a particular region, as in the case of an animal or plant, or a disease.

endoparasite—*see* **internal parasite**.

entomophagous—feeding on other insects, as opposed to **phytophagous**; predaceous or parasitic on other insects.

eonymph—early nymphal or larval stage. (*Eo*— means "early part of a period.")

epidemic—rapid spread of a disease or contagion. Improperly applied to **epizootic.**
epidermis—layer of cells underlying and secreting the cuticle. *Syn.,* **hypodermis.**
epinotum—dorsal part of the prothorax; in ants, the **propodeum.**
epipharynx—palate-like organ on inside of labrum. *See* **hypopharynx.**
epiphysis—movable process on inner side of front tibiae of some Lepidoptera.
epipleuron (*pl.,* **epipleura**)—infolded (bent under) edge of elytron.
episternum (*pl.,* **episterna**)—larger anterior sclerite on sides of the thorax.
epizootic—rapid spread of a contagion among the arthropods. *See* **epidemic.**
estivation—period of dormancy during the hot or dry season; analogous to **hibernation** (*which see*).
evaginated—turned inside out.
eversible—capable of being everted—turned outside, or inside out.
excretion—act of getting rid of the waste products of metabolism though not necessarily eliminating them; the waste products excreted. The waste matter eliminated is also called **excreta**—a term sometimes applied to **feces.**
exserted—protruded.
external parasite—one that feeds from outside the host; **ectoparasite.**
extruded—pushed or forced out.
exudation—the oozing out of drops from an opening; the substance exuded. The matter exuding is also called an **exudate.**
exuviae (*sing.,* **exuvia**)—cast-off skins from molting.
eyespot—rudimentary ocellus; any colored spot that looks like an eye.
facet—side or plane surface, as in a diamond or the compound eye of an insect.
facultative parasite—normally free-living but able to adopt a parasitic life.
falcate—sickle-like; said of a wing incurved below the point of the apex.
fascia—a broad transverse stripe or band. *See* **vitta.**
feces—waste products of digestion which have been eliminated; **excrement.**
femur (*pl.,* **femora**)—third segment of the leg, usually the stoutest, between the trochanter and the tibia; the thigh.
filariasis—a tropical disease caused by nematodes and transmitted by mosquitoes.
filiform—slender, uniform in width; threadlike.
flagellates—small organisms (Protozoa) having one or more flagella or tails.
flagellum—whip- or tail-like process; of antennae, the **funicle** (*which see*).
fontanel (**fontanelle**)—shallow depression on the head (opening of frontal pore) in termites.
fossa (*pl.,* **fossae**)—pit or deep **sulcus** (*which see*).
fovea (*pl.,* **foveae**)—deep, well-defined depression.
frass—feces (or fecula) deposited on leaves, stems, or fruit by larvae.
frenulum—bristle at the base of the hindwing of a moth which hooks into the **retinaculum** and holds the wings together in flight.
frons (**front**)—unpaired sclerite on upper anterior portion of the head.
frontal—front of the head.
frontal lunule—oval or crescentric space above the base of the antennae, bounded by the **frontal suture;** characteristic of the Cyclorrhapha, or "circular-seamed" flies.
fungi (*sing.,* **fungus**)—unicellular plants, without chlorophyll, growing in unorganized masses without roots, stems, or leaves, often microscopic; includes molds, rusts, mildews, yeast, mushrooms. Antibiotics are developed from fungi.
fungivorous—fungus-eating.

funicle (funiculus)—a small cord, or slender stalk; in the insect antenna, that part of the **clavola** between the club and **ring joints.**

furcula—forked process; a "spring" on underside of the fourth abdominal segment in Collembola, used for jumping (*see* **tenaculum**); pair of overlapping forked appendages protruding from base of anus in Orthoptera.

fusiform—spindle-shaped; broad at the middle, narrowed toward both ends.

gall—plant growth caused by irritation of tissues owing to presence of insects, fungi, bacteria or other microscopic forms (e.g., nematodes).

ganglion (*pl.*, ganglia)—a nerve center composed of a mass of cells and fibers.

gaster—enlarged part of the abdomen (the last seven or eight segments) extending beyond the petiole of ants; the abdomen.

gena (*pl.*, genae)—side of head below the eye, extending to the throat or **gular suture;** the cheek.

genal comb—*see* **ctenidium.**

geniculate—knee-jointed; sharply bent.

genitalia—external parts of the sexual organs.

gills—respiratory organ in aquatic forms which extracts dissolved oxygen and salts from the water. *See* **tracheal gills.**

girdle—silk thread wrapped around the thorax of suspended butterfly pupae.

glabrous—smooth, hairless, and without punctures or raised structures.

gonads—the female and male reproductive glands: ovaries and testes.

granulate—surface roughened by small granules.

granulosis (*pl.*, granuloses)—a virus infection of insects, producing granular bodies (or capsules) in cells of the affected tissues. *See* **polyhedrosis,** also **virus.**

gregarious—living in communities or colonies but *not* social (e.g., aphids, squash bugs, bark beetles, some caterpillars and sawflies); parasites are said to be gregarious when more than one progeny of a given parent normally develops in or on a single host. Opposite of **solitary.**

gula—the throat.

guttate—having **guttae** or small spots.

haemocoele—main body cavity, containing the blood.

halteres—knoblike rudimentary wings (in place of hindwings) occurring in the Diptera; they are on the metathorax and believed to act as balancers.

haustellate—having mouthparts adapted for sucking liquids; having a **haustellum** or sucker.

heartwood—*see* **sapwood.**

hemelytra (*sing.*, hemelytron)—forewings of the Hemiptera, with basal half thickened and leathery, apical half membranous; sometimes refers to the **tegmina** of the Orthoptera.

herbivorous—plant-feeding, as opposed to **carnivorous.**

hibernaculum (*pl.*, hibernacula)—a tentlike shelter made from debris by some insects for hibernation or seclusion.

hibernation—state of dormancy during the winter period, analogous to **estivation.** *See* **diapause.**

honeydew—anal secretions of aphids and other Homoptera; exudate of some galls.

horns—*see* **antennae.**

humeral lobe—process on the hindwing of butterflies and skippers making contact with the forewing during flight.

humerus (*pl.*, humeri)—shoulder; in the Coleoptera, the basal exterior angles of the folded elytra; in the Diptera, the anterior angles of the mesonotum; in the Orthoptera, the femora of the front legs; in the Hemiptera, the lateral angles of the prothorax; in some Hymenoptera, the subcostal vein.

hypermetamorphosis—a complex form of metamorphosis involving more than the usual number of stages, as in the blister beetles. *See* **planidium.**

hyperparasite—a parasite that attacks another parasite.

hypodermis—*see* **epidermis.**

hypognathous—having the head vertical and the mouth directed downward. *See* **prognathous.**

hypopharynx—tonguelike organ attached to upper surface of the labium (floor of mouth); serves as organ of taste. *See* **epipharynx.**

imago (*pl.,* **imagos** or **imagines**)—mature adult insect.

imbricate—arranged like scales or shingles.

immaculate—without spots or markings.

inarticulate—not jointed or segmented.

inflexed—bent inward at an angle.

infuscated—smoky gray-brown, with blackish tinge.

inquiline—an insect living habitually as a "guest" (not parasite) in the nest of another; bees, wasps, ants, termites commonly harbor guests.

insectivorous (*n.,* **insectivore**)—insect-eating; applies to animals other than insects.

instar—the form between molts; each is numbered to designate the stage, that between the egg and first molt being the *first* instar. The time interval between two molts is called the **stadium.**

integument—*see* **cuticle.**

internal parasite—one that feeds within the host; **endoparasite.**

introduced—brought by man accidentally or intentionally; not indigenous or native.

jugum—a process at the base of the forewing, at the inner margin, of certain moths (suborder Jugatae) which overlaps the hindwing during flight, holding them together. Some Trichoptera have a similar device. In the Hemiptera, two lateral lobes of the head.

labial palpi (**palps**)—pair of sensory appendages associated with the labium or lower lip; they are shorter than the **maxillary palpi**, usually three-segmented.

labium—the lower lip; floor of mouth, opposed to the **labrum**; the second **maxilla.**

labrum—the upper lip; covers the base of the mandibles, forming the roof of the mouth and opposed to the **labium.**

lamella (*pl.,* **lamellae**)—a process composed of thin, leaflike plates.

larva (*pl.,* **larvae**)—the developmental or growing stage, between the egg and pupal stages.

larviform—shaped like a larva.

larviposit—to deposit larvae rather than unhatched eggs.

lateral—toward or related to the side.

lignivorous—feeding on wood.

longitudinal—running lengthwise; opposite of **transverse.**

lunulate—having **lunules** or crescent-shaped marks.

maculation—marking of various shapes and colors.

malaxate—to knead or chew prey preparatory to feeding it to the young.

mandibles—upper pair of jaws, toothed in chewing types, sharply pointed and hollow or grooved in sucking types. *See* **maxillae.**

marginate—having elevated or flattened border. *See* **emarginate.**

maxillae (*sing.,* **maxilla**)—lower jaws of mandibulate insects. *See* **mandibles.**

maxillary palpi (**palps**)—pair of sensory appendages associated with the maxillae or lower jaws; longer than the **labial palpi,** with one to seven segments.

media—fourth longitudinal vein (M) of wing, having up to four branches, which are numbered beginning nearest the apex (*see* **costa; subcosta; radius; cubitus; anal vein**).

median—the middle.

mesonotum—top of mesothorax.

mesopleuron—lateral surface of mesothorax.

mesosternum—underside of mesothorax.

mesothorax—second (middle) segment of the thorax; it bears the middle pair of legs and the forewings.

metamorphosis—developmental changes from egg to adult, said to be *complete* when the pupal stage is inactive, and *incomplete* when no pupal stage occurs or it is active and feeds in this form.

metanotum—top of metathorax.

metapleuron—lateral surface of the metathorax.

metasternum—underside of metathorax.

metathorax—third (posterior) segment of the thorax; it bears the hind pair of legs and the hindwings (or halteres in the case of Diptera).

microsporidian—microorganism (Protozoa) causing disease in insects.

Midwest (**Middle West**)—comprises the following states: Michigan, Ohio, Indiana, Illinois, Wisconsin, Minnesota, Iowa, Missouri, Kansas, Nebraska, South Dakota, and North Dakota.

migration—movement of a group from one place to another, usually seasonal.

molt—to shed or cast off the outer skin to accommodate growth of the body. Also applied to other animals which shed hair, feathers, or horns.

monogamous—having only one mate; opposite of **polygamous.** *See* **monogynous.**

monogynous—having only one fecund female in the colony; opposite of **polygynous.** *See* **monogamous.**

monomorphic—having only one form. *See* **dimorphic** and **polymorphic.**

multiple parasitism—attack of a single host by two or more primary parasites of different species simultaneously. *See* **superparasitism.**

mutualism—*see* **symbiosis.**

mycetophagous—feeding on fungi.

myiasis—disease resulting from the presence of maggots of flies in or on the living body of man or other animal. As defined by West, "this form of parasitism may involve the skin and subcutaneous tissues, the natural apertures and adjacent cavities (such as the nasal passages and sinuses, mouth, anus, orbit of the eye, vagina), or the more internal cavities (such as the intestine, stomach, or bladder)."

myriapoda—a group of the Arthropoda comprising the centipedes, millipedes, symphylans, and pauropods; it is sometimes considered as a class, with the others as subclasses.

myrmecophilous—associated with ants.

mystax—a beardlike brush of stiff hairs on lower part of the face, just above the mouth, as in the robber flies.

naiad—aquatic nymph; young of dragonfly, damselfly, mayfly, or stonefly.

neck—stricture between the head and thorax. *See* **cervix.**

nematodes—threadlike worms ranging in size from microscopic forms to several feet long. Some are parasitic, others (called eelworms) feed on plants.

nocturnal—active or flying at night. *See* **diurnal** and **crepuscular.**

notum—the dorsal or upper part of a segment. *Syn.,* **tergum.**

nymph—the young or immature form of insect with incomplete metamorphosis.

obligate (obligatory) parasite—a parasite dependent on one host species.

obovate—inversely egg-shaped (broad at the top). *See* **ovate.**

occiput—hind part of head between **vertex** and **neck;** in Diptera, posterior surface of the head.

ocelli (*sing.,* **ocellus**)—simple eye consisting of a single lens, occurring singly or in groups; adults often have three arranged in a triangle on top of the head (moths have two; butterflies and beetles, among others, have none), larvae commonly have one to six on a side. A colored spot with sharply contrasting ring.

ommatidia (*sing.,* **ommatidium**)—*see* **compound eye.**

omnivorous—eating both plant and animal food.

ootheca—covering or case enclosing an egg mass; leathery egg capsule formed by female cockroach.

opaque—not allowing light to pass through; not transparent. *See* **translucent.**

operculum (*pl.,* **opercula**)—a lidlike cover.

osmeteria (*sing.,* **osmeterium**)—fleshy eversible processes which expose odoriferous secretions.

ova (*sing.,* **ovum**)—eggs.

ovary (*pl.,* **ovaries**)—mass of tubes, one on each side of body cavity of the female insect, in which the eggs develop.

ovate—egg-shaped (broad at the base). *See* **obovate.**

oviparous—reproducing by laying eggs. *See* **ovoviviparous** and **viviparous.**

ovipositor—tube or valves from which the female deposits (oviposits) eggs.

ovoid (oval)—egg-shaped; elliptical.

ovoviviparous—giving birth to living young, after hatching of ova in the mother. *See* **oviparous** and **viviparous.**

Pacific Northwest—Oregon, Washington, Idaho, and British Columbia.

paedogenesis—"parthenogenetic reproduction of the larvae which give birth to similar larvae" (Essig); reproduction in the immature larval stage.

pala (*pl.,* **palae**)—widened front tarsal joint as found in water boatmen (Hemiptera).

Paleozoic—third of the five *eras* of geologic time; comprises six *periods* beginning with the Cambrian (600 million years ago) and ending with the Permian (280 million years ago).

pallid—pale or wan; faint in color.

palmate—arranged like the palm of the hand with the fingers spread.

palmate hairs—in mosquito (*Anopheles*) larvae, paired fan-shaped tufts on the abdominal segments.

palpi (palps) (*sing.,* **palpus**)—*see* **labial palpi** and **maxillary palpi.**

papillae (*sing.,* **papilla**)—minute flexible projections.

parasite—*see* **predator.**

parthenogenesis—a form of asexual reproduction; development of egg without fertilization.

pectinate—comblike teeth or hairs on one side. *See* **bipectinate.**

pedicel—*see* **petiole.**

peduncle—a stalk or stem, pedicel or **petiole;** flower or fruit stem.

Pennsylvanian—*see* **carboniferous.**

penultimate—next to the last.

petiole—stem or stalk; one or two segments joining thorax and gaster in ants.

pharynx—cavity between the mouth and esophagus; back part of the mouth and upper part of the throat.

pheromone—substance emitted by an animal which causes a specific response from others of the same species (or colony in the case of honey bees).

phoresy—the association in which an insect is dispersed by riding on another.

phytophagous—feeding on plants. *See* **entomophagous.**

pile—thick, fine erect hairs, furlike and velvety in appearance.

planidium—the first-instar larva of certain parasites that undergo hypermetamorphosis, the form and habits in this stage being radically different from those in succeeding stages. The segments are heavily armored and have spines that aid in locomotion. Characteristic of parasitic larvae that search for host.

plankton—microscopic or very small plants (phytoplankton) and animals (zooplankton) drifting in the ocean or lakes, of great importance in the food chain and oxygen supply.

pleura (*sing.*, **pleuron**)—in general the sides of the body between the dorsum and sternum; lateral sclerites of the thorax.

plica (*pl.*, **plicae**)—fold or wrinkle; infold of margin.

plumose—branched, like a plume or feather, with long cilia on each side.

pollen—powdery sperm (male cells) on stamen of a flower which must reach or **pollinate** the stigma to produce seed; powdery covering of flies.

pollen basket—*see* **corbicula.**

pollinose—covered with a loose, mealy dust resembling pollen.

polyembryony—repeated division of a single egg to form numerous embryos, as occurs in some species of chalcid wasps; simple forms of polyembryony occur among braconid and platygasterid wasps.

polygamous—having two or more mates. Opposite of **monogamous.** *See* **polygynous.**

polygynous—having more than one fecund female in a colony. Opposite of **monogynous.** *See* **polygamous.**

polyhedrosis (*pl.*, **polyhedroses**)—a virus infection of insects, producing many-sided crystals in cells of the affected tissues. *See* **granulosis.**

polymorphism—having several forms. *See* **dimorphism** and **subspecies.**

polyphagous—eating many different kinds of food.

postantennal organ—a sensory organ behind base of each antenna in the Collembola, varying from "simple, elliptical depression to elaborate rosette or fern-like protuberance," sometimes discernible only with a microscope.

posterior—hinder, rear. Opposite of **anterior.**

postpetiole—second segment of the petiole in ants, where two segments occur.

predator—an animal living at the expense of others. An insect "predator" differs from a parasite in that it usually requires more than one host to complete its development; a "parasite" normally completes its development on one host, feeding in such a way as to avoid the vital organs until its own development is assured. The term "parasitoid" was suggested by Wheeler for this kind of parasite to distinguish it from lice, fleas, and other parasites which attack higher forms of animals including man. Predators are said to be predatory or predaceous.

pretarsus—terminal part of the tarsus; the claws and pad or pulvillus.

primary parasite—one that attacks other insects not themselves parasites.

primates—an order of mammals comprised of man, apes, monkeys, lemurs, and marmosets.

proboscis—extended mouthparts: rostrum or beak in the Hemiptera, labium in the Diptera, "tongue" in the Lepidoptera and "short-tongued" bees (rarely applied to the mouth of "long-tongued" bees). *See* **rostrum.**

process—any prominent extension of the body or an appendage.

prognathous—having the head directed forward, horizontal to the jaws. *See* **hypognathous.** (These are the two most common positions of the head.)

prolegs—fleshy unjointed legs of caterpillars and the larvae of some sawflies, used in clinging to surfaces and for support in locomotion.

pronotal comb—*see* **ctenidium.**

pronotum—top of prothorax.

propleuron—lateral surface of prothorax.

propodeum—in Hymenoptera, the first abdominal segment when it forms part of the thorax (the two are called **alitrunk**); epinotum in ants; median segment.

propygidium—next to the last abdominal segment, especially in beetles with short elytra. *See* **pygidium.**

prosternum—underside of prothorax.

prothoracic shield—*see* **cervical shield.**

prothorax—first (front) segment of the thorax; it bears the front pair of legs but never wings.

protozoa—single-celled microscopic animals (phylum Protozoa), "differing from bacteria in having at least one well-defined nucleus."

proximal—end nearest point of attachment or base. Opposite of **distal.**

pruinose—having frosty appearance.

pseudopod—a fleshy projection or appendage on the underside of some larvae.

pseudopupa—a quiescent stage preceding one or more larval instars and prior to the true pupal stage, occurring in **hypermetamorphosis.**

ptilium—bladder-like sac, thrust out through the frontal suture when certain flies (Cyclorrhapha) emerge from the pupa.

pulvillus (*pl.,* **pulvilli**)—soft pad between the tarsal claws.

punctate—impressed with punctures or pointed indentations.

pupa (*pl.,* **pupae**)—inactive stage of insects during which the larva transforms into the adult form. *See* **chrysalis; cocoon; puparium.**

puparium—thickened, hardened larval skin in which some insects pupate.

pupiparity—giving birth to mature larvae which are ready to pupate; the larvae are retained in the female uterus in a pouch where they are nourished by absorbing glandular secretions.

pygidium—the tergum or dorsal surface of the last abdominal segment; in the Coleoptera, the segment often left exposed by the elytra (*see* **propygidium**). In diaspid or armored scales, "a strongly chitinized unsegmented region terminating the abdomen of the adult female, following the first four abdominal segments, not to be confused with the true pygidium of other insects."

radius—third longitudinal vein (R) of the wing, having up to five branches (*see* **costa; subcosta; media; cubitus; anal vein**).

raptorial—adapted for seizing prey, as are the front legs of many predaceous species.

recumbent—lying down, reclined or flattened, often used in reference to hairs and bristles. *See* **decumbent** and **tomentose.**

reticulation—mesh or network of lines.

retinaculum (*pl.,* **retinacula**)—a straplike flap (in the male) or bristles (in the female) on the underside of the forewing of moths into which the **frenulum** (*which see*) engages.

rickettsiae—very small rod-shaped microorganisms, bacteria-like in nature and visible under an ordinary microscope; they will not grow in artificial media as bacteria do.

ring joints—the shorter proximal segments of the **clavola** (*which see*).

rostrum—snoutlike prolongation of the head, as in weevils; the **beak** or extension of the labium, as in Hemiptera; jointed sheath formed by the labium and enclosing the stylets and other mouthparts. *See* **proboscis.**

rufous—pale red; brilliant reddish yellow.

rugose—wrinkled.

ruminant—a cud-chewing hoofed animal with stomach of four chambers, one of which (the rumen) prepares vegetation for regurgitating and chewing; e.g., cattle, sheep, goat, deer, bison, antelope, camel, giraffe.

saprophagous—feeding on dead or decaying animal or vegetable matter; scavenging.

sapwood—outer wood of tree in which certain of the cells are still alive and serve to conduct water from the roots to the leaves. The central portion of the tree is called the **heartwood**; it is entirely dead and without function, is usually darker and more durable than the sapwood. The **cambium**—the growing part of the tree—is the soft layer between the wood and bark, consists of a single layer of cells which give rise to both the wood and the bark.

scape—three basal joints of the antennae in Hymenoptera; long basal joint in geniculate or elbowed antennae, as in ants.

scavenger—a saprophyte. *See* **saprophagous.**

sclerite—hardened (sclerotized) plates or portions of the insect body wall bounded by sutures.

scutellum—triangular piece of the thorax (sclerite) between the bases of the elytra in Coleoptera or of the hemelytra in Hemiptera, often exposed when the wings are folded; smaller part of wing-bearing plate (set off by a suture) in certain Hymenoptera; rounded part cut off from the mesonotum by an impressed line in Diptera. *See* **scutum.**

scutum—major part of wing-bearing plate on upper surface of the thorax. *See* **scutellum.**

secondary parasite—a parasite that attacks a **primary parasite** (*which see*); a hyperparasite.

sericeous—silky, with short, thick down.

serrate—toothed like a saw.

serrulate—minutely serrated or toothed.

sessile—closely seated, attached directly by its base or permanently fixed; abdomen and thorax attached over most or all of their width, as in the sawflies and horntails.

setae (*sing.,* **seta**)—slender hairs, believed to be sensory in function.

setose—covered with setae or stiff hairs.

sinuate—with wavy edges or margins.

social—living in organized communities or colonies, with division of labor and castes. The social insects are: termites, ants, paper-nest wasps, bumble bees, honey bees, and stingless bees.

solitary—not living in communities or aggregations; opposite of **gregarious.**

species (*sing.* and *pl.*)—"a variable population reproductively isolated from similar groups capable of freely interbreeding."

spermatheca—receptacle in the female which receives the sperm during coition and where it is stored.

spine—a thornlike process or multicellular outgrowth of the cuticle, not separated from it by a joint; a large seta with cuplike attachment to the cuticle.

spinule—a small spine.

spiracle—paired breathing pores along the sides of the thorax and abdomen through which the tracheae receive air. *Syn.*, **stigma.**

spur—a spinelike projection of the cuticle, connected to the body by a joint, usually found on the tibiae.

square—the three bracts enclosing the flower of the cotton plant. *See* **boll.**

stadium—*see* **instar.**

stem mother—a form of aphid hatching from the overwintering egg and giving rise to **agamic** (*which see*) summer broods.

sternite—"ventral piece in a ring or segment."

sternum (*pl.*, **sterna**)—generally the underside of the thorax between the coxal cavities; the entire ventral part of any segment.

stigma (*pl.*, **stigmata**)—**spiracle** (*which see*) or breathing pore, being paired—one on each side of a segment; dense part of costal margin of wing; colored spot near tip of auxiliary vein on the wing of Diptera; patch of black scales on forewing of skippers; tip of pistil in the female flower which receives the **pollen** (*which see*).

stria (*pl.*, **striae**)—fine impressed line, longitudinal and often with punctures in the Coleoptera, transverse in the Lepidoptera.

stridulate—make noise "by rubbing two ridged or roughened surfaces" together.

style or **stylus** (*pl.*, **styluses, styli**)—small, slender tubular process, usually pointed; most often applied to certain appendages at the end of the abdomen, such as the slender tube at the end of the abdomen in aphids, a long spinelike terminal process in male coccids, and the single organ below the forceps in male crane flies. Also applied to mouthparts of the Hemiptera and Diptera.

subcosta—second longitudinal vein (Sc) of the wing, unbranched, reaching the outer margin before the costa (*see* **costa; radius; media; cubitus; anal vein**).

subsocial—loosely organized in colonies, with some cooperation among individuals but without castes.

subspecies—a subdivision of a species according to geographical or host variations. Interbreeding is possible between subspecies of the same species and does occur but not as often as within the confines of each subspecies; grading of one subspecies into another is the result of incomplete isolation. It is thought that subspecies may indicate the development of new species. Not to be confused with **polymorphism** (*which see*).

sulcus (*pl.*, **sulci**)—deep furrow or groove. *See* **fovea.**

superparasitism—parasitism of the same host by more than one female of the same species; more eggs laid in a single host than can complete development. The excess is usually eliminated by cannibalism.

sutural hairs—in mosquito (*Anopheles*) larvae, hairs on upper part of the head. *See* **clypeal hairs.**

suture—infold or seam dividing parts of the body wall; the line formed by juncture of the elytra (Coleoptera), tegmina (Orthoptera), and hemelytra (Hemiptera).

symbiosis—intimate association of two dissimilar organisms; the association may benefit both and is so close in some instances that neither can exist without the other. Many insects harbor microscopic **symbiotes** that live harmlessly or **mutualistically** (to the benefit of both) in the tissues or organs of their hosts; best-known examples are termites, lice, aphids, cockroaches, certain bugs, and bark beetles. *See* **commensalism.**

tarsus (*pl.*, **tarsi**)—the "foot" or distal part of the leg bearing the claws and pad (pulvillus), attached basally to the tibia; it consists of one to five segments. *See* **pretarsus.**

tegmen (*pl.*, **tegmina**)—thickened forewing of the Orthoptera and certain Homoptera; sometimes refers to the **hemelytra** of the Hemiptera.

tegula—small scalelike sclerite at base of costa on forewing.

tenaculum—clasp or catch; two prongs on underside of the third abdominal segment of Collembola holding the **furcula** in place.

tenent hairs—sticky hairs; "guard setae" associated with claws of Collembola.

tergite—dorsal sclerite of a body segment; it usually pertains to a single sclerite. *See* **tergum.**

tergum—dorsal or upper surface of a body segment; it may have one or more sclerites. *See* **tergite.**

testaceous—having a test or hard cover; brownish yellow.

thorax—the middle part of division of the insect body bearing the legs and wings. It consists of three segments respectively called prothorax, mesothorax, and metathorax, with the last two bearing the wings.

tibia (*pl.*, **tibiae**)—fourth segment of the leg, between the femur and tarsus.

tomentose—covered with short, recumbent, matted hairs.

tracheae (*sing.*, **trachea**)—system of tubes which distribute oxygen throughout the body and terminate on the surface in the spiracles; the finer branches or endings are called **tracheoles.**

tracheal gills—flattened or hairlike processes by which aquatic insects extract oxygen from the water.

tracheoles—*see* **tracheae.**

translucent—partially transparent; light passing through but with objects on the other side indistinct. *See* **opaque.**

transparent—allowing light to pass through; clear. *See* **translucent** and **opaque.**

transverse—running across; broader than long. Opposite of **longitudinal.**

tribe—a unit of classification less than a subfamily and ending in *-ini;* a group of genera.

trichome—a hair, being a hollow projection of the cuticle. *See* **setae.**

triungulin—the active, first-instar larval form of meloid beetles and stylopids (Strepsiptera).

trochanter—second segment of the leg, between the coxa and femur, sometimes divided or fused with femur.

truncate—cut off squarely at the tip.

tubercle—a small pimple or button, which on caterpillars often bears a seta, bristle, or other process.

tubule—a small tube or minute tubular process.

tympanum (*pl.,* **tympana**)—drum or vibrating membrane; the insect ear, connected to the chordontal organs.

unguis (*pl.,* **ungues**)—a claw at the end of the tarsus (usually toothed) in the Collembola and opposed to a small claw, the **unguiculus.**

unisexual—only the female appearing, with reproduction by **parthenogenesis.**

urticating—stinging with nettles, causing itching and irritation of skin.

uterus—vaginal portion of oviduct in the female insect; sometimes refers to enlarged portion of the **vagina** (*which see*) at junction of the oviducts.

vagina—tubular structure in the female insect formed by union of the oviducts (paired tubes which conduct the eggs from the ovaries); it leads externally to the ovipositor and receives the penis during copulation.

vector—carrier or bearer; any insect that carries a pathogen or disease-causing organism from one animal or plant host to another.

venation—system of veins or **neuration** in the wings. *See* **costa** (C); **subcosta** (Sc); **radius** (R); **media** (M); **cubitus** (Cu); **anal vein** (A).

venter or **ventral side**—underside or the belly.

ventral declivity—concave posterior ventral surface of the abdomen, as found in bark beetles of the genus *Scolytus.*

ventral tube—tube on underside of first abdominal segment in Collembola; it is bilobed, with filaments which can be extruded, probably respiratory in function or for adhering to surfaces.

vermiform—wormlike.

vertex—top of head between the eyes, front and **occiput** (*which see*).

vestiture—clothing; surface covering of insects such as scales and hairs.

virus (*pl.,* **viruses**)—minute infectious agent consisting of rod-shaped or other particle which can pass through a thick porcelain filter (hence said to be "filterable") and cannot be seen under an ordinary microscope. Virus infections in insects are usually **polyhedroses** or **granuloses,** with the particles enclosed in many-sided crystals or in "granular," capsule-like bodies; "free viruses" as found in other animals or plants are rare. Unlike bacteria, viruses cannot be cultured on artificial media, can be seen only under an electron microscope.

vitta (*pl.,* **vittae**)—broad longitudinal stripe. *See* **fascia.**

viviparous—giving birth to living young. The term is often used synonymously with **ovoviviparous** (*which see*) but in strict usage applies to animals having a direct connection (placental or otherwise) between the mother and embryo.

warble—boil; lump or swelling on the back of an animal caused by a fly maggot (bot) lodged in the flesh, under the skin. The flies are called warble flies or bot flies. *See* **bot.**

West—"the West" as referred to in the U.S. is generally considered to be west of the 100th meridian, which extends roughly through the middle of Texas, Oklahoma, Kansas, Nebraska, South Dakota, and North Dakota.

xylem—vascular tissue of wood (beneath the cambium) which conducts water and minerals throughout the tree, and gives it firmness. *See* **sapwood.**

BIBLIOGRAPHY

General

Allee, W. C., et al. *Principles of Animal Ecology.* Philadelphia: W. B. Saunders Co., 1963. 837 pp., 263 figs.

Almand, Lyndon K., and Thomas, John G. *Insects and Related Pests Attacking Lawns and Ornamental Plants.* Bul. 1078, Agr. Ext. Serv., Texas A&M Univ., College Station, 1963. 29 pp., 66 figs. (40 color).

Anderson, R. F. *Forest and Shade Tree Entomology.* New York: John Wiley & Sons, 1960. 487 pp., il.

Barker, Paul C., et al. *The Study of Insects.* Agr. Ext. Serv., Univ. of California, Berkeley, 1962. 20 pp., il. (Collection and preservation of insects.)

Barker, Will. *Familiar Insects of America.* New York: Harper & Bros., 1960. 236 pp., 28 figs., 4 pls. (color).

Barnes, Martin M., and Madsen, Harold F. *Insect and Mite Pests of Apple in California.* Cir. 502, California Agr. Exp. Sta., Berkeley, 1961. 31 pp., il.

Bell, Jerold. *Grasshoppers in Nebraska.* Ext. Cir. 61–1594, Coop. Ext. Serv., Univ. of Nebraska, Lincoln, 1961. 31 pp., il.

Borror, Donald J., and DeLong, Dwight M. *An Introduction to the Study of Insects.* New York: Holt, Rinehart and Winston, 1963. 819 pp., il.

Brevoort, Harry F. (photographs), and Fanning, Eleanor I. (text). *Insects from Close Up.* New York: Thomas Y. Crowell Co., 1965. 145 pp., 131 photos.

Brown, F. Martin, Eff, Donald, and Rotger, Rev. Bernard, C. R. *Colorado Butterflies.* Denver: Denver Mus. Nat. Hist., 1957. 368 pp., 250 photos.

Buchsbaum, Ralph. *Animals Without Backbones: An Introduction to the Invertebrates.* Harmondsworth, England: Penguin Books, 1957. 2 vols.: 605 pp., 128 pls.

Canerday, T. Don, and Arant, F. S. *Control of Caterpillars Attacking Cabbage.* Cir. 163, Agr. Exp. Sta., Auburn Univ., Auburn, Alabama, 1968. 18 pp., 6 figs.

Chandler, Stewart C. *Peach Insects of Illinois and Their Control.* Cir. 43, Illinois Nat. Hist. Survey, Urbana, 1950. 63 pp., 39 figs.

Chu, H. F. *How to Know the Immature Insects.* Dubuque, Iowa: Wm. C. Brown Co., 1949. 234 pp., 631 figs.

Clarke, J. F. G. *Butterflies.* New York: Golden Press, 1963. 68 pp., 187 figs. (color).

Comstock, J. H. *An Introduction to Entomology.* Ithaca, N.Y.: Comstock Pub. Co., 1949. 1064 pp., 1228 figs.

Comstock, John Adams. *Butterflies of California.* Los Angeles: J. A. Comstock, 1927. 333 pp., 63 pls. (color), text figs.

Cott, Hugh B. *Adaptive Coloration in Animals.* London: Methuen & Co., 1957. 508 pp., 67 figs., 48 pls.

Craighead, F. C. *Insect Enemies of Eastern Forests.* Misc. Pub. No. 657, U.S. Dept. Agr., Washington, D.C., 1950. 679 pp., 197 figs.

Crowell, H. H., and Every, R. W. *Vegetable Garden Insect Pests.* Bul. 747, Coop. Ext. Serv., Oregon State Univ., Corvallis, 1968. 12 pp., il.

Cutright, C. R. *Insect and Mite Pests of Ohio Apples.* Res. Bul. 930, Ohio Agr. Exp. Sta., Wooster, 1963. 78 pp., 50 figs., 11 pls.

Dillon, Elizabeth S. and Lawrence S. *A Manual of Common Beetles of Eastern North America.* Evanston, Ill.: Row, Peterson, 1961. 884 pp., 544 figs., 81 pls.

Dowdy, A. C. *Fruit Insects of Michigan.* Ext. Bul. 372, Coop. Ext. Serv., Michigan State Univ., E. Lansing, 1960. 40 pp., 44 figs.

Dunstan, G. Gordon, and Davidson, Thomas R. *Diseases, Insects and Mites of Stone Fruits.* Pub. 915, Canada Dept. Agr., Ottawa, 1967. 48 pp., 30 figs.

Ebeling, Walter. *Subtropical Fruit Pests.* Div. Agr. Sci., Univ. of California, Berkeley, 1959. 436 pp., 160 figs., 8 pls. (color).

Eden, W. G., and Yates, Harold. *Corn Earworm Control on Sweet Corn in Alabama.* Bul. 326, Agr. Exp. Sta., Auburn Univ., Auburn, Alabama, 1960. 30 pp., il.

Ehrlich, Paul R. and Anne H. *How to Know the Butterflies.* Dubuque, Iowa: Wm. C. Brown Co., 1961. 262 pp., 525 figs.

Elton, Charles S. *The Ecology of Invasions by Animals and Plants.* London: Methuen and Co., Ltd., 1958. 181 pp., 51 figs., 50 pls.

English, L. L. *Illinois Trees and Shrubs: Their Insect Enemies.* Cir. 47, Illinois Nat. Hist. Survey, Urbana, 1968. 92 pp., 59 figs.

————— and Turnipseed, G. F. *Control of the Major Pests of Satsuma Orange in South Alabama.* Bul. 248, Agr. Exp. Sta., Alabama Polytech. Inst., Auburn, 1940. 48 pp., 17 figs.

—————. *Insect Pests of Azaleas and Camellias . . .* Cir. 84, Agr. Exp. Sta., Alabama Polytech. Inst., Auburn. 20 pp., 13 figs.

Essig, E. O. *College Entomology.* New York: The Macmillan Co., 1951. 900 pp., 308 figs.

—————. *Insects of Western North America.* New York: Macmillan, 1926. 1036 pp., 766 figs.

Evans, Howard E. *Wasp Farm.* Garden City, N.Y.: Natural History Press, 1963. 178 pp., 16 figs., 25 photos.

Fabre, J. Henri. *The Insect World of J. Henri Fabre.* Ed. by Edwin Way Teale. Apollo Edition. New York: Dodd, Mead & Co., 1949. 333 pp.

Fenton, F. A. *Field Crop Insects*. New York: The Macmillan Co., 1952. 405 pp., 224 figs.

Ferree, Roy J., et al. *Pecan Pest Control*. Cir. 484, Coop. Ext. Serv., Clemson Univ., Clemson, S.C., 1967. 30 pp., 11 figs.

Flink, Paul R. *Field Key to Michigan Forest Insects*. Forestry Div., Michigan Dept. of Conservation, Lansing, 1961. 14 pp., 61 figs.

Foote, Richard H., and Cook, David R. *Mosquitoes of Medical Importance*. Agr. Handb. No. 152, U.S. Dept. Agr., Washington, D.C., 1959. 159 pp., 68 figs.

Ford, E. B. *Butterflies*. London: Collins, 1957. 368 pp., 176 pls. (87 color).

————. *Moths*. London: Collins, 1955. 266 pp., 56 pls. (32 color).

Frisch, Karl von. *The Dancing Bees*. Harvest Book. New York: Harcourt, Brace & Co., 1953. 182 pp., 61 figs., 30 pls.

Frost, S. W. *General Entomology*. New York: McGraw-Hill Co., 1942 (reprint: Dover Pubs., New York). 526 pp., il.

Gaines, J. C. *Cotton Insects*. Bul. 933, Agr. Ext. Serv., Texas A & M Univ., College Station, 1965. 16 pp., 28 figs.

Garman, Philip, and Townsend, J. F. *Control of Apple Insects*. Bul. 552, Connecticut Agr. Exp. Sta., New Haven, 1952. 84 pp., 71 figs.

Garth, John S., and Tilden, J. W. *Yosemite Butterflies*. Arcadia, Calif.: The Lepidoptera Foundation, 1963. 96 pp., 6 figs., 6 pls. (4 color).

Gates, Dell E., and Peters, Leroy L. *Insects in Kansas*. Ext. Div., Kansas State Univ., Manhattan, 1962. 307 pp., 458 figs., 5 pls. (color).

Grant, Vern E. and K. A. *Flower Pollination in the Phlox Family*. New York: Columbia Univ. Press, 1965. 180 pp., 46 figs., 3 pls. (color).

Gunderson, Harold, and Lambe, R. C. *Insects and Diseases in the Family Vegetable Garden*. Pm. 230, Coop. Ext. Serv., Iowa State Univ., Ames, 1968. 8 pp., il.

———— and Stockdale, H. J. *Control of Soil Insects Which Attack Iowa Corn*. IC–368, Coop. Ext. Serv., Iowa State Univ., Ames, 1967. 8 pp., il.

Hagmann, L. E., and Jobbins, D. M. *The Story of the Mosquito and Its Control*. Cir. 585–B, Agr. Exp. Sta., Rutgers Univ., New Brunswick, N.J., 1968. 15 pp., il.

Hanson, A. J., et al. *Biology of the Cabbage Seedpod Weevil in Northwestern Washington*. Bul. No. 498, Agr. Exp. Sta., State Col. of Washington, Pullman, 1948. 15 pp., 10 figs.

Haskell, P. T. *Insect Sounds*. London: H. F. & G. Witherby, 1961. 189 pp., 97 figs.

Haufe, W. O. *Control of Cattle Lice*. Pub. 1006, Canada Dept. Agr., Ottawa, 1962. 15 pp., 3 figs. (color).

Hawkins, J. H., et al. *Wireworms Affecting the Agricultural Crops of Maine*. Bul. 578, Agr. Exp. Sta., Univ. of Maine, Orono, 1958. 40 pp., 24 figs.

Helfer, Jacques R. *How to Know the Grasshoppers, Cockroaches and Their Allies*. Dubuque, Iowa: Wm. C. Brown Co., 1962. 351 pp., 579 figs.

Holland, W. J. *The Butterfly Book*. Garden City, N.Y.: Doubleday & Co., 1931. 423 pp., 198 figs., 77 pls. (73 color).

————. *The Moth Book*. New York: Doubleday, Page & Co., 1914. (reprint: Dover Pubs., New York). 479 pp., 263 figs., 48 pls. (color).

Howard, L. O. *The Insect Book*. New York: Doubleday, Page & Co., 1914. 429 pp., 264 figs., 48 pls.

Jaeger, Edmund C. *The North American Deserts.* Stanford, Calif.: Stanford Univ. Press, 1961. 308 pp., 355 figs.

—————. *A Sourcebook of Biological Names and Terms.* Springfield, Ill.: Charles C Thomas, 1962. 323 pp., il.

James, Ray L., et al. *Christmas Tree Insect Control.* Ext. Bul. 353, Coop. Ext. Serv., Michigan State Univ., E. Lansing, 1967. 28 pp., il.

Jaques, H. E. *How to Know the Beetles.* Dubuque, Iowa: Wm. C. Brown Co., 1951. 372 pp., 865 figs.

—————. *How to Know the Insects.* Dubuque, Iowa: Wm. C. Brown Co., 1951. 205 pp., 411 figs.

Jeannel, R. *Introduction to Entomology.* London: Hutchinson, 1960. 344 pp. 150 figs., 46 pls. (11 color).

Johansen, Carl. *Beekeeping.* PNW Bul. 79, Washington State Univ., Pullman, 1966. 27 pp., 17 figs.

Keen, F. P. *Insect Enemies of Western Forests.* Misc. Pub. No. 273, U.S. Dept. Agr., Washington, D.C., 1952. 280 pp., 111 figs.

Kerr, T. W., and Hansen, H. L. *Insects of Alfalfa and Their Control.* Misc. Pub. 56, Agr. Exp. Sta., Univ. of Rhode Island, Kingston, 1959. 11 pp., 5 figs.

Klots, Alexander B. *A Field Guide to the Butterflies of North America, East of the Great Plains.* Boston: Houghton Mifflin Co., 1960. 349 pp., 232 photos, 247 figs. (color).

—————, and Klots, E. B. *Living Insects of the World.* Garden City, N.Y.: Doubleday & Co., 1965. 304 pp., 277 photos (152 color).

Lathrop, Frank H. *Apple Insects in Maine.* Bul. 540, Agr. Exp. Sta., Univ. of Maine, Orono, 1955. 88 pp., 23 figs.

Leonard, Justin W. and Fannie A. *Mayflies of Michigan Trout Streams.* Cranbrook Inst. of Sci., Bloomfield, Mich., 1962. 139 pp., 6 pls. (color).

Linn, M. B., and Luckmann, W. H. *Tomato Diseases and Insect Pests.* Cir. 912, Coop. Ext. Serv., Univ. of Illinois, Urbana, 1967. 55 pp., 34 figs.

Lutz, Frank E. *Field Book of Insects.* New York: G. P. Putnam's Sons, 1948. 510 pp., 100 pls. (23 color).

Madsen, H. F., and Arrand, J. C. *The Recognition and Biology of Orchard Insects and Mites in British Columbia.* British Columbia Dept. Agr., Victoria, 1966. 41 pp., 19 figs., 4 pls. (color).

—————— and Barnes, M. M. *Pests of Pear in California.* Cir. 478, Agr. Exp. Sta., Univ. of California, Berkeley, 1959. 40 pp., il.

—————— and McNelly, L. B. *Important Pests of Apricots.* Bul. 783, Agr. Exp. Sta., Univ. of California, Berkeley, 1961. 40 pp., 36 figs.

Maeterlinck, Maurice. *The Life of the Bee.* New York: Mentor Books, 1954. 168 pp.

Mallis, Arnold. *Handbook of Pest Control.* New York: MacNair-Dorland Co., 1960. 1132 pp., 237 figs.

Matheson, Robert. *Entomology for Introductory Courses.* Ithaca, N.Y.: Comstock Pub. Co., 1951. 629 pp., 500 figs.

Merkley, Don R. *Insect Pests of Ornamental Plants in Montana.* Bul. 606, Agr. Exp. Sta., Montana State Univ., Bozeman, 1966. 43 pp., 24 figs.

Merrill, L. G., Jr., et al. *Economic Insects of New Jersey.* Bul. 293, 295, 296, 305, 306, 321, 354, Ext. Serv., Rutgers Univ., New Brunswick, 1956–61. 112 pp., il. (color).

Metcalf, C. L. and Flint, W. P.; rev. by R. L. Metcalf. *Destructive and Useful Insects.* New York: McGraw-Hill Book Co., 1962. 1087 pp., il.

Michener, Charles D. and Mary H. *American Social Insects.* New York: D. Van Nostrand Co., 1951. 267 pp., 109 photos (30 color).

Milliron, H. E. *Economic Insects and Allied Pests of Delaware.* Bul. No. 321, Agr. Exp. Sta., Univ. of Delaware, Newark, 1958. 87 pp.

Needham, J. G., and Westfall, M. J. *A Manual of the Dragonflies of North America.* Berkeley: Univ. of California Press, 1955. 615 pp., 341 figs.

Neiswander, R. B. *Insect and Mite Pests of Trees and Shrubs.* Res. Bul. 983, Ohio Agr. Res. and Dev. Center, Wooster, 1966. 54 pp., 85 figs.

Nettles, W. C., et al. *Insects and Diseases of Ornamentals.* Cir. 502, Coop. Ext. Serv., Clemson Univ., Clemson, S.C., 1968. 36 pp., 18 figs.

————. *Soybean Insects and Diseases.* Cir. 504, Coop. Ext. Serv., Clemson Univ., Clemson, S.C., 1968. 24 pp., 14 figs., 2 pls. (color).

Neunzig, H. H., and Falter, J. M. *Insect and Mite Pests of Blueberry in North Carolina.* Bul. 427, Agr. Exp. Sta., North Carolina State Univ., Raleigh, 1966. 34 pp., 29 figs.

Newcomer, E. J. *Insect Pests of Deciduous Fruits in the West.* Agr. Handb. No. 306, U.S. Dept. Agr., Washington, D.C., 1966. 57 pp., 67 figs.

Newton, Weldon H., et al. *Insects Attacking Vegetable Crops.* Bul. 1079, Agr. Ext. Serv., Texas A & M Univ., College Station, 1964. 28 pp., 56 figs.

Oatman, E. R., and Ehlers, C. G. *Cherry Insects and Diseases of Wisconsin.* Bul. 555, Agr. Exp. Sta., Univ. of Wisconsin, Madison, 1962. 43 pp., 88 figs.

————. *Wisconsin Apple Insects.* Bul. 548, Agr. Exp. Sta., Univ. of Wisconsin, Madison, 1960. 29 pp., 89 figs.

Oldroyd, Harold. *Collecting, Preserving and Studying Insects.* London: Hutchinson, 1958. 327 pp., il.

————. *The Natural History of Flies.* London: Weidenfeld & Nicholson, 1964. 324 pp., 40 figs., 32 pls.

Painter, Reginald H. *Insect Resistance in Crop Plants.* New York: The Macmillan Co., 1951. 520 pp., il.

Papp, Charles S. *Scientific Illustration: Theory and Practice.* Dubuque, Iowa: Wm. C. Brown Co., 1968. 318 pp., 1345 figs.

Peairs, L. M. *Insect Pests of Farm, Garden, and Orchard.* New York: John Wiley & Sons, 1948. 549 pp., 648 figs.

Peterson, Alvah. *Entomological Techniques: How to Work with Insects.* Ann Arbor, Mich.: Edwards Bros., 1959. 435 pp., il.

Pfadt, Robert E. *Key to Wyoming Grasshoppers.* Mimeo. Cir. 210, Agr. Exp. Sta., Univ. of Wyoming, Laramie, 1965. 25 pp., 37 figs.

Portman, Roland W., and Manis, H. C. *Idaho Potato Insect Handbook.* Bul. 505, Agr. Ext. Serv., Univ. of Idaho, Moscow, 1969. 20 pp., il.

Puckering, D. L., and Post, R. L. *Butterflies of North Dakota.* Dept. Agr. Ent., North Dakota Agr. Col., Fargo, 1960. 32 pp., il.

Quist, John A. *Approaches to Orchard Insect Control.* Bul. 517S, Agr. Exp. Sta., Colorado State Univ., Fort Collins, 1966. 95 pp., 27 figs.

Randolph, N. M., and Garner, C. F. *Insects Attacking Forage Crops.* Bul. 975, Agr. Ext. Serv., A & M Col. of Texas, College Station, 1961. 26 pp., 44 figs.

Richards, O. W. *The Social Insects.* Harper Torchbook. New York: Harper & Row, 1961. 219 pp., 12 figs., 51 pls.

Rings, Roy W., and Neiswander, R. B. *Insect and Mite Pests of Strawberries in Ohio.* Res. Bul. 987, Ohio Agr. Res. and Dev. Center, Wooster, 1966. 19 pp., 27 figs.

Ross, Edward S. *Insects Close Up: Pictorial Guide for Photographer and Collector.* Berkeley: Univ. of California Press, 1953. 81 pp., 125 figs. (7 color).

Ross, Herbert H. *How to Collect and Preserve Insects.* Cir. 39, Illinois Nat. Hist. Survey, Urbana, 1966. 71 pp., 79 figs.

———. *A Textbook of Entomology.* New York: John Wiley & Sons., 1965. 539 pp., 401 figs.

Scott, H. E., and Brett, C. H. *Vegetable Insects of North Carolina.* Cir. 313, Coop. Ext. Serv., North Carolina State Univ., Raleigh, 1968. 24 pp., il.

Smith, Leslie M., and Stafford, Eugene M. *Grape Pests in California.* Cir. 445, Agr. Exp. Sta., Univ. of California, Berkeley, 1951. 63 pp., il.

Snyder, T. E. *Our Enemy the Termite.* Ithaca, N.Y.: Comstock Pub. Co., 1948. 257 pp., 84 figs.

Storer, Tracy I., and Usinger, Robert L. *Sierra Nevada Natural History.* Berkeley: Univ. of California Press, 1963. 374 pp., 23 figs., 89 pls. (24 color).

Summers, Francis M. *Insect and Mite Pests of Almonds.* Cir. 513, Agr. Exp. Sta., Univ. of California, Berkeley, 1962. 16 pp., il.

Swain, Ralph B. *The Insect Guide.* Garden City, N.Y.: Doubleday & Co., 1948. 261 pp., 175 figs. (159 color).

Swan, Lester A. *Beneficial Insects.* New York: Harper & Row, 1964. 429 pp., 68 figs.

Teale, Edwin Way. *The Strange Lives of Familiar Insects.* New York: Dodd, Mead & Co., 1962. 208 pp., 26 photos.

———. 1946. *The Golden Throng: A Book About Bees.* Ibid. 208 pp., 85 photos.

———. 1944. *Near Horizons: Story of An Insect Garden.* Ibid. 319 pp., 160 photos.

Thomas, John G., et al. *Peach and Plum Insects.* MP–685, Agr. Ext. Serv., Texas A & M Univ., College Station, 1967. 16 pp., 23 figs.

Van den Bosch, Robert, and Hagen, Kenneth S. *Predaceous and Parasitic Arthropods in California Cotton Fields.* Bul. 820, Agr. Exp. Sta., Univ. of California, Berkeley, 1966. 32 pp., 20 figs.

Wallis, Robert C. *Common Connecticut Flies.* Bul. 650, Connecticut Agr. Exp. Sta., New Haven, 1962. 23 pp., 14 figs.

Wallner, W. E. *Insects Affecting Woody Ornamental Shrubs and Trees.* Ext. Bul. 530, Coop. Ext. Serv., Michigan State Univ., E. Lansing, 1969. 45 pp., 67 figs.

Werner, Alfred, and Bijok, Josef. *Butterflies and Moths.* Ed. by Norman Riley. New York: The Viking Press, 1965. 126 pp., 39 pls. (color).

Wheeler, W. M. *Ants.* New York: Columbia Univ. Press, 1910. 663 pp. 284 figs.

Whitcomb, W. H., and Bell, K. *Predaceous Insects, Spiders, and Mites of Arkansas Cotton Fields.* Bul. 690, Agr. Exp. Sta., Univ. of Arkansas, Fayetteville, 1964. 84 pp., 20 figs.

Wigglesworth, V. B. *The Life of Insects.* Cleveland, Ohio: World Pub. Co., 1964. 359 pp., 164 figs., 36 pls. (12 color).

Williams, C. B. *Insect Migration.* New York: The Macmillan Co., 1958. 235 pp., 48 figs., 24 pls. (8 color).

Wingo, Curtis W. *Poisonous Spiders and Other Venomous Arthropods in Missouri.* Bul. 738, Agr. Exp. Sta., Univ. of Missouri, Columbia, 1969. 12 pp., il.

Wright, J. M., and Apple, J. W. *Common Vegetable Insects.* Cir. 671, Coop. Ext. Serv., Univ. of Illinois, Urbana, 1959. 38 pp., il.

Wyman, Leland C., and Bailey, Flora L. *Navaho Indian Ethnoentomology.* Albuquerque: Univ. of New Mexico Press, 1964. 158 pp., 5 pls.

Insects: Yearbook of Agriculture 1952, U.S. Dept. Agr. Supt. of Documents, U.S. Govt. Print. Office, Washington, D.C. 952 pp., 80 pls. (72 color), text figs.

Insects: Price List 41. Supt. of Documents, Govt. Printing Office, Washington, D.C. See *List* for other references, mostly pubs. of the U.S. Dept. Agr.

Technical

Alexander, Charles P. *The Crane Flies of Maine.* Bul. T-4, Agr. Exp. Sta., Univ. of Maine, Orono, 1962. 24 pp.

Allen, H. W. *The Oriental Fruit Moth.* Agr. Inf. Bul. No. 182, U.S. Dept. Agr., Washington, D.C., 1958. 28 pp., 11 figs.

_____. *Parasites of the Oriental Fruit Moth.* Tech. Bul. No. 1265, U.S. Dept. Agr., Washington, D.C., 1962. 139 pp., 28 figs.

_____. "Two-winged Flies of the Tribe Miltogrammini," *U.S. Natl. Mus. Proc.,* 68, Art. 9:1–106 (1926), il.

Arant, F. S. *Cotton Insects and Their Control* . . . Cir. 106, Agr. Exp. Sta., Alabama Polytech. Inst., Auburn, 1951. 36 pp., 19 figs.

_____. *Life History and Control of the Cowpea Curculio.* Bul. 246, Agr. Exp. Sta., Alabama Polytech. Inst., Auburn, 1938. 34 pp., 14 figs.

Arnett, Ross H., Jr. *The Beetles of the United States: A Manual for Identification.* Washington, D.C.: Catholic Univ. of Amer. Press, 1963. 1112 pp., il.

Arnold, J. W. *Blood Circulation in Insect Wings.* Mem. Ent. Soc. of Canada No. 38, 1964. 48 pp., 72 figs.

Baker, Charles W. *Larval Taxonomy of the Troginae in North America with Notes on Biologies and Life Histories (Coleoptera: Scarabaeidae).* Bul. 279, U.S. Natl. Mus., 1968. 79 pp., 59 figs.

Ball, E. D., et al. *The Grasshoppers and Other Orthoptera of Arizona.* Tech. Bul. 93, Agr. Exp. Sta., Univ. of Arizona, Tucson, 1942. 373 pp., 11 figs., 4 pls.

Bellinger, P. F. *Studies of Soil Fauna with Special Reference to the Collembola.* Bul. 583, Connecticut Agr. Exp. Sta., New Haven, 1954. 67 pp., 18 figs.

Bennett, Stelmon E. *A Decade with the Alfalfa Weevil in Tennessee.* Bul. 446, Agr. Exp. Sta., Univ. of Tennessee, Knoxville, 1968. 33 pp., 15 figs.

Bequaert, J. *The Hippoboscidae or Louse Flies (Diptera) of Mammals and Birds.* Ent. Amer., Vols. 32–36, 1952–57. 1053 pp., 104 figs.

_____. "A Monograph of the Melophaginae, or Kedflies, of Sheep, Goats, Deer and Antelopes," *Ent. Amer.,* 22(3):1–210 (1942), 19 figs.

Berner, L. "The Mayflies of Florida," Univ. of Florida Pubs. Biol. Sci. Serv., 4(4):1–267 (1950), il.

Betten, C. *The Caddisflies or Trichoptera of New York State.* N.Y. State Mus. Bul. 292, Albany, 1934. 576 pp., 61 figs., 67 pls.

Blatchley, W. S. *Heteroptera or True Bugs of Eastern North America.* Indianapolis: The Nature Publishing Co., 1926. 1116 pp., il.

————. *Orthoptera of Northeastern North America with Special Reference to the Fauna of Indiana and Florida.* Indianapolis: The Nature Publishing Co., 1920. 784 pp., il.

———— and Leng, C. W. *Coleoptera or Beetles Known to Occur in Indiana.* Indianapolis: The Nature Publishing Co., 1910. 1386 pp., il.

————. *Rhynchophora or Weevils of Northeastern North America.* Indianapolis: The Nature Publishing Co., 1916. 682 pp., il.

Blickenstaff, C. C., and Huggans, L. J. *Soybean Insects and Related Arthropods in Missouri.* Res. Bul. 803, Agr. Exp. Sta., Univ. of Missouri, Columbia, 1962. 51 pp.

Blickle, R. L., and Morse, W. J. *The Caddisflies (Trichoptera) of Maine Excepting the Family Hydroptilidae.* Bul. T-24, Univ. of Maine, Orono, 1966. 12 pp.

Boyd, J. P., et al. *Control of the Nantucket Pine Tip Moth . . .* Bul. B-661, Agr. Exp. Sta., Oklahoma State Univ., Stillwater, 1968. 39 pp., 3 figs.

Brezner, Jerome. *Biology, Ecology, and Taxonomy of Insects Infesting Acorns.* Res. Bul. 726, Agr. Exp. Sta., Univ. of Missouri, Columbia, 1960. 40 pp., 15 figs.

Bright, Donald E., Jr. "Review of the Tribe Xyleborini in America North of Mexico (Coleoptera: Scolytidae)," *Canad. Ent.*, 100:1288–1323 (1968), 19 figs.

Brooks, A. R., and Kelton, L. A. *Adult Elateridae of Southern Alberta, Saskatchewan, and Manitoba (Coleoptera).* Supp. 20, Canad. Ent., 1960. 92 pp., 138 figs.

————. *Aquatic and Semiaquatic Heteroptera of Alberta, Saskatchewan, and Manitoba.* Mem. Ent. Soc. of Canada, No. 51, 1967. 63 pp., 94 figs.

Brown, Leland R., and Eads, Clark O. *Insects Affecting Ornamental Conifers in Southern California.* Bul. 834, California Agr. Exp. Sta., Riverside, 1967. 72 pp., 103 figs.

————. *A Technical Study of Insects Affecting the Elm Tree in Southern California.* Bul. 821, California Agr. Exp. Sta., Riverside, 1966. 24 pp., 27 figs.

————. *A Technical Study of Insects Affecting the Oak Tree in Southern California.* Bul. 810, California Agr. Exp. Sta., Riverside, 1965. 105 pp., 145 figs.

————. *A Technical Study of Insects Affecting the Sycamore Tree in Southern California.* Bul. 818, California Agr. Exp. Sta., Riverside, 1965. 38 pp., 49 figs.

Brown, W. J. "A Revision of the Forms of *Coccinella* L. Occurring in America North of Mexico (Coleoptera: Coccinellidae)," *Canad. Ent.*, 94:785–808 (1962), il.

Brues, Charles T., Melander, A. L., and Carpenter, Frank M. *Classification of Insects: Keys to the Living and Extinct Families of Insects, and to the Living Families of Other Terrestrial Arthropods.* Cambridge, Mass.: Bul. Mus. Compar. Zool. at Harvard Col., 1954. 917 pp., 1219 figs.

————. *Insect Dietary.* Cambridge, Mass.: Harvard Univ. Press, 1946. 466 pp., 68 figs., 22 pls.

Burks, B. D. "The Mayflies or Ephemeroptera of Illinois," *Illinois Nat. Hist. Survey Bul.*, 26:1–216 (1953), 395 figs.

Byers, G. W. "Notes on North American Mecoptera," *Ent. Soc. Amer. Ann.*, 47:485–510 (1954).

Byers, R. A. *Biology and Control of a Spittlebug,* Prosapia bicincta (Say), *on Coastal Bermudagrass.* Tech. Bul. N.S. 42, Agr. Exp. Sta., Univ. of Georgia, Athens, 1965. 26 pp., 12 figs.

Campbell, W. V. *Stored Grain Insects and Their Control in the Middle Atlantic States.* Bul. No. 75, South. Coop. Series, North Carolina Agr. Exp. Sta., Raleigh, 1969. 30 pp., 32 figs.

Capriles, J. Maldonado. *The Miridae of Puerto Rico (Insecta: Hemiptera).* Tech. Paper 45, Agr. Exp. Sta., Univ. of Puerto Rico, Rio Piedras, 1969. 133 pp., 37 figs.

Carpenter, Frank M. "Biology of the Mecoptera," *Psyche*, 38:41–55 (1931).

_____. "Revision of the Nearctic Raphidiodea," *Amer. Acad. Sci. Proc.*, 71:89–157 (1936), 13 figs., 2 pls.

Childers, C. C., and Wingo, C. W. *Genus Culicoides (Diptera: Cerstopogonidae) in Central Missouri.* Res. Bul. 934, Agr. Exp. Sta., Univ. of Missouri, Columbia, 1968. 32 pp., 51 figs.

Chillcott, J. G. *A Revision of the Nearctic Species of Fanninae (Diptera: Muscidae).* Supp. 14, Canad. Ent. 1960. 295 pp., 289 figs.

Claassen, P. W. "A Catalog of the Plecoptera of the World," *Mem. Cornell Univ. Agr. Exp. Sta.*, 232:1–235 (1940).

_____. *Plecoptera Nymphs of North America (North of Mexico).* Springfield, Ill.: Charles C Thomas, 1931. 199 pp., 238 figs.

_____ and Needham, J. G. "Monograph of the Plecoptera or Stoneflies of America North of Mexico," Thos. Say Foundation, Ent. Soc. Amer., *Pub.*, 2:1–297 (1925).

Clausen, C. P. *Biological Control of Insect Pests in the Continental United States.* Tech. Bul. No. 1139, U.S. Dept. Agr., Washington, D.C. 1956. 151 pp.

_____. *Entomophagous Insects.* New York: Hafner Pub. Co., 1940. 688 pp., 257 figs.

Cole, Frank R. "Cyrtidae of North America," *Amer. Ent. Soc. Trans.*, 45:1–88 (1919), 15 pls.

_____. "Revision of the Family Therevidae," *U.S. Natl. Mus. Proc.*, 62(4): 1–140 (1923), il.

_____ and Schlinger, Evert I. *The Flies of Western North America.* Berkeley: Univ. of California Press, 1969. 693 pp., 360 figs.

Connell, W. A. *Nitidulidae of Delaware.* Pub. No. 318 (Tech.), Agr. Exp. Sta., Univ. of Delaware, Newark, 1956. 67 pp., 13 figs.

Criddle, Norman. "Habits of Some Manitoba Tiger Beetles," *Canad. Ent.*, 39:105–115 (1907); 42:9–15 (1910), 2 figs.

Curran, C. H. *The Families and Genera of North American Diptera.* Woodhaven, N.Y.: Henry Tripp, 1965. 515 pp., il.

Darby, Rollo E. "Midges Associated with California Rice Fields, with Special Reference to Their Ecology (Diptera: Chironomidae)," *Hilgardia*, Univ. of California, 32:1–206 (1962), 214 figs.

Daugherty, D. M., and Brett, C. H. *Nitidulidae Associated with Sweet Corn in North Carolina.* Tech. Bul. No. 171, Agr. Exp. Sta., North Carolina State Univ., Raleigh, 1966. 40 pp., 3 pls.

Davis, Donald R. *A Revision of the American Moths of the Family Carposinidae.* Bul. 289, U.S. Natl. Mus., 1968. 105 pp., 122 figs.

_____. *A Revision of the Moths of the Subfamily Prodoxinae (Lepidoptera: Incurvariidae).* Bul. 255, U.S. Natl. Mus., 1967. 170 pp., 155 figs.

Davis, J. J. "A Contribution to the Knowledge of the Natural Enemies of Phyllophaga," *Illinois Nat. Hist. Survey Bul.*, 13:53–138 (1919), 46 figs., 13 pls.

Dean, R. W., and Chapman, P. J. *Biology and Control of the Apple Redbug*. New York State Agr. Exp. Sta., Cornell Univ., Geneva, 1946. 42 pp., 12 figs.

DeBach, Paul (ed.), and Evert I. Schlinger (ass't. ed.). *Biological Control of Insect Pests and Weeds*. New York: Reinhold Pub. Corp., 1964. 844 pp., 123 figs.

DeLong, D. M. "The Leafhoppers or Cicadellidae of Illinois (Eurymelinae—Balcluthinae)," *Illinois Nat. Hist. Survey Bul.*, 24:97–376 (1948), 514 figs.

Dethier, Vincent G. *Chemical Insect Attractants and Repellents*. Philadelphia, Pa.: The Blakiston Co., 1947. 289 pp., 69 figs.

Dillon, Lawrence S. and Elizabeth S. "The Nearctic Components of the Tribe Acanthoconini," *Ent. Soc. Amer. Ann.*, 49:134–167, 207–235, 332, 355 (1956).

————. "The Tribe Dorcaschematini," *Amer. Ent. Soc. Trans.*, 73:173–298 (1947).

————. *The Tribe Monochamini (Cerambycidae) in the Western Hemisphere*. Reading Public Mus. and Art Gallery, Sci. Pub. No. 1, 1941. 135 pp., 5 pls.

————. *The Tribe Onciderini*. Reading Public Mus. and Art Gallery, Sci. Pubs. Nos. 5 and 6, 1945, 1946. 413 pp.

Dos Passos, Cyril F. *A Synonymic List of the Nearctic Rhopalocera*. New Haven, Conn.: The Lepidopterists' Society, 1964. 145 pp.

Duncan, Carl D. *A Contribution to the Biology of North American Vespine Wasps*. Stanford Univ. Pub. Biol. Sci., Vol. 8, No. 1, 1939. 272 pp., 255 figs.

Dunn, Henry A. *Cotton Boll Weevil* (Anthonomus grandis *Boh.*): *Abstracts of Research Publications 1843–1960*. Misc. Pub. No. 985, U.S. Dept. Agr., 1964. 194 pp.

Dupree, M. *Insecticidal and Cultural Control of the Lesser Cornstalk Borer*. Mimeo. Ser. N. S. 197, Agr. Exp. Sta., Univ. of Georgia, Athens, 1964. 21 pp.

Ebeling, Walter. *Termites: Identification, Biology, and Control of Termites Attacking Buildings*. Man. 38, California Agr. Exp. Sta., Berkeley, 1968. 74 pp., il.

Eckert, John E., and Bess, Henry A. *Fundamentals of Beekeeping in Hawaii*. Ext. Bul. 55, Agr. Ext. Serv., Univ. of Hawaii, Honolulu, 1952. 59 pp., 29 figs.

Evans, Howard E. *The Comparative Ethology and Evolution of the Sand Wasps*. Cambridge, Mass.: Harvard Univ. Press, 1966. 526 pp., il.

Eyer, J. R. *The Mexican Bean Beetle and Its Control in Southern New Mexico*. Bul. 377, Agr. Exp. Sta., New Mexico Col. Agr. & Mech. Arts, Las Cruces, 1953. 20 pp., 7 figs.

Falcon, Louis A., et al. *Light Traps and Moth Identification: An Aid for Detecting Insect Outbreaks*. Leaf. 197, Agr. Exp. Sta., Univ. of California, Berkeley, 1967. 16 pp., 4 pls. (color).

Felt, Ephraim P. *Plant Galls and Gall Makers*. Ithaca, N.Y.: Comstock Pub. Co., 1940 (reprint: Hafner Pub. Co., New York). 364 pp., 344 figs.

Ferris, G. F. *The Sucking Lice*. San Francisco, Calif.: Mem. Pac. Coast Ent. Soc., Vol. 1, 1951. 320 pp., 124 figs.

Flint, Oliver S., Jr. *The Caddisflies (Trichoptera) of Puerto Rico*. Tech. Paper 40, Agr. Exp. Sta., Univ. of Puerto Rico, Rio Piedras, 1964. 80 pp., 19 figs.

Forbes, W. T. M. *The Lepidoptera of New York and Neighboring States*. Mem. 68, 274, 329, 371, Cornell Univ. Agr. Exp. Sta., Ithaca, N.Y., 1923–60 (Mem. 68 reprint: Ent. Reprint Specialists, Los Angeles, Calif.). 1613 pp., il.

Fox, Irving. *Fleas of Eastern United States*. Ames: Iowa State Col. Press, 1940 (reprint: Hafner Pub. Co., New York). 191 pp., 166 figs.

Frick, Kenneth E., et al. *Bionomics of the Cherry Fruit Flies in Eastern Washington*. Tech. Bul. 13, Agr. Exp. Sta., State Col. of Washington, Pullman, 1954. 66 pp., 34 figs.

Frison, T. H. "Descriptions, Records, and Systematic Notes Concerning Western North American Stoneflies," *Pan-Pac. Ent.*, 19:9–16, 61–73 (1942).

————. "Fall and Winter Stoneflies of Illinois," *Illinois Nat. Hist. Survey Bul.*, 18:345–409 (1929).

————. "Stoneflies or Plecoptera of Illinois," *Illinois Nat. Hist. Survey Bul.*, 20: 281–471 (1935).

————. "Studies of North American Plecoptera," *Illinois Nat. Hist. Survey Bul.*, 22:235–355 (1942).

Gaud, Silverio Medina. *The Thysanoptera of Puerto Rico*. Tech. Paper 32, Agr. Exp. Sta., Univ. of Puerto Rico, Rio Piedras, 1961. 160 pp., 5 pls.

Gjullin, C. M., et al. *The Mosquitoes of Alaska*. Agr. Handb. No. 182, U.S. Dept. Agr., Washington, D.C., 1961. 98 pp., 81 figs.

Glass, E. H., and Chapman, P. J. *The Red-banded Leaf Roller and Its Control*. Bul. No. 755, New York State Agr. Exp. Sta., Geneva, 1952. 42 pp., 10 figs.

Graham, Kenneth. *Concepts of Forest Entomology*. New York: Reinhold Pub. Corp., 1963. 388 pp., 64 figs.

Graves, R. C. "Cicindelidae of Michigan," *Amer. Midland Nat.*, 69:492–507 (1963), 21 figs.

Gray, E. G. "The Fine Structure of the Insect Ear," Roy. Soc. London, *Trans.*, Ser. B, 243:75–94 (1960), 9 figs.

Halffter, Gonzalo, and Matthews, Eric G. *The Natural History of Dung Beetles of the Subfamily Scarabaeinae (Coleoptera: Scarabaeidae)*. Sociedad Mexicana de Entomología, Apartado Postal 31–312, Mexico 7, D.F., 1967. 312 pp., il.

Hardwick, David F. *The Corn Earworm Complex*. Mem. Ent. Soc. of Canada, No. 40, 1965. 247 pp., 146 figs. (49 color).

————. *The Genus Euxoa (Lepidoptera: Noctuidae) in North America*. Mem. Ent. Soc. of Canada, No. 67, 1970. 175 pp., 326 figs. (72 color).

Hatch, Melville, H. *The Beetles of the Pacific Northwest*. Univ. of Washington Pubs. Biol., Vol. 16 (4 pts.). Seattle: Univ. of Washington Press, 1953–65. 1514 pp., 169 pls.

Heinrich, Gerd H. *Synopsis of Nearctic Ichneumoninae Stenopneusticae with Particular Reference to the Northeastern Region (Hymenoptera)*. Supps. 15, 18, 21, 23, 26, 27, 29, Canad. Ent. 1958–62. 925 pp., il. (See Peck, Oswald.)

Herms, William B., and James, Maurice T. *Medical Entomology*. New York: The Macmillan Co., 1966. 616 pp., 185 figs.

Hine, James S. "Robberflies of the Genus *Erax*," *Ent. Soc. Amer. Ann.*, 12:103–154 (1919), 3 pls.

Hobbs, G. A. "Ecology of Species of *Bombus* Latr. (Hymenoptera: Apidae) in Southern Alberta," *Canad. Ent.*, 96:1465–1470; 97:120–128, 1293–1302; 98:33–39, 288–294 (1964–66), il.

Holland, George P. *Contribution Towards a Monograph of the Fleas of New Guinea*. Mem. Ent. Soc. of Canada, No. 61, 1969. 77 pp., 216 figs.

————. *The Siphonaptera of Canada*. Canada Dept. Agr., Pub. 817, Tech. Bul. 70, 1949. 306 pp., 350 figs.

Hopping, G. R. "North American Species of *Ips* DeGeer (Coleoptera: Scolytidae)," *Canad. Ent.*, 95: Nos. 1, 2, 5, 10, 11: 96: No. 7; 97: Nos. 2, 4, 5, 8 (1963–65). il.

Horsfall, William R. *Biology and Control of Mosquitoes in the Rice Area.* Bul. 427, Agr. Exp. Sta., Univ. of Arkansas, Fayetteville, 1942. 46 pp., 4 figs.

Howden, Henry F. "Biology and Taxonomy of North American Beetles of the Subfamily Geotrupinae with Revisions of the Genera *Bolbocerosoma, Eucanthus, Geotrupes* and *Peltotrupes* (Scarabaeidae)," *U.S. Natl. Mus. Proc.* 104:151–319 (1955).

————. *The Geotrupine of North and Central America.* Mem. Ent. Soc. of Canada, No. 39, 1964. 91 pp., 88 figs.

————. *A Revision of the New World Species of* Thalycra *Erichson, with a Description of a New Genus and Notes on Synonymy (Coleoptera: Nitidulidae).* Supp. 25, Canad. Ent., 1961. 61 pp., 106 figs.

Howell, J. F. *Biology of* Zodion obliquefasciatum *(Macq.) (Diptera: Conopidae), a Parasite of the Alkali Bee,* Nomia melandri *Ckll. (Hymenoptera: Halictidae).* Tech. Bul. 51, Agr. Exp. Sta., Washington State Univ., Pullman, 1967. 33 pp., 25 figs.

Hubbard, C. A. *Fleas of Western North America: Their Relation to Public Health.* Reprint. New York: Hafner Pub. Co., 1947. 533 pp., il.

Huber, L. L. *Some Aspects of Corn Ecology.* Bul. 679, Agr. Exp. Sta., Pennsylvania State Univ., University Park, 1961. 55 pp., 15 figs.

Huckett, Hugh C. *The Muscidae of Northern Canada, Alaska, and Greenland (Diptera).* Mem. Ent. Soc. of Canada, No. 42, 1965. 369 pp., 280 figs.

Hurd, Paul D., Jr., and Linsley, E. Gorton. "The Squash and Gourd Bees—Genera *Peponapis* Robertson and *Xenoglossa* Smith—Inhabiting America North of Mexico," *Hilgardia,* Univ. of California, 35:375–477 (1964), 18 figs.

Jacobson, Martin. *Insect Sex Attractants.* New York: John Wiley & Sons, 1965. 154 pp.

Jones, C. R. *A Contribution to Our Knowledge of the Syrphidae of Colorado.* Bul. 269, Colorado Agr. Exp. Sta., 1922. 72 pp., 8 pls.

————. "New Species of Colorado Syrphidae," *Ent. Soc. Amer. Ann.,* 10: 219–231 (1917).

Karlin, E. J., and Naegele, J. A. *Biology of the Mollusca of Greenhouses in New York State.* Mem. 372, Cornell Univ. Agr. Exp. Sta., Ithaca, 1960. 35 pp., 16 figs.

Kerr, T. W., Jr. *The Arborvitae Weevil,* Phyllobius intrusus *Kono.* Bul. 305, Agr. Exp. Sta., Univ. of Rhode Island, Kingston, 1949. 30 pp., 8 figs.

King, W. V., et al. *Handbook of Mosquitoes of the Southeastern United States.* Agr. Handb. No. 173, U.S. Dept. Agr., Washington, D.C., 1960. 188 pp., il.

Knight, Harry H. "The Plant Bugs, or Miridae, of Illinois," *Illinois Nat. Hist. Survey Bul.,* 22:1–234 (1951), 181 figs.

Knull, Josef N. "The Long-horned Beetles of Ohio (Coleoptera: Cerambycidae)," *Ohio Biol. Survey Bul.,* 39:133–354 (1946), 29 pls.

Kosztarab, Michael. *The Armored Scale Insects of Ohio (Homoptera: Diaspididae).* Bul. Ohio Biol. Survey, New Series, Vol. II, No. 2, Columbus, 1963. 120 pp., 57 figs.

_____, and Hoffman, Richard L. *The Insects of Virginia:* No. 1. Res. Div. Bul. 48, Virginia Polytech. Inst., Blacksburg, 1969. 62 pp., il.

Krombein, Karl V., and Burks, B. D. *Hymenoptera of America North of Mexico.* Agr. Monog. No. 2, Second Supp., U.S. Dept. Agr., Washington, D.C., 1967. 584 pp. First Supp., 1958. 305 pp. See Muesebeck, C. F. W.

LaBerge, W. E., and Webb, M. C. *The Bumblebees of Nebraska.* Res. Bul. 205, Agr. Exp. Sta., Univ. of Nebraska, Lincoln, 1962. 38 pp., 12 figs.

Lange, W. H., Jr. "Biology and Systematics of Plume Moths of the Genus *Platyptilia* in California," *Hilgardia,* Univ. of California, 19:561–668 (1950), 8 figs., 16 pls.

Leng, C. W. *Catalogue of Coleoptera of North America North of Mexico.* New York: John D. Sherman, 1920. 470 pp.

_____ and A. J. Mutchler. *Catalogue of Coleoptera* . . . First Supp., 1919–24. New York: Sherman, 1927. 78 pp.

_____. *Catalogue of Coleoptera* . . . Second and Third Supps., 1925–32. New York: Sherman, 1933. 112 pp.

Leonard, D. E. *Biosystematics of the "Leucopterus Complex" of the Genus* Blissus (*Heteroptera: Lygaeidae*). Bul. 677, Connecticut Agr. Exp. Sta., New Haven, 1966, 47 pp., il.

_____. *Differences in Development of Strains of the Gypsy Moth* Porthetria dispar (*L.*). Bul. 680, Connecticut Agr. Exp. Sta., New Haven, 1966. 31 pp., il.

Liljeblad, Emil. *Monograph of the Family Mordellidae (Coleoptera) of North America North of Mexico.* Misc. Pub. Mus. Zool. Ann Arbor: Univ. of Michigan, 1945. 229 pp., 7 pls.

Lindroth, Carl H. *The Ground Beetles (Carabidae, excl. Cicindelinae) of Canada and Alaska.* Pts. 1 to 6: Supp. 20, 24, 29, 33, 34, and 35, Opusc. Ent. Zool. Inst., Univ. of Lund, Sweden, 1961–69. 1192 pp., il.

Linsley, E. Gorton. *The Cerambycidae of North America.* Univ. of California Pubs. Ent., Vols. 18–22, 1961–64. 788 pp., 202 figs., 37 pls.

_____. "Ecology of Solitary Bees," *Hilgardia,* Univ. of California, 27:543–599 (1958).

Luginbill, P., Sr., and Painter, H. R. *May Beetles of the United States and Canada.* Tech. Bul. No. 1060, U.S. Dept. Agr., Washington, D.C., 1953. 102 pp., 78 pls.

Mackay, Margaret Rae. *Larvae of the North American Olethreutidae (Lepidoptera).* Supp. 10, Canad. Ent., 1959. 338 pp., il.

_____. *Larvae of the North American Tortricinae (Lepidoptera: Tortricidae).* Supp. 28, Canad. Ent., 1962. 182 pp., 85 figs.

_____. *The North American Aegeriidae (Lepidoptera): A Revision Based on Late-instar Larvae.* Mem. Ent. Soc. of Canada, No. 58, 1968. 112 pp., 50 figs.

Mann, John. *Cactus-feeding Insects and Mites.* Bul. 256, U.S. Natl. Mus., 1969. 158 pp., 8 pls.

Marston, Norman. *A Revision of the Nearctic Species of the* Albofasciatus *Group of the Genus* Anthrax Scopoli (*Diptera: Bombyliidae*). Tech. Bul. 127, Agr. Exp. Sta., Kansas State Univ., Manhattan, 1963. 79 pp., 6 pls.

_____. *Revision of New World Species of* Anthrax . . . *Other than* Anthrax albofasciatus *Group.* Smithsn. Contrib. Zool. No. 43, 1970. 146 pp., il.

Massey, C. L., and Wygant, N. D. *Biology and Control of the Engelmann Spruce*

Beetle in Colorado. Cir. 944, Forest Serv., U.S. Dept. Agr., Washington, D.C., 1954. 35 pp., 12 figs.

Mathewson, J. A. "Nest Construction and Life History of the Eastern Cucurbit Bee, *Peponapis pruinosa,*" *Kansas Ent. Soc. Jour.,* 41:255–261 (1968), 1 fig.

Maynard, E. A. *The Collembola or Springtails of New York.* Ithaca, N.Y.: Comstock Pub. Co., 1951. 339 pp., il.

McGregor, S. E., et al. *Beekeeping in the United States.* Agr. Handb. No. 335, U.S. Dept. Agr., Washington, D.C., 1967. 147 pp., il.

Metcalf, C. L. "Syrphidae of Maine," *Maine Agr. Exp. Sta. Bul.* (ser. 2), 253: 193–264 (1916), 10 figs.; 263:153–176 (1917), 5 figs.

————. "Syrphidae of Ohio," *Ohio Biol. Survey Bul.,* 1:7–122 (1913), 3 figs., 11 pls.

Michelbacher, A. E., et al. *Control of Household Insects and Related Pests.* Cir. 498, Agr. Exp. Sta., Univ. of California, Berkeley, 1961. 40 pp., il.

———— and Ortega, J. C. *Controlling Melon Insects and Spider Mites.* Bul. 749, Agr. Exp. Sta., Univ. of California, Berkeley, 1955. 46 pp., il.

————. *A Technical Study of Insects and Related Pests Attacking Walnuts.* Bul. 764, Agr. Exp. Sta., Univ. of California, Berkeley, 1958. 86 pp., 81 figs.

Miller, C. D. F. *Taxonomy and Distribution of Nearctic Vespula.* Supp. 22, Canad. Ent., 1961. 52 pp., 84 figs.

Miller, J. M. and Keen, F. P. *Biology and Control of the Western Pine Beetle.* Misc. Pub. 800, Forest Serv., U.S. Dept. Agr., Washington, D.C., 1960. 381 pp., il.

Miller, W. E., and Neiswander, R. B. *Biology and Control of the European Pine Shoot Moth.* Res. Bul. 760, Ohio Agr. Exp. Sta., Wooster, 1955. 31 pp., 7 figs.

Mills, H. B. *The Collembola of Iowa.* Ames: Collegiate Press. 1934. 143 pp., il.

Morrison, H. E. *Controlling the Garden Symphylan.* Bul. 816, Coop. Ext. Serv., Oregon State Univ., Corvallis, 1965. 12 pp., il.

Muesebeck, C. F. W., et al. *Hymenoptera of America North of Mexico: Synoptic Catalog.* Agr. Monog. No. 2, U.S. Dept. Agr., Washington, D.C., 1951. 1420 pp. See Krombein, Karl V.

Needham, J. G., et al. *The Biology of Mayflies, with Systematic Account of North American Species.* Ithaca, N.Y.: Comstock Pub. Co., 1935. 759 pp., 169 figs., 40 pls.

Newsom, L. D., et al. *The Tobacco Thrips: Its Seasonal History and Status as a Cotton Pest.* Tech. Bul. 474, Agr. Exp. Sta., Louisiana State Univ., Baton Rouge, 1953. 36 pp., 5 figs.

Niemczyk, H. D., and Guyer, G. E. *The Distribution, Abundance and Economic Importance of Insects Affecting Red and Mammoth Clover in Michigan.* Tech. Bul. 293, Agr. Exp. Sta., Michigan State Univ., E. Lansing, 1963. 38 pp., 7 figs.

Ode, P. E., and Matthysse, J. G. *Bionomics of the Face Fly,* Musca autumnalis DeGeer. Mem. 402, Agr. Exp. Sta., Cornell Univ., Ithaca, 1967. 91 pp., 39 figs.

Okumura, George T. *Identification of Lepidopterous Larvae Attacking Cotton, with Illustrated Key.* Bureau Ent. Spec. Pub. No. 282, Dept. Agr., State of California, Sacramento, 1961. 80 pp., 52 figs.

————. "Warehouse Beetle a Major Pest of Stored Food," *National Pest Control Oper. News,* 32 (1): 4, 1972. 4 figs.

Onsager, Jerome A., and Mulkern, Gregory B. *Identification of Eggs and Eggpods of North Dakota Grasshoppers (Orthoptera: Acrididae)*. Bul. 446 (Tech.), Agr. Exp. Sta., North Dakota State Univ., Fargo, 1963. 48 pp., 7 pls.

Painter, Reginald H., and Hall, Jack C. *A Monograph of the Genus* Poecilanthrax *(Diptera: Bombyliidae)*. Tech. Bul. 106, Agr. Exp. Sta., Kansas State Univ., Manhattan, 1960. 132 pp., il.

Papp, Charles S. "Checklist of Tenebrionidae of America North of the Panama Canal," *Opuscula Entomologica* (Lund, Sweden), 26:97–140 (1961).

————. "The Cleridae of North America," Pt. I: "The Geographic Distribution of Cleridae of North America North of the Panama Canal." *Bul. Southern California Acad. Sci.*, 59:76–88 (1960).

————. "An Illustrated and Descriptive Catalogue of the Ptinidae of North America," *Deutsche Entomologische Zeitschrift* (Berlin), New Series, 9:367–423 (1962), 35 figs.

————. "The Hispinae of America," *Portugaliae Acta Biologica* (Lisbon) (B), 4:1–147 (1953).

————, and Okumura, George T. "A Preliminary Study of the Ptinidae of California," *Bul. Dept. Agr.*, State of California, 48:228–248 (1959), 32 figs.

Peck, Oswald. *A Catalogue of the Nearctic Chalcidoidea (Insecta: Hymenoptera)*. Supp. 30, Canad. Ent., 1963. 1092 pp.

————. See Heinrich, Gerd H. Pt. VIII. Mem. Ent. Soc. of Canada No. 35, 889–925, 1964.

Peters, D. C., et al. *The Biology and Control of the European Corn Borer in Missouri*. Res. Bul. 757, Agr. Exp. Sta., Univ. of Missouri, Columbia, 1961. 26 pp., 3 figs.

Peterson, Alvah. *Larvae of Insects*. Columbus, Ohio: A. Peterson, 1948. 2 vols.: 731 pp., il.

Peterson, B. V. *The Prosimulium of Canada and Alaska (Diptera: Simuliidae)*. Mem. Ent. Soc. of Canada No. 69, 1970. 216 pp., 158 figs.

Pinto, J. D., and Selander, R. B. *The Bionomics of Blister Beetles of the Genus* Meloe *and a Classification of the New World Species*. Illinois Biol. Monog. 42, Univ. of Illinois, Urbana, 1970. 222 pp., 198 figs.

Pletsch, D. J. *The Potato Psyllid,* Paratrioza cockerelli *(Sulc.), Its Biology and Control*. Bul. 446, Agr. Exp. Sta., Montana State Col., Bozeman, 1947. 95 pp., 36 figs.

Polivka, J. B. *Distribution and Control of the Japanese Beetle in Ohio*. Res. Cir. No. 4, Ohio Agr. Exp. Sta., Wooster, 1950. 15 pp., 10 figs.

Race, S. R. *Western Flower Thrips,* Frankliniella occidentalis, *on Seedling Cotton*. Bul. 497, Agr. Exp. Sta., New Mexico State Univ., Las Cruces, 1965. 26 pp., il.

Ribble, D. W. *A List of Recent Publications on the Alkali Bee,* Nomia melanderi, *with Notes on Related Species of Bees*. Sci. Monog. 11, Agr. Exp. Sta., Univ. of Wyoming, Laramie, 1968. 18 pp.

Richards, W. R. *The Callaphidini of Canada (Homoptera: Aphididae)*. Mem. Ent. Soc. of Canada, No. 44, 1965. 149 pp., il.

————. *The Sminthuridae of the World (Collembola)*. Mem Ent. Soc. of Canada, No. 53, 1968. 54 pp., il.

————. *A Synopsis of the World Fauna of Myzocallis (Homoptera: Aphididae)*. Mem. Ent. Soc. of Canada, No. 57, 1968. 76 pp., 186 figs.

Ross, Edward S. "Revision of the Ebioptera of North America," *Ent. Soc. Amer. Ann.*, 33:629–676 (1940), il.

————. "A Revision of the Embioptera or Web Spinners of the New World," *U.S. Natl. Mus. Proc.*, 94(3175):401–504 (1944), il.

Ross, Herbert H. "The Caddis Flies or Trichoptera of Illinois," *Illinois Nat. Hist. Survey Bul.*, 23:1–326 (1944), 957 figs.

————. "Descriptions of Nearctic Caddis Flies (Trichoptera)," *Illinois Nat. Hist. Survey Bul.*, 21:101–183 (1938), 128 figs.

————. "A Review of Nearctic Lepidostomatidae (Trichoptera)," *Ent. Soc. Amer. Ann.*, 39:265–290 (1946), 37 figs.

————. *A Synopsis of the Mosquitoes of Illinois.* Illinois Nat. Hist. Survey Biol. Notes No. 52, 1965. 50 pp., 231 figs.

————. and Frison, T. H. "Nearctic Alderflies of the Genus *Sialis*," *Illinois Nat. Hist. Survey Bul.*, 21:57–78 (1937), 63 figs.

Roth, L. M., and Eisner, T. "Chemical Defenses of Arthropods," *Ann. Rev. Ent.*, 7:107–136 (1962).

Sabrosky, Curtis W. "A Further Contribution to the Classification of the North American Spider Parasites of the Family Acroceratidae (Diptera)," *Amer. Midland Nat.*, 39:383–430 (1948), 2 pls.

————. "A Revision of the American Spider Parasites of the Genera *Ogcodes* and *Acrocera* (Diptera: Acroceridae)," *Amer. Midland Nat.*, 31:385–413 (1944), 1 pl.

Say, Thomas. *The Complete Writings of Thomas Say on the Entomology of North America.* Ed. by J. L. LeConte. New York: Ballière Bros., 1859. 2 vols., 1226 pp., il.

Schlinger, Evert I. "A Review of the Genus *Eulonchus* Gerstaecker," Pt. I: "The Species of the Smaragdinus Group (Diptera: Acroceridae)," *Ent. Soc. Amer. Ann.*, 53:416–422 (1960), 2 figs., 1 pl.

————. "A Revision of the Genus *Ogcodes* Latreille with Particular Reference to Species of the Western Hemisphere," *U.S. Natl. Mus. Proc.* 111:227–336 (1960), 9 figs., 13 pls.

Selander, R. B. *Bionomics, Systematics, and Phylogeny of* Lytta, *a Genus of Blister Beetles.* Illinois Biol. Monog. 28, Univ. of Illinois, Urbana, 1960. 295 pp., 350 figs.

————, and Mathieu, J. M. *Ecology, Behavior, and Adult Anatomy of the Albida Group of the Genus* Epicauta. Illinois Biol. Monog. 41, Univ. of Illinois, Urbana, 1969. 168 pp., 60 figs.

Shands, W. A., and Landis, B. J. *Potato Insects: Their Biology and Biological and Cultural Control.* Agr. Handb. No. 264, U.S. Dept. Agr., Washington, D.C., 1964. 61 pp., 61 figs.

Smart, John. *A Handbook for the Identification of Insects of Medical Importance.* London: British Mus. (Nat. Hist.), 1965. 303 pp., 178 figs., 13 pls.

Smith, Marion R. *House-infesting Ants of Eastern United States: Their Recognition, Biology, and Economic Importance.* Tech. Bul. No. 1326, U.S. Dept. Agr., Washington, D.C., 1965. 105 pp., 50 figs.

Snodgrass, R. E. *Principles of Insect Morphology.* New York: McGraw-Hill Book Co., 1935. 667 pp., 319 figs.

Sorenson, Charles J., and Cutler, Lowell. *The Superb Plant Bug* Adelphocoris superbus *(Uhler): Its Life History and Its Relation to Seed Development in*

Alfalfa. Bul. 370, Agr. Exp. Sta., Utah State Agr. Col., Logan, 1954. 20 pp., 7 figs.

———— and Gunnell, Farrell H. *Biology and Control of the Peach Twig Borer* (Anarsia lineatella *Zeller*) *in Utah.* Bul. 379, Agr. Exp. Sta., Utah State Agr. Col., Logan, 1955. 19 pp., 7 figs.

Spencer, Kenneth A. *The Agromyzidae of Canada and Alaska.* Mem. Ent. Soc. of Canada, No. 64, 1969. 311 pp., 552 figs.

Stehr, Frederick W., and Cook, Edwin F. *A Revision of the Genus* Malacosoma Hubner *in North America* (*Lepidoptera: Lasiocampidae*): *Systematics, Biology, Immatures and Parasites.* Bul. 276, U.S. Natl. Mus., 1968. 321 pp., 399 figs.

Steinhaus, Edward A. (ed.). *Insect Pathology: An Advanced Treatise.* New York: Academic Press, 1963. 2 vols., 1350 pp., il.

————. *Principles of Insect Pathology.* New York: McGraw-Hill Book Co., 1949 (reprint: Hafner Pub. Co., New York). 757 pp., il.

Stone, Alan, and Snoddy, E. L. *The Black Flies of Alabama* (*Diptera: Simuliidae*). Bul. 390, Agr. Exp. Sta., Auburn Univ., Auburn, Ala., 1969. 93 pp., 306 figs.

———— et al. *A Catalog of the Diptera of America North of Mexico.* Agr. Handb. No. 276, U.S. Dept. Agr., Washington, D.C., 1965. 1696 pp.

Stone, M. W. *Biology and Control of the Lima-bean Pod Borer in Southern California.* Tech. Bul. No. 1321, U.S. Dept. Agr., Washington, D.C., 1965. 46 pp., 15 figs.

Sutherland, Douglas W. S. *Biological Investigations of* Trichoplusia ni (*Hubner*) *and Other Lepidoptera Damaging Cruciferous Crops on Long Island, New York.* Mem. 399, Cornell Univ. Agr. Exp. Sta., Ithaca, 1966. 99 pp., 25 figs.

Sweetman, Harvey L. *The Principles of Biological Control.* Dubuque, Iowa: Wm. C. Brown Co., 1958. 560 pp., 328 figs.

Taboada, Oscar. *Medical Entomology.* Bethesda, Md.: Naval Medical School, National Naval Medical Center, 1967. 395 pp., 198 figs., 44 pls.

Tashiro, H., et al. *Development of Blacklight Traps for European Chafer Surveys.* Tech. Bul. No. 1366, U.S. Dept. Agr., Washington, D.C., 1967. 52 pp., 18 figs.

Teskey, H. J. *Larvae and Pupae of Some Eastern North American Tabanidae* (*Diptera*). Mem. Ent. Soc. of Canada No. 63, 1969. 147 pp., 148 figs.

Thomas, J. B. "The Immature Stages of Scolytidae: The Genus *Dendroctonus* Erichson," *Canad. Ent.*, 97:374–400 (1965).

Timberlake, P. H. *A Revisional Study of the Bees of the Genus* Perdita *F. Smith, with Special Reference to the Fauna of the Pacific Coast* (*Hymenoptera: Apoidea*). Univ. of California Pubs. Ent., 1954–68. 7 pts.: 931 pp., 1471 figs.

Torre-Bueno, J. R. de la. *A Glossary of Entomology,* with Supp. A ed. by George S. Tullock. Brooklyn, N.Y.: Brooklyn Ent. Soc., 1962. 372 pp., 9 pls.

Townes, Henry and Marjorie. *Ichneumon-Flies of America North of Mexico.* Bul. 216, U.S. Natl. Mus., 1959–62. 3 pts., 1596 pp., 974 figs.

————. *Nearctic Wasps of the Subfamilies Pepsinae and Ceropalinae.* Bul. 209, U.S. Natl. Mus., 1957. 286 pp., 161 figs., 4 pls.

Turner, Neely (ed.). *Effect of Defoliation by the Gypsy Moth.* Bul. 658, Connecticut Agr. Exp. Sta., New Haven, 1963. 30 pp., 10 figs.

Udvardy, M. D. F. *Dynamic Zoogeography: With Special Reference to Land*

Animals. New York: Van Nostrand Reinhold Co., 1970. 445 pp., il.

Underhill, G. W. *Blackflies Found Feeding on Turkeys in Virginia* (Simulium Nigroparvum *Twinn and* Simulium Slossonae *Dyar and Shannow*). Tech. Bul. 94, Agr. Exp. Sta., Virginia Polytech. Inst., Blacksburg, 1944. 32 pp., 15 figs.

Urquhart, F. A. *The Monarch Butterfly.* Toronto: Univ. of Toronto Press, 1960. 361 pp., 79 figs.

Usinger, R. L. (ed.). *Aquatic Insects of California with Keys to North American Genera and California Species.* Berkeley: Univ. of California Press, 1963. 508 pp., il.

Van Duzee, M. C., et al. *The Dipterous Genus* Dolichopus *in North America.* Bul. 116, U.S. Natl. Mus., 1921. 304 pp., il.

Vaurie, Patricia. "Notes on the Habitats of Some North American Tiger Beetles," *Jour. N.Y. Ent. Soc.,* 58:143–153 (1950).

Vockeroth, J. R. *A Revision of the Genera of the Syrphini (Diptera: Syrphidae).* Mem. Ent. Soc. of Canada No. 62, 1969. 176 pp., 100 figs.

Walker, Edmund M. *The Odonata of Canada and Alaska.* Toronto: Univ. of Toronto Press, 1953. 2 vols., 610 pp., 108 pls.

Wallis, J. B. *The Cicindelidae of Canada.* Toronto: Univ. of Toronto Press, 1961. 84 pp., 4 figs., 4 color pls.

Weaver, C. R., and King, D. R. *Meadow Spittlebug.* Res. Bul. 741, Ohio Agr. Exp. Sta., Wooster, 1954. 99 pp., 17 figs.

West, L. S. *The House Fly.* Ithaca, N.Y.: Comstock Pub. Co., 1951. 584 pp., 176 figs.

White, R. E. *A Review of the Genus* Cryptocephalus *in America North of Mexico (Chrysomelidae: Coleoptera).* Bul. 290, U.S. Natl. Mus., 1968. 124 pp., 140 figs.

Wigglesworth, V. B. *Principles of Insect Physiology.* London: Methuen, 1965. 741 pp., 407 figs.

Wilcox, John A. "Leaf Beetles of Ohio (Chrysomelidae: Coleoptera)," *Ohio. Biol. Survey Bul.,* 43:353–506 (1954). 36 pls.

Wirth, W. W. "The Heleidae of California," *Univ. of California Pubs. Ent.,* 9:95–266 (1952). 33 figs.

Wolcott, George N. *The Insects of Puerto Rico.* Jour. Agr. Univ. Puerto Rico, Vol. 32, Nos. 1 to 4. Agr. Exp. Sta., Rio Piedras, 1948. 975 pp., il.

Wood, S. L. "New Synonymy in the Platypodidae and Scolytidae (Coleoptera)," *Great Basin Nat.,* 26:17–33 (1966).

————. "A Revision of the Bark Beetles Genus *Dendroctonus* Erichson (Coleoptera: Scolytidae)," *Great Basin Nat.,* 23:1–117 (1963).

Wylie, W. D. *The Oriental Fruit Moth as a Peach Pest in Arkansas.* Bul. 709, Agr. Exp. Sta., Univ. of Arkansas, Fayetteville, 1966. 38 pp., il.

————. *The Plum Curculio on Peaches in Arkansas.* Bul. 542, Agr. Exp. Sta., Univ. of Arkansas, Fayetteville, 1954. 46 pp., il.

Guide to the Insects of Connecticut VI: The Diptera or True Flies. State Librarian, Hartford: Conn. Geol. and Nat. Hist. Survey, 1942–60. 9 fascicles, 1438 pp., 228 figs., 88 pls.

Common Names of Insects Approved by *The Entomological Society of America.* Ent. Soc. Amer., College Park, Maryland, Dec. 1970. 36 pp.

Index of Subjects and Common Names

Braconids   521
breathing tube. *See* respiratory funnel
Brill's disease   100
Bristletails   57
Broadbeam Weevil   474
Broad-necked Root Borer   442
Bronze Birch Borer   392
Bronze Poplar Borer   392
Brush-footed Butterflies   224
Buckeye   236
Budworm, Black-headed   320
   Jack-pine   320
   Spruce   319, 530
Bugs, the True   109–10. *See* individual
   names
Burrower Bug   131
Butterflies   197, 202. *See* individual
   names

Cabbage Looper   280
Cabbageworm, Imported   210
   Southern   209
Caddisflies   191
   Large   194
   Log-cabin   195
Caddisworms   191
Cadelle   397
California Five-spined Ips   506
California Flatheaded Borer   391
California Pine Engraver   506
California Prionus   441
California Ringlet   221
California Sister   239
Camel Crickets   78
Cankerworm, Fall   292
   Spring   293
cantharidin   379
Carolina Copris   430
Carpenter Ants   561
Carpet Beetles   392, 394
   Black   394
   Varied   395
Carpet Moth   326
Carpetworm   297
Carpetworm Moths   297
Carrion Beetles   355
   American   357
   Garden   357
   Giant   356
   Gold-necked   356
   Pustulated   356
   Red-lined   357
   Spinach   357
Carrot Beetle   436
Casebearer Moths   326
Casemaking Clothes Moth   326
Caterpillar, Oak Leaf   285
   Red-humped   285
   Walnut   284
   Yellow-necked   284
Cattle Grub, Common   651
Ceanothus Silk Moth   266
Chafer, European   433
   Northern Masked   433
   Pine   435
   Rose   434
   Southern Masked   433
Chagas' disease   121
Chalcids   531–540

Charcoal Beetle   391
Changa   78
Checkered Beetles   374
   Dubious   376
   Hairy   376
   Red-blue   375
Checkerspot, Acasta   229
   Chalcedon   228
   Edwards'   229
   Harris'   229
   Leanira   229
Cheese Skipper   628
Cherry Fruitworm   313
Chigoe   659
Chinch Bug   **124,** 541
   False   125
   Hairy   125
   Southern   125
   Western   125
Chinese Seed Beetle   475
Chocolate Serica   431
chorodontal organs   30
Cicada Killer   562
Cicadas   132–136
Cigarette Beetle   416
circular-seamed flies   584
circulation   33
Citrus Blackfly   155
Citrus Whitefly   154
claws   22
Clearwing Moths   298
Click Beetles   384
   Eyed   386
   Faded   387
   Narrow-necked   388
   Small-eyed   386
   Western Eyed   386
Cloudy Wing, Northern   257
   Southern   257
Clover Hayworm   305
Clover Head Caterpillar   314
Clover Seed Chalcid   537
Clover Stem Borer   401
Coccids   **155,** 532
cocoon   16
Cockroaches   65, **66,** 540
   American   66
   Australian   67
   Brown   66
   Brown-banded   67
   German   67
   Oriental   67
   Smoky-brown   67
   Surinam   68
Codling Moth   313
Coffee Bean Weevil   475
colliphore   59
Collops, Four-spotted   373
   Two-lined   373
Colorado Potato Beetle   463
Common Alpine (butterfly)   222
Common Sooty Wing   258
Conchuela   127
conjunctivitis   630, 631
Constricted Flowerbug   384
Contort Dryinid   568
Cooley Spruce Gall Aphid   144
Copper   244
   American   250

Copper (*cont'd*)
  Bog 251
  Bronze 250
  Great 251
  Purplish 251
Corn Earworm 279, 532
Corn Rootworm, Northern 466
  Southern 466
  Western 467
Corn-seed Beetle 341
Cotton Fleahopper 118
Cotton Leaf Perforator 327
Cotton Leafworm 281
  Brown 281
Cotton Square Borer 245
Cotton Stainer 131
  Arizona 131
Cotton Stem Moth 325
Cottonwood Borer 454
Cowpea Weevil 474
coxa 22
Cranberry Fruitworm 309
Cranberry Girdler 306
Cranberry Rootworm 462
Cranberry Toad Bug 141
Crane Flies 585
  European 586
  Range 585
crawlers 155, 161
Crawling Water Beetles 348
cremaster 200
Crescent, Bates' 230
  Field 230
  Mylitta 231
  Painted 231
  Pearl 230
  Phaon 231
Cricket, Black-horned Tree 77
  Cave 78
  California Camel 79
  Coulee 75
  Field 77
  Four-spotted Tree 77
  House 76
  Jerusalem 79
  Mormon 75
  Northern Mole 78
  Pigmy Mole 78
  Snowy Tree 77
  Southern Mole 78
  Spotted Camel 79
crochets 22, 198
Cucubano 385
Cucumber Beetle, Banded 466
  Spotted 466
  Striped 467
  Western Spotted 466
  Western Striped 467
Curculio, Apple 489
  Cabbage 493
  Cambium 495
  Cowpea 494
  Grape 493
  Plum 494
  Quince 495
  Rhubarb 492
  Rose 478
  Walnut 495

Currant Borer 299
cuticle 16
Cutworm, Army 278
  Black 277
  Clay-backed 277
  Dingy 278
  Glassy 276
  Pale-sided 278
  Pale Western 278
  Spotted 277
  Variegated 276
  Western Bean 278
  Yellow-headed 274
Cynipids 519
Cypress Beetle 436
Cypress Twig or Bark Moth 316

dactyls 77, 78
Dagger Moth, American 273
  Cottonwood 274
  Smeared 274
Damsel Bug 119
Damselflies 170, 174
  Black-wing 175
  Broad-winged 175
  Narrow-winged 175
  Ruby Spot 175
Darkling Beetles 417, 423, 424
  Coastal 420
  Costate 420
  False 425
  Horned 421
  Maculated 421
  Mayhew's 420
  Puffed 419
  Red-necked 421
  Round 418
  Sculptured 418
  Shouldered 417
Darling Underwing 282
Darner, Big Green 173
Deathwatch 95, 415
Deer Fly 604, 605
defensive mechanisms 32
dengue 588, 589
Dermestid Beetles 393
Desert Locust 73
Destructive Pruneworm 309
diapause 17
digestion 33
Digger Wasp 548
Diving Beetle 351
  Fasciated 350
  Giant 351
  Striped 350
Dobsonflies 182
Dog Face, California 214
  Southern 215
Dogwood Twig Borer 457
Douglas-fir Beetle 504
Douglas-fir Engraver 500
Dragonflies 170, 172–174
  Ten-spot 172
  Widow 172
Dried-fruit Beetle 399
Drugstore Beetle 415, 416
dry flies (lures) 82, 167
Dull Canthon 429

Moths (*cont'd*)
Luna 267
Mediterranean Flour 310
Nantucket Pine Tip 314
Oriental Fruit 313
Pandora 265
Pea 315
Pine Cone 315
Pine Tube 321
Pitch-pine Tip 314
Polyphemus 266
Promethea 268
Phycitid 308
Pyralid 304
Pyraustid 301
Raisin 310
Regal (Royal) 270
Satin 288
Sequoia Pitch 300
Silver-spotted Tiger 270
Spotted Tussock 271
Spruce Seed 315
Strawberry Crown 299
Tiger 270
Tobacco 310
Western Pine Tip 315
Winter 292
Yucca 328
Mourning Cloak 235
mousies 618
Mud Daubers 562
Black and Yellow 564
Blue 564
Pipe Organ 563
mummies 146, 535
murine (endemic) typhus 100, 654
mutilation 532, 533, **536**
myiasis 634, 636, 638, 641, 642, 643, 644, 653

Nantucket Pine Tip Moth 314
Narrow Flower Scarab 433
Nautical Borer 448
Navel Orangeworm 311
Negro Bug 131
Net-winged Beetles 368
Banded 368
Eastern 369
Flat 369
Golden 369
Northern Corn Rootworm 466
notum 21

Oak Timberworm 477
Oak Twig Pruner 452
Oakworm, California 283
Orange-striped 269
Pink-striped 269
Spiny 269
ocellus 32
Oil Beetles 381
Old-house Borer 447
Omnivorous Leaf Roller 321
Omnivorous Looper 294
Orange Dog 206
Orange Tortrix 320
Orchidfly 538
Oriental Beetle 435

Oriental Fruit Moth 313
osmeteria 34
Owlet Moths and Underwings 272

Pacific Flatheaded Borer 392
paedogenesis 18, 600
Painted Beauty 237
Painted Lady 236
Pallid Scolops 141
Palm Borer 426
Parnassians 203, 207
parthenogenesis 18, 69, 95, 103, 105, 145, 167, 198, 294, 481, 484, 508, 512, 514, 522, 527, 568, 583
Pea Moth 315
Pea Weevil 473
Peach Bark Beetle 502
Peach Tree Borer 299
Lesser 300
Western 300
Peach Twig Borer 323
Pear Slug 512
California 512
Pearly Eye 221
Creole 221
Pecan Catocala 282
Pecan Leaf Casebearer 308
percussion disc 83
Periodical Cicada 133
broods 134, 135
pheromones 16, 572
Phycitid Moths 308
Phylloxeras 141, 145
Pickleworm 301
Picnic Beetles 398
Pigeon Tremex 518
Pill Beetles 396
Pine Bark Aphid 145
Pine Beetle, Arizona 504
Eastern 503
Mountain 503
Red-winged 502
Southern 504
Southwestern 503
Western 503
Pine Butterfly 211
Pine Engraver 506
California 506
Monterey 507
Pine Leaf Chermid 144
Pine Needle Miner 324
Pine-stump Borer 444
Pine Tip Moth, Southwestern 315
Western 315
Pine Tube Moth 321
Pink Bollworm 324
Pink Glowworm 370
Pistol Casebearer 326
Pitch-pine Tip Moth 314
plague 654, 656, 658
Plant Bugs 117
Four-lined 118
Superb 118
Tarnished 118
Yucca 119
Planthoppers 141
pleura 21
Plum Catocala 282

Walkingsticks 65, 69, 70
  Arizona 70
Warehouse Beetle 395
Wasp-like Clerid 375
Wasps 508
  Burrowing 567
  Cuckoo 541
  Digger 548
  Eastern Sand 566
  Evaniid 540
  Larrid 563
  Mason 546
  Mossy-rose-gall 520
  Pacific Gall 521
  Platygasterid 541
  Polistes 529, 544
  Potter 545
  Sand 562, 565, **567**
  Scelionid 540
  Sphecid 562
  Spider 546
  Trigonalid 542
  Vespid 542
  Weevil 568
  Western Sand 566
Water Boatman 110
Water Bug, Giant 113
Water Scavenger Beetles 353
  Giant 354
  Minute 354
  Narrow 355
Waterscorpions 111
Water Striders 113
Water Treader 114
wax (from coccids) 156
Webbing Clothes Moth 315
Webspinners 85
Webworm, Alfalfa 304
  Beet 303
  Bluegrass 306
  Cabbage 304
  Cross-striped Cabbage 304
  Garden 303
  Striped Sod 306
Weevils 477–498
  Alfalfa 483
  Apple Flea 490
  Black Elm Bark 485
  Black Oak 491
  Black Vine 481
  Boll 487
  Broad-nosed Grain 496
  Bronze Apple Tree 486
  Cabbage Seedpod 493
  Carrot 484
  Clover Head 483
  Clover Leaf 483
  Clover Root 483
  Clover Seed 483
  Cocklebur 497
  Cranberry 489
  Filbert 491
  Granary 497
  Hazelnut 491
  Hister-like 479
  Hollyhock 479
  Knotweed 492
  Large Chestnut 491
  Lesser Clover Leaf 483

Weevils (*cont'd*)
  Monterey Pine 485
  Mullein 491
  Northern Pine 485
  Obscure 482
  Pales 486
  Pea Leaf 483
  Pecan 490
  Pepper 487
  Pine 486
  Pine Reproduction 492
  Pine Root Collar 486
  Rice 497
  Rice Water 486
  Rough Strawberry Root 482
  Small Chestnut 491
  Square 479
  Stem-boring 492
  Strawberry 489
  Strawberry Root 482
  Sweet Clover 483
  Sweetpotato 478
  Vegetable 484
  White Pine 485
  Yosemite Bark 485
West Coast Lady 237
Western Bloodsucking Conenose 122
Western Corsair 122
Western Twig Borer 426
wet flies (lures) 82, 167
Wheat Jointworm 537
Wheat Straw-worm 537
Wheel Bug 121
Whirligig Beetles 351
white blast 107
Whiteflies 154
  Banded-wing 155
  Citrus 154
  Citrus Blackfly 155
  Cloudy-winged 155
  Greenhouse 155
  Woolly 155
White-fringed Beetle 481
White Peacock 236
Whites 208
  Becker's 209
  California 209
  Great Southern 210
White-spotted Sawyer 453
Wireworms 384
  Community 388
  Eastern Field 387
  False 419
  Great Basin 388
  Gulf 386
  Pacific Coast 387
  Plains False 419
  Prairie 388
  Puget Sound 388
  Southern Potato 386
  Sugar-beet 387
  Tobacco 386
  Western Field 387
  Wheat 387
Wood Borers, Flatheaded 389
  Metallic 389
  Roundheaded 440
Wood Nymphs 219
  Clouded 220

Index of Scientific Names

Numbers in **boldface** refer to major entries.

Aleochara bilineata 365
  bimaculata 364
Aleocharinae 364
Aleurocanthus woglumi 155
Aleurothrixus floccosus 155
Aleyrodidae 154
Alphitobius diaperinus 424
Alphitophagus bifasciatus 422
Alsophila pometaria 292
Altica carduorum 469
  chalybea 469
  ignita 468
  sylvia 469
Alticini 467
Alypia octomaculata 273
Amara impuncticollis 342
Amarini 342
Amathes c-nigrum 277
Amauromyza 632
Amblycheila 332
Amblyscirtes vialis 262
Ametastegia glabrata 513
  pallipes 514
Amitermes 93
Ammobia ichneumonea 565
Ammophila 565
Ampedus collaris 386
  nigricollis 386
Amphicerus bicaudatus 426
  cornutus 426
  hamatus 426
Amphicyrta dentipes 396
Amphimallon majalis 433
Anabrus simplex 75
Anaea aidea floridalis 240
   morrisoni 240
  andria 240
  portia 240
Anaedus brunneus 424
Anagasta kuehniella 310
Anagyrus antoniae 160
Anarhopus sydneyensis 160
Anarsia lineatella 323
Anartia jatrophae 236
Anasa tristis 126
Anastatus disparis 536
Anastoechus barbatus 613
Anastrepha ludens 627
  mombinpraeoptans 628
  suspensa 628
Anatis quindecimpunctata 410
Anax junius 173
  walsinghami 173
Ancylis comptana fragariae 312
Ancyloxypha arene 259
  numitor 259
Andrena carlini 575
  regularis 575
Andrenidae 575
Andricus californicus 520
Anisodactylus nigerrimus 341
Anisoptera 172
Anisota rubicunda 270
   alba 270
  senatoria 269
  stigma 269
  virginiensis 269

Anobiidae 415
Anomala oblivia 435
  orientalis 435
Anomalini 435
Anopheles albimanus 589
  crucians 589
  freeborni 589, **594**
  maculipennis 594
  punctipennis 594
  quadrimaculatus 594
Anoplodera cordifera 450
  vittata 450
Anoplura 99
Antheraea polyphemus 266
Anthicidae 383
Anthicus cervinus 384
  floralis 384
Anthocharis genutia 212
  sara 211
   stella 212
Anthocoridae 119
Anthocoris melanocenus 120
Anthomyiidae 632
Anthonomini 487
Anthonomus eugenii 487
  grandis 487
  musculus 489
  scutellaris 487
  signatus 489
Anthophora abrupta 578
  bomboides 578
  furcata terminalis 578
  occidentalis 547, **578**
Anthrax analis 613
  tigrinus 613
Anthrenini 394
Anthrenus coloratus 396
  scrophulariae 395
  verbasci 395
Anthribidae 475
Anthribus 475
Anticarsia gemmatalis 282
Antigonus 257
Antonina graminis 160
Antron douglasii 520
Anuraphis rosea 147
Aonidiella aurantii 162
  citrina 162
  taxus 536
Apamea amputatrix 274
Apanteles carpatus 523
  congregatus 263, **523**
  flaviconchae 524
  glomeratus 522, **523**
  harrisinae 523
  lacteicolor 523
  medicaginis 524
  melanoscelus 523
  militaris 522
  solitarius 523
Apaturinae 240
Aphelinus diaspidius 534
  mali 534
Aphididae 145
Aphidius matricariae 526
  smithi 526
Aphis fabae 150
  forbesi 150

Aphis (cont'd)
  gossypii 149
  illinoisensis 150
  maidiradicis 148
  nasturtii 150
  pomi 149
  spiraecola 149
Aphodiinae 430
Aphodiini 430
Aphodius distinctus 430
Aphoebantus 613
Aphrastus taeniatus 480
Aphrophora saratogensis 136
Aphycus helvolus 535
  lounsburyi 536
  stanleyi 536
Aphytis chrysomphali 533
  lepidosaphes 533
  lingnanensis 533
  maculicornis 533
  proclia 534
Apidae 578
Apinae 579
Apiocera haruspex 609
Apioceridae 609
Apiomerus crassipes 122
Apion fuscirostre 479
  impunctistriatum 479
  longirostre 479
  porcatum 479
  ulicis 479
Apioninae 479
Apis mellifera 573, **582**
Aplopus mayeri 69
Apocrita 508, 509, **519**
Apodemia mormo 244
    virgulati 244
  nais 243
Apoidea 569
Apterobittacus apterus 190
Apterygota 55
Araecerus fasciculatus 475
Archilestes californica 176
  grandis 176
Archips argyrospilus 319
  cerasivoranus 319
Archytas californiae 647
  marmoratus 648
Arctiidae 270
Arctocorixa alternata 110
Argynninae 225
Argynnis 225
Argyresthia thuiella 322
Argyrotaenia citrana 320
  franciscana 530
  pinatubana 321
  velutinana 321
Arilus cristatus 121
Aristotelia fragariae 323
Arrhenodes minutus 477
Asarcopus palmarum 141
Ascia monuste 210
  phileta 210
Ascogaster quadridentata 313, 522
Aseminae 444
Asemini 444
Asemum atrum 444
  striatum 444

Asidinae 418
Asilidae 610
Aspidiotus ancylus 164
  forbesi 164
  hederae 165
  juglansregiae 164
  nerii 165
  ostreaeformis 164
  perniciosus 164
Astenus longiusculus 362
Asterocampa celtis 240
  alicia 241
  montis 241
  clyton 241
  leilia 241
Astrotus contortus 419
  regularis 419
Ataenius cognatus 430
Atalopedes campestris 260
Atanycolus charus 525
Atherix variegata 607
Atima confusa 444
    dorsalis 445
    maritima 445
Atimini 444
Atlides halesus 244
Attagenini 394
Attagenus megatoma 394
  pellio 394
  piceus 394
Attalus circumscriptus 374
  morulus smithi 374
  scincetus 374
Atta texana 88, **556**
Attelabinae 479
Attelabus bipustulatus 479
  nigripes 479
Aulicus 375
Autochton cellus 257
Autographa californica 281
Automeris io 268
Autoserica 431

Baccha clavata 619
Bacillus popilliae 438
Bacillus thüringiensis 210, 213, 280, 301,
    310, 313
Badister pulchellus 345
Baeocera falsata 365
Baetidae 167
Baetinae 168
Baetis bicaudatus 168
  insignificans 168
  posticatus 168
  spinosus 168
Baliosus ruber 471
Barini 492
Basilona 270
Bathyplectes 526
  curculionis 484
Battus philenor 205
    hirsuta 205
  polydamus 205
Beauveria bassiana 125, 291, 313, 482,
    487, 512
Belonia 172
Belostoma 113
Belostomatidae 112

Foxella ignota   657
Frankliniella   **106**, 108
    fusca   106
    occidentalis   106
    tritici   106
Frenatae   199, **262**
Fucellia costalis   633
    rufitibia   632
Fulgoridae   141

Galerita   344
Galeritula bicolor   344
    janus   344
    lecontei   344
Galerucella   464, 465
Galerucinae   464
Galleria mellonella   308
Galleriidae   308
Gargaphia solani   124
    tiliae   124
Gasterophilidae   639
Gasterophilus haemorrhoidalis   640
    intestinalis   639
    nasalis   639
Gelastocoridae   115
Gelastocoris oculatus   115
Gelechiidae   322
Geocoris atricolor   125
    decoratus   125
    pallens   125
    punctipes   125
Geometridae   200, **291**
Geotrupes splendidus   430
Geotrupinae   430
Geotrupini   430
Gerridae   113
Gerris comatus   114
    marginatus   114
    remigis   113
Gerydinae   244, **250**
Gibbium psylloides   412
Gibbobruchus mimus   474
Glaucopsyche lygdamus   254
Glipa   378
Glischrochilus fasciatus   398
    quadrisignatus   398
    sanguinolentus   398
Glossina   634
Glycobius speciosus   448
Gnathocerus cornutus   423
Gnorimoschema   322
Goes pulcher   452
    tesselatus   452
Gomphidae   173
Gonatopus contortulus   568
Gonia porca   650
Gonicotes gallinae   97
Gossyparia spuria   159
Gracillaria azaleella   325
    syringella   325
Gracillariidae   325
Graphisurus   454
Graphium marcellus   207
Graphognathus leucoloma   481
Grapholitha interstinctana   314
    molesta   313
    packardi   313
    prunivora   313
Gratiana pallidula   472

Gryllacrididae   78
Gryllidae   76
Gryllotalpa gryllotalpa   78
    hexadactyla   78
Gryllotalpidae   77
Gryllus   76, 77
Gymnetini   437
Gymnetron antirrhini   491
    tetrum   491
Gynembia tarsalis   86
Gyrinidae   351
Gyrinus borealis   353
    minutus   353
    ventralis   353

Habrocerinae   363
Habrocerus capillaricornis   363
    magnus   363
    schwartzi   363
Habrodais grunus   245
Haematobia irritans   638
Haematopinus asini   102
    eurysternus   102
    quadripertusus   102
    suis   102
Haematosiphon inodorus   117
Hagenius brevistylus   173
Halictidae   573
Halictus confusus   574
    farinosus   573
    rubicundis   574
    zonulum   574
Haliplidae   348
Haliplus ruficollis   349
    triopsis   349
Halisidota argentata   270
    caryae   271
    maculata   271
    tessellaris   271
Halobates sericus   114
Halticotoma valida   119
Halticus bractatus   119
Haplothrips faurei   104
Harmolita   537
Harpalini   341
Harpalus caliginosus   341
    compar   341
    pennsylvanicus   341
Harrisina americana   297
    brillians   297
    metallica   297
Hebridae   115
Hebrus sobrinus   115
Hedriodiscus   607
Heleidae   594
Heliconiidae   223
Heliconius charitonius   223
Helicoverpa   279
Heliopetes ericetorum   257
    domicella   258
Heliothis virescens   280
    zea   279
Heliothrips haemorrhoidalis   105
    rubrocinctus   105
Hellula rogatalis   304
Helochares   354
Helophorus   354
Hemerobiidae   184
Hemerobius pacificus   184

Microbembex monodonta 547
Microcentrum retinerve 75
  rhombifolium 75
Microcoryphia 57
Microlarinus lareyniei 492
  lypriformis 492
Micropelinae 361
Micropeplus cribratus 361
Microphotus angustus 370
Milesiinae 618, 619
Minois 219
Miridae 117
Mitoura gryneus 248
Molorchini 449
Molorchus bimaculatus 449
  eburneus 449
  semiustus 449
Monarthrum mali 504
Monobia quadridens 546
Monocesta coryli 465
Monochamini 452
Monochamus maculosus 453
  notatus 453
  oregonensis 453
  scutellatus 453
  titillator 454
Monocrepidius 386
Monodontomerus montivagus 578
Monoleptini 467
Monomorium minimum 554
  pharaonis 553
Monophadnoides geniculatus 513
  rubi 513
Mordella albosuturalis 378
  atrata 378
  marginata 378
  octopunctata 378
Mordellidae 377
Mordellistena aspersa 379
  comata 379
  pustulata 379
  scapularis 379
  trifasciata 379
Murgantia histrionica 129
Musca autumnalis 635
  domestica 634
Muscidae 634
Muscina assimilis 638
  stabulans 637
Mutilla 547
Mutillidae 547
Mydas clavatus 609
  heros 609
  maculiventris 610
Mydidae 609
Mylabris 473
Myodopsylla insignis 660
Myopa 621
Myrmecocystus mexicanus 560
  hortideorum 560
Myrmeleon immaculatus 186
Myrmeleontidae 185
Myrmicinae 552
Myzus cerasi 147
  persicae 146

Nabidae 119
Nabis ferus 119
Nasonia vitripennis 533

Nathalis iole 212
Nearctaphis bakeri 148
Necrobia ruficollis 376
  rufipes 377
  violacea 377
Necrodes surinamensis 357
Necydalini 451
Neididae 127
Nemapogon granella 326
Nematocera 584, **585**
Nematus ribesii 513
Nemestrinidae 613
Nemomydas pantherinus 610
Neodiprion burkei 516
  lecontei 515
  pinetum 516
  pratti banksianae 515
  sertifer 514
  swainei 516
  taedea linearis 516
  tsugae 516
Neokolla circellata 141
Neolecanium cornuparvum 157
Neonympha 221
Neophasia menapia 211
Neopyrochroa femoralis 383
  flabellata 383
Neorhynchocephalus sackenii 613
Neotermes castaneus 91
Nepa 112
Nepidae 111
Nepytia canosaria 293
Neriidae 624
Neuroptera 181
Neuroterus saltatorius 521
Nezara viridula 128
Nicrophorus americanus 356
  investigator 356
  marginatus 356
  pustulatus 356
    melsheimeri 356
    nigritus 356
  tomentosus 356
  vespilloides 356
Niptus hololeucus 414
  ventriculus 415
Nitidula bipunctata 398
Nitidulidae 397
Nitidulinae 398
Noctuidae 272
Nodonota puncticollis 461
Nomia howardi 575
  melanderi 574
Norape cretata 297
Nosopsyllus fasciatus 657
Notodontidae 283
Notonecta insulata 111
  undulata 111
Notonectidae 111
Notoxus constrictus 383
  monodon 383
  talpa 384
Noviini 406
Novius 406
Nycteribiidae 640
Nygmia phaeorrhoea 288
Nymphalidae 200, **224**
Nymphalinae 232
Nymphalis antiopa 235

Nymphalis (cont'd)
  californica 235
  j-album 235
  milberti 234
  vau-album 235
Nysius ericae 125
  raphanus 125

Oberea bimaculata 456
  myops 457
  tripunctata 457
Odacanthini 344
Odonata 170
Odontoloxozus longicornis 624
Odontomyia communis 606
Oebalus pugnax 129
Oecanthus 65, 77
  fultoni 77
  nigricornis 77
  niveus 77
  quadripunctatus 77
Oecetis cinerascens 196
  inconspicua 195
Oeciacus vicarius 117
Oecophylla smaragdina 551
Oedionychus circumdata 468
  miniata 468
Oeme rigida 446
Oeneis jutta 222
  nevadensis nevadensis 222
    gigas 222
  taygete 222
Oestridae 651
Oestrus ovis 652
Ogcodes adaptatus 615
Oidini 464
Olethreutidae 312
Oligonychus punicae 406
Oligota oviformis 364
Oligotoma humbertiana 86
  nigra 86
  saundersii 86
Oligotomidae 86
Olla abdominalis 410
  plagiata 411
Omalinae 361
Omophron americanum 348
  ovale 348
  tessellatum 347
Omophronini 347
Omorgus monachus 439
  suberosus 438
Omphrale 608
Omus californicus 332
  dejeani 333
  lecontei 333
Onchocerca volvulus 598
Oncideres cingulata 452
Onciderini 452
Oncopeltus fasciatus 125
Onthophagini 429
Onthophagus hecate 429
  incensus 430
Ooencyrtus anasae 129
  kuwanai 536
Operophtera bruceata 292
  brumata 292
Ophyra leucostoma 637
Opisocrostis bruneri 658

Opius melleus 525
Opsimini 445
Opsimus quadrilineatus 445
Orchopeas howardi 659
  wickhami 659
Orgyia antiqua 286
Orius insidiosus 119
  tristicolor 120
Orsodacninae 459
Orthoptera 65
Orthorrhapha 584
Orthosoma brunneum 442
Oryctini 436
Oryzaephilus mercator 401
  surinamensis 401
Oscinella frit 631
Osmoderma eremicola 438
Osoriinae 363
Ostomidae 397
Ostrinia nubilalis 302
  obumbratalis 303
Otiorhynchinae 480
Otiorhynchus 481
Otitidae 624
Oulema melanopus 459
Oxyporinae 363
Oxyporus major 363
Oxytelinae 361

Pachybrachini 460
Pachybrachis bivittatus 461
  obsoletus 461
  othonus 461
  tridens 460
  trinotatus 461
Pachypsylla celtidismamma 142
  celtidisvesicula 142
Pachysomoides fulvus 529
Paederinae 361
Paederus 359
Palaminus testaceus 361
Paleacrita vernata 293
Palorus ratzeburgi 424
  subdepressus 424
Pamphiliidae 516
Pangaeus bilineatus 131
Panonychus ulmi 104
Panorpa nebulosa 188
Panorpidae 188
Pantomorus godmani 480
  cervinus 480
Papaipema nebris 274
Papilio ajax 204
  bairdi bairdi 205
    brucei 205
  cresphontes 206
  daunus 203
  eurymedon 203
  glaucus 202
    canadensis 202
  multicaudata 203
  oregonius 204
  palamedes 206
  polyxenes asterius 204
  rutulus 203
  thoas 207
  troilus 206
  zelicaon 204
Papilionidae 200, **202**

Reduvius personatus 121
Reticulitermes 88, 92, 93
 flavipes 88
 hageni 90
 hesperus 90
 tibialis 90
 virginicus 90
Rhabdopterus picipes 462
Rhagio dimidiatus 607
Rhagionidae 607
Rhagoletis cingulata 626
 completa 626
 fausta 626
 indifferens 626
 mendax 626
 pomonella 625
Rhagovelia obesa 114
Rhamphomyia scolopacea 616
Rhapidia bicolor 187
Rhapidiidae 187
Rhapidiodea 181, 186
Rhaphiomidas acton 609
Rhinotermitidae 88
Rhizobius ventralis 407
Rhodobaenus tredecimpunctatus 497
Rhopalocera 199, 202
Rhopalosiphum fitchi 149
 maidis 152
Rhyacionia buoliana 314
 bushnelli 315
 frustrana 314
 neomexicana 315
 rigidana 314
Rhyacophila 192
 fenestra 193
Rhyacophilidae 193
Rhynchaenus pallicornis 490
Rhynchites bicolor 478
Rhynchitinae 478
Rhynchitini 478
Rhyssa 526
 lineolata 528
Rhyzopertha dominica 427
Rickettsia mooseri 654
 prowazeki 100, 654
 typhi 654, 656
Riodinidae 242
Rodolia cardinalis 406
 koebelei 406
Rosalia funebris 449
Rosaliini 449
Rutelinae 435
Rutelini 436

Sabulodes caberata 294
Saissetia coffeae 156
 hemisphaerica 156
 oleae 156
Salda buenoi 116
Saldidae 116
Samia cynthia 266
Sanninoidea exitiosa 299
 graefi 300
Saperda calcarata 455
 candida 455
 cretata 456
 tridentata 455
 vestita 456
Saperdini 455

Saprininae 366
Saprinus lugens 366
 mayhewi 366
 oregonensis 366
 pennsylvanicus 366
Sarcophaga aldrichi 645
 cooleyi 644
 falciformis 645
 haemorrhoidalis 644
 kellyi 645
 sarraceniae 645
Sarcophagidae 644
Sathrobrota rileyi 318
Saturniidae 265
Satyridae 219
Satyrium 246, 247, 248
Scaeva pyrastri 619
Scambus detritus 529
 tecumseh 528
Scaphidiidae 365
Scaphinotus elevatus 340
 germari 340
Scapteriscus acletus 78
 vicinus 78
Scarabaeidae 428
Scarabaeinae 429
Scarabaeus sacer 429
Scarites subterraneus 347
Scaritini 347
Scelionidae 129, 540
Sceliphron caementarium 564
Scenopinidae 608
Scenopinus fenestralis 608
 glabrifrons 609
Scephidae 562
Schistocerca americana 74
 gregaria 74
 paranensis 74
 shoshone 74
 vaga 74
Schizaphis graminum 154
Schizura concinna 285
Sciomyzidae 584
Sciopithes obscurus 482
Scirtothrips citri 107
Scolia dubia dubia 548
 haematodes 548
 monticola 548
 manilae 548
Scoliidae 548
Scolops pallidus 141
Scolytidae 498
Scolytinae 499
Scolytus multistriatus 500
 quadrispinosus 500
 rugulosus 499
 unispinosus 500
 ventralis 500
Scopaeus exiguus 361
Scudderia furcata 74
Scymnini 405
Scymnus amabilis 407
 americanus 405
 ardelio 407
 binaevatus 405
 bisignatus 407
 caurinus 407
 collaris 407
 coniferarum 407